ENGINEERING

ECONOMIC AND

COST ANALYSIS

THIRD EDITION

ENGINEERING

ECONOMIC AND

COST ANALYSIS

THIRD EDITION

COURTLAND A. COLLIER, P. E.

University of Florida

Associate Professor Emeritus

CHARLES R. GLAGOLA, P. L. O., P. E.

University of Florida

Assistant Professor

ADDISON-WESLEY

An imprint of Addison Wesley Longman, Inc.

Menlo Park, California • Reading, Massachusetts • Harlow, England
Berkeley, California • Don Mills, Ontario • Sydney • Bonn • Amsterdam • Tokyo • Mexico City

Acquisitions Editor: Michael Slaughter
Assistant Editor: Susan Slater
Production Manager: Pattie Myers
Senior Production Editor: Teri Hyde
Art and Design Supervisor: Kevin Berry
Composition: Interactive Composition Corporation
Cover Design: Yvo Riezebos
Cover Image: Yvo Riezebos
Text Design: R. Kharibian & Associates
Text Printer and Binder: Maple-Vail Book Manufacturing Group
Cover Printer: Phoenix Color

Library of Congress Cataloging-in-Publication Data

Collier, Courtland A.
    Engineering economic and cost analysis / Courtland A. Collier,
  Charles R. Glagola. — 3rd ed.
      p. cm.
    Includes index.
    ISBN 0-673-98394-3
    1. Engineering economy.   I. Glagola, Charles R.   II. Title.
TA177.4.C65   1998
658.15—dc21   98-15681
                                        CIP

Instructional Material Disclaimer
The programs presented in this book have been included for their instructional value. They have been tested with care but are not guaranteed for any particular purpose. Neither the publisher or the authors offer any warranties or representations, nor do they accept any liabilities with respect to the programs.

The full complement of supplemental teaching materials is available to qualified instructors.

ISBN 0–673–98394–3

1 2 3 4 5 6 7 8 9 10—MV—02 01 00 99 98
Addison Wesley Longman, Inc.
2725 Sand Hill Road
Menlo Park, California 94025

CONTENTS

6   Geometric Gradient: The Constant Percentage Increment   137

PART II

Comparing Alternative Proposals   163

7   Present Worth Method of Comparing Alternatives   165

**8   Annual Payments Method for Comparing Alternatives
(Equivalent Series of Uniform Payments)   208**

9   Future Worth Method of Comparing Alternatives   243

PART III

Preface

This text is especially designed for those engineering students and professional practitioners who are new to the study of engineering economics. The language, explanations, examples, and illustrative figures are carefully selected to present the subject matter in a straightforward manner that will enable the reader to comprehend and use the material in the solution of practical everyday problems. In so doing, we have included 215 illustrative figures, 170 example problems fully worked and explained, together with 519 problems available for homework assignments or practice. About one-half of these problems have answers included with the problem. Solutions to all the problems are available to the teaching faculty in the Solutions Manual available from the publisher.

This third edition is in response to the enthusiastic reception afforded the previous editions, which were adopted as a practice-oriented text in many colleges and universities around the nation. The authors gratefully appreciate these adoptions and wish to thank all of those who made it possible.

NEW TO THIS EDITION

As in most professional subject areas, the scope of applications of engineering economy is widening continually, and this third edition has been updated to reflect many of these new applications. The first six chapters, which deal with an introduction to time-value-of-money concepts, are broadened with updated examples and homework problems. Since there is usually a choice of alternative courses of action, Chapters 7 through 14 demonstrate a variety of basic methods for comparing alternative proposals.

To deal competently with private-sector cost analysis problems the income tax consequences must be taken into account. Therefore, Chapter 15 begins a series of chapters dealing with a variety of after-tax situations likely to be encountered by the practicing engineer. The authors realize that engineering students and practitioners may not be working with a familiar subject area when they consider tax matters. Therefore, these chapters are written carefully in language familiar to engineers. This third edition has been updated to reflect the current Tax Reform Act. In addition to the new features of the act, other traditional features of tax law are included, both for a sense of historical perspective, as well as in anticipation that many of these traditional features likely will be reintroduced into U.S. tax law within the foreseeable future. Chapters 18 and 19 explain how to determine the profitability of investments involving real estate, when to sell, and when to keep. The text is well illustrated with specific examples. Chapter 20 presents straightforward solutions to the problems of determining the optimum replacement time for a currently owned piece of equipment, and what sort of life-cycle time to expect from the replacement equipment.

Since computers are readily available to both students and professional practitioners, several examples of spreadsheet programs have been included to facilitate problem solving.

Chapter 21 on double gradients contains updated material dealing with escalating cost which is often an overlooked problem area that has frustrated engineers for many years. For example, in doing a cost analysis of an excavator, when the productivity (in cubic yards per hour) is decreasing with age, but the price of the excavation (in dollars per cubic yard) is also changing with time, the double gradient can eliminate potentially significant errors that are inherent in older methods of solution.

ACKNOWLEDGMENTS

As in the previous editions, the authors again acknowledge with deep gratitude the very helpful input to this text from alert, inquiring students, as well as conscientious faculty and professional practitioners. Those of you who have worked seemingly endless hours on favorite projects know that endeavors like this are made possible in large part by the patience and loving support of wife and family to whom we owe so much in helping bring this edition to fruition. A sincere word of thanks is due also to the understanding and hardworking editor and staff at Addison Wesley Longman, Inc.

Courtland A. Collier
Charles R. Glagola

PART ONE

Basic Principles

CHAPTER 1

Quantifying Alternatives for Easier Decision Making

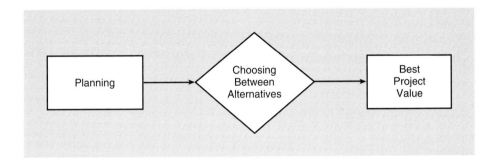

THE DECISION PROCESS

Life is full of choices. In matters small and large we are constantly confronted with the need to make selections from arrays of alternative courses of action concerning how best to spend our time, where to live, which job to take, how to determine spending priorities for our money, and so forth. In short, how can we best invest our limited supply of resources such as time and money? Decisions, decisions, decisions! The decision-making process would be greatly simplified if the alternatives could be quantified objectively and some numerical value attached to each. Then an optimum return on our limited resources could be assured by simply selecting the alternative with the highest numerical value from each array of choices.

Not all alternatives can be quantified, but a surprising number can. For instance, the question of whether or not to install a traffic light at an intersection can be reduced to a summation of costs versus benefits to all concerned. The costs of stopping a vehicle, costs of waiting time, costs of damage from accidents, and even of air pollution from idling engines can be assigned some numerical value. These, in contemporary applications, are *user costs,* which are applied in highway feasibility studies. Later chapters explain how to equate costs and benefits that occur in different quantities and at different times, so that alternatives can be compared directly and the better choice made evident. One of the primary purposes of this text is to enable the reader to make intelligent selections from among a wide variety of alternatives.

As in other areas of life, the practice of engineering requires a considerable amount of decision making. The decision-making process may vary somewhat depending upon the needs of the project sponsor. Most engineering works are sponsored either by the government, as public works projects, or by private enterprise, as profit-making ventures.

Determining Public Works Priorities

The selection, planning, funding, construction, and operation of public works projects typically involve the following process:

1. List current and future needs of the public that can be met by public works projects.
2. List the alternative proposals for meeting those needs.
3. Compare the alternatives.
   a. Quantify and numerically compare *tangible values,* such as the costs, benefits, effectiveness, or income of each alternative in meeting the need.
   b. Compare the *intangible values,* such as quality, appearance, effect on neighborhood image, pride and respect, the comparative needs of the beneficiaries, and the distribution of the benefits.
4. Establish priorities based on ratios of benefits to costs or equivalents, with proper consideration given to intangibles.
5. Make recommendations for funding, and hold public hearings to obtain approval and funding.
6. Construct and operate the project. Evaluate accomplishments compared to previous estimates.

Since public works planning involves decision making in the spending of the public's tax money, it should be done with the public's sense of priority in mind. The conscientious engineer performs a valuable public service in preparing and presenting comparative numerical data for use by the public when facing decisions between competing alternatives.

Determining Private Enterprise Priorities

Planning for privately owned facilities differs from public works planning basically in the accountability for funds, both investment and return. In public works, the funds must be fairly distributed to reflect the needs of citizens, their sources of revenue, and other factors of concern to them. In the private sector, the investor has wide discretion concerning the allocation and amount of funds to be expended, within the broad limits set by law. Usually the private investor is primarily concerned with profitability. As a result, the private investor, like the public works planner, is vitally concerned with the needs of the public, since the return on investment usually depends on the patronage of the consuming public as well as on the satisfaction of individual demands. Therefore, rather than rely on intuition and hope, planners of privately owned projects need factual numerical comparisons in order to make intelligent decisions.

Dollar Value: The Common Denominator

In planning for either public or private projects, all significant aspects of each project are analyzed, and viable alternatives are listed, quantified, and compared. Whatever can be quantified should be, since decision makers find it easier to compare numbers than other qualities. In order to quantify and compare as many aspects of a proposal as possible, a common denominator for most purposes is the dollar value. For this reason, most of the significant attributes of a project are reduced to equivalent dollar values wherever possible.

INTEREST: THE RENT PAID FOR USE OF MONEY

Money, like the goods and services it represents, can be owned, or it can be rented (borrowed or owed). When money is borrowed, owners usually expect some compensation for the inconvenience of doing without their money for the specified period of time, so there is usually an agreed upon rental fee charged for the loan. Of course, the rental fee is more commonly called *interest,* but the function is the same as other rental fees; that is, the rent (or interest) is paid by the borrower in order to acquire the use of the owner's money for a stipulated period of time. By the same reasoning, the borrower who has the use of the money expects to pay rent or interest. For instance, a contractor-builder often needs short-term money and takes a construction loan to cover the cost of labor and materials put into the project until reimbursed by the project owner. A project owner usually needs a long-term loan (mortgage) on the project, which can be repaid with interest in equal monthly installments out of current income from the project.

Interest Rates

Owners of engineering projects, whether public or private, may finance the project with their own money, or with borrowed money, or with a combination of both owned and borrowed money.

Owned Money   If the owners of the project have money of their own available, they may choose between investing the money in the project or lending it to others to generate interest income. If they invest in the project, the money is not available for loan, and the interest income is lost. The lost interest income therefore should be charged to the project as one of the costs of the project.

Borrowed Money   If the owners choose to finance the project with borrowed money, the interest paid on the borrowed money is also one of the costs of the project.

Combination of Owned and Borrowed Money   On private projects, most lenders will lend only from about 60 to 80 percent of the value of a project. Therefore, most projects require owners to finance from 20 to 40 percent of the project with their own money.

Since most owners expect to make a return on their projects greater than the interest rate charged by money lenders, they usually prefer to borrow as much as possible and use a minimum of their own funds. Borrowing is sometimes called "leveraging," and many owners find it profitable to leverage as much as possible. On public works projects owned by agencies of the government, 100 percent financing can be obtained commonly, since the risk of default is usually small. This type of financing is often done through bond issues.

Determination of Interest Rates

Interest rates are determined by a number of factors, prominent among which are the following:

1. The risk of not getting the loaned money back. A high risk warrants a high interest rate. For instance, if the normal interest rate is 12 percent on a one-year loan with ordinary risk, and a loan application is received which has an additional risk probability of default of 1 in 50, add about 2 percent to the 12 percent normal interest to cover this risk. (Since the 98 good loans have to make up the 12% interest for all 100 loans, the required interest rate can be calculated as $112/98 - 1 = 14.286$ percent, so 2.286 percent would have to be added to the normal 12 percent interest rate to compensate for this added risk.)

2. The supply of and demand for funds. A high demand for loans usually drives interest rates up. For instance, the high interest rates of the early 1980s were caused, in part, by the federal government's borrowing to cover budget deficits that at times exceeded \$200 billion per year. A typical supply-demand curve is illustrated in Figure 1.1. For the particular money market and point in time shown, an amount equal to V dollars will be loaned at i interest rate.

3. Overhead costs of bookkeeping, collection fees, and so forth, and length of time of loan. High collection costs, high accounting costs for small weekly payments, low dollar amount of loan, and short-term loans with turnaround costs all result in higher interest charges.

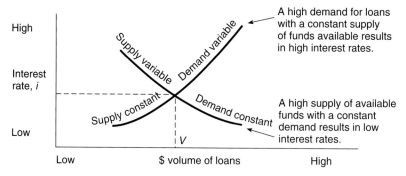

Figure 1.1 Effect on interest rates of supply and demand for loan funds.

4. Government regulations. Usury laws put ceilings on certain types of loans. Also some federally chartered banks are limited in the interest they may pay on deposits, and the discount rate at which banks borrow certain funds is regulated by the Federal Reserve Board.

MARR, Minimum Attractive Rate of Return

Both in the public works sector and in the private enterprise area, the following question frequently arises: "What interest rate should I use for a feasibility study?" Interest rates that are unrealistically low magnify the importance of future costs and benefits, and may cause the project to falsely appear too costly or too beneficial. On the other hand, interest rates that are too high tend to diminish the effect of future costs and benefits and likewise can give a false impression of the true worth of the project. To reflect as realistically as possible the actual interest rate at which the project should be evaluated, the concept of *Minimum Attractive Rate of Return* (**MARR**) is commonly used. As the name implies, the MARR is the minimum rate of return at which the owner is willing to invest. Investment opportunities yielding less than the MARR are considered not worthwhile. In general the MARR will be near the higher of the following two values:

1. Cost of borrowed money.
2. Opportunity cost. This refers to the rate of return on alternative investments.

For example, assume that an investment opportunity arises to build or buy project A. The funds to invest may come from two usual sources.

1. Borrowed funds If the funds to finance project A are borrowed, then the interest rate of return from the investment in project A obviously should exceed the interest paid on the borrowed funds. For example, to borrow $100,000 at 12 percent interest (paying $12,000 per year interest) and invest it in a project that yields only 8 percent return ($8,000 per year return) is a bad investment. Therefore the MARR for the investment should exceed the cost of borrowed money (unless there is some offsetting intangible or other benefit).

2. Owned funds If the funds to finance the investment are already in possession of the owner (firm, individual, or government agency), then these funds alternatively could be used to (a) pay off existing debt, or (b) fund alternative investments. Therefore, once again project A is competing with the cost of borrowed funds, as well as other alternative investment opportunities (projects B, C, etc.).

For most public works projects, the MARR usually may be taken to be somewhere between the taxpayer's cost of funds and the interest rate paid on bond funds used to finance the project. For private projects, the MARR is usually determined by competitive opportunities to invest (taking into account risk, liquidity, timing of cash flow, size of investment, etc.), since the interest rate on borrowed funds is typically lower than the interest rate of return on available alternative investment opportunities. Therefore, the MARR for private enterprise is usually assumed to be the opportunity cost (the rate of return obtainable on alternative investment opportunities).

SIMPLE AND COMPOUND INTEREST

Simple interest   is the term used to describe interest usually paid to the lender as soon as it is earned. If the simple interest payment is not claimed by the lender when due, no interest accrues on that interest payment no matter how long the lender leaves it unclaimed.

Compound interest   is interest which is not paid out of the investment as soon as it is earned, but is considered as an increased increment of principal and thus earns additional interest with time. Compound interest implies that each subsequent interest payment is calculated on the total of principal plus all accumulated interest to date.

Simple interest example ($i = 10\%$):	
Principal (original investment)	$1,000
Interest earned and paid at end of year 1 @ 10%	100
Interest earned and paid at end of year 2 @ 10%	100
New total of principal plus *simple* interest	$1,200
Compound interest example ($i = 10\%$):	
Principal (original investment)	$1,000
Interest earned, end of year 1 @ 10%	100
New principal plus accumulated interest	$1,100
Interest earned, end of year 2 @ 10%	110
New total of principal plus *compounded* interest	$1,210

Nominal Interest Rate

The **nominal interest rate** is the *annual* interest rate, disregarding the effects of any compounding that may occur during the year. In the marketplace, interest rates usually are expressed as the nominal annual rate, even though interest actually may be paid out or compounded at any convenient period, such as quarterly, monthly, or daily. For instance, 8 percent nominal interest compounded semiannually on a $1,000 investment would yield ($8/2 = 4$) 4 percent × $1,000 = $40 credited to the lender at the end of the first six-month period, plus an additional 4 percent × $1,040 = $41.60 credited at the end of the twelfth month. Even though the interest earned sums up to $40 + $41.60 = $81.60, which on a $1,000 deposit yields an actual return of $81.60 / $1,000 = 8.16 percent, the nominal interest is nonetheless calculated as 2 × 4 percent = 8 percent, and expressed as 8 percent nominal.

If the interest is not compounded but paid directly to the lender twice a year, the $40 interest earned each six months is still referred to as 8 percent nominal annual interest, but there is usually some indication that the interest is simple interest and not compounded. More details are presented in the following chapters.

Percent Growth Similar to Interest

The concepts of simple interest and compound interest are readily applicable to nonfinancial subjects as well. For instance, population growth is often expressed in terms of simple or compound percentage. If a city that had a population of 100,000 a year ago now has a population of 110,000, the growth rate may be calculated as 10 percent for that year. If the same rate of growth continues during this year, the population next year will grow by another 10 percent (or 11,000 people) and the total will reach 121,000. Thus the growth rate would be 10 percent compounded. Sometimes the growth rate is expressed in terms of simple percentage, similar to simple interest. For instance, if a state had a population of 4,000,000 ten years ago and 5,000,000 now, some commentators would calculate the increase of 1,000,000 in 10 years, or an average of 100,000 per year on a base of 4,000,000. The rate of increase therefore is 2.5 percent per year (100,000 / 4,000,000 = 0.025 or 2.5%). The growth rate of many numbers important to all of us are commonly measured in percentages. A partial list includes the following:

- Changes in Consumer Price Index (cost of living)
- Population growth of a neighborhood, county, state, nation, continent, planet Earth
- Changes in Construction Cost Index
- Changes in unemployment and other sociological indexes, such as the crime rate (Index of Crime)
- Increase in traffic count on a given road or at a particular intersection

PROBLEMS FOR CHAPTER 1, QUANTIFYING ALTERNATIVES

Problem Group A

Find or apply the percent increase in several commonly encountered indexes over a one-year period.

A.1. Find a recent issue of *Engineering News Record* (*ENR*) magazine and look up the annual rise in the Construction Cost Index expressed in percent (see the Market Trends page). Assume a construction project last year was bid at $10,000,000, but the award was postponed for one year due to high prices. How much would it cost now, one year later, if the Construction Cost Index adequately reflects the change in construction costs? (Note the issue and page numbers of the *ENR* on which you find the information.)

A.2. Find the current Consumer Price Index and the value of one year ago. Then derive the percent increase over the past year. (See monthly *U.S. Bureau of Labor Statistics Bulletin* [*BLS*], or other financially-oriented periodicals.)

A.3. A small retirement village had a population of 1,750 one year ago. During the past year, 14 people died and 52 people moved into the village. Based upon the population one year ago, answer the following questions.

(a) What was the percent death rate during the year?

(*Ans.* 0.80%)

(b) What was the percent *net* growth rate during the year?

(*Ans.* 2.17%)

Problem Group B

Find current interest rates used by credit institutions.

B.1.   Find what current interest rates are being paid or charged on any two types of loans from the list below. Name the source of your information and briefly summarize the salient features of the loan, such as, whether repayment is in installments or a lump sum, and when it occurs.
(a) Passbook savings.
(b) Finance company on a personal signature loan (unsecured).
(c) Major credit card company on the unpaid balance.
(d) New auto loan.
(e) First mortgage loan on an owner-occupied residence.
(f) Prime rate.
(g) Certificate of deposit.
(h) U.S. savings bond.

B.2.   Call a local finance company and find what interest rate is charged on a $1,000 personal (unsecured) loan for any time period of your choice. Is the interest simple or compounded? Show name of company and date called.

B.3.   Call a local credit union and find the interest rate and compounding period (a) for unsecured loans to members and (b) for savings or share accounts of members.

B.4.   What annual and monthly interest rates are charged by any one of the major credit card companies on credit accounts (Visa or Mastercard)?

Problem Group C

Consider the problem of quantifying some typical tangibles and intangibles.

C.1.   When considering whether or not to install a traffic signal at a street intersection, how would you approach the problem of finding dollar-value equivalents to the following:
(a) Waiting time for motorists now trying to cross the intersection without a signal as well as waiting time for motorists stopped by the signal if one is installed.
(b) Safety without the signal compared to safety with the signal.

C.2.   When considering whether to specify brick facing for the outside of a proposed school building or whether to leave the outside as painted concrete block, how do you go about evaluating the numerical value of appearance and aesthetics? List the items you would consider. Briefly discuss your approach to evaluating each item.

C.3.   Assume you need an automobile (new or used), and that you have $20,000 cash to spend either as payment in full or as a cash down payment on a car costing more than $20,000. Describe the features you would consider (including cost) when shopping for this car and describe at least 10 factors that would influence your final selection. Indicate beside each factor listed whether it is "tangible" or "intangible."

C.4.   List five tangible and five intangible factors influencing your decision to attend the college of your choice.

Problem Group D

Find the MARR.

D.1.   A firm is considering an investment involving the design and construction of project A. The question arises of what interest rate to use in the feasibility study. Assume that project A is competitive in all respects with projects B and C. Suggest a logical MARR for the firm, using the following data:
Investment opportunity, project B $= 14\%$
Investment opportunity, project C $= 16\%$
Cost of borrowed money, loan D $= 12\%$
Cost of borrowed money, loan E $= 10\%$

D.2.   A young engineer has just received a $2,000 tax rebate from the IRS. The engineer's financial situation appears as follows:
Car loan balance, $2,432 @ 14.5%
Credit card balance, $2,150 @ 18%
Investment opportunity, bank CDs (low risk) 6%
Investment opportunity, mutual funds (medium risk) 13%
(a)   Suggest a logical value for the engineer's MARR.
(b)   Where could the $2,000 be invested most profitably?
(c)   If the tax rebate were $4,000, would the MARR change? Explain.

CHAPTER 2

The Value of a Single Payment Now Compared to a Single Payment in the Future (*P/F, F/P*)

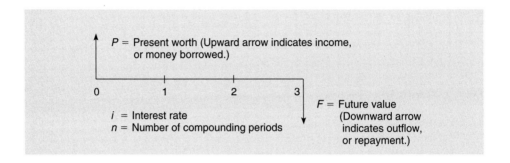

P = Present worth (Upward arrow indicates income, or money borrowed.)

0   1   2   3

i = Interest rate
n = Number of compounding periods

F = Future value
(Downward arrow indicates outflow, or repayment.)

KEY EXPRESSIONS IN THIS CHAPTER

P = Present value of all payments under consideration. Present value* may be viewed as a single lump sum cash payment now, at time zero, the *beginning* of the first of n time periods. No interest has accrued.

F = Future value. A single lump sum cash payment occurring in the future at the *end* of the *last* of n time periods. The future value equals the present value plus all the interest payments that have accrued at interest rate i at the end of each of n compounding periods.

i = Interest rate per compounding period. This is the interest rate paid on the accumulated balance of funds in the account. In nonfinancial problems, i is the rate of growth or decline of the base number. It is assumed as annual and compounded unless otherwise indicated.

n = Number of compounding periods. Frequently n is expressed in years but also can occur in terms of quarters, months, days, minutes, or any other convenient period of time, provided that it is adjusted to suit the time period.

i = Interest rate per interest period

i_n = Nominal *annual* interest rate

---

* Note: The words *value, worth, sum, amount,* and *payment* are used interchangeably. For instance, present value, present worth, present sum, etc., all mean the same.

i_e = Effective interest rate (usually annual but can be for longer or shorter compounding periods)

m = The number of compounding periods in *one* year

e = The base of natural logarithms, 2.71828+

EOY = An abbreviation for *End Of Year*, usually followed by a number indicating
which year (e.g., EOY 3 indicates End Of Year 3)

Expressions used to describe the time at which payments are made.

BOY = Beginning of the year

BOM = Beginning of the month

EOY = End of the year

EOM = End of the month

INVESTMENT FOR PROFIT

Most investments are made with the hope of some financial gain or profit. When money is
invested (deposited) in a savings account, we expect to get more out than we put in. The
value of the savings account should increase as interest payments accumulate in the account.
In describing the investment in more general terms, we can state that the future value (designated as F) of the account is always higher than the present value (designated as P), provided that the interest rate (i) is greater than zero and that the number of time periods (n)
during which interest is paid is equal to or greater than one. The size of the increase in the
balance in the account will depend upon the interest rate i, and the number of interest payment periods n. A measure of the size of the increase is given by the ratio F/P. This is the
ratio of the dollar balance in the account, F, at the end of a series of n compounding periods
compared to the original deposit, P, made at the beginning of the first compounding period.

CASH FLOW

As the name implies, cash flow occurs whenever cash or its equivalent "flows" from one
party to another. When you pay money for a cup of coffee or the morning newspaper, cash
flows from you to the vendor. If the morning paper comes by monthly subscription, the
paper carrier has earned a little more with each delivery but cash does not flow until you
pay the monthly bill.

The cash flow exchange may occur by using currency, checks, a transfer through bank
accounts, or some other means, providing the transaction is readily convertible to cash.
Cash flow *in* (income) occurs when you receive payment, and cash flow *out* (cost) occurs
when you pay out. Some other examples of cash flow follow:

1. A savings bond is purchased for $75 ($P$) that will mature in seven years and can be
   cashed in at that time for $100 ($F$). Each year of the seven years (n), the bond
   increases in value and could be cashed in at any time. But there is no cash flow
   until the bond is actually exchanged for money.

2. As an engineer, you agree to design a structure and charge a fee of 7 percent (i) of the cost of the structure, payable upon satisfactory completion of the design. When your design work is half completed, you may feel you have earned one-half of your fee (some accountants might even add one-half of the fee to your statement of earned income at this time), but the actual cash flow does not occur until the payment is made upon the completion of the design.

3. As an investor, you purchase a choice piece of real estate for $10,000. You pay 20 percent down (cash flow out of $2,000), with the balance due in a lump sum in five years (no cash flow during the five-year period). At the end of five years you pay off the remaining 80 percent due (cash flow out) and sell the property for a $5,000 down payment ($5,000 cash flow in) plus a mortgage that entitles you to $2,000 per year for five years ($2,000 cash flow in at the end of each of the next five years).

Cash Flow Diagram

Cash flow is illustrated graphically by use of a line diagram, a type of graph used extensively because of its simplicity and ease of construction. It consists of two basic parts, (1) the horizontal time line, and (2) the vertical cash flow lines. The horizontal time line is subdivided into n periods, with each period representing whatever time duration is appropriate for the problem under consideration, such as a year, month, day, and so forth. The vertical lines represent cash flow and are placed along the time line at points corresponding to the timing of the cash flow. The vertical lines are not necessarily to scale, although a large cash flow is usually represented by a longer line than a small cash flow. If you borrow $1,000 for one year and agree to pay 8 percent interest upon repayment, the cash flow diagram representing this transaction from your (borrower's) point of view is illustrated in Figure 2.1. Note that n represents the number of completed periods. Therefore, n does not equal 1 in Figure 2.1 until the full period is completed at the right end of the diagram.

In Figure 2.1, the use of the double arrow point designates a derived equivalent value, an unknown value to be found. In this case $1,080 after one year is derived as the equivalent to $1,000 now by simply adding the accrued interest of $80 to the original $1,000 loan.

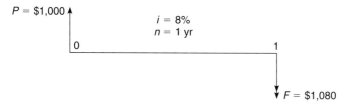

Figure 2.1   Borrower's cash flow line diagram.

Normally, receipts or income are represented by upward arrows since they represent cash flow in and an increase in cash available. Conversely, disbursements or expenditures are usually downward arrows indicating a decrease in available cash due to a cash flow out.

Notice that when the cash flow diagram is drawn from the *lender's* point of view, the arrows point in the opposite directions. In loaning you the $1,000, the lender has a cash flow out (arrow downward), and upon repayment the lender has a cash flow in (arrow upward); see Figure 2.2.

If the $1,000 is borrowed for a four-year period ($n = 4$) at 8 percent interest per year, the borrower and lender might select one of the following four alternative plans of repay-

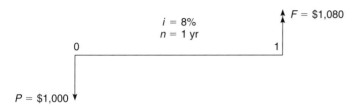

Figure 2.2   Lender's cash flow line diagram.

ment. These four plans illustrate some of the more common repayment plans but are by no means a complete list of all the alternatives available for repaying $1,000 in four years with 8 percent interest on the balance.

The cash flow diagrams for this example are drawn from the borrower's point of view (borrowed funds are a cash flow in, with arrow upward), and each plan provides for 8 percent interest on the unpaid balance.

Plan A pays interest only on the $1,000 loan at the end of each year. At the end of the fourth year the loan is repaid, together with interest due for that year (Figure 2.3).

Plan B repays an equal fraction of principal each year (one-fourth, or $250 in this case) plus interest on the amount of principal owed during the year. Thus, at the end of

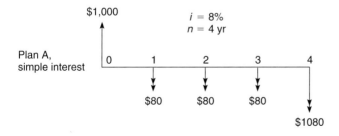

Figure 2.3   Borrower's cash flow diagram, plan A.

year 1, the interest owed for the use of $1,000 for one year at 8 percent is $80. This $80 interest payment plus repayment of $250 of the principal results in a total cash flow of $330 at the end of year 1. Payments for subsequent years are calculated in a similar manner (Figure 2.4).

Plan C provides for *equal annual* payments, as called for in most standard installment loan contracts. At the end of year 1, $80 is due as interest for the $1,000 owed during the

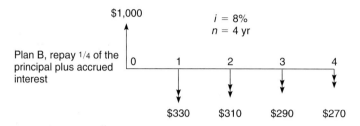

Figure 2.4   Borrower's cash flow diagram, plan B.

year. The remainder of the payment, $301.92 − $80 = $221.92 pays the balance of principal due, so that only $1,000 − $221.92 = $778.08 is owed during the second year. The interest due at the end of the second year then is only $0.08 \times \$778.08 = \$62.25$. The amount of payment is calculated by methods shown in later chapters so that the final payment pays all remaining principal due plus the final interest payment (Figure 2.5).

Plan D accumulates interest until the end of the four years. The balance owed at the end of the first year is $1,080. At the end of the second year the total amount borrowed is

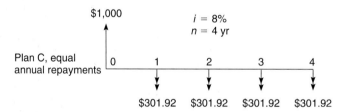

Figure 2.5   Borrower's cash flow diagram, plan C.

considered to be $1,080 instead of the original $1,000, and interest is calculated on this whole amount as $0.08 \times \$1,080 = \$86.40$. Interest calculated in this manner is said to be compounded (Figure 2.6).

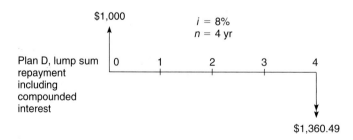

Figure 2.6   Borrower's cash flow diagram, plan D.

EQUIVALENCE

The four preceding plans are termed *equivalent* to each other. Equivalent means that one sum or series differs from another only by the amount of accrued interest at the stated interest rate, i, accumulated during the intervening n compounding periods. In other words, a lump sum of $1,000 borrowed today is equivalent to three end-of-year interest payments of $80 each plus a fourth payment at EOY 4 of $1,080 (interest plus principal, as in plan A). The $1,000 borrowed differs from the total of $1,320 repaid (4 × $80 + $1,000 = $1,320) only by the amount of $320 accumulated interest. Plan A is also equivalent to the other payment plans shown by plans B, C, and D, providing that i remains at 8 percent. In all four plans if the lender reinvests all the returns as they are received (including all payments of both principal and interest) at 8 percent, then the original $1,000 will grow to a total sum of $1,360.49 at the end of four years. If a borrower has a good credit rating and borrows $1,000 from a bank, the bank may permit repayment of the $1,000 using *any* one of the preceding four plans because they are equivalent, and the bank receives 8 percent interest on the full balance owed in every case. This equivalence concept is the basis for comparing alternative proposals in terms of equivalent dollar values, and thus a clear understanding of the equivalence concept is fundamental to understanding the remainder of this text.

INTEREST USUALLY PAID IN ARREARS

In day-to-day money matters, some payments are customarily paid in advance of receiving the goods or services, while some payments are made in arrears, only after the goods or services have been delivered. For instance, rent payments, insurance premiums, and theater admissions are customarily paid in advance at the beginning of the time period (before the services are received), while interest payments, restaurant checks, and property taxes are usually paid at the end of the period or service. Sometimes when a credit agreement is drawn up, the lender and borrower may ignore custom and agree that interest shall be paid in advance (prepaid interest). This is uncommon, but perfectly acceptable. Of course prepaying the interest increases the return to the lender, and consequently increases the effective interest rate, since the lender (rather than the borrower) has the use of the interest payment money during the term of the loan.

SINGLE PAYMENT COMPOUND AMOUNT FACTOR (*F/P*)

If interest is compounded, the interest is not paid out of the account but accumulates in the account, increasing the balance in the account and earning further interest. Thus, on a loan of $1,000 at 10 percent per year interest, the interest payment of $100 earned at the end of the first year (EOY 1) is added to the loan at EOY 1 and treated as a loan of $1,100 for the entire second year. Then at EOY 2, the 10 percent interest is due on a loan of $1,100 rather than on just the original $1,000. Thus, the interest due is added to the original loan (or invested capital) and in turn earns interest. Calculating interest on the original principal plus accumulated interest is called compounding.

EXAMPLE 2.1

The sum of $1,000 is borrowed at 10 percent compounded annually, to be repaid in one lump sum at the end of three years. How much should the lump sum repayment amount to at the end of the three-year period?

Solution. At EOY 1 the borrower owes the principal ($1,000) plus 10 percent interest ($0.10 \times \$1,000 = \$100$) which yields a total of interest plus principal owed of $1,100. If the interest payment is not paid out but simply credited to the $1,000 loan, then at the end of two years the amount owed is $1.10 \times \$1,100 = \$1,210.00$. At the end of the full three years, the borrower should repay a total of $\$1,210.00 \times 1.10 = \$1,331.00$.

In terms of the original principal (P) and interest (i), this sequence of events may be written as shown below, where F is the future sum of all principal and interest accumulated at the end of each time period shown.

$$
\begin{array}{ll}
\text{end of first year} & F = P = iP = P(1+i) \\
\text{end of second year} & F = P(1+i) + iP(1+i) \\
& \quad = P(1+i)(1+i) = P(1+i)^2 \\
\text{end of third year} & F = P(1+i)^2 + iP(1+i)^2 = P(1+i)^3 \\
\text{end of } n\text{th year} & F = P(1+i)^n \qquad\qquad (2.1)
\end{array}
$$

Thus, in Example 2.1, for $1,000 at 10 percent compounded for three years the total future sum, F, may be found very simply by

$$
F = \$1,000(1 + 0.10)^3 = \underline{\underline{\$1,331.00}}
$$

Note that the factor $(1 + i)^n$ represents the ratio of F/P and is sometimes referred to as the *Single Payment Compound Amount Factor* (SPCAF). Then to find F, given P, multiply P by this F/P factor. For convenience this factor may be designated by the notation

$$
F/P = (F/P, i, n)
$$

Using this notation Equation 2.1 would be expressed as

$$
F = P(F/P, i, n) \qquad\qquad (2.1a)
$$

This notation is *not* a formula, although it is functionally correct in that, when the *P* values are canceled, both sides of the equation revert to *F*.

$$F = P(F/P, i, n)$$

To find *P*, given *F*, simply use the inverse of Equation 2.1:

$$P = F\frac{1}{(1+i)^n} \tag{2.2}$$

The *P/F* factor $1/(1 + i)^n$ is often termed the *Present Worth Compound Amount Factor* (PWCAF) and can be expressed in its functional form as $(P/F, i, n)$. Using this form, Equation 2.2 becomes

$$P = F(P/F, i, n) \tag{2.2a}$$

As an aid to remembering the sequence of characters, note that when the *F* values are canceled, both sides of the equation revert to *P* (therefore the sequence of *P* and *F* is the same both inside and outside the parentheses).

For Equations 2.1 and 2.2 the interest rate, *i*, is the interest rate per *period*, and the time, *n*, is *n* compounding *periods*. Thus, *i* and *n* are linked by an identical number of compounding periods. If *i* is expressed as a rate per quarter, then *n* must also be expressed using the quarter as a unit of time and represents the number of quarters over which the compounding takes place. These notations (2.1a and 2.2a) will appear frequently in later chapters.

Solution Using the Tables in Appendix A.   The expression inside the parentheses is called the **Functional Notation.** As an aid in solving engineering economy problems, values for the Functional Notation are provided in tables for various values of interest rate (*i*) and periods (*n*). As an example, the value of the Functional Notation $(F/P, i, n)$ is the calculated value of the expression $(1 + i)^n$ for given values of interest (*i*) and period (*n*). Thus, the equation $F = P(F/P, i, n)$ can be solved for *F*, knowing the value of *P*, and looking up the value of the Functional Notation in the tables when given the values for *i* and *n*. While solving problems by use of the equations is a required skill, for many problems using the tables saves time. Each table contains values for only one interest rate *i*. The time periods *n* are shown in the extreme right-hand and left-hand columns of each page. Example 2.1 could be solved by proceeding through the following steps.

REPEAT EXAMPLE 2.1   **(Given *P*, *i*, *n*, find *F*)**

If $1,000 is borrowed for three years at 10 percent per year, how much is due at the end of three years? (See Figure 2.7.)

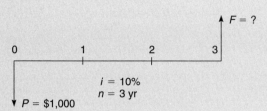

Figure 2.7   Cash flow diagram for Example 2.1.

Solution

1. Find the table in Appendix A for $i = 10$ percent.
2. Proceed down the n column to $n = 3$.
3. Go horizontally to the right along the $n = 3$ line to the F/P column, finding $F/P = 1.3310 = (F/P, 10, 3)$.
4. Since $F/P = 1.3310$, then $F = P \times 1.3310$. Since P is given as $1,000, then $F = \$1,000 \times 1.3310 = \underline{\$1,331}$, the future amount resulting from $1,000 invested principal plus compound interest at 10 percent after three years.

Note the answer is correct to five significant digits *only,* because the tables in Appendix A are complete to only five significant digits. Thus, some round-off has been introduced through the tables, and the reader is cautioned not to include more digits in an answer than are significant.

Examples of Single Payment *F / P* and *P / F* Relationships

These examples include single payments deposited (invested) or borrowed, accumulating interest compounded, and accruing to a larger future sum. If any three of the four variables are given, the fourth may be found, as illustrated in the following examples.

EXAMPLE 2.2   (Given *P, i, n,* find *F*)

If $1,000 is deposited now at 8 percent compounded annually, what is the accumulated sum after 10 years? (See Figure 2.8.)

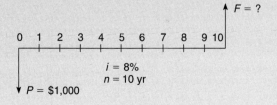

Figure 2.8   Cash flow diagram for Example 2.2.

Solution.   This problem may be solved by either of two methods.

Alternate Solution 1.   By equation:

$$F = P(1+i)^n = \$1,000(1+0.08)^{10} = \$1,000 \times 2.1589 = \underline{\underline{\$2,159}}$$

or $1,000 deposited now at 8 percent will accumulate to $2,159 in 10 years.

Alternate Solution 2.   By use of the tables in Appendix A. From Appendix A, $(F/P, 8\%, 10)$ is found as 2.1589. Then,

$$F = P(F/P, 8\%, 10) = \$1,000 \times 2.1589 = \underline{\underline{\$2,159}}$$

EXAMPLE 2.3   (**Given** F, i, n, **find** P)

The sum of $10,000 must be available 10 years from now (at EOY 10) to pay off a debt that will be coming due in one lump sum. How much should be invested *now* at 8 percent compounded interest in order to accumulate $10,000 in 10 years?

Solution.   This problem may be solved by either of two methods.

Alternate Solution 1.   By Equation 2.2.

$$P = \frac{F}{(1+i)^n} = \frac{\$10,000}{(1+0.08)^{10}} = \frac{\$10,000}{2.1589} = \underline{\underline{\$4,632}}$$

That is, $4,632 invested now at 8 percent will accumulate to $10,000 in 10 years.

Alternate Solution 2.   By use of the tables in Appendix A. From Appendix A the $(P/F, 8\%, 10)$ factor is found as 0.46319. Therefore

$$P = F(P/F, 8\%, 10) = 10,000(0.46319) = \underline{\underline{\$4,632}}$$

EXAMPLE 2.4   (**Given** P, F, i, **find** n)

How many years must $1,000 stay invested at 6 percent compounded in order to accumulate to $2,000? (See Figure 2.9.)

Solution.   This problem may be solved by either of two methods.

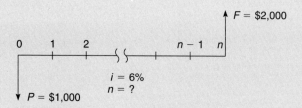

Figure 2.9   Cash flow diagram for Example 2.4.

Alternate Solution 1.   By a form of Equation 2.1.

$$F/P = (1 + i)^n$$
$$2,000/1,000 = (1 + 0.06)^n$$

A direct solution is obtained by taking the natural logarithm of both sides of the equation to find

$$\ln 2 = n \ln 1.06$$

Therefore, $n = \ln 2 / \ln 1.06 = 11.90$ years.

$$n = \underline{\underline{11.9 \text{ yr}}}$$

Alternate Solution 2.   By interpolation of the F/P tables in Appendix A. From the problem, determine that

$$F/P = 2,000/1,000 = 2.0$$

Then in the $i = 6$ percent tables of Appendix A, find values of F/P that bracket the value $F/P = 2.0$ (see Figure 2.10). Therefore, find

$$(F/P, 6\%, 11) = 1.8983 \quad \text{and} \quad (F/P, 6\%, 12) = 2.0122$$

Then by rectilinear (straight-line) interpolation* with $i = 6$ percent,

---

* Caution: The compound interest relationships are not rectilinear but curvilinear. Therefore solutions by rectilinear interpolation are approximations. Interpolated between narrow limits, the results usually are adequate for most practical purposes. However, if the limits are too widely separated, or if large values of n or i are used, the results can contain significant errors.

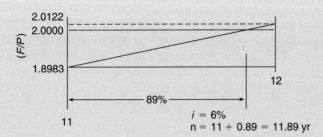

Figure 2.10   Interpolation graphic for Example 2.4.

n	F/P	F/P
12 yr	2.0122	
n (unknown)		2.0000
11 yr	−1.8983	−1.8983
	0.1139	0.1017

Then the interpolated value is determined as

$$(12 \text{ yr} - 11 \text{ yr}) \times (0.1017/0.1139) + 11 \text{ yr} = \underline{\underline{11.89 \text{ yr}}}$$

The linear interpolation can be represented graphically. This graphic shows that we have five known values and one unknown value. All values for F/P are known, so that we can determine the unknown value of *n* by seeing that this value lies the same distance from the end of the graphic as its corresponding F/P value of 2.0000. Considering the distance from either end as 100%, the F/P value of 2.0000 is 89% down the line from the left end of the graphic. Thus, the value for *n* is also 89% from the left end, or 89% of the difference between 12 yr and 11 yr.

Instinctively we could conclude that if $1,000 is invested at 6 percent, it will accumulate to $2,000 at the end of 11.9 years. On further reflection, however, we remember that this equation assumes interest payments occur only at the *end* of each compounding period, so that no final interest payment is available at the end of 11.9 years (since it is not the end of a full year). The correct conclusion then is that when the interest *is* paid at the end of year 12, the balance of principal plus accrued interest will exceed $2,000. If desired, the exact balance at EOY 12 can be calculated, as $F = \$1{,}000(F/P, 6\%, 12) = \$2{,}012.20$.

EXAMPLE 2.5   (Given P, F, n, find i)

Assume that 10 years ago you invested \$1,000 in a stamp collection. Today you are offered \$2,000 for the collection. What is the equivalent annual compounded interest rate? (See Figure 2.11.)

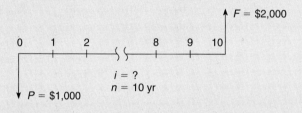

Figure 2.11   Cash flow diagram for Example 2.5.

Solution.   This problem may be solved by any of three methods.

Alternate Solution 1.   Use exponential roots.

$$F = P(1 + i)^n$$

Therefore,

$$i = (F/P)^{1/n} - 1 = (2,000/1,000)^{1/10} - 1 = 1.07177 - 1 = 0.07177$$
$$i = 0.0718 = \underline{7.18\%}$$

Alternate Solution 2.   Use natural logarithms.

$$\ln\left(\frac{2,000}{1,000}\right) = 10\ln(1 + i)$$

$$\ln(1 + i) = \frac{\ln 2}{10}$$

$$i = 0.071 = \underline{7.18\%}$$

Alternate Solution 3.   Use rectilinear interpolation of the tables in Appendix A to find the interpolated i value corresponding to $(F/P, i, 10) = 2.0000$.

i	F/P
8%	= 2.1589
i unknown	= 2.0000
7%	= 1.9671

$$i = 7\% + \frac{2.0000 - 1.9671}{2.1589 - 1.9671} \times 1\% = 7.172\% = \underline{7.17\%}$$

NOMINAL AND EFFECTIVE INTEREST

To clarify the relationship between the types of interest rates discussed in this chapter, the following nomenclature is used:

i = interest rate per interest period

i_n = nominal *annual* interest rate (often termed r)

i_e = effective interest rate (usually annual but can be for longer or shorter compounding periods)

n = number of compounding periods (often years, but not always)

m = number of compounding periods in *one* year

e = the base of natural logarithms, 2.71828+

Throughout most of this text and other literature on this subject, the subscripts to i are not used to differentiate between the three types of interest rates. The reader is usually able to determine the type of interest under discussion by the context in which it is used. It is important, therefore, that the distinction between these types of interest be carefully noted, since most discussions regarding interest rates assume readers can discern the differences on their own.

1. The *actual* interest rate, i, is the rate at which interest is compounded at the end of each period, regardless of how many compounding periods occur in one year. Thus, interest compounded monthly at 1 percent per month implies an actual interest rate of 1 percent per month.

2. The *nominal* interest rate, i_n, is the *annual* interest rate *disregarding* the effects of any compounding that may occur during the year, as explained in Chapter 1. In the financial press, interest rates are often expressed in terms of the nominal annual rate, even though the interest may be paid or compounded quarterly, monthly, daily, or otherwise. For instance, a nominal interest rate of 12 percent compounded semiannually on a $1,000 investment would yield 6 percent interest every six months. Thus at the end of the first six months this investment would yield 6 percent × $1,000 = $60 credited to the $1,000 balance, plus at the end of the second six months an additional 6 percent × $1,060 = $63.60 credited at the end of one year. Even though the interest earned sums up to $60 + $63.60 = $123.60, which on a $1,000 investment yields an actual return of $123.60/$1,000 = 12.36 percent return, the nominal interest is nevertheless calculated as 2 × 6% = 12%, and expressed as 12 percent. In equation form this appears as

$$i_n = im \tag{2.3}$$

or a nominal interest rate of

$$12\% = 6\%/\text{6-month period} \times \text{two 6-month periods/yr}$$

If the interest is not compounded but paid directly to the investor at the end of each six months, the $60 interest earned each six months is still referred to as 12 percent nominal annual interest, but there is frequently some indication that the interest is simple interest (e.g., a footnote stating "paid semiannually") and not compounded. Many debt instruments (e.g., bonds, mortgages, etc.) use the nominal interest rate, followed by the number and type of compounding periods per year. Thus, 6 percent compounded quarterly means $i_n = 6$ percent and $m = 4$ (number of quarters per year); therefore, interest payments of 1.5 percent are made every quarter (every three months). When dealing with all real-life financial calculations and negotiations, the reader is cautioned to carefully document the actual interest intended, since serious misunderstandings can occur when the interest rate is only implied and not thoroughly detailed in writing.

3. The *effective* interest rate, i_e, is the *annual* rate *including* the effect of compounding at the end of periods shorter than one year. Thus, an actual rate of 1 percent per month implies an effective rate of $1.01^{12} - 1 = 0.1268$, or 12.68 percent per year. In equation form using i,

$$i_e = (1 + i)^m - 1$$

$$i_e = 1.01^{12} - 1 = 0.1268 \quad \text{or} \quad 12.68\%/\text{yr} \tag{2.4}$$

or, using i_n,

$$i_e = \left(1 + \frac{i_n}{m}\right)^m - 1$$

$$i_e = \left(1 + \frac{0.12}{12}\right)^{12} - 1 = 0.1268 \quad \text{or} \quad 12.68\%/\text{yr} \tag{2.5}$$

Note: Effective interest rate, i_e, is, by convention, usually the *annual* interest rate, although it is sometimes used as the equivalent interest rate per payment period where there are several compounding periods between payments. The actual interest rate, i, may or may not be for a one-year period. Also, note that, when $m = 1$, $i = i_n = i_e$.

Alternatively, where only one compounding period separates P and F, and the known data include the amount, P, deposited at the beginning of the period, and the resulting balance, F, at the end of the period, then i can be found as

$$i = F/P - 1 \tag{2.6}$$

The following examples illustrate nominal and effective interest rates.

EXAMPLE 2.6   (Given P, F, i, i_n, find n)

How many months must $1,000 stay invested at a nominal interest rate of 6 percent compounded monthly to accumulate to $2,000? (Compare with Example 2.4.)

Solution.   The nominal interest rate is given as $i_n = 6$ percent per year, so the actual interest rate is found by Equation 2.3 as

$$i = \frac{i_n}{m} = \frac{6}{12} = \frac{1}{2}\%/\text{month} = 0.005$$
$$n = \text{number of } \textit{monthly} \text{ periods}$$

Then the value n in terms of months may be determined from the equation

$$F/P = (1+i)^n$$
$$n = \frac{\ln(F/P)}{\ln(1+i)} = \frac{\ln 2}{\ln 1.005} = \underline{\underline{139.0 \text{ months}}}, \text{ or } 11.58 \text{ yr}$$

(compared to 11.90 yr found in Example 2.4 at a lower interest rate)

Alternate Solution.   An alternate solution involves finding the effective annual interest rate i_e and solving by the same equation for n in terms of years. The i_e is found as

$$i_e = (1+i)^n - 1 = 1.005^{12} - 1 = 0.0617$$

Then

$$n = \frac{\ln(F/P)}{\ln(1+i)} = \frac{\ln 2}{\ln 1.0617} = \underline{\underline{11.58 \text{ yr}}}$$

This answer of course coincides with the previous answer, since 139 months divided by 12 months per year = 11.58 years.

CONTINUOUS INTEREST

Up to this point interest payments have been paid at the end of finite compounding periods, at say monthly, quarterly, annual, or other intervals. This is known as *discrete compounding*. More broadly, discrete compounding implies that there is one interest payment for each compounding period, and that the compounding period is a finite (measurable) interval of time. Interest is usually paid at the end of each compounding period, but it can be paid at any other agreed upon point in the compounding period.

On the other hand, *continuous compounding* implies that the compounding periods become so short that they are considered infinitely short, and thus compounding becomes mathematically continuous. To derive an equation for continuous compounding, let $m =$ the number of compounding periods in a year. If compounding occurs every six months then $m = 2$. As the compounding periods get shorter m increases. For monthly compounding, $m = 12$; for weekly compounding, $m = 52$; for daily, $m = 365$. As m increases and approaches infinity, the reciprocal, $1/m$ becomes smaller and approaches zero. In mathematical terms this can be expressed as

$$\text{if } 1/m \rightarrow 0, \qquad \text{then } m \rightarrow \infty$$

Since $i_e = [1 + (i_n/m)]^m - 1$, as m gets infinitely large, the limiting value of $[1 + (i_n/m)]^m - 1$ can be shown to be

$$e^{i_n} - 1$$

Thus, for continuous compounding,

$$i_e = e^{i_n} - 1 \tag{2.7}$$

Combining this with Equation 2.5 yields an equation for determining a future value of a present sum with interest compounding continuously.

$$F = Pe^{i_n n} \qquad \text{or} \qquad P = F/e^{i_n n} \tag{2.8}$$

If n or i are the unknown variables, then the equation is simply transposed to read

$$n = \frac{\ln(F/P)}{i_n} \tag{2.8a}$$

or

$$i_n = \frac{\ln(F/P)}{n} \tag{2.8b}$$

For example, a nominal interest rate of 12 percent per year compounded continuously yields an effective annual interest rate of

$$i_e = e^{0.12} - 1 = 0.1275 \qquad \text{or} \qquad 12.75\%$$

If the sum of $1,000 is invested for ten years and compounded continuously at a nominal interest rate of 12 percent per year, the resulting future balance at the end of 10 years will be (from Equation 2.8)

$$F = Pe^{i_n n} = \$1,000 \times e^{0.12 \times 10} = \$3,320$$

Alternatively, how long will it take for a deposit of $1,000 to double in value if compounded continuously at $i_n = 12$ percent? Find

$$n = \frac{\ln(F/P)}{i_n} = \frac{\ln 2}{0.12} = 5.8 \text{ yr}$$

The concept of continuous compounding is useful as an approximation to compound increases of income, expenditures, population, traffic counts, or similar events that are spread more or less evenly throughout the year instead of concentrated at year's end.

OTHER APPLICATIONS OF COMPOUND INTEREST FORMULAS

In addition to growth rates of financial investment, growth rates of many other kinds follow the same pattern as money drawing interest. For example, consumer-demand growth predictions are often estimated by using compound interest formulas.

EXAMPLE 2.7   (Find *n*, given *P, i, F*. Then find *F*, given *P, i, n*)

A city's sewage treatment plant has capacity enough to serve a population of 70,000. The current population is 50,000 and growing at the rate of 5 percent. (See Figures 2.12 and 2.13.)

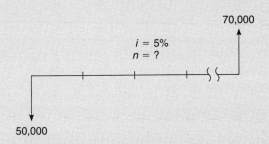

Figure 2.12   Illustrating Example 2.7, Solution 1.

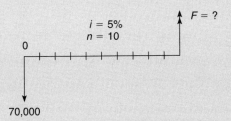

Figure 2.13   Illustrating Example 2.7, Solution 2.

1. How many years before the plant reaches capacity (a) if compounding occurs continuously; (b) if compounding is calculated at the end of each year (EOY)?

2. Additions to the plant should be planned to serve for a 10-year period. What increment of population should the next addition serve (a) if compounding is calculated at EOY?

Solution 1.   Find n, given $P = 50{,}000$, $F = 70{,}000$, $i = 5$ percent.

  (a) Continuous compounding.

$$n = \frac{\ln(F/P)}{i} = \frac{\ln(70{,}000/50{,}000)}{0.05} = \underline{\underline{6.7 \text{ yr}}}$$

  (b) EOY compounding.

$$F = P(1+i)^n$$
$$70{,}000 = 50{,}000(1 + 0.05)^n$$
$$1.4 = 1.05^n$$
$$\ln 1.4 = (n)\ln 1.05$$
$$n = \frac{\ln 1.4}{\ln 1.05} = \underline{\underline{6.9 \text{ yr}}}$$

In either case the population will reach 70,000 about seven years from now if growth continues at the same rate. For relatively small values of n, the difference between continuous and end-of-period compounding is usually small. For large values of n, the results of continuous compounding can easily double the value for end-of-period compounding.

Solution 2.   Find F, given $P = 70{,}000$, $i = 5$ percent, $n = 10$.

  (a) Continuous compounding.

$$F = Pe^{i \times n} = 70{,}000e^{0.05 \times 10}$$
$$= 115{,}410 \text{ future population}$$
$$\underline{-70{,}000} \text{ present capacity}$$
$$45{,}410 \text{ capacity of new addition}$$

  (b) EOY compounding.

$$F = 70{,}000(1 + 0.05)^{10}$$
$$= 114{,}022 \text{ population 10 yr after city reaches 70,000}$$
$$\underline{-70{,}000} \text{ population served by present plant}$$
$$44{,}022 \text{ increment of population served by}$$
$$\text{addition to sewage plant}$$

A population of about 45,000 needs to be served by the addition to the sewage plant.

Comparative future values sometimes are of some concern and need to be found, as illustrated in the next example.

EXAMPLE 2.8 (Given P_1, i_1, P_2, i_2, find n so that $F_1 = F_2$)

An older port city, J, in your state ships 10,000,000 tons of cargo per year, but shipments are diminishing at a rate of -0.7 percent compounded annually. Another port, T, ships 8,000,000 tons per year but is increasing at 5.2 percent annually. How long before the two ports ship equal amounts, if the percent changes in shipping tonnages continue at the same rate (1) compounding annually, and (2) compounding continuously?

Solution 1. Let $F_1 = F_2$, so $P_1(1 + i_1)^n = P_2(1 + i_2)^n$; then

$$\ln\left(\frac{P_1}{P_2}\right) = n \times \ln\left(\frac{1 + i_2}{1 + i_1}\right)$$

or

$$\ln\left(\frac{10}{8}\right) = n \times \ln\left(\frac{1.052}{0.993}\right)$$

$$n = \frac{0.223}{0.0577} = \underline{\underline{3.9 \text{ yr}}}$$

At that time the amount shipped would be

$$F = P(1 + i)^n$$
$$F = 8 \text{ million } (1 + 0.052)^{3.9}$$
$$F = 9.7 \text{ million tons}$$

Solution 2. Compounding continuously with a negative i_1 value brings up a dilemma. If i_2 is taken as -0.007, you discover you cannot take the logarithm of a negative number, so simply substituting values into the formulas does not yield an answer! The problem can be solved in two ways.

Alternate Solution 2A

$$F_1 = F_2, \quad \text{so} \quad P_1 e^{ni_1} = P_2 e^{ni_2}$$

Therefore

$$P_1/P_2 = e^{ni_2}/e^{ni_1}$$

Then

$$\ln(P_1/P_2) = (n \times i_2) - (n \times i_1) = n(i_2 - i_1)$$

and

$$n = \frac{\ln(P_1/P_2)}{i_2 - i_1} = \frac{\ln(10/8)}{0.052 - (-0.007)}$$

$$n = 3.78 = \underline{\underline{3.8 \text{ yr}}}$$

Alternate Solution 2B. Convert continuous interest to annual effective interest and solve the same as for Solution 1; that is,

$$i_{1e} = e^{0.052} - 1 = 0.05338$$
$$-i_{2e} = -(e^{0.007} - 1) = -0.00702$$

Then,

$$\ln \frac{P_1}{P_2} = n \times \ln\left(\frac{1 + i_2}{1 + i_1}\right)$$

$$\ln \frac{10}{8} = n \times \ln\left(\frac{1.05338}{1 - 0.00702}\right)$$

$$n = 3.78 = \underline{\underline{3.8 \text{ yr}}}$$

Thus, the continuous compounding slightly shortens the time required to achieve equal tonnages in this example. As values of n increase, the differences between the results of continuous compounding and periodic compounding can increase to sizable amounts.

RESALE OR SALVAGE VALUE

Often the incomes or costs of a facility will be separated by a period of time. For example, some income may be anticipated due to resale of a facility at some future date. The present value of this future income should be considered when calculating the net present value of any investment, as illustrated in the following example.

EXAMPLE 2.9

A public works facility has an initial estimated cost of $100,000. At the end of 10 years the facility will be replaced and have a resale or salvage value estimated at $50,000. If $i = 6$ percent, what is the net present value of the facility, considering both the present cost and the anticipated income from resale of the facility? (See Figure 2.14.)

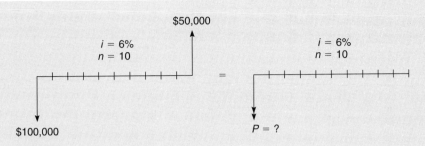

Figure 2.14 Illustrating Example 2.9.

Solution. The present value of the facility consists of the initial cost ($100,000) less the present value of the $50,000 income from salvage 10 years from now.

$$P_1 = -\$100,000$$
$$P_2 = F(P/F, i, n) = \$50,000 \underbrace{(P/F, 6\%, 10)}_{0.55839} = \$27,920$$

Since P_2 is an income, it has an opposite sign from the cost. Therefore, the net present value is the difference between P_1 and P_2. This yields $(-\$100,000 + 27,920) = \underline{-\$72,080}$ net present value (cost in this case).

Another way of viewing the proposal follows.

1. The city raises $72,080 cash to pay for the facility.

2. In addition the city gets a loan for the remaining $27,920 at 6 percent, with interest compounded for 10 years and repayment due in a single lump sum at the end of that time.

3. At the end of 10 years, the loan has now accumulated principal and interest to total $50,000, and the facility is sold for $50,000 to pay off the loan.

4. Total cash outlay (present value) for the city at the start of the project is $72,080.

COMPARISON OF ALTERNATIVES

When confronted with alternatives involving costs or income, the usual approach is to compare the alternatives on a common basis and select the least costly, or most profitable. The same approach is used when the alternatives involve a choice between building a project all at one time or in stages. The following example illustrates one logical method of solution.

EXAMPLE 2.10   (Given P_1, i, n, P_2, F_2, compare PW_1 versus PW_2)

A subdivision developer asks your opinion on whether to construct roads all at once or in stages. He finds he can put in the base course and pavement complete now for $210,000. As an alternative, the county engineer will permit him to install only the base course now (cost estimated at $120,000), with the paving installed two years from now (cost estimated at $100,000). The developer lends and borrows at 10 percent (so use $i = 10\%$ for this example). Which do you recommend as the more economical alternative?

Solution.   The present worth (cost in this case) of each alternative may be determined and the alternative with the lowest cost selected. (See Figure 2.15.)

$$\text{present cost of future pavement cost, } P_3 = F(P/F, i, n)$$
$$P_3 = \$100,000\underbrace{(P/F, 10\%, 2)}_{0.8264} = -82,640$$

present cost of base course $= -120,000$

total present cost of alternative $B = -\$202,640$

In other words, the developer can either allocate $210,000 (alternative A) for the base and pavement installation all at once now, or allocate $202,640 (alternative B) for stage construction. If stage construction is selected (alternative B), $120,000 is spent now for the base, while $82,640 is invested at 10 percent to accumulate to $100,000 in two years, enough to pay for the pavement installation at that time.

Conclusion.   The developer saves ($210,000 − $202,640) = $7,360 by utilizing stage construction, if all goes according to plan. This concept is fully developed in Chapter 7.

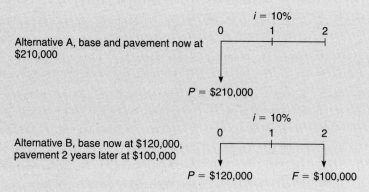

Figure 2.15   Illustrating Example 2.10.

Note: As a practical matter the developer may also consider other factors, such as the following:

1. The worry and risk of future price increases may not be worth the $7,360 saving. This consideration favors alternative A.
2. The extra two years of pavement life due to the two-year delay in installation favors alternative B.

DOUBLING TIME AND RATE: THE "RULE OF 70"

The "Rule of 70" is a useful method for approximating the time or rate required to double the value of a present sum that is increasing at a compound interest rate. Given the compound interest rate in percent, divide that percent into 70 and find the approximate time required to double the initial value. Conversely, given the doubling time, the interest rate may be approximated by dividing the doubling time into 70. For example,

QUESTION: How long does it take for the population of a country to double if the growth rate is 3.5 percent compounded?

APPROXIMATE ANSWER: $70/3.5 = 20$ years (actually, $n = 20.15$ yr).

QUESTION: If a $1,000 investment made 5 years ago is worth $2,000 today, what has been the rate of return?

APPROXIMATE ANSWER: $70/5$ years $= 14$ percent (actually, $i = 14.87\%$).

SUMMARY

In this chapter the following equivalence factors were developed between P and F:

$$F = P(1+i)^n \quad \text{or} \quad F = P(F/P, i, n)$$

$$P = \frac{F}{(1+i)^n} \quad \text{or} \quad P = F(P/F, i, n)$$

These factors are dependent only upon i and n and thus are independent of the particular values of F or P. Knowing any three of the four variables (F, P, i, n), the fourth can be easily determined. Note that P occurs at the beginning of the first of n compounding periods, while F occurs at the end of the last of n compounding periods, and i is the interest rate per period.

Also introduced in this chapter are the concepts of nominal and effective interest. The nominal annual interest, i_n, is

$$i_n = im$$

and the effective annual interest, i_e, is

$$i_e = (1+i)^m - 1 = \left(1 + \frac{i_n}{m}\right)^m - 1$$

For the special case when interest is compounded continuously, the effective annual interest, i_e, is

$$i_e = e^{i_n} - 1$$

The basic F/P equations for discrete and continuous compounding may be expressed in the following forms:

Discrete Compounding	Continuous Compounding
$F = P(1+i)^n$	$F = Pe^{i_n n}$
$P = \dfrac{F}{(1+i)^n}$	$P = \dfrac{F}{e^{i_n n}}$
$n = \dfrac{\ln(F/P)}{\ln(1+i)}$	$n = \dfrac{\ln(F/P)}{i_n}$
$i = (F/P)^{1/n} - 1$	$i_n = \dfrac{\ln(F/P)}{n}$

PROBLEMS FOR CHAPTER 2 (F/P, P/F)

Note: Assume all payments are made at the end of the period and that compounding is discrete unless otherwise noted in the problem. EOY is an abbreviation for end of year. Throughout the remainder of this text some problems may contain extraneous information not needed to solve the problem. This is to simulate the real world, where a myriad of useless information is often available that confuses the real problem.

Problem Group A

Simple and compound interest. Construct cash flow diagrams.

A.1. Which of the following are examples of simple interest, and which are examples of compound interest?
   (a) Jones buys a bond for $1,000, receives a check for $60 every year until the bond's maturity, and then receives $1,000.
   (b) Jones deposits $1,000 in a savings account bearing 6 percent interest. He makes no withdrawals for a period of 10 years. At the end of this period he closes the account and receives $1,791.

A.2. Construct a cash flow diagram illustrating the following cash flows, from (a) the borrower's viewpoint; (b) the lender's viewpoint.

EOY
0   Loan of $5,000 @ 10% interest
1   Pay $500 interest
2   Repay $5,000 plus $500 interest

A.3. Construct a cash flow diagram illustrating the cash flows involved in the following transaction from the borrower's viewpoint. The amount borrowed is $2,000 at 10 percent for five years.

(a) Year-end payment of interest only. Repayment of principal at the end of the five years.

(b) Year-end repayment of one-fifth of the principal ($400) plus interest on the unpaid balance.

(c) Lump sum repayment at EOY 5 of principal plus accrued interest compounded annually.

(d) Year-end payments of equal installments as in a standard installment loan contract (by methods developed in later chapters, the amount of each installment will be $527.60).

Problem Group B

Given three of four variables, find P, F, or i.

B.1. A contractor offers to purchase your old D-9 tractor dozer for $10,000 but cannot pay you the money for 12 months. If you feel $i = 1$ percent per month is a fair interest rate, (a) what is the present worth to you today of the $10,000 if it is paid 12 months from now? (How much should be deposited today into an account bearing interest at 1 percent per month in order to accumulate to $10,000 in 12 months from now?)

(*Ans.* $8,874)

B.2. If the starting salaries of engineering graduates are expected to increase at the rate of 12 percent per year, using the most recent average starting salary, what will be the average starting salary (a) next year, (b) 5 years from now, and (c) 25 years from now?

B.3. A firm wants to lease some land from you on a 20-year lease and build a warehouse on it. As your payment for the lease, you will own the warehouse at the end of the 20 years, estimated to be worth $20,000 at that time.

(a) If $i = 8$ percent, what is the present worth of the deal to you?

(*Ans.* $P = $4,291$)

(b) If $i = 2$ percent per quarter, what is the present worth of the deal to you?

(*Ans.* $P = $4,102$)

B.4. A new branch of the Interstate is expected to open a new area for tourist trade five years from now. Your client expects to construct a $1,000,000 facility at that time. She has funds available now that can be invested at 12 percent compounded annually. How much should she so invest in order to have the $1,000,000 ready at the end of five years?

B.5. In payment for engineering services, a client offers you a choice between (a) $10,000 now and (b) a share in the project, which you are fairly certain you can cash in for $15,000 five years from now. With $i = 10$ percent, which is the most profitable choice?

B.6. The increase in new purchase price of a certain model of a grader has averaged an annual rate of 9 percent compounded. If this model can be purchased right now for $12,000 new, what is the expected price for a replacement in five years if the price increase curve continues on the same trend?

(*Ans. F* = $18,460)

B.7. Your client has 1,000 acres of "average" residential-development type land now worth $3,000 per acre. What should it be worth in five years if it appreciates at the rate of 8 percent compounded (a) annually? (b) quarterly? (c) monthly? (d) continuously?

B.8. A reliable client owes you $10,000 for professional services you have rendered. He is short of cash, but offers you an IOU payable in one lump sum five years from now, with interest compounding at 15 percent. How much would you receive at EOY 5?

(*Ans. F* = $20,110)

B.9. A sum of $1,000 is deposited in a savings account.
(a) If the account earns 8 percent nominal interest compounded quarterly, how much interest is earned (i) after one year? (ii) after five years?
(b) If the account earns 9 percent interest compounded continuously, how much interest would be earned (i) after one year? (ii) after five years (answer in terms of $)?

Problem Group C

Find i (actual, nominal, effective) and also P and F.

C.1. Assume an Indian head nickel, saved when new 60 years ago, will sell now for $2.00. What compound annual interest rate has been earned?

(*Ans. i* = 6.34%)

C.2. In 11 years the Consumer Price Index (a commonly used measure of inflation) climbed from 100 to 193.
(a) What was the annual compound growth rate?
(b) If it continued compounding at the same annual rate, how high would the index be in 11 more years?
(c) Why cannot the increase for the first 11 years simply be doubled to find the increase for the 22-year period?

C.3. Twenty years ago laborers in a certain city were earning $2.00 per hour. Now they earn $20.00 per hour.
(a) What is the compound annual growth rate?

(*Ans. i* = 12.2%)

(b) If they are able to sustain the same growth rate in the future, what pay will they earn 20 years from now?

(*Ans. F* = $200.00/hr)

C.4. As a contractor, you need to buy a dragline. The equipment dealer knows you well and offers to take either $200,000 cash now, or a note for $300,000 payable in one lump sum five years from now. What annual compound interest rate would the note be equivalent to?

C.5. Assume that 22 years ago the sum of $10,000 was invested in a certain mutual fund, that all dividends were reinvested, and that the total investment is worth $104,153 today. What compound rate of increase is this?

(*Ans. i* = 11.24%)

C.6. A rapidly growing city with a current population of 50,000 wishes to control its growth rate so that it will not exceed 100,000 population in 20 years. The regulation will occur by limiting the number of building permits for dwelling units. They now have 15,625 dwelling units, which yields an average of (50,000/ 15,625 =) 3.2 persons per dwelling unit.

    (a) Assuming discrete EOY compounding, what should be the annual planned rate of growth expressed in percent?

    (b) How many building permits should be issued this year?

    (c) At the end of the tenth year what will the population be?

    (d) How many permits should be issued during the eleventh year?

C.7. A type of calculator which sold new for $130 two years ago has declined to a selling price new of $59.95 today. What negative percent rate of change per year does this decline in price represent?

(*Ans. i* = −32.1%)

C.8. A $1,000 investment earns $20 in interest each quarter. This principal remains intact.

    (a) What actual rate of interest is earned?

    (b) What nominal annual rate of interest is earned?

    (c) What annual effective rate of interest is earned?

C.9. A major credit card company charges 1.5 percent interest on the unpaid balance each month.

    (a) What nominal annual interest rate is charged?

    (*Ans. i_n* = 18%)

    (b) What effective annual interest rate is charged?

    (*Ans. i_e* = 19.56%)

Problem Group D

Find *F, P, i,* and *n* with *discrete* compounding.

D.1. Assume a medium-sized town now has a peak electrical demand of 105 megawatts increasing at an annually compounded rate of 15 percent. Assume the generating capacity is now 240 megawatts.

    (a) How soon will additional generating capacity be needed on-line?

    (b) If the new generator is designed to take care of needs five years past the on-line date, what size should it be? Assume the present generators continue in service.

D.2. Assume that the population of a certain country is now 218,000,000 and will increase this coming year by 1,700,000.

    (a) What is the current rate of increase?

    (*Ans. i* = 0.78%)

    (b) If the increase continues to compound annually at that rate, how long before it reaches 300,000,000?

    (*Ans. n* = 41.1 yr)

D.3. If the birthrate in a certain country is 35 per 1,000 population, and the death rate is 12 per 1,000, how many years does it take for the population to double, assuming no net loss to emigration?

D.4. A city now has a water treatment plant with a capacity of 10 million gallons per day (MGD). The peak actual consumption this year is recorded as 8.7 MGD. Each year the peak consumption is expected to increase by 7.2 percent compounded annually. A consultant recommends construction of an addition to the existing plant that would increase capacity to a total of 30 MGD. Find how many years it will be before this 30-MGD capacity is fully utilized if growth in peak demand continues at this annual rate.

(*Ans.* $n = 17.8$ yr)

D.5. A county school district now has two high schools with a total student enrollment capacity of 4,000 students. The actual enrollment this year is 2,200 students. The school enrollment is expected to increase by 3.80 percent compounded annually. How long will it be before this total capacity is fully utilized?

D.6. The population of a certain sunbelt state is estimated at 9,540,000 this year, with 4,500 new residents added each week.
(a) What nominal annual growth rate in percent is this?
(b) What effective annual growth rate in percent is this?
(c) How long before the population reaches 10.5 million if the growth continues at the same rate?
(d) How long until reaching 15,000,000?

D.7. Assume a local government agency asks you to consult regarding acquisition of land for recreation needs for the urban area. The following data are provided:

urban population 10 yr ago $= 49,050$
urban area population now $= 89,920$
desired ratio of recreation land in acres per
   1,000 population $= 10$ acres/1,000
actual acres of land now held by local government
   for recreational purposes $= 803$ acres

(a) Find the annual growth rate in the urban area population by assuming growth has compounded discretely at year's end over the past 10 years.

(*Ans.* $i = 6.13\%$)

(b) How many years from now will it be before the desired ratio of recreation land per 1,000 population is exceeded if no more land is acquired and the population continues to grow at its present rate.

(*Ans.* $n = 1.71$ yr)

(c) The local government is planning to purchase more land to supply the recreational needs for 10 years past the point in time found in (b) above. How many acres of land should they purchase in order to maintain the desired ratio, assuming the population growth continues at the same rate?

(*Ans.* $F = 653$ acres more)

D.8. The school board asks your help in determining the size and timing of a proposed increase in classroom space over the coming years. Assume end-of-year compounding of growth. ($ft^2 =$ square feet of floor area)

classroom space needed per student $= 25$ ft^2/student
present number of students $\qquad = 21,286$
number of students 4 yr ago $\qquad = 18,142$
classroom space in existing buildings $= 600,000$ ft^2

(a) Find the discrete end-of -year growth rate in number of students over the past four years.

(*Ans.* $i = 4.08\%$)

(b) If the number of students continues to grow at the present rate, how long before the design capacity of the existing classrooms is reached?

(*Ans.* $n = 3.00$ yr)

(c) The plans for the new classrooms should allow enough space to accommodate the projected growth for the five-year period following the date of completion. How many square feet of new classrooms will be needed assuming the growth in number of students continues at the same rate?

(*Ans.* $F = 132,550$ ft^2 added)

D.9. A university asks your help in determining the size and timing of a proposed increase in bicycle parking facilities on campus over the coming years. Assume end-of-year compounding of growth.

bicycle parking space needed per student $= 0.25$ spaces/student
present number of students $\qquad = 14,200$
number of students 5 yr ago $\qquad = 10,350$
parking spaces now existing $\qquad = 4,100$ spaces

(a) Find the growth rate in number of students over the past five years in terms of the discrete end-of-year compounding rate, i.

(b) If the number of students continues to grow at the present rate, how long before the capacity of the existing parking facilities is reached?

(c) The new parking facilities should contain enough space to accommodate the growth over a five-year period from the date of completion. How many new spaces will be needed, assuming the growth in number of students continues at the same rate? The date of completion of the new parking facilities is scheduled as the same date on which the capacity of existing facilities is reached.

Problem Group E

Find F, P, i, and n with *continuous* compounding.

E.1. Eighty-five years ago the world's population was estimated as 1.0 billion people. Ten years ago the population was estimated at 4.0 billion. Now the population is estimated at 5.0 billion.

(a) Compare the rate of increase for the earlier 75-year period to the recent 10-year period (use continual compounding).

(b) If the i found in part (a) for the most recent period remains constant, how long will it take for the population to double?

(c) There are about 52 million square miles of land area on the earth. Assume about one-half of this area is arable and that with current technology an average of 1 acre is required to grow food for each person. How long before the population reaches the limit that can be sustained by the available food, using current technology, and assuming that the rate of increase continues as found in part (a) (640 acres = 1 square mile)?

E.2.   You are called upon to consult with a nearby municipality concerning plans for a future expansion of their waste-water treatment plant. The following data are provided (MGD = million gallons per day).

present population	= 87,452
waste-water treatment demand 5 yr ago	= 13.2 MGD
waste-water treatment peak demand at present	= 18.2 MGD
design capacity of present waste-water	
treatment plant	= 23.2 MGD

Assume that growth in peak demand is compounding continuously each year.

(a) Find the rate of growth in peak demand for the past five years.

(*Ans. i = 6.42%*)

(b) If the peak demand continues to increase at the present rate, how long before the present plant reaches design capacity?

(*Ans. n = 3.78 yr*)

(c) To supplement (not replace) the existing plant the municipality wants to plan for a new plant that will provide adequate peak demand capacity for 10 years after the present plant reaches design capacity. For what capacity should the new supplemental plant be designed, assuming the same growth rate continues.

(*Ans. F = 20.9 MGD added*)

E.3.   The airport passenger terminal for a certain city will soon need expansion. The following data are provided. Assume all growth is continuously compounded (ppd = passenger per day).

peak passenger count 5 yr ago	= 1,760 ppd
peak passenger count now	= 2,952 ppd
capacity of existing passenger terminal	= 3,400 ppd
present population of the city	= 162,111

(a) Find the growth rate in peak passenger count over the past five years in terms of the continuously compounding rate, i.

(*Ans. i = 10.3%*)

(b) Assuming the peak passenger count continues to grow at the same rate, how long before the capacity of the existing terminal is reached?

(*Ans. n = 1.37 yr*)

(c) If the new addition to the existing terminal is planned to accommodate the needs of the city for 10 years following the date at which the capacity of the existing terminal is reached, how much of a peak passenger count should the new addition (only) be designed for in terms of ppd?

(*Ans. F = 6,163 ppd*)

E.4.   The electrical generating plant for a certain city will soon reach capacity demand load and will require a new addition due to anticipated growth in demand. The following data are provided. Assume all growth is continuously compounded (MW = megawatts, a measure of electrical demand).

peak electric demand 5 yr ago   = 150 MW
peak electric demand now   = 210 MW
capacity of existing generating plant = 240 MW
present population of the city   = 62,111

(a) Find the growth rate in peak demand over the past five years in terms of the continuously compounding rate, i.

(b) Assuming the peak demand load on the plant continues to grow at the same rate, how long before the capacity of the existing plant is reached?

(c) If the new addition to the existing plant is planned to accommodate the needs of the city for 10 years following the date at which the capacity of the existing plant is reached, how much load should the new addition be designed for, in terms of MW?

Problem Group F

Continuous and discrete compounding: Find F, i, n.

F.1. Assume a certain two-lane road has a traffic volume of 10,000 cars per day, increasing at a compound rate of 10 percent per year. It will be two years before a new road can be built, and the new facility should be designed to be adequate for an additional 15 years.

(a) If the simplifying assumption is made that each two-lane pair (one lane in each direction) can carry 15,000 cars per day, how many lanes should be provided?

(b) How many lanes are needed if the compounding is continuous at 10 percent rather than annual?

(*Ans.* (a) $F = 3.36$, use 4 pairs, (b) $F = 3.64$, use 4 pairs)

F.2. Assume that 362 years ago Peter Minuit, governor of the Dutch West India Company, bought Manhattan Island from the Indians for $24 worth of beads, cloth, and trinkets.

(a) If the Indians had insisted on $24 cash and invested the $24 at 6 percent, what would the compounded amount be worth now?

(b) If the compounding were continuous instead of annual, how much would the $24 investment be worth now?

(c) Assume the actual value of Manhattan Island, land only (no improvements), is estimated to be $28 billion. What is the average percent increase in value with annual compounding each year?

F.3. If a city's current population is 65,000 and compounding at the rate of 5 percent per year, (a) what will the projected population be in 10 years? (b) How long before it reaches 200,000 at this rate? (c) What will the projected population be in 10 years if the compounding is continuous rather than annual?

(*Ans.* (a) $F = 105,900$, (b) $n = 27.0$ yr, (e) $F = 107,200$)

F.4. The county engineer asks your help on sizing a replacement for a two-lane highway bridge. The following information is provided.

present daily traffic count at bridge   = 5,000
daily traffic count at bridge 5 yr ago   = 4,100
design capacity of bridge (daily)   = 6,200
number of vehicles registered in county = 32,000

Assume continuous compounding of traffic growth.

(a) Find the growth rate of daily traffic over the past five years in terms of continuously compounding rate, i.

(b) If the traffic count continues to increase at the rate found above, when will the design capacity of the bridge be reached?

(c) The replacement bridge will be designed to accommodate the growth in traffic over a 20-year period from the date the design capacity on the present bridge is reached. Assuming the same rate of growth, what should the design capacity be in terms of traffic count?

F.5. Assume you are called to consult with a nearby municipality concerning plans for a future expansion of their library.

number of books in the library	$= 237{,}411$
city population 5 yr ago	$= 87{,}120$
city population now	$= 124{,}760$
recommended ratio of square feet of library floor space	
to population	$= 0.5 \text{ ft}^2/\text{citizen}$
size of present library	$= 75{,}000 \text{ ft}^2$

(a) Find the continuously compounded growth rate, i, in the city population over the past 5 years.

(*Ans.* $i = 7.18\%$)

(b) Assume the city maintains the recommended ratio of library floor space to population. How many years from now will the city reach the capacity of their present library?

(*Ans.* $n = 2.57$ years)

(c) To supplement (not replace) the existing library, the city wants to plan for an addition that will provide adequate capacity for 15 years after the present library reaches design capacity. How many square feet should be added, assuming the same rate of continuously compounding growth?

(*Ans.* $F =$ add $145{,}200 \text{ ft}^2$)

Problem Group G

Given P_1 and P_2, both increasing at different rates i_1 and i_2, find the n at which $F_1 = F_2$.

G.1. A mass transit terminal for a certain city will soon reach capacity and requires expansion. The following data are provided (ppd $=$ passengers per day).

peak passenger count now	$= 6{,}995$ ppd
current rate of growth of passenger count	$= 6.21\%$ discrete EOY
capacity of terminal, now	$= 9{,}000$ ppd

Due to increased operating efficiencies, it is estimated that the capacity of the existing terminal can be increased by 1.25 percent per year each year for the foreseeable future.

(a) How long before the increasing peak passenger count reaches the increasing capacity?
(*Ans.* $n = 5.26$ yr)

(b) In terms of ppd, what will the peak passenger count be at that point in time?
(*Ans.* $F = 9{,}603$ ppd)

G.2. The population of India was once estimated at 667,326,000, increasing at a rate of 2.01 percent per year. On the same date, the population of China was estimated at 1,027,000,000, increasing at a rate of 1.71 percent per year. Assume continuous compounding. If the growth in both countries continues to compound at the current rate:

(a) How long will it be until the increasing population of India equals the increasing population of China?

(b) What will be the population of each country at that time?

Problem Group H

Rule of 70.

H.1. Five years ago a water treatment plant was treating 2 MGD average, and now it is treating 4 MGD.

(a) Using the Rule of 70, approximately what rate of annual increase, i, is the plant experiencing?

(b) Use the F/P equation to find the actual annual discrete i value, and compare with the result in (a).

H.2. A property purchased 14 years ago for $50,000 is now selling for $200,000.

(a) Use the Rule of 70 to find the approximate rate of return in terms of annual interest, i, earned on this investment.

(b) Use the F/P equation to find the actual i, and compare with (a).

(*Ans.* (a) $i =$ approx. 10%, (b) $i = 10.41\%$)

H.3. Create a table of values. In the first column list interest rates in i percent. Begin with a figure of 1 percent, then 5 percent, then list values increasing by 5-percent increments to 20 percent. Then increase by increments of 10 percent from 20 to 50 percent. In the second column calculate by the equation the actual doubling time that corresponds with each of the i values listed in column 1. In the third column list the product found by multiplying the doubling time in column 2 by the i percent in column 1. For what i values does the Rule of 70 give an answer within (a) 5 percent of the correct answer; (b) within 10 percent; (c) not within 20 percent?

H.4. A city population is increasing at a rate of 3.5 percent per year. If the same growth rate continues, use the Rule of 70 to find how long before the city (a) doubles in size. (b) If the city now has 100,000 inhabitants and growth continues at the same rate, how long before it reaches 6,400,000?

(*Ans.* $n = 120$ yr)

CHAPTER 3

The Value of One Future Payment Compared to a Uniform Series of Payments (*F/A, A/F*)

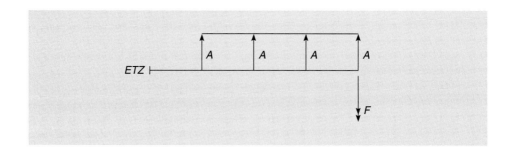

KEY EXPRESSIONS IN THIS CHAPTER

A = Uniform series of n end-of-period payments or receipts, with interest compounded at rate i, on the balance in the account at the end of each period.

ETZ = Equation Time Zero: with reference to a particular payment series, the time at which $n = 0$.

PTZ = Problem Time Zero, the present time according to the statement of the problem narrative.

THE FUTURE SUM OF PERIODIC DEPOSITS

In planning for future expenditures of a city, a private company, or a single person, the need for making large, future lump sum cash outlays can often be foreseen. A city soon may need new buildings, parks, or transportation facilities. A company may need production equipment, a new plant, or store, and a person may be looking ahead to making a down payment on a new car or home, or to providing a comfortable nest egg for retirement. One method of saving for these future lump sum outlays is to periodically deposit uniform amounts into an interest-bearing account. Many individuals have savings accounts to serve this purpose, while businesses and governments often use interest-bearing sinking funds (savings accounts) to accumulate capital.

A simple equation may be derived to relate the future lump sum, F, that will accumulate from a uniform series of n end-of-period deposits, A, with interest compounding at the end of every period at rate i, on all sums on deposit. (As a memory aid, the A used to designate periodic deposits can be associated with Annual deposits, even though periods of other durations are frequently used.)

Generalized Problem

Find the total future lump sum, F, that will accrue by the end of the nth period resulting from n end-of-period deposits, each of amount A, plus interest compounded at the end of every period at rate i.

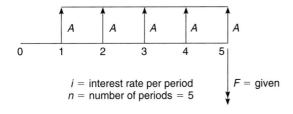

i = interest rate per period
n = number of periods = 5

F = given

Figure 3.1 General form for uniform series cash flow diagrams.

Generalized Solution, Development of F/A Equation

If *no* money is deposited at the *beginning* of the first period, and A_1 dollars are deposited at the *end* of the first period, the future sum F accumulated at the end of that one period is simply $F = A_1$, since A_1 has not had time to earn any interest.

If A_1 dollars are deposited at the end of the first period, and the same amount (A_2 dollars) is deposited at the end of the second period, the future sum F accumulated at the end of the two periods is $F = A_1 + A_1 i + A_2$. Since $A_1 = A_2$ the subscripts are omitted, and this may be written as $F = A[1 + (1 + i)]$. When the interest is paid at the end of each period, the accumulated balance on deposit is multiplied by $(1 + i)$, representing the principal plus interest accruing at the end of the period. In addition, as soon as the interest is paid, another periodic deposit of amount A is made. Thus if A dollars are deposited at the end of each period for n periods, accumulating i interest at the end of each period, the future amount F accumulated at the end of each period for n periods is shown in the following table:

End of period	Accumulated amount at the end of each period
1	$F_1 = A$
2	$F_2 = A(1 + i) + A = A[1 + (1 + i)]$
3	$F_3 = A[1 + (1 + i)](1 + i) + A = A[1 + (1 + i) + (1 + i)^2]$
$\vdots$	
n	$F_n = A[1 + (1 + i) + (1 + i)^2 + \cdots + (1 + i)^{n-1}]$

The general form of this equation for n periods is

$$F = A[1 + (1+i) + (1+i)^2 + \cdots + (1+i)^{n-1}] \tag{3.1}$$

The reduction of the general form to a simpler, more practical form is performed by the following sequence:

1. Multiply both sides by $(1+i)$.

$$F(1+i) = A[(1+i) + (1+i)^2 + \cdots + (1+i)^n] \tag{3.2}$$

2. Subtract Equation 3.1 from Equation 3.2.

$$Fi = A[-1 + (1+i)^n] = [(1+i)^n - 1]$$

3. Multiply both sides by $1/i$ to get the equation relating F, A, n, and i.

$$F = A\left[\frac{(1+i)^n - 1}{i}\right] \quad \text{or} \quad F/A = \frac{(1+i)^n - 1}{i} \tag{3.3}$$

Note once again that the factor $[(1+i)^n - 1]/i$ is independent of the amount of either F or A. This factor is sometimes termed the uniform-series, compound-amount factor (USCAF).

Graph of *F/A* Versus *n*

A graph of the equation $F/A = [(1+i)^n - 1]/i$ is shown in Figure 3.2. Note the rapid increases in future value, F, with increases in either n or i.

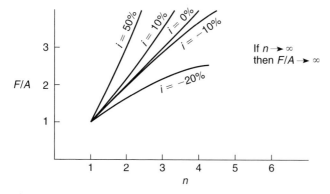

Figure 3.2   Graph of the equation F/A.

Graph of the Balance in the Account, *F* Versus *n*

The balance-in-account graph is shown in Figure 3.3, which illustrates the timing of the cash flows of *A* related to *F*. Note that the first deposit of *A* occurs at the *end* of the first time period. Therefore, the first interest is not earned until the end of the second time period by definition. At the end of the last (*n*th) period there occurs both a final deposit *A* and a final interest payment. It is important to remember that the *F/A* series begins at a point in time that is one time period before the first periodic payment *A* is made. In other words, equation time zero for the *F/A* series occurs one time period *before* the first periodic payment *A* occurs.

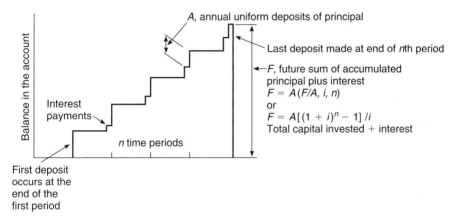

Figure 3.3 Balance-in-account graph showing the timing of cash flows.

Notation

Rather than write out the equation each time, a briefer notation format is frequently used:

$$F = A(F/A, i, n) \tag{3.3a}$$

Cash Flow Diagrams

The general form of the cash flow diagram illustrating the *F/A* relationship is shown in Figure 3.4. *The double arrow notation indicates an unknown value to be found (unknown dependent variable).* If any three of the four variables are known or determinable, the fourth can be found. Therefore, each of the four variables may at some time appear as the unknown dependent variable.

Note that the first periodic deposit, *A*, occurs at the *end* of the first period. Also note that the future value, *F*, occurs *simultaneously* with the last periodic deposit, *A*. Students

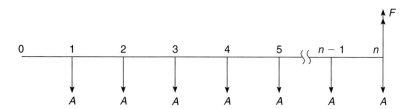

Figure 3.4   Equivalent future value (F) from a series of uniform periodic amounts (A).

are cautioned against common errors illustrated in the incorrect F/A cash flow diagrams that follow (Figures 3.5 and 3.6).

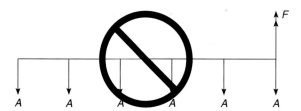

Figure 3.5   Incorrect representation of F/A relationship, with extra (initial) deposit.

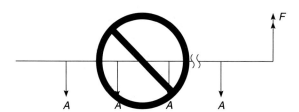

Figure 3.6   Incorrect representation of F/A relationship, with missing (final) deposit.

A/F, or Uniform Series Sinking Fund Factor (USSFF)

The reciprocal of the F/A is A/F, sometimes referred to as the "Uniform Series Sinking Fund Factor" (USSFF) and written as

$$A/F = \left[\frac{i}{(1+i)^n - 1}\right]$$

(3.4)

In functional notation, this is designated by

$$A = F(A/F, i, n) \tag{3.4a}$$

Note that the expression inside the parentheses, the $(F/A, i, n)$ factor in Equation 3.3a and its reciprocal factor $(A/F, i, n)$ in Equation 3.4a, are dependent on the values of i and n *only,* and are independent of the values for F or A. For convenience, values of $(A/F, i, n)$ and $(F/A, i, n)$ have been tabulated in Appendix A for commonly used combinations of i and n.

SIGN OF *F* VERSUS SIGN OF *A*

In most cash flow diagrams the future value arrow for F is drawn on the opposite side of the horizontal time line from the arrows representing uniform periodic payments, A. If A represents the deposits and F the lump sum withdrawal, this opposite sign is correct. Similarly, if an upward A arrow represents periodic borrowing, then a downward F arrow depicts the lump sum repayment (or balance owed) at the end of the series. However, sometimes the F arrow and the A arrow may appear on the same side of the horizontal time line. This may occur to show a lump sum future value that is *equivalent* to an A series of deposits or withdrawals, and has the same sign as the periodic payments, A. In such cases it is correct to show all arrows on the same side of the cash flow diagram horizontal time line. Example 3.1 illustrates this.

EXAMPLE 3.1   (Find *F*, given *A, i, n*)

Assume your banker is willing to accept equivalent quarterly payments (one payment every three months) on a home mortgage instead of the usually stipulated monthly payments of $500 per month. If the interest rate is 1 percent per month, what would be the equivalent quarterly payments? The cash flow diagram for this problem is shown in Figure 3.7.

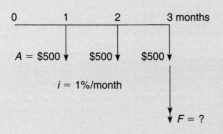

Figure 3.7   Cash flow diagram for Example 3.1. Note that the arrow for the equivalent future value F points in the same direction as the periodic values A.

Solution.　Using the equation, we have

$$F = A[(1 + i)^n - 1]/i$$
$$F = 500[(1.01)^3 - 1]/0.01 = 500 \times 3.0301 = \$1,515$$

Conclusion.　The payment of $1,515 per quarter gives the banker the equivalent of three monthly payments of $500 per month.

APPLICATIONS FOR THE *F/A* AND *A/F* EQUATIONS

F/A problems involve the four variables, F, A, i, n. Whenever any three of the variables are known, the fourth may be determined. The following four examples illustrate typical situations involving each of the four variables, in turn, as the unknown.

EXAMPLE 3.2　(**Find *F*, given *A*, *i*, *n***)

Aiming to replace their mainframe computer five years from now, a firm of consulting engineers is depositing $10,000 per year into an investment fund earning interest compounded at 10 percent annually. Find the balance in the fund (including deposits and accumulated interest) at the end of five years. Figure 3.8 shows the cash flow diagram.

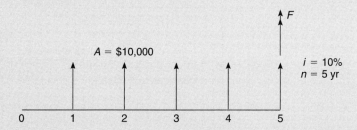

Figure 3.8　Cash flow diagram for Example 3.2.

Alternate Solution 1.　Using the equation:

$$F = A[(1 + i)^n - 1]/i$$
$$F = \$10,000[(1.10)^5 - 1]/0.10$$
$$F = \$10,000 \times 6.1051 = \$61,051$$

Alternate Solution 2.   Using the tables: For many problems a solution using the tables for F/A and A/F in Appendix A saves time. The value of $[(1+i)^n - 1]/i$ is represented by F/A, and the inverse is represented by A/F. Each table contains values for only one interest rate, i. The time periods, n, are shown in the left and right outer columns of each table.

   Example 3.2 could be solved by the following step-by-step procedure:

1. Find the table in Appendix A for $i = 10$ percent.

2. Proceed down the n column to find the row where $n = 5$.

3. At the $n = 5$ row proceed horizontally to the F/A column, finding $F/A = 6.1051$.

4. Since $F/A = 6.1051$, then $F = A \times 6.1051$. Since $A = \$10,000$, then $F = \$10,000 \times 6.1051 = \underline{\underline{\$61,051}}$.

EXAMPLE 3.3   **(Find A, given F, i, n)**

A young engineer, now 22, expects to retire in 40 years at age 62. He antici- pates that a lump sum retirement fund of \$400,000 will see him nicely through his sunset years. How much should be deposited annually into an investment fund earning 9 percent compounded for the next 40 years to accumulate a \$400,000 retirement fund? (See Figure 3.9.)

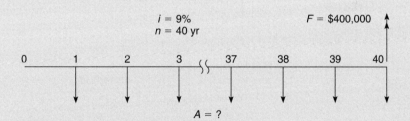

Figure 3.9   Cash flow diagram for Example 3.3.

Solution.   Using the 9% table in Appendix A:

$$A = F(A/F, i, n)$$
$$A = \$400,000(A/F, 9\%, 40)$$
$$A = \$400,000(0.00296)$$
$$A = \underline{\underline{\$1,184/\text{yr}}}$$

EXAMPLE 3.4   (Find n, given A, F, i)

A city finances its park acquisitions from a special tax of $0.02 a bottle, can, or glass on all beverages sold in the city. It derives $100,000 a year from this tax. The city has an option to buy a 40-acre lakefront park site for $1,000,000 as soon as sufficient funds accumulate. How soon will this be if the tax receipts are invested annually at the end of each year at 6 percent?

Alternate Solution 1.   By logarithms:

$$F/A = [(1+i)^n - 1]/i$$

Therefore

$$n = \frac{\ln[(Fi/A)] + 1}{\ln(1+i)}$$

$$n = \frac{\ln\left(\dfrac{1,000,000 \times 0.06}{100,000} + 1\right)}{\ln(1 + 0.06)} = 8.07 \text{ yr}$$

The same result may be obtained by an alternative method.

Alternate Solution 2.   By interpolation of the tables in Appendix A (see Figure 3.10):

$$A = F(A/F, i, n)$$
$$100,000 = 1,000,000(A/F, 6\%, n)$$

Then $(A/F, 6\%, n) = 100,000/1,000,000 = 0.1000$. Find the following in the 6 percent table:

$$
\begin{array}{lll}
(A/F, 6\%, 8) = & 0.1010 & 0.1010 \\
(A/F, 6\%, n) = & -0.1000 & \\
(A/F, 6\%, 9) = & \underline{\hphantom{000000}} & \underline{-0.0870} \\
& 0.0010 & 0.0140
\end{array}
$$

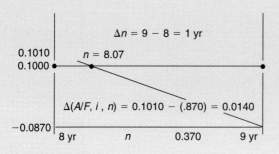

$\Delta n = 9 - 8 = 1$ yr

0.1010
0.1000 $n = 8.07$

$\Delta(A/F, i, n) = 0.1010 - (.870) = 0.0140$

−0.0870

8 yr n 0.370   9 yr

Figure 3.10   Graphic interpolation for Example 3.4.

Interpolation yields

$$n = 8 + \frac{0.0010}{0.0140} \times 1 = \underline{\underline{8.07 \text{ yr}}}$$

EXAMPLE 3.5   (Find *i*, given *A, F, n*)

Referring to Example 3.4, the Recreational Advisory Board feels that the park land should be acquired no later than seven years from now, and that a financial consultant should be engaged who can increase the interest income on invested special tax funds. How much would the interest have to increase in order for the fund to reach $1,000,000 by EOY 7?

Solution.   Since *i* appears twice in the equation, the answer can more readily be found by interpolation of the tables.

$$A = F(A/F, i, n)$$
$$100{,}000 = 1{,}000{,}000(A/F, i, 7)$$
$$(A/F, i, 7) = 0.100$$

Find the two tables where $(A/F, i, 7)$ brackets 0.100. These turn out to be the 11 percent table and the 12 percent table, with values listed as shown.

$$
\begin{array}{lll}
(A/F, 11\%, 7) = & 0.10222 & 0.10222 \\
(A/F, i, 7) & = -0.10000 & \\
(A/F, 12\%, 7) & & = -0.09912 \\
\hline
& 0.00222 & 0.00310
\end{array}
$$

Interpolation yields

$$i = 11\% + \frac{0.00222}{0.00310} \times 1\% = 11.72\% = \underline{\underline{11.7\%}}$$

Note: Rectilinear interpolation of this curved function results in a slight error (should be $i = 11.71$ percent), so the answer is rounded as shown.

WHEN PROBLEM TIME ZERO DOES NOT CORRESPOND WITH EQUATION TIME ZERO (PTZ≠ETZ)

As explained previously, the beginning point in time for the *F/A* equation is one compounding period of time before the first payment in the series, as shown in Figures 3.3 and 3.4. Many problem situations occur where the beginning time for the problem (*Problem*

Time Zero, or PTZ) does not coincide with the beginning time for the equation (*Equation Time* Zero, or ETZ). The general approach is to subdivide the problem into simple components, solve each component, and sum the results. The following examples illustrate the method.

EXAMPLE 3.6 (PTZ≠ETZ, series of $n = 20$ Monthly Payments, problem $n = 30$)

A young engineer decides to save for a down payment on a new car to be purchased 30 months from now. He decides to deposit $50 per month into a savings account which earns 6 percent nominal interest compounded monthly (use $i = 6\%/12 = 0.5\%/$ month). He plans to make a series of 20 regular monthly deposits, with the first deposit scheduled for one month from today. After EOM 20, the account will receive no more deposits but will receive monthly interest compounded on the balance at the end of each month for another 10 months. How much will be in the account at the end of the 30-month period?

Solution. A cash flow line diagram illustrates the cash flow of the deposits. The problem is divided into two parts and solved, and the results summed (see Figure 3.11).

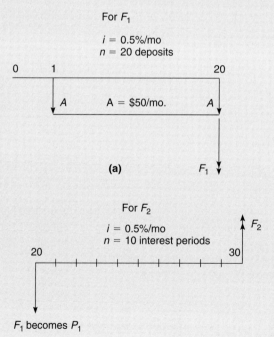

Figure 3.11 Cash flow diagrams for Example 3.6.

1. First find the lump sum future amount, F_1, in the account at the end of the first 20-month period using Equation 3.1 (or the 0.5 percent table).

$$F_1 = A\left[\frac{(1+i)^n - 1}{i}\right] = 50\left[\frac{(1.005)^{20} - 1}{0.005}\right] = \$1,049$$

2. Second, find the future amount, F_2, that will accumulate as a result of just the lump sum F_1 balance at EOM 20, plus 10 additional monthly interest payments (without any more deposits for the final 10-month period). For this equation, F_1 becomes the initial deposit, P_1 at ETZ.

$$P_1 = F_1$$
$$F_2 = P_1(1+i)^n = \$1,049(1.005)^{10} = \underline{\underline{\$1,103}}$$

Looking back at the solution of this example, it is evident that the procedure can be greatly simplified by combining the three lines of equations that were used into one line of calculations as follows:

$$F_1 = \$50(F/A, 0.5\%, 20)(F/P, 0.5\%, 10) = \underline{\underline{\$1,103}}$$

Another frequently encountered type of problem where problem time zero does *not* correspond with equation time zero (PTZ $\neq$ ETZ) occurs when the initial payment in a series occurs at the *beginning* of the first period rather than the *end* of the first period. This case is easily handled by locating the equation time zero (ETZ) one period to the left of problem time zero (PTZ); that is, find ETZ for the F/A equation at one compounding period before the first deposit is made. Then n for the equation is counted from that ETZ, as in Example 3.7.

EXAMPLE 3.7   **(PTZ $\neq$ ETZ, and the first payment in the series occurs at PTZ)**

Assume a proud father bought his son a $100 bond on the day he was born and continued buying an additional $100 bond on every birthday. How much is the total value of the investment on his twenty-first birthday after the twenty-second bond has been received? Assume all the bonds accumulate 6 percent interest compounded annually.

Solution.   First draw a cash flow diagram as shown in Figure 3.12. Since the initial deposit occurs at PTZ, then ETZ is found one year earlier, at a time one year before the baby is born. The cash flow diagram showing both problem

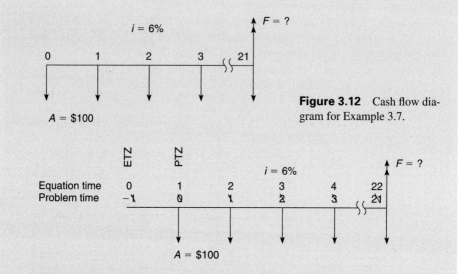

Figure 3.12   Cash flow diagram for Example 3.7.

Figure 3.13   Cash flow diagram for Example 3.7 showing ETZ and PTZ.

time and equation time is shown in Figure 3.13. Counting from the new ETZ, $n = 22$ when the twenty-first birthday occurs. The equation becomes

$$F = A(F/A, i, n) = 100(F/A, 6\%, 22) = \underline{\underline{\$4,339}}$$

TWO OR MORE SEQUENTIAL SERIES

Most problems that at first seem complex can be broken down into relatively simple components. For instance, a problem may be composed of more than one series of periodic payments. These may be treated very simply as two or more separate series (split series). Each series is solved independently of the other, and the results are summed.

EXAMPLE 3.8   (**Two sequential series with different *i* values**)

An owner of a building expects the present roof to last nine more years and then need replacement at EOY 9. Anticipating this expenditure, five years ago the owner began depositing $1,000 per year into an investment account bearing interest at 6 percent per year. It is now EOY 5, and the fifth annual deposit has just been made. The bank announces that beginning today, all funds on deposit will bear interest at 7 percent. The owner, realizing that costs are increasing, decides to raise the amount of the annual deposits to $2,000 per year, beginning with the deposit due one year from today. As advisor to the owner, you are requested to quickly calculate how much the account will total at the end of nine more years, assuming the deposits remain at $2,000 per year and interest remains at 7 percent.

Solution. The cash flow line diagram illustrating this example is shown in Figure 3.14 . The problem may be divided into three parts:

1. The amount on deposit at the end of five years is F_1.

$$F_1 = \$1,000(F/A, 6\%, 5) = \$5,637$$

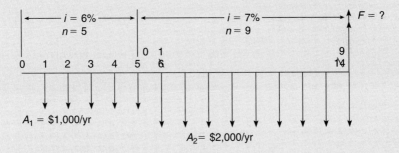

Figure 3.14 Cash flow diagram for Example 3.8.

2. This amount, $5,637, may now be considered as the present value, P_1, of a lump sum deposit that begins drawing interest at 7 percent from EOY 5 until EOY 14, or $n = 9$. The future amount that accumulates at EOY 14 from this one source is

$$F_2 = \$5,637(F/P, 7\%, 9) = \$10,364$$

3. The annual deposits are raised to $2,000 from EOY 6 until EOY 14, with $i = 7$ percent on the accumulated balance. This results in the accumulation of

$$F_3 = \$2,000(F/A, 7\%, 9) = \$23,956$$

The total amount on hand at EOY 14 is

$$F_2 + F_3 = \$10,364 + \$23,956 = \underline{\$34,320}$$

Example 3.8 was divided into three different parts. Each part was brought to a common point in time (in this case the F values at EOY 14). Then the F

values representing lump sum equivalents at EOY 14 were summed, and this sum is the total amount on deposit at EOY 14. To sum any combination of F values (or P values) **each value must be transferred to a common point in time** on the cash flow diagram. This series of equations could be simplified and telescoped into three lines as follows:

$$
\begin{aligned}
F_2 &= \$1,000(F/A, 6\%, 5)(F/P, 7\%, 9) = \$10,364 \quad \text{(combining steps)} \\
F_3 &= 2,000(F/A, 7\%, 9) = 23,956 \\
F_{\text{total}} & = \$34,320
\end{aligned}
$$

NEGATIVE INTEREST RATE

Problems sometime occur where the interest rate (or equivalent growth rate) is negative. For instance, an investment that turns out poorly may have a negative net cash flow (you paid in more than you received out); thus it yields a negative rate of return. Most investors who take risks typically have experienced this type of unfortunate investment at one time or another.

Population growth patterns can also yield a negative rate of increase. For instance, consider a retirement area with a large number of older people. Due to the large proportion of older residents, the birth rate may be lower than the death rate, leading to a negative natural increase. The natural rate of increase is often simulated by compound interest, so a negative increase may be simulated by a negative interest rate. For instance, if the annual birth rate is 8 per 1,000 population and the annual death rate is 14 per 1,000 population, then the natural increase is a negative 6 per 1,000 population. Thus, *if there were no immigration* into the area from outside, the population would decline at the rate of 0.6 percent per year $(+0.8\% -1.4\%)$ compounded. However, if there *were* immigrations into the area, and if the incoming residents were similar to those already there, and if these new residents continued to move into the area at a fairly constant rate (same number of new people moving in per year), then estimates of future population may be made using the F/A series equation. The following example illustrates the method.

EXAMPLE 3.9　(Negative interest rate)

A new retirement community will need to add a second unit to the new water treatment plant when the population reaches 40,000. The community has just been built and is starting out now with a zero population. On average, 6,000 new residents are expected to move into the community each year. The estimated birthrate for the community is 8 per 1,000, and the death rate is 14 per 1,000 resulting in a negative annual increase of −0.6 percent. How long will it be before the second unit is needed for the new water treatment plant?

Solution. The population must reach 40,000 before the second unit is needed, so that $F = 40,000$. The number of new residents expected each year is 6,000, so $A = 6,000$. The residents that are there should have a natural rate of increase of −0.6 percent, so $i = -0.006$.

A diagram similar to the cash flow line diagram may be drawn for this problem, as shown in Figure 3.15. The similarity to previous problems involving deposits into an interest-bearing account should be apparent. In this case, the accumulated amount "on deposit" loses at rate i instead of gaining at rate i.

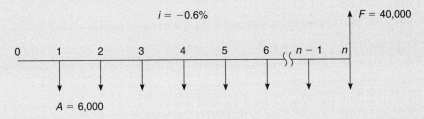

Figure 3.15   Cash flow diagram for Example 3.9.

To solve for the number of years, n, that will yield a net population increase of 40,000, use the F/A equation, as follows.

$$n = \frac{\ln\left(\dfrac{Fi}{A} + 1\right)}{\ln(1 + i)}$$

$$n = \frac{\ln\left(\dfrac{40{,}000 \times (-0.006)}{6{,}000} + 1\right)}{\ln(1 - 0.006)}$$

$$n = 6.78 = \underline{\underline{6.8 \text{ yr}}}$$

Note that since the negative i is small, and n is not large either, the number of years required to reach 40,000 is just a little more than if i were zero ($40{,}000/6{,}000 = 6.67 = 6.7$ yr).

If the retirement community in Example 3.9 already has an existing population at PTZ, the solution involves one more equation and a slightly different logarithmic equation, as follows.

EXAMPLE 3.10   (Find n, given F, P, A, i)

Assume the retirement community already has a base population of 10,000 persons. Now how long will it take to reach a population of 40,000 with all other factors the same as in Example 3.9?

Solution.   Given

$$F = 40,000 \text{ pop.}$$
$$P = 10,000 \text{ pop.}$$
$$A = 6,000/\text{yr}$$
$$i = -0.006$$

Find n.

$$F_1 = P(1+i)^n \qquad \text{for the 10,000 base population}$$

$$F_2 = A\left[\frac{(1+i)^n - 1}{i}\right] \qquad \text{for the 6,000 per year addition}$$

$$F = F_{\text{total}} = F_1 + F_2 = P(1+i)^n + A\left[\frac{(1+i)^n - 1}{i}\right]$$

$$F = P(1+i)^n + \frac{A(1+i)^n}{i} - \frac{A}{i}$$

$$(1+i)^n = \frac{F + A/i}{P + A/i}$$

$$n = \frac{\ln\left(\dfrac{F + A/i}{P + A/I}\right)}{\ln(1+i)}$$

$$n = \frac{\ln\left[\dfrac{40,000 + 6,000/(-0.006)}{10,000 + 6,000/(-0.006)}\right]}{\ln(1 - 0.006)} = 5.11 = \underline{\underline{5.1 \text{ yr}}}$$

RATES, FARES, AND TOLLS

Frequently the cost of financing capital investments is met by charging each individual user a certain amount for each use or unit consumed. The amount charged per use is usually called the fare or toll, while the amount charged per unit consumed is commonly called the rate. For instance, to accumulate funds to build a new water treatment plant, the current water rate charge may be increased by an extra $0.10 per thousand gallons of water, or a sinking fund for a proposed toll bridge may be funded by charging a toll of $0.25 per vehicle.

When the total cost of the facility is known and the rate is unknown, the customary approach is to substitute an unknown y (any letter will do) for the rate and set up an equation. The following example illustrates the procedure.

EXAMPLE 3.11   (Find the required fare increase, y, with sequential series)

A city transit authority is planning on replacing its aging bus fleet six years from now at an estimated lump sum cost of $1,000,000. To accumulate this amount, they decide to raise bus fares by whatever amount is required, with the raised fare going into effect three months from today. The increased fare increment will be deposited monthly at the end of each month of collections (the first deposit will be made four months from today) into a fund bearing interest at 0.5 percent per month, compounded monthly. The number of fares per month for the next six years is estimated in the following table:

During months	Fares per month
4 through 24	140,000
25 through 72	160,000

How much of a fare increase per passenger is necessary in order to reach the desired lump sum? The fare increase per passenger will remain constant until the end of the sixth year (seventy-second month).

Solution.   The amount collected each month equals the number of fares per month times the fare per passenger (designated by y). For the first 21 months of collections the monthly receipts should be

$$A_1 = 140,000y$$

The monthly receipts for the final 48 months are designated as

$$A_2 = 160,000y$$

The cash flow diagram for this problem is shown in Figure 3.16a. The first series involves 21 payments ($24-3 = 21$). A future lump sum F_1 equivalent to this first series is found at EOM 24 as

$$F_1 = 140,000y \underbrace{(F/A, 0.5\%, 21)}_{22.084} = 3,092,000y$$

(Figure 3.16b)

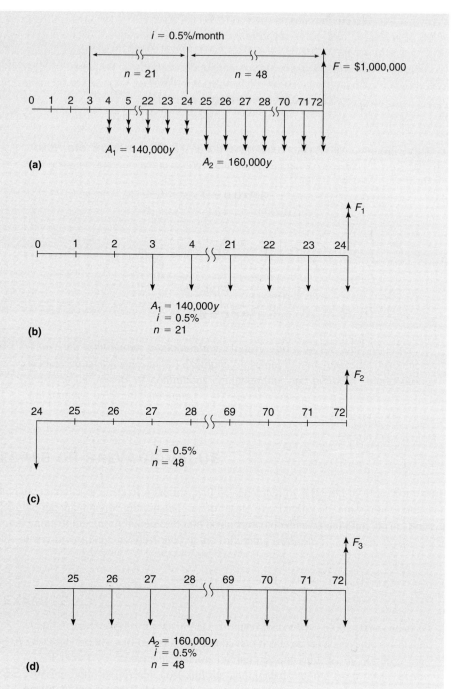

Figure 3.16   Cash flow diagrams for Example 3.11.

Then this lump sum F_1 is treated as P_1 and transferred another 48 months $(72-24=48)$ to an equivalent future lump sum F_2 at EOM 72, as

$$F_2 = 3{,}092{,}000y \underbrace{(F/P, 0.5\%, 48)}_{1.12705} = 3{,}928{,}000y \qquad \text{(Figure 3.16c)}$$

The future value F_2 of the second series of payments A_2 is found at EOM 72, as

$$F_3 = 160{,}000y \underbrace{(F/A, 0.5\%, 48)}_{54.0978} = 8{,}656{,}000y$$

$$F_2 + F_3 = 12{,}584{,}000y \qquad \text{(Figure 3.16d)}$$

The total amount needed at the end of the six-year period is $1,000,000, so $F_2 + F_3$ is equated to this amount and the equation is solved for y. Thus,

$$y = \frac{\$1{,}000{,}000}{12{,}584{,}000} = \$0.079 \text{ or,}$$

the fare must increase by $\underline{\underline{\$0.08}}$ per passenger

NOMINAL AND EFFECTIVE INTEREST RATES

EXAMPLE 3.12 (Find i_n and i_e, given i, n)

Credit card interest often is given as a percent per month. This figure, however, does not represent the effective interest rate that you will pay annually. If the interest rate quoted by a credit card company is 1 percent per month, what annual *nominal* and annual *effective* interest rate would you be paying?

Solution. From Equation 2.3, the annual nominal interest rate, i_n, is

$$i_n = im = 1 \times 12 = \underline{12\%}$$

From Equation 2.4, the annual effective interest rate, i_e, is

$$i_e = (1 + i)^n - 1 = (1.01)^{12} - 1 = 0.1268 = \underline{12.7\%}$$

FUNDS FLOW PROBLEMS

In most business transactions and economy studies, interest is compounded at the end of each compounding period, and cash flows are assumed to occur either as lump sums at the beginning or end of such periods, or as series of end-of-period payments. This will be the case for most situations in this text. However, occasionally different flows occur, as illustrated in the two types of cases described in the following paragraphs.

Case 1:   The Intervals Between Payments in a Series Consist of Several Compounding Periods (interest is compounded more than once between payments in a series)   The cash flow diagram is shown in Figure 3.17 for a typical situation involving more than one compounding period between payment deposits (e.g., monthly compounding, annual payments).

Where a regular series of payments is made at intervals of more than one interest-compounding period, then two common methods are available to find the future sum: the sinking fund method and the equivalent interest method.

Method 1.   Sinking Fund Method   Use the A/F equation (where $n = m$, and m represents the number of compounding periods between payments) to find the periodic payment at the end of every interest-compounding period that is equivalent to the less frequent actual payment.

EXAMPLE 3.13   **(The period between periodic payments is some whole multiple of the compounding period)**

The following problem and cash flow diagram represent a requirement that $1,200 be paid annually. If you were to pay monthly with compounding monthly at $i = 1$ percent per month, find the future lump sum equivalent at EOM 36.

Solution.   The cash flow diagram appears as shown in Figure 3.17. Find the monthly payment that is equivalent to the annual payment of $1,200.

$$A = F(A/F, i, n) = \$1,200(A/F, 1\%, 12) = \$94.62/\text{month}$$

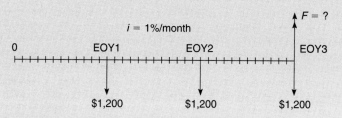

Figure 3.17   Cash flow diagram for Example 3.13.

In order to accumulate $1,200 in an account at EOM 12, you can deposit $94.62 at the end of every month for 12 months and earn 1 percent interest compounded monthly. Then, having found that $94.62/ month is equivalent to $1,200 per year, the equivalent future lump sum at EOM 36 can be determined as follows.

$$F = A(F/A, i, n) = \$94.62(F/A, 1\%, 36) = \underline{\$4,076}$$

Thus if $1,200 is deposited at the end of every 12 months for 36 months, and if interest is credited every month compounded at 1 percent per month on the balance in the account, then the balance in the account at EOM 36 is $4,076. These calculations can be reduced to one line as follows:

$$F = \$1,200(A/F, 1\%, 12)(F/A, 1\%, 36) = \underline{\$4,076}$$

Method 2. Effective Interest Method Instead of compounding the interest every month at 1 percent, the interest can be compounded every m months (corresponding to the payment interval) at a higher interest rate. This rate will be a little higher than m times i to account for the compounding. The effective interest is actually $i_e = (1 + i)^m - 1$. The following example illustrates this procedure.

EXAMPLE 3.14 (Use an equivalent effective interest rate to find F, given A, i, n)

Assume the same $1,200 annual payment and $i = 1$ percent/month, as in Example 3.13. Find F at EOM 36.

Solution. The effective interest rate for compounding every 12 months that is equivalent to compounding at 1 percent per month is

$$i_e = (1 + i)^m - 1 = 1.01^{12} - 1 = 0.12683 = 12.68\%$$

Thus the future lump sum equivalent, where $i = i_e$ and $n = 3$, is found as

$$F = A \left[\frac{(1 + i)^n - 1}{i} \right] = \$1,200 \frac{(1.12683^3 - 1)}{0.12683} = \underline{\$4,076}$$

This indicates that an interest rate of 12.68 percent compounded annually on annual deposits of $1,200 yields the same balance at EOY 3 as an interest rate of 1 percent compounded monthly on the same $1,200 annual deposits.

Case 2:   **Interest Is Compounded Periodically, While Payments Occur at Random Points in Times Within the Periods**   In this case, interest normally is earned only on funds that are on deposit for the entire term of the compounding period. If $1,000 is on deposit at the beginning of the period and $200 is withdrawn one day before the end of the period, then interest is earned on only $800. By the same token, if $400 is on deposit at the beginning of the period and $600 is deposited the very next day, then interest is earned only on the $400, providing that at least $400 remains in the account throughout the entire compounding period. For funds received after the start of the period, normally there are *no* interest earnings attributable to the time between the receipt of the funds and the end of the compounding period. The following illustrates this approach.

EXAMPLE 3.15   **(Find *F*, given several *A* payment sequences, *i*, *n*)**

Given the cash flows shown in Figure 3.18 with $i = 5$ percent per quarter, find the equivalent future lump sum F.

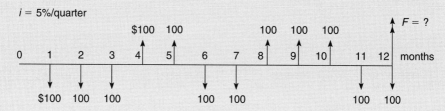

Figure 3.18   Cash flow diagram for Example 3.15.

Solution.   Since interest is earned at the close of each quarter, the simplified cash flow diagram is shown in Figure.3.19. Notice the receipts (positive cash flows) can be grouped at the beginning of each quarter in which received, while the disbursements (negative cash flows) can be grouped at the end of

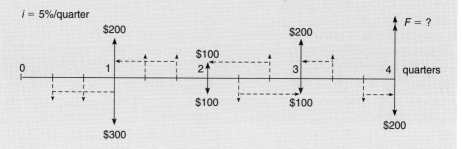

Figure 3.19   Simplified cash flow diagram for Example 3.15.

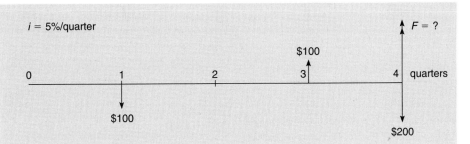

Figure 3.20   Cash flow for Example 3.15 with each quarter's totals summed.

each quarter. The cash flow diagram can be further simplified by summing each quarter's totals, as shown in Figure 3.20. The balance in the account at the end of the fourth quarter is calculated as follows:

$$F = -100(F/P, 5\%, 3) + 100(F/P, 5\%, 1) - 200 = -\$210.76$$

SUMMARY

This chapter describes the relationship between the future value, F, and a uniform series of payments (or receipts). The equations employed are

$$F = A \left[\frac{(1+i)^n - 1}{i} \right] = A(F/A, i, n)$$

$$A = F \left[\frac{i}{(1+i)^n - 1} \right] = F(A/F, i, n)$$

$$n = \frac{\ln \left(\frac{Fi}{A} + 1 \right)}{\ln(1 + i)}$$

The F/A [or $(F/A, i, n)$ factor] is sometimes termed the "Uniform Series Compound Amount Factor" (USCAF), while the A/F [or $(A/F, i, n)$ factor] is often termed the "Uniform Series Sinking Fund Factor" (USSFF). Note that the F/A factor, and all of its equivalents, is the reciprocal of the A/F and its equivalents. The cash flow diagram is shown in Figure 3.21.

Note that the *first* A payment occurs at the *end* of the first compounding period, and the last A payment and the F payment occur simultaneously.

Values for these two factors can be found in Appendix A for common combinations of i and n.

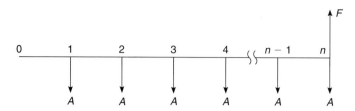

Figure 3.21   Cash flow diagram showing F/A relationship.

Finally, funds flow problems were discussed for two cases,

Case 1.   Interest is compounded more than once between payments in a series of payments. Either the sinking fund method or an equivalent effective interest rate may be used to solve this type of problem.

Case 2.   Interest is compounded periodically while payments occur at random points in time within the compounding periods. Normally, for disbursements the funds may be summed together at the end of each compounding period; for receipts the funds may be summed together at the beginning of each period.

PROBLEMS FOR CHAPTER 3 *(F/A, A/F)*

Note: Assume all payments are made at the end of the period (a year in most cases) unless otherwise noted in the problem.

EOM = End of Month
EOY = End of Year

Problem Group A

Given three of the four variables, find A, F.

A.1.   A municipality wishes to set aside enough of its annual allocation of gasoline tax funds to finance a $1,000,000 road project scheduled for contract three years from now. If the annual deposit can be invested at 8 percent, how much should be deposited at the end of each of the three years to accumulate $1,000,000 at the end of the third year?

(*Ans.* $A = \$308,030$)

A.2.   Your client has constructed a small medical office complex. It is anticipated that in 20 years the complex will need a complete renovation at a cost of $300,000.
   (a) How much rent should be charged annually just to cover this one future expense item if the rent is deposited at the end of each year into an investment account earning interest?
   (b) How much rent should be charged monthly if the rent is deposited at the end of each month in an account drawing 8 percent nominal interest compounded monthly?

A.3.   A newly opened toll bridge has a life expectancy of 25 years. Considering rising construction and other costs, a replacement bridge at the end of that time is expected to cost $5,000,000. The bridge is expected to serve an average of 300,000 toll-paying vehicles per month for the full 25 years. At the end of every month the tolls will be deposited into an account bearing interest at a nominal annual $i = 7.5$ percent, compounded monthly. The operating and maintenance (O & M) expenses for the toll bridge are estimated at $0.15 per vehicle.

(a) How much toll must be collected per vehicle in order to accumulate $5,000,000 by the end of 25 years (300 months)?

(*Ans. A* = $0.019/vehicle)

(b) What should be the total toll per vehicle in order to include both O & M costs as well as replacement costs?

(*Ans. A* = $0.169/vehicle)

A.4.   The central air conditioner compressor and air handler in a municipal auditorium have an estimated life expectancy of 12 more years. Replacement at that time is expected to cost $200,000. The city manager is considering a plan to set aside equal annual deposits into a fund bearing 8 percent interest so that the $200,000 will be on hand when needed 12 years from now. He is ready to make a deposit *right now* and requests you as consultant to determine how much the deposit should be. (Assume 13 equal annual deposits, including one at the beginning of the first period and one at the end of the last period.)

A.5.   An insurance company uses 40 percent of its premium income for overhead (OH) (salaries, rent, commissions, profit, etc.). Premium income not used for claims or OH is deposited in an account bearing 12 percent interest per year. The remaining 60 percent of premium income is used to pay death benefits to policy holders. Assume that you are doing a rough estimate of insurance charges for policy holders who are between 30 and 35 years of age. (The rough estimate does not take into account statistical variations within the group.) The tables show a remaining life expectancy for this group of an average of 44 more years.

(a) How much should the insurance company charge per year over the remaining life of each policy holder just to cover each $1,000 worth of death benefits?

(*Ans. A* = $0.825/ 1,000)

(b) How much should the total annual charge be to cover overhead as well as death benefits?

(*Ans. A* = $1.38/ 1,000)

Problem Group B

Find *A* and/or *F*. Two or more steps required.

B.1.   A young engineer decides to save $240 per year toward her retirement in 40 years.

(a) If she invests this sum at the end of every year at 9 percent, how much will be accumulated by retirement time?

(b) If by astute investing the interest rate could be raised to 12 percent, what sum could be saved?

(c) If she deposits one-fourth of this annual amount each quarter ($60 per quarter) in an interest-bearing account earning a nominal annual interest rate of 12 percent, compounded quarterly, how much could be saved by retirement time?

(d) In part (c) above, what annual effective interest rate is being earned?

B.2. A young engineer gets a $250-a-month raise and decides to save it toward a down payment on a car. He deposits $250 now and $250 a month for the next 20 months in an account earning an annual $i = 12$ percent compounded monthly on his deposit. How much will he have at EOM 20 (with 21 deposits)?

(*Ans.* $F = \$5,810$)

B.3. If $1,000 is deposited in an account at the end of every month, how much will be contained in the account after five years (60 deposits) if
(a) the interest rate is 12 percent compounded monthly?
(b) the interest rate is 12 percent compounded quarterly?
(c) the interest rate is 12 percent compounded annually?

B.4. Your client is collecting rent of $10,000 per year on a commercial building. His operating expenses are $5,000 per year, and he deposits the remaining $5,000 per year into an account that earns 10 percent. The accumulated balance in the account including principal and interest at the present time is $42,312, including a very recent year-end deposit of $5,000. Your client plans to demolish the existing structure five years from now (he will make five more deposits at $5,000 each) and replace it with a structure estimated to cost $150,000. How much cash will he need five years from now in addition to what is expected to be in the account?

(*Ans.* Need $51,330 more)

B.5. The citizens of a resort town in the mountains want to build a civic center at a cost of $1,000,000. You are asked for an opinion on the financial feasibility of the civic center. A philanthropist with a vacation home in the area promises to donate $700,000 if the town can raise the other $300,000 within three years. You estimate that the town can hold a series of benefits during the vacation months of June, July, August, and September and raise $20,000 per month. Assume it is now the end of September. The first deposit of $20,000 will be made at the end of next June. Funds on deposit bear interest at 7.2 percent per year, compounded *monthly.* The deadline for reaching the $300,000 balance in the fund is 36 months from now and will include the deposit and interest payment at the end of September, three years from now.
(a) How much will be in the fund at that time if all projections are correct?
(b) By how much should the monthly deposits be increased or decreased to come out to exactly the $300,000 amount? (Can this increase or decrease be found simply by multiplying the ratio of the actual balance to the desired balance times the monthly deposit?)

B.6. A lump sum of $100,000 is borrowed now, to be repaid in one lump sum at EOM 120, and bearing interest at 1 percent compounded monthly. No partial repayments will be accepted on this loan. To accumulate the repayment lump sum due, monthly deposits are made into an interest-bearing account. This repayment deposit account bears interest at 0.75 percent per month from EOM 1 until EOM 48. Then from EOM 48 until EOM 120, the interest rate changes to 0.5 percent. Monthly deposits of amount A begin with the first deposit at EOM 1, and continue until EOM 48. Beginning with EOM 49 the deposits are doubled to amount $2A$ and continued at this level until the final deposit at EOM 120. Find the initial monthly deposit amount A.

(*Ans.* $A = \$1,293$)

B.7. A small electric utility company borrows $1,000,000 to construct an addition to its power plant. The loan will accrue interest compounded annually at 10 percent and be repaid in one lump sum at EOY 5. **(a)** Find the amount to be repaid at EOY 5.

In order to repay the amount due at EOY 5, the company needs to raise the monthly electric rates. It decides on a two-step rate increase. The first step to level *A* will go into effect now, with monthly deposits to the repayment account of amount *A* beginning at EOM 1 and continuing through EOM 30. The increase then doubles to level 2*A*, with deposits beginning at EOM 31 and continuing through EOM 60. The repayment account earns interest at a rate of 1 percent per month from EOM 1 until EOM 30. The rate then rises to 1.5 percent effective from EOM 30 until EOM 60. **(b)** Find the first monthly deposit *A*.

B.8. A family deposits funds into an interest-bearing account for college expenses for the children. Find the balance in the account after the deposits and withdrawals shown in the following table.

(*Ans.* $11,570)

Time	Amount	Deposit or withdrawal
EOY 1 through 5	$1,000/yr	Deposit
EOY 6 through 13	1,500/yr	Deposit
EOY 10 through 13	5,000/yr	Withdrawal

The interest rate earned by the account is:

EOY 1 through 5	9%
EOY 6 through 13	14%

B.9. Several little league baseball teams hope to finance their summer activities by a series of off-season garage sales. They hope to raise $100 per month from October through May, and deposit this amount at the *end* of each month into an account earning interest rate at 0.5 percent per month. At the *beginning* of June through September, they need to withdraw $200 per month. How much is left in the account at the *end* of September?

B.10. The city manager proposes to build a new airport terminal building six years from now at an estimated cost of $2,000,000. He proposes to raise the necessary money by charging a fee for each passenger using the existing terminal facilities for the next six years, and depositing the proceeds into an interest-bearing account. The fee would go into effect two months from now and would be deposited at monthly intervals, with the first deposit made three months from now. The account will earn a nominal 6 percent annual interest to be compounded *monthly*. The number of passengers passing through the existing terminal for the next six years is estimated as shown in the following table.

During years	Passengers/month
1 through 2	25,000 passengers/month
3 through 6	30,000 passengers/month

(a) How many deposits will be made?

(*Ans.* 70)

(b) How much must be collected from each passenger in order to accumulate $2,000,000 by six years from now? The charge per passenger will be constant over the entire collection period.

(*Ans.* $0.848)

Problem Group C

Find *i*.

C.1. A young man of 25 decides he will retire at 50 with a sum of $200,000. He feels that by careful budgeting he can invest $2,000 per year toward this sum.
   (a) What annual interest rate does his investment need to earn in order to attain his goal?
   (b) What actual quarterly interest rate does his investment have to earn if he deposits $500 quarterly?
   (c) What effective annual interest rate is earned in part (b)?

C.2. A young engineer decides to save her tax rebate each year and invest it for retirement. She arranges to over-deduct and expects a $1,000 rebate at the end of each year. She is 23 years old now and would like to retire at age 58 with a $300,000 nest egg. How much annual interest must her retirement investments yield in order to reach her goal?

(*Ans.* $i = 10.47\%$)

Problem Group D

Find *n*.

D.1. To pay for replacing his dragline, a contractor is setting aside $0.10 per cubic yard moved by the dragline. The dragline is averaging 200 yd^3 per hour and 170 hours per month. The money is invested at the end of every month at the nominal annual rate of 10 percent per year. How many months will pass before he accumulates the $150,000 needed to purchase a new dragline? (Assume monthly compounding of interest.)

D.2. A new city in a retirement area is expected to have an influx of 10,000 new residents per year on the average. Due to the large number of older people attracted to the area, the anticipated birthrate is only 5 per 1,000 per year, whereas the death rate is 15 per 1,000 residents. Therefore, there will be a negative natural annual increase of 10 per 1,000. The city is just getting started, and the population of the area now is essentially zero. How long will it be before the population reaches 60,000 if growth continues to occur at the estimated rates?

(*Ans.* $n = 6.16$ yr)

D.3. A friend is starting a savings program with deposits of $100 per month into an account bearing interest at 0.5 percent per month compounded. When the balance in the account reaches $2,500, it can be transferred to another program that pays 1 percent per month.
   (a) How soon will the balance reach (exceed) $2,500 (no partial deposits are made)?
   (b) How much will be in the account at EOM 60, assuming the switch is made to the new account as soon as possible?

Problem Group E

Find the financing cost.

E.1. A contract calls for a lump sum payment of $1,000,000 to you, the contractor, when a job is completed. Estimated time to complete the job is 10 months. Your expenditures will be very close to $95,000 per month. You pay your banker 2 percent per month on the amount loaned out to you and borrow the full $95,000 per month from her at the end of every month. Find the financing cost. (The financing cost is the total amount of interest charges.)

E.2. A contractor is constructing a project for which he will be paid in one lump sum upon satisfactory completion of the project. The contractor needs cash to finance the construction (payroll and materials) and opens a line of credit at his bank (an agreement whereby the contractor borrows in increments, as needed and at his own convenience, and repays in one lump sum upon completion of the project). The construction is expected to last for 16 months, and the contractor expects to borrow $100,000 at the beginning of the first month (EOM 0) and at the end of each of the first six months, making a total borrowing of $700,000. The contractor then needs to borrow $250,000 per month at the end of month 7 through 16, for an additional $2,500,000. The owner agrees to pay the contractor the lump sum payment for the project at EOM 20, at which time the contractor will repay the bank. Interest on the contractor's outstanding loan balance is 1.5 percent per month for the first six months until EOM 6, and then 2 percent per month until EOM 20, compounded monthly.

  (a) Draw a cash flow diagram of the bank loans and repayment?

  (b) How much should the contractor add to his lump sum bid just to cover the cost of financing the job (interest only)?

    (*Ans.* $728,980)

E.3. An engineer-contractor (E-C) has a contract to build a hotel that has an estimated direct cost of $2,600,000. The E-C opens a line of credit at the bank in order to obtain funds for construction. The job will last for 16 months, and the owner will pay the E-C at EOM 18, at which time the E-C will repay the bank. The E-C's schedule for monthly borrowing is given in the following table.

EOM	Amount borrowed
1 through 6	$100,000/month
7 through 16	200,000/month
	Interest rate
1 through 7	1%/month
7 through 18	1.5%/month

Find the cost to the E-C just for financing the project. (Find how much the bank receives just for interest by EOM 18.)

Problem Group F

Periodic payments, use i_e.

F.1. Your client must make a lump sum payment of $150,000 at EOM 120. To accumulate the necessary sum by EOM 120, the client plans to make a series of quarterly deposits of amount A into an account bearing interest compounded at 1 percent per month. The first deposit is made now, at EOM 0. Each subsequent deposit is made at intervals of three months thereafter, until the last deposit at EOM 117. Find the amount A to be deposited at three-month intervals.

(*Ans. A* = $1,917)

F.2. A greenspace trust fund has been established to accumulate funds to buy parkland and recreation areas. Every three months a deposit of $100,000 is made into the fund, which earns interest at 1 percent per month compounded. The first deposit is made today. The fund trustees have an option to buy a tract of land which could be purchased today for $1,000,000. The terms of the option allow an increase in price of 0.5 percent per month for every month of delay in purchase. How soon will the trustees have enough in the fund to purchase this tract at its escalated price?

F.3. A city collects a fuel tax to finance new highway construction. This tax brings in $500,000 per year, which is deposited at the end of each year into an account earning a nominal 8 percent, compounded quarterly. If each major project empties the fund every five years, how much will have accumulated in the fund at the end of each five-year period?

F.4. A town with a population of 13,700 decides to set up a trust fund with which to finance new parks and recreation facilities. They tax each bottle or can of beverage sold in the town at $0.01. Since each resident drinks an average of one beverage per day, they raise $50,000 per year. They hope to finance one major project every five years. How much will be available to spend at the end of each five-year period if

(a) the $50,000 is deposited at the end of each year and earns annual interest of 8 percent compounded quarterly?

(*Ans. F* = $294,800)

(b) the $50,000 is deposited in 52 weekly installments, and the 8 percent interest is compounded weekly?

(*Ans. F* = $295,220)

The Value of One Present Payment Compared to a Uniform Series of Payments (*P/A, A/P*)

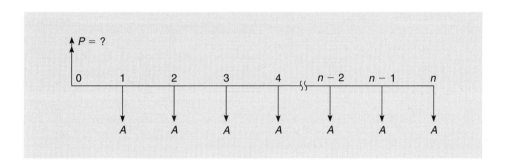

INSTALLMENT LOAN

A uniform series of payments, such as installment loan payments, is familiar to everyone in today's credit-oriented society. Appliances, cars, houses, even vacations, may be purchased on installment plans. Many engineering projects are financed in essentially the same way. Funds to construct the project are borrowed and repaid in annual (or other end-of-period) payments. The value of the periodic repayments (*A*) must equal the value of the amount borrowed (*P*) plus the interest compounded periodically at rate (*i*) on the unpaid balance over *n* periods of repayment. The following equation may be derived for this equivalence relationship.

    P dollars are borrowed now, and a uniform series of repayments, *A*, will be made at the *end* of each of *n* periods. Each payment *A* will include two parts: payment of interest on the unpaid balance, plus a repayment of part of the principal.

P/A EQUATION

To derive the *P/A* equation, simply combine two previously derived equations,

$$P = F \left[\frac{1}{(1+i)^n} \right] \tag{2.2}$$

and

$$F = A\left[\frac{(1+i)^n - 1}{i}\right] \tag{3.3}$$

Substitute the right side of Equation 3.3 for F in Equation 2.2 to obtain

$$P = A\left[\frac{(1+i)^n - 1}{i}\right]\left[\frac{1}{(1+i)^n}\right]$$

This may be combined and simplified to

$$P = A\left[\frac{(1+i)^n - 1}{i(1+i)^n}\right] \tag{4.1}$$

The factor in brackets in Equation 4.1 often is termed the Uniform Series Present Worth Factor (USPWF) and is more simply designated by

$$P = A(P/A, i, n) \tag{4.1a}$$

The correct cash flow diagram for this relationship is shown in Figure 4.1. The inverse of the USPWF commonly is termed the Uniform Series Capital Recovery Factor (USCRF) and is written as

$$A = P\left[\frac{i(1+i)^n}{(1+i)^n - 1}\right] \tag{4.2}$$

The briefer common designation is written as

$$A = P(A/P, i, n) \tag{4.2a}$$

As pointed out in Chapters 2 and 3 the P/A and A/P factors depend only upon i and n. Values for these two factors are tabulated in Appendix A for commonly used combinations of i and n.

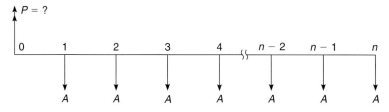

Figure 4.1   Correct cash flow diagram for P/A (USPWF).

Equations 4.1 and 4.2 may be rearranged to solve for n as follows:

$$n = \frac{\ln\left(\dfrac{1}{1 - Pi/A}\right)}{\ln(1 + i)} \tag{4.3}$$

The cash flow diagram for Equation 4.1 P/A (USPWF) is shown in Figure 4.1. Beginners sometimes make errors in constructing the cash flow diagrams, and some of the more common errors for this relationship are shown in Figure 4.2.

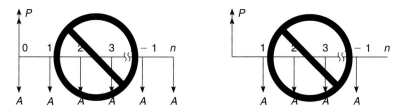

Figure 4.2   Illustrations of several commonly occurring errors in constructing cash flow diagrams for P/A (USPWF).

Graph of P/A

A graph of the equation

$$P/A = \frac{(1 + i)^n - 1}{i(1 + i)^n}$$

is shown in Figure 4.3. The graph illustrates the response of P/A to variations in i and n. This information is helpful in determining the effect of changes in input variables on problem solutions.

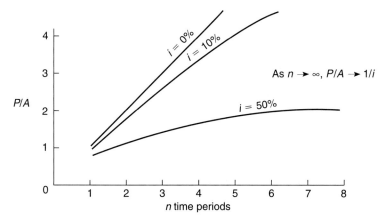

Figure 4.3   Graph of P/A.

Graph of Decrease in Principal Balance with Periodic Payments

Figure 4.4 represents the balance of an account that begins with a present amount P on the left side. At the end of each of n time periods, a payment A is made until the account is reduced to zero on the right side. Each periodic payment A consists first of interest on whatever current balance is left in the account. The remaining portion of the periodic payment A pays off another part of the balance, until finally the balance is reduced to zero. Note that the first periodic payment A is made at the *end* of the first period.

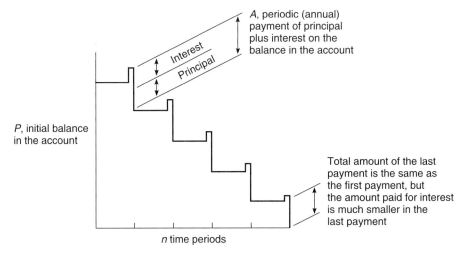

Figure 4.4   Balance-in-the-account diagram for P/A.

It is important to remember that equation time zero (beginning time for the equation) for the P/A series occurs *one period before* the first periodic payment A occurs. In analyzing problems, this information is used to find the correct value of n, the number of compounding periods.

As in the equations discussed in earlier chapters, the P/A (or A/P) equation contains four variables, P, A, i, and n. If any three of the variables are given (or can be derived from other data in the problem), then the fourth variable can be found. The next four examples demonstrate how each of the four variables may be found in turn.

EXAMPLE 4.1   (Find P, given A, i, n)

A young engineer decides to budget $600 per month for payments on a new car. The bank offers a 60-month repayment period on new vehicle loans at 1 percent per month on the unpaid balance. How much can be financed under these terms? (See Figure 4.5.)

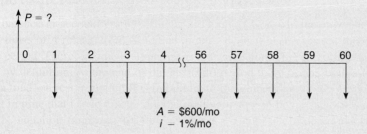

Figure 4.5 Cash flow diagram for Example 4.1.

Solution

$$P = \frac{A[(1 + i)^n - 1]}{i(1 + i)^n}$$

$$P = \frac{\$600[(1 + 0.01)^{60} - 1]}{0.01(1 + 0.01)^{60}}$$

$$P = \$600/\text{month} \times 44.95504 = \$26,974$$

EXAMPLE 4.2 (Find A, given P, i, n)

The engineer finds a very desirable car, but the balance to be financed is $18,000. Under the same terms, 1 percent per month and 60 months, what would the monthly payment be, with the first payment due one month from now? (See Figure 4.6.)

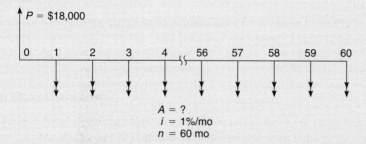

Figure 4.6 Cash flow diagram for Example 4.2.

Solution. Using the equation and solving for A yields

$$A = P\frac{i(1 + i)^n}{(1 + i)^n - 1}$$

$$A = 18,000 \, \frac{0.01(1.01)^{60}}{(1.01)^{60} - 1}$$

$$A = \$400.40/\text{month}$$

For commonly encountered values of i and n, the P/A and A/P values tabulated in Appendix A can be used, as follows:

$$A = P(A/P, i, n) = 18,000 \underbrace{(A/P, 1\%, 60)}_{0.02224} = \$400.32/\text{month}$$

The \$0.08 difference is due to the rounding of the tables.

EXAMPLE 4.3   (Find i, given P, A, n)

A contractor buys a \$40,000 truck with no down payment and makes annual payments of \$10,400 per year for five years. What interest rate is he paying?

Solution

$$A = P(A/P, i, n)$$

$$\$10,400 = \$40,000(A/P, i, 5)$$

$$10,400/40,000 = 0.2600 = (A/P, i, 5)$$

The equation with this data included appears as

$$A/P = 0.2600 = \frac{i(1+i)^5}{(1+i)^5 - 1}$$

Since the i appears three times in this equation, the solution must be found by interpolation from the tables. (In other problems, if values should fall outside the tables, bracket the solution by trial and error, and then interpolate.)

In order to interpolate from the tables, look through the interest rate tables of Appendix A, noting that each page represents a different interest rate. On each page look under the A/P column and across the $n = 5$ row until the A/P values are found which bracket 0.2600, as follows:

i	A/P	A/P
$(A/P, \underline{10\%}, 5) =$	0.2638	
$(A/P, \underline{i\%}, 5) =$		0.2600
$(A/P, \underline{9\%}, 5) =$	0.2571	0.2571
Subtract	0.0067	0.0029

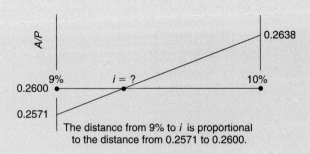

The distance from 9% to i is proportional to the distance from 0.2571 to 0.2600.

Then interpolate

$$i = 9\% + \frac{0.0029}{0.0067} \times 1\% = \underline{\underline{9.43\%}}$$

EXAMPLE 4.4   (Find n, given P, A, i)

A new civic auditorium is estimated to cost \$20,000,000 and produce a net annual revenue of \$2,000,000. If funds to construct the auditorium are borrowed at $i = 8\%$, how many annual payments of \$2,000,000 are required to retire the debt? (See Figure 4.7.)

A = \$2,000,000/yr.

$i = 8\%$
$n = ?$

P = \$20,000,000

Figure 4.7   Cash flow diagram for Example 4.4.

Alternate Solution 1.   Use Equation 4.3.

$$n = \frac{\ln\left(\dfrac{1}{1 - Pi/A}\right)}{\ln(1 + i)} = \frac{\ln\left(\dfrac{1}{1 - 10 \times 0.08}\right)}{\ln 1.08} = \underline{\underline{20.91 \text{ yr}}}$$

Of course the answer figure of 20.9 years does *not* mean that 20 payments of \$2,000,000 each will be made at the end of 20 successive years and then another \$2,000,000 will be paid in another 0.9 of a year. Customarily, payments are paid at the end of each full compounding period rather than part way through the period. Therefore, if n does not equal a whole number, a partial payment at the end of a full time period is implied (rather than a full payment paid partway through the period). For instance, for this example *approximately* $0.9 \times \$2,000,000 = \$1,800,000$ will be paid as a final payment at the end of the twenty-first year. The method of calculating the correct amount of the last odd payment is outlined later in this chapter.

Alternate Solution 2.   Interpolate from the tables. A solution also may be found by interpolating from the tables in Appendix A, as follows:

$$P/A = (P/A, i, n)$$
$$\$20,000,000/\$2,000,000 = 10 = (P/A, 8\%, n)$$

From the interest tables in Appendix A, bracket the P/A factor of 10, and interpolate to find n.

$$
\begin{array}{lrr}
(P/A, 8\%, 21) = 10.0168 & & \\
(P/A, 8\%, n) \;\; = & & 10.000 \\
(P/A, 8\%, 20) = & \underline{9.8181} & \underline{9.8181} \\
\text{Subtract} & 0.1987 & 0.1819
\end{array}
$$

$$n = 20 \text{ yr} + \frac{0.1819}{0.1987} = \times 1 \text{ yr} = \underline{\underline{20.9 \text{ yr}}}$$

AMOUNT OF THE FINAL PAYMENT IN A SERIES

As illustrated in Example 4.4, when n is the derived quantity in a problem involving repayment, n does not normally equal a whole number. When n is not a whole number, this indicates that the last payment in the series is less than the rest of the payments and will be paid at the *end* of a full compounding period. To find the odd final payment, use the following three-step procedure.

1. Given the present worth, P_1, of the loan, and the amount of the periodic repayment, A_1, find the present worth, P_2, of the series of full payments, leaving out the final odd payment.

2. Subtract P_2, from P_1 to find P_3, the present value of the last odd payment.

3. Find the future value, F_1, of P_3 at the time the last payment is due. Then F_1 is the amount due as the odd last payment.

To find the amount of the last payment due at EOY 21 on the civic auditorium of Example 4.4, the three steps are applied as follows:

Given: The amount of the loan is $P_1 = \$20,000,000$

The amount of the periodic repayment is $A_1 = \$2,000,000$.

1. Since large numbers are involved, use the equations to find the present worth P_2 of the first 20 payments. This yields

$$P_2 = \$2,000,000(P/A, 8\%, 20) = \$19,636,295.$$

2. Subtract P_2 from the original loan of P_1 D $\$20,000,000$ and find P_3. This P_3 is the small fraction of the original loan that must be paid off by the final odd payment F_1 at the end of 21 years.

$$P_3 = \$20,000,000 - \$19,636,295 = \$363,705$$

3. Find F_1, the amount of the final odd payment at the end of the twenty-first year.

$$F_1 = \$363,705(F/P, 8\%, 21) = \$1,830,831.$$

AMOUNT OF THE CURRENT BALANCE

Often it is necessary to find the unpaid principal balance still owed on a loan partway through the repayment time. For instance, suppose in Example 4.4 that at EOY 5 (of the 21-year repayment schedule) funds became available to pay off the remaining balance of the debt owed. How much of the original $\$20,000,000$ debt is still owed after making five payments of $\$2,000,000$ each?

The basic concept to remember is simply that *the current balance equals the present value of the remaining payments.* All payments on the debt consist of two parts. One part of each payment is interest (which pays the periodic rent due on the current balance), and the other is a payment of principal that serves to reduce the current balance. By taking the present value of the remaining payments, the interest portion of these payments is stripped away, and only the principal portion remains.

EXAMPLE 4.5

In Example 4.4 after five payments have been made, there are $20 - 5 = 15$ full payments remaining plus one partial payment at EOY 21. Thus the current balance due at EOY 5 (after payment 5 has been made) is

$$P_1 = A(P/A, i, n) = \$2,000,000(P/A, 8\%, 15) = \$17,118,957$$
$$P_2 = F(P/F, i, n) = \$1,830,831(P/F, 8\%, 16) = \underline{\$\quad 534,402}$$
$$\text{Current balance due at EOY 5, } P_{\text{total}} = \underline{\$17,653,359}$$

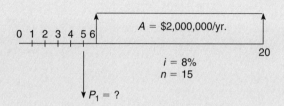

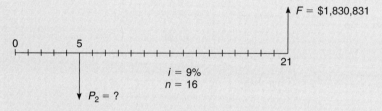

Figure 4.8   Cash flow diagram for Example 4.5.

Figure 4.9   Cash flow diagram for Example 4.5.

Thus the debt owed on the civic auditorium (the current balance at EOY 5) is found as the P_{total} shown.

Alternative Method of Determining the Remaining Balance

The remaining debt also may be determined by finding the equivalent future value, F, of the previous cash flows upon making the fifth payment at EOY 5.

$$F_1 = P(F/P, i, n) = +20,000,000(F/P, 8\%, 5) = +\$29,386,562$$
$$F_2 = A(F/A, i, n) = -2,000,000(F/A, 8\%, 5) = \underline{\ -\ 11,733,202\ }$$
$$F_{total} \qquad\qquad\qquad\qquad\qquad\qquad\qquad\quad = +\$17,653,360$$

IF PROBLEM TIME ZERO DIFFERS FROM EQUATION TIME ZERO

Problem situations sometimes occur that involve a different time base than that specified for the P/A and A/P factors. (For example, see Figure 4.2.) The general approach to solving this type of problem is to subdivide the problem into simple components, solve each component individually, and then find the sum or product of the parts. The following two examples will illustrate the method. In Example 4.6, time zero for the problem (PTZ) occurs several periods *before* the start of the series of A payments (ETZ), while in Example 4.7, problem time zero occurs *after* the A payments start.

EXAMPLE 4.6 ETZ ≠ PTZ (ETZ later than PTZ)

Your firm wants to borrow $50,000 now to buy land the company will need in the future. The proposed terms of the loan follow:

1. No cash payments on the loan (neither interest nor principal) until EOY 5, although interest accrues.
2. The loan will be paid off in eight equal annual installments, with the first installment due at EOY 5.
3. Interest is earned on the outstanding balance at 9 percent compounded annually on the loan, including, of course, the period during which no payments are made.

You are requested to find the size of the annual repayments, A, required from years 5 through 12. A cash flow diagram of the problem appears in Figure 4.10.

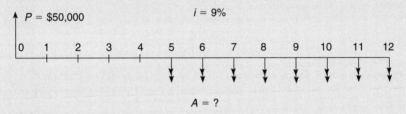

Figure 4.10   Cash flow diagram for Example 4.6.

Solution.   This type of problem may be solved by dividing it into two parts. The first few years are periods during which the interest accumulates as in an F/P problem. The second series of years involves an A/P series during which the loan is paid off by a series of equal annual payments, A. A graph of the loan balance appears in Figure 4.11.

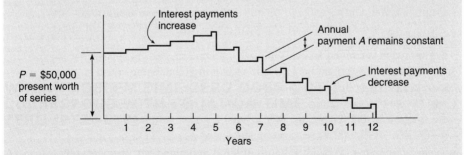

Figure 4.11   Balance-in-the-account diagram for Example 4.6.

At this point in their studies, students sometimes encounter difficulty in finding which year in the problem corresponds to time zero for the equation (where is ETZ with respect to PTZ?), and a consequent difficulty in determining the correct value of n. Such difficulties should be overcome (by further review, practice, and consultation) before proceeding further, since the determination of proper n values is essential to the solution of many problems throughout the remainder of the text. By referring to the cash flow diagram of the repayment series (Figure 4.10) and remembering the ETZ is one time period before the first payment for normal series equations, the ETZ for the A/P equation can be found as EOY 4 (one period before the first payment at EOY 5). After this ETZ point is found, the problem can be solved in two steps as follows:

1. Find F_1, the amount owed at the ETZ at EOY 4. The cash flow diagram is shown in Figure 4.12.

$$F_1 = P(F/P, i, n) = \$50,000(F/P, 9\%, 4) = \$70,580$$

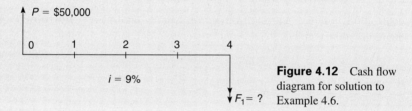

$P = \$50,000$

$i = 9\%$

$F_1 = ?$

Figure 4.12   Cash flow diagram for solution to Example 4.6.

2. Find A, the annual payment required to pay out the loan in eight years. The value of F_1 found in step (1) is considered as the P_1 value of the loan to be paid off ($P_1 = F_1$). The cash flow diagram is shown in Figure 4.13.

$$A = P_1(A/P, i, n) = \$70,580(A/P, 9\%, 8)$$
$$= \$12,753 = \$12,800/\text{yr}$$

These three lines of equations may be simplified and combined into one line, as follows:

$$A = \$50,000(F/P, 9\%, 4)(A/P, 9\%, 8) = \$12,800/\text{yr}.$$

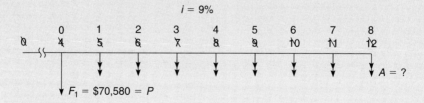

$i = 9\%$

$F_1 = \$70,580 = P$

$A = ?$

Figure 4.13   Cash flow diagram for solution to Example 4.6.

EXAMPLE 4.7   **(ETZ ≠ PTZ; ETZ earlier than PTZ: Find the unpaid balance)**

A year ago your firm bought a computer and financed it with a $20,000 loan, repayable in 36 monthly installments, with 1 percent interest per month charged on the unpaid balance. Now the firm has some extra cash and would like to pay off the loan. You are requested to determine (1) the balance due on the loan, and (2) the total amount of interest that would be paid as part of the last 24 payments. The twelfth monthly installment has just been paid.

Solution 1.   The cash flow diagram for this problem is shown in Figure 4.14. The graph of the loan balance appears in Figure 4.15. Referring to the cash flow diagram and the graph of the loan balance, we can subdivide the problem into two parts: the series before PTZ and the series after PTZ. We need to find the payoff amount, which is the same as the present worth of the series after problem time zero. Since PTZ occurs at the end of the twelfth month, the value of n for the last series is $36 - 12 = 24$. With this information, the problem may be solved in two steps as follows:

     Step 1.   Find the amount, A, of the monthly payments.

$$A = P(A/P, i, n)$$
$$A = \$20,000(A/P, 1\%, 36)$$
$$A = 20,000 \times 0.0332 = \$664 \text{ monthly payment}$$

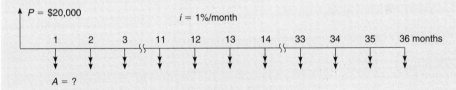

Figure 4.14   Cash flow diagram for Example 4.7.

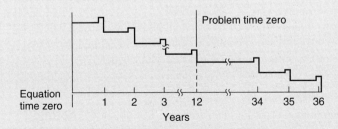

Figure 4.15   Balance-in-the-account diagram for Example 4.7.

Step 2.   Find the present worth, P, of the 24 remaining monthly payments after the twelfth payment has been made. The cash flow diagram is shown in Figure 4.16.

$$P = A(P/A, i, n)$$
$$P = 664(P/A, 1\%, 24) = \underline{\underline{\$14,106}}$$

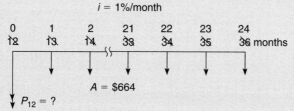

Figure 4.16   Cash flow diagram for the solution to Example 4.7.

Solution 2.   Find the amount of interest only, apart from principal, on the last 24 payments. Notice that the amount of cash required to pay off the loan ($14,106) is considerably smaller than the total cash flow if the loan were to be paid using monthly installments (24 × $664 = $15,936). The difference is the amount of interest that would be paid ($15,936 − $14,106 = $1,830) as part of the last 24 payments. (Remember that if the tables in Appendix A are used, then the numerical answer is only significant to the number of digits used in the factors from Appendix A.)

CAPITALIZED COSTS FOR INFINITE LIVES
($P/A \rightarrow 1/i$ as $N \rightarrow \infty$)

Occasionally problem situations occur involving practically infinite values of n. For instance a government department of transportation (DOT) may need land for a right of way (R/W). The land will require maintenance (mowing, etc.) for as long as the DOT owns it, and the DOT intends to own the land for a very long time, or perpetually for all practical purposes. If the annual cost of maintenance is valued at A, then the present worth of that annual cost charged at the end of every year forever, from Equation 4.1, is

$$P = A\left[\frac{(1+i)^\infty - 1}{i(1+i)^\infty}\right] = A\left[\frac{(1+i)^\infty}{i(1+i)^\infty} - \frac{1}{i(1+i)^\infty}^{\,0}\right]$$

$$P = A\left(\frac{1}{i}\right) = \frac{A}{i}, \quad \text{or} \quad A = Pi \quad \text{for} \quad n \rightarrow \infty \qquad (4.4)$$

EXAMPLE 4.8   **(Given $P_1, A_1, i, n \to \infty$, find P_2, A_2)**

A state DOT requests you to calculate the present value of acquiring and maintaining a new R/W that costs \$300,000 to purchase and \$7,200 per year to maintain. They intend to hold the R/W in perpetuity and ask for equivalent costs both in terms of (1) present worth (P), and (2) annual worth (A). Use $i = 6$ percent.

Solution 1.   The equivalent total present cost in terms of present worth, P, is

$$\text{cost of acquisition}, P_1 = \$300,000$$
$$\text{cost of maintenance}, P_2 = A/i = \$7,200/0.06 = \$120,000$$
$$\text{total equivalent present worth}, P_t = \underline{\underline{\$420,000}}$$

(P_t is sometimes referred to as capitalized value or capitalized cost.)

Comment: The \$120,000 equivalent present worth of maintenance may be visualized as the amount of money needed in a fund drawing 6 percent interest so that the interest alone is sufficient to pay all of the annual costs.

Solution 2.   The equivalent annual cost is

$$\text{cost of acquisition}, A_1 = Pi = \$300,000 \times 0.06 = \$18,000/\text{yr}$$
$$\text{cost of maintenance}, A_2 = \underline{\quad 7,200 \quad}$$
$$\underline{\underline{\$25,200/\text{yr}}}$$

Comment: The \$18,000-per-year equivalent annual cost of acquisition may be viewed as the cost of annual interest if the \$300,000 is borrowed and not repaid. In other words, the \$300,000 could be borrowed, exchanged for the land, and then repaid whenever the land is resold at the same purchase price. The only cost of ownership would be the interest cost on the borrowed money.

MORE THAN ONE COMPOUNDING PERIOD OCCURS BETWEEN PAYMENTS IN A SERIES

Sometimes a project will require a series of expenditures at intervals of time greater than the compounding period, such as a dam that needs dredging of accumulated silt every five years. The series may be delayed in starting. The capitalized value may be found by any of several methods, two of which are shown in the following example.

EXAMPLE 4.9   (Find P, given $i, n \rightarrow \infty$, a series of payments with the period between payments being some multiple of the interest compounding period, and a delayed start)

A dam will be completed three years from now, and water impounded. Five years after that at EOY 8 and every subsequent five years, the accumulated silt will be dredged out at a cost of $500,000 for each dredging. This series will continue for a long time to come ($n \rightarrow \infty$). Find the present worth if $i = 10$ percent.

Solution.   Method 1:   Sinking fund savings account.   Find the annual series amount A required for deposit into an account that will build up to an F of $500,000 every five years. Then find the present value of this A series with $n = \infty$. The ETZ for this series is EOY 3, so a further P/F transfer to EOY 0 is required (see Figure 4.17).

$$A = F(A/F, 10\%, 5) = 500{,}000 \times 0.1638 = \$81{,}900/\text{yr.}$$
$$P_1 = A(P/A, 10\%, \infty) = \frac{\$81{,}900}{0.10} = \$819{,}000 \text{ at EOY 3}$$

Then P_1 becomes F' on our time line and

$$P_2 = F(P/F, 10\%, 3) = \$819{,}000 \times 0.7513 = \underline{\underline{\$615{,}300}}$$

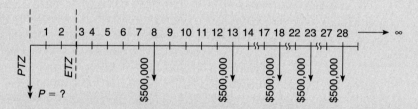

Figure 4.17   Cash flow diagram for Example 4.9.

Conclusion.   If a lump sum (capitalized value) of $615,300 is deposited now into an account bearing interest compounded annually at 10 percent, then $500,000 may be withdrawn at EOY 8, and every five years thereafter forever.

   Method 2:   Effective interest.   Find the effective interest for a five-year compounding period that is equivalent to 10 percent compounded per year; solve the problem as a regular periodic series.

$$i_e = (1 + i)^m - 1 = 1.1^5 - 1 = 0.6105 \quad (61.05\%/5 \text{ yr} = 10\%/ \text{ yr for 5 yr})$$

At EOY 3, $P_1 = A(P/A, i_e, \infty) = A/i_e$.

$$P_1 = \$500{,}000/0.6105 = \$819{,}000 \text{ at EOY 3}$$

Then the P_1 value needs to be moved three years to the left, so it becomes the F value with respect to P_2.

$$P_2 = F(P/F, 10\%, 3) = \$819{,}000 \times 0.7513 = \underline{\$615{,}300}$$

Conclusion: The capitalized value is \$615,300, which is the same as for method 1.

SUMMARY

This chapter develops equivalence factors relating a lump sum present value, P, to a uniform series of values, A. The equations and their designations are

$$P = A\frac{(1+i)^n - 1}{i(1+i)^n} = A(P/A, i, n)$$

$$A = P\frac{i(1+i)^n}{(1+i)^n - 1} = P(A/P, i, n)$$

$$n = \frac{\ln[1/(1 - Pi/A)]}{\ln(1+i)}$$

The cash flow diagram for these relationships is shown in Figure 4.18.

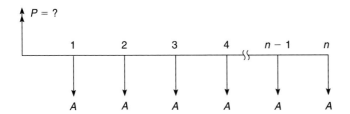

Figure 4.18   Cash flow diagram for P/A relationship.

When an investment has a long life, the calculations may be simplified by assuming $n \to \infty$. The present worth is expressed as

$$P = \frac{A}{i} \quad (\text{for } n \to \infty)$$

The uniform series is expressed as

$$A = Pi \quad (\text{for } n \to \infty)$$

PROBLEMS FOR CHAPTER 4 (P/A, A/P)

Note: Assume that all payments are made at the end of the period unless otherwise noted in the problem.

Problem Group A

Given three or four variables, find A or P. One step.

A.1. A contracting firm needs a new addition to its equipment maintenance shop. The firm estimates that the maximum repayment it can afford to make is $8,000 per year. How much can they borrow for the shop if

(a) Loan money is available at 15 percent for 15 years?

(*Ans.* $P = \$46,780$)

(b) Loan money is available at 20 percent for 15 years?

(*Ans.* $P = \$37,400$)

A.2. A backhoe is purchased for $200,000. The terms are 10 percent down and 2 percent per month on the unpaid balance for 60 months.

(a) How much are the monthly payments?

(b) What annual effective interest rate is being charged?

A.3. A city estimates that for each new family that moves into town it must eventually add about $3,200 worth of new improvements (water, sewer pipes and treatment facilities, vehicle R/W and lane capacity, schools, recreation, library, enlarged government buildings, etc.). In order to assess this cost directly to the new residents, it is suggested that a "hook-on" charge be made for utility services to newly constructed residential units. Assuming this payment would be included in the mortgage, how much would the annual mortgage payments be increased to cover the $3,200 added charge if $i = 9$ percent and $n = 25$ years?

(*Ans.* $A = \$325.79/\text{yr}$)

A.4. Your firm needs concrete aggregate and has an opportunity to buy a five-year lease on a rock quarry that is expected to yield a net profit of $58,000 per year (paid at EOY) for the next five years. How much can your firm pay for this lease (lump sum payment now) and earn 14 percent? The lease expires worthless at EOY 5. In other words, how large an investment could be paid off with payments of $58,000 per year with 14 percent interest on the unpaid balance?

Problem Group B

Find i, or whether a given i is exceeded.

B.1. A contractor invests $1,500,000 in a scraper fleet with an expected life of 10 years with zero salvage value. He expects to move an average of 100,000 yd^3 of earth per month at $1.00 per yd^3, of which $0.80 is the cost of operating the fleet, leaving $0.20 per yd^3 to pay off the principal plus interest on the investment in the machines. Using $n = 120$ months,

(a) What is the actual monthly rate of interest earned on the investment in this equipment?

(*Ans.* $i = 0.85\%$/month)

(b) What is the annual nominal interest earned?

(*Ans.* $i = 10.24\%$/yr)

(c) What is the annual effective interest earned?

(*Ans.* $i_e = 10.72\%$/yr)

B.2.   Your firm owns a large earthmoving machine and has contracts to move earth for $1 per yd³. For $100,000 this machine may be modified to increase its production output by an extra 10 yd³ per hour, with no increase in operating costs. The earthmoving machine is expected to last another eight years, with zero salvage value at the end of that time. Find whether or not this investment meets the company objective of earning at least a 15 percent return. Assume the equipment works 2,000 hours per year.

Problem Group C

Find n, final payment, and i_e.

C.1.   A proposed toll parking structure is estimated to cost $1,000,000. Loan funds are available at 6 percent. Net income (after other expenses) available to pay principal and interest is anticipated at $100,000 per year.

(a) How long will it take to pay off the loan?

(*Ans.* $n = 15.7$, so 15 full payments plus 1 partial payment at EOY 16 are required)

(b) Find the amount of the final payment.

(*Ans.* $F = \$73,100$)

C.2.   A left-turn lane can be installed at an intersection at a cost of $125,000. The turning lane is expected to save motorists an average of $0.01 (stopping, delays, and accident costs) for each vehicle traversing the intersection. Traffic is estimated at 10,000 vehicles per day every day of the year. Assuming $i = 7$ percent, how many years will it take for the savings to amortize (pay out) the cost of the improvement?

C.3.   A developer client of yours purchased some property for $75,000. The terms were $20,000 down and a loan for $55,000, payable at the rate of $650 per month, compounded monthly, at a nominal annual interest rate of 12 percent on the unpaid balance.

(a) How long will it take to pay off the loan?

(*Ans.* 188 full payments plus 1 partial payment at EOM 189)

(b) What annual effective interest rate is being charged?

(*Ans.* $i_e = 12.68\%$)

C.4.   Five years ago a contracting firm bought a concrete batch plant. To finance it, they borrowed $100,000 on a 20-year mortgage bearing interest at 1 percent per month on the unpaid balance. Repayment is scheduled in 240 equal monthly installments. The firm now has money available to pay off the loan and asks you to calculate the balance due. The sixtieth monthly payment has just been made.

Problem Group D

ETZ ≠ PTZ (equation time zero does not equal problem time zero). Find A, current balance.

D.1. Your firm wants to purchase a $50,000 computer, no money down. The $50,000 will be paid off in 10 equal end-of-year payments with interest at 8 percent on the unpaid balance.

    (a) What are the annual end-of-year payments?

        (*Ans.* A = $7,451/yr)

    (b) What hourly charge should be included to pay off the computer, assuming 2,000 hours work per year, credited at end of year?

        (*Ans.* $3.73/hr)

    (c) Assume that five years from now you would like to trade this one in and purchase a new computer. You expect a 5 percent increase in price each year. What would the new computer cost at EOY 5?

        (*Ans.* $63,814)

    (d) What is the unpaid balance on the current computer after five years?

        (*Ans.* $29,752)

D.2. Your firm asks your help in establishing a revolving student loan fund to help children of the firm's employees finance their college education. The provisions of the fund follow:

    1. Each student is eligible for a loan of $5,000 per year at the *beginning* of each year of the four college years.

    2. Interest accrues at 0.5 percent per *month* on the outstanding balance of the loan, compounded at the end of each *month,* until repaid in full.

    3. Repayments are to be made in 10 equal end-of-*year* payments, with the first payment due two years *after* graduation.

        (a) Draw a cash flow diagram for the program, using only one student as an example.

        (b) Find the amount of annual repayment, A.

D.3. Your hometown bank offers to loan you money to finance your last two years in college. They will loan you $4,000 today, and another $4,000 one year from today. They will accept repayment in 10 equal annual installments with 15 percent interest on all unpaid balances. The first annual repayment is due at the end of your first full year of work, three years from today. What is the amount of each installment payment?

    (*Ans.* A = $1,970)

D.4. A transportation authority asks you to check on the feasibility of financing for a toll bridge. The bridge will cost $2,000,000 at PTZ. The authority can borrow this amount to be repaid from tolls. The bridge will take two years to construct, and then be open for traffic at EOY 2. Tolls will be accumulated throughout the third year and will be available for the initial annual repayment at EOY 3. In subsequent years the tolls are deposited at the end of each year. The authority is expecting an average of 10,000 cars per day every day of the year. How much must be charged to each car in order to repay the borrowed funds in 20 equal annual installments (first installment due at EOY 3), with 8 percent compound interest on the unpaid balance?

D.5. Your client is trying to arrange financing for the construction of a new $1,200,000 office building. The building will take 12 months to construct. In order to make monthly payments to the contractor, your client needs to borrow $100,000 per month at the end of each of the 12 months during which construction is underway. He agrees to repay the entire amount owed in 240 monthly payments, with the first repayment at EOM 14. The lender charges a nominal 12 percent interest compounded monthly on the balance owed at all times, including the construction period.

(a) Draw a line diagram illustrating the cash flow for this problem.

(b) How much does your client owe the lender at the end of the twelfth month after making the last payment to the contractor?

(*Ans.* $1,268,250)

(c) How much are the monthly repayments?

(*Ans.* $14,102)

D.6. A wealthy father whose daughter is just getting married wishes to set up a college education fund for his grandchildren-to-be that will pay $10,000 per year for 10 years commencing 20 years from today (10 payments, with first payment at EOY 20). What annual deposit must be made, starting with the first deposit today and continuing for the next 19 years (20 deposits), if $i = 9$ percent?

D.7. Assume that five years ago the county borrowed $10,000,000 for a road construction program. The existing loan is being repaid in 240 equal monthly installments, with interest on the unpaid balance at 0.5 percent per month compounded monthly. The sixtieth payment has just been made. The source of the repayment funds is a gas tax, which currently yields an average of $200,000 per month.

(a) How much of the $200,000 per month gas tax income is needed to pay off the existing loan?

(*Ans.* $A_1 = $71,600/month)

(b) If state law allows you to pledge no more than 80 percent of the average monthly income toward repayment of loans, how much is available (in dollars per month) to pledge to repay a new loan?

(*Ans.* $A_2 = $88,400)

(c) If cash is needed now for more construction, how much new money could be borrowed based on the monthly income available in (b) if $i = 0.75\%$ per month and $n = 240$ for the new loan? ($P_1 = $9,825,000)

(d) If the full 80 percent of $200,000 were used to repay a new loan ($i = 0.75\%/$ month, and $n = 240$), and part of the cash received from the new loan were used to pay off the old loan in one lump sum now, how much net cash lump sum would be available for new construction? [*Hint:* **(i)** Find the lump sum cash balance due on the existing loan. **(ii)** Find the new loan amount that can be financed with the 80 percent of $200,000 per month. **(iii)** Subtract (i) from (ii).]

(*Ans.* $P = $9,297,800)

(e) Which yields the most cash, plan (c) or plan (d)?

(f) If the interest on the new loan were 0.5 percent per month, which would yield the most cash? (No calculations needed.) Why?

D.8. A prosperous consulting engineer decides to aid student engineers by establishing a student loan fund. The provisions of the loan follow:

1. Each student is eligible for a loan of $5,000 per year at the *beginning* of each year of the four college years.

2. Interest accrues at 4 percent per *year* on the outstanding balance of the loan, compounded at the end of each, *year,* until repaid in full.

3. Graduation occurs at EOY 4, and repayments are to be made in 10 equal end-of-year payments, with the first payment due two years *after* graduation.

   (a) Draw a cash flow diagram for the program, using only one student loan as an example.

   (b) Find the equal annual series repayment amounts.

   (c) If the consultant borrows at 15 percent, find the present value of the subsidy for each loan. (In terms of PV, how much does each loan cost the consultant?)

D.9. An equipment-leasing company leases earthmovers (rubber-tired scrapers) at a monthly rate. Drive-wheel tires for the scraper last about six months and cost about $12,000 a pair. The company buys tires at EOM 0 and each six months thereafter for the 10-year life of the scraper (no expenditure at EOM 120). How much in terms of dollars per month should the leasing company charge just to cover the cost of the tires if they borrow and lend at 1.5 percent per month? The first lease payment is received at EOM 1, and the last at EOM 120.

(*Ans.* $2,106/month)

D.10. Assume your firm is negotiating a loan of $100,000, with interest compounded monthly on the unpaid balance at 1.5 percent per month. Instead of the usual monthly payments, your firm offers to repay the loan in 20 equal *annual* installments of principal plus interest over a 20-year period. The first payment is due at EOM 12.

(a) Find the amount of the annual payment.

(b) If the firm wanted to pay off the balance due on the loan immediately after making the fifth payment, how much would be due?

D.11. Assume your firm borrowed $400,000 just 36 months ago at 1 percent per month interest and is repaying the loan at the rate of $10,000 per month.

(a) From the time the money was borrowed, how many months will it take to pay off the loan?

(*Ans.* 51 full payments plus 1 partial at EOM 52)

(b) How much is the last payment?

(*Ans.* $3,387)

(c) The firm wants to pay off the balance in one lump sum now, just after the thirty-sixth payment has been made. How much do they owe?

(*Ans.* $147,179)

D.12. Your firm is conducting a feasibility study on a proposed water treatment plant which will require major maintenance at eight-month intervals over the 20-year life of the plant. The first major maintenance is scheduled for the EOM 8, and the last for EOM 232. Each major maintenance is expected to cost $10,000. The interest rate compounded monthly is $i = 1$ percent per month. Find the present worth of the periodic series of expenditures for major maintenance.

Problem Group E

Perpetual ($n \rightarrow \infty$) or very long lives. Find P, A, or i.

E.1.   A grateful college graduate wishes to set up a $1,000-a-year perpetual scholarship in his name at his alma mater. If he can receive 8 percent on this investment, how much money will be required to set up the scholarship?

(*Ans.* $12,500)

E.2.   A dam costing $17,000,000 is expected to have annual maintenance expenses of $2,000 per year. If $i = 7$ percent and the expected life is 100 years,

(a) What is the equivalent present worth of the annual maintenance expenses over the 100-year period?

(b) What is the equivalent present worth of the annual maintenance expenses if the dam lasts forever?

(c) What is the equivalent annual worth of the purchase price of the dam over the 100-year period?

(d) What is the equivalent annual worth of the purchase price of the dam if the dam lasts forever?

E.3.   A major real estate investor is considering the purchase of a land site near downtown Houston for $10,000,000. He plans to construct a 50-story building that is estimated to cost $80,000,000. He assumes the building will last 50 years with zero salvage value. For $i = 15$ percent:

(a) What equivalent annual cost will be incurred for the building and land, assuming the land will not change in value during the 50-year period and can be sold for $10,000,000 at EOY 50?

(*Ans.* $13,511,100)

(b) How far off would the estimate of the annualized cost be if **(i)** the resale value of the land were zero?

(*Ans.* annual cost increased about $1,000/yr); **(ii)** the resale value of the land increased fivefold (worth $50,000,000 in 50 years)?

(*Ans.* annual cost decreased about $6,000/yr from (a)).

E.4.   Sky boxes have just been constructed in a football stadium at a cost of $5,000,000 paid at EOM 0. In addition to the construction, heavy maintenance is expected to occur every 24 months thereafter for a long time to come ($n \to \infty$) and cost $100,000 for every occurrence. The first heavy maintenance will occur at EOM 24, and then again at EOM 48, 72, and so on. Assume $i = 1$ percent per month.

(a) Find the present worth (capitalized cost) of the project.

(b) Assume income from the skyboxes occurs monthly for 3 months out of each 12. Find the monthly income required to pay for all costs mentioned above if the first income occurs at EOM 8, 9, 10 and every 12 months thereafter (e.g., EOM 20, 21, 22, then 32 33, 34 for a very long time, so $n \to \infty$).

E.5.   The construction of south-end zone seating in a football stadium is scheduled to occur 12 months from now at a cost of $10,000,000. In addition to the construction cost, heavy maintenance is expected to occur every 24 months thereafter for a long time to come ($n \to \infty$) and cost $100,000 for every occurrence. Assume $i = 0.75$ percent per month (first heavy maintenance at EOM 36, then EOM 60, 84, 108, etc.).

(a) Draw a representative section of the cash flow diagram.

(b) Find the capitalized cost (present worth) of this project.

(*Ans.* $9,607,800)

(c) Assume income occurs monthly for 3 months out of each 12. Find the monthly income required to pay for all costs mentioned above if the first income occurs at EOM 15, 16, and 17 and every 12 months thereafter (e.g., EOM 27, 28, 29, then 39, 40, 41, etc.).

(*Ans.* $309,520)

E.6. Assume you are the consultant to the county, which asks your help in finding money for a much-needed county road and bridge construction program. The county currently has an income of $10,000 per month to use for roads and bridges, but the whole amount is being used every month to pay off an existing loan. The existing loan was made 10 years ago; the amount borrowed was $1,395,808, which is being repaid in 240 monthly installments of $10,000 each with interest on the unpaid balance at 0.5 percent per month compounded. The one hundred-and-twentieth payment has just been made. To raise new money for the new road and bridge program, you suggest refinancing the loan, involving the following steps:

1. Pledge the $10,000 per month to pay off a new loan. (Money is available at 0.75 percent per month for 300 months.)
2. Use part of the new loan to pay off the old loan.
3. Use the remainder of the new loan to finance the new road and bridge program.
    (a) What is the balance due on the old loan after the one-hundred-and-twentieth payment is made?
    (b) How much could be borrowed with repayments of $10,000 per month for 300 months? (Do not subtract the old loan at this point.)
    (c) Under this plan, how much new money would be available for the new road and bridge program after paying off the existing loan?

E.7. Assume that a drainage project is under study that would require heavy maintenance costing $10,000 every six months. The first heavy maintenance occurs at EOM 6. With $i = 1$ percent per month, find the present worth of the heavy maintenance expenditures if $n = \infty$.

(*Ans.* $162,600)

CHAPTER 5

Arithmetic Gradient, *G*:
The Constant Increment to
a Series of Periodic Payments

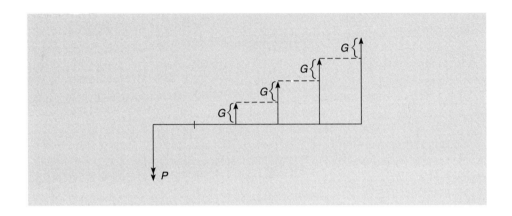

KEY EXPRESSIONS IN THIS CHAPTER

G = Arithmetic gradient increase in funds flow at the end of each period for n periods.

P & i = Principal and Interest, the two parts of each periodic payment for an amortization schedule.

PAYMENT SERIES WITH A CONSTANT INCREMENT

Engineers frequently encounter situations involving periodic payments that increase or decrease by constant increments from period to period. For example, annual maintenance costs on a proposed engineering project are estimated as shown in the following table:

Year	Maintenance
1	$ 0
2	1,000
3	2,000
4	3,000
etc.	etc.

The growth increment G in this example is $G = \$1,000$ per year and is termed a "gradient." Since the increment of the gradient is a *constant amount* at $\$1,000$, the progression is described as arithmetic (compared to geometric if the growth increases by the same *percent* each year, as in Chapter 6).

This situation occurs often enough to warrant the use of special equivalence factors relating the arithmetic gradient to other cash flows. In this chapter we will be determining equivalency of an arithmetic gradient G to a single future value F, a single present worth P, and a uniform series A.

FUTURE VALUES OF GRADIENT AMOUNTS (F/G)

As with a uniform series, gradients are end-of-year disbursements. Notice that the maintenance cost table above starts with a value of $\$0$ at the end of year 1 and then increases by a G amount of $\$1,000$ per year. The cash flow for such a series is shown in Figure 5.1.

The gradient amount, or per-term increase, is designated as G in the generalized problem and can be either positive or negative, but always increases in absolute value with increasing time (the values grow larger from left to right on the cash flow diagram), as shown in Figure 5.2.

Generalized Derivation of the Equation Relating G to Future Value F (Find F, given G, i, n)

An investor decides to invest on an arithmetic gradient schedule as shown in the table below:

End of year	Amount invested
1	$0G$
2	$1G$
3	$2G$
$\vdots$	$\vdots$
n	$(n-1)G$

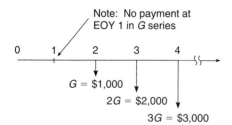

Note: No payment at EOY 1 in G series

$G = \$1,000$

$2G = \$2,000$

$3G = \$3,000$

Figure 5.1   Arithmetic gradient series.

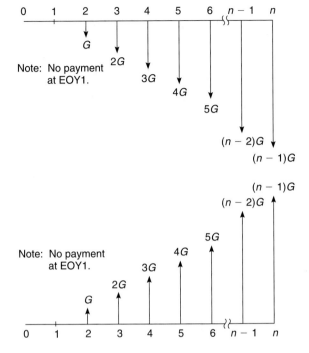

Figure 5.2 Cash flow diagrams for arithmetic gradient (*G*).

If the investor earns an interest rate of *i* on the investment, what is the future value of this investment at the end of year *n*?

Solution. In order to arrive at a simpler equation, begin by considering an arithmetic gradient to be the sum of an overlapping group of uniform series of annual payments, each of value $A = G$, but for decreasing time periods beginning with $n - 1$, as shown in Figure 5.3. Thus,

$$F = G(F/A, i, n - 1) + G(F/A, i, n - 2) + \cdots + G(F/A, i, n - n + 1)$$

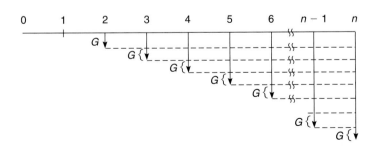

Figure 5.3 Overlapping group of uniform series of annual payments.

From Equation 3.2

$$(F/A, i, n) = \frac{(1+i)^n - 1}{i}$$

$$F = G\left[\frac{(1+i)^{n-1} - 1}{i} + \frac{(1+i)^{n-2} - 1}{i} + \cdots + \frac{(1+i) - 1}{i}\right] \qquad (5.1)$$

Simplifying

$$F = \frac{G}{i}[(1+i)^{n-1} + (1+i)^{n-2} + \cdots + (1+i) - (n-1)] \qquad (5.2)$$

Equation 3.1 is

$$F = A[(1+i)^{n-1} + (1+i)^{n-2} + \cdots + (1+i)^2 + (1+i) + 1] \qquad (3.1)$$

Subtracting 3.1 from 5.2 yields

$$\frac{Fi}{G} - \frac{F}{A} = -n$$

or

$$F = \frac{G}{i}\left(\frac{F}{A} - n\right) \qquad (5.3)$$

But $F/A = [(1 + i)^n - 1]/i$ (from Equation 3.3). Therefore,

$$F = \frac{G}{i}\left[\frac{(1+i)^n - 1}{i} - n\right] \qquad (5.4)$$

or, in functional notation form

$$F = G(F/G, i, n) \qquad (5.4a)$$

Values for F/G are listed in Appendix A for commonly occurring values of i and n. The use of this equivalence factor is demonstrated in the following examples.

EXAMPLE 5.1   (Find F, given G, i, n)

A student scheduled to graduate this year is planning on saving for a major purchase and decides to invest on the following schedule. The investment will earn 6 percent compounded annually. How much will he have saved after five years?

Solution. The cash flow diagram is shown in Figure 5.4, and the problem is solved using the general equation as follows:

$$F_2 = G(F/G, i, n) = \frac{200}{0.06}\left[\frac{(1 + 0.06)^5 - 1}{0.06} - 5\right] = \underline{\underline{\$2,124}}$$

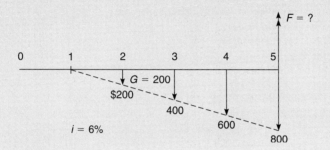

Figure 5.4 Cash flow diagram for Example 5.1.

Graph of Increase in *F* as *n* Increases

Figure 5.5 illustrates the timing of the cash flows of G, as well as the relationships between G, F, i, and n, as illustrated by Example 5.1. Note that the first deposit of G ($200 in this example) occurs at the *end* of the *second* period from ETZ for an arithmetic gradient. Therefore the first interest is not earned until the end of the third period. Gradient deposits are made at the end of every year except the first year, so there are a total of $(n - 1)$ deposits in the G series (as contrasted to n deposits in the A series). Deposits are increased by the amount G ($200 in this case) for each period (year for this example). Therefore, the amount of the last deposit is $G(n - 1)$, or in this example $200(5 - 1) = \$800$. The ability to determine the correct value of n is essential in solving many problems commonly encountered by the professional engineer. Since there are always $(n - 1)$ gradient deposits in the G series, the value of n may be determined as the number of gradient deposits plus one.

Graph of *F/G* and *G/F*

A graph of the equation

$$F/G = \frac{1}{i}\left[\frac{(1 + i)^n - 1}{i} - n\right]$$

and its inverse is shown in Figure 5.6.

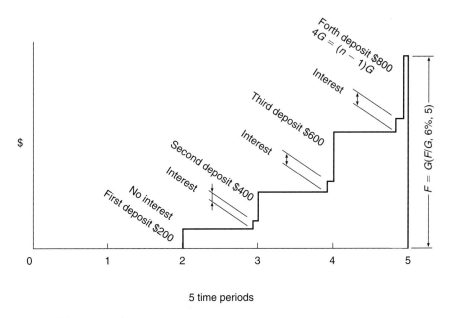

Figure 5.5   Timing of the cash flows of $F = \$200\ (F/G, 6\%, 5)$.

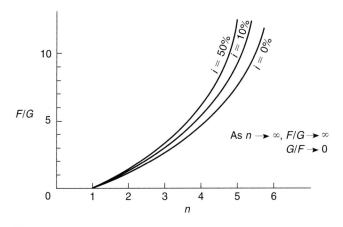

Figure 5.6   Graph showing the F/G relationship.

DECREASING ARITHMETIC GRADIENT

With each succeeding payment, the arithmetic gradient can only *increase* in absolute value (increase in size, whether it increases in positive size or negative size). Therefore for problems involving a *decreasing* gradient an annual series must be subtracted from the gradient series. For example, consider a series that begins with an initial payment of

$1,000 at the EOY 1. Each subsequent payment decreases by $100, so that at EOY 2 the payment is $900, at EOY 3 it is $800, and so on.

The future value of this decreasing gradient series is obtained by finding the future value of an annual series of $1,000 payments and then subtracting the future value of the arithmetic gradient series ($G = -\$100$) from it. Example 5.2 illustrates the procedure.

EXAMPLE 5.2   (Find *F*, given *A*, *G*, *i*, *n* with a decreasing gradient, *G*)

Your client is a contractor who needs to save $500,000 to replace worn out equipment upon completion of a five-year contract for a mining company. To accumulate the needed savings, the bid on the contract includes an extra $100,000 the first year to be deposited at EOY 1 in a fund earning 10 percent compounded annually. Due to declining productivity as the equipment grows older, only $90,000 is included the second year, $80,000 the third year, and so on. Determine whether or not the full $500,000 will be accumulated in the fund by the end of the fifth year.

Solution.   The problem is solved by separating it into two parts (the *A* series, and the *G* series), solving for the respective future values, *F*, and summing. The cash flow diagram for this problem is shown in Figure 5.7.

The declining gradient is solved by breaking it into two components: 1) a uniform series, and 2) an increasing gradient indicated by the shaded area Y. As can be seen in the figure, the uniform series *A* will have as its value the highest value of the declining gradient. We solve the series for an equivalent value F_1 and then solve the increasing gradient indicated by the shaded area, for its value F_2. Subtracting F_1 from F_2 results in an equivalent future value of *F* for the declining gradient.

Notice the use of signs to indicate a gradient *G* in the opposite direction from the uniform series *A*. The $100,000 annual series payments are shown as downward arrows and negative, in the calculations, while the $10,000 gradient *G* series is shown as upward arrows and positive.

$$A = -\$100,000/\text{yr}, \quad i = 10\%, \quad n = 5 \text{ for both } A \text{ and } G$$
$$G = +\$10,000/\text{yr}$$

The future value of an annual deposit of $100,000 per year is

$$F_1 = A(F/A, i, n) = -\$100,000 \underbrace{(F/A, 10\%, 5)}_{6.1051} = -\$610,510$$

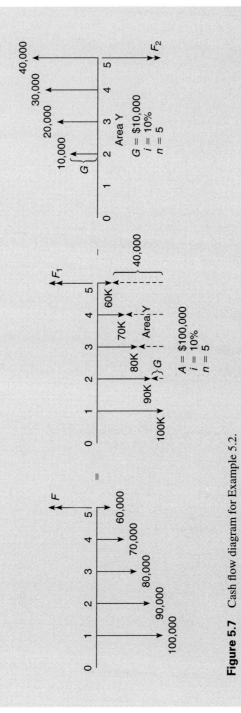

Figure 5.7 Cash flow diagram for Example 5.2.

Then the future value of the positive gradient of +$10,000 per year is

$$F_2 = G(F/G, i, n) = +\$10,000 \times 11.051 = +\$110,510$$
$$F = F_1 + F_2 = -\$610,510 + \$110,510 = \underline{\underline{-500,000}}$$

The total future value of the deposits is $500,000. The answer is yes, this plan will provide the $500,000 by the end of the five years if each deposit is invested at 10 percent interest compounded annually.

EXAMPLE 5.3   **(Decreasing gradient: Find *G* and *A* in terms of dollars per cubic yard, given *F*, *i*, *n*, with *G* and *A* in terms of cubic yards per hour)**

Your contractor client forsees a need for a total of $160,000 10 years from today in order to replace a dragline. The dragline is expected to move about 100,000 yd^3 of earth this year, but due to age and increasing "down time," this amount will decrease by about 5,000 yd^3 per year each year over the 10-year period. Other operating and maintenance (O & M) costs run about $1.00 per yd^3. Determine the additional charge per cubic yard required to accumulate the $160,000 in 10 years if all end-of-year deposits earn 10 percent interest compounded on the amount invested.

Solution.   First, identify the variables:

$$F = \$160,000$$
$$n = 10 \text{ yr}$$
$$A = (100,000 \text{ yd}^3/\text{yr})(y/\text{yd}^3)$$
$$G = (-5,000 \text{ yd}^3/\text{yr})(y/\text{yd}^3)$$
$$i = 10\%$$

The "other O & M costs" of $1.00 per yd^3 is extraneous information which is often available in practice, but is not needed to solve the problem. The inclusion of such information here simulates the real world, where a myriad of useless information is often available to confuse the problem.

One basic problem here is translating cubic yards per year into dollars per year. This may be done by letting *y* represent the extra charge in terms of dollars per cubic yard. Then find how many dollars per cubic yard times cubic yards per hour are needed to accumulate the $160,000.

$$A = (100,000 \text{ yd}^3)(y), \text{ the annual base income}$$
$$G = (-5,000 \text{ yd}^3)(y), \text{ the declining gradient}$$
$$F = A(F/A, i, n) - G(F/G, i, n)$$

$$\$160,000 = (100,000y)\underbrace{(F/A, 10\%, 10)}_{15.937} - 5,000y\underbrace{(F/G, 10\%, 10)}_{59.374}$$

$$\$160,000 = 1,593,700y - 296,870y$$

$$y = \frac{160,000}{1,296,830} = \$0.0617 = \underline{\underline{\$0.124/yd^3}}$$

By charging an extra \$0.124 per yd^3 and investing the resulting funds at the end of each year at 10 percent compounded, the contractor will accumulate \$160,000 at the end of 10 years to purchase a new dragline.

Examination of Examples 5.1, 5.2, and 5.3 reveals that a uniform series A was involved with the gradient G. This is often the case in real-life problems in which an investment, or disbursement, is initially made and then decreased (or increased) by a constant amount in succeeding years.

EQUIVALENT UNIFORM SERIES VALUES OF GRADIENT AMOUNTS (*A/G*)

A problem often requires an answer in terms of a uniform series A, rather than a simple value such as F or P, that is equivalent to a gradient series G. The cash flow diagram for this type of problem is given in Figure 5.8.

An equation relating A and G can be developed as follows.

Referring to Equation 5.3,

$$F = \frac{G}{i}\left(\frac{F}{A} - n\right)$$

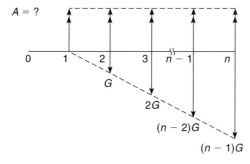

Figure 5.8   Uniform series values equivalent to arithmetic gradient series.

and dividing both sides by G, then

$$\frac{F}{G} = \frac{1}{i}\left(\frac{F}{A} - n\right) \qquad (5.5)$$

Multiplying both sides by A/F yields

$$\frac{A}{F} \times \frac{F}{G} = \frac{1}{i}\left[\frac{A}{F} \times \frac{F}{A} - \frac{A}{F}(n)\right]$$

$$\frac{A}{G} = \frac{1}{i}\left[1 - \frac{A}{F}(n)\right] \qquad (5.6)$$

Substituting the equation for A/F (Equation 3.4) yields

$$A = G\left[\frac{1}{i} - \frac{n}{(1+i)^n - 1}\right] \qquad (5.7)$$

In functional notation this relationship becomes

$$A = G(A/G, i, n) \qquad (5.7a)$$

Values for $(A/G, i, n)$ are tabulated in Appendix A for commonly occurring combinations of i and n.

Graph of *A/G*

A graph of the equation

$$A/G = \frac{1}{i} - \frac{n}{(1+i)^n - 1}$$

is shown in Figure 5.9. Knowing how A/G values respond to changes in i and n will prove helpful in evaluating the effect on problem solutions of possible variations in i and n.

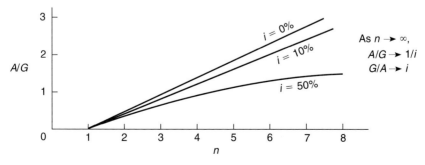

Figure 5.9 Graph of A/G relationship.

Account-Balance Graph of *A* Equivalent to *G*

Figure 5.10 illustrates an *A* series that is equivalent to a *G* series. The graph shows a balance in the account over a five-year period resulting from annual deposits of the *A* series and withdrawals of the *G* series. The *A* payments are considered as a cash flow into the account balance (graph moves upward), and the *G* payments are cash flow out (down on the graph). Interest is paid at the end of each period on whatever balance has been in the account for the duration of that period. Note that the first *A* payment occurs at the end of the first period, whereas the first *G* payment does not occur until the end of the second period. At the end of the final period three payments occur: (1) an interest payment, (2) a payment of *A,* and (3) a *G* payment amounting to $(n - 1)G$. Since the *A* series is exactly equivalent to the *G* series, the graph ends with a zero balance in the account at the end of the final period.

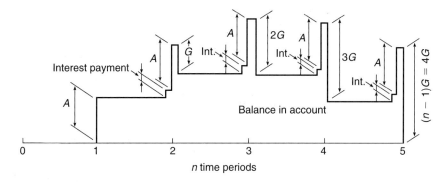

Figure 5.10   Graph of balance in the account for an *A* series of deposits combined with an equivalent *G* series of withdrawals.

The examples that follow illustrate the A/G relationship. Example 5.4 shows a negative annual income, *A*, coupled with a positive gradient, *G*. The method of solution is to divide the problem into simple components and solve as usual.

EXAMPLE 5.4   (Given *A, G, i, n,* find equivalent *A*)

A bridge authority collects $0.50 per vehicle and deposits the revenues once a month in an interest-bearing account. Currently 600,000 vehicles a month cross the bridge, and the traffic count is expected to increase at the rate of 3,500 vehicles per month added for each succeeding month. What equivalent uniform monthly revenue would be generated over the next five years if the interest in the account were (1) 6 percent nominal interest compounded monthly (0.5 percent per month), and (2) 6 percent compounded annually?

Solution for Part 1, with Monthly Compounding.   The monthly costs for the problem in terms of *A* and *G* are determined as follows:

$$A = 600,000 \text{ vehicles/month} \times \$0.50/\text{vehicle} = \$300,000/\text{month}$$
$$G = 3,500 \text{ vehicles/month} \times \$0.50/\text{vehicle} = \$1,750/\text{month}$$
$$n = 5 \text{ yr} \times 12 \text{ months/yr} = 60 \text{ months}$$

The cash flow diagram is shown in Figure 5.11.

$$A_1 = \$300,000/\text{month} = \$300,000/\text{month}$$
$$A_2 = 1,750(A/G, 0.5\%, 60) = \$ \ 49,000/\text{month}$$
$$A = A_1 + A_2 = \underline{\$349,000/\text{month}}$$

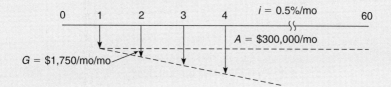

Figure 5.11   Cash flow diagram for Example 5.4, Solution 1.

Solution for Part 2, with Annual Compounding.   Since interest in part 2 is earned only once each year, the payments can be grouped together at the end of each year, negating any benefit from depositing at monthly periods (see Chapter 4 for a discussion of this situation). Thus, the base annual income deposited at EOY 1 is

$$A_1 = [600,000 \text{ vehicles/month} \times 12 + 3,500 \text{ vehicles/month/month}$$
$$\times (1 + 2 + \cdots + 11) \ \$0.50/\text{vehicle}] = \$3,715,500/\text{yr}$$

The gradient income deposited at the end of each subsequent year is

$$G = 3,500 \text{ vehicles/month/month} \ (1 + 2 + 3 + \cdots + 12)\text{yr}$$
$$\times \$0.50/\text{vehicle} = \$136,500/\text{yr/yr}$$

See Figure 5.12 for the cash flow diagram.
  Solving for the *annual A* yields

$$A_1 = \qquad\qquad\qquad\qquad\qquad \$3,715,500/\text{yr}$$
$$A_2 = \$136,500 \underbrace{(A/G, 6\%, 5)}_{1.8836} = \qquad \underline{257,312}$$

$$\text{annual } A = \qquad\qquad\qquad\qquad \$3,972,600/\text{yr}$$

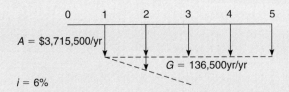

Figure 5.12   Cash flow diagram for Example 5.4, Solution 2.

Since interest is earned only at the end of each *annual* period, the monthly equivalent is simply one-twelfth of the annual equivalent. Thus

$$\text{monthly } A = \$3,972,600/12 = \$331,000/\text{month}$$

(Compare to $A = \$349,000/\text{month}$ in Solution 1.)

This answer is significantly different from Solution 1, even though the interest rate does not appear to be too different. (For Solution 1 the annual $i_e = (1.05)^{12} - 1 = 6.17\%$ versus 6% in Solution 2.) The reason for this difference is the simplifying assumption that receipts that occur between the points in time at which compounding occurs (intermediate funds flow) earn *no* interest until the *next* complete compounding period ends. Thus, the precise terms of interest-compounding periods and earnings on intermediate funds flow become very important in actual practice.

EXAMPLE 5.5   (Find G, given P, A, i, n)

A young engineer decides to go into business and open a consulting office. The anticipated income will be low to begin with but will gradually increase. In order to provide cash to start up the office, a bank will lend $200,000 at 12 percent for 10 years, to be amortized in 120 monthly payments. If the repayment of the loan requires equal monthly payments, the engineer would have to pay

$$A = \$200,000(A/P, 1\%, 120) = \$2,870/\text{month}$$

However, payments of this size would be a big financial drain for the new business during the early months, so the lender is willing to take graduated gradient payments as long as the principal balance outstanding earns 12 percent. They agree to begin with a payment of $1,000 at EOM 1, and then raise the payment amount each month throughout the 120-month repayment period. Find G, the amount of increase needed each month in order to pay off the loan by EOM 120.

Solution.   The monthly payments to amortize the $100,000 loan are found as

$$A_1 = \$200,000(A/P, 1\%, 120) = \$2,870/\text{month}$$

By agreement, the initial monthly payment is

$$A_2 = \$1,000/\text{month}$$

The equivalent monthly payment deficit to be made up by the periodic increase is

$$\text{deficit} = A_1 - A_2 = \$2,870 - \$1,000 = \$1,870/\text{month}$$

The required monthly gradient increase in payment is found as

$$G = \$1,870/\underbrace{(A/G, 1\%, 120)}_{47.834} = \underline{\underline{\$39.09}}$$

Check:

$$P = A(P/A, i, n) + G(P/G, i, n)$$
$$\$200,000 = 1,000(P/A, 1\%, 120) + 39.09(P/G, 1\%, 120)$$
$$\$200,000 = \qquad \$69,701 \qquad + \qquad \$130,342$$

Conclusion.   Instead of a series of 120 equal repayments of $2,870 per month, the loan may be repaid with 120 payments beginning with $1,000 at EOM 1, and then increasing by $39.09 per month ($1,000 at EOM 1, $1,039.09 at EOM 2, $1078.18 at EOM 3, etc.). No doubt the engineer would be concerned about the growing size of the payments as time goes on. The amount of the last payment at EOM 120 is found as

$$\text{last payment} = A + G(n - 1)$$
$$= \$1,000 + \$39.09 \times (120 - 1) = \$5,650 \text{ at EOM } 120$$

EQUIVALENT PRESENT WORTH VALUES OF GRADIENT AMOUNTS (*P/G*)

This *P/G* series provides for solution of problems involving a payment (cost or income) that changes (increases or decreases) by $G per period for *n* periods. Interest is earned on any balance that remains in the loan or deposit. The equation is derived by finding the present sum required to finance this *G* series for *n* periods (given *G*, *i*, *n*, find *P*).

Derivation of the *P/G* Equivalence Factor

It was previously found from Equation 5.3 that

$$\frac{F}{G} = \frac{1}{i}\left(\frac{F}{A} - n\right)$$

Both sides may be multiplied by P/F as follows:

$$\frac{F}{G} \times \frac{P}{F} = \frac{1}{i}\left[\frac{F}{A} \times \frac{P}{F} - n\left(\frac{P}{F}\right)\right]$$

The F's cancel out to provide the general equation relating P, G, i, and n.

$$\frac{P}{G} = \frac{1}{i}\left[\frac{P}{A} - n\left(\frac{P}{F}\right)\right] \quad \text{or} \quad P/G = \frac{1}{i}\left[\frac{(1+i)^n - 1}{i(1+i)^n} - \frac{n}{(1+i)^n}\right] \qquad (5.8)$$

In notation form,

$$P = G(P/G, i, n) \qquad (5.8a)$$

Values for P/G are listed in Appendix A.

Graph of *P/G*

A graph of the equation

$$P/G = \frac{1}{i}\left[\frac{(1+i)^n - 1}{i(1+i)^n} - \frac{n}{(1+i)^n}\right]$$

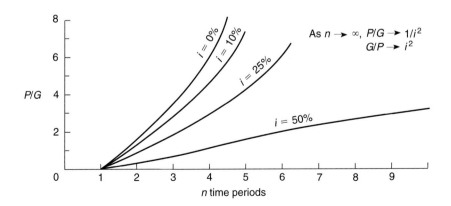

Figure 5.13   Graph of the P/G relationship.

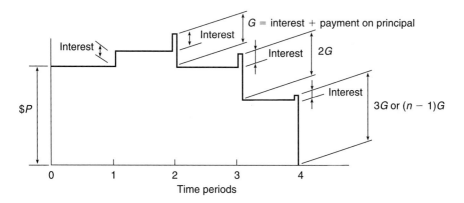

Figure 5.14   Graph of balance of investment *P* with (*n* − 1) repayments periodically increasing by amount *G*.

is shown in Figure 5.13. Figure 5.14 illustrates a graph of the balance of investment *P* with (*n* − 1) repayments periodically increasing by an amount *G*. Example 5.6 illustrates an application of the *P*/*G* equation in finding a present value, *P*, given a gradient, *G*.

EXAMPLE 5.6   **(Find *P*, given *G*, *i*, *n*)**

The repair costs on a water works pump are expected to be zero the first year and increase by $1,000 per year over the five-year life of the pump. A fund bearing 6 percent on the remaining balance is set up in advance to pay for these costs. How much should be in the fund? Costs are billed to the fund at the end of each year.

Solution.   Using the equation:

$$P = \frac{G}{i}\left[\frac{(1+i)^n - 1}{i(1+i)^n} - \frac{n}{(1+i)^n}\right]$$

$$P = \frac{1,000}{0.06}\left[\frac{(1+0.06)^5 - 1}{0.06(1+0.06)^5} - \frac{5}{(1+0.06)^5}\right]$$

$$P = 1,000 \times 7.935 = \underline{\underline{\$7,935}}$$

Solution.   Using tables in Appendix A:

$$P = G(P/G, 6\%, 5) = \$1,000(7.9345) = \underline{\underline{\$7,934}}$$

The slight difference in the answers is due to rounding off of the factors.

The following example demonstrates applications of all three arithmetic gradient equations to the same problem situation. Each application is then explained to help clarify the mental image of each solution.

EXAMPLE 5.7　(Find P, A, F, given G, i, n)

To operate, maintain, and repair a motor grader costs $2,000 the first year and increases by $1,000 per year thereafter, so that the second year's costs are $3,000, the third year's, $4,000, and so forth, for five years; $i = 8$ percent. Find (1) the present worth of these costs; (2) the equivalent uniform annual worth of these costs; (3) the future worth of these costs.

Solution 1

$$P = A(P/A, i, n) + G(P/G, i, n)$$
$$P = 2,000(P/A, 8\%, 5) + 1,000(P/G, 8\%, 5)$$
$$P = \$7,986 + \$7,372 = \underline{\underline{\$15,358 \text{ present worth}}}$$

Comment: If $15,358 were put into an account drawing 8 percent interest on the balance left in the account, and $2,000 were drawn out at the end of the first year to pay for operating and maintenance, $3,000 the second year, and so forth, the account would be drawn down to zero at the end of the fifth year.

Solution 2

$$A = A + G(A/G, i, n)$$
$$A = \$2,000 + 1,000(1.8465)$$
$$A = \$2,000 + 1,846 = \underline{\underline{\$3,846/\text{yr annual worth}}}$$

Comment: If $3,846 were paid into an account drawing 8 percent interest every year, $2,000 could be drawn out the first year, $3,000 the second year, and so forth, and the fifth year the account balance would be zero.

Solution 3

$$F = A(F/A, i, n) + G(F/G, i, n)$$
$$F = \$2,000(5.8666) + 1,000(10.832)$$
$$F = \$11,734 + 10,832 = \$22,566 = \underline{\underline{\$22,566 \text{ future worth}}}$$

Comment: If $2,000 were borrowed at 8 percent at the end of the first year to pay for operating and maintenance expenses, $3,000 borrowed for the second year, and so forth, at the end of the fifth year, the amount owed would be $22,566.

LATE START FOR GRADIENT SERIES (ETZ ≠ PTZ)

The arithmetic gradient series equation assumes, by definition, that the first payment occurs at the end of the second period after ETZ. In a problem where the ETZ does not coincide with PTZ (if the first gradient payment occurs at some point in time other than EOY 2 of the problem at hand), then an additional calculation is needed in order to find the equivalent present worth or annual worth, as illustrated by Example 5.8.

EXAMPLE 5.8 **(ETZ ≠ PTZ: Find *P*, given *G*, *i*, *n*)**

A county engineer is thinking of lining a certain drainage ditch in order to save on maintenance costs. The ditch lining is expected to save the following amounts over its 11-year life. With $i = 12$ percent, find the present worth of the savings.

EOY	Income
1 through 4	$ 0
5	1,000
6	2,000
7	3,000
8	4,000
9	5,000
10	6,000
11	7,000

Assume there is no salvage value and that after the eleventh year the ditch lining will be worthless.

Solution.   To help clarify the problem, a cash flow diagram is sketched as shown in Figure 5.15. This problem may be solved by means of a four-step procedure as follows.

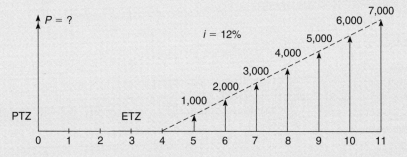

Figure 5.15   Cash flow diagram for Example 5.8.

Step 1. Find ETZ. The arithmetic gradient series equation assumes that the ETZ is located at a point two periods before the first gradient payment. In this problem the first gradient payment occurs at EOY 5. Two periods before EOY 5 is EOY 3, so the ETZ occurs at EOY 3.

Step 2. Find n. Either of two methods can be used to find n.

Alternative Method A. Subtract the n that corresponds to the ETZ from the n that represents the last payment in the gradient series. In this problem subtract EOY 3 (found in step 1) from EOY 11 and find $n = 8$.

Alternative Method B. Recall that $n =$ the number of gradient payments plus 1. In this problem the number of gradient payments is 7, so $n = 7 + 1 = 8$.

Step 3. Find PW at ETZ. The PW of the gradient is found at EOY 3 as

$$P_1 = G(P/G, 12\%, 8) = \$1{,}000 \times 14.471 = \$14{,}471$$

Step 4. Find PW at PTZ. This step consists of moving the PW (P_1) from EOY 3 to EOY 0, which requires P_1 to become F_1 for purposes of moving it back in time. Therefore,

$$F_1 = P_1$$
$$P_2 = F_1(P/F, 12\%, 3) = \$14{,}471 \times 0.71178 = \underline{\underline{\$10{,}300}}$$

Conclusion. The savings have a present worth of $10,300.

PERPETUAL LIFE WITH GRADIENT INCREASES

Some perpetual-life problems involve a gradient increase in cost. For instance, assume the maintenance costs on each mile of right-of-way increase by $100 per year over a perpetual life. To find the present worth of this arithmetic gradient, simply solve Equation 5.8, letting n approach infinity.

$$P/G = \frac{1}{i}\left[\frac{(1+i)^n - 1}{i(1+i)^n} - \frac{n}{(1+i)^n}\right]$$

Consider the elements of the equation one at a time. As n approaches infinity, both the numerator and the denominator of both parts of the equation become very large. To analyze, use L'Hospital's rule; as n approaches infinity, $(1 + i)^n$ becomes very large and also approaches infinity.

As $n \to \infty$

$$\frac{n}{(1+i)^n} = \frac{d(n)/dn}{d(1+i)^n/dn} = \frac{1}{n(1+i)^{n-1}} = 0$$

Therefore

$$\frac{n}{(1+i)^n} \to 0 \quad \text{as} \quad n \to \infty$$

and

$$\frac{(1+i)^n - 1}{(1+i)^n} \to 1 \quad \text{as} \quad n \to \infty$$

so that

$$P/G \to \frac{1}{i}\left(\frac{1}{i} - 0\right) \quad \text{as} \quad n \to \infty$$

or, for $n = \infty$,

$$P/G = \frac{1}{i^2} \tag{5.9}$$

Applying this solution to the foregoing right-of-way problem (assuming $i = 6\%$), we find the present worth of the perpetually increasing maintenance cost of $100 per year as

$$P = \frac{G}{i^2} = \frac{\$200}{0.06^2} = \$55,556$$

In other words, if a trust fund of $55,556 earns 6 percent interest, then withdrawals of $200 may be commenced at EOY 2 and increased by $200 per year each year forever.

Perpetual Life for Other Time-Value Equations

As $n \to \infty$, the other time-value equations are evaluated in a manner similar to the evaluations in the preceding paragraph. The results are as follows:

$$A/F = 0$$
$$A/P = i$$
$$A/G = 1/i$$
$$P/F = 0$$
$$P/A = 1/i$$
$$P/G = 1/i^2$$
$$F/P = \infty$$
$$F/A = \infty$$
$$F/G = \infty$$

GRADIENT PAYMENTS OCCURRING AT PERIODIC INTERVALS LONGER THAN THE COMPOUNDING PERIOD

Many projects include a periodic series of increasing payments (representing major maintenance or similar costs or income) spaced at intervals longer than the interest-

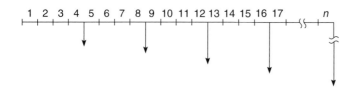

Figure 5.16   Typical periodic series of increasing payments.

compounding period, as illustrated in Figure 5.16. The usual characteristics of these long-interval gradient series are the following:

1. The payments are scheduled at regular intervals.
2. The intervals between payments are longer than the compounding periods, and are some whole multiple of the compounding period.
3. The payments increase by regular arithmetic gradient increments (possibly due to inflation, higher maintenance costs with age, or other factors).

The method of solution treats the series as a whole and requires the application of an equivalent interest rate. This equivalent interest rate has a compounding period equal in length to the interval between payments, and an interest rate increased sufficiently to compensate for the lengthened compounding period. Thus the equivalent interest rate will yield the same final balance in the account as if the account were compounded at the actual rate and period.

To differentiate between the two time periods and interest rates, the following nomenclature is used.

n_a = actual number of compounding periods between payments
n_p = number of payments in the entire series
i_a = the actual interest rate at which the account is periodically credited with compound interest
i_e = the equivalent interest rate compounding at the end of each interval consisting of n_a periods

The **equivalent interest,** i_e, is found as

$$i_e = (1 + i_a)^{n_a} - 1$$

To find the equivalent present, future, or annual value of a series of gradient payments occurring at some interval n_a periods apart, simply substitute i_e in place of i, and n_p for n in the appropriate equation, as illustrated below.

EXAMPLE 5.9

Find the present worth of heavy maintenance payments scheduled at the end of every four years. The first maintenance payment is $20,000 occurring at EOY 4, and each subsequent payment (at four-year intervals) is expected to increase by $4,000. Use $i = 10$ percent per year compounded at the end of each year, and assume the life of the project is 40 years.

Solution

1. To clarify the problem, draw the cash flow diagram as shown in Figure 5.17.

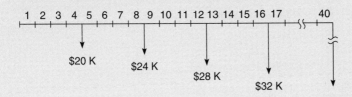

Figure 5.17 Cash flow diagram for Example 5.9.

2. Find the number of compounding periods between payments. In this problem the interval is given as four years, so, $n_a = 4$.

3. Find the number of payments in the series, given as 40 years divided by four years between payments, so $n_p = 10$.

4. The equivalent interest rate i_e is found as

$$i_e = (1 + i_a)^{n_a} - 1 = 1.10^4 - 1 = 0.4641 = 46.41\%$$

5. The payments are subdivided as follows:
   a. A periodic series of 10 payments of $20,000, each paid at the end of every four years.
   b. A gradient series increasing by $4,000 every four years.

Substituting i_e and n_p in the appropriate equations yields

$$P_1 = A \left[\frac{(1+i)^n - 1}{i(1+i)^n} \right] = \$20,000 \left[\frac{1.4641^{10} - 1}{0.4641 \times 1.4641^{10}} \right] = \$42,142$$

$$P_2 = \frac{G}{i} \left[\frac{(1+i)^n - 1}{i(1+i)^n} - \frac{n}{(1+i)^n} \right]$$

$$= \frac{4,000}{0.4641} \left[\frac{1.4641^{10} - 1}{0.4641 \times 1.4641^{10}} - \frac{10}{1.4641^{10}} \right] = \$16,256$$

$$P_{\text{total}} = P_1 + P_2 = \$42,142 + \$16,256 = \$58,398 = \underline{\underline{\$58,400}}$$

By way of illustration, a deposit of $58,400 (the PW found above) in an account drawing interest at 10 percent compounded annually is sufficient to supply payments from the account of $20,000 at the end of year 4, plus nine subsequent payments at four-year intervals, with each payment increasing by $4,000 over the previous one. Each withdrawal from the account reduces the balance, and the last withdrawal draws the balance in the account down to zero.

If the project were perpetual ($n \rightarrow \infty$), such as a road R/W, the present value could be calculated as shown in Example 5.10.

EXAMPLE 5.10

Find the PW of a series of periodic payments. The first payment of $20,000 occurs at EOY 4. Each subsequent payment occurs at each four-year interval thereafter and increases by $4,000 over the previous payment. ($24,000 at EOY 8, $28,000 at EOY 10, etc.); $i = 10$ percent, and $n \rightarrow \infty$.

Solution. The payments are divided into a uniform periodic series plus a gradient periodic series consisting of

1. A perpetual periodic series of $20,000 every four years.

2. A perpetual gradient series increasing by $4,000 every four years.

Substituting i_e and n_p in the equations for $n \rightarrow \infty$ yields

$$P_1 = A/i_e = \$20,000/0.4641 = \$43,094$$
$$P_2 = G/i_e^2 = \$4,000/0.4641^2 = \underline{18,572}$$
$$P_{\text{total}} = \qquad\qquad\qquad\qquad \$61,666$$

Thus if the sum of $61,666 were deposited in an account bearing interest at 10 percent compounded annually, the account balance would be sufficient to permit withdrawals of $20,000 at EOY 4, with subsequent withdrawals increasing by $4,000 at the end of each succeeding four years *forever* without exhausting the account. Note that the difference in PW due to the change of n from 40 years to ∞ is not large for $i = 10$ percent. Would the difference diminish or increase as i increases?

(*Ans.* diminish)

When all else fails, a simplistic P/F method can always be devised to determine the equivalent present, future, or annual values. This method requires just two steps:

1. Find the present or future value of each individual irregular payment by multiplying each payment by the appropriate P/F or F/P factor.

2. Sum the results to find the total present or future value of the entire irregular series.

Where many irregular payments are anticipated, this method can be programmed for computer application to reduce the time requirements.

PERIODIC SERIES OF *G* PAYMENTS WITH ETZ ≠ PTZ

Sometimes gradient payments will begin at some odd time interval from problem time zero. To solve this feature of the problem, simply find the PW of the series at the ETZ for the series. Then transfer the PW from the ETZ to the PTZ. This procedure is illustrated in the next example.

EXAMPLE 5.11   **(Find *P*, given *A*, *G*, *i*, *n*)**

The net income for a proposed project begins with a *negative* cash flow of $18,000 at EOY 2. Payments are received from the project at four-year intervals thereafter. Each subsequent payment is $10,000 more profitable than the one before (−$8,000 at EOY 6, +$2,000 at EOY 10, etc.) as shown in Figure 5.18. Find the equivalent PW with $n \to \infty$, and $i = 10$ percent per year.

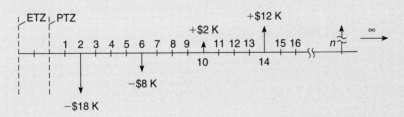

Figure 5.18   Cash flow diagram for Example 5.11.

Solution.   For $n \to \infty$,

$$P = A/i$$
$$P = G/i^2$$
$$i_e = 1.10^4 - 1 = 0.4641$$

At ETZ,

$$P_1 = -\frac{\$18K}{i_e} + \frac{\$10K}{i_e^2}$$

$$P_1 = -\frac{\$18K}{0.4641} + \frac{\$10K}{(0.4641)^2}$$

$$P_1 = \$7,643$$

At PTZ,

$$F_1 = P_1(F/P, 10\%, 2)$$
$$F_1 = \$7,643(1.2100)$$
$$F_1 = \$9,248$$

SUMMARY

In this chapter the equivalence factors equating the arithmetic gradient G to future worth F, annual worth A, and present worth P, are developed. The equations for relating G to A, F and P are

$$A = G\left[\frac{1}{i} - \frac{n}{(1+i)^n - 1}\right] = G(A/G, i, n)$$

$$F = \frac{G}{i}\left[\frac{(1+i)^n - 1}{i} - n\right] = G(F/G, i, n)$$

$$P = \frac{G}{i}\left[\frac{(1+i)^n - 1}{i(1+i)^n} - \frac{n}{(1+i)^n}\right] = G(P/G, i, n)$$

As $n \to \infty$, then $A/G \to 1/i$, and $P/G \to 1/i^2$.

PROBLEMS FOR CHAPTER 5, ARITHMETIC GRADIENT

Note: Assume all costs and income are incurred at the end of the year designated, unless otherwise specified in the problem.

Problem Group A

Given three variables, find the fourth.

A.1. Five years from now, how much will there be in an account that receives the deposits listed below, and bears interest at a rate of 6 percent compounded annually on the balance on deposit?

EOY	Deposit
1	$ 0
2	300
3	600
4	900
5	1,200

(*Ans.* $3,185)

A.2. What is the equal annual equivalent to the following series of costs if $i = 7$ percent?

EOY	Cost
1	$ 0
2	1,000
3	2,000
4	3,000

A.3. What is the present worth of the following series of income payments if $i = 9$ percent?

EOY	Income
1	$ 0
2	2,500
3	5,000
4	7,500
5	10,000
6	12,500

(*Ans.* $25,230)

Problem Group B

Find *i* or compare *i* to a given minimum standard. (*Hint:* In order to determine if the *i* of the problem meets a certain minimum standard *i*, simply find the PW using the standard *i* and see if the PW is positive or negative.)

B.1. A client of yours is interested in buying a five-year franchise for the sale and construction of patented prefab metal buildings if a 20 percent return can be assured. Since the system is new, sales are expected to be slow at first and pick up in future years. The estimated profit at the end of each year is shown in the table.

(a) If the selling price for the franchise is $50,000, with no salvage value at the end of five years, will it yield 20 percent?

(b) What is the maximum your client could pay for the franchise and still make 20 percent?

EOY	Net profit
1	$ 0
2	10,000
3	20,000
4	30,000
5	40,000

B.2. A tract of land, which can be subdivided and sold as building lots, is available for $120,000. The net return after expenses is projected as follows:

EOY	Net income
1	$ 0
2	10,000
3	20,000
4	30,000
5	40,000
6	50,000
7	60,000
8	70,000
9	80,000

At EOY 9, all the lots will be sold, and the investment will have no further value. A developer is interested in the property if it will yield a 15 percent return and asks you to calculate whether it will or not.

B.3. A state regulatory commission is examining an application by a cable TV firm to provide exclusive service to a community. One stipulation is that the firm's rate of return shall not exceed 15 percent. The firm proposes to charge $100 per customer per year, credited at EOY. The franchise expires worthless at EOY 10. Does the data below indicate that this rate will be more or less than 15 percent?

EOY	Number of customers	Expenses
0		$200,000
1	600	84,000
2 through 10	Increase of 200 customers per year	84,000, same through year 10

(*Ans.* $i > 15\%$)

B.4. A construction contractor needs to make at least a 20 percent return on all funds invested in equipment. The purchase of a new trenching machine is now under consideration, and he estimates the costs and incomes as shown in the following table.

EOY	Cost	Income
0	Purchase $100K	0
1	30K	$60K
2 through 8	*Increase* by 2K/yr/yr	*Decrease* by 2K/yr/yr
8	44K	46K
8	Sell trencher 0	15K

Find whether this investment meets the contractor's goal of $i \geq 20$ percent.

Problem Group C

Find *n*.

C.1. A fund of $100,000 was set up, earning 8 percent on the balance, to pay for maintenance of a facility at the rate of $14,903 per year over a 10-year period. The first year's maintenance did cost $14,903 per year, but maintenance the second year rose to $15,903 per year. Due to inflation, it is now estimated the costs will rise by $1,000 per year each year. If the costs continue to increase by $1,000 per year, how long will the fund last?

(*Ans.* $n = 7.71$ yr. The fund will sustain seven full payments plus a portion of an eighth payment.)

Problem Group D

These may require two or more equations to find *A*, *P*, comparisons, rates.

D.1. The operating cost for a proposed waste-water treatment plant is estimated as $500,000 per year, increasing by $100,000 each year after the first. Assume $i = 6$ percent. Find the following:

    (a) Equivalent annual cost of all operating costs, for a life of 25 years.

       (*Ans.* $1,407,200/yr)

    (b) Present worth of all operating costs, for a life of 25 years.

       (*Ans.* $17,988,800)

    (c) Equivalent annual cost of all operating costs, for a perpetual life.

       (*Ans.* $2,166,700/yr)

    (d) Present worth of all operating costs, for a perpetual life.

       (*Ans.* $36,111,000)

D.2. The plant in Problem D.1 is estimated to cost $10,000,000 to build. After 25 years it will have a $2,000,000 salvage value. Assume the same $n = 25$ and $i = 6$ percent, and that the money to construct is borrowed. Find the following:

    (a) The total equivalent annual cost of **(i)** P & I payments, plus **(ii)** annual operating costs, plus **(iii)** salvage value.

(b) This plant will serve 25,000 housing units. What annual charge per year per unit is required if the sewer rate remains constant for 25 years?

(c) What annual rate per housing unit is required to cover only the first year's operating costs plus P & I payment on the capital cost (deduct annual equivalent to salvage value for P & I payment)?

(d) What annual increment G is necessary per housing unit to cover the increasing operating costs?

(e) If it is necessary because of competition to begin with an annual rate of $36 per unit per year, what annual increment, G, is necessary to reach the breakeven point (on operating costs plus P & I, plus salvage) over the period of 25 years?

D.3. A friend of yours, just turned 23, wants to become a millionaire by age 63. Regular deposits will be made into an investment account that will continue to bear interest compounded at 10 percent. Whatever the amount of the initial investment deposit, each subsequent deposit will increase by $200 per year throughout the 40-year period. Calculate the proper amount for the initial investment on the twenty-fourth birthday.

(*Ans.* $440.00)

D.4. An engineering firm wishes to establish a trust fund to provide one engineering scholarship every year for a deserving student. The amount of the first scholarship to be awarded at the end of the first year is $1,000. However, due to the historical trend of generally rising costs, it seems wise to provide for an increase in the award of $100 per year. Therefore the award will be $1,100 at the end of the second year, $1,200 at the end of the third year, and so on. The trust fund bears interest at 7 percent on the unspent balance. The firm asks you to find how much would be needed to set up a trust fund for each of the following options.

(a) To provide for annual awards for the next 20 years.

(b) To provide annual awards forever ($n \rightarrow \infty$).

(c) Assuming the amount needed in part (a) is available to set up the trust fund, and that this amount could be invested at 9 instead of 7 percent, how much could the initial award be raised to, assuming the gradient is still $100 and $n = 20$?

D.5. An intersection now has a traffic count of 10,000 cars daily, expected to increase by 1,000 cars per day each year. A new traffic signal system for the intersection is estimated to cost $50,000 and will save each car passing through an average of $0.01 worth of fuel, wear, and time. With $i = 6$ percent and a 10-year life, compare the present worth of the savings versus the cost of the proposed system. (For simplicity, assume the traffic count increases, and savings to motorists occur at EOY each year.)

(*Ans.* savings PW = $376,700 versus cost PW = $50,000)

D.6. Your client, a toll turnpike authority, is considering the feasibility of a 100-mile extension to the present turnpike at a cost of $2,000,000 per mile. The traffic is estimated to average 25,000 cars per day (365 days per year) for the first year. Each year thereafter the traffic is expected to increase by an additional 2,500 cars per day for the full 20-year life of the project. (Use $n = 20$). The money to build the road can be borrowed on road bonds at 7 percent. Sixty percent of the toll receipts will be needed to pay operating and maintenance costs, and the remaining 40 percent will be available to retire the bonds. The authority asks you to determine how much toll must be charged each car in terms of dollars per mile in order to pay O & M costs as well as retire the bonds.

D.7. Your firm needs a new computer. Two competing brands are under consideration. All other factors have been checked and balance out, except the estimated O & M costs, which are listed below. You are asked to determine which of the following two schedules of costs yields the lesser cost ($i = 9\%$).

EOY	O & M costs, brand A	O & M costs, brand B
1	$40,000	$62,000
2	45,000	62,500
3	50,000	62,500
4	55,000	62,500
etc.	etc.	etc.
10	85,000	62,500

(*Ans.* $AW_A = \$58,990$; $AW_B = \$62,500$)

D.8. A manufacturing firm selling a new type of form tie is expected to lose $4,000 the first year, lose $2,000 the second year, and just break even the third year. Every year thereafter the firm is expected to earn $2,000 more than the previous year for the life of the manufacturing equipment (10 years). For $i = 15$ percent, find the present value of the total earnings over the 10-year period.

D.9. The cost for a handheld calculator is $139 this year. Due to technological advances, the costs are expected to decrease by $14 each year for the next five years. An engineering firm estimates it will purchase eight calculators each year for the next five years. If $i = 15$ percent, what is the present value of the total purchase of calculators? (Assume each group of calculators is purchased at the end of each year.)

(*Ans.* $3,119.34)

D.10. A contractor is considering the purchase of a new concrete-pumping machine. The machine will save $5,000 the first year over the costs of placing concrete by the present method. Savings should increase by $500 per year every year thereafter. The estimated life of the pump is nine years with negligible salvage value at that time. The contractor aims at a 15 percent return on all invested capital. What is the highest price the contractor can pay for the concrete pump and still earn 15 percent?

Problem Group E

ETZ $\neq$ PTZ.

E.1. A graduating class of 100 engineers wants to present a gift of $500,000 to the university at their twenty-fifth class reunion. They set up a trust fund that earns 8 percent per year compounded. They plan to start with a small donation of G dollars to the fund on the first anniversary of their graduation and increase the amount by regular increments every year ($2G, 3G, \ldots, nG$). If they put nothing into the fund at the time of their graduation, but continue increasing their donations by a regular amount, G, each year, find how much G amounts to in terms of the donation each classmate should give on his or her first anniversary.

(*Ans.* $G = \$7.41/\text{yr}/\text{yr}$)

E.2. Find which of the two alternative sewer designs has the lowest present worth. The expected life of each is 20 years, and $i = 8$ percent per year. All annual payments are end-of-year.

Design A.   Gravity flow, deep installation
Cost new, $780,000
O & M, $1,000/yr
Salvage value, none

Design B.   Lift station, shallow installation
Cost new, $500,000
O & M, $12,000 for years 1 through 5, increasing by $1,200/yr/yr thereafter,
($13,200 at EOY 6, $14,400 at EOY 7, etc.)
Salvage value, $100,000 at EOY 20

(a) Draw the cash flow diagram for *design B only.*
(b) Find the present worth of each alternative.

E.3. A small municipality has just received competing bids from two utility companies that want to pay the municipality for a franchise to serve the area exclusively for the next 20 years. The amounts they are willing to pay are shown in the table below. Calculate which is the higher bid if $i = 7$ percent.

EOY	Company A will pay	Company B will pay
0	$100,000	0
1 through 5	20,000/yr	$10,000/yr
6 through 20	20,000/yr	increase $5,000/yr each year,
		(EOY 6, pay $15,000; EOY 7, pay 20,000,
		etc.)

(*Ans.* NPW$_A$ = $311,880, NPW$_B$ = $325,380)

E.4. A city engineer is thinking of mechanizing a routine construction operation currently performed by a crew of laborers. The estimated costs of each are shown below. With $i = 7$ percent, calculate which is the lower cost alternative. The machine will have zero salvage value at the end of year 10.

EOY	Machine	Crew
0	$30,000	0
1	15,000	$24,000
2 through 10	15,000	Increase $2,000/yr each year,
		($26,000 for year 2, 28,000 for year 3, etc.)

E.5. Two types of road surfacing are under consideration. They both have the same first cost of installation, but the maintenance costs are different, as shown in the table below. Calculate which has the lower cost, using $i = 0.5$ percent per month. After four years there is no salvage value for either pavement.

EOM	Type A	Type B
1 through 24	$1,000/month	$500/month
25 through 48	1,000/month	Increase $50/month each month over the previous month (twenty-fifth month costs $550, twenty-sixth month costs $600, etc.)

(*Ans.* AW$_A$ = $1,000/month, AW$_B$ = $788.22/month)

E.6. A contractor will need a hydraulic crane for a period of 24 months and has a choice of either leasing one for $2,100 per month or buying one fully financed, with no down payment. If purchased, the monthly payments plus maintenance costs are estimated as follows:

EOM	Cost
1 through 6	$1,000/month
7	$1,200
8 through 24	Increase $200/month each month from previous month
24	Income of $10,000 resale value

The total cash flow for the purchase and maintenance of the crane is $48,200. If the crane is leased, the cash flow is 24 months × $2,100 per month = $50,400. Since the purchase alternative is the least costly in terms of cash flow, the contractor is tempted to purchase the crane rather than lease it. Determine which actually is the least costly alternative by calculating the monthly equivalent cost. Assume $i = 1$ percent per month, and lease payments are made at the *end* of each month.

E.7. The County Road Department is considering the adoption of a new system of road maintenance. The estimated annual costs for the county are listed below. Since the total cash flow for the present method is less than the total cash flow for the new method, the department has serious doubts about the new method. Find which is least costly using $i = 9$ percent.

EOY	Present method	Proposed new method	
1	$ 7,500	$ 5,000	
2	7,500	5,000	
3	7,500	5,000	
4	7,500	6,000	
5	7,500	7,000	
6	7,500	8,000	
7	7,500	9,000	
8	7,500	10,000	
9	7,500	11,000	
10	7,500	12,000	
	$75,000	$78,000	Total cash flow

(*Ans.* AW$_{present}$ = $7,500/yr; AW$_{proposed}$ = $7,215/yr)

Problem Group F

Periodic gradient, finite n value.

F.1. An urban tree-planting project is proposed that would involve the following costs:

Cost of trees and planting, $1,000,000.

Periodic Maintenance (PM), $80,000 every 3 months

First PM occurs at EOM 1.

Gradient PM, *decrease* by $4,000 each subsequent PM event.

Life of project, 48 months (after EOM 48, maintenance is assumed by others and not charged to this project).

$i = 1$ percent per month.

(a) Find the PV of the proposed project.

(*Ans.* $1,671,500)

Funding: To induce the city to undertake the project, a philanthropic foundation is willing to sponsor the project, provided that the city pays one-half of the cost. The city will be billed at EOM 0 for one-half the cost found in (a). Due to a limited budget, the city proposes to pay off the bill by making a series of annual *end-of-year* payments, beginning with an initial payment of $100,000 now, and increasing the amount by $100,000 each subsequent year ($200K at EOY 1, $300K at EOY 2, etc.).

(b) If the foundation agrees to finance the balance of the debt at an actual interest rate of 12 percent compounded annually, how long will it take the city to pay off the bill?

(*Ans.* Three full payments plus one of $37,320)

F.2. An investment opportunity will yield either $5,000 per month for 30 months (paid at EOM), or the periodic gradient payments listed below. Find the equivalent equal monthly payments of those tabulated below and compare. Use $i = 1$ percent per month.

EOM	Payment
0	$20K
4	23K
8	26K
12	29K
16	32K
20	35K
24	38K

F.3. A local utility system (US) is planning a package plant waste-water system to serve a large new development. Use $i = 1$ percent per month, $n = 120$ months.

(a) Find the amount of equal monthly EOM payments to be made by residents in the development in order to adequately reimburse the expenses to the US over the 120-month life of the project.

EOM	US cost
0	$1,000K
6	100K
10	120K
14	140K
18	160K
etc.	etc.

The gradient continues to increase by $20K every four months until the last cost incurred at EOM 118.

(*Ans.* *A* = $91,550/month)

F.4.   A small city is considering starting a city bus system (CBS) and running it for a 60-month trial period. Use *i* = 1 percent per month.

(**a**) Find the amount of equal monthly EOM fare receipts that the CBS should receive in order for the system to break even over the 60-month period.

EOM	CBS cost
0	$1,000K new business
2	100K
7	110K
12	120K
etc.	etc.

The gradient continues to increase by $10K every five months until the last cost incurred at EOM 57. At EOM 60, the buses have a resale value of $400K.

Problem Group G

Periodic gradient, *n* → ∞.

G.1.   A proposed flood-protection works has the following estimated costs:

Cost new, $1,000,000.

Periodic Maintenance (PM) every three years, $200,000.

First PM occurs at EOY 2.

Gradient PM, increases by $25,000 for each subsequent PM event.

Life of flood works, a long time, so *n* → ∞.

i = 10 percent

(**a**) Find the PV (capitalized cost) of the proposed project.

(*Ans.* $1,915,700)

Funding: The project would be constructed by the federal government, *but* the local government is required to pay one-half the cost. The local government will be billed at EOY 0 for one-half the cost found in (a). The local government can only afford to pay $100,000 on this bill right now (EOY 0), but in each subsequent year they can increase the amount of the payment by $100,000 (pay $200K at EOY 1, $300K at EOY 2, etc.).

(b) If the federal government is willing to finance the balance at 10 percent, how long will it take the local government to pay off the debt using the proposed gradient payment schedule.

(*Ans.* Three full payments plus one payment of $140,300)

G.2. A proposed highway will require periodic heavy maintenance expenditures as shown below. Note that the first expenditure occurs at EOY 2, while the infinite series of increasing expenditures occurs at five-year intervals thereafter. Find the equivalent annual cost with $n \rightarrow \infty$ and $i = 10$ percent per year.

EOY	Expenditure
2	$200K
7	300K
12	400K
17	500K
etc. to ∞	etc. to ∞

G.3. A proposed geothermal energy project has an initial cost of $1,000K. In addition it will require an infinite series of periodic maintenance expenditures at five-year intervals following the first maintenance expenditure at EOY 2. Find the equivalent annual cost with $n \rightarrow \infty$ and $i = 88$ percent per year.

EOY		Expenditure
0	Initial cost	$1,000K
2	Periodic maintenance	140K
7	Same	200K
12	Same	260K
17	Same	320K
etc. to ∞		etc. to ∞

(*Ans.* NAW = $137,510K/yr)

CHAPTER 6

Geometric Gradient: The Constant Percentage Increment

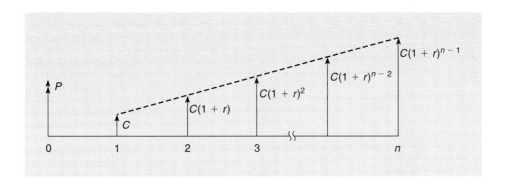

KEY EXPRESSIONS IN THIS CHAPTER

C = First end-of-period payment of a series changing by a constant percentage; base payment.

r = Rate of growth (or decline) of the series. [For example, a growth rate of 15 percent is shown as $r = 0.15$, and a decline rate (decay rate or negative growth) of 3 percent is shown as $r = -0.03$.]

w = Derived quantity as shown below.

$i, n, P, F,$ and A all have the same meanings as in previous chapters.

A PAYMENT SERIES CHANGING BY A CONSTANT PERCENTAGE

A geometric gradient series is a series of end-of-period payments with each payment increasing (or decreasing) by a fixed percentage.

Change has become an ordinary feature of cost analysis problems encountered in everyday practice. Costs are subject to periodic change, and incomes change along with a great array of changes in other variables. The two basic tools available for evaluating the effects of these changes are the two sets of gradient equations, (a) the *arithmetic* gradient discussed in Chapter 5, and (b) the *geometric* gradient developed in this chapter.

The arithmetic gradient in Chapter 5 dealt with constant gradient amounts, G, such as an annual increase in the cost of maintenance of $1,000 per year. The arithmetic gradient G does *not* include any annual base amount A, which is accounted for separately. This chapter deals with geometric increments, where the percentage change is constant (but the resulting numerical amount of change varies), such as an annual increase in the cost of maintenance of 10 percent per year, or increase in traffic count of 8 percent per year. The base amount C is *included* in the geometric gradient instead of being excluded, as in the arithmetic gradient. As an example of a geometric gradient, if maintenance costs for a particular project are $10,000 per year the first year and increase 10 percent per year thereafter, the costs for each year are as shown in the following table.

End of year	Maintenance cost increasing 10% each year	Maintenance cost each year in terms of base year cost C (C in this case is $10,000)	Maintenance cost each year in terms of rate of growth r (r in this case is 0.10)	
1	$10,000 = C	$C \times 1$	C	$= C(1 + r)^0$
2	11,000	$C \times 1.1$	$C + Cr$	$= C(1 + r)^1$
3	12,100	$C \times 1.21$	$C(1 + r) \times (1 + r) = C(1 + r)^2$	
4	13,310	$C \times 1.331$		$= C(1 + r)^3$
etc.	etc.	etc.	etc.	etc.
n	$10^4 \times (1.1)^{n-1}$	$C \times (1.1)^{n-1}$		$= C(1 + r)^{n-1}$

Graphic Representation

Figure 6.1 shows a graph of the balance in the account showing the present worth of a series of geometrically increasing annual payments that start at first year level C and

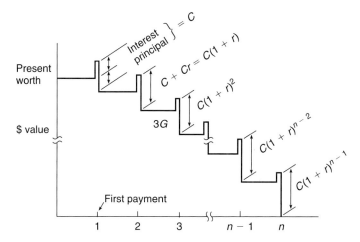

Figure 6.1   A graph of the balance in the account with initial balance P and a series of geometrically increasing payments, beginning with an initial payment C.

increase by r each period for n periods. The payment for each subsequent period is $(1 + r)$ times the payment for the previous period. Notice that the first payment occurs at the end of the first period, as in the A series, rather than at the end of the second period, as in the G series.

Derivation of Equations for the Present Worth

Following is the derivation of an equation to determine the present worth P of the series of payments with an initial payment C, and increasing at a constant rate r (or decreasing at a constant rate, $-r$) for n periods with interest at i on the balance remaining in the fund.

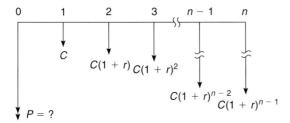

Figure 6.2   Cash flow diagram for geometric gradient.

The cash flow diagram for this situation is given in Figure 6.2. In the extreme case where $n = 1$, then only one payment C is made at the end of the first period. In this case C is a single future sum, and its present worth may be found by

$$P = \frac{F}{(1 + i)^n} = \frac{C}{1 + i}$$

If $n = 2$, the present worth is the sum of the two individual future sums

$$P = \frac{F_1}{(1 + i)^{n-1}} + \frac{F_2}{(1 + i)^n} = \frac{F_1}{(1 + i)^1} + \frac{F_2}{(1 + i)^2}$$

The two successive payments, F_1 and F_2, can be replaced respectively by C and $C(1 + r)$. Then the same present worth, P, is now expressed in terms of the initial payment C and the growth rate r.

$$P = \frac{C}{1 + i} + \frac{C(1 + r)}{(1 + i)^2}$$

With this example of $n = 2$ in mind, the series logically may be expanded to accommodate larger values of n, as follows.

$$P = C\left[\frac{1}{1 + i} + \frac{1 + r}{(1 + i)^2} + \cdots + \frac{(1 + r)^{n-1}}{(1 + i)^n}\right]$$

To condense the equation into a more manageable form, the following procedures are employed.

Multiply the right-hand side by $(1 + r)/(1 + r)$ and obtain

$$P = \frac{C}{1+r}\left[\frac{1+r}{1+i} + \frac{(1+r)^2}{(1+i)^2} + \cdots + \frac{(1+r)^n}{(1+i)^n}\right] \tag{6.1}$$

Case 1. $r > i$ When $r > i$, let $1 + w = (1 + r)/(1 + i)$ and substitute into Equation 6.1 to obtain

$$P = \frac{C}{1+r}[(1 + w) + (1 + w)^2 + \cdots + (1 + w)^n]$$

The term $(1 + w)$ may be extracted from the brackets, as follows:

$$P = \frac{C}{1+r}(1 + w)[1 + (1 + w) + \cdots + (1 + w)^{n-1}]$$

Then, multiply both sides by $(1 + i)/w$, subtract the new equation from the old, and multiply both sides by w/i to find the final standard form relating P, C, i, and n.

$$P = \frac{C}{1+i}\left[\frac{(1 + w)^n - 1}{w}\right] \quad \text{for} \quad r > i, \quad \text{and} \quad w = \frac{1+r}{1+i} - 1 \tag{6.2}$$

As a further convenience, Chapter 3 showed that

$$F/A = \frac{(1+i)^n - 1}{i}$$

and that the right-hand side may be designated by $(F/A, i, n)$.

Therefore, using w in place of i to indicate a derived equivalent,

$$P = \frac{C}{1+i}(F/A, w, n) \quad \text{for} \quad r > i \tag{6.2a}$$

where

$$w = \frac{1+r}{1+i} - 1 \quad \text{for} \quad r > i$$

Case 2. $r < i$ When $r < i$, let

$$\frac{1}{1+w} = \frac{1+r}{1+i}$$

Then substituting $1/(1 + w)$ for $(1 + r)/(1 + i)$ in Equation 6.1 yields

$$P = \frac{C}{1+r}\left[\frac{1}{1+w} + \frac{1}{(1+w)^2} + \cdots + \frac{1}{(1+w)^n}\right]$$

Since

$$\left[\frac{1}{1+w} + \frac{1}{(1+w)^2} + \cdots + \frac{1}{(1+w)^n}\right] = \frac{(1+w)^n - 1}{w(1+w)^n}$$

then

$$P = \frac{C}{1+r}\left[\frac{(1+w)^n - 1}{w(1+w)^n}\right] \quad \textbf{when} \quad \boldsymbol{r < i} \quad \text{and} \quad \frac{1+r}{1+i} = \frac{1}{1+w} \tag{6.3}$$

Chapter 4 showed that

$$P/A = \frac{(1+i)^n - 1}{i(1+i)^n}$$

The right-hand side can be designated by $(P/A, i, n)$. Therefore, substituting w in place of i yields

$$P = \frac{C}{1+r}(P/A, w, n) \quad \text{for} \quad r < i \tag{6.3a}$$

where

$$w = \frac{1+i}{1+r} - 1 \quad \text{for} \quad r < i$$

Case 3. $\boldsymbol{r = i}$ If $r = i$, a simplified equation results. Examining Equation 6.1, if $r = i$, then

$$P = \frac{C}{1+r}(1^1 + 1^2 + \cdots + 1^{n-1} + 1^n)$$

or

$$P = \frac{Cn}{1+r} \quad \text{for} \quad r = i \tag{6.4}$$

Applications of Geometric Gradient

The simplest problem situation occurs where $r = i$, as in the following example.

EXAMPLE 6.1   (Find P for $r = i$, given C, r, i, n)

A consulting engineering firm is acquiring a new computer with operating costs estimated at \$60,000 the first year, increasing by 10 percent per year thereafter, until EOY 3. Assume all costs are debited at EOY. With the firm's interest rate at 10 percent, find the present worth of the operating costs for the three-year life.

Solution

$$r = i = 10\%$$
$$C = \$60,000$$
$$n = 3$$

Since $r = i$, use Equation 6.4.

$$P = \frac{Cn}{1 + r} = \frac{\$60,000 \times 3}{1.10} = \$163,636$$

The present worth of the three annual cost payments is \$163,636. The equivalence can be represented graphically as shown in Figure 6.3.

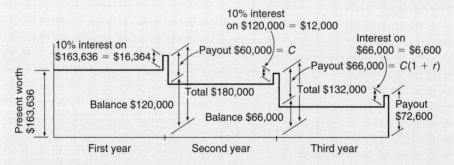

Figure 6.3   Balance-in-the-account diagram illustrating Example 6.1.

To help visualize the concept, a lump sum amount of \$163,636 could be deposited now in an account earning interest at 10 percent. Then a series of three end-of-year withdrawals could be made to pay operating costs, starting with \$60,000 at EOY 1 and increasing the size of the withdrawal by 10 percent each year (\$66,000 at EOY 2, \$72,600 at EOY 3). The third withdrawal at EOY 3 will draw the account balance down to zero.

FINDING THE EQUIVALENT FUTURE LUMP SUM OR EQUIVALENT UNIFORM SERIES PAYMENTS OF A GEOMETRIC GRADIENT SERIES

The future worth or annual worth of the geometric gradient can be determined by first finding the present worth, and then applying the appropriate F/P or A/P factor, as illustrated in the following continuation of Example 6.1.

EXAMPLE 6.1   (continued)

Assume that due to the capabilities of the new computer the consulting firm expects to earn an extra profit of $60,000 at EOY 1, increasing by 10 percent per year for the three-year life of the computer. They decide to invest these funds in an account that earns 10 percent in order to have cash on hand to buy the new computer model when it becomes available at EOY 3. How much is in the account at EOY 3?

Solution.   The equivalent future worth lump sum can be obtained directly from the present worth (of $163,636 found previously) by applying the F/P equation as follows.

$$F = P(F/P, i, n)$$
$$F = \$163,636 \underbrace{(F/P, 10\%, 3)}_{1.331} = \underline{\underline{\$217,800}}$$

There would be $217,800 in the account at the end of three years.

The relationships may be portrayed graphically in two ways:

1. Assume the equivalent cash is on deposit now and will receive interest at the rate of 10 percent per year, as shown in Figure 6.4.

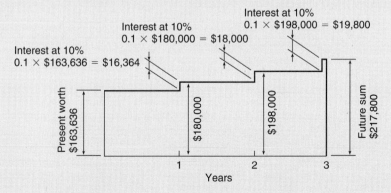

Figure 6.4   Balance-in-the-account diagram illustrating P/F relationship.

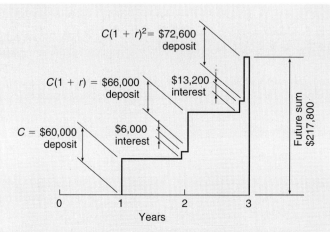

Figure 6.5   Balance-in-the-account diagram illustrating F/C relationship of Example 6.1.

2. Assume the payments will be deposited as made and earn interest at 10 percent from the date of deposit, as shown in Figure 6.5.

RATE OF GROWTH IS GREATER OR LESS THAN THE INTEREST RATE

The next two examples illustrate solutions to problems involving rates of growth r that are not the same as the interest rate i.

EXAMPLE 6.2   (Find P for $r < i$, given C, r, i, n)

The operating costs for an air conditioning system currently cost an owner of a commercial building \$18,000 per year. The costs are expected to increase at the rate of 5 percent per year. If interest is 6 percent, find the present worth of the operating costs over a 10-year period.

Solution.   Since $r < i$, use Equation 6.3, as in the second line below.

$$w = \frac{1+i}{1+r} - 1 = \frac{1.06}{1.05} - 1 = 0.009524 = 0.9524\%$$

$$P = \frac{C}{1+r}\left[\frac{(1+w)^n - 1}{w(1+w)^n}\right] = \frac{18,000}{1.05}\left[\frac{1.009524^{10} - 1}{0.009524 \times 1.009524^{10}}\right]$$

$$= \$162,781$$

EXAMPLE 6.3   (Find P for $r > i$, given C, r, i, n)

A traffic intersection improvement is under study. The improvement is estimated to cost $200,000. The benefit to the motoring public is estimated at an average saving of $0.010 per vehicle if the improvement is made. The current traffic count averages 8,000 cars per day for 365 days per year and is expected to increase 10 percent per year over the 10-year anticipated life of the installation. Money to finance the improvement comes from road bond funds bearing an interest rate of 5.77 percent. What is the present worth of the benefits to the public?

Solution.   Identify the variables.

$$\text{benefit, } C = 8{,}000 \text{ vehicles/day} \times 365 \text{ days/yr} \times \$0.010/\text{vehicle}$$
$$= \$29{,}200/\text{yr}$$
$$r = 10\%$$
$$i = 5.77\%$$

Since $r > i$, use Equation 6.2.

$$w = \frac{1+r}{1+i} - 1 = \frac{1.10}{1.0577} - 1 = 0.0400 \quad \text{or} \quad 4.000\%$$

Thus the present worth of the benefit is

$$P = \frac{C}{1+i} \left[\frac{(1+w)^n - 1}{w} \right] = \frac{29{,}200}{1.06} \left(\frac{1.0400^{10} - 1}{0.0400} \right) = \$330{,}600$$

DECLINING GRADIENT ($r < 0$)

Sometimes a decline in income will occur, due to declining productivity, increasing competition, or costs increasing more rapidly than income. The following example illustrates a situation with $r < 0$.

EXAMPLE 6.4   (Find C, given P, F, i, r)

Due to increasing age and down time, the productivity of a contractor's excavator is expected to decline with each passing year. Using the data below, calculate the price in terms of dollars per cubic yard that the contractor must charge to cover the cost of buying and selling the excavator.

$$
\begin{aligned}
\text{cost new} &= \$200{,}000 \\
\text{resale value at EOY 6} &= \$\ 40{,}000 \\
\text{contractor borrows at } i &= \quad 10\% \\
\text{production declines at } r &= \quad -6\% \\
\text{production for year 1} &= \ 200{,}000 \ \text{yd}^3/\text{yr}
\end{aligned}
$$

Assume all funds are credited at EOY.

Solution

Step 1.   The PW of the income from excavation equals the PW of the cost of the machine, so the first step is to find the PW of the cost of the machine.

$$
\begin{aligned}
\text{cost new } P_1 &= \qquad\qquad -\$200{,}000 \\
\text{PW of resale value } P_2 &= +40{,}000(P/F,\ 10\%,\ 6) = \ \$\ 22{,}578 \\
\text{net PW of cost } P_{1+2} &= \qquad\qquad = -\$177{,}422
\end{aligned}
$$

Step 2.   The income is found as

$$
C\ \$/\text{yd}^3 \times 200{,}000\ \text{yd}^3/\text{yr} = 200{,}000C\ \$/\text{yr}
$$

for the first year in the series. The equation for the PW of this geometric gradient income is

$$
\text{income, } P_3 = \frac{C}{(1+r)}\left[\frac{(1+w)^n - 1}{(1+w)^n - w}\right]
$$

where $r = -0.06$
$i = +0.10$

Since $r < i$, use Equation 6.3. Therefore,

$$
w = \frac{1+i}{1+r} - 1 = \frac{1.10}{0.94} - 1 = 0.1702
$$

Substituting the values for this problem into the equation yields

$$
\text{income, } P_3 = \frac{200{,}000C}{0.94}\left[\frac{1.1702^6 - 1}{0.1702 \times 1.1702^6}\right] = 763{,}280C
$$

Step 3.   Income equals cost, so the P_3 value is set equal to the $P_1 + P_2$

values, and solved for C. Thus,

$$763,280C = \$177,422$$

and

$$C = 177,422/763,280 = \underline{\$0.232/\text{yd}^3} \quad \text{must be charged to cover costs.}$$

The cash flow diagram for this problem is shown in Figure 6.6.

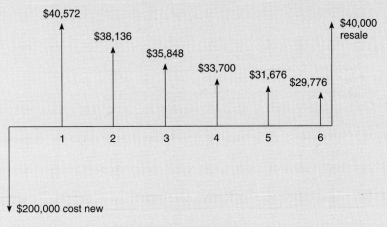

Figure 6.6   Cash flow diagram for Example 6.4.

Geometric Gradient Passing Through Zero

Ordinarily the geometric gradient never passes through zero, since any number increased or decreased by an exponential factor never can quite reach zero.

An exception can occur when a constant is added or subtracted to the gradient. The following example illustrates this type of problem.

EXAMPLE 6.5   **(Geometric gradient passing through zero. Find P, r, C, given the cash flow series, n, i)**

Assume that the income for a certain project increases as a geometric gradient, but the costs remain constant. As a result the net income (shown below as income minus cost) could begin as a negative value dominated by high costs, but subsequently rise through zero into the positive range as income improves. Find the present worth of the net income, with $i = 10$ percent.

Year	Net income in $1,000's (income—costs)
1	− $50
2	− 35
3	− 17.75
4	+ 2.09
5	+ 24.90
⋮	⋮
25	+2712.52

Solution. Since it is known that this growth is geometric (or a reasonable approximation thereto), but the base amount of cost, income, and growth rate are unknown, the constants r and C may be found as follows.

1. List the net income for the first three years.
2. Find the change in net income for the years 1 to 2, and from 2 to 3, and list these under Δ net as shown.
3. Find the change in the change and list it under $\Delta\Delta$ net as shown.
4. The growth rate $r = \Delta\Delta/\Delta$.
5. The first value, C, in the geometric gradient series is found as $C = \Delta/r$. In this case C is the first year's income.

Year	Net income	Δ net	$\Delta\Delta$ net
1	−$50		
		15	
2	−35		2.25
		17.25	
3	−17.75		

Then $r = \Delta\Delta/\Delta = 2.25/15 = 0.15$, and $C = \Delta/r = 15/0.15 = 100$. From this information the income and cost columns can be reconstructed as follows:

Year	Income ($1,000's)	Cost ($1,000's)	Net income ($1,000's)
1	$100	−$150	− $50
2	115	− 150	− 35
3	132.25	− 150	− 17.75
4	152.09	− 150	+ 2.09
5	174.90	− 150	+ 24.90
⋮	⋮	⋮	⋮
25	$100 \times 1.15^{n-1}$	− 150	+2712.52

The next step is to find the present worth of the net income with $i = 10$ percent. This is done by adding the present worth of the income, P_1, to the present worth of the cost, P_2. The sum of $P_1 + P_2$ is the present worth of the net income for the project.

Since $r > i$, use Equation 6.2.

$$w = \frac{1+r}{1+i} - 1 = \frac{1.15}{1.10} - 1 = 0.04545$$

and

$$P_1 = \frac{C}{1+i}\left[\frac{(1+w)^n - 1}{w}\right] = \frac{100}{1.10}\left(\frac{1.04545^{25} - 1}{0.04545}\right) = +\$4,076K$$

$$P_2 = -150\ \underbrace{P/A, 10, 25}_{9.077} \qquad\qquad\qquad\qquad = -\ 1,362K$$

$$\overline{+\$2,714K}$$

At EOY 0 the income from the project is worth $P_1 + P_2 = \$2,714,000$.

PERIODIC GEOMETRIC GRADIENT WITH PTZ $\neq$ ETZ

Frequently the periodic interval between series payments will be several times as long as the compounding period. For example, a pump for a waste-water system may need maintenance every four months, whereas interest is compounded at the end of each month. In addition the first maintenance event may occur at some point in time other than EOM 4. The periodic intervals can be handled easily by using the effective interest rate to find the actual effective compounding rate, i_e, between payments. The solution to the offset time base situation is to (1) find the PW at ETZ, and then (2) use P/F or F/P to bring the PW to the PTZ. Example 6.6 illustrates the procedure.

EXAMPLE 6.6 **(Periodic geometric gradient with PTZ $\neq$ ETZ)**

A new grit pump for a waste-water treatment plant is being considered for purchase. The pump will need major maintenance every four months, with the first maintenance occurring at EOM 6 and then again at EOM 10, 14, 18, and so on. The cost of the maintenance at EOM 6 is $4,000, and the cost increases 5 percent each time it is performed. The life of the pump is 120 months, and $i = 0.75$ percent per month. Find the PW of the periodic maintenance over the life of the pump.

Solution

$$i = 0.75\%$$
$$i_e = (1.0075)^4 - 1 = 0.030339$$
$$n = (120 - 2)/4 = 29.5$$

The major maintenance occurs every four months, so 29 of these events will occur; therefore

$$n = 29$$
$$r = 5\%$$

$$r > i_e, \qquad \text{therefore} \qquad w = \left(\frac{1+r}{1+i_e}\right) - 1 = 0.01908$$

$$C = \$4,000$$

$$P = \frac{C}{(1+i_e)}\left[\frac{(1+w)^{29} - 1}{w}\right](P/F, 0.75\%, 2)$$

$$P = -148,533 \times 0.98517 = \$146,330 \qquad\qquad \text{(Equation 6.2)}$$

Conclusion.   The PW of the major maintenance costs for the grit pump is $146,330.

GEOMETRIC GRADIENT WITH $n \rightarrow \infty$

As n grows larger and approaches infinity, only one of the three geometric gradient equations survives.

For $r > i$, the $C(1 + r)^n$ value of the expenditure is growing more rapidly than the interest income on the balance in the account, so as n approaches infinity, so likewise does the P/C value. Therefore, the P/C equation cannot be used for an infinite series where $r > i$.

For $r = i$, the equation for these values of r and i is $P/C = Cn/(1 + r)$, so that as n approaches infinity, P/C is growing correspondingly large and also approaches infinity. Therefore, the P/C equation cannot be used for an infinite series where $r = i$.

For $r < i$, the $C(1 + r)^n$ value of the expenditure is growing less rapidly than the interest accumulating on the principal balance in the account. Therefore, a finite value of P/C is obtainable for all values of n, including $n \rightarrow \infty$. Actually, the equation becomes somewhat simpler, so for problem situations of long duration and reasonably high interest rates, the infinite series equation can be employed to simplify the calculations. The interest rate is a factor, since the higher the rate, the greater the discount on the effect of future payments. The equation for $r < i$ and $n \rightarrow \infty$ becomes simply $P/C = 1/(1 + r)w$.

Fortunately, most real-life problem situations are found in the $r < i$ category. The increase in expenditures, r, will usually parallel the rate of inflation, whereas the investment rate, i, is normally a few percentage points higher than the rate of inflation. Thus this equation has broad areas of potential application.

EXAMPLE 6.7   **(Periodic geometric gradient, ETZ $\neq$ PTZ, $n \rightarrow \infty$**

A proposed bridge design is being evaluated. This bridge will require periodic heavy maintenance every five years. The first maintenance will occur at EOY 3, with subsequent maintenance at five-year intervals (EOY 3, 8, 13, 18, etc.). The first maintenance at EOY 3 will cost $80,000, and each subsequent maintenance will increase in cost by 25 percent from the previous time. The bridge is expected to last a long time, so use $n \rightarrow \infty$, and $i = 12$ percent per year. Find the PW of the infinite series of periodic maintenance expenditures.

Solution

$$C = \$80,000$$
$$r = r_e = 25\%$$
$$i = 12\%$$
$$i_e = (1+i)^n - 1 = 1.12^5 - 1 = 0.7623 \text{ (or } 76.23\%)$$

Since $r < i$,

$$w = \left(\frac{1+i}{1+r}\right) - 1 = \left(\frac{1.7623}{1.25}\right) - 1 = 0.4098$$

$$P_1 = \frac{C}{(1+r)w} = \frac{80,000}{1.25 \times 0.4098} = 156,174$$

$$P_2 = P_1(F/P, 12\%, 2) = 156,174 \times 1.2544 = \underline{\$195,904}$$

The PW of the infinite series of periodic maintenance expenditures increasing by 25 percent every five years is $195,904.

COMPARISON OF THE GEOMETRIC GRADIENT TO THE ARITHMETIC GRADIENT

The two types of gradients have a number of similarities as well as differences that need to be clearly understood. They are listed in the following table:

Arithmetic gradient	Geometric gradient
1. Changes by a constant *amount*, so the percent change varies. As the G series increases, the percent of change decreases.	Changes by a constant *percent,* so the actual amount of change varies. As the C series increases, the amount of the change increases.
2. The first G payment occurs at EOY 2.	The first C payment occurs at EOY 1.
3. G does *not* include an initial payment at EOY 1 but may be added to an underlying series, A.	C does include the full amount of the initial payment at EOY 1.
4. n = number of G payments + 1.	n = number of C payments.
5. G series increases to the right. May decrease only if added to an A of the opposite sign.	C series increases or decreases to the right depending on $+$ or $-$ r value.
6. May pass through zero depending on the underlying A series.	Asymptotic to zero, but does not pass through zero unless added to a constant A series.
7. Always the same equation (plus one more for $n \to \infty$).	Three different equations depending on i and r (plus one more for $n \to \infty$).

SPREADSHEET APPLICATIONS

The following repeats Example 6.6, using a spreadsheet.

	A	B	C	D	E	F
1		EXAMPLE 6.6: (Periodic geometric gradient with PTZ not equal to ETZ)				
2						
3		A new grit pump for a waste-water treatment plant is being considered for				
4		purchase. The pump will need major maintenance every four months,				
5		with the first maintenance occurring at EOM 6 and then again at EOM				
6		10, 14, 18, and so on. The cost of the maintenance at EOM 6 is $4,000,				
7		and the cost increases 5 percent each time it is performed. The life of the				
8		pump is 120 months, and $i = 0.75$ percent per month. Find the PW of the				
9		periodic maintenance over the life of the pump.				
10						
11						
12	SOLUTION:					
13						
14		$i = 0.75\%$ per period = 0.0075/period				
15						
16		$i_e = (1.0075)^4 - 1 = 0.030339$				
17						
18		$n = (120 - 2)/4 = 29.5$		-or-		
19						
20		$n = [(120_{\text{month total}} - 6_{\text{month maintenance starts}})/4_{\text{month interval}}] + 1 = 29.5$				

(Spreadsheet continued)

	A	B	C	D	E	F
21		(where " $+1$ " accounts for first maintenance event at EOM 6)				
22						
23		The major maintenance occurs every four months, so 29 of these events				
24		will occur; therefore:				
25						
26						
27		n	29		$n =$ number of periods interest will be compounded	
28		m	4		$m =$ number of compounding subperiods	
29		i	0.75%		$i =$ interest rate per period	
30		i_e	3.03%		$i_e =$ effective interest rate per compounding subperiod	
31		i	5.00%		$r =$ rate of growth or decline	
32		w	1.91%		$w =$ derived quantity	
33		C	$4,000		$C =$ 1st end-of-period pmt. of series changing by constant %	
34						
35		PW	($146,330)		Note: For $r = i_e$, PW $= (C^* n/(1 + r)$	
36						
37						
38	CELL CONTENTS: The following cell contents will allow us to evaluate the problem with a					
39			spreadsheet.			
40		cell C31:	$= (1 + cC30) \wedge C29 - 1$			
41						
42		cell C33:	$=$ IF(C32>C31,((1 + C32)/(1 + C31) $-$ 1,((1 + C31)/(1 + C32)) $-$ 1)			
43						
44		cell C36:	$=$ PV(C30,C28,,((C34/(1 + C31)))*(((1 + C33)^C28 $-$ 1)/C33)),0)			
45			where **Excel's** present value function: $=$ PV(rate,nper,pmt,fv,type)			
46						
47	CONCLUSION:					
48						
49		The PW of the major maintenance costs for the grit pump is $146,330.				

SUMMARY

This chapter describes the use of the geometric gradient. The geometric gradient is a series of gradient payments that increase (or decrease) in sequence by a fixed percentage from the first end-of-period payment C.

The equations for the geometric gradient are the following:

1. When $r > i$, then

$$w = \frac{1 + r}{1 + i} - 1 \qquad\qquad\qquad \text{as } n \to \infty$$

$$P = \frac{C}{1 + i}\left[\frac{(1 + w)^n - 1}{w}\right] = \frac{C}{1 + i}(F/A, w, n) \qquad P/C \to \infty$$

2. When $r < i$, then

$$w = \frac{1+i}{1+r} - 1 \qquad\qquad\qquad \text{as } n \to \infty$$

$$P = \frac{C}{1+r}\left[\frac{(1+w)^n - 1}{w(1+w)^n}\right] = \frac{C}{1+r}(P/A, w, n) \qquad P/C \to \frac{1}{(1+r)w}$$

3. When $r = i$, then

$$P = \frac{Cn}{(1+r)} \qquad\qquad\qquad P/C \to \infty$$

The future worth or annual worth of the geometric gradient can easily be determined by first finding the present worth through the use of the above equations and then applying the appropriate F/P or A/P factor using i (not r).

PROBLEMS FOR CHAPTER 6, GEOMETRIC GRADIENT

Note: Assume all payments are made at the end of the year unless otherwise noted in the problem.

Problem Group A

Given C, i, r, n, find P using the basic equations.

A.1. Find the present worth of the geometric gradient series shown in the table below, using the geometric gradient equations with $i = 6$ percent.

EOY	Maintenance costs increasing 10%/yr
1	$10,000
2	11,000
3	12,100
4	13,310
5	14,641

(*Ans.* $P = \$50,870$)

A.2. A contractor is considering purchase of a new concrete-pumping machine. The machine will save $5,000 the first year over costs of placing concrete by the present method, and savings should increase by 10 percent per year every year thereafter. The estimated life of the pump is nine years with negligible salvage value at that time. The contractor aims for a 15 percent return on all invested capital. What is the highest price that can be paid for the concrete pump and still earn 15 percent?

A.3. A municipal power plant will earn an annual net profit of $1,000,000 this year, and this annual amount is expected to increase by 10 percent per year for the next 20 years. Use $i = 6$ percent.

(a) What is the present worth of this series of increasing annual net profits over the next 20 years?

(*Ans.* $27,440,000)

(b) If current long-term debt now stands at $10,000,000, how much additional borrowing could be secured by the estimated present worth of the net profit if lenders will lend on 100 percent of that value?

(*Ans.* $17,440,000)

A.4. Annual maintenance costs on a certain road are $2,000 per mile this year and are expected to increase 4 percent per year. $i = 7$ percent.

(a) What is the present worth of the maintenance costs for the next five years (including the cost incurred at the end of the fifth year), assuming that all costs are billed at the end of the year?

(b) What is the annual equivalent (equivalent equal annual costs) of these increasing costs?

A.5. A homeowner asks your help in calculating costs versus savings for home insulation. The following amounts of energy are consumed annually:

cooling: 60×10^6 Btu/yr from electricity at $12.00/$10^6$ Btu

heating: 25×10^6 Btu/yr from oil at $8.00/$10^6$ Btu

Additional insulation, weather stripping, and storm windows will cost $2,000 and have a life of 20 years, and are estimated to save 20 percent of the BTUs now consumed. Energy costs are expected to rise at 12 percent per year compounded. The homeowner borrows and invests at $i = 15$ percent annual rate.

(a) What is the amount of anticipated saving for the first year only? (Assume the savings accrue during the year and are credited at the end of each year).

(*Ans.* $184/yr)

(b) Find the present worth, *P*, of the savings in energy costs over the full 20-year period.

(*Ans.* $2,518)

(c) What is the present worth, *P*, of the cost of additional insulation?

(*Ans.* $2,000)

(d) Compare the present worths of the insulation cost versus energy savings and tell whether the fuel savings justify the cost of insulation.

(*Ans.* $518)

A.6. Your client is considering purchase of a new solar energy heater and cooler (SEHAC) for use in a large building and asks you to determine whether it is worth the extra cost if your client's $i = 15$ percent. The following cost information is provided:

cost new of SEHAC $= $150,000

expected life of installation $= 25$ yr

savings in energy costs the first year $= $18,000 at EOY 1

Expected increase in energy costs (and savings) over the 25-year life of the installation is 8 percent per year.

(a) By PW, find which system is more economical.

(b) Find the PW of the savings.

Problem Group B

Find P, C, A, or F involving two or more steps.

B.1. A student scheduled to graduate at age 23 is making plans to retire by age 48. He plans to begin by investing $2,400 per year ($200/month) the first year, an additional $2,640 the second year, and increase the annual amount added to the investment by 10 percent every year to $2,904 the third year, and so on. He hopes by careful management to average a 10 percent return compounded on reasonable investments.

(a) How much would he accumulate by age 48 (25 years)?

   (*Ans.* $591,000)

(b) If he could raise his average interest rate of return on the investment to 12 percent, what would the amount be by age 48?

   (*Ans.* $739,800)

(c) How much has the annual amount required to be invested per year grown in 25 years?

   (*Ans.* $23,640)

B.2. Your client, the Metro Solid Waste Authority, asks you to design a sanitary land fill with capacity to receive all the solid waste from the area for the next ten years. Collections are averaging 300 yd^3 of land fill per week, and the amount is increasing on the average of 5 percent per year. You calculate that you need about 1 acre of disposal site for each 10,000 yd^3 of solid waste received. What area in acres will you recommend to the authority as being adequate for the next 10 years, just for the disposal site? (This is a special case where $i = 0$.)

B.3. A rate study is underway to determine water rates to pay for a new water supply well field and treatment plant financed by $2,000,000 in revenue bonds at 6 percent. It is assumed customers would rather pay less now and more later, since incomes in the community are expected to rise on the average. Assume that rates can be raised 4 percent per year and the pay-out period is 20 years.

(a) How much should the first payment on the bonds be at the end of the first year?

   (*Ans.* $126,300)

(b) How much will the last payment be at the end of the twentieth year?

   (*Ans.* $266,000)

B.4. An engineering student decides to accumulate $500,000 by her sixty-fifth birthday. She expects to start by investing a certain amount, C, on her twenty-third birthday and then increase the payment by 10 percent each year. She feels she can safely invest her funds at 12 percent compounded.

(a) How much should her initial investment, C, be?

(b) n for this problem is _____ years.

B.5. A young engineer on his twenty-fifth birthday decides to accumulate the equivalent of $100,000 by his sixty-fifth birthday, but is concerned about the effects of inflation on the purchasing power of $100,000.

(a) If he assumes that inflation will increase the price of the goods and services he normally buys (thus reducing the purchasing power of the dollar) by 6 percent compounded annually, how much money will he have to accumulate in order to have the same purchasing power 40 years from now that $100,000 has at the present time?

(*Ans.* $1,029,000)

(b) If he starts his savings program with a deposit now on his twenty-fifth birthday, and continues making equal annual deposits on each birthday including his sixty-fifth, how much should he deposit annually in order to accumulate the amount found in (a) above if his savings draw $9\frac{1}{4}$ percent interest compounded annually?

(*Ans.* $2,599)

(c) The young engineer reasons that his salary should be increasing as time goes on, and it makes more sense for him to start depositing a small amount in the earlier years, with increasing deposits in later years. He estimates that his salary will increase by 8 percent per year, and decides to increase the amount of his deposits by that percentage each year also. How much should he deposit the first year in order to reach the sum found in (a) above?

(*Ans.* $908.90)

(d) Same as (c), but now find the amount of the last deposit made on his sixty-fifth birthday?

(*Ans.* $19,750)

B.6. Find the present worth of the infinite stream of payments tabulated below.

EOY	Payment	i (%)	
1 through 24	$100K/yr	12	
25	90	10	(*Note:* change in interest rate at EOY 25)
26	81	10	
27	72.9	10	
⋮	Decreases at the same % rate	Same	
∞			

B.7. The school board asks your help in comparing alternative methods of providing schools in a newly developed suburban area. Find the equivalent annual costs of each of the following alternatives, using $i = 10$ percent, over a 20-year period.

Build a new school

cost new for school and equipment = $2,000,000
O & M costs for the first year of operation = $ 400,000/yr
O & M cost gradient for subsequent years = increase 8%/yr gradient
salvage value at EOY 20 = $ 200,000

Contract for busing pupils to nearest existing school

cost the first year = $300,000/yr
arithmetic gradient increase each subsequent year = $ 30,000/yr/yr

(*Ans.* Build, A = $952,000/yr, bus, A = $495,200/yr)

B.8. The electric power supply for a new industrial complex is being planned, and you are asked to compare the cost of constructing a completely new electric generating system to the cost of buying power from a nearby utility already in existence. Find the equivalent *annual* cost of each alternative.

In either case, $i = 12$ percent, and the system will last 25 years. All annual costs are charged at the end of the year.

Alternative A. Construct a new system.

cost of new construction = $25,000,000

            O & M cost = $3,000,000/yr for the first 3 yr, increasing by 6%/yr for
                          every year thereafter ($3,180,000 at EOY 4,
                          $3,370,800 at EOY 5, etc.)

salvage value at EOY 25 = $5,000,000

Alternative B. Purchase electric power from an existing nearby electric company on a long-term, 25-year contract for $6,000,000 per year, with no escalation clauses. All necessary construction is paid for by the neighboring utility.

Find the equivalent annual cost of alternative A, and compare with alternative B.

Problem Group C

Involves comparisons and unit costs.

C.1. An overpass and traffic interchange is proposed for a certain intersection. The traffic this year will average 10,000 vehicles per day and is expected to increase at the rate of 10 percent per year. This growth rate is expected to continue for the next 20 years. The interchange is estimated to cost $2,000,000. Maintenance costs are estimated to be about the same with the interchange as without it. The savings in time, fuel, wear and tear, accidental property damage, and personal injury are expected to average about $0.02 per vehicle. Assume all savings accrue at EOY. The project can be financed at 7 percent interest on a loan and road bond money. Is the project justifiable on the basis of the information given?

(*Ans.* Savings, $P = \$1,796,000$)

C.2. A fleet of earthmoving equipment may be purchased for $2,000,000 cash. In an average year the fleet is expected to move 5,000,000 yd^3 of the earth. The O & M costs are currently running about $0.60/yd^3$ but are expected to increase about 8 percent per year for the next five years. Earthmoving of this type is currently being successfully bid at about $0.90/yd^3$ but is expected to increase at the rate of about 5 percent per year. Overhead (supervision, clerical, home and field office, etc.) costs are running about $500,000 per year and are expected to remain fairly constant at that level. It is estimated that at the end of five years the fleet can be sold for $600,000. A prospective buyer says he will buy the fleet if it will return at least 15 percent on his invested capital. Should he buy the fleet? Show all the calculations.

C.3. Solar hot water heaters are available for $600 installed. No maintenance is expected for the first 15 years. After 15 years they are expected to have a value of $100. A developer asks your advice on whether to install the solar heaters in all the new houses in a development in place of conventional gas hot water heaters costing $100 installed. Fuel costs are expected to average $8 per month at the start, with an increase of 5 percent per year over the 15-year period. Maintenance costs on the gas heaters are expected to be negligible, and the resale value after 15 years will be $10. Mortgage money is used to finance either installation, at 8 percent. Assume end-of-year payments.

(a) Which is the most economical alternative for the home buyer, if the house with the solar heat costs an extra $500?

(*Ans.* Solar, $P = \$568.50$; gas, $P = \$1,200$)

(b) In your opinion, do you think customers will pay an extra $500 for a house with the solar heater? Why?

C.4. A dealer is trying to lease you a new dozer, rented and operated for $40,000 per year on a five-year contract. Your present dozer will bring $40,000 on the used equipment market. Operating and maintenance costs are $30,000 a year and are increasing 10 percent per year. You can borrow money from your bank at 8 percent. Estimated life of the dozer is five more years, with a resale value at that time of $10,000. Compare the equivalent annual cost of owning and operating your present dozer with the dealer's offer.

C.5. Your firm asks you to calculate how much should be charged per ton of rock processed by the rock crusher in order to have sufficient funds on hand to replace the rock crusher in eight years. At the end of every year the extra amount charged during the year will be placed in an account bearing interest at 7 percent. Due to generally increasing prices, it is expected that the extra charge can be increased by 5 percent per year. The rock crusher is expected to process 6,000 tons of rock per year. The estimated replacement cost eight years from now is $182,000.

(*Ans.* $2.52/ton)

C.6. A contractor has a contract to construct a 12,000-ft-long tunnel in 30 months. He is trying to decide whether to do the job with his own forces or subcontract the job. He asks you to calculate the equivalent *monthly cost* of each alternative. His $i = 1$ percent per month. Production under both alternatives will be 400 ft per month.

Alternative A.   Buy a tunneling machine and work with contractor's own forces.

cost of tunneling and machine = $2,000,000 paid now
salvage value of machine at EOM 30 = $400,000
cost of labor and materials = 320/ft for the first 10 months, increasing
by 0.5%/month at EOM each month
thereafter ($321.60/ft at EOM 11,
$323.208/ft at EOM 12, etc.)

Alternative B.   Subcontract the work. Cost $800 per foot of advance of the tunnel heading, for the full 12,000 ft.

C.7. Calculate the amount of the toll per car for the first year to pay for a toll bridge with data as indicated. Assume EOY payments.

cost new of toll bridge = $20,000,000
traffic = 20,000 vpd*, no increases
life expectancy of toll bridge = 25 yr, with zero salvage value
interest rate = 12%

Toll: assume the toll starts out at a comparatively low value ($*C*/vehicle) and is increased every year by 5 percent per year compounded.

(*Ans.* $0.240/vehicle)

*vpd = vehicles per day for 365 days per year.

C.8.   A construction firm is for sale. It is expected to break even the first two years and then make a profit of $100,000 at EOY 3. The profit should increase by 10 percent per year every subsequent year for 19 additional EOY payments. Then at EOY 22 the firm will be sold for $500,000. If your MARR is 15 percent, what is the most you can afford to pay for the firm?

C.9.   A new type of insulated window has just come on the market and your client asks you to determine whether it is worth the extra cost. He wants to use it in a new office building and develops the following cost information:

$$\text{extra cost of insulated window} = \$50,000$$
$$\text{expected life of installation} = 40 \text{ yr}$$
$$\text{savings in energy costs the first year} = \$6,000 \text{ at EOY 1}$$
$$\text{expected increase in energy costs (and}$$
$$\text{savings over the 40-yr life of the installation)} = 5\%$$

Mortgage money is available to finance the extra cost of the insulated window installation at 14 percent.

(a) Find the present worth of the cost.

   (Ans. $50,000)

(b) Find the present worth of the savings.

   (Ans. $64,190)

(c) Are the windows a recommended purchase?

C.10.   Assume that at age 25 you have decided to become a millionaire by age 65. You decide to invest regularly at the end of every year for 40 years into a fund earning 12 percent interest. The initial investment deposit at EOY 1 will be relatively small, but you expect to increase the amount of each subsequent deposit by 8 percent every year thereafter.

(a) Find the amount of the *initial* investment (deposit at EOY 1).

(b) Find the amount of the *final* investment (deposit at EOY 40).

C.11.   A cable TV installation is needed for a new residential development, and you are asked to compare the cost of constructing a completely new system to the cost of buying the service from an existing TV cable company located nearby. Find the equivalent *annual* cost of each alternative. Use $i = 13$ percent, and assume the system will last 25 years. All annual costs are charged at the end of the year.

Alternative A.   Construct new system.

cost of construction = $4,000,000
         O & M = $400,000/yr for the first 5 yr increasing by 5% for each
                 year thereafter ($420,000 at EOY 6,
                 $441,000 at EOY 7, etc.)

Salvage value at EOY 25 is estimated at $1,000,000.

Alternative B.   Purchase the service from the existing TV cable company at an equivalent annual cost of $1,200,000 per year.

(Ans. $A_A = \$1,030,000/\text{yr}$; $A_B = \$1,200,000/\text{yr}$)

C.12.   The water supply for a new development is being planned, and you are asked to compare the cost of constructing a completely new water system to the cost of buying water from a nearby utility already in existence. Find the equivalent *annual* cost of each alternative. In either case $i = 11$ percent, and the system will last 25 years. All annual costs are charged at the end of the year.

Alternative A.   Construct new system.

cost of new construction = $3,000,000

O & M = $300,000/yr for the first 3 yr, increasing by 6%/yr for every year thereafter ($318,000 at EOY 4, $337,080 at EOY 5, etc.)

salvage value at EOY 25 = $0.

Alternative B.   Purchase from the existing nearby utility at an equivalent annual cost of $600,000 per year. All necessary construction will be paid for by the neighboring utility.

C.13. As consultant to the county engineer you are requested to compare the cost of purchasing an asphalt plant versus continuing to purchase asphalt on a contract from an independent supplier. The comparative costs are listed below. Use $i = 10$ percent per year. The county expects to use 7,000 tons per year.

Purchase an asphalt plant

cost new = $800,000

O & M costs, including materials the first year = $100,000/yr

each subsequent year = 8% geometric gradient increase in costs

salvage value at EOY 10 = $80,000

tonnage output per year = 7,000 tons/yr

Purchase asphalt from an independent supplier

contract price for first year = $40.00/ton

arithmetic gradient increase each subsequent year = $4.00/ton/yr

(*Ans.* $P_{\text{purch.}} = \$1,607,000$; $P_{\text{contr.}} = \$2,361,000$)

C.14. Compare the equivalent *annual* costs of the two alternative central heating systems for buildings, using the data below. Both systems are expected to last 20 years. Use $i = 10$ percent.

Alternative A.   Conventional oil-fired central heat.

cost new = $4,000

fuel and maintenance = $1,200 the first year, increasing by 6% for each year thereafter ($1,272 the second year, $1,348.32 the third year, etc.)

All fuel and maintenance costs are charged *in advance* at the *beginning* of each year. Salvage value at EOY 20 is zero.

Alternative B.   Solar central heat.

cost new = $12,000

maintenance costs = $100/yr

increasing by = $10/yr/yr ($110/yr at EOY 2, $120/yr at EOY 3, etc.)

All maintenance costs are charged at the *end* of the year. Salvage value at EOY 20 is $2,000.

C.15. A new automated traffic signal has just come on the market, and your municipal client asks you to determine whether it is worth the extra cost.

extra cost of each new signal = $13,200

expected life of installation = 40 yr

savings to motorists the first year

for each new installation = $600 at EOY 1

expected increase in motorists' savings

over the 40-yr life of the installation = 4%/yr

Assume $i = 9$ percent.

(*Ans.* Cost, $P = \$13,200$; savings, $P = \$10,160$; reject)

C.16. An electric power and light utility company asks your help in evaluating the cost of converting boiler furnaces to use solid waste as boiler fuel in an electric generating plant. The plant currently consumes the following amount of fuel annually: coal: 1,000,000 tons per year at $60 per ton for the first year. The conversion to burn solid waste will cost $40,000,000 and have a life of 20 years, and is estimated to save 10 percent of the fuel now consumed by substituting solid waste (delivered free) for 10 percent of the fuel now consumed. Salvage value of the conversion at EOY 20 equals $5,000,000. Coal costs are expected to rise at 8 percent per year compounded, and the annual savings will rise proportionally. Use $i = 12$ percent.

    (a) What is the amount of anticipated fuel cost saving for the first year only? (Assume the expenditure for the conversion occurs right now, and the savings accrue during each year and are credited at the end of each year.)

    (b) Find the present worth, P, of the savings in fuel costs over the full 20-year period.

    (c) Compare the present worth of the conversion costs versus fuel savings and tell whether the fuel savings justify the cost of the conversion.

Problem Group D

Periodic geometric gradient, PTZ $\neq$ ETZ, $n \to \infty$.

D.1. Navigation buoys in a certain ship channel are expected to require periodic expenditures as tabulated below. Note that the first expenditure occurs at EOY 3, with subsequent expenditures at five-year intervals, increasing by 20 percent for each expenditure. Find the equivalent annual cost with $n \to \infty$ and $i = 10$ percent per year.

EOY	Maintenance
3	$200,000
8	240,000
13	288,000
etc.	etc.

(*Ans.* $58,950/yr)

D.2. A new flood control project is expected to involve expenditures for periodic heavy maintenance as tabulated below. Note that the first expenditure occurs at EOY 2, with subsequent expenditures at four-year intervals, increasing by 20 percent for each expenditure. Find the equivalent annual cost with $i = 15$ percent per year and $n \to \infty$.

EOY	Expenditure
2	$250,000
6	300,000
10	360,000
etc.	etc.

PART TWO

Comparing Alternative Proposals

Present Worth Method of Comparing Alternatives

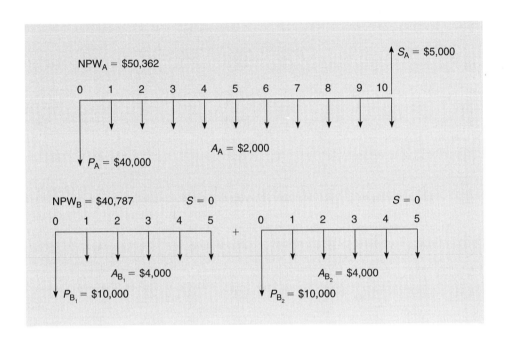

KEY EXPRESSIONS IN THIS CHAPTER

NPW = Net Present Worth, the equivalent lump sum value now after summing the present worths of all incomes, costs, and savings.

P & I = Principal and Interest, the two parts of each amortizing repayment on a loan. To amortize is to liquidate a debt by periodic payments to a creditor.

O & M = Operating and Maintenance costs, the two general categories of cost involved in the utilization of facilities and equipment.

REVIEW AND PREVIEW

The previous chapters dealt with the concept of equivalence involving the time value of money. Equations involving an interest rate i per period and a number of periods n were developed to relate the various forms of cash flow (P, F, A, G, C). Those relating to present worth are summarized as follows:

$$(P/F, i, n) = \frac{1}{(1+i)^n}$$

$$(P/A, i, n) = \left[\frac{(1+i)^n - 1}{i(1+i)^n}\right]$$

$$(P/G, i, n) = \frac{1}{i}\left[\frac{(1+i)^n - 1}{i(1+i)^n} - \frac{n}{(1+i)^n}\right]$$

$$(P/C, i, n) = \frac{1}{1+i}\left[\frac{(1+w)^n - 1}{w}\right] \quad \text{for} \quad r > i \quad \text{and} \quad w = \frac{1+r}{1+i} - 1$$

$$(P/C, i, n) = \frac{1}{1+r}\left[\frac{(1+w)^n - 1}{w(1+w)^n}\right] \quad \text{for} \quad r < i \quad \text{and} \quad w = \frac{1+i}{1+r} - 1$$

$$(P/C, i, n) = \frac{n}{1+r} = \frac{n}{1+i} \quad \text{for} \quad r = i$$

A clear understanding of the use of these equations is essential to the comprehension of the concepts in this section. Once such an understanding is achieved, the reader should have little difficulty in moving readily through the remainder of the text.

As discussed in Chapter 1, there are usually several ways to solve an engineering problem. For instance, a bridge can be built of either steel or concrete or some combination thereof. After the most promising solutions have been outlined, the problem becomes one of selecting the optimum design alternative from among those available.

COMPARING ALTERNATIVES

In selecting among alternatives, the simplest type of input to consider is that which comes in *tangible* terms that can be translated into dollars. These tangible terms include units such as cubic yards of concrete, number of vehicles per day, millions of gallons of water per day, and so on. As long as the conversion factor is available (in terms of dollars per unit), the equivalent dollars can be readily determined. Sometimes a second type of input of the category termed *intangibles* will enter the problem situation, and these can become difficult to evaluate. These are expressed in terms such as aesthetics, color, texture, location, compatibility, quality, comfort, style, durability, and many other types of value judgments. These can be very important considerations and have frequently been deciding factors in determining the fate of large projects. For example, the citizens of Boston, New Orleans, and San Francisco have at some time in the past objected to various interstate highway projects that were already partially completed in their respective cities. These citizens and their objections, which mostly dealt with aesthetics and related matters, were so persuasive that the projects in all three cities were subsequently abandoned at a cost of millions of dollars. For the most part this text deals with considerations that can be expressed in economic terms (dollars, interest rates, time), but intangibles should never be overlooked and will be discussed where appropriate.

The usual goal of engineering economy studies is to minimize costs or maximize profits or savings. These studies usually fall into either of the following two problem situations:

Situation	Criterion
Alternatives involve costs only.	Select the alternative with the lowest equivalent cost.
Alternatives involve both costs and benefits (income or savings)	Select the alternative that maximizes the equivalent net benefit or worth (the do-nothing alternative should always be considered, if available).

METHODS OF COMPARING ALTERNATIVES

In comparing alternatives several methods are available, including the following:

Present worth analysis	(Chapter 7)
Annual worth analysis	(Chapter 8)
Future worth analysis	(Chapter 9)
Rate of return analysis	(Chapters 10 and 11)
Benefit/cost analysis	(Chapter 14)

Each of these methods has advantages and limitations, which will be discussed, with examples, to assist the student in determining the preferred method of analysis for a wide variety of problem formats.

The Present Worth Method of Comparing Alternatives

In this approach alternatives are compared on the basis of equivalent present worth. To make such a comparison, the interest rate, i, must be known or assumed. The value of i should be selected carefully, since the resulting PW is usually sensitive to changes in i value. Changes in the i value will cause changes in the PW, and at times these changes are large enough to reverse the selection of the lowest cost alternative.

The length of life, n, for each alternative also should be considered carefully. To make a valid PW comparison, the same time periods must be used for each alternative. There are three types of life-span situations encountered when comparing two or more alternatives by PW.

1. Compare alternatives where each has the *same life span.* Simply find the PW of each alternative and compare.

2. Compare alternatives where each has a *different life span.* In order to satisfy the requirement for equal time periods for each alternative, multiple lives must be found

by calculating the lowest common product of each respective life span times some whole number multiplier.

3. The lives of both alternatives are very long ($n \to \infty$). Since the alternatives are each compared over an equivalent infinite life span, the requirement for equal lives is satisfied.

Examples of each of these situations follow.

COMPARISON OF ALTERNATIVES WITH EQUAL LIFE SPANS

EXAMPLE 7.1   (Alternatives with equal lives. Find the NPW of each and compare, given P, A, i, n)

An engineer is in need of an automobile for business purposes, and finds that he can either lease or purchase, on the following terms:

a. Lease a car for $550/month, paid monthly at the end of each month for twenty-four months.

b. Purchase the same car for $20,000 now and sell it in two years for $12,000.

In both cases, the engineer pays all operating, maintenance, and insurance costs. Find the least costly alternative, assuming interest is compounded at 1 percent monthly.

Solution.   Since the life spans of both alternatives are equal (two years), the alternatives can be compared directly by finding the net present worth of each and selecting the one with the least cost.

a. *Lease.* The NPW of the lease alternative is the PW of the 24 payments of $550 per month.

$$P_1 = -550(P/A, 1\%, 24) = -550 \times 21.243 = \underline{-\$11,684}$$

b. *Purchase.* The NPW of the purchase alternative is the PW of the purchase price minus the PW of the income from the resale at the end of two years.

$$
\begin{aligned}
P_2 &= -\$20{,}000 \text{ purchase price} \\
P_3 &= 12{,}000(P/F, 1\%, 24) \\
&= 12{,}000 \times 0.78757 &= \quad 9{,}451 \text{ resale} \\
P_2 + P_3, \text{ NPW of purchase alternative} &= \underline{-\$10{,}549} \text{ net}
\end{aligned}
$$

Comparing the net present worths, it is apparent that the purchase alternative is much less expensive than the lease alternative. In the above example, alternatives involving payments of different amounts made at different points in time are compared by reducing each series, or lump sum payment, to an equivalent PW and then simply comparing the equivalent NPW of each alternative. (Of course each equivalency derived is valid only for the one interest rate used. If the interest rate changes, the present worth of any future payments also changes.) While the NPW of the monthly lease rate is higher than the purchase alternative, in order to find the difference in terms of equivalent dollars per month, it is only necessary to find the monthly equivalent to the purchase alternative and compare as follows.

Monthly equivalent of purchase alternative,

$$A = 10,549(A/P, 1\%, 24)$$
$$= \$496.86/\text{month, the monthly equivalent to the purchase alternative}$$

Thus, if the monthly lease rate were lowered to $496.86 per month, the two alternatives would have equal cost, and a break-even point is reached. The lump sum cash flow out of $20,000 now together with the cash flow in of $12,000 at the end of two years is equivalent to a periodic series cash flow out of $496.86 per month for 24 months, provided that the interest rate is 1 percent.

Alternatives with Equal Life Spans and Gradient Payments

As modern technology makes mechanization and automation cheaper and more dependable, economic comparisons will determine what work can be done more efficiently by machine, and what should be left to manual labor. As costs are projected into the future, many payments will rise (but possibly at different rates of increase) while some may stay constant or decline. For those that rise or decline, gradients are an indispensable tool for determining the equivalent PW values necessary for making economic comparisons. The following example is typical of such situations.

EXAMPLE 7.2   **(Compare the NPW of two alternatives with equal lives. Find NPW, given P, A, G, C, n)**

In order to improve solid waste pickup, a new mechanical arm is proposed. This arm extends from the truck, picks up garbage cans at curbside, and dumps them into the truck. The mechanical arm is estimated to last 10 years. Compare the cost of this arm with the cost of conventional pickup.

The estimated costs are:

$$
\begin{aligned}
\text{new} &= \$80{,}000 \\
\text{salvage value in 10 yr} &= \$\ 5{,}000 \\
\text{O \& M first year} &= \$\ 2{,}000/\text{yr} \\
\text{increase in O \& M per year} &= \$\quad 400/\text{yr/yr}
\end{aligned}
$$

The mechanical arm will replace one man on the crew who now draws $16,000 per year with anticipated increases of $1,200 per year per year.

Find the NPW of each alternative and compare for $i = 10$ percent.

Solution

$$
\begin{aligned}
\text{mechanical arm}, P_1 &= -\$\ 80{,}000 \\
\text{salvage}, P_2 = +\ 5{,}000(P/F, 10\%, 10) &= +\quad 1{,}928 \\
\text{O \& M}, P_3 = -\ 2{,}000(P/A, 10\%, 10) &= -\quad 12{,}289 \\
\text{gradient}, P_4 = -\quad 400(P/G, 10\%, 10) &= -\quad 9{,}156 \\
\text{NPW of mechanical arm}, P_{\text{total}} &= -\$\ 97{,}517
\end{aligned}
$$

$$
\begin{aligned}
&\text{manual loading} \\
&\text{worker}, P_5 = -16{,}000(P/A, 10\%, 10) = -\$\ 98{,}312 \\
&\text{gradient}, P_6 = -\ 1{,}200(P/G, 10\%, 10) = -\quad 27{,}469 \\
&\qquad\qquad \text{NPW of manual loading} = -\$125{,}781
\end{aligned}
$$

Using these costs and a projected life of 10 years, the mechanical arm appears more economical. In actual practice, before accepting the results, a check of all important variables should be made.

Sensitivity Check

A sensitivity check involves ranging of all the input variables over their most probable range of values and observing the effect on the resulting answer. In this case, for instance, a sensitivity check reveals that the solution is quite sensitive to changes in the estimated life of the mechanical arm. Changing the life from 10 years to 6 years reverses the outcome.

EXAMPLE 7.2   (continued)

Compare the NPW of each alternative with $i = 10$ percent and n changed to 6 years.

Solution

$$\begin{aligned}
\text{mechanical arm}, P_1 &= -\$80,000 \\
\text{salvage}, P_2 &= +\ 5,000(P/F, 10\%, 6) = +\quad 2,822 \\
\text{O \& M}, P_3 &= -\ 2,000(P/A, 10\%, 6) = -\quad 8,710 \\
\text{gradient}, P_4 &= -\quad 400(P/G, 10\%, 6) = -\quad 3,874 \\
\text{NPW of mechanical arm} &= -\$89,762
\end{aligned}$$

$$\begin{aligned}
\text{manual loading worker}, P_5 &= -16,000(P/A, 10\%, 6) = -\$69,683 \\
\text{gradient}, P_6 &= -\ 1,200(P/G, 10\%, 6) = -\quad 11,621 \\
\text{NPW of manual loading} &= -\$81,304
\end{aligned}$$

If the mechanical arm lasted only 6 years (instead of 10 years), then manual loading would be more economical. Further analysis could reveal the break-even points for life n, interest rate i, cost new, and salvage value.

COMPARISON OF ALTERNATIVES WITH UNEQUAL LIFE SPANS

Frequently, two or more alternatives occur that do *not* have identical life expectancies. These cannot be compared directly by means of present worth. For instance, alternative A with NPW of $20,000 and a life of 10 years is undoubtedly less costly than alternative B with an NPW of $10,000 and a life of 1 year. Even though alternative A has a higher NPW, the much longer life of A more than compensates for the higher cost (assuming a normal range of interest rates). Thus the life expectancy is an important factor and must be taken into account. To account for the differences in life expectancy, two methods are available:

1. Assume the shorter-lived alternative will be replaced as many times as necessary to obtain equal life spans. In the case of a 10-year life for A and a 1-year life for B, assume B will be replaced every year with a similar installation. This compares one 10-year A with 10 sequential 1-year B alternatives. If A were 10 years and B were 4 years, then two 10-year A's (20 years) could be compared to five sequences of B at 4 years each (also 20 years).

2. The second method involves shortening the comparison period to the expected life span of the shorter alternative and attributing a salvage value to the longer lived alternative. For instance, in comparing the 10-year A to the 1-year B, the life of A would be shortened to 1 year, and some estimated salvage value would be placed on A at the end of 1 year. Then this salvage value is treated as a lump sum income at the end of 1 year and credited to A in the comparison.

EXAMPLE 7.3   (Unequal life spans; compare NPW, given P, A, i, n for each, where $n_1 \neq n_2$)

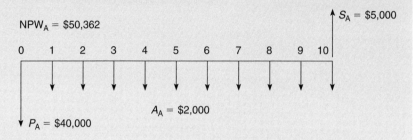

NPW$_A$ = $50,362

A_A = $2,000

P_A = $40,000

S_A = $5,000

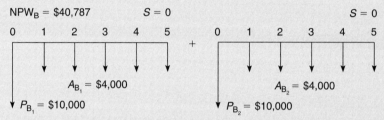

NPW$_B$ = $40,787 $S = 0$ $S = 0$

A_{B_1} = $4,000 A_{B_2} = $4,000

P_{B_1} = $10,000 P_{B_2} = $10,000

A county engineer has a choice of paving a short section of county roadway with either type A pavement or type B pavement. Type A pavement has a life expectancy of 10 years, after which part of the material can be salvaged and reused. Type B pavement only lasts 5 years but is much less expensive. Which is the better alternative?

	Pavement cost	
	Type A	Type B
Cost new	$40,000	$10,000
Annual maintenance	2,000	4,000
Estimated life	10 yr	5 yr
Salvage value at end of life	5,000	0
i	10%	10%

Solution.   Two sequential 5-year installations of type B are compared to one 10-year life of type A. Find the NPW of each.

Type A	Present worth
Cost new	−$40,000
Annual maintenance $P_1 = 2,000 \times (P/A, 10\%, 10)$	− 12,290
Less salvage $P_2 = 5,000 \times (P/F, 10\%, 10)$	+ 1,928
NPW of pavement type A	−$50,362

Type B	Present worth
Cost new first application	−$10,000
Second application $P_1 = 10,000 \times (P/F, 10\%, 5)$	− 6,209
Annual maintenance $P_2 = 4,000 \times (P/A, 10\%, 10)$	− 24,578
NPW of pavement type B	−$40,787

Comparing the two alternatives, we see that type B is less expensive. Notice that this method assumes that the second application at EOY 5 costs the same as the first application at EOY 0. In actual practice this is usually not the case due to price increases caused by inflation and other market forces. If prices are expected to increase by 6 percent compounded every year, the price of the second application at EOY 5 could be computed as

$$F = \$10,000(F/P, 6\%, 5) = \$13,382$$

The present worth of the second application then becomes

$$P_2 = 13,382(P/F, 10\%, 5) = -\$8,310$$

and the total NPW for type B becomes $42,888, still the lower of the two options in this case.

If the annual maintenance costs are expected to increase, for example, by $400 per year each year for type A pavement, and by $1,000 per year for type B, the present worth of the savings of A over B may be found as

$$P = [-400 - (-1,000)](P/G, 10\%, 10) = \$13,734$$

Since before consideration of the increasing maintenance cost gradient, B was favored by ($50,362 − 42,888 =) $7,474, the newly calculated saving of $13,734 attributed to A reverses the lower cost selection from type B to type A pavement.

Some Guidelines on Replacement Costs When Comparing Unequal Lives

When comparing different life spans, and assuming that replacement costs are the same as the original cost new, we need to recognize certain features inherent in the comparison system. For instance, if item A has one-half the life and one-half the cost of B, then the NPW of A will be less (for all $i > 0$). Why? Because there is no investment in the second A (and therefore no interest is paid on the second A) until it is purchased at the end of the life of the first A. By contrast, interest on the funds representing the second half of B's life is paid over the entire life span of B. On the other hand, if the replacement cost is not equal to the cost new but is rising at the same rate as the interest rate ($r = i$), then a break-even situation exists. The cost of A will equal the cost of B. Therefore, another rule of thumb can be formulated as, if the replacement cost of A is rising at a lower rate than i, then A is still the lower cost alternative (with A at one-half the life and one-half the cost of B).

CAPITALIZED COST OF PERPETUAL LIVES, $n \to \infty$

Capitalized cost has two popular meanings. Accountants use capitalized cost to describe expenditures that may be depreciated over more than one year, as contrasted to expenditures that may be written off (expensed) entirely in the year they were made. The second meaning for capitalized cost is found in the traditional literature of engineering economy. Here capitalized cost refers to the equivalent lump sum amount required to purchase and maintain a project in perpetuity (net present worth for $n \to \infty$). Capitalized cost problems may involve one or more of the following categories of payments:

a. An infinite series of uniform payments.

b. An infinite series of gradient payments (arithmetic or geometric).

c. Lump sum first cost of purchase or construction.

d. A series of lump sum costs of future replacements required for perpetual service or periodic major maintenance (the interval between payments is some multiple of the interest-compounding period).

e. Periodic gradients to replacement costs or periodic maintenance.

a. Where only an infinite series of uniform costs is involved, the capitalized cost is simply the lump sum amount required now in an interest-bearing account that will produce an infinite series of interest payments sufficient to meet all uniform costs. For instance, the capitalized cost, P, equals the annual cost, A, divided by the interest rate, i, or, the capitalized cost $= P = A/i$. An example follows.

EXAMPLE 7.4   (Find the capitalized cost)

The annual cost of maintaining a certain right-of-way is $10,000 per mile. Find the capitalized cost with $i = 8$ percent.

Solution.   Finding the capitalized cost means find the PW with $n \to \infty$.

$$\text{capitalized cost} = \$10{,}000/0.08 = \underline{\$125{,}000}$$

Mental picture: If funds could be invested at 8 percent, how much should be invested so that the interest alone would pay for the right-of-way maintenance?

(*Ans.* $125,000)

Note that where lives are comparatively long, the capitalized cost closely approximates the PW. For example, if the time in the example above had been 100 years instead of infinity, then the PW would only be reduced to

$$\text{PW} = \$10{,}000(P/A, 8\%, 100) = \underline{\$124{,}940}$$

compared to the $125,000 found for an $n \to \infty$.

b. Where an arithmetic gradient is involved, simply apply the equation for PW of an arithmetic gradient for $n \to \infty$, $P = G/i^2$. For instance, in Example 7.4, if the maintenance cost increases by \$1,000 per year per year, the capitalized cost increases by

$$P = \$1,000/0.08^2 = \$156,250$$

If instead of an arithmetic gradient, a geometric gradient occurs, then apply the P/C equation (providing that $r < i$), in which case

$$\text{capitalized cost, } P = \frac{C}{(1+r)w}$$

For example, if the maintenance costs increase by 5 percent per year, then

$$\text{capitalized cost } P = \frac{10,000}{1.05 \times 0.02857} = \underline{\$333,333}$$

c. Where a first cost is involved in addition to the PW of uniform and gradient series payments, simply add the sums to obtain the total capitalized cost of the project. For instance, in Example 7.4, if the first cost (cost of acquisition) of the right-of-way were \$47,000, then the total capitalized cost of the project would be the sum of the following costs:

> \$125,000 (for maintenance), plus
> \$156,250 (for arithmetic gradient maintenance), plus
> $\underline{\$\ 47,000}$ (for first cost)
> $\underline{\underline{\$328,250}}$ capitalized cost (NPW)

d. Where a facility is needed for perpetual service, but needs replacement from time to time, additional annual interest income must be available that will accumulate to the replacement cost at the appropriate time. For instance, in the above example assume drainage structures in the right-of-way have an estimated life of 20 years and cost \$120,000 to construct now, and \$100,000 to replace at the end of every 20-year period. The cost of \$120,000 to construct now is simply added to the capitalized cost as in (b) above. The \$100,000 cost to replace at the end of every 20 years is accumulated by annual payments, A, deposited into a sinking fund (savings account bearing compound interest, i). The amount required for the annual deposit to accumulate to \$100,000 at the end of every 20 years is calculated as

$$A = \$100,000(A/F, 8\%, 20) = \$2,190$$

The amount of capitalized cost required to generate this annual income forever at 8 percent is

$$P = A/i = \$2,190/0.08 = \underline{\$27,400}$$

e. When the replacement costs increase by either an arithmetic or geometric gradient, the equivalent capitalized cost may be calculated using the effective interest rate, $i_e = (1 + i)^m - 1$. For instance, assume the cost of the replacement drainage structure previously described in (d) doubles every 20 years ($r_{20} = 100\%$). Then the effective interest rate for compounding once every 20 years (instead of every year at 8 percent) is found as

$$i_e = 1.08^{20} - 1 = 3.661 \text{ or } 366.1\%$$

Since $r_e < i_e$, then

$$P = C(P/A, w, n)/(1 + r) \tag{6.3a}$$

where

$$w = (1 + i)/(1 + r) - 1 = 4.661/2 - 1 = 1.330$$
$$P = \$100{,}000/(1.330 \times 2) = \underline{\underline{\$37{,}580}}$$

Thus, the total capitalized cost of the project for parts (a) through (e) is

capitalized cost of	(a) annual maintenance	$125,000
	(b) arithmetic gradient, maintenance	$156,250
	(c) first-cost R/W	$ 47,000
	(d) first-cost drainage	$120,000
	(e) geometric gradient, 20-yr	
	replacement of drainage	$ 37,580
	Total capitalized cost	$485,830

This is the total lump sum amount required now in an account bearing 8 percent interest in order to pay all the anticipated costs of the project for an infinite period of time.

Unfortunately, the term *capitalized cost* when applied to engineering economy sometimes gives rise to confusion, since it is more frequently encountered in the accounting sense. The engineering use of the term probably could be laid aside and neglected by engineering students were it not for its frequent appearance on licensing exams.

PRESENT WORTH WHEN A SERIES OF PAYMENTS STARTS AT THE PRESENT, PAST, OR FUTURE

In order to optimize the return on their resources, many astute people spend considerable time and effort on analyzing, managing, operating, buying, selling, or trading property of one kind or another. This property may consist of machines, vehicles, structures, real estate, or other property. Property ownership and operation usually involve cash flows of income and cost, both uniform and gradient series as well as lump sums occurring at various points in time. To optimize the return on the property investment, it frequently is necessary to calculate an NPW, NAW, NFW, or rate of return. Some problems can become somewhat complex and challenging. The basic approach to solving complex problems is to separate the problem into simple components, solve each component, transform each component into an equivalent present worth, and then sum the results to determine the net present worth, or equivalent, of the entire investment. In applying this method, single lump sum payments seldom pose any difficulty, since they can be reduced readily to present worth at PTZ by the simple P/F relationship. Uniform and gradient series, however, sometimes occur in more complex surroundings.

The nomenclature used to discuss these noncoincident time problems follows:

PTZ = problem time zero, the present time now, according to the statement of the problem

ETZ = equation time zero, with reference to a particular payment series, the time at which n = zero. Where the beginning or end of a series does not conveniently correspond with PTZ, a separate ETZ may be used to designate the beginning point in time for the particular series under consideration. For a uniform series of equal payments, the ETZ is always either one period before the first payment occurs (P/A), or at the time of the last payment (F/A).

Problems involving the present worth of a uniform series can be divided conveniently into three groups, as listed below:

Group 1. Uniform series of payments begins at the present time. The first payment in the series occurs one compounding period from now. Therefore, PTZ = ETZ.

Group 2. The ETZ for the uniform series of payments occurred one or more periods in the past. Therefore, the ETZ is earlier than the PTZ.

Group 3. The ETZ for the uniform series of payments will occur sometime in the future. Therefore, the PTZ is earlier than the ETZ.

The cash flows for these groups are shown in Figures 7.1, 7.2, and 7.3.

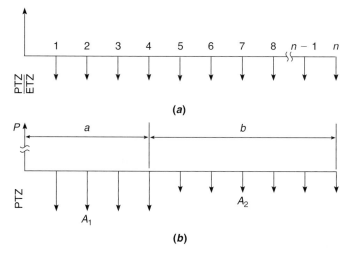

Figure 7.1   Group I. The ETZ for the uniform series of payments occurs now, at the present time, PTZ = ETZ. (a) Uniform series coincides with economic life. $P = A(P/A, i, n)$, (b) Uniform series is divided into two or more parts, a and b. $P_a = A_1(P/A, i, a)$. $P_b = A_2(P/A, i, b)(P/F, i, a)$.

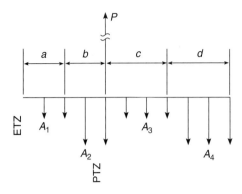

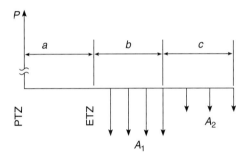

Figure 7.2   Group II. The ETZ for the uniform series of payments occurred in the past; the ETZ comes before the PTZ.

Figure 7.3   Group III. The ETZ for the uniform series of payments occurs at some future date; the PTZ comes before the ETZ.

Notes: The letters *a*, *b*, *c*, *d* represent some stipulated number of time periods, as shown on the cash flow diagrams. For the uniform series, the number of payments is always equal to the number of compounding time periods, and the ETZ is always found either one period before the first payment in the *P/A* series, or at the time of the last payment for the *F/A* series.

PRESENT WORTH OF A SERIES OF UNIFORM AND ARITHMETIC GRADIENT PAYMENTS AT INTERVALS EQUAL TO SOME MULTIPLE OF THE COMPOUNDING PERIOD

At times a series of periodic payments may involve repetitive but long intervals. For instance, heavy construction equipment may require periodic overhaul every six months while the compounding period is monthly, or a certain structure may require heavy maintenance at five-year intervals while the compounding period is annual. In addition, the payments may increase each time the event occurs as costs continue to rise. To find the equivalent PW for these long-interval gradient payments, the *equivalent interest rate, i_e,* is used to transform the long-interval series into an equivalent normal gradient series, as illustrated in the following example.

EXAMPLE 7.5   **(Periodic overhaul every six months with gradient, compounded monthly)**

Find the cost of excavation in terms of dollars per yd^3 (where $yd^3 =$ cubic yard) over the 60-month life of the excavator whose data are listed below. Use $i = 1$ percent per month.

$$
\begin{aligned}
\text{excavator cost new} &= \ \$300{,}000 \\
\text{resale value at EOM 60} &= \ \$\ 75{,}000 \\
\text{O \& M} &= - \ \ 6{,}000/\text{month} \\
\text{gradient in O \& M} &= \ \ 75/\text{month/month} \\
\text{periodic overhaul every 6 months} &= - \ \ 18{,}000/\text{cach} \\
\text{gradient in periodic overhaul} &= - \ \ 1{,}200/\text{each/6 months} \\
\text{expected production} &= \ \ 20{,}000 \ yd^3/\text{month}
\end{aligned}
$$

The first periodic overhaul occurs at EOM 6 and the last at EOM 60.

Solution.   The equivalent interest rate for a six-month interval is determined first as

$$
i_e = (1.01)^6 - 1 = 6.152\%
$$

Then the present worth of each cost item is calculated and summed as follows:

$$
\begin{aligned}
\text{cost new}, P_1 &= &&= -\$300{,}000 \\
\text{resale}, P_2 &= +75{,}000(P/F, 1\%, 60) &&= + \ \ 41{,}284 \\
\text{O \& M}, P_3 &= -6{,}000(P/A, 1\%, 60) &&= - \ \ 269{,}730 \\
\Delta \text{ O \& M}, P_4 &= -75(P/G, 1\%, 60) &&= - \ \ \ \ 87{,}460 \\
\text{periodic overhaul}, P_5 &= -18{,}000(P/A, 6.152\%, 10) &&= - \ \ 131{,}535 \\
\Delta \text{ periodic OH}, P &= \ \ \ 1{,}200(P/G, 6.152\%, 10) &&= -\$ \ 35{,}160 \\
&\text{NPW} = P_{\text{total}} &&= -\$784{,}601
\end{aligned}
$$

$$
\begin{aligned}
\text{monthly equivalent}, A &= \ \ \ 784{,}601(A/P, 1\%, 60) = \$17{,}450/\text{month} \\
\text{unit cost} &= \frac{17{,}450/\text{month}}{20{,}000 \ yd^3/\text{month}} = \$0.873/yd^3
\end{aligned}
$$

The unit cost for excavation that accounts for all of the given cost items is $0.873/yd^3. For example, if the contractor received $0.873/yd^3 for the 60-month period, all the given cost items could be paid, and any account balances either positive or negative would earn (in case of a positive balance) or pay (in case the account was temporarily in debt) interest at 1 percent on the balance at the end of each month.

Periodic Payment with ETZ ≠ PTZ

Sometimes the first periodic payment will occur at an odd time, followed at repetitive intervals by a normal periodic series. For instance, see the continuation of Example 7.5.

EXAMPLE 7.5   (continued)

Assume that the first periodic overhaul still costs $18,000 but occurs at EOM 2 instead of EOM 6. The subsequent periodic overhauls follow at regular six-month intervals and still increase by $1,200 each time (periodic overhauls now occur at EOM 2, 8, 14, . . . , 50, 56). Find the present worth of this periodic series.

Solution.   The actual compounding period is monthly at $i = 1$ percent per month. The interval between payments is six months, so the equivalent interest rate is $i_e = 6.152$ percent. The ETZ is located six months before the first payment. Since the first overhaul occurs at EOM 2, the ETZ is located at EOM minus 4. This moves the last event four months earlier, from EOM 60 to EOM 56, so there are still 10 payments within the 60-month term of ownership, and $n = 10$. The lump sum equivalent PW is found at the ETZ at EOM -4 and moved back to PTZ as follows:

$$\text{periodic OH, } P_5 = -18,000(P/A, 6.152\%, 10)(F/P, 1\%, 4) = -\$137,280$$
$$\Delta \text{ periodic OH, } P_6 = -1,200(P/G, 6.152\%, 10)(F/P, 1\%, 4) = -\$\ 36,585$$
$$\text{total } P_5 + P_6 \text{ with first OH at EOM } \underline{2} = \underline{-\$173,865}$$

Similarly, if the first periodic overhaul were to occur at EOM 10 (instead of EOM 2) and every six months thereafter, the ETZ would occur at EOM 4 (six months before the first payment), and, since the last (tenth) event would be pushed four months later (to EOM 64), which is beyond the term of ownership, the number of overhauls within the 60-month ownership term would be reduced to 9, so $n = 9$. The equations for finding the PW with an ETZ at EOM 4 are

$$\text{periodic OH, } P_5 = -18,000(P/A, 6.152\%, 9)(F/P, 1\%, 4) = -\$116,880$$
$$\Delta \text{ periodic OH, } P_6 = -1,200(P/G, 6.152\%, 9)(P/F, 1\%, 4) = -\$\ 28,080$$
$$\text{total } P_5 + P_6 \text{ with first OH at EOM } \underline{10} = \underline{-\$144,960}$$

Present Worth of Periodic Geometric Gradient Payments

At times, series of payments made at some multiple of the compounding period are observed to rise or fall *geometrically* (by a fixed percentage from payment to payment). For instance, if a generator or pump is overhauled once every year for 15 years, the costs may increase, say 10 percent each year, while the interest may compound monthly. This type of periodic series can be accommodated by finding the effective interest rate for the interval between payments, and then treating the periodic series as an ordinary geometric series. The following example illustrates the procedure.

EXAMPLE 7.6 **(Geometric gradient in periodic payments)**

A water utility department is considering the purchase of a new pump that will require an overhaul at the end of every year of service. The overhaul costs are expected to increase by 10 percent from year to year, as indicated on the cash flow diagram in Figure 7.4. With a life of 15 years, and $i = 1$ percent per month, find the present worth of these costs.

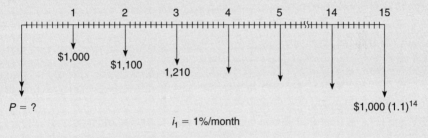

Figure 7.4 Cash flow diagram for Example 7.6

Solution. Find the effective interest rate as

$$i_e = 1.01^{12} - 1 = 0.1268 = 12.68\%$$

Then the present worth is calculated using the appropriate P/C equation. Since $r = 10$ percent, and $i_e = 12.68$ percent, $r < i$ and

$$w = \frac{1.128}{1.10} - 1 = 0.02439$$

Thus

$$P = \frac{C}{1+r}\left[\frac{(1+w)^n - 1}{w(1+w)^n}\right] = \frac{1,000}{1.10}\left[\frac{1.02439^{15} - 1}{0.02439(1.02439)^{15}}\right]$$

$$P = \$11,300$$

If the first overhaul occurs at EOM 6 (instead of EOM 12), with subsequent overhauls occurring every 12 months thereafter, then ETZ occurs at EOM minus 6 (EOM 6 − 12 = EOM − 6). The value of n remains at 15, since the last overhaul now occurs at EOM 174 (180 − 6). Therefore

$$P = \frac{C}{1+r}(P/A, w, 15)(F/P, 1\%, 6) = \frac{1,000}{1.1}(12.4372)(1.06152) = \underline{\$12,000}$$

FINANCING A LOAN, WHERE THE LIFE OF THE LOAN ≠ LIFE OF THE PROPERTY

For loans secured by property, the banks or other lending institutions usually insist that the payout period for the loan be shorter than the expected life of the property. Naturally, the lenders are concerned with the security of their loans. If the collateral for the loan has been fully depreciated before the loan is paid out, the risk of default is increased, since the loan can no longer be foreclosed against property of equal or greater value than the balance of the loan. For instance, on new cars with an average life expectancy of 12 years, lenders typically allow a maximum four-year repayment period. For used cars, a three-year maximum is common. On buildings with a life expectancy of 40 to 50 years and more, mortgage lenders usually permit 25- to 30-year repayment schedules. Frequently, however, the borrower does not own the property for the full life of the loan. The owner will often decide to sell the property, pay off the principal balance on the loan (or let the new buyer assume the loan), and buy something else before the loan is paid out. In the case of houses, the average homeowner moves about every five years. The former residence is sold, and the new buyer either may take over payments on the old loan contract, or arrange a new loan (refinance) so that the balance of the old loan contract can be paid off. Thus, it is important to be able to calculate both the amount that the loan has been paid down, as well as the balance due at any point in time on the loan contract.

EXAMPLE 7.7 **(Find the current balance owed on a partially amortized loan)**

For a typical example, suppose that 18 months ago you borrowed $5,000 at 9 percent (nominal annual) with equal monthly repayments for 48 months. Now, after the eighteenth payment has been made you need to calculate: (a) how much is the principal balance due if paid in cash now; (b) how much of the loan has been paid off; and (c) how much of the total amount paid was attributable to interest.

Solution to (a). This example involves a uniform series of payments, where the equation time zero (ETZ) (18 months ago) began before the problem time zero (PTZ), which occurs right now after the eighteenth payment. It is the type of situation shown in group 2 previously described in this chapter.

   Step 1. Find the amount of the monthly payment, A, using $i = \frac{9}{12}$ percent = 0.75 percent.

$$A_1 = P(A/P, i, n) = 5,000(A/P, 0.75\%, 48) = 124.425 = \$124.43/\text{month}$$

Note that the payment amount is rounded up to the nearest whole cent. This will cause the last payment at EOM 48 to be a few cents lower than the other monthly payments. If the payment were rounded down, the last payment would be a few cents higher than the other monthly payments.

Step 2.   Calculate the balance due by finding the present worth of the remaining 30 payments (48–18).

$$P_1 = A_1(P/A, i, n) = 124.43(P/A, 0.75\%, 30) = \underline{\$3,331.62 \text{ loan balance}}$$

(Soln. 7.7a)

a. Note the $0.22 difference in the two results. Which is correct? Remember the first method assumes 30 more payments of $124.43 each, whereas actually the thirtieth payment will be a few cents less. The second method assumes that 18 payments of $124.43 each have already been made, which is a correct assumption. Therefore, the second method yields the more accurate balance. In this example, as in actual practice, the differences in the balance caused by rounding up the payment are normally inconsequential from a practical standpoint, but are interesting from an academic and accounting viewpoint.

b. Note that, alternatively, the balance after 18 payments could be found by subtracting the FV of the 18 payments from the FV of the original loan amount, as follows:

$$\text{loan balance}, F = \$5,000(F/P, 0.75\%, 18) - \$124.43(F/A, 0.75\%, 18)$$
$$F = \$5,719 - \$2,388 = \underline{\$3,331 \text{ loan balance}}$$

(alternate Soln. 7.7a)

Solution to (b).

Step 3.   The amount of the loan already paid off is found as the original amount less the balance remaining

$$P_2 = P_0 - P_1 = \$5,000 - \$3,331 = \underline{\$1,669} \quad \text{(Soln. 7.7b)}$$

Solution to (c).

Step 4.   Frequently it is necessary to find the dollar amount of interest that has been paid on a loan, since under many circumstances the amount of interest paid is an expense item that is deductible when calculating income tax. The amount of interest paid is found as the sum of all the payments made (total cash flow) less the amount of those payments attributed to principal.

$$\text{total amount paid}, P + 1 = \$124.43/\text{month} \times 18 \text{ months} = \$2,240$$
$$\text{amount of principal} = -1,669$$
$$\text{amount of interest (subtract to find)} = \underline{\$ \quad 571}$$

(Soln 7.7c)

Find Uniform Rates ($ per Month, Hour, etc.) or Unit Costs ($ per yd³, etc.) by the NPW Method

Rental rates or unit costs often need to be calculated for construction equipment or other property used in a productive enterprise. For instance, an owner may rent a dragline out by the month, and a rental fee must be determined that will cover all costs plus margin of profit. Or a contractor may need to determine costs for excavating equipment on a unit price basis ($/yd³). One method of obtaining these answers involves the use of present worth.

The steps involved follow:

Step 1. First, each component cost item is reduced to an equivalent present worth.

Step 2. Second, all the components are summed to obtain an equivalent lump sum single payment net present worth.

Step 3. Third, a uniform equivalent (monthly, weekly, daily, hourly) rate is determined by the use of A/P factors.

Step 4. Then to find a unit cost, this periodic rate (e.g., $/hr) can be divided by the production rate (e.g., yd³/hr) to obtain costs per unit of production (e.g., $/yd³).

The following example illustrates the method.

EXAMPLE 7.8   **(Find the unit costs; the life of the financing loan is shorter than the ownership life of the property)**

A contractor has just bought a dragline for $500,000. He paid $100,000 down and will pay the balance of $400,000 in 60 monthly installments with 1 percent per month interest on the unpaid balance. He expects to keep the dragline for *seven years* (84 months) and then sell it. His installment payment contract on the machine runs for *five years* (60 months). He asks you to compute how much he must charge per cubic yard of earth moved in order to cover the following costs.

P & I payments on the financing loan.

O & M costs of $20,000 per month.

Resale value at the end of seven years, $150,000.

The contractor expects to earn, $i = 1.5\%$ per month on invested capital. Production is estimated to average 100,000 yd³ of material per month.

Solution.   Find the present worth of all costs. Then find the equivalent monthly cost and divide by the monthly production to obtain dollars per cubic yard. The monthly payments for the dragline in 60 months are

$$A_1 = \$400,000(A/P, 1\%, 60) = \$8,896/\text{month}$$

Note: The 1 percent rate is used only to calculate the monthly payments on the loan. Thereafter the 1.5 percent rate is used for all calculations.

The present worth of all costs and the resale value with $i = 1.5$ percent per month are

$$
\begin{aligned}
\text{down payment } P_1 &= -\$\ 100{,}000 \\
\text{monthly payments } P_2 = \$\ \ 8{,}896(P/A,\ 1.5\%,\ 60) &= -\$\ 350{,}327 \\
\text{O \& M } P_3 = \$\ 20{,}000(P/A,\ 1.5\%,\ 84) &= -\$\ 951{,}404 \\
\text{resale } P_4 = \$150{,}000(P/F,\ 1.5\%,\ 84) &= +\$\ \ \ 42{,}948 \\
\text{NPW of given payments} &= -\$1{,}358{,}783
\end{aligned}
$$

monthly equivalent $A_2 = \$1{,}358{,}783(A/P,\ 1.5\%,\ 84) = \$28{,}562/\text{month}$

The charge per cubic yard is obtained as

$$
\frac{\$28{,}562/\text{month}}{100{,}000/\text{yd}^3/\text{mo}} = \$0.286/\text{yd}^3
$$

To provide a simplified mental picture of these payments, imagine the contractor as having the sum of $1,358,783 invested in a savings account that pays 1.5 percent per month on whatever balance is in the account at the end of each month. At the end of each month withdrawals are made to make the O & M and P & I payments as they come due. At the end of the seven years he has paid all his bills, but is overdrawn by (has borrowed) $150,000. The resale of the dragline provides the $150,000 income needed to bring the account back up to zero.

For an alternate mental picture, imagine the contractor borrows the $1,358,783 at PTZ at 1.5 percent per month. To repay the loan, the contractor will make a series of 84 equal monthly payments in order to reduce the balance owed to zero at EOM 84. The amount of the monthly payment is the same amount that the contractor has to charge each month to cover all of the expenses above.

Conclusion. If the contractor charges $0.286/\text{yd}^3$ for excavation by this machine, he will recover his costs as listed above plus 1.5 percent per month on all EOM balances of funds invested in the machine. Notice also that the contractor makes 1.5 percent on the balance of the money that is borrowed at 1 percent. This is normal, since the contractor takes more risk than the lender.

TRADE-INS

When an older machine is replaced by a new one, the older machine may be disposed of by selling it outright on the competitive market (termed *salvage value* or *resale value*), or by trading it in on the new machine. Salesmen sometimes promote the sale of their new machine by offering an inflated trade-in allowance for the older machine. The salesman's loss from this part of the deal must then be recouped by extra charges of other types such as charging a higher price for the new machine, or charging extra for financing charges, credit investigation, dealer's preparation fee, transportation charges, special add-on features, and so on. Therefore, the trade-in value may not be the same as the salvage value or the resale value. A variety of methods for the acquisition of new machines is available and the buyer must carefully weigh each alternative to determine which is the better total cost. Three common methods of acquiring the use of machines follow:

1. Cash purchase
2. Trade-in, and/or finance with equal periodic repayments
3. Lease

In evaluating the alternatives, two basic concepts should be kept in mind:

a. Account for all of the cash flow or equivalent trade-in value that is invested into each alternative.
b. Trade-ins are worth the best cash offer (or equivalent) on the open competitive market, not the fictitious trade-in allowance offered by the salesman.

The following example illustrates a comparison between the three common methods of acquiring new equipment.

EXAMPLE 7.9   **(Illustrates: (a) use of trade-in value; (b) use of actual cash flow resulting from dealer's calculations)**

Your firm uses a small fleet of cars and pickups. As a matter of company policy all vehicles are replaced upon reaching three years of age. This year your firm needs to replace several three-year-old vehicles whose current cash market value totals $15,000. Specifications for the new vehicles have been distributed, and three dealers have submitted proposals, which you are asked to evaluate. It is estimated that three years from now the new fleet will have a cash market resale value of $17,500. Your firm can borrow money to finance the acquisition of this fleet at 9 percent interest (nominal annual).

Proposal A.   Lease the replacement vehicles for 36 months at $1,400 per month, paid in advance monthly (this fee includes insurance). All other operating and maintenance (O & M) costs are paid by the customer (your firm).

Proposal B.   Purchase for cash. The total list price for all the replacement vehicles is $60,000. For your firm, this dealer is offering a 15 percent discount. In addition, insurance costs $75 per month, payable at the beginning of each month.

Proposal C.   Trade in the existing three-year-old vehicles. The dealer will allow $18,000 total and give financing for 33 months at 7 percent, plus insurance, calculated as follows:

(Dealer's calculations, Beware!)

$$\text{list price} = \$60{,}000$$
$$\text{less trade-in allowance} = \underline{\$18{,}000}$$
$$\text{balance owed} = \$42{,}000$$
$$\text{dealer's preparation charges} = \$\ 2{,}300$$
$$\text{required insurance at \$90/month} \times 33 \text{ months} = \underline{\$\ 2{,}970}$$
$$\text{balance to be financed} = \$47{,}270$$
$$7\% \text{ Interest} \times \frac{33 \text{ months}}{12 \text{ month/yr}} = 19.25\%$$

(Dealer's own method of calculating interest!)

$$\text{interest charges } 19.25\% \times \$47{,}270 = \underline{\$9{,}099}$$
$$\text{total amount owed} = \$56{,}369$$

Total monthly payments for 33 months, $\$56{,}369/33 = \$1{,}708$ per month.

Note: Regardless of how the dealer calculates the payments, if no fraud is involved and the firm signs the contract, the contract is probably enforceable.

Solution. Since all three proposals have the same ownership life, they can be compared by finding the present worth of all of the cash flow (or equivalent trade-in) into each alternative over the 36-month anticipated life of the vehicles. So the solution consists of simply identifying the amount of each cash flow, and then finding the NPW of each at PTZ.

Proposal A. Lease. The cash flows consist of 36 beginning-of-month payments, which is the same as one payment now and 35 more at the end of each succeeding month.

$$\text{beginning of first month } P_1 = \$\ 1{,}400$$
$$35 \text{ additional EOM } P_2 = 1{,}400(P/A, 9/12\%, 35) = \underline{\$42{,}956}$$
$$\text{lease NPW } P_{\text{total}} = \$44{,}356$$

Alternatively, this could be solved by finding the PW of a series of 36 payments with ETZ at EOM minus 1 and then moving ETZ to PTZ, as follows:

$$P = 1{,}400(P/A, 9/12\%, 36)(F/P, 9/12\%, 1) = \$44{,}356$$

Proposal B. Purchase for cash (with a 15 percent discount).

$$\text{cash price } P = 0.85 \times \$60{,}000 = -\$51{,}000$$
$$\text{insurance paid monthly in advance}$$
$$P = \$75(P/A, 9/12\%, 36)(F/P, 9/12\%, 1) = -\$\ 2{,}376$$
$$\text{resale value at end of 3-yr life}$$
$$P = +\$17{,}500(P/F, 9/12\%, 36) = +\$13{,}513$$
$$\text{total NPW of purchase for cash } \underline{-\$39{,}863}$$

Proposal C.   Trade-in.

cash market value of trade-in
(*not* the salesman's inflated figure), $P_1 = -\$15,000$
Monthly payments for 33 months,
$P_2 = -\$1,708(P/A, 9/12\%, 33) = -\$49,776$
Proposal C includes only 33 months insurance, so the cost
of the last 3 months insurance at \$75/month paid
in advance (first payment at EOM 33, so
ETZ is at EOM 32) is found as
$P_3 = -\$75(P/A, 9/12\%, 3)(P/F, 9/12\%, 32) = -\$\quad 175$
Resale value at end of 3-yr life,
$P_4 = +\$17,500(P/A, 9/12\%, 36) = +\$13,513$
total PW of trade-in and finance $= -\$51,428$

Discussion: Comparison of the three alternatives indicates that proposal
B, purchase for cash, is the more economical in this instance. Note that the
\$15,000 cash market value of the existing vehicles is *not* credited to proposal
A or proposal B. To do so would reduce the apparent cost of proposals A and B
by a PW of \$15,000. In actual practice these fleet costs frequently are recov-
ered by charges to the clients or to other departments in the firm. Artificially reducing the fleet costs would result in undercharging and failing to recover
\$15,000. The value of the existing vehicles is an asset of the firm just like any
other asset, and may be retained or spent as the firm desires. In proposal C this
asset is being *spent* as a part payment on the new vehicles. Accepting either
proposal A or B will probably cause the presently owned vehicles to become
surplus, and thus allow them to be converted to cash. However, the conversion
of a presently owned asset to cash is *not new income and does not increase the
net worth of the firm*. The firm simply converts \$15,000 worth of already
owned steel and rubber into \$15,000 worth of paper.

Value Attributed to Trade-In

When considering trading in some asset of value, such as a used car, remember that the
value traded in is just as much an expenditure of assets as any cash that would have to replace the trade-in. For instance, assume a new car can be purchased for either \$25,000
cash, or \$20,000 cash plus your present car worth \$5,000. The total expenditure for the
new car in either case is \$25,000, not \$20,000. If the present car is truly worth \$5,000 it
represents \$5,000 worth of assets that can be sold on the open market (e.g., through an ad
in the newspapers) and converted into \$5,000 cash. Therefore, trading in the present car
is equivalent to trading in (or spending) \$5,000 cash. By the same reasoning, in any trade-
in or replacement-decision problem, the asset traded in constitutes an expenditure needed
to acquire the replacement alternative.

EXAMPLE 7.10

A contractor has an old bulldozer worth $40,000. A dealer offers to take the old dozer in trade plus $100,000 for a new dozer. Regardless of the dealer's figures, the contractor is being offered a new dozer for $140,000. Another dealer may tell the contractor he will allow $60,000 for the trade-in and then mark up the price on the new dozer to $160,000 in order to compensate. But if the old dozer is really worth $40,000 on the open market then the contractor is still spending $140,000 worth of his assets in the form of $40,000 worth of bulldozer plus $100,000 worth of cash to purchase the new bulldozer.

Where a replacement results in the release of an existing facility for conversion to cash, a similar outlook prevails. A facility, such as a bridge or a building, usually represents money invested in a certain location. If the existing facility has any cash salvage value this cash usually is available only if the new replacement facility is selected. Therefore, if the existing facility remains in place, the salvage value remains invested in the existing facility. It is *not* an income credited to the selection of the new replacement facility.

EXAMPLE 7.11

An existing bridge with a present salvage value of $100,000 may be renovated at a cost of $500,000 and left in service, or the existing bridge may be sold for salvage value and replaced with a new bridge costing $550,000. If both alternatives provide equal service, which is the least costly?

Solution.   The existing bridge could be converted into $100,000 cash if salvaged, so leaving it in service will cost that $100,000 cash plus another $500,000 for renovation for a total of $600,000. The new bridge costs only $550,000 and is therefore less costly. If the new bridge alternative is selected, the $100,000 cash now available from salvage of the old bridge is not new earned income, but simply a conversion of assets already owned, from $100,000 worth of bridge to $100,000 worth of cash.

The basic approach to trade-in problems is to treat the existing asset as equivalent cash. If a car worth $5,000 is traded in, the trade-in transaction is equivalent to spending $5,000 cash. If an existing bridge currently owned by the DOT could be sold for $30,000 salvage, the DOT is foregoing the cash in order to leave the bridge in service. Therefore, if the DOT selects the alternative of leaving the existing bridge in service, that alternative costs them $30,000. Further, in the event that the existing bridge costs $20,000 to demolish (it has a negative $10,000 salvage value), then leaving the bridge in service actually saves $10,000 during the period of time over which the bridge is retained in service. In effect the expenditure is postponed until the bridge finally is demolished. As a consequence interest charges are saved for the extra years of service.

SUNK COSTS

Usually there is a difference between cost and value. Cost is the price paid for an item, whereas value is a measure of the benefit received. The value-to-cost ratio should be greater than 1 to warrant an investment or purchase. When people make a purchase, in effect they are reasoning that this purchase item has greater value to them than the money they are paying for it. Therefore, they trade their money for the purchased item. Some time after the purchase the item normally will change in value. Such items as construction equipment, tools, automobiles, clothes, and so on, typically depreciate over time, and decrease in value after purchase, so the original cost is usually higher than the current value. A few categories of investments, such as real estate, frequently *appreciate* in value, so that the original cost is often *below* the current market value. In either case, once the cost has been paid and the transaction is completed, the cost becomes past history and is usually irretrievable. Any drop from cost to current market value is lost and cannot be regained; it is referred to as "sunk cost." For instance, suppose a new pickup truck is purchased for $20,000. Shortly thereafter it is wrecked without insurance, and the junk yard offers $100 for the wrecked remains. The difference of $19,900 ($20,000 cost less $100 salvage value) is termed "sunk cost." If the pickup were financed so that the payments continue on despite the absence of the pickup itself, the $19,900 is still sunk cost. By the same reasoning, if a bridge were completed just yesterday at a cost of $1,000,000 and if today it has a current market (salvage) value of $100,000, then the value of the bridge is $100,000, and the $900,000 drop in value is sunk cost. If the bridge is financed by bonds that are not yet paid off, the bridge is still worth just today's salvage value. If the bridge is destroyed by flood or earthquake, the bond debt lives on and must be repaid. One advantage of the low market value is that, when comparing the NPW of the existing bridge against other alternatives, the low capital cost of the existing bridge (or other similar facilities) gives it a decided cost advantage (at least for the capital cost portion of the NPW) over any new proposed replacements.

Replacement Cost

The value of the existing bridge is *not* the replacement value, since the replacement bridge is actually another alternative. Using the replacement value as the actual present worth of the existing bridge erroneously favors the new proposed alternative, as illustrated in the following example, simplified to clarify the point.

EXAMPLE 7.12   **(Comparing alternative bridges using the current market value and the replacement cost of the existing bridge)**

The state DOT is considering the removal of an existing bridge A, and may replace it with either proposed bridge B or C. The traffic-carrying capacity, maintenance costs, and future salvage values are the same for all three bridges. Only the capital costs are different; they are listed below. Determine which is the lower cost alternative of the three.

	Existing bridge A	Proposed bridge B	Proposed bridge C
Present market (salvage) value	$ 100K		
Replacement cost of existing bridge	$1,500K		
Cost of new bridges B and C		$1,200K	$1,000K

Solution. If the existing bridge A, is valued at the replacement cost, $1,500,000, then proposed bridge C at $1,000,000 is the lower cost selection. Actually (assuming the existing bridge is still serviceable), the only capital cost involved in retaining existing bridge A is the loss of $100,000 of present market (salvage) value. So retaining A really only costs $100,000, not $1,500,000. Therefore, the lower cost alternative is to retain existing bridge A at a true cost of $100,000, rather than construct proposed bridge C at a cost of $1,000,000.

Examples of Sunk Costs, Current Market Value, and Periodic Gradient

EXAMPLE 7.13   **(Existing bridge with current salvage value, periodic maintenance and gradient, loan life ≠ project life)**

Due to an unexpected growth in traffic counts, an existing DOT-owned bridge must either be widened or replaced with a new bridge.

Existing bridge.   The existing bridge was built just 12 years ago at a cost of $1,000,000, financed by money borrowed at $4\frac{1}{2}$ percent and being repaid in 30 equal annual installments (there are 18 more annual payments yet to come due). The current maintenance costs are $25,000 per year (charged at the end of this year), with an anticipated escalation rate of $2,000 per year each year. If removed now, the existing bridge would yield a salvage value of $150,000. The replacement cost of the existing bridge is estimated as $2,300,000 (for design and construction of a new bridge).

Alternative 1: Widen the existing bridge.   If the existing bridge is widened, a federal government loan of $1,500,000 is available to pay the full construction cost of widening. The loan must be repaid in 10 equal end-of-year payments, with 3 percent interest on the unpaid balance, with the first payment due at EOY 1. None of the loan money may be used for any other costs except construction for the widening. Other costs of the widened bridge are as shown below. All of these costs are paid from DOT funds with $i = 10$ percent.

O & M (40 EOY payments, with the first payment
charged at EOY 1) These O & M costs completely
replace the current maintenance costs for the
existing bridge. = $ 50,000/yr
gradient increase in O & M = 5,000/yr/yr
periodic heavy maintenance every 4 yr, first
periodic maintenance at EOY 4, last at EOY 36 = 100,000/4 yr
gradient in periodic maintenance = 40,000/each/4 yr
salvage value at EOY 40 = 20,000

Alternative 2: New bridge

design and construction costs of new bridge = $2,300,000
O & M cost (equivalent annual cost,
including gradients
and periodic maintenance) = 60,000
salvage value at EOY 40 = 100,000

No federal loan is available for the replacement bridge.

Compare the NPWs of each at $i = 10$ percent. Assume both alternatives have life spans of 40 more years from now.

Solution.   The $1,000,000 cost of the existing bridge 12 years ago, together with the financing and repayments, is sunk cost and, together with replacement cost and some other data, has no effect on the problem. This type of data is included here to simulate the myriad of useless background information that normally accompanies the necessary data on projects of this type in actual practice.

The NPW of each alternative is found as (all figures are in 1,000s):

Alternative 1: Widen the existing bridge

current cash value, $P_1 = -\$ 150K = -\$ 150K$
PW of government loan repayments
$P_2 = -\$1,500K(A/P, 3\%, 10)(P/A, 10\%, 10) = - 1,080K$
O & M $P_3 = -\$50K(P/A, 10\%, 40) = - 489K$
Δ O & M $P_4 = -\$5K(P/G, 10\%, 40) = - 445K$
periodic maintenance every 4 yr
$P_5 = \$100K(A/F, 10\%, 4)(P/A, 10\%, 36) = - 208K$

Δ periodic maintenance, $i_e = (1+i)^n - 1 = 1.1^4 - 1 = 0.4641$

$$P_6 = \frac{G}{i}\left[\frac{(1+i)^n - 1}{i(1+i)^n} - \frac{n}{(1+i)^n}\right]$$

$$P_6 = \frac{-40}{0.4641}\left[\frac{1.4641^9 - 1}{0.4641 \times 1.4641^9} - \frac{9}{1.4641^9}\right] = -\quad 155K$$

$$\text{salvage } P_7 = \$20(P/F, 10\%, 40) = \underline{\qquad 0K}$$

$$\text{NPW of alternative 1 } P_t = -\$2,527K$$

Alternative 2: New bridge

$$\text{cost new } P_1 = -\$2,300K$$
$$\text{O \& M } P_2 = -60K(P/A, 10\%, 40) = -\quad 587K$$
$$\text{salvage } P_3 = 100K(P/F, 10\%, 40) = +\quad 2K$$
$$\text{NPW of alternative 2} = P_{\text{total}} = -\$2,885K$$
$$\text{compare to NPW of alternative 1 at } \underline{-\$2,527K}$$

Conclusion: *Alternative 1 is less costly than alternative 2.*

SUBSIDY BY MEANS OF LOW INTEREST RATES

The federal government sometimes aids needy groups (such as those in disaster areas, low-income groups, or developing foreign nations) by making low-interest loans available. When the government borrows money at market interest rates and lends it out at low, subsidized interest rates, the government loses money. The lost interest constitutes a subsidy (a grant of money) because the present value of the amount loaned by the government is greater than the present value of the amount paid back by the needy group. A similar situation occurs in private enterprise whenever the borrowing rate is less than the investment rate. The following example illustrates the method of determining the value of the subsidy.

EXAMPLE 7.14   (Find the NPW of interest rate subsidy)

Farmers in a certain drought-stricken area are entitled to government loans at 3 percent with repayments in equal annual installments over a 20-year period. The first installment is due one year from now. If $10,000,000 is loaned at 3 percent under this program, and the government is borrowing at 7 percent, what is the present worth of the subsidy?

Solution. The amount of the subsidy is determined by a simple three-step procedure.

Step 1. Find the actual cash flow. In this case the cash flow from the government's viewpoint consists of the following:

a. The cash flow out of the $10,000,000 lump sum loan paid out at PTZ.

b. The annual repayment amount, A, that the lender (government) will receive in annual installments.

$$A = \$10,000,000(A/P, 3\%, 20) = \$672,160/\text{yr}$$

Step 2. Find the present worth to the lender (government) of this series of 20 annual repayments at the borrowing rate of 7 percent used by the lender. The i used in this step is the i at which the lender borrows money to obtain the funds used in this program. Thus, i in this equation is 7 percent.

$$P = \$672,160/\text{yr}(P/A, 7\%, 20) = \$7,120,900$$

Step 3. To find the cost of the program (the amount of the subsidy), subtract the present worth of the repayment income from the present worth of the total loaned under the program.

$$
\begin{aligned}
\text{PW of total amount loaned} &= -\$10,000,000 \\
\text{PW of all repayments} &= +\ \ 7,120,900 \\
\text{NPW of the program (program subsidy)} &= -\$\ 2,879,100
\end{aligned}
$$

Conclusion. To fully fund this program, the government will need to appropriate the equivalent of $2,879,100 in a lump sum now, in addition to borrowing an additional $7,120,900 at 7 percent. The $7,120,900 (plus interest at 7%) will eventually be repaid in full by the repayment income from the borrowers, but the $2,879,100 is a gift by the government to the program borrowers.

Comment. Other features are sometimes added to such programs. For instance, suppose the repayments were deferred for five years with no additional interest charges. The approach to solving for the value of the subsidy is basically the same. The present worth of the 20 repayments that begin with the first repayment at EOY 6, with no interest charged to the needy borrower up to EOY 5 is calculated as

$$P = \$672,160/\text{yr}\ (P/A, 7\%, 20)(P/F, 7\%, 5) = \$5,077,200$$

The amount of the subsidy is found as before:

$$PW \text{ loaned} = -\$10,000,000$$
$$PW \text{ repaid} = + 5,077,200$$
$$PW \text{ subsidy} = -\$ 4,922,800$$

The amount of the subsidy increased in this case due to the extension of time at no interest cost to the borrower.

A variety of similar provisions can be added to the terms of such programs, depending on the aims and desires of the legislators. Whenever possible, these provisions should be translated into present-worth costs, so that these grants of taxpayers' dollars can be weighed directly against the value of the benefits envisioned.

Annual Increases in Monthly Rates

In an environment of changing costs or productivity, sometimes an owner will charge a monthly rate that will need to be increased periodically. The owner does not want to raise the rate every month (which conveniently would fit the gradient equation pattern), but will change the rates once a year (or at any other regular interval). In this case the owner needs to know how much to raise the rate once a year in order to equal the results of a monthly increase. Example 7.13 illustrates such a problem.

EXAMPLE 7.15

Start with a low monthly rate, and then increase annually. Find the annual increase, $G1$, required in rate $B2$ \$/yd^3 that is equivalent to rate $B1$ \$/yd^3 throughout the life cycle, with D yd^3/month remaining constant.

The contractor of Example 7.8 found that he must receive the equivalent of \$0.286/yd^3 on 100,000 yd^3/month for 60 months in order to cover his ownership and operating costs. However, he finds that the competition is bidding \$0.15/yd^3 and he will have to lower his price to that level for the next 12 months. He asks you to calculate how much the price in dollars per cubic yard must be raised at the end of each 12 months in order to make the equivalent of \$0.286 per yd^3 over the 60-month period. Assume $i = 1.5$ percent per month.

Solution. One simple approach is outlined below:

1. Find the PW of the deficit amount to be made up.
2. Find the periodic gradient amounts required at the end of each year.
3. Find the monthly equivalent of these end of year amounts.

1. The deficit to be made up is found as

$$P = (\$0.286 - 0.15)/\text{yd}^3 \times 100,000 \text{ yd}^3/\text{month} \times (P/A, 1.5\%, 60)$$
$$P = \$535,568$$

2. The periodic gradient amount required at the end of each 12-month period is found by using the i_e, as

$$i_e = (1 + 0.015)^{12} - 1 = 0.195618 = 19.562\%/12 \text{ months}$$

Then

$$G = P(G/P, 19.562\%, 5) = \$535,568/4.9750 = \$107,652$$

If the income is increased by a gradient of $G = \$107,652/\text{yr/yr}$, then the deficit amount will be recovered.

3. This EOY gradient is equivalent to a monthly amount for the preceding 12 months of

$$A = F(A/F, 1.5\%, 12) = \$107,652 \times 0.07668 = \$8,255/\text{month}$$

Since the excavation rate is 100,000 yd^3/month, the unit charge works out to

$$\frac{\$8,255/\text{month}}{100,000 \text{ yd}^3/\text{month}} = \underline{\$0.083/\text{yd}^3}$$

Conclusion. The contractor can obtain an income equivalent to $\$0.286/\text{yd}^3$ by the following schedule of increases:

EOM	Price ($/yd³)
1–12	0.15
13–24	0.15 + 0.083 = 0.233
25–36	0.233 + 0.083 = 0.316
etc.	etc.

SPREADSHEET APPLICATIONS

	A	B	C	D	E	F	G
1	EXAMPLE 7.2 (continued): Compare the NPW of two alternatives with equal lives.						
2	Find the NPW, Given P, A, G, C, n						
3							
4	Compare the NPW of using a mechanical arm to pick up solid waste and the conven-						
5	tional method of pickup. Compare each with $i = 10$ percent, and vary n from 5 to 10						
6	years.						
7	**SOLUTION:**						
8							
9		**Mechanical Arm:**					
10				Interest rate, $i =$		10%	
11				New Cost $=$		$80,000	
12				Salvage value in 10 yr. $=$		$5,000	
13				O & M first year $=$		$2,000	
14				Gradient increase in O & M per year $=$		$400	
15							
16		n	Salvage	O & M	Gradient	NPW	
17		(yr.)	PW($)	PW($)	PW($)	($)	
18		5	$3,105	($7,582)	($2,745)	($87,222)	
19		6	$2,822	($8,711)	($3,874)	($89,762)	
20		7	$2,566	($9,737)	($5,105)	($92,276)	
21		8	$2,333	($10,670)	($6,411)	($94,749)	
22		9	$2,120	($11,518)	($7,769)	($97,166)	
23		10	$1,928	($12,289)	($9,157)	($99,518)	
24							
25							
26		**Conventional Method of Manual Loading:**					
27							
28				Interest rate, $i =$		10%	
29				Worker's Annual Wage $=$		$16,000	
30				Gradient Increase in wage per year $=$		$1,200	
31							
32		n		Wage	Gradient	NPW	
33		(yr.)		PW($)	PW($)	($)	
34		5		($60,653)	($8,234)	($68,887)	
35		6		($69,684)	($11,621)	($81,305)	
36		7		($77,895)	($15,316)	($93,210)	
37		8		($85,359)	($19,234)	($104,593)	
38		9		($92,144)	($23,306)	($115,450)	
39		10		($98,313)	($27,470)	($125,783)	
40							
41							
42	**CELL CONTENTS:** The following cell contents will allow us to evaluate the problem with a spreadsheet.						
43		cell **C18:**	= PV($F$10,B18,,−$F$12)				
44		cell **D18:**	= PV($F$10,B18,$F$13)				
45		cell **E18:**	= −($F$14/$F$10)*(((1 + $F$10)^B18 − 1)/($F$10*(1 + $F$10)^B18) − (B18/(1 + $F$10)^B18))				
46		cell **F18:**	= −$F$11 + C18 + D18 + E18				
47		cell **D34:**	= PV($F$28,B34,$F$29)				
48		cell **E34:**	= −($F$30/$F$28)*(((1 + $F$28)^B18 − 1)/($F$28*(1 + $F$28)^B18) − (B18/(1 + $F$28)^B18))				
49		cell **F34:**	= D34 + E34				

50	**Conclusion:**				
51					
52		The solution is quite sensitive to changes in the estimated life of the mechanical			
53		arm. Changing the life from 10 years to 6 years reverses the outcome of which			
54		method has the greatest NPW.			

SUMMARY

The general objective of engineering cost analyses is to maximize profits or savings, or minimize costs. One widely applicable method involves comparison of the equivalent net present worths. In this chapter alternative engineering projects or equipment acquisitions are analyzed by comparing their net present worths.

One restriction on the use of NPW is that the alternatives must be compared over identical time periods. When comparing these time periods, three variations occur.

1. When the useful lives of each alternative are the same, each alternative's new present worth is calculated and compared directly.

2 **a.** When the useful lives of each alternative are *not* the same length, and the replacement costs *are* the same as the original costs new, then select the lowest common multiple of the lives involved to compare each alternative over the same time span of multiple lives. Or a salvage value may be assigned to the longer-lived alternative at the end of the life of the shorter-lived one and the respective present worths compared.

   b. When the useful lives are *not* the same length, and the replacement costs are *not* the same as the costs new, then use escalating replacement costs.

   i. With varying replacement costs, bring each back to PTZ as a lump sum using P/F.

   ii. With predictable gradients (either arithmetic or geometric), bring the series back to an equivalent PW at PTZ by using an effective interest rate i_e to account for multiple compounding periods between periodic payments.

3. When the useful lives of all the alternatives are very long, the net present worths can be compared by using $n \rightarrow \infty$. The NPW for investments with infinite lives are often termed "capitalized cost."

Present worth techniques are widely utilized in many engineering, business, and personal finance situations, but present worth techniques are only one of several methods used to compare alternatives. Other comparable techniques are presented in subsequent chapters.

PROBLEMS FOR CHAPTER 7, PRESENT WORTH ANALYSIS

Problem Group A

Given three or more of the variables P, A, F, i, n, find the present worth of each proposal. Compare alternative proposals.

A.1. Parking meters are being considered for a certain district. Two types are available with the following estimated costs:

	Type A	Type B
Estimated life	5 yr	5 yr
Cost new	$100/each	$200/each
Annual O & M	$300/yr	$250/yr
Resale value at end of life	0	$ 75

Using present worth analysis, determine which type of parking meter should be selected for $i = 12$ percent.

(*Ans.* $P_A = \$1,180$; $P_B = \$1,060$)

A.2. Same as Problem A.1, except type B has an estimated life of 10 years.

A.3. A small utility company is for sale. The estimated net income is $40,000 per year for the next 40 years, after which its franchise expires, and the system is virtually worthless. If an interest rate of 15 percent is agreed upon as adequate, what should the selling price of the utility be at the present time (or what is the present worth)?

(*Ans.* $P = \$265,670$)

A.4. A contractor needs to choose between two makes of motor grader. Money can be borrowed at 18 percent. Which is the most economical?

	Brand A	Brand B
Cost new	$20,000	$15,000
Repair costs	2,000/yr	3,000/yr
Resale value in 10 yr	10,000	5,000

A.5. Same as Problem A.4, except brand B will last five years (same resale value).

(*Ans.* $P_A = \$27,080$; $P_B = \$27,530$)

A.6. A tree spade is proposed for a public works department to replace hand labor. Estimated costs for the next five years are given in the following table:

	Tree spade	Hand labor
Cost new	$20,000	—
Annual O & M	10,000/yr	$15,000/yr
Resale in 5 yr	5,000	—

Using present worth analysis, determine which alternative should be chosen for $i = 7$ percent.

(*Ans.* $P_{TS} = \$57,400$; $P_{HL} = \$61,500$)

A.7. A contractor needs a new truck, which can be acquired by either of the following plans:
(a) Pay $50,000 cash.
(b) Pay nothing down and $8,500 per year for nine years.

What is the more advantageous plan for the contractor? Compare by present worth analysis, with $i = 6$ percent.

(*Ans.* $P_A = \$50,000$; $P_B = \$57,810$)

A.8. Three different types of outside covering are available for a concrete frame and block building. If $i = 7$ percent and the building life is 20 years, which is the lowest cost covering? Use present worth analysis.

	Cast stone facing	Stucco	Paint
Cost new	−$15,000	−$5,000	−$1,000
Add to resale value of building in 20 yr	+ 5,000	+ 1,000	0
Add to rental income for 20 yr	+ 1,200/yr	+ 600/yr	0
Maintenance cost (including repainting)	− 100/yr	− 200/yr	− 300/yr

A.9. Same as Problem A.8, except $i = 15$ percent.

(*Ans.* $P_1 = \$2,055$; $P_2 = \$504$; $P_3 = \$4,178$)

A.10. A representative selection of roof beam spans for a store are available as follows. If $i = 12$ percent and the building life is 30 years, which is lowest cost?

	25-ft spans	50-ft spans	100-ft spans
Cost new	−$2.50/ft^2	−$2.74/ft^2	−$3.25/ft^2
Add to resale value of building in 30 yr	0	+ 0.15/ft^2	+ 0.50/ft^2
Add to rental income for 30 yr	0	+ 0.10/ft^2/yr	+ 0.20/ft^2/yr

A.11. Find which of the two types of floor covering has the lowest NPW if $i = 15$ percent.

	Type A	Type B
Estimated life	8 yr	4 yr
Maintenance	$3,000/yr	$2,000/yr
Cost new	$150,000	$100,000

(*Ans.* $P_A = \$163,462$; $P_B = \$166,150$)

A.12. A municipal parking lot to be owned and operated by the city government is proposed for downtown. Costs and income are estimated to be

$$\text{land cost} = \$100,000$$
$$\text{improvement costs} = \$15,000 \quad \text{(cost of construction)}$$
$$\text{maintenance} = \$50/\text{space}/\text{yr}$$
$$\text{parking lot capacity} = 40 \text{ cars}$$
$$\text{revenue} = \$0.10/\text{hr}/\text{space}$$
$$\text{occupancy} = 80\% \times 9 \text{ hr/day} \times 250 \text{ days/yr}$$
$$n = 20$$
$$i = 6\%$$
$$\text{resale value} = \$100,000 \text{ at end of life}$$

Using present worth analysis determine whether or not this lot should be built. Since the city will be taking over land now privately held, there will be a loss of tax revenue of $3,000 per year. This property has held a constant value for some time, and for purposes of estimating lost income, assume the tax loss would remain constant over the 20-year project life.

A.13. Same as Problem A.12, except that $n = 40$ years.

(*Ans.* $P = \$-45,090$. No)

A.14. Same as Problem A.12, except that $n = 30$ years, and $i = 8$ percent.

A.15. You have an opportunity to construct a small neighborhood shopping center for an estimated cost of $100,000. The property would consist of three small stores. The estimated income is $12,000 per year. Average taxes, insurance, and so forth you estimate to be about $2,700. You estimate the building can be sold in 10 years for $50,000 (its depreciated value). You need to earn at least 10 percent on your investment. Should you construct this building?

(*Ans.* $P = \$-23,580$. No)

A.16. Comparing alternatives for commercial building design, one alternative calls for extensive landscaping, costing $100,000 but resulting in an increased rental income of $10,000 per year over the 20-year life of the building. Assume zero salvage value. If $i = 6$ percent, is the landscaping a good investment? Compare using present worth analysis.

A.17. Same as Problem A.16, except that $i = 12$ percent.

(*Ans.* $P = \$-25,310$. No)

A.18. As an engineer you can purchase a computer under any of the following options. Which costs the least in terms of present worth ($i = 12\%/\text{yr}$, compounded monthly)?
(**a**) $10,000 cash.
(**b**) $200 per month for 60 months, no down payment, first payment due EOM 1.
(**c**) $3,000 cash and $133 per month for 60 months, first payment due EOM 1.

A.19. A building owner wants you to build a $1,000,000 building and gives you a choice of methods of payment. You can build it on a cost plus 5 percent contract whereby he will finance the job and pay you $50,000 profit when the building is completed at EOY 1. Or you can take your profit in ownership shares expected to yield $5,000 per year for 8 years (first payment at EOY 1), after which the building will be sold with your share worth an estimated $60,000. If money is worth 12 percent to you, which is the most profitable in terms of present worth?

(*Ans.* $P_A = \$44,640$; $P_B = \$49,070$)

A.20. A sanitary sewer may be run either through a deep cut or a lift station, with costs estimated as follows:

	Deep cut	Lift station
Extra cost of CI pipe	$50,000	
Extra cost of lift station		$20,000
Annual operating expenses	$ 500/yr	$ 2,000/yr
Expected life	40 yr	20 yr
i	6%	6%

Which alternative should be selected, using present worth analysis?

A.21. Two types of roofing materials are available to cover a building scheduled for demolition in 20 years. The building has no salvage value.

	Type A	Type B
Estimated life	10 yr	20 yr
Maintenance	$3,000/yr	$2,000/yr
Cost new	$50,000	$100,000

With the interest rate on borrowed money at $i = 17$ percent, which alternative should be selected? Use present worth analysis.

(*Ans.* $P_A = \$77,280$; $P_B = \$111,300$)

A.22. Flooring. Assume $i = 10$ percent. Find the lowest present worth.

	Type A	Type B
Estimated life	12 yr	20 yr
Maintenance	$0.10/ft^2/yr	$0.08/ft^2/yr
Cost new	$0.30/ft^2	$0.50/ft^2

A.23. Paint. Assume $i = 12$ percent. Find the lowest present worth.

	Type A	Type B
Estimated life	3 yr	5 yr
Cost materials (paint)	$0.03/ft^2	$0.045/ft^2
Cost labor and equipment	$0.06/ft^2	$0.06/ft^2

(*Ans.* for $n = 15$; $P_A = \$0.255/ft^2$; $P_B = \$0.198/ft^2$)

A.24. You have an opportunity to invest in a hamburger establishment to be constructed adjacent to a campus. The investment would be $10,000, for which you should receive annual receipts of $1,650 for 10 years. If you can earn 8 percent interest from a certificate of deposit, should you invest in this venture? Assume zero salvage value.

A.25. An equipment rental firm has $30,000 to invest and is considering the addition of a backhoe to its rental inventory. If the firm uses a 15 percent return on investment, which (if either) of the following alternatives should be selected? Explain.

	Alternative A	Alternative B
First cost	−$20,000	−$30,000
Salvage value, 5 yr	+ 8,000	+ 10,000
Annual maintenance	− 5,000	− 6,000
Annual rental income	+ 9,000	+ 14,000

Solve by present worth analysis.

(*Ans.* $P_A = -\$2,615; P_B = +\$1,789$)

Problem Group B

This group involves gradients.

B.1. A dragline is being considered for a long-term canal drainage project. Estimate the cost of ownership in terms of present worth ($i = 15\%$).

$$\text{cost of dragline} = \$400,000$$
$$\text{resale value in 5 yr} = \$200,000$$
$$\text{annual O \& M} = \$100,000/\text{yr increasing } \$10,000/\text{yr}$$

(*Ans.* $P = \$693,500$)

B.2. Find the present worth of a dozer with $i = 12$ percent.

$$\text{original cost} = \$20,000$$
$$\text{annual O \& M increasing at a rate of } \$2,000/\text{yr} = \$10,000/\text{yr/first yr}$$
$$\text{resale value in 5 yr} = \$ \ 5,000$$

B.3. A bus company and franchise with ten years of service life remaining is up for sale. Net income the first year is estimated at $50,000 per year. Each year thereafter, the net is expected to drop $5,000 per year.

Year	Net income
1	$50,000
2	45,000
3	40,000
etc.	etc.

Resale value at the end of 10 years is estimated at $20,000. You are looking for at least a 10 percent return on your investment. What is your maximum bid for this company?

(*Ans.* $P = \$200,500$)

B.4. Forty acres of land is on the market, for sale to the highest bidder. Your client asks you to calculate what his highest bid should be under the following conditions. Hold the land for several years and then sell 5-acre tracts with anticipated net income as scheduled below. Expenses are negligible.

EOY	Income or cost
0	Purchase price (find)
1	0
2	0
3	0
4	20,000
5	40,000
6	60,000
7	80,000
8	0

Your client requires a 20 percent rate of return ($i = 20\%$). What is the maximum bid amount that will yield the desired return?

B.5. A transportation authority is trying to finance the construction of a toll bridge. They plan to sell bonds with repayments of principal and interest coming from the bridge tolls. How much in bonds can be financed now and be paid off by the end of the twenty-fifth year if net receipts available are as listed below, and $i = 7$ percent?

EOY	
1–2	Bridge under construction
3	$20,000
4 through 8,	Increase by 20,000/yr/yr
8	120,000
9 through 25,	Increase by 5,000/yr/yr
25	205,000

(*Ans.* $P = \$1,167,000$)

Problem Group C   Geometric Gradients

C.1. Same as Problem B.1, except cost of maintenance and repairs increases by 10 percent per year compounded.

(*Ans.* $P = \$69,900$)

C.2. A contractor currently owns dragline A and is thinking of replacing it with new dragline B. Costs over the next five years are estimated in the table below, and $i = 10$ percent.

	Existing dragline A	New dragline B
Present market value	$100,000	$200,000
Estimated market value, 5 yr from now	50,000	100,000
Annual maintenance and repairs	40,000	20,000
Increase in O & M costs each yr	5%	5%
Annual costs of insurance, taxes, storage		
as a % of present market value	4%	4%

Compare the NPW of each.

C.3. A young high school graduate just turned 18 is contemplating future career alternatives based on the probable salaries to be earned. Which of the alternatives yields the higher present worth (assume $i = 10\%$, expenses occur at BOY, while income occurs at EOY)?

  (a) Go directly to work and continue until age 65. Starting annual salary $20,000 per year, increasing 5 percent per year.

  (b) Invest two years and $12,000 per year in a junior college education. Work the remaining years until age 65. Starting annual salary of $28,000 per year, increasing 6 percent per year.

  (c) Invest four years and $12,000 per year in a four-year college education. Work remaining years until age 65. Starting annual salary of $60,000 per year, increasing 8 percent per year.

  (d) Invest five years and $12,000 per year for a graduate degree and earn a starting salary of $66,000 per year, increasing 10 percent per year.

  (*Ans.* $P_a = \$355,000$; $P_b = \$446,400$; $P_c = \$1,076,400$; $P_d = \$1,514,600$)

C.4. The county engineer asks you to calculate the cost of maintaining a portion of the county's graded roads by using county equipment compared to contracting with a private contractor. Use $i = 1$ percent per month.

Alternative A.   Use county equipment and personnel.

    buy equipment = $200,000 paid now
   resale at EOM 60 = $30,000
    labor & materials = $20,000/month for the first 10 months
    equipment O & M = increasing by 0.5%/month for each subsequent month

Alternative B.   Contract the work for $30,000 per month with no increases over the 60-month period.

  Find the equivalent present value of each alternative and compare.

C.5. A new solar energy heater and cooler (SEHAC) is proposed for use in a large building. Determine whether it is worth the extra cost if $i = 12$ percent. Assume all annual costs are paid at EOY.

       extra cost new of SEHAC compared
            to conventional = $150,000
      expected life of both installations = 15 yr
          extra salvage value = 0
     SEHAC savings in energy costs the
            first year = $18,000 at EOY 1

expected increase in energy costs (and savings
over the 15-yr life of the installation) = 3%/yr increase at EOY 2
through EOY 4.
Then a 6%/yr increase
for each subsequent year
through EOY 15

(*Ans.* P_{cost} = −$150,000; $P_{savings}$ = +$157,300; P_{net} = +$7,300)

C.6. The city engineer asks you to compare the cost of installing water pipe with city forces to the cost of contracting it out. They need to lay 3,000 lineal feet of pipe per month for 48 months. Use i = 1 percent per month.
    Alternative A.

buy equipment = $1,300,000 paid now
resale at EOM 48 = $200,000
equipment O & M, labor, materials = $10/lineal foot for the first 12 months
increasing by 0.5%/month for each
subsequent month

    Alternative B. Contract the work for $25/lineal foot with no increases over the 48-month period.
    Find the equivalent present value of each alternative and compare.

Problem Group D   Arithmetic and Geometric Gradients

D.1. Your client asks you to check on the feasibility of constructing an apartment house with data as follows:

current income = $100,000/yr
gradient increase in income = $ 10,000/yr/yr
current costs = $ 60,000/yr
gradient increase in costs = $ 5,000/yr/yr

The resale value is expected to equal the current construction costs now, escalating at a rate of 2 percent per year compounded until resale in 10 years. What can he afford to construct the apartments for now if he requires a 20 percent return on investment?

(*Ans.* P = $289,000)

D.2. An electric power dam is proposed with the following estimated data:

cost of new dam = $12,000K
cost of new generator = $ 5,000K

The generator must be replaced every 15 years, and the cost escalates 4 percent per year.

annual O & M costs at EOY 1
through EOY 5 = $500K/yr
O & M cost at EOY 6 = $600K/increasing $10K/yr/yr thereafter
salvage value of dam at end of life = $1,000K
life = permanent
interest rate i = 10%

Find the capitalized cost of the electric power dam.

D.3. Find the equivalent present worth of a new pump for a water treatment plant. The anticipated costs for the pump are

$$\text{cost new} = \$100,000$$

Major maintenance is required every four months, with the first at EOM 6, then again at EOM 10, 14, 18, and so on. Cost of major maintenance at EOM 6 is $4,000. The cost of major maintenance increases 5 percent each time it is performed. Life of pump is 120 months, at which time the salvage value is zero; $i = 0.75$ percent per month.

(*Ans.* $P = \$246,300$)

D.4. To help in renovating a blighted downtown area, construction of a new downtown plaza is under consideration.

$$\text{cost of plaza, acquisition, and construction} = \$1,000,000$$
$$\text{periodic renovation every 5 yr (first at EOY 7,}$$
$$\text{second at EOY 12, etc.)} = \$200,000 \text{ at EOY 7}$$
$$\text{geometric gradient in periodic renovation} = 8\%/\text{yr compounded annually}$$

The plaza is constructed with funds borrowed at 10 percent per year, and $n \to \infty$. Find the capitalized cost (present worth) of the project.

D.5. The county is considering the purchase of land for a nature preserve for passive recreation.

$$\text{cost of nature preserve, purchase land}$$
$$\text{and place in service} = \$5,000,000$$
$$\text{periodic maintenance every 12 months}$$
$$\text{(first at EOM 15 and second}$$
$$\text{at EOM 27, etc.)} = \$20,0000 \text{ at EOM 15}$$
$$\text{geometric gradient in period maintenance} = 8\%/12\text{-month period}$$
$$(\$20,000 \text{ at EOM}$$
$$15, +21,600 \text{ at EOM 27, etc.})$$

The county borrows at $i = 1$ percent per month. The preserve will be used a long time, so $n \to \infty$. Find the capitalized cost (present worth) of the project.

(*Ans.* $P = \$5,415,000$)

Problem Group E   Subsidized Loans

E.1. The federal government sometimes makes low-interest loans in some areas for specified public purposes (to farmers for erosion control, aid for disaster areas, etc.). Assume that $100,000,000 in loans is available to a certain hurricane disaster area. Qualified applicants can borrow up to specified limits and repay in equal annual installments over 30 years with interest on the unpaid balance at 2 percent. Assume that the federal government obtains the $100,000,000 by borrowing at 7 percent. What is the equivalent present worth of this program if the full $100,000,000 is loaned out under these terms (what is the net cost to the government for this program)?

(*Ans.* $44,590,000)

CHAPTER **8**

Annual Payments Method for Comparing Alternatives (Equivalent Series of Uniform Payments)

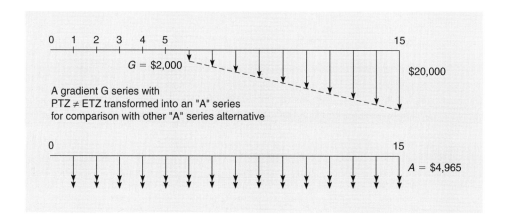

G = $2,000

$20,000

A gradient G series with
PTZ ≠ ETZ transformed into an "A" series
for comparison with other "A" series alternative

A = $4,965

KEY EXPRESSIONS USED IN THIS CHAPTER

NAW = Net Annual (periodic) Worth, the equivalent annual (or other uniform series) worth after summing the equivalent worth of all costs, incomes, and savings.

COMPARING EQUIVALENT UNIFORM SERIES

When deciding which of several alternative construction projects to fund or other investments to make, whether in government or private enterprise, there are often a number of competing opinions for dividing up the limited resources and spending the money. Each point of view usually has its own dedicated sponsors and attractive features. The selection process can be simplified somewhat if the competing proposals can be reduced to a numerical comparison between dollar benefits versus dollar costs. One common method of comparing competing alternative proposals is to reduce all cash flows involved in each proposal to an equivalent series of uniform payments. For example, assuming equal benefits, alternative **A** with a cost equivalent to $10,000 per month can be compared to alternative **B** with a cost of $13,000 per month. Alternative **A** is the least costly, and may be the better selection, provided that benefits such as productivity, safety, appearance, durability, and other features of alternative **A** are at least equal to those of competing alternative **B**. The compounding period indicated for this example is monthly, but any suitable interval that fits the problem will do (yearly, quarterly, weekly, daily, etc.).

The equivalent series of uniform payments method may be used for simple or complex analysis. However, since equivalent uniform payments are easy to comprehend and visualize, it is a particularly useful method to employ when making presentations to the public or to other decision-making groups with widely varied and often limited background in time-value-of-money concepts. While present worth might be a difficult concept to present without lengthy preparation, a simple comparison of equivalent annual payments required to support project **A** versus those for project **B** usually is grasped readily by even the most casual audience.

The major advantage of the annual worth (AW) method of comparing alternatives on the basis of periodic payments is that the complication of unequal lives of competing alternatives is automatically taken into account without any extra computations. If the lives of the competing investment alternatives are unequal, then this automatic feature is predicated on the assumption of *equal replacement costs*. With this assumption, the AW technique will yield the same selection decision as the PW technique (this is logical, since the two values are related by a constant factor).

Basic Approach

The recommended approach to solving even the most complex problems involves three steps that cannot be over-emphasized.

1. Subdivide the problem into simple components.
2. Solve each component separately.
3. Sum the component solutions to obtain a solution to the original complex problem.

The key to applying this basic procedure to annual worth problems involves mainly the ability to recognize the standard types of cash flow typically encountered. These cash flows occur in three basic types:

1. Single payment lump sum.
   a. Occurring at PTZ (ETZ = PTZ)
   b. Occurring at some other time (ETZ ≠ PTZ); see Figure 8.1.
2. Uniform series of equal payments.
   a. ETZ ≠ PTZ
   b. $n_1 \neq n_2$ (unequal lives)
   c. $i_1 \neq i_2$ (unequal interest rates)
   d. Period between payments does *not* equal the compounding period (periodic series where $i_e \neq i$).

3. Periodic series of unequal payments.
   a. Arithmetic gradient
   b. Geometric gradient
   c. Irregular amounts

CONVERTING THE THREE TYPES OF CASH FLOW INTO EQUIVALENT UNIFORM SERIES

1. A single-payment lump sum converted into an equivalent uniform series.
   a. Lump sum payment occurs at the beginning of the series, so that problem time zero equals equation time zero (PTZ = ETZ).

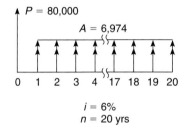

EXAMPLE 8.1 (Given P, i, n, find AW)

A short segment of road pavement costing $80,000/mi is expected to last 20 years with no salvage value. If $i = 6$ percent what is the equivalent annual cost per mile?

Solution

$$A = P(A/P, i, n) = \$80,000 \underbrace{(A/P, 6\%, 20)}_{0.08718} = \$6,974/\text{yr}$$

   b. Time base variations. If the lump sum payment occurs at any time other than the beginning or end of the uniform series, (PTZ $\neq$ ETZ), the lump sum amount is
      i. Converted to an equivalent lump sum payment at the beginning or end of the desired uniform series by means of the P/F or F/P factors.
      ii. There the new lump sum amount is converted into an equivalent uniform series as in (a) above by using the A/P factor or the A/F factor. Of course, the two steps may be combined, as illustrated in the equations below.

Typical Cash Flow Diagrams

A typical problem-solution equation for converting lump sum cash flow *not* at PTZ into equivalent annual payments is shown in Figure 8.1.

$$A = F(P/F, i, n_1)(A/P, i, n_2)$$

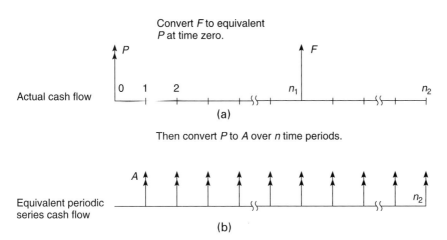

Figure 8.1(a) and (b)   Cash flow diagrams from single payments.

An example of the conversion of such a lump sum into an equivalent uniform series payment is shown in Example 8.2.

EXAMPLE 8.2   **(Given P, i, n, ETZ $\neq$ PTZ, find AW)**

A bulldozer is scheduled for a major overhaul costing $20,000 at the end of the third year of its seven-year economic life. What is the equivalent annual cost over the seven-year period, where $i = 10$ percent?

Solution.   The cash flow diagram for this problem is shown in Figure 8.2. The example problem can be solved by converting the lump sum to an equivalent lump sum at the ETZ for the uniform series, and then converting again to a uniform series, as follows:

$$A = F(P/F, i, n_1)(A/P, i, n_2)$$
$$A = \$20,000 \underbrace{(P/F, 10\%, 3)}_{0.75131} \underbrace{(A/P, 10\%, 7)}_{0.20541}$$
$$= \$3,086/\text{yr equivalent annual cost}$$

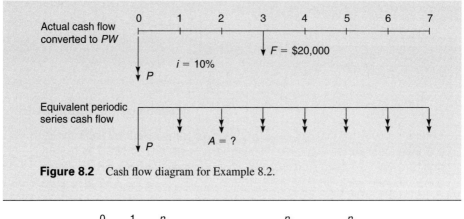

Figure 8.2   Cash flow diagram for Example 8.2.

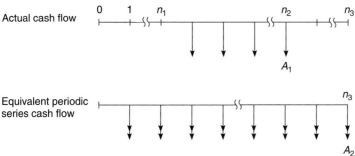

Figure 8.3   Cash flow diagrams—two equivalent periodic series.

2. One uniform series converted into another equivalent uniform series. Sometimes one component of a problem involves a series of equal payments over a time period that does *not* coincide with the time period for the entire problem. Either the beginning or end (or both beginning and end) of the series occurs at a time period different from that of the main problem. When this occurs, the following solution sequence is suggested:

 a. The odd series may be transformed into an equivalent lump sum at either n_1 using (P/A) or at n_2 using (F/A).

 b. Then the lump sum is translated to either the beginning (time zero) or end (time n_3) of the problem time.

 c. Then the new lump sum is converted into an equivalent series of periodic payments, using the proper problem time n, and using either (A/F) from n_3 or (A/P) from time zero.

These steps may be combined into one, as illustrated by the following equation, Example 8.3 and Figure 8.3.

$$A_2 = A_1(P/A, i, n_2 - n_1)(P/F, i, n_1)(A/P, i, n_3)$$

To determine the proper value of n, find where the ETZs occur and find the number of compounding periods between them. This method is illustrated in the following example.

EXAMPLE 8.3   **(Given A, i, n, ETZ $\neq$ PTZ, $n_1 \neq n_2$, payments occur at BOY; find AW)**

A marine dredge has an expected life of 12 years. The estimated maintenance costs for years 3 through 6 are estimated at $25,000 per year, payable in this case at the *beginning* of the year. What is the equivalent annual cost over the 12-year life of the dredge? $i = 9$ percent.

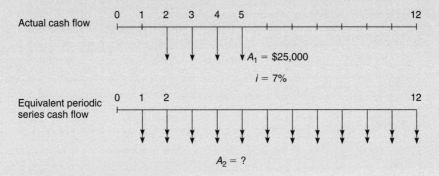

Figure 8.4   Cash flow diagrams for Example 8.3.

The cash flow diagrams for this problem are shown in Figure 8.4. The problem may be solved by converting the uniform series to an equivalent lump sum at the beginning (or end) of the series, then converting the lump sum to another equivalent lump sum at the beginning (or end) of the desired equivalent uniform series, and then converting that last lump sum into the desired equivalent uniform series with n equal to the proper n for the whole problem. Notice that with payment at the *beginning* of years 3 through 6, the ETZ for the series is at EOY 1. The solution is derived by the following calculations.

$$A_2 = A_1(P/A, i, n_1)(P/F, i, n_2)(A/P, i, n_3)$$
$$A_1 = \$25,000(P/A, 9\%, 4)(P/F, 9\%, 1)(A/P, 9\%, 12)$$
$$= \$10,377/\text{yr, annual equivalent over 12-yr period}$$

FIND THE EQUIVALENT SERIES WHEN i_1 <OR> i_2

Usually the investment rate is different from the borrowing rate. (No one purposely borrows funds at 15% in order to invest at 10%.) In addition the repayment time on a loan ordinarily is different from the life of the investment. Both of these situations are illustrated in the following example.

EXAMPLE 8.4   (Given P, i_1, i_2, n_1, n_2, find A)

Your client needs a new car and is trying to decide whether to lease or buy. The following terms are available:

If the car is purchased, cost new = $25,000, there is no down payment, and 100 percent financing by dealer at i_{n_1} = 6 percent, 48 months.

If the same car is leased, the monthly lease fee paid at EOM is $550 per month.

At the end of the 60-month lease period the client has an option to purchase the car for $1.00. (*Note:* For all practical purposes, a purchase price of $1.00 is essentially a zero purchase price. However, contract law typically differentiates between a purchase for any sum, even $1.00, and a donation or gift where no money is received in exchange for the car.) The client ordinarily invests and borrows at i_n = 12 percent, and expects to keep the car for 60 months. Find the equivalent monthly cost of the fully financed car purchase and compare to the monthly lease fee.

Solution.   Notice that the car payments are for 48 months ($n_1 = 48$) while the life of the investment is 60 months ($n_2 = 60$). Furthermore, the i values are not equal. To clarify the problem:

1. Draw the cash flow diagram.

2. Determine the amount of cash flow loan repayments. These are calculated as $A_1 = \$25,000(A/P, \frac{1}{2}\%, 48) = \$587.25$/month.

3. Find the PW of this cash flow at the client's usual i rate,

$$P_1 = \$587.25(P/A, 1\%, 48) = \$22,300$$

4. Find the equivalent monthly AW over the 60-month life of the investment,

$$A_2 = \$22,300(A/P, 1\%, 60) = \underline{\underline{\$495.96}}$$

Compare: The purchase alternative will cost the equivalent of $495.96 per month, and the client owns the resale value of the car at EOM 60. The lease alternative will cost $550 per month, and the client also will own the resale value at EOM 60. Therefore, in this case the client incurs less cost under the purchase alternative.

Variation.   ETZ<or>PTZ. Suppose the dealer offers an additional bonus feature under the purchase offer whereby the first payment is not due until EOM 4. This should result in an advantage to the client, since in effect the subsidized loan is being extended an additional three months. Delivery of the car is made now, and interest accrues on the balance owed. The 48 monthly payments are calculated on the amount owed at EOM 3, the ETZ for the repayment series. To solve, simply preface the previous set of equations with a calculation to find the lump sum amount owed at EOM 3, and then move that lump sum to PTZ before continuing with the process used above. The solution is found as

$$A = 25{,}000 \underbrace{(F/P, 0.5\%, 3)}_{1.0150} \underbrace{(A/P, 0.5\%, 48)}_{0.02349} \underbrace{(P/A, 1\%, 48)}_{37.974} \underbrace{(P/F, 1\%, 3)}_{0.97059}$$

$$\times \underbrace{(A/P, 1\%, 60)}_{0.02224}$$

$$A = \$488.59/\text{month}$$

As expected, the client's cost is slightly reduced by the three-month extension of the term of the subsidized financing loan.

3. Gradient or irregular series of nonequal payments converted into an equivalent uniform series. A gradient series, either arithmetic or geometric, may be converted into an equivalent uniform series in a manner similar to that employed in item 2. For an arithmetic series, the major difference is in locating the ETZ for the series. The first payment in the arithmetic gradient series occurs at the end of the second period (instead of at the end of the first period, as in the geometric gradient and equal payment series). If the series involves regular periods but irregular payment amounts (or regular payments but irregular times), then as a last resort the payments can always be treated as individual lump sum payments, as in Example 8.2.

Typical Cash Flow Diagrams

A typical problem-solution equation (arithmetic gradient) for the cash flow shown in Figure 8.5 is

$$A = G(P/G, i, n_2 - n_1)(P/F, i, n_1)(A/P, i, n_3)$$

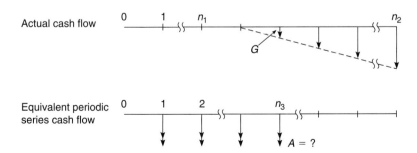

Figure 8.5   Cash flow diagrams—equivalent series.

EXAMPLE 8.5   **(Given G, i, n, ETZ $\neq$ PTZ, find AW)**

The pavement in a new city parking lot is expected to require no maintenance costs for the first five years of its life. At EOY 6, the first maintenance is expected to cost $2,000, with a gradient increase of $2,000 per year each year thereafter, until a scheduled resurfacing sometime after the annual maintenance is performed at EOY 15. What is the equivalent uniform annual maintenance cost for the first 15 years of pavement life if $i = 9$ percent?

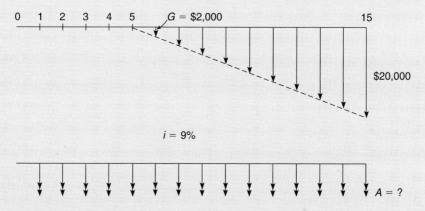

Figure 8.6   Cash flow diagrams for Example 8.5.

Solution.   The cash flow diagrams for this problem are shown in Figure 8.6. The equation for the solution of the cash flows shown in Figure 8.6 is

$$A = 2,000 \underbrace{(F/G, 9\%, 11)}_{72.892} \underbrace{(A/F, 9\%, 15)}_{0.03406}$$

$$A = \$4,965/\text{yr equivalent annual cost}$$

EXAMPLE 8.6   **(Given geometric gradient C, r, i, n, ETZ $\neq$ PTZ, find AW)**

Operating and maintenance costs for one segment of a traffic signal system are expected to occur at the end of years 3 through 8. The first expenditures will be $10,000 at the end of the third year, increasing by 10 percent per year thereafter for five additional years. The entire system has an expected life of 20 years. With $i = 10$ percent, find the equivalent annual cost of this part of the system for the entire 20-year period.

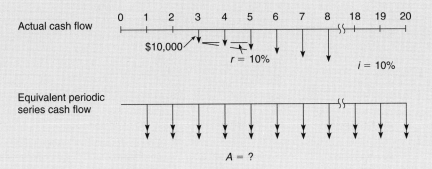

Figure 8.7   Cash flow diagrams for Example 8.6.

Solution.   The cash flow diagram for this example is shown in Figure 8.7. The example problem illustrated in Figure 8.7 may be solved by converting the geometric gradient series into an equivalent lump sum at the ETZ at EOY 2, then converting the lump sum into another equivalent lump sum at PTZ at EOY 0 (same as BOY 1), and then converting the last lump sum into an equivalent uniform annual series over the 20-year life of the installation. Since $r = i$, then $P = Cn/(1 + r)$. Then

$$A = \frac{Cn}{1+r}(P/F, i, n_2)(A/P, i, n_3)$$

$$A = \frac{10{,}000 \times 6}{1.10} \underbrace{(P/F, 10\%, 2)}_{0.8264} \underbrace{(A/P, 10\%, 20)}_{0.1175}$$

$$A = \$5{,}296/\text{yr equivalent annual cost over the 20-yr life}$$

PERPETUAL LIVES

Some investments, such as land, typically do not *de*preciate, but *ap*preciate in value. For investment study purposes, the land underlying a project is often viewed (conservatively) as merely maintaining its original value. The annual cost of such a land investment may be pictured as the annual amount required to pay interest on a loan used to acquire the land. If at some future time the land is sold for the same amount as the purchase price, then the loan is paid off, and the only costs have been the interest payments. (If the land or other investment is expected to sell at an increased price, the increment of increase can be separated from the purchase price and the A/F equation used to find the equivalent annual income payment.)

EXAMPLE 8.7   (Given $P, i, n = \infty$, find AW)

Some right-of-way land is needed for a highway that will be in use for a very long time ($n \rightarrow \infty$). If the land costs $200,000, and $i = 8$ percent, find the equivalent annual cost of the land.

Solution

$$A_1 = i\,P = 0.08 \times \$200,000 = \underline{\underline{\$16,000/\text{yr}}}$$

Mental picture: Borrow the money to buy the land. The only funds needed are those to pay the annual interest payment of 8 percent, or $16,000 per year on the $200,000 loan. It is assumed that at any time in the future, should the need arise, the land could be sold for $200,000 and the loan repaid.

A SERIES OF EQUAL EXPENDITURES OCCURRING AT PERIODIC INTERVALS SEVERAL COMPOUNDING PERIODS APART

Frequently a series of expenditures is required at repetitive intervals several years apart, while interest is compounded yearly. For instance, on a given building, painting may be needed every 4 years, new HVAC units every 12 years, and so on. The annual costs equivalent to these lump sums may be found as illustrated in the following example.

EXAMPLE 8.8   (Given P, periodic payments with the period between payments equaling some whole number of compounding periods, F, i, n; find NAW)

A civic center with an expected life of 40 years is proposed at an estimated cost of $14,000,000, including $2,000,000 for the land, and $12,000,000 for construction of the center. The following maintenance costs are anticipated.

paint every 4 yr = $80,000/each painting
replace air conditioner system
every 10 yr = $1,000,000/each replace
replace wood floors and seats
every 20 yr = $3,000,000/each replacement

Resale value of the salvageable components at the end of 40 years is $1,000,000. The civic center will be financed with 40-year revenue bonds at 9 percent, paid out in equal annual installments. How much annual revenue must be obtained to finance the project? The land value remains constant at $2,000,000 (a common assumption in engineering cost-analysis studies). This problem may be solved by any of several approaches.

Solution 1: Sinking Fund Method. Set up a savings account (sinking fund) for each major cost. The first cost of the center is $12,000,000, and the annual equivalent payment to pay off this amount in 40 years is shown below as A_1. In addition, interest payments on a loan to buy the land are provided as shown by A_2. Then an equal amount $(A_3 + A_4 + A_5)$ is deposited annually so that $80,000 will accumulate every 4 years for painting, $1,000,000 will accumulate every 10 years for replacing the air conditioning unit, and $3,000,000 will accumulate every 20 years for replacing the wood floors and seats. This takes care of the expenses as they arise, but also gives a balance in the account at the end of 40 years ready to be spent for new paint, air conditioning, and floors and seats just when the structure is scheduled for demolition. This balance may then be treated as additional salvage value and counted as equivalent annual income in A_6.

first cost of civic center	$A_1 = -\$12,000,000(A/P, 9\%, 40)$	$= -\$\ 1,116,000$
interest on value of land	$A_2 = -2,000,000 \times 0.09$	$= -\$\quad 180,000$
paint	$A_3 = -80,000(A/F, 9\%, 4)$	$= -\$\quad 17,500$
air conditioning	$A_4 = -1,000,000(A/F, 9\%, 10)$	$= -\$\quad 65,800$
wood floor and seats	$A_5 = -3,000,000(A/F, 9\%, 20)$	$= -\$\quad 58,500$
salvage value plus balance in accounts	$A_6 = +(1,000,000 + 80,000 \\ \quad + 1,000,000 \\ \quad +3,000,000)(A/F, 9\%, 40)$	$= +\$\quad 15,240$
net annual worth	$A_1 + A_2 + A_3 + A_4 + A_5 + A_6$	$= -\$1,422,560/\text{yr}$

An annual revenue of $1,422,560 is needed to finance the project. This will result in a debt of $5,080,000 just before liquidation of the last accounts and salvage value, which are needed to finally bring the account down to zero.

Solution 2: Amortize over the Periodic Interval. Each periodic lump sum amount may be separated from the original cost of the project and amortized over each interval between periodic payments.

The advantage of this method is that all accounts (except salvage value) are paid down to zero at the end of 40 years. Assume the contract for construction is divided among four contractors. The painting contractor gets $80,000, the air conditioning contractor gets $1,000,000, the wood floors and seats contractor gets $3,000,000, and the remaining $7,920,000 goes to the general contractor. All of the $12,000,000 construction money is borrowed at one time but paid off in separate accounts over the life of each account. For instance, the $80,000 borrowed for painting is paid off over four years. Then another $80,000 is borrowed every fourth year to pay for the repainting required at the end of the fourth year, the eighth year, and so on. At the end of 40 years the painting loan account is again paid down to zero, as it has been at the end of every fourth year.

paint	$A_1 = -\$80,000 \underbrace{(A/P, 9\%, 4)}_{0.3087}$	$= -\$\ \ 24,700/\text{yr}$
interest on land	$A_2 = -2,000,000 \times 0.09$	$= -\$\ 180,000$
air conditioning	$A_3 = -1,000,000 \underbrace{(A/P, 9\%, 10)}_{0.1558}$	$= -\$\ 155,800$
wood floors and seats	$A_4 = -3,000,000 \underbrace{(A/P, 9\%, 20)}_{0.1095}$	$= -\$\ 328,500$
remainder of civic center	$A_5 = -7,920,000 \underbrace{(A/P, 9\%, 40)}_{0.0930}$	$= -\$\ 736,560$
salvage value	$A_6 = +1,000,000 \underbrace{(A/F, 9\%, 40)}_{0.0030}$	$= +\$\ \ \ \ \ 3,000$
net annual worth	$A_1 + A_2 + A_3 + A_4 + A_5 + A_6$	$= -\$1,422,560\ \text{yr}$

PERIODIC GRADIENT SERIES

EXAMPLE 8.8   (continued)

What happens if the first repainting required at EOY 4 costs more than the $80,000 cost of the original paint job at EOY 0? Experience indicates that prices do tend to rise as time goes by, so assume that in this case painting prices are going to either: (a) rise by the same *amount* every four years, as in an arithmetic gradient, or (b) rise by the same *percentage* every four years, as in a geometric gradient. Find the equivalent annual worth of the periodic series of paintings, including the first painting when the structure is built, assuming the first painting costs $80,000, and the building is repainted every four years throughout the 40-year life of the building, and that after the first one, each subsequent painting increases in price by (a) $10,000 (paint at EOY 0 costs $80,000; at EOY 4 it costs $90,000, at EOY 8 it costs $100,000, etc.).

Solution to (a).   The payments may be divided into two series. First is the underlying $80,000 series every four years, and second the gradient increment series of $10,000 every four years. The annual equivalent of the underlying series may be found by either of the two methods shown in Example 8.8, or by a third method, using the equivalent interest rate over a four-year interval, i_e. The steps involved in this equivalent interest rate method are as follows:

1. Find the i_e for the four-year period.

$$i_e = (1+i)^n - 1 = (1.09)^4 - 1 = 0.41158 = 41.158\%$$

2. The series of paint payments made once every four years may be treated as a periodic series of 10 payments (the first at EOY 0, and the last at EOY 36) with interest compounded once every four years at 41.158 percent. Then the present worth of this series is found using the P/A equation. Since the first payment in the series occurs at EOY 0, the ETZ for the equation is located four years earlier at EOY minus 4. Therefore, the PW lump sum equivalent at ETZ must be moved from EOY 4 to EOY 0, by using the F/P equation with 9 percent and four years. With this accomplished, the equivalent annual series of payments over the 40-year life of the building may be found with the A/P equation. In equation form these calculations may be shown as

$$A = 80,000 \underbrace{(P/A, 41.158\%, 10)}_{2.3523} \underbrace{(F/P, 9\%, 4)}_{1.4115} \underbrace{(A/P, 9\%, 40)}_{0.09296}$$

$$= 24,690/\text{yr}$$

The $10,000 gradient in painting payments is treated in similar fashion, using the same steps but substituting P/G in place of P/A, as follows:

$$A = 10,000 \underbrace{(P/G, 41.158\%, 10)}_{4.9417} \underbrace{(F/P, 9\%, 4)}_{1.4115} \underbrace{(A/P, 9\%, 40)}_{0.09296} = 6,484/\text{yr}$$

Conclusion.   The periodic arithmetic gradient increase of $10,000 in painting costs occurring every four years increases the equivalent annual cost by $6,484 per year.

GEOMETRIC GRADIENT

EXAMPLE 8.8   (continued)

If the painting costs increase by a certain percentage every four years (instead of by a fixed amount), then the effective interest rate i_e still may be used. For instance, assume that the original paint job at EOY 0 still costs $80,000, but that painting costs are expected to increase by 4 percent per year. The PW may be found at ETZ (again located at EOY 4) by means of the P/C equation. Then the PW lump sum is moved to PTZ at EOY 0, and then distributed across the 40-year life of the building with the A/P equation as before. The equations appear as

$$\text{find } w = \frac{1 + i_e}{1 + r_e} - 1 = \frac{1.41158}{1.16986} = 0.20663 = 20.663\%$$

Then

$$A = [C \underbrace{(P/A, 20.663\%, 10)}_{80,000(3.50462)} /(1 + r_e)] \underbrace{(F/P, 9\%, 4)}_{(1.4115)} \underbrace{(A/P, 9\%, 40)}_{(0.09296)}$$
$$= 36{,}788/\text{yr}$$

Conclusion.   If the paint costs $80,000 for the first application at EOY 0 and increases in price by 4 percent per year (16.986 percent between paintings), the equivalent annual cost over the 40-year life of the building will be $36,788 per year. This represents an annual equivalent increase of $(36{,}788 - 24{,}694 =)$ $12,094 per year over the zero-increase estimate.

     What happens if the first period in a periodic series is an odd interval? For example, suppose that instead of occurring at EOY 4, the second application of paint occurs at EOY 3, and every four years thereafter (EOY 7, 11, 15, etc.)?

Solution.   Separate the first odd interval from the series. Then for the remaining nine payment series of periodic payments simply find the ETZ at four years before the first payment in the new series. Since the paint work at EOY 0 is no longer part of this series keep that $80,000 payment separate and add it on after completing the series calculations. Omitting the paint at EOY 0, the first payment in the new series occurs at EOY 3, so ETZ is located four years earlier at EOY minus 1. The cost of paint is increasing by 4 percent per year from PTZ at EOY 0, so the first payment at EOY 3 costs $C = \$80{,}000(1.04)^3 = \$89{,}990$. The PW of the nine-payment series only is found at PTZ as

$$P = [C/(1 + r_e)](P/A, w\%, n)(F/P, 9\%, 1)$$

where

$$w = (1 + i_e)/(1 + r_e) - 1 = 0.2066, \text{ as before}$$
$$P = [89,990/(1.16986)]\{[(1.20669^9) - 1]/0.2066 \times 1.2066^9\}1.09$$
$$= \$330,948$$

After finding the PW of the series, the cost of the first painting at EOY 0 is added back in, as

$$P = 330,948 + 80,000 = \$410,948$$

Having found the PW of the full series of paintings, the annual equivalent is easily determined by using the A/P equation.

$$A = 410,948 \underbrace{(A/P, 9\%, 40)}_{0.09296} = \underline{\underline{\$38,202/\text{yr}}}$$

Conclusion.   The change in the painting schedule of just one year raised the equivalent annual cost of painting by the equivalent of $(38,202 - 36,788 =)$ $\$1,414$ per year for 40 years.

GIVEN THE ANNUAL SERIES, FIND THE GRADIENT

Many practical situations occur where the present level of income is not adequate to cover the payments required to amortize the capital cost. Whenever the present worth is greater or less than the equivalent annual series, a gradient may be found to make up the difference. This type problem is illustrated in the following example.

EXAMPLE 8.9   **(Given P, A, G, periodic lump sum payments, F, i, n; find NAW, G)**

A 100-unit university married-housing project (that uses state bond funds at about 6 percent) is expected to incur the following costs and income.

$$\text{cost new} = \$2,000,000$$
$$\text{maintenance} = \$200,000/\text{yr increasing } \$4,000/\text{yr}$$
$$\text{major overhaul every 10 yr} = \$400,000$$
$$\text{salvage value at the end of 40 yr} = \$200,000$$

    a. What annual income is required for this project to break even?

    b. The initial annual rental income is \$3,000 per unit per year (\$300,000/yr for the entire project). An annual increase in rent is contemplated, since this initial income level is not sufficient to cover all costs. What annual arithmetic gradient increase will be necessary every year so that income over the 40-year life will be sufficient to cover all costs (provided the books need not be balanced until the end of the 40 years).

Solution

cost new	$A_1 = \$2,000,000 \underbrace{(A/P, 6\%, 40)}_{0.0665}$	$= -\$133,000$
maintenance	$A_2 = 200,000$	$= -\$200,000$
maintenance gradient	$A_3 = 4,000 \underbrace{(A/G, 6\%, 40)}_{12.359}$	$= -\$\ 49,440$
major renovation	$A_4 = 400,000 \underbrace{(A/F, 6\%, 10)}_{0.0759}$	$= -\$\ 30,360$
balance in the renovation account at EOY 40 plus salvage	$A_5 = (400,000 + 200,000) \underbrace{(A/F, 6\%, 40)}_{0.0065}$	$= +\$\ \ \ 3,900$
net annual worth		$= -\$408,900$

    a. The annual income required for breakeven is $-\$408,900$.

    b. The required gradient is found as:

$$-408,900 \text{ annual income required}$$
$$+300,000 \quad \text{actual income}$$
$$-108,900 \text{ deficit per year}$$

$$\text{required gradient } G = 108,900 \underbrace{(G/A, 6\%, 40)}_{1/12.359} = \$8,812$$

The project could be supported with an initial annual income of \$300,000 the first year and an increase of \$8,812 each year thereafter.

COMBINED VARIETY OF PAYMENT TIMES AND SEQUENCES CONVERTED TO EQUIVALENT ANNUAL WORTH

A wide variety of payment times and sequences can be combined to produce an equivalent annual series. In each case the problem is divided into basic components. The components are then converted into terms of an equivalent series over the life of the project and then summed. Care must be taken when summing that income is counted as plus and cost is minus.

EXAMPLE 8.10   (Given P, A, G, F, ETZ $\neq$ PTZ, find NAW)

A backhoe is expected to incur the following costs and income:

$$\begin{aligned}
\text{cost new} &= -\$120{,}000 \\
\text{annual O \& M} &= -\$\ 14{,}000 \quad \text{the first year} \\
\text{with increases of} &= -\$\ \ 2{,}000 \quad \text{each year} \\
\text{major overhaul after 5 yr} &= -\$\ 40{,}000 \\
\text{extra repair costs in years 7--10} &= -\$\ \ \ 3{,}000 \\
\text{resale value after 10 yr} &= +\$\ 20{,}000 \quad \text{(income)}
\end{aligned}$$

What must be the equivalent annual income to offset these cost and income items if $i = 20$ percent?

Solution

cost new	$A_1 = -\$120{,}000(A/P, 20\%, 10)$	$= -\$28{,}620$
annual O & M	$A_2 = -14{,}000/\text{yr}$	$= -\$14{,}000$
annual increase	$A_3 = -2{,}000 \underbrace{(A/G, 20\%, 10)}_{2.0739}$	$= -\$\ 6{,}148$

major overhaul $A_4 = -40{,}000 \underbrace{(P/F, 20\%, 5)}_{0.4019} \underbrace{(A/P, 20\%, 10)}_{0.2385} = -\$\ 3{,}834$

extra repair cost years 7–10 $A_5 = -3{,}000 \underbrace{(P/A, 20\%, 4)}_{2.5887} \underbrace{(P/F, 20\%, 6)}_{0.3349}$

$$\times \underbrace{(A/P, 20\%, 10)}_{0.2385} \qquad\qquad = -\$\quad 620$$

resale value $A_6 = +20{,}000 \underbrace{(A/F, 20\%, 10)}_{0.0385} \qquad\qquad = +\$\ 2{,}770$

annual income required to break even with costs, $\sum A_{1-6}$ $\quad = \$\ \ 52{,}452$

SELECTING THE BETTER ALTERNATIVE BY COMPARISON OF EQUIVALENT ANNUAL WORTHS

As discussed in Chapter 7, the comparison of alternatives by various methods should yield the same answer regardless of the method used since the A, P, and F factors are related to one another by constants. Thus the discussion in Chapter 7 dealing with types of alternatives and criteria for mutually exclusive alternatives is equally relevant here.

Annual worth comparisons are often the quickest and easiest to perform of all the techniques, because the problem of unequal lives is *automatically* handled without further calculations. However, this fact carries the serious implication that *equal replacement costs are inherently implied.* (For any specific problem under consideration, if replacement costs actually do increase, then additional calculations are needed.) The following example illustrates the concept that annual worth calculations imply equal replacement costs.

EXAMPLE 8.11   **(Given two alternatives with P, A, F, i, n, find and compare NAW)**

The problem statement as given in Example 7.3 is repeated here.

REPEAT OF EXAMPLE 7.3   **(Unequal life spans; compare NPW, given P, A, i, n, for each, where $n_1 \neq n_2$)**

A county engineer has a choice of paving with either type A pavement or type B pavement. Type A pavement has a life expectancy of 10 years, after which part of the material can be salvaged and reused. Type B pavement only lasts 5 years but is much less expensive. Which is the better alternative? Two sequential 5-year installations of type B are compared to one 10-year life of type A.

	Pavement cost	
	Type A	Type B
Cost new	$40,000	$10,000
Annual maintenance	2,000	4,000
Estimated life	10 yr	5 yr
Salvage value at end of life	5,000	0
i	10%	10%

Solution.   In this example the life of type A pavement is 10 years, while the life of type B pavement is 5 years. In comparing the equivalent annual costs in typical textbook problems, frequently no adjustment is made for escalating replacement costs at the end of the shorter life. This implies that the replacement cost at EOY 5 is identical to the first cost at PTZ. Since real-world problems often *do* involve escalating replacement costs, later in this text several problems will illustrate methods of dealing with escalating replacement costs.

Type A

A_1 = annualized
    purchase cost $= 40,000 \underbrace{(A/P, 10\%, 10)}_{0.16275}$ $= -\$6,510/\text{yr}$

A_2 = annualized maintenance cost $= -\$2,000$
A_3 = annualized salvage $= 5,000 \underbrace{(A/F, 10\%, 10)}_{0.06275}$ $= +\$\ \ 314$

net annual worth of type A $= A_1 + A_2 + A_3$ $= -\$8,196/\text{yr}$

Type B

A_1 = annualized
    purchase cost $= 10,000 \underbrace{(A/P, 10\%, 5)}_{0.26380}$ $= -\$2,638$

A_2 = annualized maintenance cost $= -\$4,000$
net annual worth of type B $= A_1 + A_2$ $= -\$6,638/\text{yr}$

The answer is that *type B* has the lowest net annual worth of costs.

This is the same decision reached when working the problem by the present worth method (Example 7.3) even though the unequal lives were not considered in the annual worth method (or were they?). To check the answer, the present worth of costs calculated in Example 7.3 is converted to annual worth over the 10 year period for *both* types as follows:

$$\text{NAW (type A)} = 50,362 \underbrace{(A/P, 10\%, 10)}_{0.16275} = -8,196/\text{yr}$$

$$\text{NAW (type B)} = 40,788 \underbrace{(A/P, 10\%, 10)}_{0.16275} = -6,638/\text{yr}$$

Since the results by the conversion of the NPW solution correspond to the results by the NAW method, this supports the conclusion that the unequal lives *were* accounted for by assuming equally valued replacements.

The computational advantages of comparisons by the annual worth method over the present worth method become obvious when unequal lives such as 7 and 5 years or 10 and 11 years are being considered. In the first case the lowest common multiple of 7 and 5 is 35. In the second case the lowest common multiple of 10 and 11 is *110*. Calculations based on present worth would be tedious and time-consuming. The advantage of annual worth comparisons for different life spans is illustrated in the following example.

EXAMPLE 8.12   **(Given three alternatives with $P, A, F, i, n_1 \neq n_2 \neq n_3$; find and compare NAW)**

Three different artificial turfs are available for covering the playing field in a college football stadium. The costs associated with each are tabulated as follows (assume $i = 15\%$).

	Turf King	Turf Ease	Turf Magic
Cost new (installed) ($)	540,000	608,000	467,000
Annual maintenance cost ($)	2,300	1,600	2,500
Expected life (yr)	12	15	10
Salvage value ($)	54,000	57,000	40,000

Solution.   Since different lengths of lives are involved, the lowest common multiple of 10, 12, and 15 would be 60. With multiples like this, the AW method has a distinct advantage over the PW method.

$$\text{NAW (King)} = -540,000 \underbrace{(A/P, 15\%, 12)}_{0.18448} - 2,300$$

$$+54,000 \underbrace{(A/F, 15\%, 12)}_{0.03448} \qquad = -\$100,070$$

$$\text{NAW (Ease)} = -608,000 \underbrace{(A/P, 15\%, 15)}_{0.17102} - 1,600$$

$$+57,000 \underbrace{(A/F, 15\%, 15)}_{0.02102} \qquad = -\$104,370$$

$$\text{NAW (Magic)} = -467,000 \underbrace{(A/P, 15\%, 10)}_{0.19925} - 2,500$$

$$+40,000 \underbrace{(A/F, 15\%, 10)}_{0.04925} \qquad = -\$\ 93,600$$

The results indicate that *Turf Magic* is the lower cost selection since it represents the lowest net annual worth.

Remember that the resulting values of both PW and AW are heavily dependent on the value of interest used (see Chapters 10 and 11). A different value of *i* will result in a different NAW and sometimes will reverse the order of precedence.

In the preceding problem the assumption of equal replacement costs was assumed valid. If this assumption is incorrect, then the problem must be worked differently, as follows.

EXAMPLE 8.13   **(Given two alternatives with *P*, rising replacement costs, $i, n_1 \neq n_2$; find and compare NAW)**

Two roofs are under consideration for a building needed for 20 years. Their anticipated costs and lives are given in the table below:

	Roof C	Roof D
Cost new ($)	50,000	25,000
Replacement cost ($)	—	rise 10%/yr
Life of roof (yr)	20	10
Salvage value at 20 yr ($)	0	0
Interest rate (%)	12	12

Solution

Roof C

$$A_1 = -\$50,000 \underbrace{(A/P, 12\%, 20)}_{0.1339} = -\$6,695/\text{yr}$$

Roof D

$$A_2 = -\$25,000 \underbrace{(A/P, 12\%, 20)}_{0.1339} = -\$3,348/\text{yr}$$

$$A_3 = -\$25,000 \underbrace{(F/P, 10\%, 10)}_{2.5937} \underbrace{(P/F, 12\%, 10)}_{0.3220} \underbrace{(A/P, 12\%, 20)}_{0.1339}$$

$$= -2,796/\text{yr}$$

net annual worth roof $D = A_2 + A_3$ $= -\$6,144/\text{yr}$

The two $25,000 roofs still cost less, since the replacement cost is rising at a slower rate (10%/yr) than the interest (12%) that would be tied up in the investment in the second 10-year half of the 20-year life of the $50,000 roof. In the practical situation, account should also be taken of possible business interruption and other inconveniences associated with installing a new roof.

The previous examples dealt with costs only. The same procedures are involved if incomes as well as costs occur.

EXAMPLE 8.14   **(Given two alternatives with P, income, A, F, i, $n_1 \neq n_2$; find and compare NAW)**

A consulting firm has a choice of two word processing machines. Each machine will permit a reduction in clerical staff. Estimated costs, lives, and salary savings are tabulated below. Assume $i = 15$ percent.

	Processor A	Processor B
Cost new ($)	−8,000	−17,000
Salary savings/yr ($)	+3,500	+ 4,800
Life (yr)	5	6
Salvage value ($)	0	+ 4,000

Solution.   Whenever the "do nothing" alternative is possible, it should always be considered. In this case the procedure is to calculate the net annual worth for $i = 15$ percent and select the alternative with the greatest net annual worth (select "do nothing" if the NAW of the two alternatives is negative).

$$\text{NAW(A)} = -8,000 \underbrace{(A/P, 15\%, 5)}_{0.2983} + 3,500 = \underline{\underline{+\$1,114}}$$

$$\text{NAW(B)} = -17,000 \underbrace{(A/P, 15\%, 6)}_{0.2642} + 4,800$$

$$+4,000 \underbrace{(A/F, 15\%, 6)}_{0.1142} = \underline{\underline{+\$\quad 765}}$$

The results indicate that *processor A* has the greater positive net annual worth.

SPREADSHEET APPLICATIONS

	A	B	C	D	E	F	G	H
1	Example 8.12: (Given two alternatives with P, rising replacement costs, i,							
2	n_1 not equal to n_2; find and compare NAW)							
3								
4		Two roofs are under consideration for a building needed for 20 years. Their						
5		anticipated costs and lives are given in the following table:						
6								
7						Roof C	Roof D	
8		Cost new ($)				50,000	25,000	
9		Replacement cost ($)				—	rise 10%/yr	
10		Life of roof (yr)				20	10	
11		Salvage value at 20 yr ($)				0	0	
12		Interest rate (%)				12	12	
13								
14								
15	SOLUTION:							
16								
17		*Roof C:*				*Roof D:*		
18								
19		P ($)	($50,000.00)			P ($)	($25,000.00)	
20		i (%)	12.00%			i (%)	12.00%	
21		n (yr.)	20			replacement cost$_{\%\ rise}$	10.00%	
22						$n_{life\ of\ roof}$ (yr.)	10	
23		A_1 ($/yr.)	$6,693.94			$n_{total\ span\ of\ analysis}$ (yr.)	20	
24						A_2 ($/yr.)	$3,346.97	
25						F'	$64,843.56	
26						P'	($20,877.89)	
27						A_3 ($/yr.)	$2,795.11	
28								
29						A_{total} ($/yr.)	$6,142.08	
30								
31	**CELL CONTENTS:** The following cell contents will allow us to evaluate the problem with a spreadsheet.							
32								
33		cell G24:	=PMT(G20, G23, G19,,0)					
34		cell G25:	=FV(G21, G22,,G19,0)					
35		cell G26:	=PV(G20, G22,,G25,0)					
36		cell G27:	=PV(G20, G23, G26,,0)					
37		cell G29:	=G24+G27					
38								
39	CONCLUSION:							
40								
41		The two $25,000 roofs still cost less, since the replacement cost is rising at a slower rate (10%/yr)						
42		than the interest (12%) that would be tied up in the investment in the second 10-year half of the						
43		20-yr life of the $50,000 roof. In the practical situation, account should also be taken of possible						
44		business interruption and other inconveniences associated with installing a new roof.						
45								
46		The previous examples dealt with costs only. The same procedures are involved if incomes as well						
47		as costs occur.						

SUMMARY

Methods are developed for translating a variety of different types of cash flows into an equivalent series of periodic payments. Since many alternative engineering investment decisions can be represented by their respective cash flows, these alternatives can be reduced to equivalent series of periodic payments and selections made of the better alternatives on a simple numerical basis. For alternatives involving costs *only,* the least equivalent annual cost alternative is selected. For alternatives involving incomes as well as costs, the alternative with the highest net annual worth (AW of incomes minus AW of costs) is selected. Where circumstances permit, the "do nothing" alternative is considered, and it is selected if none of the proposed alternatives yield a positive net annual worth at the stated value of i.

Annual worth analysis often involves fewer calculations than present worth analysis if differing lives are under consideration, because annual worth implicitly assumes equal replacement values (unless otherwise noted) and the least common multiple of the different lives without extra calculations. Annual worth analysis is often more easily understood, especially where rates or unit costs are being determined. For example, calculation of the annualized cost of a toll bridge provides the information necessary to calculate the toll rate for vehicles using the toll bridge.

The word *annual* is used merely for convenience in considering periodic cash flow conditions. The periods of time involved may be monthly, quarterly, daily, or any other. Remember that the compounding period for interest is the same as the period used to measure n, the life of the proposal.

PROBLEMS FOR CHAPTER 8, ANNUAL WORTH ANALYSIS

Problem Group A

Given three or more of the variables P, A, F, i, n, find the annual equivalent cost or income. Where alternatives are proposed, compare and indicate which is better.

A.1. Plans are under consideration to acquire a public boat ramp and park at a lake. There is an immediate need for 5 acres with a need in another 10 years for an additional 5 acres. The first 5 acres may be acquired for about $40,000 and the second 5 acres may be purchased for an additional $40,000. If purchased when needed, 10 years from now, it will probably have increased in value about $1,500 each year between now and then.
   (a) If money can be borrowed at 6 percent, which is the most economical alternative?
   (b) What intangible considerations might cause you to recommend the other alternative?

   (*Ans.* $A_1 = $10,870/yr; $A_2 = $9,608/yr)

A.2. **(a)** A parking structure is proposed for on-campus use. It is estimated to cost $2,500 per space plus $50 per year for maintenance. Annual parking decals will be sold to pay off the structure. With $n = 20$, $i = 6$ percent, what must be the annual charge for a decal?
   (b) As an alternative, an hourly parking fee is proposed. In addition to the above costs, an attendant is required at $50 per space per year. Assuming an average of 10 hours of parking per day, 220 days parking per year, 80 percent occupancy rate, what must be the

hourly parking rate? (Calculate the cost on an annual basis and divide by the number of hours of use per year.)

A.3. A swimming pool operated by the recreation department is in need of major repairs. The pool cost $120,000 10 years ago. To overhaul now and extend its life another 6 years would cost about $40,000. A new pool will cost $140,000 and have an estimated life of 12 years. Annual maintenance on the existing pool will be about $10,000 per year if overhauled, while the new pool will cost about $5,000 per year to maintain. Salvage value of both pools is estimated at about $20,000 regardless of age. With $i = 15$ percent, which should be selected?

(*Ans.* $A_{OH} = \$23,600/\text{yr}; A_{new} = \$30,140/\text{yr}$)

A.4. A steel bridge on a county highway near the ocean cost $3,500,000 12 years ago. Although the average annual maintenance costs of $72,000 per year have seemed excessive to the county supervisors, these costs, mainly for painting, have been necessary to prevent severe corrosion in this particular location.

A consultant offers to design and oversee the construction of a reinforced concrete bridge to replace this steel bridge. He estimates the total first cost of the concrete bridge to be $4,200,000 and the net salvage from the steel bridge either now or any time in the next 38 years to be $200,000. He reasons that the required net outlay is therefore $4,000,000, and since the county receives $5,000,000 a year for county highway improvement from state gas tax funds, this new bridge could be financed without borrowing. His economic comparison follows:

Annual cost of proposed concrete bridge

$$\text{depreciation (based on 50-yr life)} = \frac{\$4,000,000}{50} = \$\ 80,000$$

$$\text{annual maintenance cost} = \underline{\quad 2,000}$$

$$\text{total annual cost} = \$\ 82,000$$

Annual cost of present steel bridge

$$\text{depreciation (based on 50-yr life)} = \frac{\$3,500,000}{50} = \$\ 70,000$$

$$\text{annual maintenance cost} = \underline{\quad 72,000}$$

$$\text{total annual cost} = \$142,000$$

Comment on the consultant's analysis. Show the calculations and results you would use to present the comparative annual costs of the two bridges. Use $i = 6$ percent.

A.5. A van was purchased three years ago for $16,000. Even though the van needs $5,000 in repairs right now, it has a current market resale value "as is" of $4,400. If repaired, the van will last four more years and will have a salvage value of $600 at EOY 4. Operating and maintenance expenses are $3,800 per year.

A new van with an estimated life of eight years can be purchased for $20,000. The operating and maintenance costs are estimated to be $3,000 per year, and the resale value at EOY 8 is $1,800. A trade-in of $4,400 will be allowed on the existing van in its unrepaired condition.

Should the existing van be repaired and continued in service, or should a new van be purchased? Assume $i = 10$ percent

(*Ans.* $A_1 = \$6,637/\text{yr}; A_2 = \$6,591/\text{yr}$)

A.6. A school crossing guard is currently employed at $7,200 per year. The guard can be replaced by a pedestrian-actuated traffic light costing $24,000 initially, plus $600 per year for maintenance. Assume a life of 10 years and no salvage value for the light, and $i = 6$ percent. How does the annual cost of the light compare with the annual cost of the guard? List other bene-fits of having a crossing guard on duty. List other benefits of the light. Which are you inclined to recommend as a matter of personal opinion (not graded)?

A.7. A computer can be acquired for $10,000 under either of two plans. Which results in the lower annual cost?

(a) Borrow $10,000 at the bank and repay in equal annual installments over a five-year period at 8 percent on unpaid balance.

(b) Borrow a $2,000 down payment and repay over five years at 10 percent. Pay off the remaining balance owed the dealer at $2,000 per year for five years.

(*Ans.* $A_a = \$2,505/\text{yr}$; $A_b = \$2,527.60/\text{yr}$)

A.8. A company is considering the following investments. If their MARR is 15 percent, which investments, if any, should they select?

	P	Q
Initial investment ($)	−40,000	−30,000
Annual net income ($)	+6,500	+5,000
Life (yr)	20	15
Salvage value ($)	0	0

A.9. A toll bridge is being considered for design and construction. It is estimated the bridge will cost $73,000,000 and require annual operating and maintenance costs of $2,000,000. If 110,000 vehicles are expected to cross the bridge each day, what toll rate should be charged if $i = 7$ percent?

(a) Assume 50-year life with no salvage value.

(b) Assume perpetual life.

(*Ans.* $A_1 = \$0.182/\text{vehicle}$; $A_2 = \$0.177/\text{vehicle}$)

A.10. Same as Problem A.9, except that $i = 10$ percent.

Problem Group B

Involves unit costs, $n_1 \neq n_2$, ETZ $\neq$ PTZ.

B.1. Compare the equivalent NAW of the two excavators in the table below for an MARR of 8 percent.

	Excavator A	Excavator B
Purchase price	$50,000	$80,000
Cash down payment	$10,000	$50,000
Loan amount amortized over 5 yr	$40,000	$30,000
Interest on loan	7%	6%
Est. annual production at $0.25/yd^{3a}	200,000 yd^3	250,000 yd^3
Est. resale value/life	$8,000/8 yr	$16,000/10 yr
Average annual maintenance and operating cost	$40,000	$45,000

^{a}Use this figure to find gross income.

(*Ans.* $A_A = \$2,234/\text{yr}$; $A_B = \$6,915/\text{yr}$)

B.2. An earthmoving machine with an estimated life of 10 years will need a scheduled major overhaul at the end of the fifth year, costing $10,000. If $i = 10$ percent, what is the equivalent annual cost over the 10-year life?

B.3. The repair costs on an earthmoving machine are expected to average $5,000 at the end of years 1 through 4 and $8,000 at the end of years 5 through 7. If $i = 10$ percent, what is the equivalent annual cost over the 7-year period?

(*Ans. A* = $6,047/yr)

B.4. A commercial parking firm is considering construction of a parking lot as an interim use of a 4-acre tract in Metro City's CBD. The firm uses a 15 percent minimum attractive rate of return. The tract can be leased for 10 years at $8,000 per year. Construction costs are estimated to be $40,000; 20 percent of the cost comes from cash on hand, while the remaining 80 percent by a 10 percent, five-year loan. At termination of the interim use as a parking lot, all improvements must be removed; estimated cost $6,000. Operating costs are expected to be $30,000 per year. What is the equivalent annual cost?

B.5. Your client needs a new car and is trying to decide whether to (1) buy and pay cash, (2) buy fully financed, or (3) lease. The following terms are available:

If the car is purchased for cash:

Cost new, $24,000

Factory rebate available immediately, $2,000

If car is purchased fully financed:

Cost new, $24,000

No down payment

100 percent financing by dealer at $i_n = 6$ percent, 36 monthly repayments

If the same car is leased:

Monthly lease fee paid in *advance* at BOM, $450 per month (the first payment is due now).

At the end of the 60-month lease period the client has an option to purchase the car for $1.00.

The client ordinarily invests and borrows at $i_n = 12$ percent compounded monthly, and expects to keep the car for 60 months.

(**a**) Find the equivalent monthly cost over the life of the car of each of the three alternatives and select the lower cost. (Disregard resale value at EOM 60, since it is the same in all three cases.)

(**b**) The dealer offers an additional bonus feature with the fully financed purchase offer whereby the first payment is not due until EOM 4. Delivery of the car is made now, and interest at 6 percent accrues on the balance owed. The 36 monthly payments are calculated on the amount owed at EOM 3. Find the equivalent monthly cost over the life of the car of acquiring the car under this alternative.

(*Ans.* (a) A_{cash} = $489.28/month; $A_{finance}$ = $488.85/month; A_{lease} = $454.50/month; (b) A_{delay} = $469.05/month)

Problem Group C

Arithmetic gradients

C.1. Operating and maintenance costs for a package sewage treatment plant are expected to be $10,000 per year the first year and increase $1,000 per year thereafter for five years. If $i = 10$ percent, what is the equivalent annual cost?

(*Ans. A* = $11,810)

C.2. A large recreation center is proposed at an estimated $10,000,000 cost and a life of 40 years. The following maintenance costs are anticipated.

Clean and paint every 4 years. The first painting at EOY 4 costs $60,000 and increases by $12,000 every 4 years.

Replace mechanical systems every 10 years. The first replacement at EOY 10 costs $800,000 and increases by $400,000 every 10 years.

Replace wood floors and seats after 20 years at $1,000,000.

The resale value of salvageable components at the end of 40 years is estimated at $2,000,000. The recreation center will be financed with a 40-year loan at 6 percent repaid in equal annual installments. How much annual revenue is needed to finance this project?

C.3. A contractor has just borrowed $100,000 from the bank at $i = 1$ percent per month, and purchased a dragline. He expects to keep it for 120 months. The following operating and maintenance costs are anticipated:

$$
\begin{aligned}
\text{monthly operator's salary, maintenance,} \\
\text{lube, and filters} &= \$\ \ 2{,}000/\text{month} \\
\text{expected gradient increase in monthly costs each month} &= \$\ \ \ 100/\text{month/month} \\
\text{replace cables and cutting edges at the end} \\
\text{of every 12 months} &= \$\ \ 3{,}000/12\ \text{month} \\
\text{complete overhaul of engines and tracks} \\
\text{at the end of every 30 months} &= \$10{,}000/\text{each time}
\end{aligned}
$$

At the end of the 120-month period the dragline should sell for about $20,000 in "as is" condition, without a final overhaul or replacement of cables. The dragline is expected to move 50,000 yd^3 the first month. Due to continued wear, age, and down time, the amount will decrease by 100 yd^3 per month each month. How much should the contractor charge in dollars per cubic yard to recover all his dragline costs plus $0.10/$yd^3$ for overhead and profit (the dollars per cubic yard charge remains constant for the five-year period)?

(*Ans.* ($0.19 + $0.10) = $0.29/$yd^3$)

C.4. A bridge is needed at a certain location, and designs in both steel and concrete are available with the following cost estimates. The steel bridge is estimated to cost $2,000,000 now with a salvage value of $300,000 at the end of 20 years.

Maintenance costs on the steel bridge are estimated at $10,000 per year for the first three years, and then they increase by $2,000 per year for the next 17 years. ($12,000 the fourth year, $14,000 the fifth year, etc.)

The concrete bridge is estimated to cost $1,800,000 with a negative salvage value of $200,000. (It will cost $200,000 to demolish the structure.) Maintenance costs on the concrete bridge are estimated at $20,000 per year for the full 20 years. Assume that maintenance costs for both bridges are paid in advance at the beginning of each year. For example, the maintenance cost for the first year is paid out at the *beginning* of year 1. Compare the NAW of the bridges, using $i = 7$ percent.

C.5. Same as Problem A.9 except maintenance costs will increase by $200,000 per year per year ($2,000,000 EOY 1, $2,200,000 EOY 2, etc.).

(*Ans.* [a] $A = \$0.244$/vehicle; [b] $A = \$0.248$/vehicle)

C.6.   A water management district finds that they could acquire their pickup trucks either by leasing or by purchase. Find the equivalent monthly cost of each and compare. Use $i = 1$ percent per month, and $n = 60$ months. The data shown are for each truck. Each truck is driven, 1,000 miles per month.

Alternative A.   Leasing costs $600 per month paid *in advance* at the *beginning* of each month.

Alternative B.   Purchase.

$$\text{Cost new} = \$15,000$$
$$\text{O \& M} = \$0.20/\text{mi}$$
$$\Delta\text{O \& M} = \$0.002/\text{mi/month}$$
$$\text{resale at EOM } 60 = \$5,000$$

C.7.   A municipal bus company asks your help in determining what fare they should charge per passenger to pay the owning and operating costs of their buses. They borrow and lend at $i = 10$ percent.

The buses run from 6:30 A.M. until 6:30 P.M., six days per week, 52 weeks per year (3,744 hours per year). They average 10 mph and pick up an average of two fares per mile. Assume all fares are credited at the end of the year. The buses last an average of 20 years.

A typical tabulation of cash flow for the costs for one bus is shown in the following table:

Year			
0	$150,000 purchase new	10	25.00 break in gradient
1	$17.00/hr O & M costs	11	26.50
2	17.00	12	28.00
3	18.00 first gradient payment	13	29.50
4	19.00	14	31.00
5	20.00	15	32.50
6	21.00	16	34.00
7	22.00	17	35.50
8	23.00	18	37.00
9	24.00	19	38.50
		20	40.00/hr
		20	salvage value $10,000

(a) Find the equivalent dollars per hour cost for the 20-year period (including purchase cost and salvage value).

(*Ans.* $A = \$27.93/\text{hr}$)

(b) How much fare should be charged each customer to cover all costs and income listed above, assuming there are no fare increases over the 20-year period?

(*Ans.* $1.40/passenger)

(c) How much fare should be charged each passenger during the first year to cover all costs and income in the table above, assuming the fares are increased $0.05 per passenger per year for each year after the first year?

(*Ans.* $1.08/passenger)

Problem Group D

Geometric gradients (some also include arithmetic gradients)

D.1. A proposed civic center is estimated to cost $10,000,000 and have a life of 45 years. The maintenance costs are estimated as follows:

Paint and normal repairs: $20,000 per year at EOY 1, increasing 10 percent per year over the 45-year life.

Replace air conditioner, floor tile, and seat upholstery every 15 years. The cost 15 years from now is estimated at $2,000,000 and the cost is expected to double for the replacement required at the end of the thirtieth year.

After 45 years it is estimated that the entire civic center will have a salvage value of $500,000.

(a) If the civic center will be financed with 45-year revenue bonds at 8 percent to be paid off in equal annual installments, how much annual revenue is required to pay off the bonds plus paint, repairs, and replacements as listed above?

(b) A group of citizens feel that revenue from the civic center itself should amount to $200,000 the first year (use EOY receipt) and increase 10 percent per year thereafter for the 45-year life. Under the 8 percent, 45-year financing of (a), how much profit will be earned after the P & I payments, maintenance, and other costs listed above?

(*Ans.* $A_a = \$1,015,000/\text{yr}$; $A_b = \$44,070/\text{yr}$)

D.2. A proposed airport terminal is expected to cost $4,000,000 with a life of 40 years. The operating and maintenance costs are estimated as follows:

operations	1–40 yr, begin at $100,000/yr, rising $5,000/yr/yr
maintenance	1–20 yr, begin at $50,000/yr, rising 10% the second year and 10% more each year thereafter
end-of-year maintenance	10 yr, $20,000 major rehabilitation
	21–40 yr, begin at $70,000/yr, rising $1,000/yr the second year and $1,000 more each year thereafter

All the costs of the terminal, including construction, will be financed at 6 percent and will be repaid by charging each passenger a fee.

(a) Assuming 1,000,000 passengers per year, what fee should be charged per passenger?

(b) Assuming 70,000 passengers the first year and an additional 100,000 passengers each year thereafter, what fee should be charged?

D.3. Same as Problem D.2, except the 1–20 year maintenance increases at the rate of $5,000 per year per year.

(*Ans.* [a] $0.514/passenger; [b] $0.394/passenger)

D.4. A contractor has a contract to construct a 12,000-ft-long tunnel in 30 months, and is trying to decide whether to do the job with his own forces or subcontract the job. Calculate the equivalent *monthly cost* of each alternative. Use $i = 1.75$ percent per month. Production under both alternatives will be 400 ft per month.

Alternative A.   Buy a tunneling machine and work with contractor's own forces.

cost of tunneling machine = $500,000 paid now
salvage value of machine at EOM 30 = $100,000

cost of labor and materials = $80/ft for the first 10 months, increasing by 0.5%/month at EOM each month thereafter ($80.40/ft at EOM 11, $80.802/ft at EOM 12, etc.)

Alternative B. Subcontract the work. Cost $200 per foot of advance of the tunnel heading, for the full 12,000 feet.

D.5. Same as Problem D.4, except that the costs of labor and materials increase by $0.50 per month at EOM each month thereafter ($80.50/ft at EOM 11, $81.00/ft at EOM 12, etc.).

(*Ans.* A_A = $52,090/month; A_B = $80,000/month)

Problem Group E

Periodic gradients.

E.1. Find the cost in terms of dollars per mile over the 48-month life of the car whose data are listed below. Use $i = 1$ percent per month.

> cost new = $9,000
> resale value at EOM 48 = $3,000
> operating and maintenance (O & M) = $0.10/mi
> gradient in O & M = $0.001/mi/mo

periodic major overhaul (OH) every 4 months, with the

> first at EOM 4 = $150/each
> gradient in periodic major OH = $10/each 4 months
> expected use = 3,000 mi/month

The last periodic overhaul is done at EOM 48.

(*Ans.* A = $0.201/mile)

E.2. A sports authority wants to construct a sports coliseum with borrowed money that will be repaid through sale of tickets to events in the coliseum. Calculate the required charges in dollars for each ticket in order to pay all the costs listed below. Use $i = 1$ percent per month.

> coliseum cost new = $20,000,000
> salvage value at EOM 360 = $20,000,000
> O & M = $40,000/month
> O & M gradient = $500/month/month

periodic major maintenance (MM) every 12 months

> (last MM at EOM 360) = $200,000/each
> MM gradient = $10,000/each/12 months
> admission tickets per month = 300,000/month

E.3. A progressive community is studying the feasibility of installing a solid waste energy conversion plant. Find what the community must charge per 1,000 Btu of energy output in order to break even over the 240-month life of the plant. Use $i = 1$ percent per month.

$$
\begin{aligned}
\text{solid waste energy conversion cost new} &= \$8,000,000 \\
\text{salvage value at EOM 240} &= \$0 \\
\text{O \& M cost} &= \$200,000/\text{month} \\
\text{gradient in O \& M} &= \$1,000/\text{month/month} \\
\text{periodic overhaul every 10 months} & \\
\text{(first at EOM 10, last at EOM 240)} &= \$600,000/\text{each} \\
\text{gradient in periodic overhaul} &= \$30,000/\text{each}/10 \text{ months} \\
\text{expected energy output of plant} &= 5 \times 10^{10}\, \text{Btu/month}
\end{aligned}
$$

The last periodic overhaul is done at <u>EOM 240</u>.

(*Ans.* $A = \$0.00883/1,000$ Btu)

E.4. The county engineer is considering the purchase of a concrete transit mix truck, and asks you to calculate the equivalent cost per cubic yard of concrete for owning and operating the transit mix truck. Use $i = 0.75$ percent per month.

$$
\begin{aligned}
\text{cost new} &= \$200,000 \\
\text{resale value at EOM 120} &= \$20,000 \\
\text{O \& M costs (including concrete mix materials)} &= \$30,000/\text{month} \\
\text{O \& M gradient} &= \$100/\text{month/month} \\
\text{periodic major overhaul every 8 months} & \\
\text{(last at EOM 120)} &= \$15,000/\text{each} \\
\text{periodic overhaul cost gradient} &= \$1,000/\text{each}/8 \text{ months} \\
\text{monthly production of concrete} &= 1,000 \text{ yd}^3/\text{month}
\end{aligned}
$$

E.5. A monorail shuttle is being planned between the airport and downtown. Find the break-even fare to charge per passenger. Use $i = 1.0$ percent per month.

$$
\begin{aligned}
\text{cost to construct monorail and place in operation} &= \$10,000,000 \\
\text{resale value at EOM 360} &= 0 \\
\text{O \& M costs} &= \$400,000/\text{month} \\
\text{O \& M gradient} &= \$40,000/\text{month/month} \\
\text{major overhaul every 12 months} & \\
\text{(last major at EOM 360)} &= 1,000,000/\text{each} \\
\text{major overhaul gradient} &= \$100,000/\text{each}/12 \text{ months} \\
\text{passengers boarded per month} &= 800,000
\end{aligned}
$$

(*Ans.* $A = \$5.1283/\text{passenger}$)

E.6. Student government on a certain campus is considering a new shuttle bus service to be supported solely by the fares charged to shuttle bus riders. Find the charge per customer to break even over the 120-month life of the bus. Use $i = 0.75$ percent per month

$$
\begin{aligned}
\text{cost of shuttle bus new} &= \$60,000 \\
\text{salvage value at EOM 120} &= 0
\end{aligned}
$$

$$O \ \& \ M \ cost = \$2,000$$
$$gradient \ in \ O \ \& \ M = \$20/month/month$$
periodic overhaul every 12 months
$$(first \ at \ EOM \ 12, \ last \ at \ EOM \ 120) = \$5,000$$
$$gradient \ in \ periodic \ overhaul = \$500/each/12 \ months$$
$$expected \ shuttle \ bus \ passengers/month = 16,000$$

E.7. As a construction contractor, you need to calculate the equivalent hourly cost of operating and owning a new bulldozer. Use $i = 1$ percent per month.

$$A_1 \ bulldozer \ cost \ new = \$60,000$$
$$A_2 \ resale \ value \ at \ EOM \ 120 = \$12,000$$
$$A_3 \ O \ \& \ M \ costs = \$5,500/month$$
$$A_4 \ gradient \ O \ \& \ M \ costs = \$60/month/month$$
$$A_5 \ periodic \ overhaul \ every \ 8 \ months$$
$$(last \ overhaul \ at \ EOM \ 120) = \$7,500/overhaul$$
$$A_6 \ gradient \ overhaul \ costs = \$800/overhaul/8 \ month$$
$$hours \ used \ each \ month = 120 \ hr$$

(*Ans.* $A = \$88.50/hr$)

E.8. A commuter airline wants to provide service between Gainesville and Tallahassee. Given the costs below, what should the ticket cost be per passenger? Use $i = 1$ percent per month.

$$cost \ of \ new \ aircraft = \$200,000$$
$$resale \ value \ at \ EOM \ 120 = \$0$$
$$O \ \& \ M \ costs \ (including \ overhead \ and \ profit) = \$25,000/month$$
$$O \ \& \ M \ gradient = \$200/month/month$$
periodic major overhaul every 12 months
$$(first \ at \ EOM \ 12, \ last \ at \ EOM \ 120) = \$10,000/each$$
$$periodic \ overhaul \ cost \ gradient = \$1,000/each/12 \ months$$
$$passengers \ per \ month = 1,000$$

E.9. A proposed flood-control project involves periodic heavy maintenance payments as listed below. With $i = 8$ percent, and $n \to \infty$, find the equivalent annual cost.

EOY	Maintenance payment
2	$10,000
6	12,000
10	14,000
etc.	etc.

Every four years the payment increases by $2,000.

(*Ans.* $A = \$4,025/yr$)

E.10. An airport authority wants to construct a new airport with borrowed money that will be repaid through charges to airlines for each passenger boarded through the new airport. Calculate the required charge in dollars for each passenger boarding in order to pay all the costs listed below. Use $i = 12$ percent per year, $n \to \infty$.

$$\text{airport cost new} = \$20,000,000$$
$$\text{O \& M the first year, payable at EOY 1} = \$1,000,000/\text{yr}$$
$$\text{O \& M gradient} = 6\%/\text{yr compounded}$$
$$\text{periodic major maintenance (MM) every 5 yr,}$$
$$\text{with the first at EOY 7} = \$2,000,000 \text{ at EOY 7}$$
$$\text{periodic MM gradient} = 6\%/\text{yr compounded}$$
$$\text{passenger boardings} = 400,000/\text{yr}$$

CHAPTER 9

Future Worth Method of Comparing Alternatives

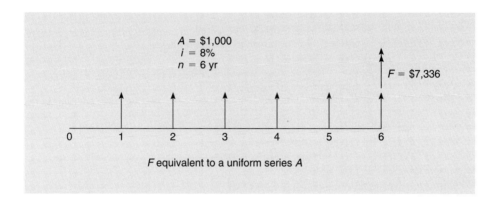

$$A = \$1,000$$
$$i = 8\%$$
$$n = 6 \text{ yr}$$
$$F = \$7,336$$

F equivalent to a uniform series *A*

KEY EXPRESSIONS USED IN THIS CHAPTER

NFW = Net Future Worth, the equivalent lump sum value at a specified future date, determined by summing the equivalent future worths of all related costs, incomes, and savings.

COMPARING FUTURE WORTHS

Frequently problems occur where we need to find the future worth (FW) of investments or other quantities that grow at some reasonably predictable rate. For instance, in order to make a future lump sum payment, an investor needs to know the future balance in a sinking fund savings account accumulating interest at a known rate. Given a good estimate of the appreciation rate for a particular real estate market, an owner needs to estimate the future selling price for a building or other property at some future date in order to determine profitability. The future worth is simply the amount of cash or equivalent value that the owner can expect at that future date.

Also, future worth concepts are commonly encountered in estimates of nonmonetary future values, such as population, traffic count, consumer's demand for water, electric power, and other products, goods, and services. When highways, water works, sewer lines, or other facilities are constructed to serve a growing population, some estimate must be made of how large a facility should be constructed and how long will it last until it reaches capacity and needs further expansion.

The method of application of future worth is similar to the methods used for present worth and annual worth. The values are readily convertible among all three by multiplication by the appropriate constants. The two basic rules when comparing alternative future worths are, like present worth comparisons, that (1) the comparisons must be made at a common future date, and (2) the life spans for comparison purposes must be identical in length.

Summary of Three Types of Cash Flow Converted to Equivalent Future Value

Cash flows previously have been classified into the following three basic types:

1. Single payment lump sum
2. Periodic series of equal payments
3. Periodic series of nonequal payments
   a. Arithmetic gradient
   b. Geometric gradient
   c. Irregular amounts and/or times

The future value of each of the above cash flows depends upon the time of occurrence of the flow, as well as the interest rate. Each type is discussed in turn in the following paragraphs.

1. *Single payment lump sum, converted into an equivalent future value.*

   a. Lump sum payment occurs at some time prior to the point in time for which the future value is calculated. Interest at rate i is earned on the payment from the time the payment is made. See Figure 9.1.

The typical problem-solution equation is

$$F = P(F/P, i, n)$$

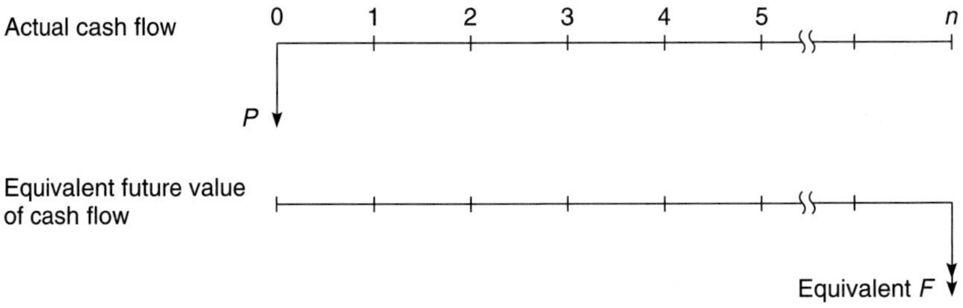

Figure 9.1   Diagram of actual present cash flow versus equivalent future cash flow.

EXAMPLE 9.1   **(Given *P*, *i*, *n*, find *F*)**

If $10,000 is invested now at 10 percent compounded, how much will accumulate (principal and interest) in 12 years?

Solution

$$F = \$10,000(F/P, 10\%, 12) = \underline{\underline{\$31,384}}$$

2. *Periodic series of equal payments, converted into an equivalent future value.*

   a. The periodic series ends at the same point in time for which the future value is calculated. Interest at rate *i* is earned on all payments made. See Figure 9.2.

The typical problem-solution equation is

$$F = A(F/A, i, n)$$

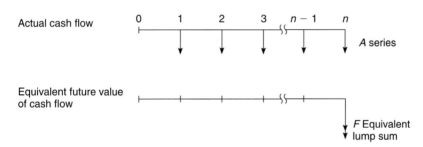

Figure 9.2   Diagram of actual periodic series cash flow versus equivalent future cash flow.

EXAMPLE 9.2   **(Given *A*, *i*, *n*, find *F*)**

If $1,000 per year is invested at the end of each year at 6 percent for 12 years, what will be the total amount of principal and interest accumulated at the end of the twelfth year after the twelfth payment?

Solution.   The cash flow diagram (Figure 9.2) for this example is similar to that above.

$$F = \$1,000(F/A, 10\%, 12) = \underline{\underline{\$21,384}}$$

b. The periodic series ends at some point in time *other* than the point in time for which the future value is calculated. See Figure 9.3.

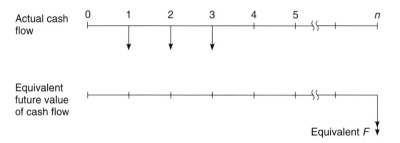

Actual cash flow

Equivalent future value of cash flow

Equivalent F

Figure 9.3 Diagram of a short series cash flow versus equivalent future cash flow.

The typical problem solution equation is shown below in combined form.

$$F = A(P/A, i, n_1)(F/P, i, n_2)$$

The first expression, $A(P/A, i, n_1)$, converts the series A to a present worth P at problem time zero. The second expression, $(F/P, i, n_2)$, converts P to an equivalent future value F at $n = 10$ years.

EXAMPLE 9.3 (Given A, i, n, ETZ $\neq$ PTZ, find F)

An income of $1,000 per year is expected from a certain enterprise at the end of years 1 through 6. These funds will be invested at 10 percent as soon as received and will remain invested at that interest until EOY 10. How much will the account be worth at EOY 10?

Solution. The cash flow diagram for this problem is shown in Figure 9.4. The problem can be solved as follows.

$$F = A(P/A, i, n_1)(F/P, i, n_2)$$
$$F = \$1,000\underbrace{(P/A, 10\%, 6)}_{4.3553}\underbrace{(F/P, 10\%, 10)}_{2.5937}$$
$$F = \$11,296 \text{ future value of the series}$$

---

3. *Periodic series of nonequal payments, converted into an equivalent future value.* If the nonequal payments form an arithmetic series or a geometric gradient series, the conversion to future value proceeds very similarly to the equal payment series

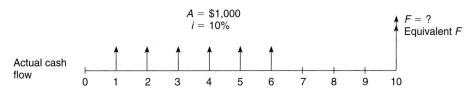

Figure 9.4   Cash flow diagram for Example 9.3.

described in paragraph 2 above. (Remember, however, that the arithmetic gradient first payment occurs at the end of the second period rather than the first, as in the geometric gradient and the equal payment series.) See Figure 9.5.

The typical problem-solution equation for an arithmetic gradient is

$$F = G(P/G, i, n_1)(F/P, i, n_2)$$

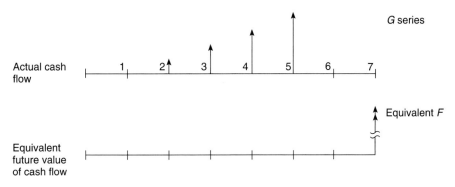

Figure 9.5   Typical cash flow diagrams.

EXAMPLE 9.4   (Given G, i, n, ETZ $\neq$ PTZ, find F)

A contracting firm is arranging financing for a project about to go under construction. They need to borrow $1,000 at the beginning of the first month, $2,000 at the beginning of the second month, and increase the amount borrowed by $1,000 per month through the beginning of the tenth month. They will not need to borrow any more, and expect payment for the entire job from the owner at the end of the twelfth month. The bank agrees to loan the money at 1 percent interest compounded monthly on the amount borrowed. How much will the contracting firm have to repay at the end of the twelfth month?

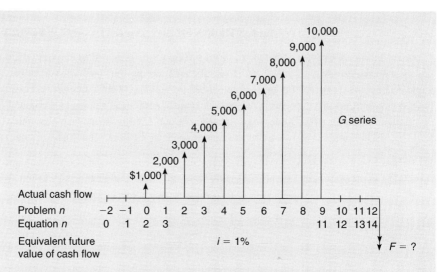

Figure 9.6   Actual cash flow diagram for Example 9.4.

Solution.   See Figure 9.6. The problem can be solved by shifting the ETZ back to -2, and using

$$F = G(P/G, i, n_1)(F/P, i, n_2)$$
$$F = \$1,000\underbrace{(P/G, 1\%, 11)}_{50.8068}\underbrace{(F/P, 1\%, 14)}_{1.1495}$$
$$F = \$58,400 \text{ future value of this gradient series}$$

EXAMPLE 9.5   **(Given C, r, i, n, find F)**

Make a series of 10 annual deposits beginning with $1,000 at the end of the first year and increase the amount deposited by 10 percent per year, with $i = 10$ percent on all sums on deposit. Find the accumulated principal and interest at the end of 12 years.

Solution.   See Figure 9.7. The problem can be solved as follows.

$$\text{since } i = r, \quad P = \frac{Cn}{1+r}$$
$$F = \frac{Cn}{1+r}(F/P, i, n)$$
$$F = \frac{\$1,000 \times 10}{1.10}(F/P, 10\%, 12)$$
$$F = \$28,530 \text{ future value of this gradient series}$$

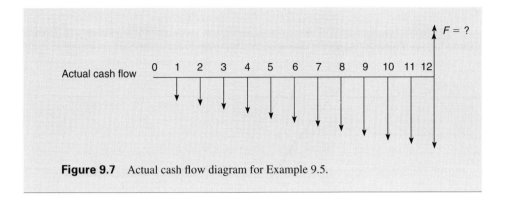

Figure 9.7   Actual cash flow diagram for Example 9.5.

COST OF FINANCING

Most construction projects are built with borrowed money. Contractors usually get paid in one of the following ways for work completed:

1. Periodically (e.g., monthly) according to the amount of work completed in each period.
2. By stages of completion (e.g., 15% on completion of foundation, 30% more on completion of framing, etc.).
3. Payment in full only upon satisfactory completion of the entire job.

In order to pay for labor and material before receiving payment from the owner, contractors normally borrow money from a bank. The bank lends the money on the good faith expectation that the job will be completed in a timely and workmanlike manner and that the owner will pay the contractor, who will then repay with interest the loan at the bank. The owner on his part is likely borrowing from a mortgage lender whose loan is secured by either a construction loan mortgage during construction, or a permanent mortgage after the work is completed and accepted. In all of these cases interest is paid on the amount borrowed, and constitutes a significant cost of the project. To determine the amount of this financing cost alone requires a two-step process.

Step 1.   Find the future worth of the accumulated principal plus interest.
Step 2.   Subtract the amount of principal borrowed. The amount remaining is the total amount of interest paid in terms of dollars.

The procedure is illustrated in the following examples.

EXAMPLE 9.6   (Given A, G, i, n, find dollar amount of interest cost)

A contractor is bidding on a construction project on which the owner requires completion of the work by EOM 12. The owner will pay the contractor in one lump sum at EOM 14. To pay for the day-to-day labor and material costs on the project, the contractor will require a loan from the bank of the amounts shown below. The bank charges interest at $i = 1$ percent per month on the outstanding balance.

EOM	Amount borrowed
1 through 12	$100,000/month

Thus at the EOM 12 the contractor will have borrowed from the bank 12 payments of $100,000 each for a total of $1,200,000 and will owe the bank the $1,200,000 plus accumulated interest. The owner is scheduled to pay the contractor at EOM 14, at which time the contractor will repay the bank. How much should the contractor include in the bid just to cover the cost of financing the job? See Figure 9.8.

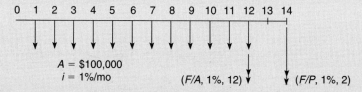

Figure 9.8   Cash flow diagram for Example 9.6.

Solution.   The amount of interest can be found by finding the future worth of interest plus principal, and then subtracting the amount of the principal borrowed.

$$\text{FW at EOM 14}, F = \$100,000 \underbrace{\frac{(F/A, 1\%, 12)}{12.682}}_{} \underbrace{(F/P, 1\%, 2)}_{1.0201} = \$1,293,700$$

$$\text{amount borrowed} = 1,200,000$$
$$\text{amount of interest only} = \$\underline{93,700}$$

Conclusion.   In order to account for the interest costs, the contractor will have to add to his bid a financing cost of $93,700.

Financing Costs Where Monthly Progress Payments to the Contractor Do NOT Equal the Paid-out Costs of the Contractor

Commonly the owner does not require the contractor to carry all the costs of financing the project during construction, but will make monthly (or other periodic) payments to the contractor. However, there is usually a retainage by the owner of a portion of the payment in order to guarantee completion, so the contractor still needs to borrow some funds in order to construct the project. To calculate the amount of the financing cost borne by the contractor, the same basic approach is employed. First, find the future worth of accumulated principal plus interest, then subtract the principal portion. The following example illustrates the process.

EXAMPLE 9.7 **(Given A, G, i, n, find the net amount of interest cost in dollars to the contractor)**

A contractor is bidding on a construction project with the anticipated cash flows shown in the table below. He borrows and lends at 1 percent per month.

	EOM	Income to contractor from owner	Cost to contractor paid out for labor and materials
	0	$ 0	$ 100,000
	1	80,000	110,000
	2	90,000	120,000
	3	100,000	130,000
Gradient	4	120,000	140,000
G=10,000/mo.	⋮	⋮	level
	11	180,000	140,000
	12	190,000	140,000
	Total	$1,620,000	$1,720,000

To arrive at the total contract bid price, in addition to the direct labor and materials costs of $1,720,000 listed above, this contractor expects to add

 a. The cost of financing.

 b. 30 percent of all costs to cover supervision, overhead, and profit (OH & P).

At EOM 12, the contractor is entitled to receive whatever balance is due from the owner on the contract, in addition to the owner's payments of $1,620,000 listed above. Find the contractor's bid amount, which consists of

 a. Direct labor and materials $1,720,000

 b. Contractor's financing costs   Find

 c. 30 percent markup for supervision, OH & P   Sum above × 1.3

Solution. First, find the contractor's financing costs for this project. Follow the approach suggested above:

 Step 1. Find the amount of interest income the contractor could earn if the payments received from the owner were deposited in an account earning 1 percent interest per month.

Step 2.   Find how much the contractor pays for interest costs for the amount the contractor borrows.

Step 3.   Subtract the contractor's interest income from the interest costs.

Future worth of income:

	FW at 1%	Actual income
$F_1 = \$80,000\underbrace{(F/A, 1\%, 12)}_{12.6826}$	$= \$+1,014,610$	$\underbrace{\$960,000}_{(80,000 \times 12 \text{ mo})}$
$F_2 = \$10,000\underbrace{(F/G, 1\%, 12)}_{68.251}$	$=\quad\quad 682,510$	$\underline{660,000}$
total FW	$=\quad \$1,697,120$	$\$1,620,000$
subtract amount of income	$=\quad -1,620,000$	
interest earned	$=\quad \$\quad 77,120$	

Future worth of costs:

amount borrowed	FW at 1%	
$F_3 = \$100,000\underbrace{(F/A, 1\%, 13)}_{13.809}$	$=\quad \$1,380,900$	$\$1,300,000$
$F_4 = 10,000\underbrace{(F/G, 1\%, 5)}_{10.100}\underbrace{(F/P, 1\%, 8)}_{1.0828}=$	$\quad 109,360$	$100,000$
$F_5 = 40,000\underbrace{(F/A, 1\%, 8)}_{8.2856}$	$=\quad\quad 331,420$	$\underline{320,000}$
total FW	$=\quad \$1,821,680$	$\$1,720,000$
subtract amount borrowed	$=\quad -1,720,000$	
cost of interest on borrowed funds	$=\quad \$101,680$	

Next, the net financing cost of the contractor is found as the cost of interest less the interest earned.

$$
\begin{aligned}
\text{cost of interest} &= \quad \$101,680 \\
\text{interest earned} &= - \quad 77,120 \\
\text{net financing cost to contract} &= \quad \$\ 24,560
\end{aligned}
$$

Conclusion.   The contractor should add $24,560 to the bid to cover the net costs of the contractor's share of financing the job. The contractor's bid then consists of:

a.	Direct labor and materials	$1,720,000
b.	Contractor's financing cost	24,560
	Subtotal	$1,744,560
c.	Markup	× 1.3
	Contractor's bid price	$2,267,930
	Sum of owner's monthly payments	1,620,000
	Final payment due contractor at EOM 12	$ 647,930

Comment.   In actual practice the cash flows often do not follow uniform or gradient series. In these cases the cost of financing can be determined using the same approach but treating each payment as a lump sum if necessary, finding the equivalent FW, and following the solution process above to obtain a correct result.

PROBLEMS OF GROWTH

Some of our greatest problems as well as our greatest opportunities revolve around population growth. As the population grows, we need more of almost everything. As population grows, the whole infrastructure must grow with it, or bottlenecks, congestion, and shortages will appear. Need for the various elements of the infrastructure is usually dictated by two factors: (a) the need or consumption per person, and (b) the number of people. For instance, electrical energy consumption is generally increasing several percent per year per person simply because we are all buying more electric and electronic equipment. In areas where there are growing numbers of people, the demand for electric energy is growing due to increasing consumption per person as well as increasing number of persons. To predict the future needs for all sorts of infrastructure elements, including roads, water plants, apartment houses, schools, and electric power generators, the following approach may be employed.

EXAMPLE 9.8   (Given P, A, G, F_1, i, find n, F_2)

As consultant to a rural electric system, you are asked to determine what size the next new electric power generator should be. You are given the following criteria and data.

$$\text{present number of customers on the system} = 8,000$$
$$\text{expected new customers added to system} = 400/\text{yr}$$
$$\text{average peak demand per customer} = 4 \text{ kW}$$
$$\text{expected increase in peak demand per customer} = 3\%/\text{yr}$$
$$\text{capacity of existing generator} = 40,000 \text{ kW, or 40 MW}$$

The consumption characteristics for the new customers are expected to be the same as for the existing customers.

a. When will the capacity of the existing generator be reached by the increasing peak load?

b. Assume the new generator should be sized so that together with the existing generator they should be adequate to handle the peak load for a period of 10 years beyond the startup date of the new generator. What size should the new generator be in terms of kilowatts?

Solution

a. F_1 represents the peak demand needs of the existing 8,000 customers, which is growing by 3 percent per year, and is like a savings account with no new deposits that earns 3 percent interest (increase) compounding each year.

$$F_1 = 4 \text{ kW} \times 8,000(F/P, 3\%, n)$$

F_2 represents the peak demands of the 400 new customers per year, and is like a series of deposits into a savings account, with the rising new balance earning 3 percent interest (increase) each year.

$$F_2 = 4 \text{ kW} \times 400(F/A, 3\%, n)$$

The sum of $F_1 + F_2$ represents the total demand for electrical energy. When the demand of these two groups totals 40 MW, a new generator needs to go on line.

$$F_1 + F_2 = 40,000 \text{ kW, or 40 MW.}$$

The appropriate values are substituted into the equations, and they can be solved directly for n as shown below.

$$32 \text{ MW}(1.03^n) + 1.6 \text{ MW}(1.03^n - 1)/0.03 = 40 \text{ MW}$$
$$n = \ln 1.0938 / \ln 1.03 = 3.03 \text{ yr}$$

until the capacity of the existing generator is reached

b. The second question requires us to find the peak demand at a time 10 years past the time when the existing generator reaches capacity, or at EOY $n = \text{EOY}(10 + 3.03) = \text{EOY } 13.03$.

$$F_1 = 32 \text{ MW}(1.03)^{13.03} = 47.035 \text{ MW}$$
$$F_2 = 1.6 \text{ MW}(1.03^{13.03} - 1)/0.03 = 25.058$$
$$F_1 + F_2 = \text{(capacity required at EOY 13.03)} = 72.093$$
$$\text{less existing capacity} = 40.000$$
$$\text{new capacity required at EOY 13.03} = 32.093 \text{ MW}$$

A similar approach may be employed to solve a wide range of problems involving planning for other future needs.

Forecasting Growth

Since growth occurs in many areas that affect both our personal as well as professional lives, growth normally must be taken into account when planning for the future. Admittedly, forecasts very far into the future typically are difficult to make with any great degree of accuracy. However, it is usually far more profitable to work with *some* estimate of future growth no matter how inaccurate than to try to ignore growth entirely. Thus some values for growth can be derived from past history, and from examining the history of other situations similar to the one under study. Projection of past history can be hazardous if not modified by the changing pressures of current and future influences. In addition, of course, the best projections of experts on growth in the area under study should be consulted. Where future projections are required, it is often prudent to project a range of values, including the following:

1. A high, "optimistic," projection, but perhaps with a common sense caveat that there is still a 10 percent (or other appropriate number) probability that even this high projection may be exceeded.

2. An expected projection, if everything goes according to the best predictions.

3. A low, "pessimistic," projection with a 10 percent probability that even that projection will not be met.

Using these estimates of growth will enable us to better plan and prepare for whatever growth does occur. It has almost always proven more profitable to make some estimate rather than no estimate at all.

COMPARING ALTERNATIVES

As with present worth and annual worth, comparisons between alternatives may be made using future worth equivalents. Two types of alternatives commonly occur: those involving costs only and those involving incomes (or benefits) as well as costs. For problems involving costs only, the alternative with the lowest equivalent future worth (cost) is usually preferable. For alternatives involving incomes as well as cost, the alternative with the greatest net future worth (NFW) is usually better, with due consideration being given to the "do nothing" alternative. The following example illustrates the method.

EXAMPLE 9.9   (Given competing alternatives with $P, A, i, n_1 \neq n_2$, find and compare the NFW)

A construction company needs to buy a new front-end loader, and is trying to select one from the two whose data are listed below.

a. If $i = 8$ percent, which loader, if any, should be selected?

b. If $i = 20$ percent, which loader, if any, should be selected?

	Front-end loader A	Front-end loader B
First cost ($)	− 110,000	− 40,000
Annual net income ($)	+ 16,000	+ 13,000
Useful life (yr)	10	5
Replacement cost escalation	NA	10%/yr compounded
Salvage value ($)	+ 10,000	+ 10,000

Solution

a. For an i of 8 percent,

$$\text{NFW(A)} = -110,000 \underbrace{(F/P, 8\%, 10)}_{2.1589} + 16,000 \underbrace{(F/A, 8\%, 10)}_{14.4866} + 10,000$$

$$= \underline{+\$4,307}$$

In order to compare loader B, which only has a five-year life span, a second loader B is purchased at EOY 5 so that the two loaders both have a 10-year service life span.

$$\text{NFW (B)} = -40,000 \underbrace{(F/P, 8\%, 10)}_{2.1589} - 40,000 \underbrace{(F/P, 10\%, 5)}_{1.6106} \underbrace{(F/P, 8\%, 5)}_{1.4693}$$

$$+13,000 \underbrace{(F/A, 8\%, 10)}_{14.4866} + 10,000 + 10,000 \underbrace{(F/P, 8\%, 5)}_{1.4693}$$

$$= \underline{+32,005}$$

Solution a. With $i = 8$ percent, <u>Loader B</u>, with the greatest positive NFW is preferred.

b. For $i = 20$ percent,

$$\text{NFW(A)} = -110,000 \underbrace{(F/P, 20\%, 10)}_{6.1917} + 16,000 \underbrace{(F/A, 20\%, 10)}_{25.958} + 10,000$$

$$= \underline{-\$255,759}$$

$$\text{NFW(B)} = -40,000 \underbrace{(F/P, 20\%, 10)}_{6.1917} - 40,000 \underbrace{(F/P, 10\%, 5)}_{1.6106} \underbrace{(F/P, 20\%, 5)}_{2.4883}$$

$$+13,000 (F/A, 20\%, 10) + 10,000 + 10,000 (F/P, 20\%, 5)$$

$$= \underline{-\$35,637}$$

Solution b.   In part (b), using $i = 20$ percent, both alternatives yield a negative NFW. This result indicates that both investments will yield a return of *less* than 20 percent. If the "do nothing" alternative is available, then *no* investment should be made if a 20 percent return is required.

THE "DO NOTHING" ALTERNATIVE

When considering an investment involving income, frequently the investor has the option of either investing in one of several alternatives or not investing at all (known as the "do nothing" alternative). However, in some instances the "do nothing" alternative is not available even when income is expected. The investment under consideration may be a part of a larger operation that would not function properly without this investment as an integral part. For instance, a construction contractor may spend relatively large sums on copy machines and word processors. This equipment may not make any profit on its own but could contribute significantly to the profitability of the firm. Similarly a department store may operate a lunch counter or a branch post office at a loss in order to attract more customers into the store. Thus, even when income is involved, the "do nothing" alternative may or may not be available depending upon the particular proposal under consideration.

In part (2) of Example 9.9, the "do nothing" alternative is the better selection (if it is available), since the company can make more money by investing the purchase price of either loader in an investment earning 20 percent [if the company is borrowing at $i = 20\%$, then just repaying some of their debts represents an investment that saves (yields) 20% rate of return] rather than in investing in either piece of equipment.

This leads naturally to the question, What rate of return will these two possible alternatives yield the company? Obviously they both will yield greater than 8 percent because in part (1) the NFW's are positive at 8 percent, and they both yield less than 20 percent because in part (2) the NFW's are negative at 20 percent. The methodology of finding the actual rate of return for each alternative is presented in the next chapter.

SPREADSHEET APPLICATIONS

	A	B	C	D	E	F	G	H
1	EXAMPLE 9.9: (Given competing alternatives with P, A, i, n_1 not equal to n_2, find and							
2	compare the NFW)							
3								
4		A construction company needs to buy a new front-end loader and is trying						
5		to select one from the two whose data are listed below.						
6								
7		a. If $i = 8$ percent, which loader, if any, should be selected?						
8		b. If $i = 20$ percent, which loader, if any, should be selected?						

(Spreadsheet continued)

	A	B	C	D	E	F	G	H
9								
10								
11						Front-end	Front-end	
12						loader A	loader B	
13		First cost ($)				−110,000	−40,000	
14		Annual net income ($)				+16,000	+13,000	
15		Useful life (yr.)				10	5	
16		Replacement cost escalation				NA	10%/yr compounded	
17		Salvage value ($)				+10,000	+10,000	
18								
19								
20	SOLUTION:							
21								
22		*Front-end loader A:*			*Front-end loader B:*			
23								
24		i	8.00%		i	8.00%		
25		n	10		$n_{\text{life of loader}}$ (yr)	5		
26		P	$110,000.00		$n_{\text{total span of analysis}}$ (yr)	10		
27		A	($16,000.00)		P	$40,000.00		
28		S	$10,000.00		A	($13,000.00)		
29					S	$10,000.00		
30		NFW	$4,303.25		replace cost% rise/yr	10.00%		
31					P'	($64,420.40)		
32								
33					NFW	$32,006.89		
34								
35								
36	CELL CONTENTS: The following cell contents will allow us to evaluate the problem with a spreadsheet.						
37								
38		cell C26:	+110000	Note: Excel expects payments, or costs, to be entered as positive numbers.				
39		cell C27:	−16000	and therefore incomes must be entered with negative numbers.**???**				
40		cell C30:	=FV(C24,C25,,C26,0)+FV(C24,C25,C27,,0)+C28					
41		cell F31:	=FV(F30,F25,,F27,0)					
42		cell F33:	=FV(F24,F26,,F27,0)+FV(F24,F26,F28,,0)+F29−FV(F24,F25,,F29,0)−FV(F24,F25,,F31,0)					
43								
44								
45	SOLUTION:							
46								
47		With i = 8 percent, Loader B with the greatest positive NFW is preferred.						
48		With i = 20 percent, both alternatives yield a negative NFW, and the "do nothing" alternative should						
49		be chosen if available.						

SUMMARY

In this chapter, the methods for finding and comparing future worths of a variety of cash flow combinations are presented. These methods are similar in many respects to the methods used for finding present worths and annual worths. Future worth is useful for finding the future value of an investment or for comparing alternatives on the basis of net future worth at some specified rate of return. As with present worth, the lives of various alternatives *must be equal* in order for the comparison to be valid.

PROBLEMS FOR CHAPTER 9, FUTURE WORTH

Problem Group A

Given three or more variables, find F; one-step operation or similar.

A.1. Assume that 30 years ago a trust fund was set up for your benefit with an initial investment of $10,000. The trustees had to choose between investing the funds in (a) 1,000 shares of a certain stock selling at $10 per share, (b) a selected mutual fund, or (c) a savings account bearing interest at 5 percent compounded annually. Which alternative would have yielded the greatest total amount of principal plus interest by now, 30 years later?

(a) Estimate the increase in dollar value of the stock investment if the stock now sells for $44.50 per share. In addition, dividends averaged 4.1 percent of the current stock value and are assumed to have been reinvested as received. (*Hint*: For a close approximation, find i for the increase in value of the stock and add 4.1% to it to obtain a new i to use in finding the future value of the $10,000 investment in stock.)

(b) The selected mutual fund advertises that $10,000, if invested in their average fund 30 years ago, would amount to $94,008 by now, 30 years later.

(*Ans.* Stock, $F = \$140,260$; Savings, $F = \$43,220$)

A.2. A developer client of yours purchased some property for $50,000. The terms were $10,000 down and a note for $ 40,000, payable at the rate of $5,000 per year with $8\frac{1}{2}$ percent interest on the unpaid balance.

(a) When is the last payment due?

(b) How much is the last payment?

A.3. Traffic on First Street is currently 1,500 cars per day and increasing at the rate of 6.3 percent per year. When the traffic reaches 4,000 cars per day, new traffic signals will be needed. How soon will that be if the present rate of increase continues?

(*Ans.* 16.1 yr)

A.4. The main trunk line of a sanitary sewer is being planned for an area experiencing a steady rate of growth. The initial flow through the pipe is expected to be 10,000 gallons per day for the first year. As the pipe ages, a small amount of infiltration is expected to occur, amounting to about 5 percent of the sanitary sewage flow each year. Additional development of the area is expected to add about 300 gallons per day for each new dwelling unit, and about 50 new dwellings per year are expected on the line every year for the next 10 years. At the end of 10 years the development will be completed, but the infiltration is expected to continue at the same 5 percent compounded per year. For simplicity, assume all new dwellings are connected at EOY.

(a) What is the expected flow at the end of 20 years?

(*Ans.* 333.8 kgal)

(b) How much of the flow at the end of 20 years is sanitary sewage, and how much is infiltration?

(*Ans.* sewage $= 160.0$ kgal, infiltration $= 173.8$ kgal)

Problem Group B

Arithmetic gradient, comparisons.

B.1. A retirement plan requires a 5 percent annual salary contribution from an employee aged 25 who expects to retire at age 60. He now makes $36,000 per year and expects annual raises averaging $3,000 per year from now until retirement. The retirement 5 percent of his salary ($0.05 \times \$36,000 = \$1,800$ the first year) is placed in a trust fund at the end of every year ($n = 35$) along with a matching 5 percent from his employer, and the fund draws 4 percent interest compounded. The plan is voluntary, and as an alternative, he realizes he can invest his own 5 percent (no employer matching funds) in mortgages and securities at an estimated 12 percent return. At age 60, how much would there be in each alternative fund?

(*Ans.* alternative 1, $F = \$555,000$; 2, $F = \$1,273,000$)

B.2. A friend of yours purchased some land 10 years ago for $10,000. At the end of the first year she paid taxes of $120. Every year thereafter the taxes have increased by $20 per year each year. Four years ago she had some drainage and fill work done that cost $1,500. A potential buyer is interested in the property now and your friend wants to know what selling price to ask in order to realize 15 percent compounded on all cash invested in the property.

B.3. An engineer uses a car extensively for company business and has a choice between using the company car or buying his own car and receiving a mileage allowance from the company. Compare the equivalent future value of each alternative and find which is preferred. Assume all payments are credited at EOY, and $i = 10$ percent.

EOY	0	1	2	3	4
Purchase price	$10,000				
Fuel (cents/mi)		8.0	8.9	9.8	10.7
Tires, repairs (cents/mi), maintenance, insurances, etc.		5.0	5.6	6.2	6.8
Resale value					$3,000
Mileage reimbursement paid by the company (cents/mi)		22.0	24.0	26.0	28.0

(a) Assume the car is driven 10,000 miles per year.

(*Ans.* NFW $= -\$7,162$)

(b) Assume the car is driven 20,000 miles per year.

(*Ans.* NFW $= -\$2,645$)

B.4. Deposit $1,000 at the end of the second year, $2,000 at the end of the third, and increase the amount deposited by $1,000 per year thereafter until the end of the tenth year, with $i = 10$ percent. Find the accumulated principal and interest at the end of 10 years.

B.5. A small irrigation project is expected to yield the incomes (+) and expenditures (−) tabulated below. Interest on the funds is 15 percent from EOY 1 until EOY 6, and then it changes to 10 percent from EOY 6 until EOY 12. (All payments are in thousands of dollars.) Draw the cash flow diagram and find the net future worth at EOY 12.

EOY	Payment ($)	EOY	Payment ($)
1	0	7	+29
2	0	8	+18
3	+10	9	+7
4	+20	10	−4
5	+30	11	−15
6	+40	12	−26

(*Ans.* +$240.8)

B.6. Your client asks your advice on a building project that has the estimated cash flows tabulated below. (Note, all payments are in thousands of dollars.) Draw the cash flow diagram and find the NFW at EOY 12.

EOY	Income ($)	Costs ($)
1	0	0
2	+40	0
3	+35	−10
4	+30	−14
5	+25	−18
6	+20	−22
7	+15	−26
8	+10	−30 (notice break in gradient)
9	+10	0
10	+10	0
11	+10	0
12	+10	0

The interest rates are

EOY 0 to EOY 3, $i = 6\%$

EOY 3 to EOY 8, $i = 10\%$

EOY 8 to EOY 12, $i = 12\%$

B.7. Your firm of consulting engineers is asked to design a complex project for a reliable client. Under the terms of the proposed agreement, the completed plans will be delivered to the client at EOM 14 from now, and your firm will be paid in one lump sum at that time. Meanwhile, your firm will borrow from the bank the amounts listed below in order to pay the direct payroll and computer costs incurred in work on this project. All amounts borrowed are in thousands of dollars.

EOM	Borrowed this month ($)	EOM	Borrowed this month ($)
0	100	7	190
1	120	8	160
2	140	9	130
3	160	10	100
4	180	11	70
5	200	12	40
6	220 (break gradient)		

The bank charges 1.5 percent per month on the compounded balance. Find how much the firm must add to their design fee in order to cover the cost of financing (how much interest is paid to the bank).

(*Ans.* $246,400)

Problem Group C

Geometric gradient.

C.1. A graduating class of civil engineers decides they will present the engineering college with a gift of $100,000 at their twenty-fifth reunion. There are 50 in the class and they can invest at 8 percent. They plan to start out giving a small amount at the end of the first year and increase their gifts to the class fund by 10 percent each year. Find the amount each graduate would donate at the end of the first year in order to reach their class goal of $100,000 at the end of 25 years.

(*Ans.* $C = $10.03)

C.2. A paper company with large land holdings asks your help in evaluating the cost of converting boiler furnaces to the use of peat as fuel. The plant currently consumes the following amount of fuel annually: oil: 2,000,000 barrels per year at $35 per barrel the first year. The conversion to burn peat will cost $40,000,000 and have a life of 20 years, and is estimated to save 50 percent of the cost of the oil now consumed. Salvage value of the conversion at EOY 20 equals $5,000,000. Oil costs are expected to rise at 9 percent per year compounded, and the annual savings will rise proportionately. Use $i = 15$ percent.
  (a) What is the amount of anticipated fuel-cost saving for the first year only? (Assume the expenditure for the conversion occurs right now, and the savings accrue during the year and are credited at the end of each year.)
  (b) Find the future worth, F, of the savings in fuel costs over the full 20-year period.
  (c) Compare the net future worth at EOY 20 of the conversion cost (including salvage value) versus fuel savings and tell whether the fuel savings justify the cost of conversion.

C.3. A firm of consultants needs computer services. They can either (A) purchase a computer and operate it themselves, or (B) lease the computer services from an existing firm. Find the equivalent future net worth at EOM 60 for each alternative. In either case $i = 1.5$ percent per month, and the system will last 60 months. All costs are charged at the EOM.

Alternative A.   Purchase computer.

cost of purchase = $300,000

operating and maintenance cost for the first 6 months = $10,000/mo

O & M costs increase by 1%/mo each month after the first 6 mo.

($10,100 at EOM 7, $10,201 at EOM 8, $10,303 at EOM 9, etc.)

salvage value at EOM 60 = $50,000

Alternative B. Lease computer services from an existing firm at an equivalent monthly cost of $20,000 per month.

(*Ans.* alternative A, F = $1,877 K; alternative B, F = $1,924 K)

C.4. Calculate the equivalent NFW of the cost of insulated windows and compare to the equivalent NFW cost of energy savings for a proposed new office building.

$$\text{extra cost of insulated windows} = \$302,000$$

$$\text{extra annual maintenance cost for insulated windows} = \$5,000/\text{yr EOY 1}$$

$$\text{gradient, annual maintenance over the 40-yr life} = \$500/\text{yr/yr}$$

$$\text{expected life of installation} = 40 \text{ yr}$$

$$\text{savings in energy costs the first year} = \$36,000 \text{ at EOY 1}$$

$$\text{expected increase in annual energy costs (and savings)}$$

$$\text{over the 40-yr life of the installation} = 4\tfrac{1}{2}\%/\text{yr}$$

Assume i = 14 percent.

Problem Group D

Both arithmetic and geometric gradients; comparison.

D.1. A contractor is considering two alternate plans to set aside a certain amount every year to replace a tractor. Plan 1 is to set aside $1,000 per year beginning at the end of the second year and continue increasing the amount $1,000 per year until the tractor is scheduled for trade-in at the end of the tenth year (nine payments).

   Plan 2 is to set aside $2,000 now, at the *beginning* of the first year and increase this amount by 25 percent every year until the end of the ten years (11 payments). In both cases the funds are invested at 10 percent interest. How much would be in each fund at the end of the tenth year?

(*Ans.* F_1 = $59,370; F_2 = $116,900)

D.2. A building project is brought to you for analysis. The costs and income are estimated as follows.

		End of year
Construction	$1,000,000	0
O & M	80,000/yr	1 through 20, increasing $5,000 the second year and $5,000/yr each year thereafter
Gross income	200,000/yr	1 through 20, increasing 8% the second year and 8% more each year thereafter
Resale value	800,000	20

Find the net gain or loss for the project at the end of 20 years by summing the future values of all costs and income. Assume the owner can lend and borrow at 10 percent interest.

D.3. A contractor client of yours set up a sinking fund (savings account) to finance purchases of new equipment. The fund now has $5,758 in it, and the contractor is earning a nominal $7\frac{1}{2}$ percent interest compounded monthly on the balance in the account. He intends to set aside $0.10 for every cubic yard of earth moved in order to build up the fund. He is currently moving $50,000$ yd^3 of earth a month and makes deposits at the end of every month.
  (a) At the end of another 24 months, how much will be in the account?

  (*Ans.* $135,700)

  (b) If the deposits are increased by 10 percent per month (geometric gradient), how much will the account contain in 24 months?

  (*Ans.* $470,100)

Problem Group E

PTZ ≠ ETZ.

E.1. If $1,000 per year is deposited at the end of years 1 through 5 at 6 percent and $2,000 per year is deposited at the end of years 3 through 7 at 8 percent, how much will the investments be worth at the end of year 10 if each deposit continues to bear interest compounded at the same rates as obtained when deposited?

  (*Ans.* $22,320)

E.2. As a consulting engineer, you plan to build your own office complex in a few more years for your own occupancy and use. You have just heard of a choice lot in a new office park that is available at a bargain price of $30,000 cash payable right now. You call your local banker and he offers to loan you the money at 12 percent (annual rate) compounded monthly. No repayments need be made until the building is constructed and a mortgage taken out.

  You plan on starting construction 35 months from now on a building that will cost $100,000. To finance the construction, your banker offers to loan an additional $20,000 per month for five months with interest also at 12 percent (annual rate) compounded monthly until repaid.

  At the end of the five-month construction period you expect to take out a mortgage in an amount sufficient to repay the banker the entire principal plus accumulated interest to date on the original $30,000 for the land, plus the $100,000 for the building. The mortgage will have a 9 percent interest rate and will be repaid in 240 equal monthly payments.
  (a) How much do you owe the banker at the end of the construction period?
  (b) How much are the monthly payments on the mortgage?

E.3. Your client, the utility board, asks that you determine the rate to charge customers in a new residential development in order to pay for a new waste-water treatment plant over the next 20 years. The following data are provided.

  Plant cost is $1,000,000. This cost will be repaid to the utility board through a special annual end-of-year assessment to each customer. The number of customers in the development is 1,000, and interest on the unpaid balance is $i = 7$ percent. To encourage the developer to build in their area, the board members agreed to charge as little as possible the first year and then increase the rate by 4 percent per year over the balance of the 20-year period.

(a) Find the annual charge per customer that must be charged the first year.

   (*Ans.* $69.22 at EOY 1)

(b) Find the annual charge per customer charged for the twentieth year.

   (*Ans.* $145.84 at EOY 20)

Problem Group F

Periodic payments, gradients.

F.1.   A water treatment plant needs a new pump with costs anticipated as follows.

cost new = $100,000

major maintenance (MM) is required every four months, with the first MM at EOM 6. Subsequent MM will occur at EOM 10, 14, 18, and so on. Cost of MM at EOM 6 is $4,000. The cost of MM increases by 5 percent each time it is performed. The life of the pump is 120 months, at which time the salvage value is $5,000; $i = 0.75$ percent per month. Find the NFW at EOM 120.

(*Ans.* NFW = $598.8)

F.2.   You are asked to estimate the cost of financing the construction of a proposed office building. The amounts tabulated below will be borrowed from the bank at the scheduled times in order to pay the contractor for work done as construction progresses. When the job is complete, the bank will be repaid from proceeds from the permanent mortgage financing. The amounts shown are in thousands of dollars.

EOM	Amount borrowed this time	EOM	Amount borrowed this time
1	$100	9	$180
3	120	11	200
5	140	13	220
7	160		

The bank lends at 1.5 percent per month, and the permanent mortgage money will be available to repay the bank at EOM 16. Find the cost of the construction financing for this project (find the total amount of interest paid to the bank).

Problem Group G

Find *n*.

G.1.   A married couple want to buy a house priced at $100,000 and ask your advice. They would like to stay out of debt, so they plan on getting a long-term option to buy, and then saving up enough cash to pay for the house in one lump sum. They can save $1,000 per month in a savings account that compounds interest at 0.5 percent per month compounded monthly. Meanwhile, due to inflation and increasing demand, the terms of the option provide that the price of the house will escalate 0.5 percent month.

(a) How long will it take before the amount in the savings account equals the rising price of the house?

(b) The couple now live in a trailer costing $200 per month, and if they bought the house now with no down payment and a 9 percent mortgage, they could use the total $1200 ($1000 + $200) to pay off the mortgage. Which method pays for the house more quickly, (a) save up and pay cash, or (b) buy now, move in with nothing down, and pay off the mortgage at $1200 per month with a mortgage at 9 percent (0.95%/month)?

(*Ans.* buy now, $n = 166$ months)

CHAPTER 10

Rate of Return Method (ROR)

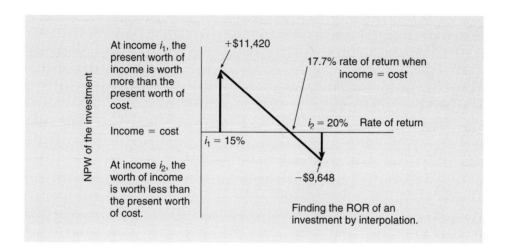

At income i_1, the present worth of income is worth more than the present worth of cost.

Income = cost

At income i_2, the worth of income is worth less than the present worth of cost.

NPW of the investment

+$11,420

17.7% rate of return when income = cost

$i_2 = 20\%$ Rate of return

$i_1 = 15\%$

−$9,648

Finding the ROR of an investment by interpolation.

KEY EXPRESSIONS IN THIS CHAPTER

ROR = Rate of Return is the equivalent interest rate earned or paid on the principal balance of an investment or loan.

BASICS OF ROR

The rate of return (ROR) on an investment is a concept that is already familiar to most people. A return of $10 interest per year on a deposit of $100 is easily understood to imply a rate of return of 10 percent. Therefore, those who make decisions on where to spend the money usually appreciate the ease, speed, and convenience of comparisons presented in terms of rates of return. The project with a return of 18.2 percent obviously is more desirable than an alternative project with a return of only 9.3 percent (providing the risk and other considerations are comparable). Because the results are easily understood, the rate of return method together with the companion incremental rate of return method are powerful tools for comparing alternatives and providing decision makers with the information required to obtain a better investment of the public or private dollar, and more efficient allocation of other resources of all types.

In order to obtain a positive rate of return, an investment must yield a positive net income; that is, the total income cash flow must exceed the total cost cash flow. For example, an investment of $1,000 today that will yield an income of $1,100 one year from today has a rate of return of 10% (calculated as ROR, $i = (F/P)^{1/n} - 1 = 1,100/1,000 - 1 = 0.10 = 10\%$). If the $1,000 investment only yields $1,000 after one year, then the rate of return is zero, whereas if the income is less than $1,000, say $900, then the rate of return is a negative 10% [$i = (900/1,000) - 1 = -10\%$].

In general terms the rate of return is the interest rate, i, at which the equivalent cash flow in equals the equivalent cash flow out: ("equivalent" here means that the time value of money has been accounted for, and future payments are appropriately discounted.) More specifically, the rate of return is the rate, i, at which:

the PW of all costs = PW of all incomes, and NPW = 0
the AW of all costs = AW of all incomes, and NAW = 0
the FW of all costs = FW of all incomes, and NFW = 0

If incomes are viewed as "benefits," while payments out are termed "costs" (both of which are expressed in equivalent terms), then the rate of return is the interest rate at which the sum of all benefits equals the sum of all costs. This condition is satisfied when any of the following occur (assume that the "benefits" and "costs" refer to the PW, or AW, or FW of each benefit and cost):

$$benefits = costs$$
$$benefits - cost = 0$$
$$benefits/costs = 1$$

EXAMPLE 10.1

A $1,000 cost results in a lump sum payment of $2,500 at the end of 10 years. What is the rate of return (ROR) on this investment?

Solution.   The solution is obtained by formulating the income and cost into *equivalent* amounts by PW (finding the FW or AW of both could be done with equal ease) and then solving for the value of i that makes the NPW $= 0$.

$$PW\ benefits = 2,500(P/F, i, n) = 2,500/(1 + i)^{10}$$
$$PW\ costs = 1,000$$
$$NPW = 2,500/(1 + i)^{10} - 1,000 = 0$$
$$1/(1 + i)^{10} = 1/2.5$$
$$(1 + i)^{10} = 2.5$$
$$i = \sqrt[10]{2.5} - 1$$
$$i = 0.09596 \quad or \quad \underline{9.60\%}$$

The problem could also be solved using NFW, or NAW.

Calculators

After the initial analysis and sketching of the cash flow diagram, simple rate-of-return problems can be solved quickly by use of inexpensive hand held calculators programmed for time value problems. Before relying on these calculators, the student should be thoroughly familiar with the step-by-step methods outlined below.

STEP-BY-STEP PROCEDURE FOR FINDING ROR

The foregoing simple Example 10.1 could be solved directly because there was only one unknown value of i. More commonly the unknown value, i, appears several times in several equations, and the most practical general method of solution is by successive approximation (trial and error). The following step-by-step procedure is recommended. As happens in the usual case, this procedure assumes that the investment cost occurs before the income returns. Therefore, the graphic plot of net worth versus i value curves downward to the right, and higher values of i are associated with lower values of net worth. If the situation were the reverse and the income came before the cost (such as when a borrower borrows money and later repays), the graph would curve upward to the right.

Step 1. Make a guess at a trial rate of return, i.

Step 2. Count the costs as negative ($-$) and the income or savings as positive ($+$). Then use the trial i value to find the equivalent net worth of all costs and income. Use either present worth, annual worth, or future worth.

Step 3. Using the trial i value, if the income from the investment (counted as positive value) is worth more than the cost of the investment (negative value), then the equivalent net worth is positive at this trial i value. When the cost precedes the income (usual case), the graphic plot of net worth versus i value curves downward to the right. When the trial i yields a positive net worth, we can surmise that the trial i is too low, and the correct i value is higher than the trial rate, i.

Step 4. Adjust the trial i value upward, and proceed with steps 2 and 3 again until one value of i is found that results in a positive ($+$) equivalent net worth, and another value of i is found with a negative ($-$) equivalent net worth.

Step 5. Interpolate between the two trial i values (one yielding a positive net worth and the other a negative), to solve for the correct value of i. Remember that the graph of NPW versus i is a *curve,* so the two trial values of i must be reasonably close in order for the interpolation to yield accurate results.

An example involving two lump sum cash flows as well as a periodic series cash flow will serve to illustrate the procedure.

EXAMPLE 10.2

Assume a bond sells for $9,500. The bondholder will receive $600 per year interest as well as $10,000 (the face amount of this bond) at the end of 10 years. Find the rate of return.

Solution.   Find the interest rate at which the present worth of the income ($600 per year for 10 years, plus $10,000 at the end of 10 years) equals the present worth of the cost (−$9,500). The steps outlined above are applied as follows.

Step 1.   Make a trial selection of 7 percent for a trial rate of return. (The yield is $600 per year on an investment of $9,500, plus an extra $500 at the end of 10 years, so the i probably will be a little above $600/9,500 = 0.063$ or 6.3 percent.)

Step 2.   Find the present worth of all costs and income using $i = 7$ percent. The income consists of $600 per year for 10 years, plus $10,000 at the end of year 10. The cost is one $9,500 payment due right now.

$$\text{income } P_1 = \$600/\text{yr} \underbrace{(P/A, 7\%, 10)}_{7.024} = \quad \$4,214.40$$

$$\text{income } P_2 = \$10,000 \underbrace{(P/F, 7\%, 10)}_{0.5083} = \quad 5,083.00$$

$$\text{cost } P_3 = -\ 9,500.00$$

$$\text{net present worth} = -\$\ 202.60$$

Step 3.   This graph curves downward to the right. Since the net present worth is negative, the trial rate, 7 percent, must be too high. The present worth of income ($9,297.40) is less than the present worth of costs ($9,500), or at 7 percent you are offered an opportunity to pay $9,500 for an income worth $9,297.40.

Step 4.   Adjust the estimate downward to, say, 6 percent. Then, the net present worth of the income minus cost is

$$P_1 = \$600/\text{yr}(P/A, 6\%, 10) = \$600 \times 7.360 = \quad \$4,416.00$$

$$P_2 = \$10,000(P/F, 6\%, 10) = \$10,000 \times .5584 = \quad 5,584.00$$

$$P_3 = -\ 9,500.00$$

$$\text{net present worth} = +\$\ 500.00$$

Two values of i have been found, each resulting in a different sign (+ or −) thus bracketing the value of NPW $= 0$. An interpolation to the correct value of i (the value of i at which the NPW equals zero) is found as follows.

i value	NPW
6%	+$500
ROR i	0
7%	− 202.60

$$\text{rate of return, } i = 6\% + \frac{500}{500 + 202.60} \times 1\% = \underline{\underline{6.7\%}}$$

The 6.7% return is called the bond's "yield to maturity" because it includes redemption income when the bond matures. (The "current yield" is $600/$9,500 = 6.3% and is frequently used in financial circles with reference to bonds whose maturity date is fairly distant. Bonds are discussed in more detail in Chapter 17.)

CONCURRENT INCOME AND COSTS

The rate of return method can be used where both costs and income occur during the same periods. To save a step, use the net payment for those periods, as in the following example.

EXAMPLE 10.3

An entire fleet of earthmoving equipment may be purchased for $2,400,000. The anticipated income from the equipment is $920,000 per year with direct expenses of $420,000 per year. The market value after five years is expected to be $1,000,000. What is the rate of return?

Solution

Step 1.   Assume $i = 10$ percent for first trial.

Step 2.   Find present worth of all income and costs.

income	$920,000/yr
expenses	− 420,000/yr
	$500,000/yr net annual income

The present worth of this net income for five years is

$$P_1 = \$500,000/\text{yr} \underbrace{(P/A, 10\%, 5)}_{3.7907} = \$1,895,350$$

Market value in five years is $1,000,000. The present worth of $1,000,000 income expected five years from now is

$$P_2 = \$1,000,000(P/F, 10\%, 5) = \$620,920 \text{ present worth of market value}$$

Total present worth of all income at 10 percent is

$$
\begin{aligned}
P_1 &= \quad \$1,895,350 \\
P_2 &= + \quad 620,920 \\
\overline{P_{1+2}} &= \quad \$2,516,270 \text{ total}
\end{aligned}
$$

This means that if a 10 percent rate of return is acceptable, you can afford to pay as high as $2,516,270 cash for this project. You would then receive $500,000 per year (part of which is interest and part is principal) for five years plus $1,000,000 at the end of the fifth year, and this income yields 10 percent return on the balance of all funds remaining in the investment at the end of each year. Actually this fleet is being offered for only $2,400,000. So the actual rate of return is higher than 10%.

Step 3.   Conclusion: Assuming a trial i of 10 percent, the income (with a PW of $2,516,270 at 10 percent) is worth more than the cost (with PW of $-\$2,400,000$), resulting in a positive NPW of $(2,516,270 - 2,400,000 =)$ $+\$116,250$, so the trial i estimate is too low.

Step 4.   Revise the trial to 12 percent and repeat steps 2 through 3. The PW at 12% is found as

$$P_3 = \qquad \text{net income } \$500,000/\text{yr } \underbrace{(P/A, 12\%, 5)}_{3.6047} = \$1,802,350$$

$$P_4 = \text{market value in 5 yrs } \$1,000,000 \underbrace{(P/F, 12\%, 5)}_{0.56743} = \underline{\quad 567,430}$$

$$P_{3+4} = \qquad \text{total PW of all income at } 12\% = \$2,369,780$$

Conclusion.   Since the PW of income at 12 percent is lower ($2,369,780) than the PW of cost ($-\$2,400,000$), resulting in a negative NPW, the rate of return is lower than 12 percent. Interpolation will give a final figure, as illustrated in Figure 10.1.

$$
\begin{array}{llll}
10\% \text{ PW income} = & \$2,516,270 & 12\% \text{ PW income} = & \$2,369,780 \\
\text{PW cost} = - & \underline{2,400,000} & \text{PW cost} = - & \underline{2,400,000} \\
& +\$ \quad 116,270 & & -\$ \quad 30,220
\end{array}
$$

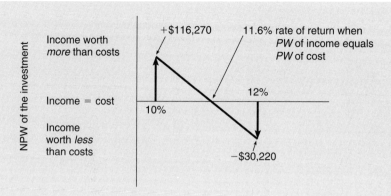

Figure 10.1 Interpolation to find rate of return.

$$i = 10\% + \frac{116,270}{116,270 + 30,220} \times 2\% = 11.59\%$$

$$i = \underline{\underline{11.6\%}}$$

This problem can also be worked by using net *annual* worth or net *future* worth.

Gradients on Income and Costs

Gradients present no unusual problems when finding the rate of return. Simply follow the standard procedure for determining the rate of return:

1. Assume a trial interest rate.
2. Find the equivalent worth of each payment item, treating gradient items the same as usual.
3. Sum the totals.
4. By trial, error, and interpolation, find the interest rate at which the sum of equivalent worths equals zero.

The next example illustrates gradients in a rate of return problem.

EXAMPLE 10.4

A bus company is for sale for $150,000. The net income this year is $50,000 but is expected to drop $6,000 per year next year and each year thereafter. At the end of 10 years the franchise will be terminated, and the company assets sold for $50,000. What is the interest rate of return on this investment?

Solution.   Try $i = 15$ percent.

$$P_1 = -\$150,000$$
$$P_2 = -G(P/G, 15\%, 10) = 6,000 \times 16.979 = -\$101,890$$
$$P_3 = A(P/A, 15\%, 10) = 50,000 \times 5.0187 = +\$250,950$$
$$P_4 = F(P/F, 15\%, 10) = 50,000 \times 0.2472 = +\$12,360$$
$$P_{total} = P_1 + P_2 + P_3 + P_4 \quad +\$11,420 \text{ for } i = 15\%$$

Try $i = 20$ percent.
$P_{total} = -\$9,648$ for $i = 20\%$

Interpolating,

$$i = 15\% + \frac{11,420}{11,420 + 9,648} \times 5\% = \underline{17.7\%}$$

Periodic Gradients, ETZ or PTZ, Financing

No special techniques are required in order to find the rate of return of an investment involving periodic gradients and financing. Simply organize the problem material in an orderly fashion and proceed with the step-by-step solution. An example follows.

EXAMPLE 10.5   **(Given P, A, G, C, i, r, n, find ROR i)**

A government agency needs more office space. They either can rent the space on the rental market or renovate an existing older office building that they already own. Find the ROR if they renovate the existing older building and count the rent saved as income over the next 40 years.

$$\text{market value of existing older building} = \$1,200,000$$
$$\text{cost of rehabilitation} = \$4,000,000$$
rehab loan available at 3% to be repaid in 10 equal annual installments, with first payment due
$$\text{at EOY } 1 = \$3,200,000$$
$$\text{rental value at EOY 1 (count as income)} = \$400,000$$
$$\text{geometric gradient in rental value} = 5\%/\text{yr}$$
$$\text{O \& M costs at EOY 1} = \$90,000$$
$$\text{geometric gradient in O \& M costs} = 5\%/\text{yr}$$

major maintenance (MM) every 5 yr, first MM
at EOY 3 = $20,000/5-yr period
gradient increase in major maintenance (MM
costs $30,000 at EOY 8, $40,000 at EOY 13, etc.) = $10,000/each/5 yr
resale value at EOY 40 = $100,000

Solution. The usual procedure is followed:

1. The equation for each item is written in terms of PW.
2. A trial i value is selected.
3. The NPW is totaled.
4. The ROR i value is sought corresponding to NPW = 0.

If the building is renovated, the $1,200,000 sale value of the existing building is committed to the renovation project and is counted as a cost of this project. Take a first trial i value of $i = 10$ percent.

		$i = 10\%$	$= 12\%$
Existing building	$P_1 = -1,200K =$	$-1,200$	$-1,200$
Renovation	$P_2 = -4,000K =$	$-4,000$	$-4,000$
Rehab loan	$P_3 = +3,200K =$	$+3,200$	$+3,200$
Repay loan	$P_4 = -3,200K(A/P, 3\%, 10)(P/A, i\%, 10) =$	$-2,305$	$-2,120$
(Rental $-$O & M)	$P_5 =$		

$[(400 - 90)/1.05](P/A, w, 40) = (310/1.05)(P/A, 4.762\%, 40) = $ 5,336
$[$where $w = (1.10/1.05) - 1 = 0.04762]$

$$P_5 =$$

$[(400 - 90)/1.05](P/A, w, 40) = (310/1.05)(P/A, 6.667\%, 40) = $ +4,094
$[$where $w = (1.12/1.05) - 1 = 0.06667]$
Periodic MM $[i_e = (1.10)^5 - 1 = 61.05\%]$
    $[i_e = (1.12)^5 - 1 = 76.23\%]$

	$P_6 = -20K(P/A, i_e\%, 8)(F/P, i\%, 2) =$	-39	-33
Gradient per MM	$P_7 = -10K(P/G, i_e\%, 8)(F/P, i\%, 2) =$	-28	-20
Resale EOY 40	$P_8 = +100K(P/F, i\%, 40) =$	$+2$	$+1$
Net present worth $P_{total} =$		$+966K$	$-78K$

Interpolating for the value of i at which NPW equals zero,

$$i = 10\% + [966K/(966K + 78K)] \times 2\% = 11.85\% = \underline{\underline{11.8\%}}$$

Conclusion. Counting the rent saved as income, the governmental agency will earn a rate of return of 11.8 percent on all funds invested in the rehabilitated building.

OTHER CONSIDERATIONS BESIDES ROR

Some organizations have employed the rate of return as the sole measure of desirability when comparing alternatives. Actually in exceptional cases the alternative with the higher ROR may *not* be the most profitable. (An investment of $1 now that returns $2 at EOY 1 yields a ROR of 100%, but only $1 of profit, whereas an investment of $10,000 now that returns $12,000 at EOY 1 yields a ROR of only 20% but a much larger profit of $2,000. Assuming the two investments are mutually exclusive alternatives, the *lower* ROR of this pair probably is the better investment for a minimum attractive rate of return (MARR) of less than 20%. If one alternative requires a higher initial investment than the other, an evaluation is needed of the rate of return on the increment of initial investment (the difference between the two initial investments). The return on this increment of investment is called the incremental rate of return (IROR) and is discussed in Chapter 11.

MULTIPLE RATES OF RETURN

For most practical problems involving an interest rate of return, there is only one solution for the ROR. However, there are a few instances where more than one ROR is possible. *This possibility is indicated when the cumulative cash flow changes signs more than once.* In the following example, a running total of the value of the cumulative cash flow is shown in the right-hand column. Note that it changes from negative to positive (first change of sign) and then from positive back to negative (second change of sign). This multiple change of sign signals the possibility (but not certainty) that more than one solution may occur.

EXAMPLE 10.6

An investment consists of the following expenditures (−) and incomes (+). Find the rate of return.

End of year	Net cash flow for this year	Cumulative cash flow
0	−$ 500	−$500
1	+ 1150	+ 650
2	− 660	− 10

Solution. The cash flow diagram is shown in Figure 10.2. The problem may be solved in the usual manner by setting the PW of the cost equal to the PW of the income, as follows.

$$\text{PW of cost} = 500 + \frac{660}{(1+i)^2} = \text{PW of income} = \frac{1150}{1+i}$$

Figure 10.2   Cash flow diagram with reversed flow.

To find the rate of return, this equation of PW cost + PW income is set equal to zero. This equation can be solved either algebraically or by trial and error. Solving algebraically gives the following results. Multiply by $(1 + i)^2$

$$P = +500(1 + i)^2 + 660 - 1150(1 + i) = 0$$
$$P = [(1 + i) - 1.1][(1 + i) - 1.2] = 0$$

This yields *two* roots,

$$i = 0.1 \quad \text{or} \quad \underline{10\%} \quad \text{and} \quad i = 0.2 \quad \text{or} \quad \underline{20\%}$$

Solving by the quadratic equation yields the same results, as does solution by trial and error.

A graph of the rate of return versus the PW of the total investment is plotted in Figure 10.3. Inspection of the graph indicates that if the initial expenditure were increased by $1 (to −$501), there would be only one solution, the tangent point of the parabola with the x axis at $i = 15$ percent. Further, if the initial investment were raised to more than $501, there would be no solution, since all total PW's of cost plus income would be negative regardless of the trial i employed.

One resolution of the dilemma posed by Example 10.6 is to report that the investment has a ROR of 15% plus a positive PW of $1.25 at that ROR. (See spreadsheet for numerical values).

In connection with the above discussion, the following observations may prove helpful:

1. Only one change or no change in the sign of the cumulative cash flow indicates there is only one solution i.

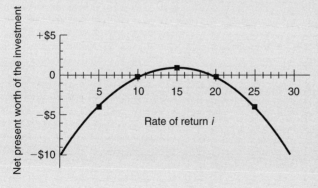

Figure 10.3   Graph of rate of return versus PW using a two-solution example.

2. It is theoretically possible to have any number of solutions, providing that the sign of the cumulative cash flow changes more than once. For example,

$$[(1+i) - 1.05][(1+i) - 1.10][(1+i) - 1.15] = 0$$

This "solution" will solve for $i = 5$ percent, $i = 10$ percent, and $i = 15$ percent. This "solution" may be multiplied out to find the equivalent cash flow, as follows:

$$+(1+i)^3 - 3.30(1+i)^2 + 3.63(1+i)^2 - 1.33 = 0$$

Any convenient multiple of these numbers will yield the same three interest rate solutions.

3. Multiple rates of return, intuitively, seem contrary to reason and common experience. However, multiple rate solutions do occur due to the implicit assumption that *all* the cash flow values (both positive income and negative costs) will earn the solution interest rate (the derived rate, i, that makes the net equivalent worth equation equal to zero). Actually, in order for any i to be valid, whatever income is taken out of the project temporarily must be invested at some externally available market rate, i, and earn interest until the income is returned to the project to pay costs when the signs change. Thus, if the maximum available *external* rate is $i = 10$ percent, and a multiple *internal* rate solution is $i = 20$ percent and 40 percent, then the solution must be modified to take into account the period of time during which the income is deposited in an external account bearing interest at the lower 10 percent rate, waiting to pay later costs on the project.

SPREADSHEET APPLICATIONS

	A	B	C	D	E	F	G	H
1	**EXAMPLE 10.6:**							
2								
3		An investment consists of the following expenditures (−) and incomes (+).						
4		Find the rate of return.						
5								
6								
7						Net cash flow	Cumulative	
8		End of year				for this year	cash flow	
9		0				− $500	−$500	
10		1				+ 1150	+ 650	
11		2				− 660	− 10	
12								
13								

(Spreadsheet continued)

	A	B	C	D	E	F	G	H
14	SOLUTION:							
15								
16		Set the PW of the cost equal to the PW of the income, as follows:						
17								
18		PW of cost = 500 + [660/(1 + i)²] = PW of income = 1150/(1 + i)						
19								
20		To find the rate of return, this equation of PW cost + PW income is set equal to zero.						
21								
22		$P = +500(1 + i)^2 + 660 - 1150(1 + i) = 0$						
23								
24		To solve for the roots by trial and error, enter a range of interest rates and inspect to see						
25		where the curve crosses the x-axis.						
26								
27		A graph of the rate of return versus the PW of the total investment is plotted below.						
28								
29		i (%)	P($)					
30		0	−$10.00					
31		5	− $3.75					
32		10	$0.00					
33		15	$1.25					
34		20	$0.00					
35		25	− $3.75					
36		30	−$10.00					
37								
38								
39								
40								
41								
42								
43	**CELL CONTENTS:**		The following cell contents will allow us to evaluate the problem with					
44	a spreadsheet.							
45		**cell C30: = − (500*(1 + B30/100)^2 + 660 − 1150*(1 + B30/100))**						
46								
47	CONCLUSION:							
48								
49		$i = 0.1$ or 10% and $i = 0.2$ or 20%						

Rate of Return vs. Present Worth (Two-solution example)

SUMMARY

The "rate of return" (ROR) on investments is defined as the interest rate for which equivalent benefits from the investment equal the equivalent costs. The cash flow analysis may be equated in either of the following forms:

net discounted benefits − net discounted costs = 0

net discounted benefits/net discounted costs = 1

where the discounting for time value of money may be done by your choice of PW, AW, or FW.

For a positive rate of return ($i > 0$), benefits (income or savings) must exceed costs. Since there are often several terms in the cash flow equation involving the unknown interest rate, i, a trial-and-error approach customarily is used.

Step 1. Make a guess at a trial rate of return.

Step 2. Solve one of the above forms of the cash flow equations.

Step 3. Compare the equivalent worth of benefits to costs. If investment cost precedes income (usual case) and equivalent worth of income is higher than the worth of costs (equivalent net worth is positive, $+$), the trial value of i is too low.

Step 4. Choose another trial rate of return and proceed with steps 2 and 3 until the value of i representing a net worth of zero is bracketed. Then solve for the correct i by interpolation.

When comparing alternatives, we must consider several factors:

1. Comparisons must be on the basis of equivalent costs and benefits. Where alternatives have different life spans and/or replacement costs, an appropriate adjustment is necessary.

2. While the alternative with the highest ROR is usually the most profitable, the ROR of the *incremental* investment must be taken into account. This subject is presented in Chapter 11.

Once in a while problems occur with more than one reversal of signs in the cash flow, and consequently more than one i-value solution to the cash flow equation is possible. In such cases, derived i values above the MARR value usually are discarded.

PROBLEMS FOR CHAPTER 10, RATE OF RETURN

Problem Group A

Given three or more of the variables, P, A, F, n, find the rate of return, i, for each proposal.

A.1. A shopping center costing $2,000,000 will have an estimated net annual income of $200,000. In 10 years it is estimated the center can be sold for $1,500,000. What is the expected rate of return?

(*Ans.* $i = 8.31\%$)

A.2. A parking lot is proposed for downtown. The land, clearing, and development costs are $150,000. The land can be sold in 10 years for $100,000. Parking revenue is expected to net $15,000 per year. What is the rate of return?

A.3. A parking garage for 160 cars is proposed for West University Avenue opposite the campus at a total cost of $250,000. The net annual income (after expenses) is expected to be $150 per car space per year for 20 years. At the end of 20 years the estimated value of the property is $200,000. What percent return would this yield on the investment?

(*Ans.* $i = 9.23\%$)

A.4. A mortgage note for $10,000 at 6 percent calls for payment of *interest only* at the end of every year ($600/yr) for 10 years. At the end of the tenth year the $10,000 will be paid off, together with the $600 interest for that year. The mortgage note is offered for sale at $7,000. What is the *actual* interest rate at this price?

A.5. Find the percent return on the investment of a $10,000 cash down payment on a $60,000 power shovel. Estimated gross income is $20,000 per year. Estimated operating and mainte-nance expenses are $8,000 per year, not including payments. Payments on the $50,000 bal-ance may be made over a 10-year period at 7 percent on the unpaid balance. The shovel can be sold in 12 years for an estimated $10,000. (*Hint:* Try 50%.)

(*Ans.* $i = 49.6\%$)

A.6. A certain bank pays interest on certificates of deposit at the time of deposit rather than wait-ing $3\frac{1}{2}$ years. A depositor of $10,000 gets a check for $2,275 at the time of deposit and $3\frac{1}{2}$ years later the original $10,000 is returned. Assume monthly compounding.
(**a**) Calculate the nominal annual rate of return on this system.
(**b**) What is the nominal annual rate of return if the $2,275 is paid at the end of $3\frac{1}{2}$ years in-stead of the beginning?

A.7. A new bus route is proposed with the following estimated costs and income:

$$2 \text{ new buses at } \$50,000 \text{ each} = \$100,000$$
$$\text{operating costs} = \$1/\text{mi}/2 \text{ buses (or } \$0.50/\text{mi/bus)}$$
$$\text{average speed including stops} = 10 \text{ mph}$$
$$\text{operating hours} = 6 \text{ A.M.–9 P.M. 7 days/week}$$
$$\text{overhead charged to this route} = \$10,000/\text{yr}$$
$$\text{average total number of riders on this route}$$
$$\text{(includes both buses) daily} = 900 \text{ at } \$0.25 \text{ each}$$
$$\text{resale value of buses after 10 yr} = \text{negligible}$$

What is the rate of return if these estimates prove correct?

(*Ans.* $i = 11.56\%$)

A.8. The public works department requests permission to purchase a new mower for $10,000. They estimate it will save $1,200 per year over the 13-year life of the machine. No salvage value is expected at the end of that time.
(**a**) What rate of return would this savings represent on the investment of $10,000?
(**b**) Funds are available for the purchase from a new $7\frac{1}{2}$ percent bond issue. Would you rec-ommend that the request to purchase be approved? Why?

A.9. The cost of heavy construction dredging equipment is rising at the rate of 6 percent com-pounded every year. A sum of $400,000 is now on hand. Two alternatives are available: (a) One $400,000 unit if purchased now should serve the needs of a growing contractor for the next 10 years. (b) One $250,000 unit may be purchased now and another identical unit purchased in five years at the price increase defined above. At what interest rate must the extra $150,000 left on hand be invested in order to provide the needed cash in five years? Ne-glect obsolescence costs and comparative operating costs.

(*Ans.* $i = 17.4\%$)

A.10. An investor friend of yours purchased some stock six years ago and kept the following cash flow record of all expenditures and income. What rate of return did the investment yield?

EOY		
0	$3,476	Purchase stock (including commission)
1	142	Dividend
2	215	Dividend
3	222	Dividend
4	375	Dividend
5	350	Dividend
6	4,013	Sale of stock (net after commission)

Problem Group B

Involves arithmetic gradients.

B.1. A contractor asks your advice on the purchase of a new dump truck that is available for $40,000. You estimate that the net profit earned by the truck the first year will be $10,000. However, with increasing maintenance and repair costs, and decreasing productivity you estimate that the net profit will decline by $600 per year each year. Thus you estimate $9,400 the second year, $8,800 the third, and so on. The contractor expects to keep the truck for 10 years and sell it for $5,000 at the end of the tenth year. Estimate the rate of return for the contractor. (*Hint:* Try 15%.)

(*Ans. i* = 16.0%)

B.2. A dragline is available for $75,000. Net income from available work is estimated at $25,000 this year, but due to rising costs, competition, and obsolescence, it is expected to decrease $3,000 per year in each succeeding year. At the end of 10 years the dragline should be worth about $25,000. What is the rate of return on this investment? (*Hint:* Try between 15 and 20%.)

B.3. A private utility company is considering the feasibility of servicing a large, new subdivision. You are provided with the following income and cost projections, and are requested to find the rate of return, *i*. (*Hint:* Try 10%.)

(*Ans. i* = 10.1%)

EOY		EOY capital cost	EOY net operating income
0	Begin design	$ 50,000.00	
1	Begin construction	300,000.00	
2	Complete construction Commence operation	500,000.00	
3	Operations underway, subdivision in growth stages		−$100,000.00

EOY	EOY capital cost	EOY net operating income
4		− 60,000.00
5		− 20,000.00
6		+ 20,000.00
7		+ 60,000.00
8		+ 100,000.00
9	Subdivision fully developed (change of gradient)	+ 140,000.00
10		+ 145,000.00
11		+ 150,000.00
.		(Increase $5,000.00 per year through thirty-fifth year)
.		
.		
35		+ 270,000.00
35	Salvage value	+ 500,000.00

Problem Group C

Involves geometric gradients.

C.1. A small shopping center can be constructed for $500,000. The financing requires $50,000 down and a $450,000, 30-year mortgage at 10 percent with equal annual payments. The gross income for the first year is estimated at $60,000, and is expected to increase 5 percent per year thereafter. The operating and maintenance costs should average about 40 percent of the gross income and rise at the same 5 percent annual rate. The resale value in 30 years is estimated at $1,000,000. Find the rate of return on this investment over the 30-year life. (*Hint:* Try between 10 and 20%.)

(*Ans.* $i = 14.1\%$.)

C.2. An engineering firm is considering trading in a used transit on the purchase of a new electronic distance measuring device (DMD). Find the rate of return on the new DMD, given the following data.

actual current market value of used transit = $1,000
list price of new DMD = $7,000
trade-in allowance by dealer for transit = $1,500
balance to be financed by dealer (amortized in 36 monthly
payments at 1%/month interest on balance) = $5,500
O & M costs for the first month, charged at EOM 1 = $1,000
monthly geometric gradient increases in O & M costs = 1.5%/month
monthly income from DMD = $1,640/month

The DMD will be used for 60 months and then sold at EOM 60 for $2,000. (*Hint:* Try 1.5%.)

Problem Group D

ETZ ≠ PTZ.

D.1. A 20-year loan to the owners of flood-damaged property provides for no interest to be paid for the first four years and for 2 percent interest to be paid from the fifth through the twentieth year. Loans in Needeep County amounted to a total of $50,000,000. The total interest payments are scheduled at $1,000,000 per year, payable at the end of the fifth through the twentieth year, with the $50,000,000 principal also due at the end of the twentieth year.

   (a) What is the true interest rate received by the lending agency?

   (Ans. 1.57%)

   (b) If this lending agency is paying 6 percent for the money it loans out, what is the PW of the subsidy?

   (Ans. P = $26,405,000)

D.2. Your client is interested in buying land and building an office park on it and operating the office park as an investment for a period of about 10 years. She asks you to determine the before-tax rate of return on the investment, using the following estimates of cost and income. *(Hint:* Find i between 10 and 20%.)

EOY		Cost	Income
		(all figures are ×1,000)	
0	Purchase land	−$ 100	$ 0
1	Complete the design	− 80	0
2	Complete construction	− 1100	0
3	Operating expenses and income	− 200	+ 100
4	Operating expenses and income	− 210	+ 150
5	Operating expenses and income	− 220	+ 200
6	Operating expenses and income	− 230	+ 250
7	Operating expenses and income	− 240	+ 300
8	Operating expenses and income	− 250	+ 350
9	Operating expenses and income	− 260	+ 400
10	Operating expenses and income	− 270	+ 450
10	Sell entire complex		+ 3,900

Problem Group E

Periodic gradient (arithmetic and geometric).

E.1. The equipment division of a city public works department is considering the purchase of an excavator, which they will charge out to the various projects that the city constructs with their own forces. To justify the purchase, the excavator must earn a rate of return of at least 0.75 percent per month.

$$\text{cost new} = \$100,000$$
$$\text{resale value at EOM } 120 = \$10,000$$
$$\text{O \& M costs} = \$8,000/\text{month}$$

gradient O & M = $100/month
periodic major overhaul every 9 months
(first at EOM 9, last at EOM 117) = $12,000/each
gradient periodic overhaul cost = $1,000/each/9 months
hours used per month = 150 hr/month
charge per hour for excavator (income) = $110/hour

Using this data, determine whether or not the excavator will earn at least 0.75 percent per month.

(*Ans.* At a trial $i = 0.75\%$, the NAW $= +\$407$/month, so the excavator earns more than 0.75%/month.)

E.2.   A shuttle bus service is proposed between downtown and the airport, and you are asked to find the rate of return, given the data below.

cost of shuttle bus new = $150,000
life of shuttle bus = 120 months
resale value at EOM 120 = $50,000
O & M cost = $2,000/month
gradient in O & M = $20/month/month
periodic overhaul every 12 months
(first at EOM 7, last at EOM 115) = $5,000 at EOM 7
gradient in periodic overhaul = 1%/month compounded
anticipated income = $5,800/month

(*Hint:* Try 1%.)

E.3.   The owners of a new planned unit development have installed a package sewage treatment plant that will be paid for by rates charged to the users.

package sewage treatment plant cost new = $1,000,000
life of the plant = 240 months
salvage value at EOM 240 = $0
O & M cost = $40,000/month
gradient in O & M = $400/month/month
periodic overhaul every 10 months
(first at EOM 4, last at EOM 234) = $100,000/each
gradient in periodic overhaul = $5,000/each/10 months
number of customers = 6,000
charge per customer = $16.00/month

The owners borrowed money at 1 percent per month interest in order to construct the plant, so while they do not expect to make a big profit, they do want the plant to earn a rate of return of at least 1 percent. Is the rate shown above sufficient to earn 1 percent rate of return?

(*Ans.* At a trial $i = 1\%$, the NAW $= +\$930$/month, so the plant earns more than 1% rate of return.)

E.4.   A concrete pump is being considered for purchase, with data estimated as shown.

$$\text{concrete pump cost new} = \$40,000$$
$$\text{life of the pump} = 120 \text{ months}$$
$$\text{resale value at EOM } 120 = \$5,000$$
$$\text{O \& M} = \$400/\text{month}$$
$$\text{gradient in O \& M} = \$8/\text{month/month}$$
$$\text{periodic overhaul every 6 months} = \$1,200/\text{each}$$
$$\text{gradient in periodic overhaul} = \$80/\text{each}/6 \text{ months}$$

The last periodic overhaul is done at EOM 120.

$$\text{expected production} = \$2,000/\text{yd}^3/\text{month}$$
$$\text{income from charges for use of pump} = \$0.85/\text{yd}^3$$

Find the rate of return earned by the pump. (*Hint*: Try 1%.)

CHAPTER 11

Incremental Rate of Return (IROR) on Required Investments

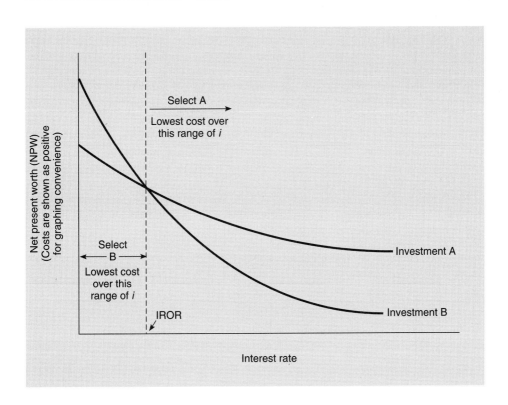

KEY EXPRESSIONS IN THIS CHAPTER

IROR = Incremental Rate of Return, the interest rate of return on each increment of added investment.

MARR = Minimum Attractive Rate of Return, the minimum acceptable rate of return. The MARR represents alternative outside investment opportunities. Investment opportunities yielding *less* than the MARR are normally considered neither competitive nor worthwhile.

RATE OF RETURN ON THE INCREMENT OF INVESTMENT

Circumstances, often beyond our control, frequently require us to make sizable expenditures that do *not* produce income or add to profit. For instance, new environmental regulations may require an electric power plant to install additional pollution-control devices, a construction contractor may receive orders from OSHA to buy new safety equipment, or a private car owner may need a new set of tires for the car. None of these expenditures produce income by themselves, but all are necessary expenditures required for the continued operation of some important function.

Competitive alternatives usually exist for fulfilling most needs in our relatively free economy. The power plant can consider several alternative methods of pollution control, the contractor can select from competing safety equipment, and the car owner can choose from a wide selection of tires. To remain competitive, the alternatives with higher first costs frequently have either longer lives, lower operating and maintenance costs, higher resale values, greater capacity, or some combination thereof. Can the car owner afford to pay $125 more in first cost for a set of tires in order to extend the expected life from 24 months to 36 months? Even though there is no direct dollar income or "return" resulting from this type investment, the incremental rate of return (IROR) method is well suited and often used to find the most economical alternative.

Instead of focusing on the rate of return earned by *every* dollar of investment, as in the rate of return method, the IROR method looks at the rate of return earned by each *additional* optional dollar of investment *above some required minimum.* Thus the IROR is the return on each optional incremental dollar of capital cost investment.

The problem now assumes the following form. If $1 *must* be invested with no measurable return expected, and an additional (incremental) $1 *may* be added to this investment with a positive return (savings in O & M, longer life, etc.) expected on the second $1, should the optional second $1 be invested in addition to the mandatory first $1? The following example illustrates this situation.

EXAMPLE 11.1 **(Given: Must purchase either A or B; find the incremental rate of return on the increment of cost)**

A car owner has worn-out tires and no alternative transportation. He can buy a set of tires expected to last two years for $200, or a set of tires expected to last four years for $325 ($125 more). Which is the better alternative?

Solution. The solution may be found through IROR. First clarify the problem situation by drawing the cash flow diagram of each alternative and the table of required expenditures A, optional additional expenditures B, and the incremental difference, B − A. For simplicity, assume that over a four-year period of time two sets of the two-year tires would be needed, and that the replacement cost for the two-year tires at EOY 2 remains at $200 and does not change. The increment of investment at PTZ is the difference in the first cost of

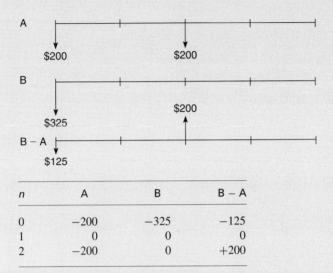

n	A	B	B − A
0	−200	−325	−125
1	0	0	0
2	−200	0	+200

the alternative sets of tires, or $325 − $200 = $125. If the extra $125 is paid out now, then there will be no need to buy another set of tires at the EOY 2, and a savings of $200 will occur at that point in time. Thus an investment of an incremental $125 now will produce a savings (or return) of $200 at EOY 2. The ROR on the incremental investment is called the IROR, and in this simplified case can be calculated as

$$F/P = (1 + i)^n$$
$$\$200/\$125 = (1 + i)^2$$
$$i = 1.6^{0.5} - 1 = 0.2649 = \underline{26.5\%} = \text{IROR}$$

An extra $125 invested in the four-year tires yields the equivalent of 26.5 percent rate of return. Thus, the IROR on the increment of investment is 26.5 percent.

Question: Could this problem be solved by the ROR, PW, AW, or FW methods?

Answer 1, ROR: It could not be solved by the ROR (rate of return) method. Since there is no income, the ROR would be infinitely negative.

Answer 2, PW, AW, or FW: The problem can be "solved" by any of these methods. However, the lower cost alternative (better buy) will be to buy the $325 tires if the interest rate is less than 26.5 percent, and then switch to the

$200 tires if an interest rate of higher than 26.5 percent is used. If exactly 26.5 percent interest is used, then the equivalent costs will be identical. For instance, examine the PW of each of the two alternatives at 10 percent.

$$\text{PW of two sets of 2-yr tires at } i = 10\%$$
$$P = -\$200 - \$200\underbrace{(P/F, 10\%, 2)}_{0.82645} =$$
$$P = -\$200 - \$165.29 = \underline{\underline{-\$365.29}}$$
$$\text{PW of one set of 4-yr tires, } P = \underline{\underline{-\$325.00}}$$

Conclusion.   The one set of *four-year* tires is the better buy by a significant margin.

Now try a solution at $i = 30$ percent.

$$\text{PW of two sets of 2-yr tires at } i = 30\%$$
$$P = \$200 + \$200(P/F, 30\%, 2) =$$
$$P = -\$200 - \$118.34 = \underline{\underline{-\$318.34}}$$
$$\text{PW of one set of 4-yr tires at } i = 30\%$$
$$P = \underline{\underline{-\$325}}$$

Conclusion.   The selection has reversed, and now the *two sets of two-year* tires are the better buy by a significant margin. If the PW were found at $i = 26.5$ percent, the PW values of the two alternatives would be equal.

Regarding the AW and FW methods, all three methods (PW, AW, FW) are related by constants. Therefore, whichever alternative costs less by one method also costs less by the other two methods. (An exception occurs when $n = \infty$, and the FW becomes infinitely large, whereas the PW and AW values normally do not.)

The *"do nothing" alternative.* Note that in Example 11.1 it was assumed that the car *must* be kept in operation and tires of some description *must* be purchased. The car owner could not afford to do nothing; therefore the "do nothing" alternative was not available. In this case if there were *no* negative consequences from doing nothing, then the "do nothing" alternative of not buying any tires at all would have been the more economical choice, since no costs are incurred.

Which set of tires should the owner buy? The better alternative depends upon the owner's Minimum Attractive Rate of Return (MARR). If the incremental $125 can be invested elsewhere at a higher rate than 26.5%, then buy the less-expensive tires and invest the $125 elsewhere at the higher rate. If the MARR is less than 26.5%, then invest the incremental $125 in the more expensive tires.

MARR, THE MINIMUM ATTRACTIVE RATE OF RETURN

In addition to the discussion of the MARR in Chapter 1, the following points should be reviewed at this time. By establishing a MARR, the following assumptions are made:

1. Sufficient funds are available to purchase the highest cost alternative under consideration, should the investment be justifiable.

2. If one of the lower cost alternatives is selected, the excess funds (the difference in cost between the selected lower cost alternative and the highest cost alternative) are assumed to be invested elsewhere earning the MARR.

Thus, the MARR involves consideration of opportunities to invest the *total available* funds, and not just the funds utilized in any particular alternative. The following paragraphs and examples illustrate these points.

OPTIONAL ORIENTATION FOR IROR SOLUTION MATRIX

Most of the IROR problems in this chapter are laid out so that each alternative investment (A, B, C, etc.) heads up a separate column, with the rows occupied by different types of payment items (cost new, O & M, etc.). Alternatively, the problems can be solved just as easily using a layout with the alternatives in the rows and the payment items in the columns.

The first part of Example 11.2 illustrates the use of alternatives placed in the columns. In Example 8.2 Continued the orientation is reversed for convenience, and the alternatives appear in the rows.

COMPARISON WITH EQUAL LIFE SPANS

The following example illustrates an IROR comparison of alternatives all having the same life span.

EXAMPLE 11.2   (Given two alternative sets of P, A, F, n, find IROR)

Assume a coal-burning electric power generating plant must select one pollution-control device from two suitable types available having the characteristics listed below. The "do nothing" alternative is not available.

Pollution-control device	A	B
Cost new	$1,000,000	$1,300,000
Operating costs	$ 300,000/yr	$ 265,000/yr
Life, n	20 yr	20 yr
Salvage value	$ 50,000	$ 60,000

Recommend the preferable alternative based on the incremental rate of return (IROR).

Solution. *Note:* Neither of these alternatives produces income for the owner except the small salvage value at EOY 20. Therefore, there is no need to determine the ROR, since it is obviously some large negative number. The problem now is to compare the extra benefits available from alternative B (save $35,000 per year in operating costs plus the small increase in salvage value) to the extra $300,000 initial cost of alternative B. To compare, first find the *incremental* rate of return (IROR) on the extra $300,000 investment in B, and then compare that IROR with the MARR.

   Comparing the two alternatives, the lowest initial investment is required by alternative A, at $1,000,000. Alternative B requires $1,300,000 capital outlay ($300,000 more than B). This raises the basic question, Since a minimum of $1,000,000 must be spent on pollution control anyway, should an extra $300,000 be invested in alternative B in order to save $35,000 per year in operating costs, plus receive $10,000 more in salvage value at EOY 20? Will the lower operating cost and higher salvage value of B more than compensate for the extra $300,000 initial cost of alternative B? To answer these questions an incremental rate of return (IROR), or rate of return on the increment of investment, is calculated. As the name implies, the method is similar to that used for rate of return, but only the *increments* of cost and saving (or income) need be considered. For this example, the columns are for the alternatives and the rows are for cost items. The IROR can be found by the following procedure, illustrated in Example 11.2.

1. Determine which alternative has the lower first cost, place the name of that alternative at the top of the left column and list all the payment items for that alternative in the same left column.

2. Find the next higher-first-cost alternative and list the corresponding payment items in a column just to the right of the previous list.

3. Subtract the cost of the left-hand column from the cost of the right-hand column. This is done by simply changing the signs in the left column and adding the figures in the left column to those in the right column. Write an equivalent-worth equation for each type of payment item in the

list, using your choice of present worth, annual worth, or future worth equations.

4. Select a trial i value for the IROR, use it in each equation, and sum the results of all the equations. If the sum turns out to be a positive value, then the trial i selected is too low (assuming the graph is normal and curves down to the right). Therefore, try a higher trial value of i, and repeat until the sum becomes negative. (If the second trial sum of the equations turns out to be a higher positive value, the graph may curve up to the right, or there may be an error in the problem setup or calculations.)

5. After finding the i values for which the equivalent-worth values bracket the zero point, interpolate to find the actual IROR value of i.

For Example 11.2, the stepwise procedure is followed below.

Step 1.   Alternative A is found to have the lower first cost of $1,000,000 compared to $1,300,000 for alternative B. The payment items for A are listed in the left-hand column.

Payment item	Alternative A
Cost new	−$1,000,000
Operating	−$ 300,000
Salvage value	+$ 50,000

Step 2.   The payment items for the next higher alternative B are listed in a column just to the right.

Payment item	Alternative A	Alternative B
Cost new	−$1,000,000	−$1,300,000
Operating	−$ 300,000	−$ 265,000
Salvage value	+$ 50,000	+$ 60,000

Step 3.   The items in column A (the left column) are subtracted from column B by changing the sign and adding. Present worth is selected for the equivalent-worth equations (either AW or FW would serve equally well) and appear as shown.

		Alternative A	Alternative B	
Cost new	$P_1 =$	+$1,000,000	−$1,300,000	=
Operating	$P_2 =$	(+ 300,000	− 265,000)$(P/A, i\%, 20) =$	
Salvage value	$P_3 =$	(− 50,000	+ 60,000)$(P/F, i\%, 20) =$	

Step 4.   Next, a trial value of i is assumed. In this case try $i = 9$ percent. This value is used in each equation and the results summed, as shown.

	Alternative A	Alternative B	B − A trial i value 9%
Cost new $P_1 =$	+ $1,000,000	−$1,300,000	= − $300,000
Operating $P_2 =$	(+ 300,000	− 265,000)$(P/A, i\%, 20) =$	+ 319,499
Salvage value $P_3 =$	(− 50,000	+ 60,000)$(P/F, i\%, 20) =$	+ 1,784
Total			+ $ 21,283

The total is a relatively small plus value, which indicates that 9 percent is a little lower than the sought-for IROR, but close. The next trial is made at $i = 11$ percent.

	Alternative A	Alternative B	B − A trial i values 9%	11%
Cost new $P_1 =$	+ $1,000,000	− $1,300,000	= −$300,000	−$ 300,000
Operating $P_2 = $	(+ 300,000	− 265,000)$(P/A, i\%, 20) =$	+ 319,499	+ 278,716
Salvage value $P_3 =$	(− 50,000	+ 60,000)$(P/F, i\%, 20) =$	+ 1,784	+ 1,240
Total			+$ 21,283	−$ 20,043

The resulting NPW now has changed from a positive to a negative value, indicating that the IROR value of i lies between 9 and 11 percent. Interpolation yields

$$i = 9\% + \frac{21,283}{21,283 + 20,043} \times 2\% = \underline{10.0\% = \text{IROR}}$$

Thus, the extra $300,000 required for alternate B over alternate A earns a rate of return (IROR) of 10.0 percent through lower operating costs and higher salvage value.

Assuming the firm has the additional $300,000 ready for investment, the decision makers must now consider whether other alternative investments would yield more than the 10.0 percent return available through investing the incremental $300,000 in alternative B. In response to this question a minimum attractive rate of return (MARR) is established. This MARR represents alternative investment opportunities of comparable risk and other attributes. Only comparable investment opportunities with a rate of return equal to or higher than the MARR are considered as worthwhile. Opportunities to invest at a rate *lower* than the MARR are usually rejected as not sufficiently profitable. If a lot of good opportunities occur to invest at the MARR rate or higher, and investment funds run short, the MARR can be raised to suit the current supply and demand for the firm's investment funds. (In addition to the rate of return, investment opportunities are evaluated for risk, compatibility with the investor's objectives, timing of cash flow, time required to service the investment, and other characteristics.) See Figure 11.1.

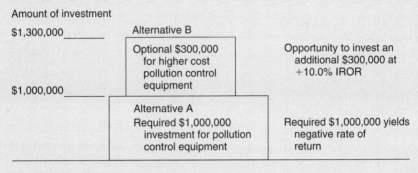

Figure 11.1   A $300,000 increment of investment earns 10.0% IROR.

COMPARING MULTIPLE ALTERNATIVES WHEN THE MARR IS KNOWN

Frequently more than two alternatives are available and need consideration. The IROR method is an excellent tool for analyzing such situations. The procedure for comparing multiple (more than two) alternatives is similar to the steps outlined previously in Example 11.2 for two alternatives. For illustration, the orientation is reversed, now with the alternatives occupying the rows, and the payment item names heading up the columns. The problem can be solved just as easily with either orientation.

   1. List all the available alternatives, and sort them by first cost. The lowest first cost alternative is at the top of the list, and the list is placed in the left column so that each alternative is at the left end of its own row.

   2. At the top of each remaining column, place a payment item name, such as Cost New, O & M, Salvage, etc.

   3. Fill in the matrix with the appropriate costs.

4. Begin with the top row, representing the alternative with the lowest first cost. Compare all the costs of this first alternative with the alternative in the next row lower (that has the second lowest first cost). Do this by assuming some i value, reducing each payment item to a common equivalent value by use of PW, FW or AW.

5. Find the IROR value for this first pair of alternatives.

6. Compare the IROR to the known MARR. If the IROR is higher, accept this second alternative. Then proceed to compare the second alternative to the third. If the IROR is lower than the MARR, reject the second alternative and compare the first to the third.

7. Continue making comparisons, rejecting any alternatives with IRORs that are less than the MARR. The last acceptable alternative is the preferred one.

EXAMPLE 11.2 (continued)

Assume that the power plant in Example 11.2 had five alternatives instead of the two listed. Assume further that each had the first costs listed below, and that the "do nothing" alternative was not a viable option in this case.

Alternative	First cost	Annual O & M costs	Salvage value at EOY 20
A	$1,000,000	$300,000	$50,000
B	1,300,000	265,000	60,000
C	900,000	325,000	40,000
D	1,150,000	270,000	60,000
E	1,400,000	260,000	65,000

The first step is to list the alternatives in order of ascending first cost, as follows:

Alternative	First cost	Annual O & M costs	Salvage value at EOY 20
C	$ 900,000	$325,000	$40,000
A	1,000,000	300,000	50,000
D	1,150,000	270,000	60,000
B	1,300,000	265,000	60,000
E	1,400,000	240,000	65,000

Then assume that additional data on operating costs, salvage values, and so forth are provided (not shown). From these values the 10 IROR values are derived (calculations not shown) and listed in matrix table form as follows:

Cost of alternative	Alternative	A	D	B	E	
$ 900,000	C	24.7% (A−C)	21.6% (D−C)	14.0% (B−C)	16.2% (E−C)	←This line shows IROR of A, D, B, E, compared to C
1,000,000	A		19.5% (D−A)	10.0% (B−A)	13.9% (E−A)	
1,150,000	D			−3.6% (B−D)	10.4% (E−D)	
1,300,000	B				2.3% (E−B)	
1,400,000	E					

The process of selecting the better alternative begins with the lowest cost alternative C at the top row of the matrix. The IROR values on the row to the right of C are examined and the largest one selected. This is the 24.7 percent that can be earned by investing $100,000(A − C) in addition to the required $900,000 expenditure for minimum first cost pollution-control device C. If the MARR is less than 24.7 percent, then alternative A is the better selection over C. If the MARR is greater than 24.7 percent, then C is the better alternative, and none of the other alternatives or rows need be examined. *Whenever the MARR exceeds the IROR between the lowest cost alternative and all other alternatives* (MARR > the highest IROR in the top row of the matrix), *the lowest cost alternative is the better investment.* Similarly, as we proceed down into the matrix, wherever the MARR exceeds all the values on any row, the alternative for any given row is the better investment. Assume that in Example 11.2 the MARR < 24.7 percent, and alternative A is temporarily accepted. Then is it profitable to invest more than the $1,000,000 needed for A?

Cost of alternative	Alternative	A	D	B	E	
$ 900,000	C	24.7%	21.6%	14.0%	16.2%	
1,000,000	A		19.5%	10.0%	13.9%	← This line shows IROR of D, B, E compared to A
1,150,000	D			−3.6%	10.4%	
1,300,000	B				2.3%	

If MARR < 24.7 percent, reject C and temporarily accept A, the higher cost alternative between A and C, and proceed down to line A. If MARR > 24.7 percent, select C and end search.

Proceed to row A and compare the IROR values to the right of A (D − A), (B − A), and (E − A). The highest value is 19.5 percent for the increment (D − A), which indicates that an additional investment of $150,000 required to purchase D over the $1,000,000 required for A would yield an IROR of 19.5 percent. If the MARR > 19.5 percent (but less than 24.7 percent found in row C), proceed no further; invest only in A at $1,000,000.

For MARR < 19.5 percent, alternative D should be accepted temporarily over both A and C, but it still must be compared with both B and E before a final choice can be made. The same procedure is followed when comparing D to B and E.

Cost of alternative	Alternative	A	D	B	E
$ 900,000	C	24.7%	21.6%	14.0%	16.2%
1,000,000	A		19.5%	10.0%	13.9%
1,150,000	D			−3.6%	10.4%
1,300,000	B				2.3%

This line shows IROR of B, E compared to D

The results may be summarized from the table as follows:

For MARR > 24.7%, use alternative C.
For 24.7% > MARR > 19.5%, use alternative A.
For 19.5% > MARR > 10.4%, use alternative D.
For MARR < 10.4%, use alternative E.

Note: There are no values of MARR under which alternative B is the better investment.

Graphically, the selection process is as shown in Figure 11.2. Since an investment in pollution-control equipment of at least $900,000 is required, would you recommend an additional investment

of $100,000 in A that yields an IROR of 24.7 percent?
of $250,000 in D that yields an IROR of 21.6 percent?
of $400,000 in B that yields an IROR of 14.0 percent?
of $500,000 in E that yields an IROR of 16.2 percent?

If the MARR is < 24.7 percent, then A is temporarily selected, since it has the greatest rate of return at 24.7 percent. Having selected A, at a cost of $1,000,000 (see Figure 11.3) would you now recommend an additional

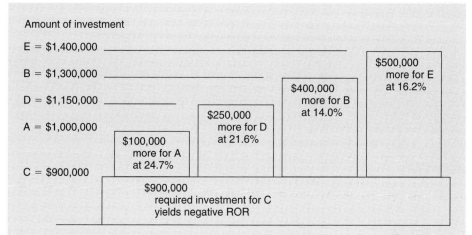

Figure 11.2 Alternative increments of investments with their respective IRORs, assuming a required initial investment in C.

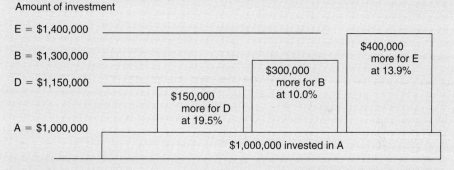

Figure 11.3 Alternative increments of investments with their respective IRORs, after committing at least $1,000,000 at level A.

investment of

> $150,000 in D that yields an IROR of 19.5 percent?
> $300,000 in B that yields an IROR of 10.0 percent?
> $400,000 in E that yields an IROR of 13.9 percent?

If the MARR is < 19.5 percent, then alternative D is the logical selection since it has the greatest rate of return of those available alternatives at 19.5 percent. Having selected D, at a cost of $1,150,000, would you now recommend an additional investment of $150,000 in B that yields an IROR of −3.6 percent or $250,000 in E that yields an IROR of 10.4 percent? (See Figure 11.4).

If the MARR < 10.4 percent then E should be selected. The total investment pattern would now appear graphically as shown in Figure 11.5.

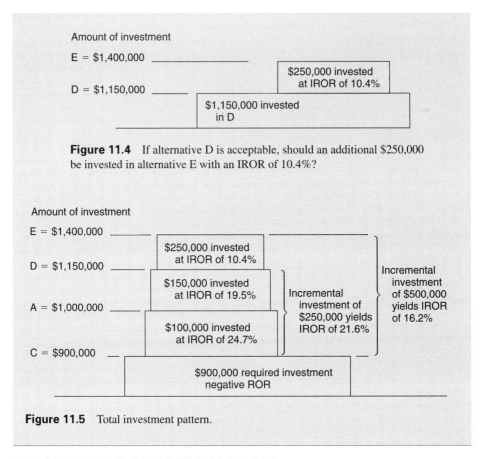

Figure 11.4   If alternative D is acceptable, should an additional $250,000 be invested in alternative E with an IROR of 10.4%?

Figure 11.5   Total investment pattern.

SOLUTION BY ARROW DIAGRAM

Where a number of alternatives are available, a graphical solution using an arrow diagram has been devised by Dr. Gerald W. Smith of Iowa State University that greatly simplifies the selection process.

- The diagram is drawn as a closed geometric figure with the same number of corners as there are alternatives. Thus, using the data from the two matrix tables on page 297, the five alternatives of Example 11.2 are drawn as a pentagon. Six alternatives would be drawn as a hexagon, and so on.

- The first corner drawn always represents the alternative requiring the lowest capital investment (C in Example 11.2).

- Proceeding clockwise around the diagram, each successive corner represents the alternative with the next higher capital cost. The sides between the corners are tipped with arrows pointing toward the next higher cost and labeled with the corresponding return on incremental investment (IROR), as shown in Figure 11.6. Note that C

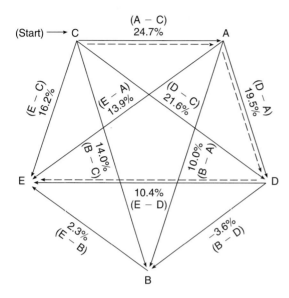

Figure 11.6 Arrow diagram for Example 11.2.

joins both A and E, since the lowest cost alternative will always be joined to two corners, the second lowest and the last (highest cost) alternative.

- To geographically determine the better alternative, start at the corner representing the lowest capital investment, in this case, C. Select the line with the highest rate of return leading from that starting corner, in this case the increment (A − C) or 24.7 percent.

- If the MARR is higher than this rate (MARR > 24.7 percent), then the lowest capital investment C, costing $900,000, is the better alternative.

- Assume for Example 11.2 that the MARR is lower (MARR < 24.7 percent). Then alternative A is temporarily the better choice, and we therefore proceed from corner A for our next comparison.

- Next, select the highest rate of return leading away from our first temporary selection (corner A), that is, 19.5 percent leading to corner D. Note that this was the better among three choices, 13.9 percent, 10.0 percent, and 19.5 percent. From D there is a choice between B, yielding −3.6 percent, and E, yielding 10.4 percent. It should also be noted here that there are only two choices from D; considering C is not a choice, because C is a lower, not a higher, cost alternative. If the MARR is lower then 10.4 percent (MARR < 10.4 percent) then accept E.

- No arrows lead away from E, and we therefore have no more comparison opportunity, indicating that E is the alternative with the highest capital investment.

- At whatever point, the IROR exceeds the MARR then temporarily accept the destination alternative at the point of the arrow. At whatever point the MARR exceeds all IRORs leading from that point, they accept the alternative represented by that point as the better selection.

Reviewing the diagram:

If the MARR is greater than 24.7 percent, select the lowest cost alternative, C.
If the MARR is between 24.7 and 19.5 percent, then A is the better selection.
If the MARR is between 19.5 and 10.4 percent, then D is the better selection.
If the MARR is less than 10.4 percent, then E is the better selection.

OPPORTUNITY FOR THE "DO NOTHING" ALTERNATIVE (NO INVESTMENT)

In Example 11.2 if the power plant is relieved of the need to install pollution-control devices, then another alternative is available: the opportunity for no investment ("do nothing"). The pollution-control devices produce no direct income for the plant (although they undoubtedly benefit the area in many ways); instead they increase the plant's cost, which in turn results in a negative rate of return when each alternative is considered independently. Thus the "no investment" opportunity is now added to the arrow diagram as the lowest first cost option, at corner 0. The comparative returns on incremental investment are as indicated along the appropriate sides of the diagram in Figure 11.7. The graphic solution now begins at corner 0, and all arrows away from 0 yield negative rates of return. The diagram simply confirms what in this case is intuitively obvious, that (unless the MARR is lower than −5.8 percent) no investment from this group of alternatives is recommended.

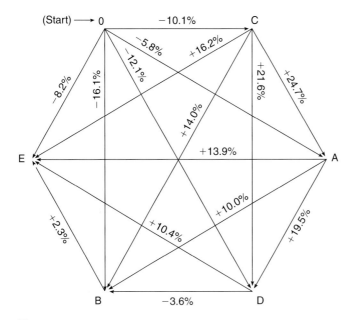

Figure 11.7　Arrow diagram with "do nothing" alternative.

COMPARING DIFFERENT LIFE SPANS

If the alternatives have different life spans, the IROR may be found by several methods.

1. *Use equivalent net annual worth.* Formulate the equation for each alternative in terms of equivalent annual worth instead of present worth. (Remember that when using equivalent annual costs for comparing alternatives with different life spans, the inference is that replacement costs equal original costs.)

2. *Assume multiple equal replacements.* Find the present worth of enough future replacements of each alternative to reach a common life span.

Example Illustrating Annual Costs

When determining the IROR, annual costs often are easier to use than present worth in cases where different life spans occur and the lowest common multiple happens to be a large or awkward figure to work with. Remember that using annual costs to compare alternatives with unequal life spans implies equal replacement costs, unless special provisions to the contrary are made. The following example illustrates this situation.

EXAMPLE 11.3

Three grades of roofing are available to replace a residential roof, and you are asked to determine which is the more economical. Data on the roofs are presented as follows:

1. 15-year roof costs $1,800 plus $100 per year for end-of-year maintenance.
2. 20-year roof costs $2,200 plus $50 per year for end-of-year maintenance.
3. 25-year roof costs $3,000 and requires no maintenance.
   a. Plot a graph of *A* versus *i* for each alternative.
   b. On the graph, show which ranges of *i* favor which alternatives.

Discussion: The least common multiple for the *n* values (lives) in this problem is 300 years, which renders the present worth method of finding IROR cumbersome and tedious. The annual cost method is much simpler.

Solution. Find *i* for the IROR in terms of annual costs. A graph of the values of annual costs versus *i* is shown in Figure 11.8. The equations for each alternative are found in terms of annual worth. Notice the use of an alternative format for this solution. Instead of the variables being listed in a separate *column* for each alternative, they are formulated in a separate *row* for each alternative. Both formats yield identical results. You may want to use only one of

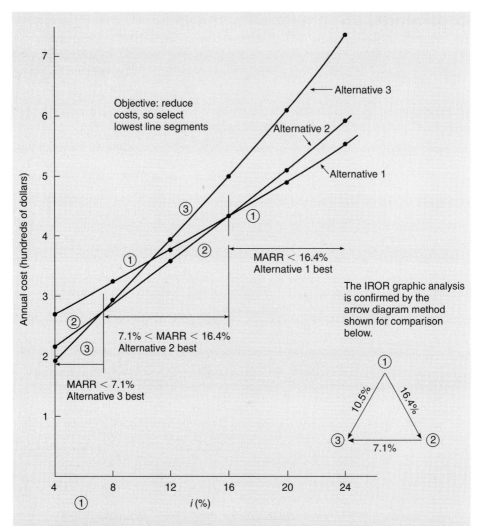

Figure 11.8   Graph of annual costs versus rate of return for Example 11.3.
Note: Costs are shown as positive values in this example.

these two formats in order to maintain consistency.

Alternative 1

$$A_1 = \$100 + \$1,800(A/P, i, 15)$$

Alternative 2

$$A_2 = \$50 + \$2,200(A/P, i, 20)$$

Alternative 3

$$A_3 = \$3,000(A/P, i, 25)$$

The solutions can be found by solving simultaneous equations in three pairs, and finding the respective three values of i for which

$$A_1 = A_2$$
$$A_1 = A_3$$
$$A_2 = A_3$$

These solutions yield the following matrix of values:

	2	3
1	16.4%	10.5%
2		7.1%

Conclusion.   The graph, arrow diagram, and matrix all indicate that

For MARR > 16.4 percent, use alternative 1.
For 16.4% > MARR > 7.1 percent, use alternative 2.
For MARR < 7.1 percent, use alternative 3.

Example Illustrating Multiple Replacements

Example 11.4 illustrates how the IROR is found by multiple replacements when life spans differ among alternatives.

EXAMPLE 11.4

A firm of engineers is in need of additional office space. A landlord offers them suitable space with three alternative leasing plans:

1. A one-year lease for $50,000 paid in advance, with option to renew for five subsequent one-year leases under the same terms.
2. A two-year lease for $90,000 paid in advance, with option to renew two more times at no increase.
3. A three-year lease for $130,000 paid in advance, with option to renew one more time at no increase.
   a. Find which ranges of MARR favor which lease terms.
   b. If the firm can invest its working capital elsewhere in the business at a 20 percent pretax rate of return, which is the better selection of lease terms?

Solution.   The one- and two-year leases are compared over the least common multiple of life spans, in this case a two-year period.

$$\text{two 1-year leases, } P_1 = -\$50{,}000 - \$50{,}000(P/F, i, 1)$$
$$\text{one 2-year lease, } P_2 = -\$90{,}000$$

The present worths of the two plans are equated and solved for i.

$$P_1 = P_2$$
$$-\$50{,}000 - \$50{,}000(P/F, i, 1) = -\$90{,}000$$
$$P/F = \frac{-\$90{,}000 + \$50{,}000}{-50{,}000}$$
$$\frac{1}{(1+i)} = 0.80$$
$$i = 0.25 = 25\%$$

In comparing the one- and two-year leases, the firm has an opportunity to spend an extra \$40,000 now in order to save spending \$50,000 a year from now. The effect is the same as investing \$40,000 now and receiving a \$50,000 return (benefit) in one year. Thus the rate of return on the incremental investment (IROR) is \$10,000/\$40,000 = 25 percent for the one-year period.

The one-year and three-year leases are compared over the least common multiple, in this case a three-year period, in the same manner as above:

$$\text{three 1-year leases, } P_3 = -\$50{,}000 - \$5{,}000(P/A, i, 2)$$
$$\text{one 3-year lease paid in advance, } P_1 = -\$130{,}000$$

Then equate P_3 to P_1 and solve as follows:

$$(P/A, i, 2) = \frac{130{,}000 - 50{,}000}{50{,}000} = 1.6000$$

Solving the P/A equation yields the factors listed below, which are then interpolated to find i.

i	$(P/A, i, 2)$
16%	1.6053
IROR	1.6000
17%	1.5853

$$i = \frac{1.6053 - 1.6000}{1.6053 - 1.5853} \times 1\% + 16\% = \underline{\underline{16.3\%}}$$

In comparing the two-year lease to the three-year lease, the shortest common time period is six years. The IROR between these alternatives is found as follows.

$$P_5 = -\$90{,}000 - \$90{,}000(P/F, i, 2) - \$90{,}000(P/F, i, 4)$$
$$P_6 = -\$130{,}000 - \$130{,}000(P/F, i, 3)$$

Equating P_5 to P_6 yields

$$P_6 - P_5 = -\$40{,}000 + \$90{,}000(P/F, i, 2) - \$130{,}000(P/F, i, 3)$$
$$+ \$90{,}000(P/F, i, 4)$$

Solving for $\Delta P = 0$, select trial i values. Try i at 8 percent.

$$\text{find } P_6 - P_5 = +\$114.10$$

This trial value of $i = 8$ percent yields a positive value of $P_6 - P_5$, so the next trial i should be higher, assuming the curve of i versus P declines to the right. Try $i = 9$ percent.

$$\text{find } P_6 - P_5 = -\$874.80$$

Interpolation yields

$$i = 8\% + \frac{114.10}{114.10 + 874.80} \times 1\% = \underline{\underline{8.1\%}}$$

The arrow diagram method provides a graphic solution process for this example, as shown in Figure 11.9.

For MARR > 25 percent, use the one-year leases.
For 25 percent > MARR > 8.1 percent, use two-year leases.
For MARR < 8.1 percent, use three-year leases.

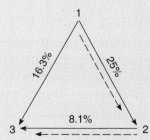

Figure 11.9   Arrow diagram for Example 11.4.

GRADIENTS

Changes in costs with the passage of time are a normal occurrence in real life and a reasonable mathematical simulation usually can be formulated by using gradients. The following example illustrates the use of gradients in IROR problems.

EXAMPLE 11.5

The state DOT has decided to construct a new road to bypass a congested area. They ask you to find the best of the three alternative routes from the data provided below.

Route	A	B	C
Initial cost ($)	1,500,000	1,300,000	1,600,000
Maintenance costs ($)	100,000/yr	120,000/yr	130,000/yr
Gradient maintenance costs ($)	12,000/yr/yr	12,000/yr/yr	12,000/yr/yr
Savings per vehicle (seconds)	25 s	15 s	40 s
Salvage value ($)	500,000	200,000	700,000
Estimated life (yr)	15	20	25

Each of the three routes is estimated to carry 5,000 vehicles per day for the first year, with a gradient of 500 vehicles per day for each year thereafter. The value of time for each vehicle is estimated at $3 per hour.

1. Find the incremental rate of return between the lowest cost route and the next lower cost alternative route.

2. The incremental rates of return for the other alternatives are as listed below. Draw an arrow diagram showing the three alternatives, including appropriate directions for each arrow between alternatives.

	A	C
B	find	16.7%
A		13.1%

3. Draw dashed arrows on the arrow diagram to illustrate the method of reaching the best alternative of the three if the MARR is 15 percent.

Solution 1.   Since the alternatives have different life spans, the equations are more easily formulated in terms of equivalent annual costs. Route B is the lowest cost and is subtracted from the next higher cost alternative route A, and the result is solved for the IROR value. Since savings are given in terms of seconds (s) rather than dollars, the first step is to find the annual dollar value of savings in time for each alternative route, calculated as

savings route A: $A_A = \$3/\text{hr} \times 5{,}000 \text{ vehicles/day} \times 365 \text{ days/yr}$
$$\times \frac{25\text{s/vehicle}}{3{,}600\text{s/hr}} = \$38{,}021/\text{yr}$$

savings route B: $A_B = \$3/\text{hr} \times 5{,}000 \text{ vehicles/day} \times 365 \text{ days/yr}$
$$\times \frac{15\text{s/vehicle}}{3{,}600\text{s/hr}} = \$22{,}813/\text{yr}$$

Then the gradient in savings per year is found as

savings route A: $G_A = \$3{,}802/\text{yr/yr}$
savings route B: $G_B = \$2{,}281/\text{yr/yr}$

The total equivalent annual costs plus savings are then formulated for each route as follows (all quantities are $\times \$1{,}000$).

A: $A_A = -1{,}500(A/P, i, 15) + 500(A/F, i, 15) - 100$
$\qquad -12(A/G, i, 15) + 38.021 + 3.802(A/G, i, 15)$

B: $A_B = -1{,}300(A/P, i, 20) + 200(A/F, i, 20) - 120$
$\qquad -12(A/G, i, 20) + 22.813 + 2.281(A/G, i, 20)$

Then trial i values are substituted into the equations until a positive and a negative result of $A_A - A_B$ are obtained.

	For $i = 15\%$	For $i = 20\%$
A_A	$= -345.421$	-408.317
A_B	$= -355.070$	-406.469
$A_A - A_B =$	$+\quad 9.649$	$-\quad 1.848$

Thus, by interpolation,

$$\text{IROR} = i = 15\% + 5\% \times \frac{9.649}{9.649 + 1.848} = \underline{\underline{19.2\%}}$$

If problems of this type are encountered often, it is well to write or acquire a computer program to save time in finding solutions and ranging variables.

Solution 2.   The three alternatives are represented by a three-sided polygon (triangle) with the lowest first cost alternative at the top (see Figure 11.10). The next higher first cost alternatives are represented by each corner in sequence, proceeding clockwise around the polygon. From every corner except the last, arrows are drawn from lower first cost to higher first cost.

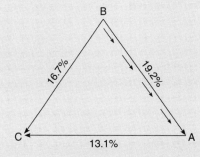

Figure 11.10   Arrow diagram for Example 11.5.

Solution 3

a. Begin at the lowest cost alternative B.

b. Find the highest IROR leading from this point (IROR = 19.2%).

c. Compare IROR with MARR (IROR = 19.2% > 15% = MARR); choose IROR and temporarily accept alternative A.

d. Find highest IROR leading from this point (IROR = 13.1%).

e. Compare IROR with MARR (IROR = 13.1% < 15% = MARR); reject alternative C, and stay with A.

Conclusion.   With the MARR at 15 percent, route A should be constructed for $1,500,000 rather than route B for $1,300,000, since the extra $200,000 will earn an IROR of 19.2 percent. Route C should be rejected, since the extra $100,000 required over route A will only earn an IROR of 13.1 percent.

COMPARING ALTERNATIVES BY THE ROR VERSUS THE IROR

The following example illustrates some of the problems encountered if only the ROR method is used for comparing alternatives. In this example the relative rankings for the preferred alternatives reverse themselves as the i values are ranged from one value to the next.

EXAMPLE 11.6

The following information has been obtained on two alternative machines:

	Machine X	Machine Y
Initial cost ($)	200	400
Annualized income ($)	75	120
End-life salvage value ($)	50	150
Life (yr)	6	12

Which alternative should be selected?

Note: There are unequal lives involved, so we must either (a) assume multiple equal investments and benefits of machine X until the 12-year period is concluded, or (b) assume some different salvage value for machine Y after 6 years. Remember that considering each alternative on the basis of equivalent annualized worth *automatically* assumes multiple equal investments and benefits.

Solution. If we make assumption (a) above and calculate net annualized worth, we obtain the following:

Machine X
$$A = 75 - 200(A/P, i\%, 6) + 50(A/F, i\%, 6) = 0$$

try 30%	0.37839	0.07839 = +3.24
try 40%	0.46126	0.06126 = −14.189

by interpolation: $i = \underline{31.9\%}$

Machine Y
$$A = 120 - 400(A/P, i\%, 12) + 150(A/F, i\%, 12) = 0$$

try 30%	0.31345	0.01345 = −3.364
try 25%	0.26845	0.01845 = +15.39

by interpolation: $i = \underline{29.1\%}$

If the maximum ROR were the correct criterion, then the decision would be to choose machine X. But if we check the net annual worth of each machine at a specific value of interest, then we may or may not arrive at the same solution.

To illustrate, let us take a value for i of 20 percent and determine the net annual worth of each machine.

Net annual worth of machine X
$$A(X) = 75 - 200\underbrace{(A/P, 20\%, 6)}_{0.30072} + 50\underbrace{(A/F, 20\%, 6)}_{0.10072} = \underline{\$19.9/yr}$$

Net annual worth of machine Y

$$A(Y) = 120 - 400\underbrace{(A/P, 20\%, 12)}_{0.225262} + 150\underbrace{(A/F, 20\%, 12)}_{0.025262} = \$33.7/\text{yr}$$

In this case machine Y has the greater net worth and thus would be selected. But this is a different answer than when the rate of return was examined, so we must conclude that utilizing the net annual worth for a specific interest rate can yield results very different from those found by simply calculating the ROR of each investment. Because different conclusions can be reached, some further analysis is needed. At this point, the major conclusion to be drawn is that choosing among multiple alternatives involves *more* than simply choosing the alternative with the highest ROR. A return of 20 percent on $10,000 yields a greater profit ($2,000) than a return of 100 percent on a $1 investment ($1 profit) (assuming the investments are mutually exclusive and nonduplicative). If machine X is chosen (earning 31.9 percent on this $200), then the remaining money available ($400 − $200 = $200 in this case) is assumed to earn some MARR (for example 20 percent). Now the complete question to be answered is whether $200 earning 31.9 percent plus $200 earning 20 percent is better or worse than $400 earning 29.1 percent (machine Y). Calculation of the net annual worth at 20 percent reveals that machine Y is the better investment, assuming an MARR of 20 percent. The IROR method considers the *differences* between the alternatives over the same number of years. Figure 11.11 shows the cash flows extended to provide a comparison over the same total life span.

$$NPW = -200 + 45\underbrace{(P/A, i\%, 12)}_{} + 150\underbrace{(P/F, i\%, 6)}_{} + 100\underbrace{(P/F, i\%, 12)}_{}$$

try 30%   3.1903   0.20718   0.04293
NPW = −21.02
try 25%   3.7252   0.26215   0.06872
NPW = +13.83
by interpolation, IROR = 27.0%

This means the $200 *additional* investment earns 27 percent. If the MARR is 20 percent, then machine X would earn 31.9 percent, and the unused additional investment would presumably earn 20 percent, while machine Y ($400 investment) would earn 29.1 percent. A $400 investment earning 29.1 percent is better than a $200 investment at 31.9 percent plus another $200 investment at 20 percent (the MARR).

The answer (decision) depends upon the MARR. For this example, if the MARR is less than 27.0 percent (the IROR in this case), then machine Y is the better alternative. If the MARR is between 27.0 and 31.9 percent, then machine X is the best alternative. If the MARR is greater than 31.9 percent, then *neither* alternative should be selected.

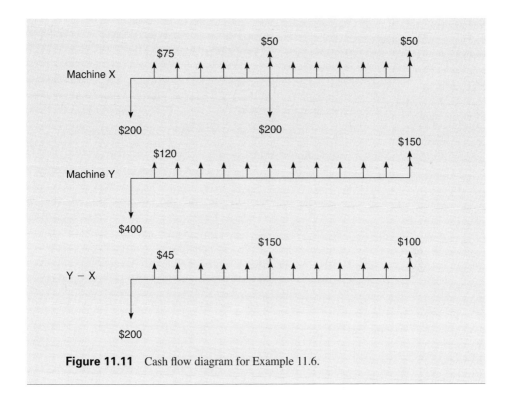

Figure 11.11 Cash flow diagram for Example 11.6.

COMPARING MULTIPLE ALTERNATIVES WHEN (1) THE MARR IS KNOWN, AND (2) THE "DO NOTHING" ALTERNATIVE IS AVAILABLE

Often there are a number of alternatives from which to choose, each being mutually exclusive but none required (the "do nothing" alternative is available). When this occurs, the time-consuming calculation of every alternative's ROR and the IROR for *every* difference between all possible combinations of alternatives becomes unnecessary *if* (1) the MARR can be established, and (2) the "do nothing" alternative is available. To analyze this type of multiple alternative, the following steps should be followed.

Step 1. List the alternatives *in order* of ascending first cost. The alternative with the lowest first cost is placed at the top of the list (or in the left column).

Step 2. Calculate the ROR for each alternative. Discard any alternative whose ROR is less than the MARR. (In those cases where there are no incomes and the "do nothing" alternative is not a possible choice, step 2 can be omitted because the ROR for all alternatives is less than zero, and the decision must be based on the IROR values.)

Step 3. Regroup the alternatives after discarding any unacceptable alternatives based on ROR.

Step 4. Stepwise determine the IROR for each difference in alternatives, beginning with the lowest cost alternative, discarding either (a) the lower cost alternative if the IROR is greater than the MARR, or (b) the higher cost alternative if the IROR is less than the MARR.

Step 5. Select the single remaining alternative after analyzing all alternatives.

The following example illustrates this method.

EXAMPLE 11.7

An engineering firm is considering five different small computers to aid the firm in running complicated and repetitive calculations. Analysis of the various machines, when applied to the firm's business, indicates the following cash flow information.

	F	G	H	I	J
First cost ($)	25,000	18,000	22,500	28,400	32,000
Annual net income ($)	5,000	3,800	4,800	6,000	6,600
Useful life (yr)	8	8	8	8	8

Which computer should be selected if the MARR is 12%?

Solution. First, arrange the alternatives in ascending order of first cost.

	G	H	F	I	J
First cost ($)	18,000	22,500	25,000	28,400	32,000
Annual net income ($)	3,800	4,800	5,000	6,000	6,600

Second, calculate the ROR of each alternative by setting the NPW = 0.

Computer G
$$\text{NPW} = -18{,}000 + 3{,}800(P/A, i\%, 8) = 0$$
$$\text{then } (P/A, i\%, 8) = 18{,}000/3{,}800 = 4.7368$$
$$\text{to find } i, \text{ try } 13\%, (P/A, 13\%, 8) = 4.7988$$
$$\text{try } 14\%, (P/A, 14\%, 8) = 4.6389$$
$$\text{then by interpolation, ROR} = 13.4\%$$

Computer H
$$NPW = -22,500 + 4,800(P/A, i\%, 8) = 0$$
$$so \ (P/A, i\%, 8) = 4.6875$$
$$then \ by \ interpolation, ROR = 13.7\%$$

Computer F
$$NPW = -25,000 + 5,000(P/A, i\%, 8) = 0$$
$$so \ (P/A, i\%, 8) = 5,000$$
$$then \ by \ interpolation, ROR = 11.8\%$$

Computer I
$$NPW = -28,400 + 6,000(P/A, i\%, 8) = 0$$
$$so \ (P/A, i\%, 8) = 4.7333$$
$$then \ by \ interpolation, ROR = 13.4\%$$

Computer J
$$NPW = -32,000 + 6,600(P/A, i\%, 8) = 0$$
$$so \ (P/A, i\%, 8) = 4.8485$$
$$then \ by \ interpolation, ROR = 12.7\%$$

Third, regroup the alternatives, after discarding those alternatives whose ROR does not meet the MARR (computer F in this case would be discarded).

	G	H	I	J
First cost ($)	18,000	22,500	28,400	32,000
Annual net income ($)	3,800	4,800	6,000	6,600
ROR (%)	13.4	13.7	13.4	12.7

Fourth, in turn determine the IROR between each combination of paired alternatives, discarding the least desirable one of each pair.

Thus computer G can be considered as the comparison between computer G and the "do nothing" alternative. Since the IROR of computer G compared to "do nothing" is 13.4 percent, and this is greater than the MARR of 12 percent, therefore discard the "do nothing" alternative. Comparing H to G, the cash flow becomes

	H–G
Incremental first cost ($)	4,500
Incremental net income ($)	1,000

$$-4,500 + 1,000(P/A, i\%, 8) = 0$$
$$(P/A, i\%, 8) = 4.500$$

By interpolation, the IROR = 14.9 percent > 12 percent MARR; therefore, discard G. Next, compare I over H.

	I–H
Incremental first cost ($)	5,900
Incremental net income ($)	1,200

$-5{,}900 + 1{,}200(P/A, i\%, 8) = 0$

$(P/A, i\%, 8) = 4.9167$

By interpolation, the IROR = 12.3 percent > 12 percent MARR; therefore, discard H. Next, compare J over I.

	J–I
Incremental first cost ($)	3,600
Incremental net income ($)	600

$-3{,}600 + 600(P/A, i\%, 8) = 0$

$(P/A, i\%, 8) = 6.000$

By interpolation, the IROR = 6.9 percent < 12 percent MARR; therefore, discard J. The one remaining alternative, computer I, should be selected.

Note that if the net present worth of each alternative is calculated at the MARR of 12 percent, the best alternative would be computer I, as determined using IROR analysis, and the same answer would be obtained using net annual worth or net future worth, as illustrated in the following table.

	F	G	H	I	J
First cost ($)	25,000	18,000	22,500	28,400	32,000
Present worth of annual net income $A(P/A, 12\%, 8)$ ($)	24,838	18,878	23,845	29,806	32,787
Sum ($)	−162	878	1,345	1,406	787

The natural question is, Why expend the effort calculating IROR when net PW or net annual worth will often yield the same result with much less effort? Two important reasons for calculating IROR follow:

1. The IROR values show the limits of i within which each selection is valid. A selection that is valid for a wide range of i values on both sides of the MARR is a more convincing selection than one whose priority would change if the MARR should move a couple of percentage points in tandem with normal market fluctuations.

2. A graph of the equivalent values versus i (which can be plotted from just the IROR values if need be) shows the amount of the differences between the equivalent worths of the competing alternatives for all values of i. This allows value judgments to be made by those who are considering buying the second or third most economical choice due to intangibles such as better dealer service, familiarity with the product, and so on. The graph shows just how much more the second or third alternatives will cost for each value of i, and this extra cost then can be weighed against the perceived intangible values.

SPREADSHEET APPLICATIONS

The calculations associated with making comparisons of investment opportunities using the incremental rate of return method are necessarily repetitive. Spreadsheets simplify the labor of making many similar calculations. Spreadsheets also offer the advantage of providing predefined formulas, such as the IRR function used in this example, which allow us to associate more than one set of values with a formula and calculate the interest rate of the incremental differences. By using simple macros, we can compare data and create text prints of decision parameters, as shown in row 41 of the following example, where we are comparing different interest-earning opportunities to our predefined minimum attractive rate of return, the MARR.

	A	B	C	D	E	F	G
1	EXAMPLE 11.7	MARR = 12%					
2	Solving our previous Example 11.7 using the EXCEL spreadsheet software						
3							
4		Order the investments, starting with the least first cost on the left side of the table.					
5	EOY	G	H	F	I	J	
6	0	−18000	−22500	−25000	−28400	−32000	
7	1	3800	4800	5000	6000	6600	
8	2	3800	4800	5000	6000	6600	
9	3	3800	4800	5000	6000	6600	
10	4	3800	4800	5000	6000	6600	
11	5	3800	4800	5000	6000	6600	
12	6	3800	4800	5000	6000	6600	
13	7	3800	4800	5000	6000	6600	
14	8	3800	4800	5000	6000	6600	
15							
16	IRR	13.4%	13.7%	11.8%	13.4%	12.7%	Calculate individual ROR.
17	Decision	Accept	Accept	Reject	Accept	Accept	Test against the MARR.
18							
19	We will temporarily accept the alternatives except for **F**, which does not provide the minimum ROR.						

(Spreadsheet continued)

	A	B	C	D	E	F	G
20	We now calculate incremental rates of returns for our investments, starting with the lowest first cost, G.						
21							
22	CELL CONTENTS: The following cell contents will allow us to evaluate the problem using a spreadsheet.						
23	Row 16 contents are typical of cell B16:				IRR(B6:B14)		
24	Row 17 contents are typical of cell B17:				IF(B16>=C1, Accept , Reject)		
25	Rows and columns for the table below are typical of cell B30:				C6-B6		
26	Row 40 contents are typical of cell B40:				IRR(B30:B38)		
27	Row 41 contents are typical of cell B41:				IF(B40>=$C$1, Accept , Reject)		
28							
29	EOY	H-G	I-H	J-I			
30	0	−4500	−5900	−3600			
31	1	1000	1200	600			
32	2	1000	1200	600			
33	3	1000	1200	600			
34	4	1000	1200	600			
35	5	1000	1200	600			
36	6	1000	1200	600			
37	7	1000	1200	600			
38	8	1000	1200	600			
39							
40	Delta IORR	14.9%	12.3%	6.9%	Calculate incremental rate of return (IROR).		
41	Decision	Accept	Accept	Reject	Test IROR against the MARR.		
42							
43	Therefore, computer I should be chosen over the other investment opportunities.						
44							

SUMMARY

In private as well as business life, many expenditures are required that do not yield a profitable return directly. Therefore, a negative rate of return often results, such as when purchasing tires for a car. However, when alternative candidates for purchase are available, an incremental rate of return (IROR) can be calculated on the increment of investment. The IROR is defined as the rate of return for an *additional* increment of investment. Finding the IRORs leads to finding which alternative is the preferred selection. To solve the IROR problems, the following procedures are employed.

Step 1. List the alternatives *in order* of ascending first cost, with the lowest first cost at the top of the list.

Step 2. Comparing each successive alternative in turn, determine the IROR for each increment of investment, beginning with the lowest cost alternative. Compare each IROR with the minimum attractive rate of return (MARR), if known. Immediately discard either (a) the lower cost alternative if the IROR is greater than the MARR or (b) the higher cost alternative if the IROR is less than the MARR.

If the MARR is not known, or if more information is desirable, then all IRORs should be calculated, and the ranges of MARR values listed that are associated with each alternative.

PROBLEMS FOR CHAPTER 11, INCREMENTAL RATE OF RETURN

Problem Group A

A series of alternative investments is given, each with one or more P, A, F values, but all alternatives have the same n values.

A.1.   Assume you are considering the following three investments. Using incremental rate of return analysis techniques, draw an arrow diagram and determine which alternative, if any, you should select (a) if MARR is 15 percent; (b) if MARR is 20 percent. The "do nothing" alternative is available.

	P	Q	R
Initial investment ($)	40,000	30,000	20,000
Annual net income ($)	7,400	5,700	3,200
Salvage value ($)	0	0	0
Life (yr)	20	20	20

[*Ans.* (a) select P; (b) select O (do nothing)]

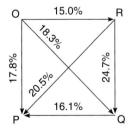

A.2.   Same as Problem A.1, except the life of each alternative is 30 years.

A.3.   A sanitary sewer can be constructed using either a deep cut or a lift station, with the following costs estimated. Using incremental rate of return analysis techniques, determine which alternative should be selected for an MARR of 7 percent.

	Deep-cut alternative	Lift-station alternative
Installation cost of pipe ($)	275,000	235,000
Installation cost of lift station ($)	—	15,000
Annual maintenance costs ($)	400	1,800
Expected life (yr)	40	40

(*Ans.* IROR = 4.73% < 7%; select lift station)

A.4. Same as Problem A.3, except the expected life of the lift station is only 10 years, while the deep-cut life remains at 40 years.

A.5. A contractor asks your advice on selecting the best of three alternative financing methods for acquiring company pickup trucks. The alternatives are listed as follows.
1. Purchase the pickups now for $15,000 each. Sell after four years for $3,000 each.
2. Lease the pickups for four years at $3,600 per year paid in advance at the beginning of each year. Contractor pays all operating and maintenance costs, just as if she owned them.
3. Purchase the pickups on time payments, with $2,250 down payment and $3,960 per year at the end of each year. Sell after four years for $3,000 each.
 (a) Draw an arrow diagram showing the three alternatives.
 (b) Show which ranges of i favor which alternatives.

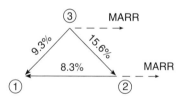

[*Ans.* (a) (b) if MARR > 15.6% select alternative 3; if 15.6% > MARR > 8.3% select alternative 2; if MARR < 8.3% select alternative 1]

A.6. A consulting firm finds it must provide reprints of a large number of items such as plans and specifications for its customers at a fixed nominal cost. It narrows the alternative means of providing this service for the next five years down to the following list of three.
1. Outside contract. A $1,000 deposit (returnable at EOY 5) is required to reserve these facilities. In addition, the fixed fee from customers is not enough to cover the cost of the work. Loss to the consultant amounts to $100 per year.
2. Lease the reproduction equipment, brand B. A deposit of $4,000 is required by the lessor, returnable upon termination of the lease. A cash flow profit of $320 per year for the reproduction work is assured under this plan.
3. Purchase brand C with guaranteed buy-back. This firm sells the reproduction equipment for $10,000 and guarantees to buy it back for 50 percent of the purchase price in five years. Calculations indicate this plan would yield a positive cash flow of $2,000 per year (not counting the purchase and resale cash flows).

In each case the study period is five years.
 (a) Find the rate of return for each alternative.
 (b) Find the incremental rate of return for each pair.

    (c) Draw two arrow diagrams: **(i)** without the "do nothing" alternative; **(ii)** with the "do nothing" alternative.

    (d) Analyze the diagrams and tell which alternative investment is best suited for each range of *i*.

    (e) If the consulting firm is earning 20 percent on capital invested in the firm but must provide the reproduction service, which is the better alternative?

    (f) If the consulting firm has the option of *not* providing the reproduction service, and it is earning 20 percent on capital, which is the better alternative?

A.7. Your client, a contractor, needs some new trucks to complete a project currently under contract. Analyze the three alternative types of trucks on the basis of the information listed below, and advise which is the best buy.

Type of truck	Cost new	Annual O & M cost	Annual increase in O & M costs	Resale value at EOY 4
A	$50,000	1.70/mi	$0.16/mi/yr	$10,000
B	56,000	1.60	0.16	15,000
C	65,000	1.50	0.09	15,000

The trucks will average 20,000 miles per year and will be sold at the end of four years for the resale values listed.

    (a) Find the incremental rate of return between A and B. (*Hint:* Try *i* for B−A between 30 and 40 percent.)

    (b) Assume the IROR of C−A is 26.7 percent and that of C−B is 23.9 percent. Draw an arrow diagram showing the IROR values and tell what ranges of MARR favor which alternatives.

    (c) Show which alternative is recommended if the contractor's MARR = 15 percent.

    [*Ans.* (a) B−A IROR = 30.7%; (b) if MARR > 30.7%, select A; if 30.7% > MARR > 23.9%, select B; if MARR < 23.9%, select C; (c) select C]

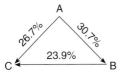

A.8. An OSHA inspector has just ruled that the lighting fixtures in your client's plant are inadequate to meet minimum standards. New lighting must be installed in order to stay in business. Three alternative installations are available with figures as tabulated below. In order to assist your client in making the best selection, develop the following information.

    (a) Find the incremental rate of return between A and B.

    (b) Assume the IROR of C−A is 17.1 percent and that of C−B is 32.4 percent. Draw an arrow diagram showing which alternative to choose for MARR = 5 percent.

    (c) Show the break points (if any) and tell which ranges of MARR favor which alternatives.

Type	Cost new	Annual operating cost	Annual increase in operating cost	Salvage value at EOY 20
A. Mercury	$30,000	$14,000/yr	$2,000/yr/yr	$ 3,000
B. Fluorescent	45,000	13,000	1,900	13,000
C. Metal halide	54,000	10,000	1,800	3,000

(*Hint:* Find *i*, B−A between 10 and 20 percent.)

A.9. Assume that you are a consulting engineer, and your company car has just been totaled in wreck. You are now selecting a replacement car. Any of the three listed below is acceptable from a functional point of view.

(**a**) Find the incremental rate of return for each pair of alternatives.

(**b**) Draw an arrow diagram showing which alternative to choose if MARR = 15 percent.

(**c**) Show the break points in the MARR (if any), and tell what ranges of MARR favor which alternatives.

Type	Cost new	Annual O & M cost	Annual increase O & M cost	Resale value at EOY 4
A. Electra	$15,000	51/mi	$0.042/mi/yr	$3,000
B. Monaco	17,100	48	0.048	4,500
C. Montego	19,500	45	0.027	4,500

The car will be driven about 20,000 miles per year and sold at the end of the fourth year. (*Hint:* Try *i* for B−A between 10 and 20 percent; C−A between 20 and 30 percent; C−B between 20 and 30 percent.)

[*Ans.* (a) − A + B 15.3%, − A + C 23.2%, − B + C 29.5%; (b) see diagram below; (c) if MARR > 23.2%, select A; if MARR < 23.2%, select C]

A.10. A friend of yours is considering the purchase of a home. The seller offers a choice of two options:

	Option A	Option B
Purchase price	$80,000	$78,000
Down payment	10,000	12,000
Mortgage	70,000	66,000
Interest rate on the mortgage	9%	9%
Payout period for mortgage	25 yr	25 yr

(**a**) What is the incremental rate of return on the difference between the two options? (*Hint:* Try between 15 and 25 percent.)

(**b**) Which option do you recommend and with what conditions (if any)?

Problem Group B

The *n* values of each alternative differ.

B.1. A subscription to a professional magazine is available at three different rates:

1 yr $100
2 yr $175
3 yr $ 250

Assume you intend to subscribe probably for the rest of your professional career. Find the before-tax incremental rate of return on:

(a) The two-year subscription compared to the one-year.

(*Ans.* 33.3%)

(b) The three-year subscription compared to the one-year.

(*Ans.* 21.5%)

(c) If you can invest your money elsewhere at 20 percent, before taxes, which alternative is the better selection?

(*Ans.* Select the two-year subscription.)

B.2. A professional society whose dues are normally $100 per year is offering life memberships for $1,250. Assume you are a young engineer of 24, considering the life membership offer. If you stay active in the society until age 65 (41 years), what rate of return will the investment in life membership yield

(a) Compared to the $100-per-year membership dues paid in advance, with the first year's dues payable right now?

(b) Compared to the $100-per-year membership with dues paid in advance, but where the dues are expected to increase by $5 per year every year ($100 the first year, $105 the second, etc.)?

B.3. The school board asks your help in determining which is the better alternative for obtaining portable classrooms.

1. One-year renewable leases, with $1,000 per month paid in advance at the first of every month.

2. Purchase with guaranteed buy-back. The portable classrooms are purchased for $100,000 each, and the vendor agrees to repurchase them from the school board at the end of three years for $57,000, if the school board should want to sell them at that time.

3. Three-year lease for $30,000 paid in advance.

Find the rate of return of:

(a) Alternative 2 compared to 1.

(*Ans.* < 0.0%)

(b) Alternative 3 compared to 1.

(*Ans.* 13.0%)

(c) If a selection of the most economical alternative is made when funds may be borrowed at 6 percent, does the selection change if interest rates rise to 9 percent?

(*Ans.* No)

B.4. Three grades of roofing are available for a proposed office building, and you are asked to determine which is the most economical. Data on the roofs are as follows. (Find *i* in terms of *A* rather than PW.)

1. Fifteen-year roof costs $18,000 plus $1,000 per year for end-of-year maintenance.
2. Twenty-year roof costs $22,000 plus $500 per year for end-of-year maintenance.
3. Twenty-five-year roof costs $30,000 and requires no maintenance.
    (a) Plot a graph of A versus i for each alternative (i is the pretax rate at which the building owner can invest available funds).
    (b) On the graph, show which ranges of i favor which alternative.

B.5. Your firm has a large fleet of cars, and tires are a major expenditure. You are asked to determine which of the three alternatives listed below is the most economical. The vehicles average 2,000 miles per month. Assume i is the monthly rate of interest and is compounded monthly.
1. 20,000-mile tire costing $63 each.
2. 30,000-mile tire costing $87 each.
3. 40,000-mile tire costing $114 each.
    (a) Find the IROR for each combination of alternatives and draw an arrow diagram.
    (b) Find which ranges of MARR favor which alternatives.

[*Ans.* (a) see diagram below; (b) for MARR > 3.6%, select 1; for 3.6 > MARR > 0.7%, select 2; for MARR < 0.7%, select 3)]

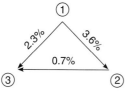

Problem Group C

Periodic and gradient payments.

C.1. Assume that 18 months ago the state highway patrol purchased a new airplane for $55,000. Now the plane needs repairs. They need to decide whether to repair the plane and keep it, or trade it in on a new one. The "do nothing" alternative is not available. Whichever plane is selected will be kept for the next 60 months and then sold. The costs of each alternative follow:

	A: Keep and repair	B: Trade in on new
Market value if sold now, as is	$40,000	
Repairs needed now	6,500	
Dealer's advertised price for new plane		$70,000
Trade-in allowance for present plane, as is		48,000
Dealer financing on the difference		22,000
New-plane amortized loan, for $22,000 at 0.75%, 48 months first payment at EOM 1		
O & M for the first month (100 hr/mo.)	$10.00/hr	$8.90/hour
Gradient O & M for each succeeding month	$0.10/hr/month	$0.10/hr/month
Periodic overhauls at 4-month intervals, with the first at EOM 6 (geo. grad.)	$ 1,000	$ 850
Each periodic OH costs 4% more than the previous one		
Resale value at EOM 60	$30,000	$44,000

(a) Use the equivalent NAW to find the IROR. (*Hint:* Try 1 percent.)

(*Ans.* 1.35%).

(b) Assume other alternatives are available, represented by C and D. Draw the arrow diagram and show which alternative is favored if MARR = 1%.

(*Ans.* Select C.)

	B	C	D
A	find	2.2%	1.2%
B		1.7%	0.9%
C			0.3%

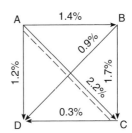

(c) Fill in the table below and show the MARR ranges favoring each alternative.

(*Ans.* > 2.2%→A; none→B;< 2.2% > 0.3%→C;< 0.3%→D)

MARR > _____ favors A
MARR < _____ but > _____ favors B
MARR < _____ but > _____ favors C
MARR < _____ favors D

C.2. A city water department is serving an urban area that is growing so large that the demand for water is about to exceed the capacity of the existing water main to supply it. The alternatives for the department include (a) installing a parallel water main, or (b) taking out the existing main and replacing it with a larger one. (The "do nothing" alternative is not available.) If the existing main is removed, it will have a net salvage value of $500,000 after all of the removal and cleanup costs. Both alternatives are expected to have 40 more years of useful life. Use the NAW method to determine which is the better alternative.

	A: Leave existing pipe in place; install parallel pipe	B: Replace with larger pipe
Cost of the one new parallel pipe only	$1,440K	
Cost of new larger pipe only		$3,000K
Government financing, available *only* for the larger pipe alternative, is a $1,800K loan at 5% amortized over 10 yr, with first payment due at EOY 1		
O & M costs	350K/yr	252K/yr
Gradient in O & M costs	10K/yr	10K/yr
Periodic maintenance at 3-yr intervals		
Each PM costs 10% more than the previous PM		
The first PM at EOY 5 costs	50K	150K
Salvage value at EOY 40	Zero	Zero

(a) Use the equivalent PW to find the IROR. (*Hint:* Try 12 percent.)

(b) Assume other alternatives, represented by C and D, are available. Draw the arrow diagram and show which alternative is favored if MARR = 10%.

	B	C	D
A	find	8.2%	9.3%
B		3.7%	8.9%
C			16.3%

(c) Fill in the table below and show the MARR ranges favoring each alternative.

MARR > _____ favors A
MARR < _____ but > _____ favors B
MARR < _____ but > _____ favors C
MARR < _____ favors D

C.3. Assume that you now own a two-year old car (present used car A) and are considering whether to keep it or buy a new car B. The data for each alternative are as follows:

	Present used car A	New car B
Current market value	$5,000	
Cost new: dealer's sticker price		$9,000
Dealer's trade-in allowance for present car		$5,800
Cash payment to dealer in addition to trade-in		$3,200
Resale value 36 months from now	$2,200	$3,000
Car will be driven	1,500 mi/month	1,500 mi/month
O & M costs (charged at EOM)	$0.14/mi	$0.09/mi
Gradient O & M costs	$0.0001/mi/month	$0.0001/mi/month
Periodic maintenance	$190/6,000 mi	$100/6,000 mi
First maintenance at EOM 4, last at EOM 36		
Gradient in periodic maintenance	$16/each/each	$6/each/each
Prestige value of new car over present used car	0	$15/month

(a) Find the IROR on the incremental investment in the new car B. Use the equivalent PW method. (*Hint:* Try $i = 2\%$ /month.)

(*Ans.* 2.50%)

(b) Assume that other alternative new cars C and D are also under consideration and that the data tabulated below is correct. Assume D has the highest first cost, and C has the next highest first cost. **(i)** Draw an arrow diagram showing the four alternatives and the direction of each arrow between alternatives. (**ii**) Show by dashed arrows the method of reaching the better alternative if the MARR is 2 percent per month.

(*Ans.* Select B.)

	B	C	D
A	find	1.5	2.2
B		1.0	1.8
C			2.7

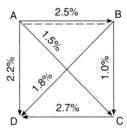

C.4. Assume you are retained as a consulting engineer to a contractor who must either overhaul or replace his presently owned bulldozer. The present dozer is just two years old and was completely financed by a loan of $60,000, which is being paid off in six equal end-of-year installment payments with interest at 10 percent on the unpaid balance. This existing dozer could be sold now for $40,000 cash. There are two replacement dozers available that could handle the anticipated work load. Analyze the costs (shown below) of the three alternatives to determine which is the better investment.

Type	Overhaul and keep A	Replace with B
Initial cost		$40,000
Cost of overhaul	$10,000	
O & M costs	$ 7,500/yr	$ 9,000/yr
Gradient O & M costs	1,000/yr/yr	1,200/yr/yr
Major maintenance every 2 yr, including EOY 10	5,000/each	6,000/each
Savings to your firm compared to existing dozer	9,000/yr	9,000/yr
Gradient savings to your firm	1,400/yr/yr	1,400/yr/yr
Increment in major maintenance	1,100/each/each	1,250/each/each
Resale value of dozer at EOY 10	5,000	2,000
Estimated life	10 yr	10 yr

(**a**) Find the incremental rate of return between the lowest cost alternative and the next higher cost alternative. Use equivalent annual costs and income. (*Hint:* Try 25 percent.)

(**b**) Assume the incremental rates of return for the other alternatives are as listed below. Draw an arrow diagram showing the three alternatives and the direction of each arrow between the alternatives.

(**c**) Show by dashed arrows the method of reaching the better alternative of the three if the MARR is 15 percent.

	A	C
B	find	12.2
A		4.8

Problem Group D

Formulate your own problem.

D.1.   Formulate a problem involving alternative methods of acquisition of goods or services of interest to you. Find the better alternative by the IROR method, and determine the ranges of i for which each alternative is feasible.

CHAPTER 12

Break-Even Comparisons

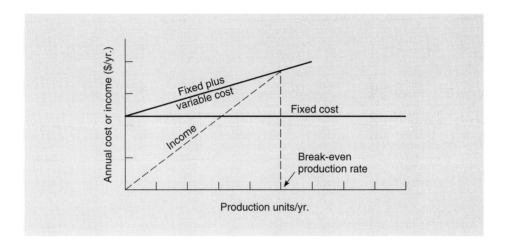

THE BREAK-EVEN POINT

Engineers frequently deal with capital investments in plant or equipment that will subsequently be operated and maintained to perform some needed function. Costs for these investments usually occur in two basic categories—the initial capital investment and the operating and maintenance (O & M) costs. Some financial managers use the following terms:

1. **Fixed costs:** those capital costs (including construction, acquisition, or installation costs) that remain relatively constant regardless of changes in the volume of production.
2. **Variable costs:** those costs that vary directly with production output.

These two types of costs can be portrayed (in simplified form) on a graph, as shown in Figure 12.1. Since the fixed costs are the costs of providing plant and equipment in a standby, ready-to-serve condition before starting production, the fixed costs are shown as a horizontal line on the graph, indicating the same constant annual cost regardless of the number of units produced. On the other hand, variable costs are usually considered as the operating costs of actually turning out a product. If each unit of the product is assumed to cost a constant amount of dollars per unit to produce, then the variable cost versus output can be plotted on the graph as an inclined straight line.

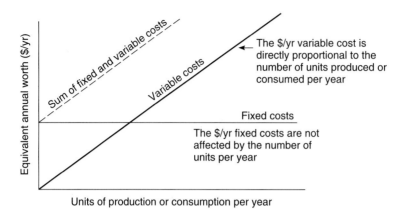

Figure 12.1 Linear fixed and variable costs.

When comparing alternative investments, often there is an inverse relationship between fixed capital cost and the variable operating expenses. In many instances a higher initial capital cost will purchase more efficient equipment, which in turn results in lower operating and production costs. Engineers are frequently confronted with decisions between higher capital cost for more efficient equipment with lower operating costs, and the converse. For example, is it more economical to purchase a new large dump truck for $85,000 with O & M costs of $0.25 per ton-mile or buy a used truck with the same total capacity for $32,000 but with O & M costs of $0.33 per ton-mile? The answer depends on how many ton-miles will be hauled. For a low number of ton-miles per year, the used truck with the lower capital costs will be less expensive. As the number of ton-miles per year increases, the lower O & M costs of the new unit become significant, so that at high ton-miles per year the new unit becomes the more economical. At some point in between, called the *break-even point,* the costs will be equal. If this point is known, the owner who expects to haul few ton-miles can benefit from the economy of the used unit in the low ton-mile range, while the owner with expectations of hauling more than the break-even amount can maximize profit by purchasing the new unit.

The break-even point may be found by the following three-step procedure:

1. Find the *annual* equivalent of the capital costs.
2. Find the independent variable (ton-miles in this case) and set up an equation for each alternative cost combination. The equations usually take the following form:

   total annual cost = annual equivalent capital cost
   + (cost/variable unit)(number of variable units/year)

   These equations graph as straight lines as in Figure 12.1.
3. Find the break-even point, which is the point of intersection of the graphed lines.

A graphical solution can be obtained by plotting the lines on a graph and scaling the points of intersection. An algebraic solution can be obtained by setting the equations equal to each other and solving. These procedures are illustrated in Example 12.1.

EXAMPLE 12.1

A contractor is thinking of selling his present dump truck and buying a new one. The new truck costs $85,000 and is expected to incur O & M costs of $0.25 per ton-mile. It has a life of 15 years with no significant salvage value. The presently owned truck can be sold now for $32,000. If kept, it will cost $0.33 per ton-mile for O & M, and have an expected life of five years, and no salvage value. Use $i = 10$ percent. Find the break-even point in terms of ton-miles per year.

Solution. The annual equivalent to the capital investment cost is

$$A_{\text{(new truck)}} = \$85,000\underbrace{(A/P, 10\%, 15)}_{0.13147} = \$11,175/\text{yr}$$

$$A_{\text{(present truck)}} = \$32,000\underbrace{(A/P, 10\%, 5)}_{0.26380} = \$8,442/\text{yr}$$

The present truck "costs" $32,000 because the contractor gives up $32,000 cash if he keeps it, just as he gives up $85,000 if he decides to purchase the new one.

The *total* annual equivalent cost for each year for each alternative is simply the annual equivalent capital cost plus the annual O & M cost, as follows:

$$\text{total annual cost}_{\text{(new truck)}} = A_1$$
$$A_1 = \$11,175/\text{yr} + (\$0.25/\text{ton-mi})(X \text{ ton-mi/yr})$$
$$\text{total annual cost}_{\text{(present truck)}} = A_2$$
$$A_2 = \$8,442/\text{yr} + (\$0.33/\text{ton-mi})(X \text{ ton-mi/yr})$$

where X is the number of ton-miles per year. The break-even value of X may be found by solving the equations simultaneously (setting them equal to each other).

$$\begin{array}{ll}
\text{general form} & y = mx + b \\
\text{for new truck} & y = 0.25X + 11,175 \\
\text{for present truck} & y = 0.33X + 8,442 \\
\text{break-even value } X = & \dfrac{11,175 - 8,442}{0.33 - 0.25} = 34,160 \text{ ton-mi/yr}
\end{array}$$

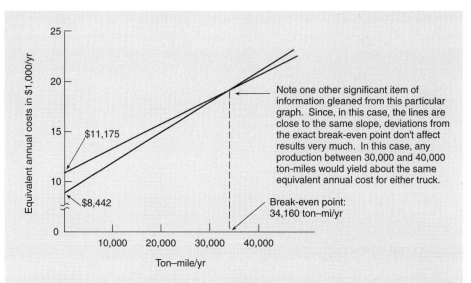

Figure 12.2   Graph of Example 12.1.

This is the break-even point. If the new truck carries an average load of 10 tons, the annual mileage at break-even is 3,416 miles per year. The graphical solution is illustrated in Figure 12.2.

From the graph it is evident that the break-even point occurs where the difference in slopes of the two lines overtakes the initial gap between the two lines. This point can be determined by the following process.

1. Find the amount of the gap between the two lines at the vertical axis ($2,733/yr at 0 ton-mi/yr);

2. Divide this amount ($2,733/yr) by the difference in the slope of the lines ($0.33 − $0.25 = $0.08/ton-mi) to obtain 34,160 ton-miles per year for the break-even point.

Note: Observe from the graph that the two lines are reasonably close together. This indicates that in this case the penalty is not great for deviating a few ton miles either way from the break-even point.

BREAK-EVEN INVOLVING INCOMES (BENEFITS) AND COSTS

Often the purpose of a capital investment is to produce a product with a direct market value. The capital investment then becomes the direct source of a cash flow income stream.

If the product can be sold for a certain number of dollars per unit produced, then the income can be depicted on the graph as another inclined straight line representing dollars

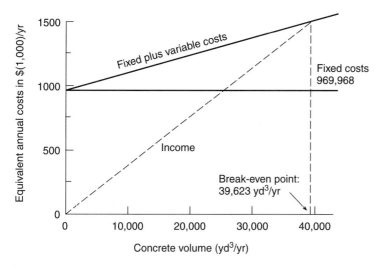

Figure 12.3   Graph of Example 12.2.

of income as a function of output volume, as shown in Figure 12.3. The concept is illustrated in the following example.

EXAMPLE 12.2

The purchase of an existing ready-mixed concrete plant in good operating condition requires a capital investment of $4,710,400 (plant, trucks, property, etc.). Operating and maintenance costs of producing concrete are $40.52/yd^3 of concrete (this includes material costs for cement and aggregate as well as O & M costs for the equipment). If the concrete is sold for $65.00/yd^3, what annual volume of concrete must the company sell to break even if their before-tax MARR is 20 percent? Assume a 20-year plant life with an estimated salvage value of $492,600.

Solution.   The equivalent annual fixed cost of the investment is

$$A = -4,710,400\underbrace{(A/P, 20\%, 20)}_{0.20536} + 492,600\underbrace{(A/F, 20\%, 20)}_{0.00536} = -\$969,968$$

The various components of costs and income are portrayed in Figure 12.3.
From the graph, the break-even point of about 39,500 yd^3 of concrete sales each year can be found.

To solve algebraically, simply set the costs equal to the income.

$$\$969,968 + \$40.52B = \$65B$$
$$B = \frac{969,968}{24.48} = \underline{\underline{39,623 \text{yd}^3}}$$

This type of analysis is helpful in determining break-even production outputs, and estimating the marginal profit or loss resulting from various production volumes. In the previous example, for every cubic yard of concrete over the break-even 39,623 yd^3, the company will earn a profit of $65.00 − 40.52 = $24.48. And conversely for every cubic yard under 39,623 yd^3 they will lose $24.48.

FOUR TYPES OF FIXED/VARIABLE COST COMBINATIONS

Four commonly encountered examples of fixed and variable cost combinations are depicted in simplified form in Figure 12.4. The first graph, which illustrates practically no fixed costs, can be exemplified by a casual day-laborer who offers his services on an hourly basis to contractors and others. He carries no tools and thus has no fixed costs except the cost of maintaining himself in good health. He gets paid only for hours worked; thus his income varies in direct proportion to those hours worked. The second graph, showing variable costs, could picture the owner of a piece of land who is receiving royalties for the removal of oil from the land. The fixed costs are the costs of the land, and the income is derived from the oil removed. The third and fourth graphs illustrate the more commonly encountered situations of combined fixed and variable costs. Many managers spend considerable time and effort looking for ways to lower the slope of the variable cost line, and they often succeed by raising the level of the fixed cost line. Example 12.3 (as well as most other examples in this chapter) illustrates this relationship.

Algebraic Relationships for Break-Even Analysis

Since many realistic investment situations can be approximated by straight-line relationships, the following simple algebraic relationships may prove useful in finding break-even points. We will use the following symbols:

B = break-even output (units/period)
F_d = equivalent fixed cost per period
R = total revenue (income) per period
V = variable cost per unit
M = profit per period (or loss)
C = total costs per period
N = output units (or time periods)
S = selling price (income/unit)
$R = NS$ (12.1)
$C = NV + F_d$ (12.2)

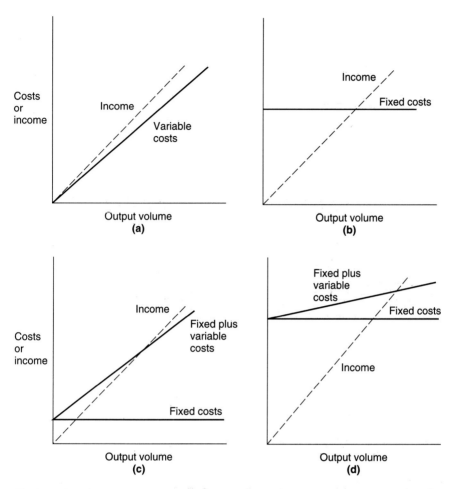

Figure 12.4   (a) Zero fixed costs. (b) Zero variable costs. (c) Low fixed costs, high variable costs. (d) High fixed costs, low variable costs.

At break-even, $R = C$. By letting $N = B$ (number of units to break even)

$$BS = BV + F_d = \text{Income per period}$$
$$B = F_d/(S - V) = \text{Number of units to break even} \qquad (12.3)$$

The quantity $(S - V)$ is sometimes referred to as the "contribution," because it is the contribution of the net profit per unit (or selling price per unit minus variable cost per unit) toward paying off the fixed costs.

$$M = R - C = N(S - V) - F_d \qquad (12.4)$$

Further Applications of Linear Break-Even

The following example illustrates a break-even analysis for hiring additional help.

EXAMPLE 12.3

A small, but profitable, consulting engineering firm is considering hiring another engineer for $38,000 salary + $35,000 (for fringe benefits, technical support, etc.) = $73,000 per year. Currently their fixed costs total $840,000 per year, and their variable costs average $19,200 per engineering contract. Their income per engineering contract averages $28,300.

1. If the new engineer is hired and all the new salary and overhead costs are fixed costs, how many *more* contracts must be acquired for the company to break even?

2. If the addition of the new engineer will permit the company to reduce its subcontracting efforts to other engineers to the point where the average cost per engineering contract is reduced to $18,200 and they expect approximately 100 contracts next year, should the new engineer be hired?

Solution 1.   The existing break-even is

$$B(\text{existing}) = \frac{\$840,000}{(28,300 - 19,200)} = 92.3 \text{ contracts}$$

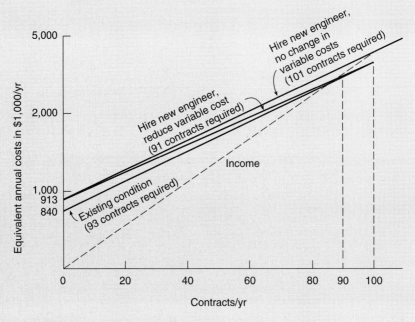

Figure 12.5   Graph of Example 12.3.

The new break-even is

$$B(\text{new}) = \frac{840,000 + 73,000}{(28,300 - 19,200)} = 100.3 \text{ contracts}$$

additional contracts $= 100.3 - 92.3 = 8.0$ contracts

Solution 2.   Here the decision would depend on the overall profit generated by the 100 contracts.

existing profit, $M = 100(28,300 - 19,200) - 840,000 = +\underline{\$70,000}$
new profit, $M = 100(28,300 - 18,200) - 913,000 = +\underline{\$97,000}$

The answer is to hire the new engineer in anticipation of an additional $27,000 profit.

Comment: The additional fixed cost (a new engineer in this case) reduced the variable costs per unit V (by $1,000) and thus raised the "contribution" $(S - V)$ that could be applied to the fixed costs. With the large number of contracts expected, the overall profit was increased. A graph illustrating this example is shown in Figure 12.5. Examination of the graph reveals that *if* the estimate of 100 contracts is *not* realized, the company could end up losing money by hiring the engineer.

LIMITATIONS OF BREAK-EVEN ANALYSIS

Problem areas commonly encountered when applying break-even analysis in actual practice include the following.

1. Difficulty in allocating all costs into purely fixed and purely variable components. Most costs have both a fixed component and a variable component. For example, the unit labor cost for erecting concrete forms may include (a) the cost of a supervisor who will be employed regardless of whether or not any forms are built and (b) the cost of carpenters and helpers, which may be totally variable. Oversimplification of the costs can result in erroneous data and conclusions.

2. The relationship between costs and outputs often is not linear. To assume that the variable cost of an activity is directly proportional to the quantity involved sometimes neglects important cost reductions available at higher production volumes, such as the discounts achieved through volume purchasing, shipping, handling, and so on.

3. The break-even chart may only be valid for a limited time because the data may change significantly with time. Therefore, break-even analysis (like many other methods) should be used with caution, particularly on long-lived investments. The real world contains many examples of preliminary studies that indicated one course toward a profitable activity, only to have circumstances change drastically, resulting in the indicated course leading to a loss rather than profit. For example, at one point in time U.S. automakers decided to continue producing large, expensive, and inefficient automobiles because of potentially larger unit profits ("contribution"). But market conditions changed abruptly when gasoline costs began escalating. Sales of large cars decreased rapidly, as sales of small foreign cars soared. The U.S. automakers who were depending on sales of large cars found themselves in serious financial difficulty because they could not achieve their break-even volume.

Used knowledgeably, break-even analysis is a useful tool for examining relationships between fixed and variable costs and incomes. It is also an aid in making rational decisions involving capital investments.

RISING RATES (GRADIENT)

Unfortunately, future expenditures usually do not remain at conveniently level rates, but typically rise at a rate the estimator must attempt to predict. The following example illustrates the effect of rising rates by comparing the break-even points using (a) constant labor rates versus (b) rising labor rates.

EXAMPLE 12.4

A parks department plant nursery is considering the acquisition of a hole auger which will dig holes for tree and plant transplants. The equipment costs $5,000 with no predictable salvage value after 10 years. The machine can dig five holes an hour and requires one operator at $8/hr. Fuel, maintenance, insurance, and overhead add another $2/hr.

At present the holes are hand dug by laborers at $5/hr at an average of two holes per person per hour.

1. How many holes per year is the break-even point if $i = 10$ percent?
2. If labor costs increase $0.50/hr per year for both laborers and equipment operators, what is the break-even point, using equivalent annual cost of the gradient, if $i = 10$ percent?

Solution 1.   Annual capitalization cost of hole auger,

$$A = \$5,000\underbrace{(A/P, 10\%, 10)}_{0.1627}$$

$$= \$813.50$$

There is zero salvage value in 10 years. Per-hole operating costs for hole auger are as follows:

$$\begin{array}{ll}
\text{operator} & \$ \ 8/\text{hr} \\
\text{fuel, etc.} & \underline{\ \ 2/\text{hr}} \\
\text{total} & \dfrac{\$10/\text{hr}}{5 \text{ holes/hr}} = \$ \ 2/\text{hole}
\end{array}$$

$$\text{total cost for auger/hole} = \$2/\text{hole} + (\$813.50/\text{yr})/(Y \text{ holes/yr})$$

$$\text{total cost for hand labor/hole} = \dfrac{\$5/\text{hr}}{2 \text{ holes/hr}} = \$2.50/\text{hole}$$

$$\text{break-even point } \$2.50 = \$2.00 + \$813.50/Y$$

$$Y = \underline{\underline{1627 \text{ holes/yr}}}$$

Solution 2. Annual equivalent of increased labor costs,

$$A = \$0.50\underbrace{(A/G, 10\%, 10)}_{3.7254} = \$1.86$$

Per-hole operating costs for hole auger are as follows:

$$\begin{array}{l}
\text{operator } \$8 + \$1.86 = \$ \ 9.86 \\
\text{fuel, etc.} = \underline{\ \ 2.00} \\
\dfrac{\$11.86}{5 \text{ holes/hr}} = \$2.37/\text{hole}
\end{array}$$

$$\text{total cost for hand labor/hole} = (5 + 1.86)/2 \text{ holes/hr}$$
$$= \$3.43/\text{hole}$$
$$\text{break-even point, } \$3.43 = \$2.37 + \$813.50/Y$$

$$Y = 813.50/1.06 = \underline{\underline{767 \text{ holes/yr}}}$$

Comment: In this case the rising labor rates significantly decrease the number of holes per year required for the break-even point.

BREAK-EVEN COSTS: FIND UNKNOWN LIFE

In ordinary algebra, as long as the number of independent equations exceeds the number of unknowns, the problem can usually be solved. Break-even problems occur under a wide variety of circumstances, and the unknown variable could be any one of the variables involved in the problem. An example of a break-even problem with an unknown service life follows.

EXAMPLE 12.5

A dredge is in need of a pump that will pump 15,000 gal/min at a 10-ft total dynamic head. The specific gravity of the dredged material will average 1.20. There are two good pumps available with data as shown ($i = 7\%$). How long must the Pumpall last to have a cost equivalent to that of the Gusher?

	Pumpall	Gusher
Cost new ($)	−15,600	−43,200
Efficiency (%)	70	78
Installation costs ($)	− 4,200	− 3,900
Maintenance and repair costs ($/yr)	− 3,850	− 3,850
Salvage value ($)	None	+10,000
Estimated life (yr)	Find	12

The pump is expected to operate 310 days per year, 24 hours per day. Power is available at $0.042/kWh. Note:

$$1 \text{ horse power (hp)} = 0.746 \text{ kW}$$

$$\text{hp output required} = \frac{\text{dynamic head} \times \text{gal/min} \times \text{specific gravity}}{3{,}960}$$

Solution.   The break-even point in terms of the estimated life of the Pumpall is found by setting the costs of the two pumps equal. The only unknown variable is the life of the Pumpall.

Pumpall

cost in place, $A_1 = -(15{,}600 + 4{,}200)\,(A/P, 7\%, n)$
$\qquad\qquad = -\$19{,}800(A/P, 7\%, n)/\text{yr}$

$$\text{operating cost, } A_2 = -\frac{\left[\begin{array}{l} 10 \text{ ft} \times 15{,}000 \text{ gal/min} \times 0.746 \text{ kW/hp} \\ \times 1.20 \times 310 \text{ days/yr} \times 24 \text{ hr/day} \\ \times \$0.042/\text{kWh} \end{array}\right]}{3{,}960 \times 0.70}$$

$$A_2 = -\$15{,}137/\text{yr}$$

Gusher

cost in place, $A_3 = -(43{,}200 + 3{,}900)\underbrace{(A/P, 7\%, 12)}_{0.1259}$

$\qquad\qquad = - \$5{,}930/\text{yr}$

salvage, $A_4 = +10{,}000\underbrace{(A/F, 7\%, 12)}_{0.0559} = +\$559/\text{yr}$

$$\text{operating cost, } A_5 = -\frac{\begin{bmatrix} 10\text{ ft} \times 15{,}000\text{ gal/min} \times 0.746\text{ kW/hp} \\ \times\ 1.20 \times 310\text{ days/yr} \times 24\text{ hr/day} \\ \times\ \$0.042\text{/kWh} \end{bmatrix}}{3{,}960 \times 0.78}$$

$$A_5 = -\$13{,}585\text{/yr}$$

$$\text{total annual cost of Gusher} = A_3 + A_4 + A_5 = -\$5{,}930 + 559 - 13{,}585$$
$$= -\$18{,}956$$

Setting the Pumpall cost equal to that of the Gusher, we have

$$-19{,}800(A/P, 7\%, n) - 15{,}137 = -18{,}956$$
$$(A/P, 7\%, n) = 0.19288$$

Checking the tables yields

$$\underline{n = \ \text{less than 7 yr}}$$

The Pumpall would need a seven-year life to equal the cost of the Gusher.

KELVIN'S LAW: COSTS THAT VARY DIRECTLY AND INVERSELY WITH ONE VARIABLE

Engineers sometimes encounter problems where an increase in the independent variable involves an increase in one type of cost and a decrease in another. For instance, in 1881 Lord Kelvin pointed out that the capital cost of conducting electricity through a wire increases as the area of the wire increases, but the cost of transmission losses due to resistance decreases with the same increase in wire area. He proposed that the best solution was the lowest total of capital costs plus transmission costs, obtained as a solution to an equation in the general form of

$$Y = AX + \frac{B}{X} + C$$

where

X = independent variable, the area of the wire in this case
AX = costs that increase with the increases in the wire area (capital costs)
B/X = costs that decrease with increases in the wire area (operating or energy costs)
C = the fixed costs incurred regardless of the wire area

The equation plots as a curve, in this case concave up, with a dip low point. The minimum point may be determined by finding the first derivative (slope equation) and setting it equal to zero (slope at tangent to low point), as follows.

$$\frac{dY}{dX} = A - \frac{B}{X^2} = 0$$

$$X = \sqrt{\frac{B}{A}}$$

Thus, the cross-sectional area at the minimum cost point is the square root of the indirectly varying costs divided by the directly varying costs. This wire area is then substituted back into the equation to determine the total cost of the installation. An example follows.

EXAMPLE 12.6

A 200-ft length of copper wire is required to conduct electricity from a transformer to an electrically powered pump at a municipal water plant. Find the most economical wire size, assuming the following data.

$$\text{pump motor size (hp)} = 100$$
$$\text{pump motor voltage (V)} = 440$$
$$\text{pump motor operating efficiency (\%)} = 80$$
$$\text{operating hours per year} = 4{,}000$$
$$\text{length of wire conductor needed (ft)} = 200$$
$$\text{installed cost of copper wire} = \$0.80/\text{lb} + \$200 \text{ for fittings}$$
$$\text{estimated life (yr)} = 20$$
$$\text{salvage value (\$/lb)} = 0.40$$
$$\text{electrical resistance of copper } (\Omega/\text{in.}^2/\text{ft length}) = 8.17 \times 10^{-6}$$
$$\text{conversion W/hp at 100\% efficiency} = 745.7$$
$$\text{cost of energy (\$/kWh)} = 0.05$$
$$\text{interest rate, } i \ (\%) = 7$$
$$\text{copper unit weight (lb/in.}^3) = 0.32$$

Solution.   The amperage required by the electric pump motor is

$$(100 \text{ hp} \times 745.7 \text{ W/hp})/(0.8 \text{ eff} \times 400 \text{ V}) = 211.8 \text{ A}$$

The electrical resistance is inversely proportional to the area of the wire cross section (a in.^{2}). The energy loss in kilowatts hours (kWh) due to resistance on the wire is calculated as

$$\frac{I^2R \times \text{number of hr/yr}}{1{,}000\text{W/kW} \times a \text{ in.}^2} = \frac{211.8^2 \times 8.17 \times 10^{-6} \times 200 \times 4{,}000}{1{,}000 \times a \text{ in.}^2}$$

$$= 293.2 \text{ kWh/}a \text{ in.}^2$$

Thus, the energy loss due to resistance in the wire amounts to 293.2 kWh/(a in.^{2}) each year. The annual cost of this lost energy is

$$A_1 = \$0.05/\text{kWh} \times 293.2 \text{ kWh/}(a \text{ in.}^2/\text{yr}) = \$14.66/a \text{ in.}^2/\text{yr}$$

Annual equivalent costs. The cost of the wire is given in terms of dollars per pound, and the weight is proportional to the area, a. Therefore, the cost is calculated as

$$P = 200 \text{ ft} \times 12 \text{ in./ft} \times 0.32 \text{ lb/in.}^3 \times a \text{ in.}^2 \times \$0.80/\text{lb} = \$614.4a$$

The annual equivalent value is

$$A_2 = \$614.4a\underbrace{(A/P, 7\%, 20)}_{0.0944} = \$58.00a/\text{yr}$$

The salvage value is $0.40 per pound, or $307.2a. The annual equivalent is

$$A_3 = \$307.2a\underbrace{(A/F, 7\%, 20)}_{0.0244} = \$7.50a/\text{yr}$$

The fixed cost of installation and fittings is $200, regardless of the area, so the annual equivalent is calculated as

$$A_4 = \$200\underbrace{(A/P, 7\%, 20)}_{0.0944} = \$18.88/\text{yr}$$

Summing, the total of all annual equivalent costs is

Total annual cost, $A_{\text{total}} = \$14.66/a + \$58.00a - \$7.50a + \$18.88$

Optimum cross section, a, for the lowest net cost. The equation for total cost derived above may be graphed as a sag curve of dollars versus area, a. The lowest cost point on the curve is found by taking the first derivative of the equation and setting it equal to zero (finding the point of zero slope). As previously explained, the low point is found at

$$X = \sqrt{\frac{B}{A}} \quad \text{or} \quad a = \sqrt{\frac{14.66}{58.00 - 7.50}} = 0.539 \text{ in.}^2$$

for circular wire, the diameter is

$$d = 2\sqrt{a/\pi} = \underline{0.828\text{-in. diameter}}$$

The weight of the wire is

$$0.539 \text{ in.}^2 \times 0.32 \text{ lb/in.}^3 \times 12 \text{ in./ft} = 2.07 \text{ lb/ft}$$

The equivalent annual cost of the entire installation is

$$A = \frac{14.66}{0.539} + (58.00 - 7.50)0.539 + 18.88 = \underline{\underline{\$73.30/\text{yr}}}$$

From an economics viewpoint, the lowest cost installation is for a 0.828-in. diameter copper cable weighing 2.07 lb/ft. Of course, electrical manufacturers' catalogues show the nearest size available, and electrical codes must be checked to ensure the resulting wire size complies with code requirements. Figure 12.6 illustrates the component costs of this problem.

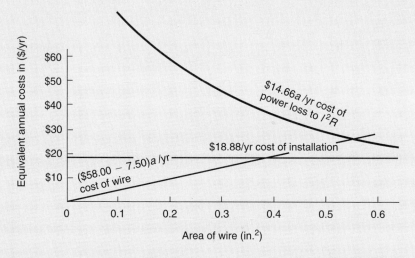

Figure 12.6   Graph of Example 12.6.

Applications to Span-Length (Column or Pier Spacing) Determination

Other problems involving a single variable with both inverse and direct relationships include the determination of column or pier spacing and the consequent span length. The longer spans save on the number of columns or piers but increase the size and cost of the spanning member. The following example illustrates this type of problem.

EXAMPLE 12.7

Assume a river is to be spanned by a highway bridge that will measure 2,000 ft from abutment to abutment, and a determination of pier spacing and span length is needed. The following data are provided. The piers are estimated to cost $150,000 each, regardless of the span design. The end abutments will cost $300,000 each, and no pier is required at the abutments (of course). The weight of the span dead load per foot is estimated as

$$w = (10s + 200)\text{lb/ft}$$

where s represents the spacing between pier centerlines in feet. The cost of the span is estimated as $0.90 per pound.

Solution.   The number of piers required is the number of spans less one, since no piers are required at the ends of the bridge where the abutments are located. This number can be expressed as $2,000/s - 1$. Therefore, the total cost of the piers is the number of piers times the cost per pier or $(2,000/s - 1)$ $150,000. Likewise, the weight of all the spans is $(10s + 200)$lb/ft $\times$ 2,000 ft. Therefore, the cost of these spans is this weight times $0.90 per pound. The total cost of the bridge T, is found by summing these parts:

$$T = \underbrace{(10s + 200)(2,000)(\$0.90)}_{\text{Spans}} + \underbrace{\left(\frac{2,000}{s} - 1\right)(\$150,000)}_{\text{Piers}} + \underbrace{2(\$300,000)}_{\text{Abutments}}$$

This is reduced to

$$T = 18,000s + \frac{3 \times 10^8}{s} + 810,000$$

Since this is the equation for the cost, the minimum cost span is found by taking the first derivative of this equation and setting it equal to zero (in order to find the equation sag point at which the slope of the line is zero).

$$\frac{dT}{ds} = 18,000 - \frac{3 \times 10^8}{s^2}$$

$$s = \sqrt{\frac{3 \times 10^8}{18,000}} = \underline{\underline{129.1\text{ft}}}$$

The theoretical minimum cost span occurs using 15.5 spans of 129.1 ft. Since the number of spans must equal a whole number, either 15 or 16 may be used. The cost of each alternative is found by substituting in the equation above.

Number of spans	Span length	$18,000s + \dfrac{3 \times 10^8}{s} + 810,000 = \text{total cost}$
15	133.3	$\$2.4 \times 10^6 + \$2.25 \times 10^6 + 0.81 \times 10^6 = \$5.46 \times 10^6$
16	125.0	$\$2.25 \times 10^6 + 2.4 \times 10^6 + 0.81 \times 10^6 = \$5.46 \times 10^6$

Since these points are equidistant from the minimum point on the curve, the resulting costs are the same in either case.

   The same solution technique may be adapted for other similar problems, such as floor spans in buildings, distance between booster pumps on a pipeline, and many other problems involving costs that vary directly with a variable X, together with costs that vary inversely with the same variable, and fixed costs.

SPREADSHEET APPLICATIONS

	A	B	C	D	E	F	G
1	**EXAMPLE 12.3**						
2							
3	A small, but profitable, consulting engineering firm is considering hiring						
4	another engineer for $38,000 salary + $35,000 (for fringe benefits, technical						
5	support, etc.) = $73,000 per year. Currently their fixed costs total $840,000						
6	per year and their variable costs average $19,200 per engineering contract.						
7	Their income per engineering contract averages $28,300.						
8							
9	1. If the new engineer is hired and all the new salary and overhead costs						
10	are fixed costs, how many *more* contracts must be acquired for the						
11	company to break even?						
12							
13	2. If the addition of the new engineer will permit the company to reduce						
14	its subcontracting efforts to other engineers to the point where the						
15	average cost per engineering contract is reduced to $18,200 and they						
16	expect approximately 100 contracts next year, should the new engineer						
17	be hired?						
18							
19							
20	**SOLUTION 1:**						
21							
22		Department	Fixed costs	Variable costs	Income	Break-even # of contracts	
23			($)	($/contract)	($/contract)	(# of contracts)	
24		Structures	840000	19200	28300	92.3	
25		Geotech	840000	19200	28300	92.3	
26		Construction	840000	19200	28300	92.3	
27		Hydrology	840000	19200	28300	92.3	
28		Transportation	840000	19200	28300	92.3	
29							
30		-or-					
31							
32		Department	Fixed costs	Variable costs	Income	Break-even # of contracts	
33			($)	($/contract)	($/contract)	(# of contracts)	
34		Existing	840000	19200	28300	92.3	
35		New	913000	19200	28300	100.3	
36							
37				Additional # of contracts required:		8.0	
38							
39							
40	**CELL CONTENTS:** The following cell contents will allow us to evaluate the problem with a spreadsheet.						
41							
42		cell F24:	=C24/(E24-D24) Copy this formula into cells F25 through F28 and F34, F35.				
43		cell F37:	=F35-F34				
44							
45	**NOTE:** Spreadsheet application not provided for "Solution #2," which changes the variable cost.						
46							
47	**CONCLUSION:**						
48	(A general solution format is presented to accommodate a variety of applications.						
49	Whatever point you want to get across, depending upon how you decide to set up the problem.)						

SUMMARY

Often an increased first investment cost will result in decreased periodic costs (such as operation and maintenance costs). These periodic costs are usually directly related to the use rate of the investment. Different initial investments can be compared on the basis of their break-even volume—the volume at which two alternatives yield the same overall cost. Knowing the break-even volume, or output, decisions can be made concerning which alternative will result in the lowest overall equivalent cost.

Both income and cost can be expressed in terms of the output volume. Under such circumstances break-even volumes can be determined on the basis of expected income versus fixed and variable costs.

PROBLEMS FOR CHAPTER 12, BREAK-EVEN COMPARISONS

Problem Group A

These are break-even problems between two or more alternatives involving costs only. Given sufficient data to determine annual costs (at a fixed production rate) and unit costs as the number of units per year increase, find break-even costs.

A.1. A tree spade is available for $30,000 with zero salvage value in 10 years. Overhead and maintenance costs are estimated at $12/hr. It can dig up and transplant an average of one tree per hour. The current method is hand labor by a crew costing $26/hr that can dig up and transplant an average of one tree every 1.5 hours. How many trees per year must be moved to justify the tree spade expenditure ($i = 8\%$)?

A.2. A trencher is available for $72,000 with an estimated resale value in 10 years of $5,000. The O & M costs are estimated at $35/hr, increasing each year by $2/hr. It is capable of digging a trench at an average rate of about 50 ft/hr. The alternate method is to buy a backhoe at $54,000 with a resale value of $5,000 in 10 years and O & M costs of $26/hr, increasing each year by $1.50/hr. It can dig a trench at an average rate of 20 ft/hr. How many feet of trench per year must be dug to justify the trencher ($i = 15\%$)? Plot the results on a graph.

(*Ans.* = 4,990 ft/yr)

A.3. A bucket truck may be rented, leased, or purchased for sign inspection by the building inspector in a small city. It rents for $40/hr, or it can be leased (including maintenance) for $10,000 per year (in advance) or purchased for $40,000 with guaranteed buy-back of $8,000 in five years. If purchased, it can be maintained for about $8 per hour. Find the most economical options with break-even points if any, in terms of hours per year of use for each option ($i = 8\%$). Plot the results on a graph.

A.4. A decision is required on what type of air conditioners should be installed in a residential project your firm is designing. Two units are under consideration, with data as follows.

	Carrier 38 G5 036	Carrier 38 SE 004
Cost installed	$750	$1,005
Efficiency rating	6.5	8.6
Btu	35,000/hr	37,000/hr

The efficiency rating is the number of btu's output per watt of electrical consumption. Find the annual cost of each unit under the following assumptions.

(a) Assume the unit will last 10 years with no salvage value ($i = 8\%$). The unit will be used an estimated 10 hours per day, six months of the year. Electricity costs an average of $0.036/kWh.

(b) Assume the hours of use per year are unknown. Graph the "dollars per year" versus "hours of use per year," and find the break-even point.

(c) Same as (b), except electrical costs increase by $0.004/kWh each year (first-year cost $0.036/kWh, second-year cost $0.04/kWh, etc.).

(d) What effect would a higher interest rate have on the solutions to (b) and (c).

A.5. A friend asks your advice on whether or not to sell his present car and buy a new one. His records indicate that his fixed costs of insurance, tag, and depreciation are currently about $750 per year. His mileage costs, including gas, tires, maintenance, and repairs, are about $0.145/mi. Authoritative figures indicate that a new car in the same class as your friend's present car would have a fixed cost of about $1,200 per year and mileage cost of $0.084/mi. Find the break-even point in terms of miles per year.

(*Ans.* = 7,380 mi/yr)

A.6. A consumer group needs information on the relative costs of each of three systems of hot-water heaters for residential use. They provide the information listed below ($i = 8\%$).

	Initial cost ($)	Service life (yr)	Operating expenses	Salvage value ($)
Solar water heater	600	20	1.00/kg	50
Electric hot water heater	100	12	2.00/kg	0
Gas hot water heater, vented through roof (kg = 1,000 gal)	200	12	1.60/kg	100

Plot a graph of kg/yr versus $/yr cost and find any break-even points that occur.

A.7. Same as Problem A.6, except that $i = 15$ percent.

A.8. A new office building is being planned, and you are requested to give an opinion on the alternative window systems listed below.

	Double pane	Single pane
Initial cost	$4.00/ft^2	$2.00/ft^2
Heat loss	10 Btu/hr/$^\circ$F/ft^2	12 Btu/hr/$^\circ$F/ft^2

Assume $i = 8$ percent; $3.00 per 1,000 Btu cost; $n = 40$ years; salvage value $= 0$. Graph the results in terms of $/yr/ft^2 versus hr/$^\circ$F/ft^2 and determine break-even points, if any.

(*Ans.* break-even $= 50.2$ hr/$^\circ$F/ft^2)

A.9. Same as Problem A.8, except $i = 15$ percent.

A.10. A designer finds that a certain building may be wired with either aluminum or copper wire. The aluminum is less expensive per foot, but the connections are more expensive. The comparative costs are shown as follows.

	Aluminum	Copper
Cost of wire per foot (cents)	12	15
Cost per connection (cents)	83	27

Assume one connection per length of wire. Find the average length of run between connections at the break-even point between aluminum and copper wire.

(*Ans.* 18.7 ft)

A.11. A buyer, in need of a used car, has several under consideration. Car A, which can be purchased for $5,595, has an EPA-rated gas consumption of 33.5 mpg. Car B can be purchased for $3,000 and has an estimated gas consumption of 16.9 mpg. The buyer intends to use either car for four years and then sell it for one-half the purchase price. Car B will require approximately $100 a year *more* maintenance expense than car A. If $i = 12$ percent, find the break-even point in miles per year, assuming gasoline costs $1.25 per gallon.

A.12. Same as Problem A.11, except that $i = 8$ percent.

A.13. Same as Problem A.11, except that gasoline costs $1.00 per gallon.

A.14. A solar heating system is under consideration for a new office building. The comparative data follow:

	Initial cost ($)	O & M costs each year ($/hr)	Estimated increase in O & M costs each year ($/hr)	Salvage value end of 10 yr ($)
Solar heat	20,000	0.10	0	$15,000
Conventional heat	8,000	1.00	0.12	0

Using $i = 10$ percent, how many hours per year must the heating system be used to reach the cost break-even point between the two systems?

A.15. Same as Problem A.14, except $i = 20$ percent.

(*Ans.* = 1,800 hr/yr)

Problem Group B

Kelvin's law. Find the least cost.

B.1. A short length of electrical conductor is needed to supply a pump at the municipal wastewater treatment plant for which you are a consultant. The pump draws 350 amps for an estimated 6,500 hours per year. Assume copper wire will be used, and the following data have been obtained.

$$\text{copper weight} = 0.32 \text{ lb/in.}^3$$
$$\text{length of conductor} = 150 \text{ ft}$$
$$\text{cost of copper wire in place} = \$0.85/\text{lb plus } \$2,500$$
$$\text{estimated life} = 25 \text{ yr}$$
$$\text{salvage value} = \$0.20/\text{lb}$$
$$\text{electrical resistance} = 8.22 \times 10^{-6} \ \Omega/\text{in.}^2/\text{ft of length}$$
$$\text{power costs} = \$0.045/\text{kWh}$$

With $i = 8$ percent, find the theoretically most economical size of wire in terms of square inches of cross section.

(*Ans.* 1.0 in.^{2})

B.2. Find the most economical span for spacing of towers for high-voltage electric lines. The intermediate towers cost $20,000 each, regardless of the spacing; the double-braced end towers every 5 mi cost $100,000, and the cost of the high-transmission wires is $0.50 per pound in place. The weight of the wire in this case is governed not only by strength requirements but also by conductance. Conductance requirements are met by a minimum weight of 50 lb/ft. Strength requirements are met by a weight per foot of $w = 0.125s + 4$. The average length of a straight-line run between double-braced end towers is 5 mi ($s =$ span between towers in feet).

Assume $i = 8$ percent, and all materials have a salvage value equal to 50 percent of the cost new at the end of a 20-year service life. Include the cost of one double-braced end tower for each 5 mi of transmission line.

(a) Find the most economical theoretical spacing.

(b) Find the nearest actual spacing that will provide a whole number of equal spans.

(c) Find the cost per mile (including $\frac{1}{5}$ of an end tower per mile).

B.3. Your client, the city, needs to determine where to locate the new solid-waste disposal landfill site and asks your recommendation. The land farther out from the city is cheaper, but transportation costs increase with distance from the city. You research the problem and come up with the following data.

Land values within a mile of the city limits average $8,000 per acre and drop off in value according to the following equation:

$$\frac{\$6{,}000}{M} + \$2{,}000, \qquad \text{where } M = \text{miles from the city limits } (M > 1)$$

The city now has a population of 50,000 and is growing at a rate of 2,000 per year. Solid-waste collections average 4 lb per person per day for 365 days per year. (To simplify calculations, assume the waste is all collected and transported at the end of each year; for 50,000 persons at the end of the first year, for 52,000 persons at the end of the second year, and so on.)

Transportation costs equal $0.10 per ton per mile of distance between the disposal site and the city limits. The compacted waste weighs an average of 1,000 lb/yd^3. The disposal site is expected to contain four lifts of compacted waste, with each lift being 3 ft high. Enough land should be purchased to last for 10 years.

The city borrows at 7 percent annual rate (1 acre $= 43{,}560$ ft^2).

(a) How much land should be purchased?

(*Ans.* 44.5 acres)

(b) How far from the city limits should it be located in order to minimize costs?

(*Ans.* 3 mi)

B.4. Find the optimum thickness of fiberglass insulation to place in a residence in an attic space above the ceiling. The resistance to heat loss is measured in units of Btu.in./hr-°F-ft^2 (termed R). Fiberglass insulation costs $0.045/ft^2-in., and has an R value of 3 Btu-in./hr-°F-ft^2. Thus the initial installation cost for the fiberglass is

$$P = \$0.045/\text{ft}^2\text{-in.} \times t \text{ in.}, \qquad \text{where } t = \text{thickness of insulation}$$

Assume that oil is used to heat the house and that 1 gal of oil emits 150,000 Btu and costs $1.12/gal the first year, with probable increases of $0.0375/gal in each subsequent year. Therefore, the price of heat is calculated as

$$1.12/150,000 \text{ Btu} = \$7.47/10^6 \text{ Btu, increasing in price by } \$0.25/10^6 \text{ Btu/yr}$$

Assume that in this particular location there is an average of 200 days per year with average attic temperature 20° below room temperature, yielding a total of

$$200 \text{ days} \times 20°\text{F} \times 24 \text{ hr/day} = 96,000 \text{ degree hr/yr}$$

Thus annual fuel costs are

$$A = 96,000°\text{F-hr/yr}(3 \text{ Btu-in./hr-°F-ft}^2)(1/t)(\text{fuel cost/Btu})$$

Assume $i = 10$ percent, and $n = 30$ years.

B.5. Same as Problem B.4, except that $i = 15$ percent and $n = 40$ years.

(*Ans.* 20 in.)

B.6. Your client plans to build a multistory apartment building and asks you to determine how many stories it should be. The cost of construction in terms of dollars per square foot increases as the number of floors increases, due to the following factors:
1. Increase in vertical services required (elevators, utilities, etc.).
2. Increased cost of foundations and structural supports.
3. Decreasing productivity of construction forces with increasing height, increased vertical transport required, and decreasing safety.

   Assume the cost of construction of a one-story building is $50 per square foot and that this cost increases according to $N^{0.1}$, where N is the number of floors. Therefore the cost of construction for a building with N floors is

$$C_1 = \$50N^{0.1}$$

On the other hand, the cost of land for each square foot of building decreases with the number of floors. The cost of land for a two-story building is one-half as much per square foot of floor area as it is for a one-story building. (On a two-story building the floor area is double; thus the land cost per square foot of floor area is just half that for a one-story building.) In this case the land cost for the entire lot is $500,000. The building will cover a ground area of 10,000 ft². Therefore, the cost of land for each square foot of floor area is

$$C_2 = \frac{\$500,000}{10,000N} = \frac{\$50}{N}$$

Maintenance costs also increase with the number of floors at about the same ratio as building costs. Thus the annual maintenance costs will equal

$$C_3 = \$1.00N^{0.1}/\text{yr plus an arithmetic gradient of } \$0.15N^{0.1}/\text{yr/yr}$$

Use $i = 10$ percent, $n = 40$ years.
(a) Find the theoretical number of floors in the building for least cost.

   (*Ans.* $= 5.3$ floors)

(b) Find the actual number of floors and the present worth per square foot.
(c) Find the annual rent in dollars per square foot which must be charged to cover the costs of land, construction, and maintenance, assuming 90 percent occupancy.

B.7. Your client, the Downtown Parking Authority, is considering a new multistory parking structure. Given the following data, find the percent of capacity the garage must operate at in order to break even ($i = 15\%$).

	Cost ($)	Economic life (yr)
Land	600,000	Infinite
Construction (capital improvements)	1,690,000	40
Fixtures	300,000	10
Salvage value	1,000,000	EOY 40
Working cash needed for entire 40-yr period	75,000	(not included in salvage)

The structure will contain 853 spaces, and customers will pay $3.65 per space per day for whatever spaces they occupy. Operating expenses are

    fixed = $230,000/yr
    variable = $0.50/day for occupied spaces only (total daily cost for the structure
        varies with the occupancy rate)

The garage will be open for business 250 days per year.

Problem Group C

Break-even analysis for profitable companies.

C.1. A firm manufacturing building supplies has the following fixed costs, variable costs, and income per unit of output. They are considering the purchase of an automated device that will change the costs as indicated. Use $i = 12$ percent.

	Existing	Automated device added
Fixed cost ($)	940,000	40,000
Life of fixed cost (yr)	17	10
Salvage value ($)	0	0
Variable costs ($/unit)	7.20	6.95
Income ($/unit)	10.20	10.20

(a) What annual volume of units must be sold under the existing system to break even?
(b) What annual volume of units must be sold to break even if the automated device is added?
(c) Prepare a break-even graph for both situations.
(d) If 75,000 units are expected to be sold each year, should the automated system be purchased?

C.2. Same as Problem C.1, except the variable costs with the addition of the automated device are $5.50 per unit.

CHAPTER 13

Probability Evaluation

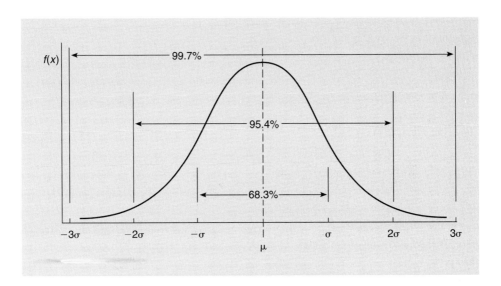

KEY EXPRESSIONS IN THIS CHAPTER

σ = Standard deviation of the population
z = Number of standard deviations from the mean

PROBABILITY DISCUSSION

Thus far in the text, engineering economic analyses of alternatives have been based on an assumed single value for costs, interest rates, and times. If specific values are assumed for P, i, and n, for example, equivalent values for A or F can be determined. Many times, however, we really do not know the precise values of P, i, or n to use. How are such situations handled? How can realistic decisions be made concerning alternatives when the values used in the calculations are subject to some variation?

Fortunately, *if* in real life the assumed values exhibit a random, or nearly random variability, we can use the power of probability and statistics to assist in understanding this variability and to analyze the expected outcome of a given set of circumstances. In this chapter some basic concepts of probability and statistics will be presented to demonstrate some applications used in engineering economic analyses. Such concepts are part of a well-developed body of knowledge, and there are several reputable texts which deal solely with this subject.

RISK

Risk involves variability for which a probability value for an outcome can be determined or estimated. Examples of risk situations include the probability of obtaining heads on the toss of a coin or the probability of rain during the next 24 hours. Probability values are known or can be estimated for a number of engineering problems. Examples include the probability of flooding in a given year, the probability of downtime on a piece of equipment during a given year, and the probability that the strength of a specific concrete test cylinder will be below a certain value. In fact, such probabilistic examples involving risk permeate our everyday life.

BASIC PROBABILITY

Probability may be regarded as the relative frequency of the occurrence of events (either discrete or continuous) in the long run. Thus, a 50 percent probability of tossing heads with a coin means that over many tosses of the coin, the average number of heads will closely approximate 50 percent of the total tosses.

Of particular interest to engineering economic analyses are events that are of one or more of the following three types: (1) independent, (2) dependent, and (3) mutually exclusive.

Independent events are events in which the occurrence or nonoccurrence of one event has no influence on the occurrence of another event. The toss of a coin has no influence on another toss of that coin. For each toss, there is a 50 percent probability of tossing a head or a tail, and thus the results of two tosses are "independent" of each other.

Dependent events are events in which the occurrence or nonoccurrence of one event depends upon the occurrence of another event. For instance, the flood water cannot reach the 10 foot level unless it has already reached the 9 foot level.

Mutually exclusive events are events that cannot occur simultaneously. Again using the toss of a coin, a result of a head or a tail can occur, but both cannot occur at the same time.

Using these terms, four time-honored axioms of probability follow:

Axiom 1   The probability (p) of an "independent" event (E) occurring $[p(E)]$ can be expressed as a number between 0 and 1.0, that is,

$$0 \leq p(E) \leq 1.0 \tag{13.1}$$

Axiom 2   The sum of the probabilities of all possible "independent" outcomes for a given event equals 1.0. (The probability of an event occurring $[p(E)]$ plus the probability of that same event not occurring $[p(\bar{E})]$ is equal to 1.0.)

$$p(E) + p(\bar{E}) = 1.0 \tag{13.2}$$

Axiom 3   The probability of either of two "independent" events occurring is equal to the sum of the probabilities of the two events; that is,

$$p(E_1 + E_2) = p(E_1) + p(E_2) \tag{13.3}$$

Axiom 4 The probability of two "independent" events occurring simultaneously (together) is equal to the product of the probabilities of the two independent events; that is,

$$p(E_1 E_2) = p(E_1) \times p(E_2) \tag{13.4}$$

These concepts can be illustrated by responses to questions about the toss of a six-sided die.

1. What is the probability of a number six turning up on the first throw of a die?

   Answer: $p(6) = 1/6$, because there are six possible outcomes (1, 2, 3, 4, 5, or 6 on any throw) and the probability of a six on one throw is one out of six possibilities, or $1/6$.

2. What is the probability of *not* throwing a six with one throw of the die?

   Answer: $p(\bar{6}) = 1 - p(6)$ [Axiom 2]

   $\qquad\quad = 1 - 1/6 = 5/6$

3. What is the probability of throwing either a five or six with one throw of the die?

   Answer: $p(E_1 + E_2) = p(E_1) + p(E_2)$ [Axiom 3]

   $\qquad\qquad\quad = 1/6 + 1/6 = 2/6$

4. What is the probability of throwing two sixes in succession with two throws of the die?

   Answer: $p(E_1 E_2) = p(E_1) \times p(E_2)$ [Axiom 4]

   $\qquad\qquad\quad = 1/6 \times 1/6 = 1/36$

Item 4 is analogous to throwing two dice simultaneously and obtaining double sixes. In this case there are 36 possibilities (the sample space) of numbers, of which only one would be a double six.

Notice the term "*independent*" in these four axioms. Throughout most of this chapter we will be dealing with "independent" events whose outcomes are generally "mutually exclusive."

Expected Value

Once situations are diagnosed as having risk-type events and outcomes (outcomes for which a probability can be estimated), the next question is, How can these situations be analyzed using engineering economy? Obviously, specific values for the various parameters of P, F, A, and so on are needed if equivalent values are to be determined.

Perhaps the simplest way to analyze these situations is to use the "expected value" (EV) of the outcome in the analysis. Although simple, this approach is also powerful.

Expected value, or weighted average $\bar{x}$, is a standard measure for assessing the most probable outcome from an event involving risk (assignable probabilities). Expected value can be determined by multiplying the possible outcomes of an event by their associated probabilities. The mathematical form is

$$EV = \sum_{j=1}^{m} p_j O_j \tag{13.5}$$

where p_j = independent probability for event j
O_j = outcomes for event j
$j = 1, 2, 3, \ldots, m$

Note:

$$\sum_{j=1}^{m} p_j = 1.0 \text{ (the sum of the probabilities must equal 1.0)}$$

To illustrate how "expected value" can be used, suppose a lottery ticket is sold for $3.00, which if drawn will pay $1,000,000, and there are 500,000 such tickets sold. If you purchase the ticket for $3.00, you will either win $1,000,000 or lose your $3.00. The probability of winning $1,000,000 (outcome $O_j = \$1,000,000$) on any one ticket is 1 in 500,000 ($p_j = 1/500,000$). However, the probability of losing the $3.00 cost of the ticket ($O_j = -\$3.00$) is $p_j = 499,999/500,000$. Therefore, for any one ticket the expected value is

$$EV = \frac{1}{500,000} \times 1,000,000 + \frac{499,999}{500,000} \times (-3.00)$$
$$EV = \$2.00 - 3.00 = \underline{\underline{-\$1.00}}$$

The $-\$1.00$ EV means that if you bought a large number of tickets you would lose, on the average, $1.00 for every ticket purchased. The expected value to the ticket purchasers for the *whole* lottery is determined as

$$EV = \sum_{j=1}^{m} p_j O_j$$

$$EV = \left[\frac{1,000,000}{500,000} - \frac{499,999}{500,000} \times (-3.00) \right] 500,000 = -\$500,000$$

The numerical value of probabilities (p) may be estimated by two basic methods.

1. **Theoretically** comparing the number of occurrences to the theoretical number of opportunities for such occurrences. Example: the probability that a flipped coin will come up heads is 0.5. This can be determined without ever flipping the coin, simply by considering the limited number of possibilities. Also in this category are probabilities of rolling a given combination with a pair of dice, drawing a certain card or combination from a deck, and so forth.

2. **Counting** the actual number of occurrences of an event and comparing it to the actual number of opportunities when a nearly infinite variety exists. Example: the

average ($p = 0.5$) percent of downtime on a given make and model of construction equipment during its first year of service must be determined experimentally. Also, the probability that a 25-year storm ($p = 0.04$) will produce a maximum runoff of 500 ft³/s at a given culvert is typically a determination based on actual field measurements.

The following examples typify the methods used in practice for deriving probabilities and expected values.

EXAMPLE 13.1

Assume a certain construction equipment manufacturer's record of the up-time availability of a given type of bulldozer during the first year in service is tabulated below.

Number of dozers reporting (Frequency, f) (a)	Percentage of time available first year (b)	Weighted dozers × % ($a \cdot b$)	Relative frequency (a/m)
1	100	100%	0.02
3	99	297%	0.07
5	98	490%	0.11
6	97	582%	0.14
5	96	480%	0.11
10	95	950%	0.23
6	94	564%	0.14
5	93	465%	0.11
2	92	184%	0.05
1	90	90%	0.02
$44 = N$		4204%	1.00

$$\text{average availability, } \bar{x} = \frac{4204}{44} = 95.5\% \text{ (the ``expected value'')}$$

This value of 95.5 percent is the "average" availability of the dozer fleet. By definition, the average availability of 95.5 percent represents a probability of 50 percent that any dozer, picked at random, will be available 95.5 percent *or more* of the time. It also means that this same dozer has a 50 percent probability of being available 95.5 percent *or less* of the time. Finally, it should be emphasized that the probability of any specific dozer being available *exactly* 95.5 percent of the time is almost zero!

These data can be represented as a relative frequency histogram, as shown in Figure 13.1. Relative frequency histograms can be a very useful tool for

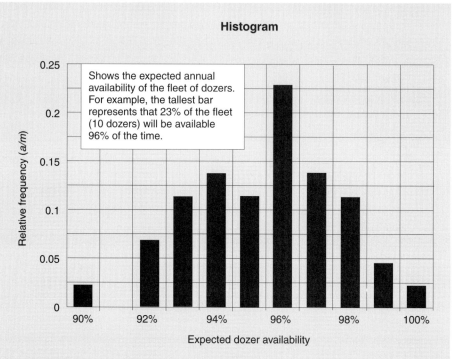

Figure 13.1   Histogram of bulldozer availability.

understanding and using data such as these. In this case, the histogram shows us that most of our equipment, dozers, have an availability of around 96%. The area under all of the bars of a relative frequency histogram represents 100% of the population. In our case, we can easily see that this would mean that 100% of our fleet is available 90% of the time or more and that only 2% of our fleet, or one dozer, can be expected to be available 100% of the time. If we were to take the area of the bars from 96% on, it would indicate that approximately 50% of our fleet would be available 96% of the time or better. This graphically represents the statements of the previous paragraph.

How are such data used? A contractor, purchasing a dozer representative of this group, can calculate expected revenues using this weighted average, transforming variable data into specific values for analysis.

By way of caution, these data tell only what has occurred in the past. The data might not repeat themselves if significant changes have occurred in any of the many variables affecting equipment availability, such as design changes, production-line variables (material or labor shortages causing substitution of material or skills), newer and more efficient automation, on-the-job maintenance, or severity of working conditions.

STANDARD DEVIATION

When analyzing data whose values vary, often it is not sufficient to know only a measure of central tendency (such as expected value), because such measures do not indicate anything about the dispersion (scatter) of the data around the mean. In the preceding example, dozer availability ranged from 90 to 100 percent. What if the range of values was from 80 to 100 percent, with the same expected value of 95.5 percent? Intuitively, this latter situation would seem more "risky" than the former situation, because the amount of data dispersion is twice as much. One measure of this dispersion is known as the "*standard deviation,*" σ, which is really the root mean square of the data scatter (squaring the numbers removes the negative signs).

For "*normally*" distributed data (a term coined for data whose histograms resemble the familiar bell-shaped curve), the "standard deviation," σ, can be expressed as

$$\sigma = \sqrt{\frac{\sum_{j=1}^{m} (x_i - \bar{x})^2}{m - 1}} \tag{13.6}$$

The larger the value of σ, the greater the dispersion of data.

The values of $\bar{x}$ and σ for a set of data provides an accurate picture of the variability of the data and probabilities can be analyzed with confidence. From the mathematics of statistics, it can be predicted that within a range of $\bar{x} \pm 1\sigma$ about 68 percent of all the data will occur, and about 95 percent of all the data will fall into a range of $\bar{x} \pm 2\sigma$, provided the data are "normally" distributed. Standard tables have been constructed that give the probabilities for normal ranges.

Using the data in the previous example, the standard deviation may be found as in the following tabulation.

Number of dozers reporting (m)	Percent availability	$(x - \bar{x})$ deviation from average $\bar{x} = 95.5$	$\Sigma (x - \bar{x})^2$
1	100	4.5	20.25
3	99	3.5	36.75
5	98	2.5	31.25
6	97	1.5	13.25
5	96	0.5	1.25
10	95	0.5	2.50
6	94	1.5	13.50
5	93	2.5	31.25
2	92	3.5	24.50
1	90	5.5	30.25
44			$205.00 = \Sigma (x - \bar{x})^2$

$$\sigma = \sqrt{\frac{\sum (x - \bar{x})^2}{m - 1}} = \sqrt{\frac{205}{44 - 1}} = 2.2$$

Thus the data indicate that there is a 68 percent probability that the dozers will be available between 95.5 ± 2.2 percent of the time or between 97.7 and 93.3 percent of the time. This same statistical theory predicts that 90 percent of the occurrences will lie within 1.64σ or $\pm 1.64 \times 2.2$ percent of the average 95.5 percent as shown in Figure 13.2. There is therefore a 90 percent probability that the availability will be between 91.9 and 99.1 percent, which means there is a 10 percent probability that the availability will be outside this range. Assuming the data are normally distributed, there will be a 5 percent probability that the availability will be less than 91.9 percent, and a 5 percent probability that the availability will be greater than 99.1 percent. Such information is valuable in analyzing risk situations, as illustrated in the next example.

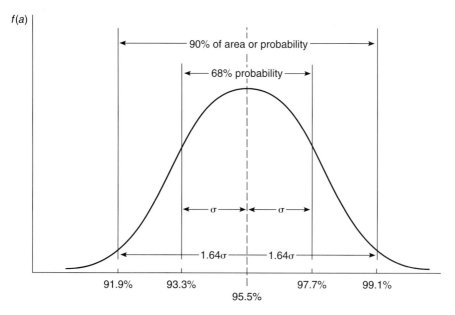

Figure 13.2   Normal distribution for Example 13.2 with $\bar{x} = 95.5\%$, and $\sigma = \pm 2.2\%$.

EXAMPLE 13.2

Each dozer in Example 13.1 is used to push-load a fleet of five scrapers. If the dozer is out of service, production stops. Production is usually at 1,000 yd³/hr with costs at $900 per hour and income at $1.20 per cubic yard. With the dozer out of service, the costs continue at $900 per hour, but production stops. A standby unit may be rented at $35 per hour. Should the unit be rented? Assume for the moment the standby unit is available 100 percent of the time.

Solution.   Assume 95.5 percent availability of the dozer (the average availability).

Without standby,

$$\text{"Expected" income } 0.955 \times \$1.20/\text{yd}^3 \times 1{,}000 \text{ yd}^3/\text{hr} = \quad \$1{,}146/\text{hr}$$
$$\text{cost} = - \quad 900/\text{hr}$$
$$\overline{\text{net} = \quad \$\ \ 246/\text{hr}}$$

With standby,

$$\text{income } \$1.20/\text{yd}^3 \times 1{,}000 \text{ yd}^3/\text{hr} = \quad \$1{,}200/\text{hr}$$
$$\text{cost } \$900/\text{hr} + \$35/\text{hr} = - \quad 935/\text{hr}$$
$$\overline{\text{net} = \quad \$\ \ 265/\text{hr}}$$

The results indicate that a standby unit would be profitable at any cost less than $1,200 − $1,146 = $54 per hour. Since this one is available at $35 per hour, it is profitable to have on hand. (This assumes the standby is 100 percent available.)

MULTIPLE PROBABILITIES

Realistically, the standby unit in the previous example would not be available 100 percent of the time. How to account for multiple probabilities is illustrated in the next example.

EXAMPLE 13.3

The standby dozer in the previous example is an older unit and has about 30 percent downtime itself. If the regular dozer has an availability of 95.5 percent (is down 4.5 percent of the time), what are the probabilities of their both being down simultaneously?

Solution.   The probability of any two independent events occurring simultaneously is $p_1 \times p_2$ (Axiom 4). The probability of their being down simultaneously is therefore

$$0.30 \times 0.045 = 0.0135 = \underline{1.35\%}$$

GRADIENT PROBABILITIES

The reliability of equipment characteristically decreases with increasing years of service. A method of accounting for gradient changes in probability is shown in the following example.

EXAMPLE 13.4

Assume a contractor needs a tractor to haul a set of compaction rollers around the fill areas on a sequence of highway jobs over the next five-year period. The availability of the tractor during the first year is expected to be 95 percent but drop about 4 percent each year thereafter. The contractor is considering putting a standby tractor on the job, since whenever the compaction stops, the whole earthmoving operation stops. When the compactor is in operation, production averages 200 yd^3/hr, and the bid price (income) averages \$1.20/yd^3. Expenses average \$100 per hour whether the compactor is running or not. The standby tractor would cost \$80,000, could be sold in five years for \$40,000, and is estimated to be available 75 percent of the time without change throughout the five-year period. Find the rate of return on the proposed investment in the standby tractor (1,000 hr/yr use).

Solution. Consider the probable loss of income without the standby tractor versus the probable loss of income if the standby is present. The expenses and net income are actually not involved in this particular part of the problem but would be needed to consider overall profitability. Gross maximum possible income from earthmoving is

$$\$1.20/\text{yd}^3 \times 200 \text{ yd}^3/\text{hr} \times 1,000 \text{ hr/yr} = \$240,000/\text{yr}$$

Probable loss of income due to downtime with *no* standby unit is tabulated below:

Year	Downtime	Cost of downtime % × max. income
1	5%	0.05 × \$240,000 = \$12,000
2	9%	0.09 = 21,600
3	13%	0.13 = 31,200
4	17%	0.17 = 40,800
5	21%	0.21 = 50,400

Probable loss of income due to downtime when both units are present but down is (the probability of two occurrences happening simultaneously is $p_1 \times p_2$)

Year	Downtime, tractor and standby
1	0.05 × 0.25 × \$240,000 = \$ 3,000
2	0.09 × 0.25 = \$ 5,400
3	0.13 × 0.25 = \$ 7,800
4	0.17 × 0.25 = \$10,200
5	0.21 × 0.25 = \$12,600

The amount saved by purchasing the standby unit is the difference between the last two sets of calculations.

Year	Loss without standby	Loss with standby	Saved by standby
1	−$12,000	−$ 3,000	+$ 9,000
2	− 21,600	− 5,400	+ 16,200
3	− 31,200	− 7,800	+ 23,400
4	− 40,800	− 10,200	+ 30,600
5	− 50,400	− 12,600	+ 37,800

Therefore the amount saved may be credited as income produced by this investment and the rate of return calculated. Notice that the amount "Saved by standby" in this case has an arithmetic gradient of $7,200 per year.

$$Try\ i = 20\%$$

PW of *income − cost*

$$\text{base annual } P_1 = \$9,000/\text{yr}\underbrace{(P/A, 20\%, 5)}_{2.991} = \quad \$26,920$$

$$\text{annual gradient } P_2 = \$7,200/\text{yr}\underbrace{(P/G, 20\%, 5)}_{4.906} = - \ 35,322$$

$$\text{resale in 5 yr } P_3 = \$40,000\underbrace{(P/F, 20\%, 5)}_{0.4019} = \quad 16,076$$

$$\text{purchase price } P_4 = \$80,000 = - \ 80,000$$

$$\text{NPW at } 20\% = -\$ \ 1,682$$

$$Try\ i = 15\ \%$$

PW of *income − cost*

$$\text{base } P_1 = \$9,000/\text{yr} \underbrace{(P/A, 15\%, 5)}_{3.352} = \quad \$ 30,168$$

$$\text{gradient } P_2 = \$7,200/\text{yr}\underbrace{(P/G, 15\%, 5)}_{5.775} = \quad 41,580$$

$$\text{resale } P_3 = \$40,000\underbrace{(P/F, 15\%, 5)}_{0.4972} = \quad 19,888$$

$$\text{purchase } P_4 = \$80,000 = - \ 80,000$$

$$\text{NPW at } 15\% = +\$11,636$$

Interpolating,

$$15\% + \frac{11,636}{11,636 + 1,682} \times 5\% = 19.37\%$$

Under the conditions stated, the "expected" rate of return on the investment in the standby tractor is 19.4 percent.

Remember the "expected," or average, availability was used as input to obtain specific values needed for the analysis, therefore the resulting rate of return output also is an "expected" rate of return. There is a probability associated with this rate of return value. Since it is the expected value, there is a 50 percent probability that the actual rate of return will be 19.4 percent *or more,* and a corresponding 50 percent probability that the actual rate of return will be 19.4 percent *or less,* as well as a practically zero percent probability of *achieving* exactly 19.4 percent on the investment. Further, to find the rate of return that would occur with a 90 percent probability, the standard deviation of the availability would have to be known. Then the solution would utilize values corresponding to the probability desired. Of course all this assumes "normal" data distribution, which often is a valid assumption.

PROBABILITIES USED TO SELECT THE BETTER ALTERNATIVE

Probabilities can also be used to select the least cost alternative from among a number of possible choices. For instance, in designing culverts, an engineer may estimate the cost of several alternative culvert sizes and types. The probability of exceeding the design capacity is calculated and an estimate made of the value of damage occurring in the event capacity is exceeded. These variables may be related as follows to determine the design with the lowest net cost.

EXAMPLE 13.5

A culvert is being designed to pass under a highway. The cost of repairs in case the highway is overtopped is estimated at $20,000. If i is estimated at 6 percent and $n = 20$ for each culvert, which of the following is the most economical?

Type of culvert	Cost new ($)	Probability (p) of overtopping in any one year
24 in. RCP	20,000	0.20
2 × 4-ft box culvert	40,000	0.10
Two 2 × 4-ft twin box	60,000	0.05

Solution.   The probability of overtopping the 24-in. RCP is given as 0.20 in any one year, for an annual equivalent of $4,000 *expected* damage. ($20,000 × 0.20 = $4,000/yr.) If the annual equivalent damage is calculated for each type of culvert and added to other annual costs, the lowest cost culvert may be determined as follows.

Type culvert	24-in. RCP	2 × 4 ft	Two 2 × 4 ft
Annual equivalent to cost new = $\underbrace{P(A/P, 6\%, 20)}_{0.08718}$	$1,743	$3,487	$5,230
Annual cost of overtopping p × $20,000	$4,000	$2,000	$1,000
Total	$5,743	$5,487	$6,230

The single 2 × 4 ft is the most economical under these circumstances.

GRADIENT IN LOSSES

In many situations the value of the loss will change depending on the year in which it occurs. For instance, if the value of the property subject to flooding increases every year due to increased development, improvement, or inflation, the loss incurred if flooding occurs in the tenth year will be greater than in the first. Either an arithmetic or geometric gradient may be used to transform gradient losses into either present worth or annual equivalents. For example, assume the probability of loss in any one year is 10 percent, or 0.10. Assume further that if the loss occurs the first year, the damage will amount to $1,000. If it occurs the second year, the damage is estimated as $1,500, and the third year at $2,000. These damages may be tabulated as follows, assuming i as 7 percent.

Year, n (col. 1)	Amount of loss, L, if occurring this year (col. 2)	Probability, p, of loss this year (col. 3)	Loss liability accounted for this year (col. 4)	PW of loss liability $Lp(P/F, 7\%, n)$ (col. 5)
1	$1,000	0.1	$100	$ 93
2	1,500	0.1	150	131
3	2,000	0.1	200	163
				$387

The sum of the present worths of the loss liability in column 5 also may be expressed as

$$P = \sum \text{col. } 5 = Lp(P/A, i, n) + \Delta Lp(P/G, i, n)$$

or

$$P = \$1,000 \times 0.1\underbrace{(P/A, 7\%, 3)}_{2.6244} + \$500 \times 0.1\underbrace{(P/G, 7\%, 3)}_{2.5061} = 262 + 125$$

$$= \underline{\underline{\$387}}$$

If the cost of losses is desired in terms of the annual equivalent, dollars per year for each of n years, the equation may be written as $A = pA + pG(A/G, i, n)$.

Comparisons Involving Gradients and Present Worth

Comparisons of either PW or AW are equally valid for determining the lowest cost alternative. In addition a gradient factor may be needed, since areas where development is occurring will probably suffer greater damage in future years. The next example illustrates both of these features.

EXAMPLE 13.6

A culvert is needed in a growing neighborhood. If the culvert is overtopped, the damages initially are estimated at $40,000 per flood. The cost and probability of overtopping of each of three alternatives are given in the following table:

Culvert	Cost of culvert	Probability of overtopping in 1 yr
A	$ 45,000	0.10
B	20,000	0.20
C	100,000	0.05

The life of the structure is estimated as 20 years, and $i = 6$ percent.

1. Which is the most economical selection?
2. A new estimate indicates that the damages from overtopping will be $40,000 if overtopping occurs the first year, but will increase $10,000 per year thereafter due to the rapid development of the area. Which is now the best choice? (Find PW of each alternative.)

Solution 1

Structure	Probability of damage p	Estimated damages if overtopped ($)	Expected annual damage, A ($)	PW of damage $A(P/A,$ 6%, 20) ($)	Cost of culvert ($)	PW of total cost ($)	
A	0.10	40,000	4,000	45,800	45,000	90,880	lowest total
B	0.20	40,000	8,000	91,760	20,000	111,760	cost
C	0.05	40,000	2,000	22,940	100,000	122,940	

Solution 2.   If the damage estimate increases $10,000 every year, then

gradient, $P_1 = G(\underbrace{P/G, 6\%, 20}_{87.230}) = p \times 10,000 \times 87.230 = p \times 872,300$

base cost $P_2 = A(\underbrace{P/A, 6\%, 20}_{11.47}) = p \times 40,000 \times 11.47 = p \times 458,800$
of damages

Struc-ture	p	$P_1 = G(P/G,$ 6%, 20) = p × 872,300 ($)	$P_2 = A(P/A,$ 6%, 20) = p × 458,800 ($)	P_3 cost of culvert ($)	PW of total cost p ($)	
A	0.10	87,230	45,880	45,000	178,110	
B	0.20	174,460	91,760	20,000	286,220	⎧ lowest
C	0.05	43,615	22,940	100,000	166,555	⎨ total
						⎩ cost

USES OF EXPECTED VALUE INVOLVING JOINT-OCCURRENCE PROBABILITIES

Often probabilities are calculated for a series of alternatives that successively reduce the probability of damage. For example, hydrology studies may report incremental flood levels from heavy rains in terms of probability, such as one in one hundred probability ($p = 0.01$) of reaching 20 ft above a certain norm, with other probabilities of reaching other levels.

Such a hydrologic tabulation may appear as follows:

Flood height above datum level (ft)	Yearly probability of exceeding height
0	1.00
2.0	0.15
4.0	0.07
6.0	0.02
8.0	0.005
10.0	0.0001

At first glance this table may appear to violate Axiom 2, in that the vertical sum of probabilities exceeds 1.00. However, further examination shows that the events are not *mutually exclusive.* A flood height exceeding 4 ft also exceeds 2 ft. The probabilities contain joint occurrences (not mutually exclusive), and Axiom 2 can only show that if the yearly probability of exceeding a flood height of 2.0 ft is 0.15, the yearly probability of *not* exceeding 2.0 ft is $1 - 0.15$ or 0.85. Also, the probability of having a flood height of between 2.0 and 4.0 ft would be $0.15 - 0.07 = 0.08$. The following example illustrates the analysis of this type of problem.

EXAMPLE 13.7

An apartment complex is subject to periodic flooding. A consultant's study shows that the complex can be protected from successive increments of flood water levels, but each higher level of protection costs more, as shown in the table.

Alternative	First cost ($)	Annual maintenance cost ($)	Provides protection against flood of depth (ft)	Probability of flood exceeding expected depth in any 1-yr period	Estimated loss when given depth is exceeded ($)
A	none	none	0.5	1.0	50,000
B	50,000	1,000	1.0	0.2	800,000
C	100,000	1,500	1.5	0.1	1,000,000
D	300,000	2,000	2.0	0.05	1,400,000
E	600,000	3,000	3.0	0.02	1,900,000
F	900,000	4,000	4.0	0.01	2,000,000

Solution. The equivalent "expected" annual costs of flood damages are calculated for each alternative, and using net annual worth analysis, the alternative with the lowest net equivalent cost is selected. Beginning with the highest cost alternative, the annual equivalent expected cost (EC) of flood damages for alternative F is simply

$$EC(F) = 2,000,000(0.01)$$
$$= \$20,000$$

The EC of alternative E is slightly more involved since the probability of a flood over 3.0 ft is $P = 0.02$, but the probability of a flood over 3.0 ft also includes a flood of over 4.0 ft (joint occurrence). Therefore, the probability of a flood occurring with a depth *between* 3.0 and 4.0 ft must be determined as $p(3 - 4 \text{ ft}) = 0.02 - 0.01 = 0.01$. Then the EC of alternative E is calculated over the range between E and F:

$$EC(E) = 1,900,000(0.02 - 0.01) + 2,000,000(0.01)$$
$$= 19,000 + 20,000 = \underline{\underline{\$39,000}}$$

In a similar manner the expected cost within the range of each alternative is

$$EC(D) = 1,400,000(0.05 - 0.02) + 39,000$$
$$= 42,000 + 39,000 = \underline{\underline{\$81,000}}$$

$$EC(C) = 1,000,000(0.1 - 0.05) + 81,000$$
$$= 50,000 + 81,000 = \underline{\underline{\$131,000}}$$

$$EC(B) = 800,000(0.2 - 0.1) + 131,000$$
$$= 80,000 + 131,000 = \underline{\underline{\$211,000}}$$

$$EC(A) = 50,000(1.0 - 0.2) + 211,000$$
$$= 40,000 + 211,000 = \underline{\underline{\$251,000}}$$

Calculation of total costs on the basis of equivalent annual costs, using an interest rate of 10% and an n value of 20 years, yields the following:

Alternative	Annualized first cost ($) $P(A/P, 10\%, 20)$ 0.1175	Annual maintenance costs ($)	Annual "expected" flood costs ($)	Total annual costs ($)
A	0	0	251,000	251,000
B	5,875	1,000	211,000	217,875
C	11,750	1,500	131,000	144,250
D	35,250	2,000	81,000	118,250 ⎧ lowest
E	70,500	3,000	39,000	112,500 ⎨ cost
F	105,750	4,000	20,000	129,750 ⎩ alternative

Based on "expected" values of damages, the lowest cost alternative is E.

INSURANCE COSTS

People buy insurance not to avoid loss but to pool the risk of loss. Problems involving insurance costs necessarily must consider the cost of the risk of loss. These are basically probability problems and can be solved as illustrated in the following example.

EXAMPLE 13.8

A new civic center is proposed with an estimated insurable value of $6,500,000. You are asked to report on the feasibility of an automatic sprinkler system costing $440,000 (which will not add to the insurable value of the building). This will reduce the annual fire insurance premium cost from 1.30 to 0.50 percent. Half of the fire insurance premium cost is used to pay the actual insured fire loss, while the other half pays overhead, sales, and claims-investigation costs and profit. The records indicate that total losses from a destructive fire will be about two and one-half times the insurable losses. The annual cost of operation and maintenance of the sprinkler system is estimated at $4,000. Assume the life of the sprinkler system is 20 years and the civic center will be financed by bond funds that are expected to cost 6 percent. Is the sprinkler system a good investment?

Solution

	Without sprinkler	With sprinkler
Annual insurance premium cost		
0.013 × $6,500,000 =	−$ 84,500	
0.005 × $6,500,000 =		−$ 32,500
Actual *insured* fire loss (estimated as one-half insurance premium cost)		
Reimbursed fire loss		
$84,500/2 =	+$ 42,250	
$32,500/2 =		+$ 16,250
Actual *total* fire loss		
2.5 × 42,250 =	−$105,625	
2.5 × 16,250 =		−$ 40,625
Annual O & M for sprinkler		−$ 4,000
First cost, sprinkler		
$A = 440,000(A/P, 6\%, 20) =$		−$ 38,359
0.08718		
	−$147,875	−$114,234

Conclusion.   The sprinkler is recommended under these circumstances.

REQUIREMENTS FOR A PROBABILITY PROBLEM

A wide variety of practical problems are susceptible to analysis if they can be defined in terms of "probability" or "percentage-of-occurrence" problems. The requirements follow:

1. There must be an exposure to a certain type of event involving a measurable loss or gain. (Example: exposure to washout of an existing storm water drainage culvert.)

2. There must be enough history of similar exposures to the event and of the resulting loss or gain so that a probability p, of future similar events occurring can be estimated (p = number of events actually occurring divided by the number of exposures to the event). (Example: history of stream indicates the existing culvert is not quite adequate for a 1 in 25 year storm or $p = 0.04$).

3. The equivalent periodic value of the event equals the probability of the event multiplied by the value of the event (loss or gain). (Example: if loss from washout is $5,000, then the equivalent periodic value is $A = 0.04 \times \$5,000 = \$200$ equivalent annual loss.)

4. If the equivalent periodic value of preventing or causing the event is less than the equivalent periodic value of the event, then the investment required to prevent or cause the event is numerically justifiable. (Example: if the equivalent annual cost of enlarging the culvert, $A = P(A/P, i, n)$ is more than $200, keep the existing culvert. If not, enlarge it.)

EXAMPLE 13.9

A contractor finds that adding an extra sack of cement per cubic yard of concrete reduced the probability of concrete cylinder test failure from 1 in 20 ($p = 0.05$) to 1 in 50 ($p = 0.02$). There is one test taken for every 20 yd^3. Each failure costs him an average of $2,500 worth of additional tests, delays, arguments, and replaced concrete. Cement costs $3 per sack. Should he add the extra cement?

Solution

$$\text{cost of failure} \times \text{probability of failure} = \text{expected cost (EC) per test}$$
$$\text{without extra cement } \$2,500 \times .05 = \$125 \text{ EC per test}$$
$$\text{with extra cement } 2,500 \times .02 = \underline{\quad 50}$$
$$\text{expected savings per test} = \$\,75 \text{ per test}$$

There are 20 yd^3 poured for each test taken, so an expected savings (ES) of $75 per test results in an ES of $3.75/yd^3. The cost of preventing the failure is one sack of cement per cubic yard, or a cost of $3/yd^3. Thus the contractor will find it profitable to add the extra sack of cement per cubic yard under the stated circumstances.

The event does not necessarily have to involve a loss. It can just as easily be a gain, as the following example illustrates.

EXAMPLE 13.10

A road-building contractor finds that infrared air photos will sometimes reveal previously unreported adverse drainage conditions, and other times will reveal good borrow sites, likewise previously unreported. He finds these events occur about 1 time in 20 ($p = 0.05$). When they do occur, they result in an average benefit of about $10,000. The cost of having each prospective job photographed is $300. Should the contractor have infrared photos made on each job?

Solution.   The benefit realized is $10,000 \times 0.05 = \$500$ average per job. The cost is $300, so under the stated conditions, the photos more than pay for themselves.

Comparing "probabilities" with "percentage-of-occurrence" problems reveals a great deal of similarity. For instance, if the probability of having a machine out of service at any given time is 1 in 20, then $p = 0.05$, and it is averaging 5 percent downtime or 95 percent availability. The two terms, *probability* and *percentage of occurrence*, can often be used interchangeably.

SPREADSHEET APPLICATIONS

	A	B	C	D	E	F	G
1	EXAMPLE 13.4						
2	Solving example 13.4 using the EXCEL spreadsheet program						
3							
4	Given:	Availability of tractor is 95% for 1st year, drop 4% year after					
5		Production average	200	yd3/hr			
6		Bid Price (income)	$ 1.20	/yd3			
7		Expenses	$ 100	/hr			
8		Purchase Price	$ 80,000				
9		Salvage Value	$ 40,000				
10		N	5	yr			
11		Usage	1,000	hr/yr			
12		Standby (100%-75%) =	0.25				
13							
14	Step 1:	Gross maximum income :	$ 240,000	/yr			
15		Cell C14 = C6*C5*C11					
16							
17	Step 2:	Loss of income due to downtime w/o standby					
18		Year	Downtime	Cost			
19		1	0.05	$ 12,000		Cost	= $C$14 * Downtime
20		2	0.09	$ 21,600		Cell B20 = B19 + 0.04	
21		3	0.13	$ 31,200		Cell B21 = B20 + 0.04	
22		4	0.17	$ 40,800			
23		5	0.21	$ 50,400			
24							
25	Step 3:	Loss of income due to downtime w/ standby					
26		Year	Downtime	Cost			
27		1	0.05	$ 3,000		Cost = $C$14* Downtime* $C$12	
28		2	0.09	$ 5,400		Cell B28 = B27 + 0.04	
29		3	0.13	$ 7,800		Cell B29 = B28 + 0.04	
30		4	0.17	$ 10,200			
31		5	0.21	$ 12,600			
32							
33	Step 4:	The amount saved by purchasing the standby unit					
34		Year	Saved by stanby	Cash Flows			
35		0		$ (80,000)		Cell C35 = -C8	
36		1	$ 9,000	$ 9,000			
37		2	$ 16,200	$ 16,200			
38		3	$ 23,400	$ 23,400			

(Spreadsheet Continued)

	A	B	C	D	E	F	G
39	4	$ 30,600	$ 30,600				
40	5	$ 37,800	$ 77,800		Cell C40 = B40 + C9		
41							
42	Step 5:	The expected rate of return (IRR) is :		19.30%			
43		Cell D42 = IRR(C35:C40)					
44							
45	Conclusion:						
46	In order to find the expected rate of return (IRR) on the proposed investment in the standby						
47	tractor, the above steps have to follow. By using IRR function in EXCEL, we do not need						
48	to do trial and error to find the IRR. The IRR function is faster than to find the IRR manually						
49	by using the i table. The IRR for this example is 19.30%						

SUMMARY

Many situations arise where a range of values is possible. Under such circumstances, if the variability is random, the mathematics of probability and statistics may be helpful. Variable data may be handled by using the "expected value," EV, of an outcome.

$$EV = \sum_{j=1}^{m} p_j O_j$$

The EV is nothing more than the weighted average, $\bar{x}$, and for random data there is a 50 percent probability that any actual value will meet or exceed the EV.

To solve engineering economy problems involving risk situations for which probabilities can be assigned, the expected outcomes are calculated and then used to calculate equivalent values for comparison purposes.

PROBLEMS FOR CHAPTER 13, PROBABILITY EVALUATION

Problem Group A

Involves probabilities.

A.1. An apartment complex is subject to periodic flooding from storms. Whenever the floods reach floor level, the cost of damage approximates $100,000. A consultant has calculated the following information.

Alternative	First cost ($)	Annual maintenance cost ($)	Probability of flood reaching floor level in any year
Do nothing	0	0	0.05
Improved storm sewer A	15,000	500	0.01
Improved storm sewer B	25,000	600	0.006

Neglecting salvage value, which of the three alternatives should be selected if $i = 10$ percent and the complex is expected to last 25 years?

A.2. A 30-unit apartment is being considered for investment purposes. For an initial investment of $350,000 for land and $850,000 for the complex, the following probabilities and net incomes are predicted.

Probability	Net income ($/yr)
0.4	100,000
0.4	130,000
0.2	180,000

(a) Assume the apartment complex will last 20 years with no salvage value, and that increases in rent will offset maintenance and inflation. Calculate the "expected" rate of return on the investment. (*Hint:* How do you treat the future value of the land?)

(*Ans.* 9.3%)

(b) What is the probability that *less* than the expected rate of return will be realized?

A.3. A construction company adds a markup on each job of 10 percent for profit. Past records indicate the company has, on any given job, a 50 percent probability of making its 10 percent profit; a 10 percent probability of making 15 percent profit; a 30 percent probability of making zero profit; and a 10 percent probability of *losing* 10 percent.
(a) On any given job what is the expected profit?
(b) The company calculates it can go to computer-based estimating, and by so doing improve its probabilities of making a profit. The new profit probabilities are estimated as follows:

making 10% profit	65% probability
making 15% profit	15% probability
making zero profit	15% probability
losing 10%	5% probability

If the computer can be used for five years and then sold for one-half the purchase price, and the MARR is 15 percent, how much can the company afford to spend on the computer for a contract volume of $70,000,000 per year [contract volume = contract cost $\times$ (1 + profit markup)]?

   (c) The company purchases a computer for $4,500,000. The new probabilities for making its profit turn out to be

making 10% profit	55% probability
making 15% profit	15% probability
making zero profit	20% probability
losing 10%	10% probability

The actual company contract volume turned out to be $55,000,000 per year. If the computer lasted ten years with no salvage value, what rate of return did the company make on the investment?

(*Ans.* 9.3%)

A.4. To protect against a burglary, you are considering the purchase of a burglar alarm. The estimated cost of a burglary would be $10,000, and you figure the probabilities of burglary are 0.01 in any given year. The cost of the burglar alarm is $550, and the sellers tell you your probability of burglary would be reduced to 0.003. Assuming the alarm would last 15 years with no salvage value, is this a good investment, if $i = 9$ percent?

A.5. There is a 0.01 probability that a certain waste-water lift pump will fail *in any given year.* If failure occurs, it will require an expenditure of $78,000. With a minimum attractive rate of return of 15 percent, what is the justifiable present expenditure to *reduce* the risk from 0.01 to 0.002 for a 25-year period?

(*Ans.* $4,034)

A.6. Same as Problem A.5, except that the period is 20 years.

A.7. A real estate developer is planning a subdivision in a valley subject to infrequent flooding. If he constructs extra drainage facilities for $34,000 and maintains them for $1,000 per year, he can reduce the probability of flood damage in any one year from 0.05 to 0.005. If the cost of a flood, should it occur, is estimated to be $178,000, and the developer expects to pay for all flood damages for a period of 15 years, should he install and maintain the extra drainage facilities? Use $i = 10$ percent.

(*Ans.* He should install the extra drainage facilities.)

A.8. Same as Problem A.7, except that the extra drainage facilities will cost $54,000, and $i = 15$ percent.

A.9. An apartment complex is subject to periodic flooding from storms. A consultant has calculated the following information:

Alternative	First cost ($)	Annual cost ($)	Probability of flood damage in year	Estimated cost if damaged by flood ($)
Do nothing	0	0	0.05	100,000
Storm sewer A	15,000	500	0.01	100,000
Storm sewer B	25,000	600	0.006	100,000

Neglecting salvage value, which alternative should be selected if $i = 12$ percent and the complex is expected to last 20 years?

A.10. Insurance to guard against water damage costs $1,000 per year, payable in advance. Water damage can be expected from severe storms, which occur with a frequency of once every 20 years. If water damage does occur, the cost will be $22,000. Assuming interest rates of 12 percent, should the insurance be purchased, if the business expects to stay in the location for the next seven years?

A.11. Income from student parking violations on a university campus generated the following incomes during the last 12 months.

Months	Monthly revenues from fines ($)
1–4	9,000
5–9	6,500
10–12	2,000

If the revenues continued in this sequence every year for a five-year period, and were deposited at the end of each month in an interest-bearing account at 0.75 percent per month, what amount would be accumulated at EOY 5?

(*Ans.* $473,000)

A.12. Same problem as A.11, except that the deposits earn an annual nominal interest of 12 percent, compounded monthly.

A.13. An enterprising student offers you $1.00 if you can toss a coin three times in succession landing on heads, provided you pay $0.25 if you fail to throw the three heads. If 1,000 people are willing to gamble, how much money should this student make (or lose)? (The probability of throwing a head is 0.5.)

A.14. A culvert is to be selected for installation where an arterial street will cross a creek. If the culvert is of insufficient size, so that water flows over the street, damages to the street are estimated to be $10,000. (Assume that the upstream and downstream damages are unaffected by the design alternative.) For interest at 6 percent and a 25-year life, which alternative would be most economical for the city to provide?

Alternative	Construction cost ($)	Probability[a] that capacity will be exceeded in any one year
A	30,000	0.25
B	45,000	0.10
C	65,000	0.04
D	95,000	0.02

[a] The set of occurrences is all possible rainfalls.

(*Ans.* Choose alternative B.)

A.15. Commercial Leasing Inc. has an option on a piece of property that they believe is an ideal location for construction of an office complex. However, much of the site is subject to occasional flooding. The Local Flood Plain ordinance permits construction in the flood zone as long as the improvements do not interfere with the free passage of water. The estimated loss to the building and contents is estimated at $300,000 if the flood exceeds the floor elevation. The cost of constructing the building is increased by $35,000 for each 1 foot the building is raised.

(a) Find the most economical floor elevation if the building life is 20 years and $i = 15$ percent.

(b) Assume the estimated loss increases by $30,000 each year.

Flood elevation (ft)	Expectancy of flood exceeding given elevation
360	1.0
361	0.2
362	0.1
363	0.05
364	0.02
365	0.01
366	0

A.16. Provincial Drilling Ltd. has been adding 10 percent to the estimated project cost (direct plus indirect costs) of each job for management, risk, and profit. The area in which the firm operates has an extremely complex geology; therefore, cost estimates are subject to random variation. An audit to the firm's operators indicates the following.

Percent profit (loss)	Percentage of jobs (probability)
10% profit	40
15% profit	15
no profit	30
(10% loss)	15

(a) What is the expected profit on a given job?

(*Ans.* 4.75%)

(b) A geological engineer calculates that additional exploration and improved bid-preparation procedures will result in the following probabilities.

Percent profit (loss)	Probability (%)
10% profit	60
15% profit	20
no profit	10
(10% loss)	10

If a job is expected to cost $2 million, how much can the firm expend on additional exploration and bid preparation and still make the profit that they previously achieved?

A.17. A contracting firm in a relatively freeze-free area is considering the question of whether or not to provide weatherproofing for its building projects. An investigation of the weather records in this area shows that during the past 20 years there have been an average of 3 weeks per year of weather cold enough to slow down the work. It is estimated on the average that this will reduce the value of the construction produced during those weeks by about 50 percent, with no reduction in costs. Average weekly gross value put in place during normal weather is estimated at $2,000 per week. The initial investment in weather-protection equipment is estimated at $15,000. The life of the weather-protection equipment is 20 years. Annual labor cost for setting up the equipment is $150. Average annual fuel cost is $100. If it is assumed that this will give complete protection against loss due to cold weather, will it be desirable to purchase and operate the weather-protection devices? Assume interest at 12 percent.

A.18. A friend wants you to do $100,000 worth of earthmoving now to launch a speculative real estate development. You would receive $200,000 cash in a lump sum in five years when the development is sold. What percent return on your investment of $100,000 would this amount to, assuming compounded interest? If you took advantage of 10 opportunities similar to this, and due to their speculative nature, 5 failed (you received nothing) and 5 succeeded (you received the $200,000 as promised) what percent return would you realize on the group of 10 (same as if the one project has a 50 percent probability of success)?

A.19. You are working for a contractor with 100 employees and an annual payroll of $2,600,000. Workmen's Compensation Insurance premiums are paid at the rate of 3 percent of the total payroll. The accident record for the firm shows an average of $40,000 worth of insured claims per year. The total loss of the company is estimated at about $120,000 (of which $40,000 is repaid by the insurance company) due to the extra paper work, lost time of other employees, and the general confusion and lack of productivity that always accompanies an accident of any kind. The insurance company is suggesting a safety training course for all employees under the following conditions.

1. Cost of the entire course paid by the contractor is $300 per employee.
2. The course is given on company time, so loss of services will amount to about 20 hours per employee at an average of $12 per hour.
3. The insurance company's experience with the course shows that accident claims will be cut 12 percent the first year, but will gradually return to their former level over a six-year period at the rate of 2 percent per year. They therefore recommend a similar safety course be offered at the end of every four years.
4. They offer to cut their insurance charges from the present rate by 12 percent the first year, 10 percent the second, 8 percent the third, and 6 percent the fourth.

   Determine whether the insurance company's proposal is a wise investment of the company resources, assuming one of these courses is held once every four years ($i = 12\%$).

A.20. A new classroom building on campus is proposed with an estimated insurable value of $10,000,000. You are asked to report on the feasibility of adding more firewalls and other fireproofing elements costing $500,000 (the insurable value of the building will also rise to $10,500,000). This will reduce the annual fire insurance premium cost from 1.00 percent of the value of the building to 0.60 percent. In either case, the fire insurance premium cost is based on an estimate that half the premium will be needed to pay actual insured fire loss, while the other half pays insurance company overhead, sales and claims-investigation costs, and profit. The records indicate that total losses from a destructive fire will be about two and one-half times the insurable losses. (Example: the fire causes $2.50 worth of losses of business, time, records, and intangibles for each $1 worth of insured loss reimbursed by the insurance company. The owner suffers a net loss of $-\$2.50 + \$1.00 = -\$1.50$ in addition to insurance premium payments.) The annual cost of maintenance of the extra fireproofing is estimated at \$3,000. Assume the life of the classroom building is 50 years with no salvage value, and that it will be built on bond funds with $i = 7$ percent. Is the fireproofing a good investment?

Problem Group B

Involves joint probabilities or gradients.

B.1. The county engineer asks your help in determining the most economical culvert design for a new county road. The costs of each alternative design are tabulated as follows.

Culvert design	Designed to be overtopped once every	Cost of culvert ($)	Annual maintenance ($/yr)	Gradient in maintenance each year ($/yr)
A	5 yr	5,000	500	50
B	10 yr	7,000	800	75
C	25 yr	12,000	1,000	100
D	50 yr	20,000	1,500	150

If the culvert is overtopped the first year, the resulting damage will approximate \$20,000. Due to increasing development in the area, the resulting damages will increase by about \$3,000 for each succeeding year. Thus if overtopping occurs during the second year, the damages will be about \$23,000, and so on. Which is the most economical design if the life of the culvert is estimated as 20 years and $i = 8$ percent?

(*Ans.* culvert C)

B.2. You are designing a parking lot for a shopping center and are concerned about pedestrian safety for people passing to and from their parked cars. Pedestrian walkways can be constructed and protected with reinforced planter boxes at a cost listed below. Without the protected walkways, an accident involving a pedestrian and an auto is expected to occur to one customer in 100,000. If the protected walkways are installed, the probability drops to 1 in 1,000,000 customers. The parking lot contains 825 parking spaces. Each space will host an average of 3.1 cars per day, and each car will contain an average of 1.5 customers. Assume 300 shopping days per year. Each accident costs an average of \$5,000 if it occurs the first year, with an increment of \$500 for each subsequent year in which the accident occurs. (If the accident occurs the second year the cost is \$5,500, the third year \$6,000, and so on.) Use $i = 15$ percent, $n = 20$ years, all costs charged at EOY. Costs of reinforced concrete planter boxes and pedestrian walkways follow:

$$\text{first cost} = \$200,000$$
$$\text{life} = 20 \text{ yr}$$
$$\text{maintenance} = \$10,000/\text{yr for the first 5 yr}$$
$$\text{increase in maintenance} = \$1,000/\text{yr/yr for the last 15 yr}$$
$$(\text{e.g., } \$11,000 \text{ for EOY 6, } \$12,000 \text{ for EOY 7, etc.})$$

(a) Find which is the more economical solution, (i) installing the planter-box protected walkways and paying for fewer accidents, or (ii) omitting the planter-protected walkways and paying for more accidents.

(b) Are there other considerations than just the economics? (If yes, list).

B.3. A contractor has a problem involving security against theft. The firm is building a new home office and shop compound for their own use. According to police records, in any one year there is a 1 in 5 chance of someone breaking and entering the compound. There will be an average of about $100,000 worth of easily portable office equipment, small, tools, supplies and cash on the premises that could be targets for the thieves. If a thief enters, there is a 50 percent probability that all the easily portable office equipment will be stolen. To protect the premises, the contractor has four alternatives: (a) electronic detection devices connected to police headquarters, (b) guard dogs, (c) security services (watchmen), or (d) theft insurance. The costs and benefits are as listed in the table. Find the equivalent annual costs of each alternative and tell which is most economical overall if $i = 10$ percent.

(*Ans.* electronic detection devices)

	Probability of breaking in	Capital cost ($)	Life (yr)	Salvage value ($)	Annual O & M ($)	Annual increase in O & M ($)
(a) Electronic detection devices	0.05	20,000	10	2,000	2,500	200
(b) Guard dogs	0.04	2,000	5	0	3,000	300
(c) Security service	0.03	0	20	0	8,000	1,000
(d) Theft insurance	0.00	0	20	0	10,000	1,000

B.4. Same as Problem B.3, except the property is expected to increase in value by about $10,000 per year as the company expands.

B.5. You are called to consult on the construction of a motel on the seafront in an area where the probability of hurricane-force winds in any one year is 1 in 12. A standard design is estimated to cost $512,000 but is not expected to survive a hurricane without severe damage. A hurricane-resistant design is estimated to cost $702,000. It is estimated that the standard design will suffer damage of 90 percent of the value current at the time of the hurricane plus loss of business costs amounting to 40 percent of the cost of the damage. Assume further that the hurricane-resistant design will survive 98 percent of the hurricanes intact, but will suffer 50 percent loss of value (including damage and loss of business) in the other 2 percent. The replacement value of both designs increases by $40,000 per year. Use $i = 9$ percent and an estimated life of 30 years for the motel. Which design is most economical and by how much in terms of annual costs?

B.6. An earthmoving contractor has a fleet of equipment consisting of ten scrapers, three tractors, one grader, and one compactor. The average income for each machine is now $40 per hour, but is expected to increase by $4 per hour each succeeding year. Each machine should earn income for an average of 1,500 hours per year, but currently about 14 percent of this time the machine is not working because of mechanical problems of some sort. A lubrication expert advises the contractor that the downtime could be cut from the present 14 percent down to 10 percent by using additional planned preventive maintenance. The plan involves purchase of a lube truck for $25,000, plus hiring another person for $12,000 per year, with the salary increasing $1,000 per year each additional year. The lube truck will cost about $4,000 per year for fuel, supplies, and operating expenses, and can probably be sold in five years for $5,000. The current interest rate for the contractor is $i = 15$ percent.

(a) Is this a good investment over a five-year period under the stated circumstances?

(b) If it turns out that the downtime is actually reduced only to 12 percent, was it a good investment?

B.7. Hydrology studies indicate that damage from flooding can be reduced by the addition of an earthen flood-control dam. Anticipated costs and probabilities for flooding are given below.

Design	First cost ($)	Maintenance cost ($)	Dam height (ft)	Probability of flooding over dam	Estimated costs when flooding occurs ($)
Do nothing	0	0	0	1.0	50,000
A	280,000	5,000	5.0	0.1	80,000
B	360,000	6,000	6.0	0.05	200,000
C	400,000	7,000	7.0	0.03	400,000
D	540,000	9,000	8.0	0.01	800,000
E	800,000	9,900	9.0	0.001	1,500,000

What is the expected net annual cost of each design, and which design should be selected if $i = 7$ percent and the dam is expected to last 40 years with no salvage value?

(*Ans.* Design A has the lowest net annual cost of $50,703 per year.)

B.8. Same as Problem B.7, except $i = 6$ percent, and the life of the dam is infinite.

B.9. A client is planning to build an office park and is trying to decide between two alternative locations. The basic difference between the two sites is the fire protection available. Site A will cost $165,000 for the land only, whereas site B can be obtained for $125,000. The office park after construction will be insured for $3,000,000. The annual insurance premium on site A is 0.60 percent of the insured value, while the premium on site B is 0.85 percent. Over the years, about one-half of these premiums will be paid back to your client as compensation for actual fire losses occurring at your park. The total cost of the fires to your client will closely approximate three times the compensation paid him by the insurance company. With $i = 12$ percent, which is the more economical site? $n = \infty$.

CHAPTER 14

Benefit/Cost Analysis

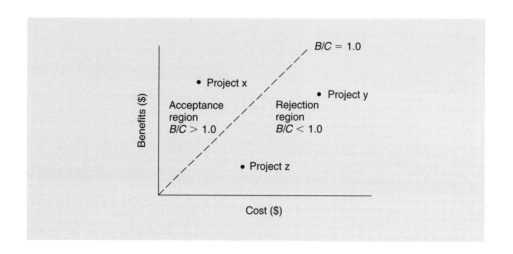

COMPARING BENEFITS TO COSTS

A commonly used method of evaluating the relative worth of a proposed project is called the benefit/cost (B/C) ratio method. As the name implies, the method consists of comparing the "equivalent net worth" of the *benefits* to the "equivalent net worth" of the *costs,* expressed as a benefit to cost (B/C) ratio. Usually the net annual worth is used, although the net present or future worth can also be used with identical results. For example, a proposal with estimated benefits of $20,000 per year and estimated costs of $10,000 per year has a ratio of $B/C = \$20,000/\$10,000 = 2.0$.

The B/C ratio method became widely used following passage of the federal government's Flood Control Act of 1936, which stated that federal funding for improvements to waterways for flood control would be justified only "if the benefits to whomsoever they may accrue are in excess of the estimated costs." Thus the B/C ratio can be used to evaluate projects involving benefits to persons other than the organization incurring the costs. The principle has been applied to many types of public projects, from airports to zoos, and is now an accepted measure of desirability for projects at most levels of government.

Today, with our expanding population and ever-growing transportation needs, new applications are being developed to determine which projects will be funded by departments of transportation and which will not. These new B/C applications include the costs, or adverse effects on the transportation user ("disbenefits"). The use of benefit/cost analysis associated with "user cost" is now being adopted by most progressive departments of transportation because of public insistence that funds are spent in the most advantageous way possible. The cost and benefit to the user is now part of the equation for determining project selection and is also becoming part of modern bidding and contracting approaches such as the $A + B$ and Lane Rental methods. Problems dealing with user costs and associated funding decisions are included in this chapter.

The B/C ratio method is similar in some respects to the present worth method, the annual worth method, or the rate of return method discussed in previous chapters. Keeping in mind that the benefits and costs must be expressed in "equivalent" terms, the B/C ratios of one or more projects are calculated. A common practice among mutually exclusive projects is to find the B/C for each of the competing projects and select the one with the highest B/C ratio. While this usually yields a valid ranking, in some cases it may lead to erroneous selections, as the following example illustrates.

EXAMPLE 14.1

Analysis of several mutually exclusive roadway alignments yields the following information ($i = 7\%$).

	X	Y	Z
Annual benefit ($)	375,000	460,000	500,000
Annual cost ($)	150,000	200,000	250,000
B/C ratio	2.5	2.3	2.0

Solution. On the basis of B/C ratio, alignment X appears the best because it has the highest B/C ratio. However, if the NAW of the costs is simply subtracted from the NAW of the benefits the following results are obtained:

	X	Y	Z
Annual benefits ($)	375,000	460,000	500,000
Annual costs ($)	150,000	200,000	250,000
Net annual worth ($)	225,000	260,000	250,000

Examining the resulting NAW values, it is evident that alternative Y has the higher NAW at $260,000 per year. This NAW is $35,000 per year higher than X; however, the *cost* of Y is $50,000 per year higher than that of X. The question then arises, "Does the extra $35,000 per year of NAW justify the additional $50,000 per year cost of Y?" To answer this question, an *incremental B/C* ratio is found (which is analogous to the incremental rate of return of Chapter 11), beginning with the lower cost pair $Y - X$. The incremental B/C ratio is determined by subtracting lower cost X from the next higher cost Y as follows:

Y minus X

$$(B/C) = \frac{460,000 - 375,000}{200,000 - 150,000} = \frac{85,000}{50,000} = 1.7 \quad B/C > 1$$

Each additional dollar invested in alignment Y over X yields a benefit of $1.70. Therefore, assuming all other aspects are equal, Y is preferred over X, and we discard X. Then compare current candidate Y to the next higher cost alignment Z as follows:

Z minus Y

$$(B/C) = \frac{500,000 - 460,000}{250,000 - 200,000} = \frac{40,000}{50,000} = 0.8 \quad B/C < 1$$

Each additional dollar invested in alignment Z over Y yields a benefit of $0.80. Therefore, Y is preferred over Z, and Z should be discarded. Select Y.

Comment: Y is the better alternative even though it has a lower B/C ratio than X (it also has a lower rate of return), because it offers the greatest benefit for the total expenditure. The comparison may be viewed more clearly by dividing alternative Y into two parts. First separate out the first $375,000 of benefits together with the first $150,000 of costs. This part of the alternative Y yields the same B/C ratio as alternative X, or $B/C = 2.5$. Then the remaining incremental part of Y yields a B/C ratio of $85,000/50,000 = 1.7$. The composite of the two parts of Y combines to yield a B/C of 2.3. Therefore if a B/C ratio of 1.7 is worthy of funding from the sponsoring agency, then project Y competes with project X and yields the same B/C on the first $150,000 worth of annual expenditures. In addition it outranks project X by providing an additional B/C of 1.7 on the remaining $50,000 annual investment.

B/C RATIOS

The usual method for computing the benefit/cost ratio is to calculate all benefits and costs on an annual basis. Thus initial construction and other capital costs are converted to equivalent annual costs by application of the appropriate $(A/P, i, n)$ factor. Similarly, future salvage values are converted to equivalent annual cost "savings" and subtracted from the construction and other capital costs. Benefits to others, in the form of cost *savings* and income, whether lump sum, annual, or gradient, are also converted to annual values. Occasionally the sign convention may be confusing. Using the conventional sign notation (benefits are positive and costs are negative), the B/C ratio sometimes turns out to be a negative number even when the project obviously has more benefits than costs. To make the B/C ratio positive under these circumstances use the following rule:

> B/C RULE: Above the line in the numerator, all benefits are positive, while all costs are negative. Below the line in the denominator, all costs are positive, and all benefits are negative.

Most texts assume the reader can discern any obvious discrepancies in sign notation and can adjust for them where appropriate.

User Benefits

When new highways or other public facilities are constructed, the user usually enjoys some savings in costs (such as reduced travel costs) when the new facility is compared to the existing one. These savings may occur due to less time required to travel a certain distance (less congestion, fewer stops or delays at traffic signals, shorter route, etc.), lower fuel consumption, less wear, fewer accidents, or other similar reasons. Savings of this type are determined by estimating the **total annual cost to the users for the present facility** (this cost is designated U_p) and subtracting from it the **total annual cost to the same number of users for the future facility** (designated U_f). Thus the **net user benefit,** designated U_n, is calculated as $U_n = U_p - U_f$. If the net costs to users of the future facility are less, the resulting U_n will be positive. The terms $(U_p - U_f)$ and U_n both designate *net* user benefits and are used interchangeably.

Owner's Costs

Owner's costs are divided into two categories: (1) capital costs, and (2) maintenance costs.

1. Capital costs are usually considered as the construction, acquisition, or other costs of investing rather than operating and maintaining a facility. If a replacement facility is being considered, the question arises of how to handle the present value of the existing facility. The present value is the cash salvage value that would be received now if the existing facility were sold or demolished. If the existing facility is maintained in place, this cash value is left invested in the existing facility. Therefore the capital cost of the existing facility is the cash salvage value. The cost of the proposed facility is *not* reduced by the worth (salvage value) of the existing facility, since the value of the existing facility is an owner's asset that can be freely kept or spent like cash or any other asset. To handle the sign notation, the following terms are defined.

$C_f =$ Equivalent capital cost of proposed (future) facility, usually expressed on an annualized basis.

$C_p =$ Equivalent capital worth of the existing facility (present salvage value), usually expressed on an annualized basis. This is the value that either can be received in cash or left invested in the existing facility.

$C_n = C_f - C_p = $ *Net* capital cost of replacing the present facility with the future facility. Note that if the worth of the present facility is greater than the cost of the proposed facility (an unusual situation), the net capital cost could be negative. Normally, using this chapter's sign convention, Cn will be positive.

2. Operation and maintenance costs are the owner's costs for operating and maintaining the facility (O & M). If a replacement facility is being considered, its O & M costs may be more, or less, than the O & M costs for the present facility. Thus, let us define

$M_f =$ Equivalent operating and maintenance costs of the future (proposed) facility, usually expressed as an annual (or annual equivalent) cost.

$M_p =$ Equivalent operating and maintenance costs of the present (existing) facility, expressed in the same terms as M_f.

$M_n = M_f - M_p =$ *Net* operating and maintenance cost of the proposed facility over the present facility. Mn may be either positive or negative.

SEVERAL METHODS FOR CALCULATING B/C RATIO

One of the major problems facing the engineer in determining B/C ratios is deciding which items to include in costs and which items to include in benefits. Several variations of methods exist for determining B/C ratios. As a result, considerable engineering judgment is required both in originating B/C studies and in evaluating B/C studies done by others.

The two most common approaches to computing the benefit/cost ratio are known as (1) the conventional B/C, or AASHTO, system,* and (2) the modified B/C method. Both of these were developed to compare a proposed facility with an existing facility. (If there is no existing facility, use the present cost of reaching the same objective. For example, if the proposed facility is a bridge over a river, and there is now no existing method of crossing the river at this point, then the present user cost is the cost of whatever existing route must be taken to reach the other side.)

---

*"Road User Benefit Analysis for Highway Improvements," American Association of State Highway Officials, Washington, D.C., 1960, pp. 27–28.

1. *Conventional B/C.* The benefits (usually annual) are considered to be only user cost savings (reduction of cost is considered a benefit). Therefore,

$$U_n = U_p - U_f$$

Most proposed projects are designed to reduce user's costs. Therefore the net benefits are equal to the user costs of the present facility *minus* the reduced user costs of the proposed (future) facility.

$C_n + M_n =$ Costs consist of the annual equivalent costs to the *owner* of the facility, including capital costs *and* maintenance.

The numerator consists of all of the user's benefits; the denominator is the sum of *all* the owner's costs. Regarding sign notation, in the denominator any increase in costs results in a net positive cost. Thus,

$$\text{conventional } B/C = \frac{\text{net savings to users}}{\text{owner's net capital cost} + \text{owner's net operating and maintenance cost}}$$

$$= \frac{U_n}{C_n + M_n} = \frac{B_n}{C_n + M_n} \tag{14.1a}$$

Where a portion of the benefits accrue to the owner as owner income (I_n), this is added to the benefit total in the numerator. For instance, if a new toll road saves a motorist $20 in terms of gas, time, and so forth, but the toll charge is $5, then the (per vehicle) user net is $U_n = \$15$, and the owner's net income $I_n = \$5$. The total net benefit is $B_n = U_n + I_n = \$15 + \$5 = \$20$. If the owner's income is included, the conventional B/C ratio becomes

$$\text{conventional } B/C = \frac{\text{net user savings} + \text{net owner income}}{\text{net owner capital cost} + \text{net owner operating and maintenance cost}}$$

$$= \frac{U_n + I_n}{C_n + M_n} = \frac{B_n}{C_n + M_n} \tag{14.1b}$$

2. *Modified B/C.* This method uses the same input data, but net operating and maintenance costs (Mn) are treated as negative benefits (or *dis*benefits) rather than as costs. Thus, they are placed in the numerator rather than in the denominator. The resulting equation is

$$\text{modified } B/C = \frac{U_n - M_n}{C_n} = \frac{B_n - M_n}{C_n} \tag{14.2}$$

The sign convention is the same as for the conventional B/C. In the numerator any net increase in benefit (or decrease in cost) is positive, and in the denominator, any net increase in cost is positive.

In Chapter 8, where the net annual worth (NAW) was shown to be

$$\text{NAW} = \text{AW benefits} - \text{AW costs}$$

the problem of treating maintenance costs (as a cost or as a *dis*benefit) did not arise. The sign convention automatically handled the problem because the *net* value was determined. When a ratio is used, however, the results can be difficult to interpret as shown in the next example.

EXAMPLE 14.2   (Comparison of conventional and modified *B/C*)

Find the conventional *B/C* for alternate routes X and Y proposed to replace an existing route between two points; $i = 8$ percent, $n = 20$ years.

	Existing route	Proposed route X	Proposed route Y
Construction cost ($)	0	100,000	100,000
Annual equivalent to construction cost $A = \$100,000\underbrace{(A/P, 8\%, 20)}_{0.1019}$	0	$ 10,190	$ 10,190
Estimated user's cost ($/yr)	200,000	165,000	195,000
Owner's operating and maintenance costs ($/yr)	250,000	270,000	240,000
Total annual cost ($)	450,000	445,190	445,190
Annual savings over existing route ($/yr)	0	4,810	4,810

Solution.   Substituting in the *conventional B/C* equation yields *Route X compared to existing route:*

$$\frac{U_p - U_f}{(C_f - C_p) + (M_f - M_p)} = \frac{200,000 - 165,000}{(10,190 - 0) + (270,000 - 250,000)}$$
$$= \frac{35,000}{30,190} = \underline{\underline{1.16}}$$

Route Y compared to existing route:

$$\frac{U_p - U_f}{(C_f - C_p) + (M_f - M_p)} = \frac{200,000 - 195,000}{(10,190 - 0) + (240,000 - 250,000)}$$
$$= \frac{5,000}{190} = \underline{\underline{26.32}}$$

By comparison, the modified B/C yields the following results.

Route X:

$$\text{modified } B/C = \frac{(U_p - U_f) - (M_f - M_p)}{C_f - C_p}$$

$$= \frac{(200,000 - 165,000) - (270,000 - 250,000)}{10,190 - 0}$$

$$= \frac{15,000}{10,190} = \underline{\underline{1.47}}$$

Route Y:

$$\text{modified } B/C = \frac{(200,000 - 195,000) - (240,000 - 250,000)}{10,190 - 0} = \frac{15,000}{10,190}$$

$$= \underline{\underline{1.47}}$$

Thus with identical capital investment, the annual savings for routes X and route Y are identical at \$4,810 per year, yet the conventional B/C ratio yields B/C of 1.16 for X and 26.32 for Y. The modified B/C ratio on the other hand yields B/C ratios of 1.47 in each case. *This and similar comparisons indicate that the modified B/C method usually yields more consistent results.*

COMPARISON OF *B/C* RESULTS WITH RATE OF RETURN

It is interesting to compare the B/C evaluation of the alternate routes with an evaluation using the rate of return method from Chapter 10.

EXAMPLE 14.3

Find the rate of return of the investment in route X or route Y (same for each) from Example 14.2.

Solution

$$(U_p - U_f) - (M_f - M_p) = C_f - C_p$$

For both cases

$$\$15,000 = \$100,000(A/P, i, 20)$$
$$0.150 = (A/P, i, 20)$$

By interpolation,

$$
\begin{aligned}
i &= 14\% & 0.1510 &= (A/P, 15\%, 20) \\
i &= ? & 0.1500 &= (A/P, i, 20) \\
i &= 13\% & 0.1424 &= (A/P, 12\%, 20)
\end{aligned}
$$

$$
i = 13 + \frac{0.0076}{0.0086} = \underline{\underline{13.9\%}}
$$

Discussion:

(a) Alternatives X and Y each have the same rate of return, as indicated by the modified B/C method, rather than the differences indicated by the conventional B/C method.

(b) The modified method indicates that for every $1 invested, there will be a benefit of $1.47, which some might interpret as a 47 percent rate of return. Careful analysis indicates the actual rate of return is only 13.9 percent.

What to Compare

B/C studies by their very nature usually involve a comparison of two or more alternatives. One of these alternatives should be the existing facility or method, or erroneous results may occur. It is possible, for instance, to compare an expensive new bridge across a bay with a more expensive new tunnel. The result could be a recommendation for construction of the bridge, simply because the existing route involving a short trip around the end of the bay was not considered in the comparison.

Sometimes there is no existing facility, as in the following instances:

1. Government regulations require construction of a sewage treatment plant where none was in service previously.

2 An interstate highway requires a river, railroad, or intersecting road crossing, and no existing crossing meets the standards.

3. A new facility is required due to population growth or growing demand for service.

Where there is no existing facility to which a comparison may be made, caution is urged both in the formulation of B/C calculations as well as in the examination of B/C studies by others.

User Benefits: User benefits in the B/C ratio are ordinarily supposed to represent the reduction in user's cost resulting from construction of the new facility. Where there is no existing facility it is difficult to establish a legitimate reduction in cost. Some practitioners have filled this void with a substitute alternative of their own selection to replace the missing existing facility. This immediately raises some question concerning whether the favorable B/C ratio resulting from such a comparison between two proposed facilities (versus a comparison of existing versus proposed) really represents a real benefit to the users. For an extreme but plausible example, a B/C comparison between a pedestrian overpass over a quiet residential street and a pedestrian tunnel under that street could be formulated to show a high B/C ratio in favor of the one or the other, but would likely be of little actual benefit to the pedestrian, who naturally prefers a street-level route. This satire on some actual B/C studies is illustrated by the following theoretically logical example.

EXAMPLE 14.4

A pedestrian overpass is estimated to cost $100,000 to build, and $40,000 per year to maintain. A pedestrian tunnel will cost $200,000 to construct and $10,000 per year to maintain. The expected life of each is 20 years, and $i = 6$ percent. User costs, before and after, are zero, since no one is expected to use either facility.

Solution. The annual equivalent of the capital costs are calculated as shown.

$$C_{\text{overpass}} = \$100,000 \underbrace{(A/P, 6\%, 20)}_{0.0872} = \$8,720$$

$$C_{\text{tunnel}} = \$200,000 \underbrace{(A/P, 6\%, 20)}_{0.0872} = \$17,440$$

The B/C calculations may then be carried out as follows.

$$\text{modified } B/C = \frac{(U_0 - U_t) - (M_t - M_0)}{C_t - C_0} = \frac{0 - (10,000 - 40,000)}{17,440 - 8,720}$$
$$= \underline{\underline{3.44}}$$

According to the example, the investment in the tunnel shows a B/C of 3.44, or an "implied" benefit of $3.44 for every $1 invested.

This example illustrates what can occur when B/C studies do not include a comparison with the existing facility or method. The B/C method is an excellent tool when properly applied, but obviously considerable care and some discretion are required in its use.

FURTHER COMMENTS ON *B/C* RATIOS

The reader should be aware that under some circumstances one or the other of the B/C ratios may not be usable at all. Considerable caution is urged when devising a B/C study, or examining one presented by others. The following cases are cited as extreme examples. However, bear in mind that as actual problem situations approach these extremes, the results may show signs of skewing toward the limits indicated by the extremes.

1. If there is a zero saving to users, then $U_n = 0$.

$$\text{conventional } B/C = \frac{U_n}{C_n + M_n} = \frac{0}{C_n + M_n} = 0$$

$$\text{modified } B/C = \frac{U_n - M_n}{C_n} = -\frac{M_n}{C_n}$$

A project should be justifiable, if the *savings* in maintenance (represented by a negative M_n) and operating costs more than offset the capital cost. This could not be demonstrated with the conventional B/C but could be by the modified B/C.

2. If the project results in a savings to users (Un is $+$) and maintenance and operating costs drop more than capital costs rise ($Cn + Mn$ is $-$), then

$$\text{conventional } B/C = \frac{U_n}{C_n + M_n} < 0$$

$$\text{modified } B/C = \frac{U_n - M_n}{C_n} => 1$$

The conventional B/C ratio becomes a negative number, whereas the modified B/C properly accounts for the benefit of lower maintenance to the owner.

3. If there is no change in the capital investment required ($C_n = 0$), then the conventional B/C can be substituted to show the relative merits of the project.

$$\text{conventional } B/C = \frac{U_n}{C_n + M_n} = \frac{U_n}{M_n}$$

$$\text{modified } B/C = \frac{U_n - M_n}{C_n} = \frac{U_n - M_n}{0} = \infty$$

The modified B/C reflects the benefits derived from capital investment only. Since in this example the added capital investment required is zero, any benefit at all from zero investment yields an infinite B/C ratio. The conventional B/C compares the user's benefits to the owner's costs. Therefore, no matter how much the owner may benefit by lower capital or operating costs, the user must benefit commensurately, or the project acquires a low B/C rating under the conventional B/C system. Two other factors affecting the validity of B/C output with regard to other variables should be examined carefully when evaluating a B/C study. These are cash flow and dollar volume.

Cash Flow

The B/C method of project comparison can be relatively insensitive to timing of cash flow if low interest rates are used. Thus the timing may require separate scrutiny apart from the B/C result itself. For instance, under certain circumstances a cost now of $100,000 resulting in a $400,000 benefit occurring 60 years from now could outrank (by means of a higher B/C) an alternative expenditure of $100,000 now that would bring a $140,000 benefit immediately. The following example illustrates how this could happen.

EXAMPLE 14.5

1. Use $i = 1$ percent. Alternative X, which involves a $400,000 benefit occurring 60 years from now, has an annual benefit equivalent to

$$B = \$400,000 \underbrace{(A/F, 1\%, 60)}_{0.0122} = \$4,880/\text{yr}$$

Assume an expenditure of $100,000 required right now to produce this $400,000 future benefit. The annual equivalent cost is

$$C = \$100,000 \underbrace{(A/F, 1\%, 60)}_{0.0222} = \$2,220/\text{yr}$$

Therefore, the B/C may be calculated as

$$\text{alternative X: } B/C = \frac{\$4,880}{\$2,220} = \underline{\underline{2.20}}$$

Assume now a competing alternative project Y involves

$$\text{benefit of \$140,000 occurring immediately} = \$140,000$$
$$\text{cost of \$100,000 occurring immediately} = \$100,000$$
$$\text{alternative Y: } B/C = \frac{\$140,000 \ (A/P, 1\%, 60)}{\$100,000 \ (A/P, 1\%, 60)} = \underline{\underline{1.4}}$$

2. For a more realistic value of i, use $i = 7$ percent.

$$\text{alternative X: } B/C = \frac{\$400,000(A/F, 7\%, 60)}{\$100,000(A/P, 7\%, 60)} = \underline{\underline{0.67}}$$
$$\text{alternative Y: } B/C = \frac{\$140,000(A/P, 7\%, 60)}{\$100,000(A/P, 7\%, 60)} = \underline{\underline{1.4}}$$

Solution 1.   For $i = 1$ percent, alternative X is preferred.

Solution 2. For $i = 7$ percent, alternative Y is preferred. Alternative X has a B/C ratio of less than 1 and thus should not be selected under any circumstances.

Evaluation of Benefits and Costs

In private enterprise the question of benefits is usually simple, since the benefits are typically measured in dollars of income. When considering public works, however, direct income frequently does not occur, and other types of benefits, both tangible and intangible, must be considered very carefully. Ultimately the decision on what benefits to include is a political decision (using "political" in its best sense). It must be made by elected political decision makers usually with input from both lay citizens and technical advisors. Technical advisors for their part undoubtedly can profit from a continuing input from their peers. Technical meetings of engineers particularly should schedule regular periodic panel discussions and encourage further consideration of methods of evaluating benefits and costs (both tangible and intangible) so that technical people may contribute more intelligently to the political decision-making process.

Returning to the phrase in the federal Flood Control Act of 1936 that reads "benefits to whomsoever they may accrue," the American people usually prefer the users of public works to pay for those works wherever feasible. For example, gasoline tax pays for roads, rate payers support sewer, water, and electric facilities, and so on. Exceptions to user taxes occur when users are difficult to determine, or are too poor to pay. Public policy prefers that public funds derived from a broad tax base should not be used to benefit a few, except in severe hardship cases. Thus the phrase ". . . benefits to whomsoever they may accrue . . ." possibly may be interpreted at times to infer broader coverage than most citizens would desire. Questions continually arise, such as the following:*

1. If a canal is constructed at government expense, is the increase in value of adjoining property a proper benefit to be counted? Arguments:
   a. Pro. The act specifies "benefits to whomsoever they may accrue," which could be interpreted as a few or many without discrimination between large landholders and others. Also, higher land values mean higher property tax payments back to the government and capital gains taxes upon sale. If the property has a higher value, isn't this evidence that it is more useful (or at least more desirable)? A nation's wealth consists of the total wealth of all its citizens. Thus, when value is added to a citizen's property, wealth is added to the nation.
   b. Con. Government funds should be used to benefit large groups of ordinary citizens. If some few landowners reap windfall profits from government projects, let them pay special taxes or assessments on it. "My taxes are not meant to make a few people wealthy" is the feeling of most.

---

*These arguments do not necessarily represent the views of the authors.

2. Are recreational and scenic benefits as valuable as direct cost savings to users? Arguments:

   a. Pro. People spend significant percentages of their income on recreation and aesthetics. The house, furniture, clothes, car, and other important items are usually chosen with appearance (or aesthetics) uppermost in mind. People have, therefore, demonstrated with their pocketbooks that scenic aesthetic benefits are important. Likewise recreation is important to the average American and every year satisfies an increasing need, as evidenced by rising annual expenditures. Benefits should be valued at the cost of traveling to and using the nearest equivalent location multiplied by the number of users.

   b. Con. Not everyone appreciates the dollar value of a tree or a fishing spot. However, a much larger number of people benefit from reduced transportation costs of consumer products. Therefore, direct savings should weigh more heavily than secondary benefits. For instance, discounting secondary benefits by 50 percent seems reasonable.

3. Should recreational and scenic locations that are destroyed due to a proposed project be counted as part of the cost? Arguments:

   a. Pro. Each asset that is destroyed is as much a part of the cost as each dollar spent. Evaluate them as fairly as possible. Lost capital assets are part of the capital costs; lost income is an annual cost. Their market value depends in part on relative scarcity, what other similar assets remain, and how accessible these remaining similar assets are.

   b. Con. Progress requires a certain amount of changes in the landscape. Nothing can be preserved forever. A practical balance must be struck between people and the environment. Property owners have a right to destroy what they have purchased if the resulting project is profitable.

Approach to Applications

Applications of B/C techniques become more complex as user costs become more difficult to determine. For instance, a recreational or public park complex may not show large, direct, tangible benefits. Most will agree, however, that the intangible benefits are very important to individuals, families, and communities. One rational approach to the determination of user benefits is to do the following:

1. Estimate how many users will use the new facility per year.

2. Assume that all of these potential users are now using an existing comparable facility somewhere else. Obviously they are not, but if they use the new facility, there is evidence of frustrated desire to use such a facility; probably the present cost in travel time, money, and effort to get there is just too high. The benefit results from lowering the cost to a level they can afford by building a facility close by.

3. Find the cost, in travel time and money, for them to use the existing comparable facility and use this for present user cost.

4. The cost of using the proposed new facility to these same people in terms of decreased travel time and money is the user cost of the future facility.

5. These costs are then compared to the maintenance and capital costs to determine the B/C ratio.

Applications of the B/C method to other projects associated with community growth are more complex yet. For instance, the addition of a new convention center to a community introduces some interesting questions. User benefits and costs in a community situation may sometimes be estimated as an average accruing to citizens within a reasonable radius affected by the development. A new convention center may have the following effects on those within the service area.

Benefits

1. Convention-goers will spend money while in town, benefiting local merchants and innkeepers.

2. Increases city revenue from sales tax receipts, and property taxes from increased property values in the vicinity of the center.

3. Brings many new visitors to the city, which increases visibility and prestige of the city and in turn attracts industry and community growth.

4. Opportunity to redevelop acreage occupied by convention center in a needy area of town.

Costs

1. Increases traffic in the area immediately around the convention center; greater congestion of existing facilities.

2. Increases cost of municipal services
   a. fire and police protection
   b. utilities services, water, sewage, electric, gas, telephone, solid waste disposal.

3. New jobs bring new people with added costs of services, schools, police, roads, fire, parks, playgrounds, and other public facilities.

4. Increases rain runoff, heat, dust (depends upon previous use of property).

For other projects, a similar type of list could be constructed. The difficulty lies in attempting to estimate the dollar's worth of benefit or cost for each of the items listed and the number of people affected. However some attempts at analysis must be made, and frequently a preliminary attempt will give rise to suggestions for a more detailed and comprehensive subsequent attempt. In any case if the community is to make wise decisions, it needs as much information as possible. Engineers have for too long been largely passive followers in these decision-making processes. Engineers by training and practice have the ability to analyze difficult and complex problems affecting the employment of both our natural and constructed environment. By virtue of that ability, engineers have an obligation to actively probe and research for answers. They should assume leadership roles. Here is an area of challenging problems affecting the enjoyment of life and, some say, the very existence of millions of people. The situation certainly warrants a great deal of creative thinking and logical analysis by engineers. Due to the press of normal professional work, this creative thinking will often be largely on the engineers' own time and of their own volition, at least at the outset.

Road Improvements

Problems requiring an evaluation of the benefits accruing to the public from the expenditure of public road-improvement funds frequently can be resolved by using B/C studies. The user's benefits usually accrue to the motoring public, while the maintenance and capital costs are incurred by the governmental agencies involved, which, of course, are financed by the public. The following example illustrates a traffic signal study.

EXAMPLE 14.6

A city is considering the installation of a traffic-activated signal system at a busy intersection now controlled by a single, preset traffic light. Traffic counts indicate that 20,000 vehicles cross the intersection daily. A major consideration is the number of accidents occurring in the intersection. The existing signal is a single traffic light with maintenance costs averaging $300 per month. The new system has an initial cost of $6,867, with monthly maintenance costs estimated at $500. The life of the new system is assumed as five years. No salvage value is assumed for either system. Current traffic studies indicate the number of vehicles stopping at the intersection is now 8,000 per day. The new traffic-activated system will stop about one half this many, or 4,000 cars per day. The average cost for a motorist to stop at this light is $0.025. Experience with similar installations indicates the new system will reduce accident costs from a current rate of $9,528 per year to an anticipated $300 per year. Find the B/C for the proposed installations (assume $i = 6\%$).

Solution.   The owner's capital cost, O & M costs, and user costs are calculated and then inserted into the B/C equation as follows.

	Annual existing	Annual future proposed
DOT owner's cost per year (C)	0	$6,867(A/P, 6\%, 5) = \$1,630$
O & M costs (M)	$ 3,600	6,000
Facility user costs (C):		
accidents	9,528	300
Stop and start $0.025/vehicle × 365 days/yr × vehicle =	73,000	36,500
Total user costs	$82,528/yr	$36,800/yr

$$\text{modified } B/C = \frac{(U_p - U_f) - (M_f - M_p)}{C_f - C_p}$$

$$\text{modified } B/C = \frac{(82,528 - 36,800) - (6,000 - 3,600)}{1,630 - 0}$$

$$= \frac{43,328}{1,630} = \underline{\underline{26.6}}$$

By way of comparison, the conventional $B/C = 11.4$ for this case.

Intangible Benefits. In addition to the readily quantifiable benefits accounted for in the equation, the following intangible benefits are anticipated.

a. Injury: reduction of pain, suffering, lost productivity, mental anguish of friends and loved ones, and so forth. Direct accident costs are accounted for in the data, but the real cost of accidents often far exceed the costs listed.

b. Air pollution: 4,000 less stops per day will enhance air quality in the area.

c. Driver fatigue and irritation will be reduced by fewer stops and starts.

APPLICATIONS OF *B/C* RATIOS

Where user costs and savings are easily determined, the application of the B/C method is simple and straightforward. This includes many traffic situations such as signalized intersections or alternate routes. A traffic count is made and projected ahead. An estimate is made of user cost and maintenance cost before and after. Capital cost is estimated. The conversion of capital cost to annual cost depends on finding valid n and i. The determination of the useful life, n, is discussed in detail in Chapter 21. Selection of i for public projects is a matter of some controversy. As noted in the examples of this and other chapters, the value of i will strongly influence the result. Very low values of i (promulgated by some public officials) will favor long-term investments yielding low benefits each year. Higher values of i (promulgated by most public officials) will favor shorter-term investments yielding more immediate benefits. Since most public agencies borrow money through the issuance of bonds or some other financial investment (see Chapter 17), they do experience the time value of money in much the same way as private companies. Hence, realistic rates of interest, at least comparable to bond rates for tax-exempt bonds, should be used.

GROWTH IN NUMBER OF USERS

A problem in arithmetic frequently arises when considering user costs. As more people are attracted to the new facility, the total user cost rises, even though the cost per individual user may drop. This can produce misleading results, as illustrated in the following example.

EXAMPLE 14.7

A new bridge is proposed to replace an existing bridge. The data below show current and projected costs. Assume $i = 6$ percent, $n = 25$.

	Old bridge	New bridge
User trips per year	2,000,000	8,000,000
User cost ($/user trip)	0.50	0.25
Annual user cost ($/yr)	1,000,000	2,000,000
Maintenance ($/yr)	200,000	200,000
Capital cost ($) (P)	0	5,000,000
Annual equivalent to capital cost ($)		
$A = P(A/P, 6\%, 25) = \$5,000,000 \times \underbrace{0.0782}_{0.0782} = $		391,000/yr

Solution. The conventional B/C ratio yields the following:

$$B/C = \frac{U_p - U_f}{(C_f - C_p) + (M_f - M_p)}$$

$$= \frac{2,000,000 \times \$0.50 - 8,000,000 \times \$0.25}{(\$391,000 - \$0) + (\$200,000 - \$200,000)}$$

$$= \underline{\underline{-2.56}}$$

The modified B/C ratio also yields the same, since the maintenance costs are a net of zero.

$$B/C = \frac{(U_p - U_f) - (M_f - M_p)}{C_f - C_p} = \frac{-\$1,000,000}{\$391,000} = \underline{\underline{-2.56}}$$

The results in both cases are negative: not only less than 1, but less than zero. Intuitively the investment appears profitable. If a person can cross the existing old bridge at a cost of $0.50 and could cross the new one for only $0.25, there should be a demonstrated saving.

The solution to the dilemma is revealed in Figure 14.1, which shows the relationship between cost and number of user trips per year. The number of users attracted to the facility increases as the cost per trip decreases. Therefore the response is said to be "price-elastic," and the increase in number of trips is assumed to be roughly proportional to the reduction in cost (varying

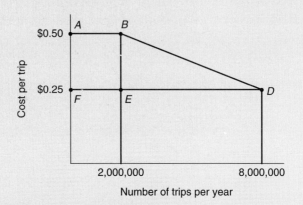

Figure 14.1   User benefits resulting from increasing number of users.

according to line *BD* in Figure 14.1). Actually the price-elastic response line could be represented by a variety of configurations between points *B* and *D*, but a straight line can usually serve as a reasonable approximation. Examining Figure 14.1 it is evident that all 2,000,000 existing user trips benefit by the full $0.25 savings, and the total savings for those user trips will amount to (2,000,000 × $0.25) = $500,000 per year. In Figure 14.1 this saving is represented by the area of the rectangle *ABEF*. If the cost were reduced only from $0.50 to $0.49 per trip, all 2,000,000 existing trips would benefit (by $0.01 per trip), but only a few new user trips would be attracted to the facility. The user benefit would be calculated as the number of existing trips times $0.01 per trip plus the *average* number of new trips times $0.01 per trip. Each added decrement in cost attracts an added increment of trips, forming an incremental *triangle* of trips times savings per trip. The new trips added with each cost decrement are assumed to be attracted at incremental points in time because previous decrements in cost did not represent a savings to them; only with the most recent decrement did a real savings occur, which then attracted that increment of trips. For Example 14.7, reduction of costs from $0.50 to $0.25 results in an added 6,000,000 trips, and the incremental benefit can be calculated as the area of the triangle *BDE*, or [($0.50 − $0.25) × (8 − 2) × 10^6]/2 = $750,000/year benefit to new users plus the $500,000/year benefit calculated previously for the existing users. Using this approach, both the conventional and the modified methods yield

$$B/C = \frac{\$500{,}000 + \$750{,}000}{\$391{,}000} = \underline{\underline{3.20}}$$

EVALUATING COSTS OF LIFE AND PERSONAL INJURY

Attempting to equate life itself with a dollar value seems crass and repugnant. However, every day many lives are lost through accident or other cause. If the cause of death involved negligence, very often the negligent party is required to compensate to some extent the heirs of the deceased. What is just compensation for the loss of a life? Difficult as a judgment may be, opinions are required, and decisions and awards are made based upon these opinions. And difficult as it may be, engineers are often required to rank proposed public works projects in order of priority, recognizing that capital is limited and not all of the proposals can be funded. For example, if funds are available to improve the skid resistance of highway X or highway Y, which one should be selected? One consideration is the savings from reduced accidents. Continuing efforts are underway to evaluate the costs of life and personal injury for use in engineering cost and feasibility studies. For example, when experts are asked to evaluate the cost of a death in order to seek compensation in a court of law, they will frequently use the following approach.

1. Loss of earnings. Assume the deceased was 30 years old, was making $30,000 per year and could expect raises of $2,000 per year until retirement at age 65. The present worth may be calculated at some reasonable investment rate, say 8 percent for illustration purposes, as

$$P_1 = 30,000 \underbrace{(P/A, 8\%, 35)}_{11.6546} + 2,000(P/G, 8\%, 35)$$

$$P_1 = 349,600 + 232,200 = \$581,800$$

Thus the jury would be requested to award the survivors the sum of around $581,800 as compensation for loss of earnings.

2. In addition the attorneys may also ask compensation for loss of (a) love and affection, and (b) physical comfort and other named benefits now lost due to the untimely death. In addition, if the death was accompanied by or caused (c) pain and suffering, and (d) mental anguish or other trauma of a similar nature, compensation may be requested for these damages. In short, almost any type of loss that can be explained and evaluated to a jury's satisfaction (and is not ruled out of order by the judge) can be listed as a compensable loss. The jury then uses its own discretion and judgment in evaluating the claim and awarding whatever compensation, if any, is justified. A similar approach may be taken by the engineer in evaluating priority of expenditure. One unfortunate difference occurs, however. The jury need not regard the ability of the defendant to pay the damages in making the award, but the engineer must always abide by the budgetary restraints of the funding agency. However, most funding agencies have some flexibility and it is quite possible that a proper presentation of the benefits versus the costs of life and injury-saving installations may wring extra funding from even the most adamant agency.

B/C Analysis for Private Investments

Although developed for public projects where benefits are difficult to determine, this method of analysis can be used for private investments. In fact, often it is a very easy method of analysis because benefits and costs are easily quantified. Use of the B/C ratio is analogous to use of the rate of return method described in Chapters 10 and 11. The steps to be followed are exactly the same, and the incremental benefits to incremental costs must be determined in order to obtain valid results. The analysis of Example 11.7 using the B/C ratio is illustrated in the following example.

EXAMPLE 14.8

An engineering firm is considering five different small computers to aid the firm in running complicated and repetitive calculations. Analysis of the various machines, when applied to the firm's business, indicates the following cash flow information (MARR = 12%).

	G	H	F	I	J
First cost ($) ($P$)	18,000	22,500	25,000	28,400	32,000
Annual net income ($)	3,800	4,800	5,000	6,000	6,600
Useful life (yr)	8	8	8	8	8
Present worth of benefits ($)	18,877	23,845	24,838	29,806	32,787
$P_B = A(P/A, 12\%, 8)$					
B/C	1.04	1.06	0.99	1.05	1.02

Solution.   Since H has the highest B/C ratio, one might be tempted to select H, but that would be wrong. At this point the only conclusion that can be made is to discard F because its B/C ratio is less than 1.0 (which means its rate of return is less than 12.0 percent). To solve the problem, each *incremental B/C* ratio must be determined. Thus,

H–G:

$$\frac{\Delta B}{\Delta C} = \frac{23,845 - 18,877}{22,500 - 18,000} = \frac{4,698}{4,500} = \underline{\underline{1.10}}$$

Since $B/C > 1.0$, discard lower cost alternative G.

I–H:

$$\frac{\Delta B}{\Delta C} = \frac{29,806 - 23,845}{28,400 - 22,500} = \frac{5,981}{5,900} = \underline{\underline{1.01}}$$

Since $B/C > 1.0$, discard lower cost alternative H.

J–I:

$$\frac{\Delta B}{\Delta C} = \frac{32,787 - 29,806}{32,000 - 28,400} = \frac{2,981}{3,600} = \underline{\underline{0.83}}$$

Since $B/C < 1.0$, discard higher cost alternative J. The remaining alternative is I. Therefore, select I. This is the same conclusion reached using incremental rate of return, net present worth, and so forth (see Example 11.2).

COST EFFECTIVENESS: UTILITY/COST METHOD OF EVALUATING INTANGIBLES

Often when comparing alternative candidates for purchase or investment, some of the desirable characteristics of the competing purchases or investments are intangible and difficult to quantify. For instance, when selecting a particular exterior surfacing for a building wall, how does the visual appearance of brick compare with the appearance of concrete block, cedar shakes, or other alternatives? When buying a car, how should an evaluation be made of comfort, noise level, styling, prestige, convenience, and other similar characteristics?

The concept of utility value can help quantify these characteristics in order to facilitate a more direct and objective comparison. The utility value itself is a derived number, obtained by comparing the actual level of performance of the candidate system to an ideal desired level of performance or performance goal for an idealized system.

The utility/cost (U/C) format may be employed on either of two levels, defined by the matrix dimension. Either a one-dimensional or two-dimensional matrix may be used. The one-dimensional matrix consists of a left-hand column listing a series of performance goals, together with a top row of column headings containing the alternative candidates to be evaluated against the performance goals. The two-dimensional matrix employs a separate matrix for each candidate. The left-hand column now contains a list of desirable system features generally comparable to the features exhibited by each of the candidate systems, while the top row of column headings contains the list of performance goals.

The step-by-step process for each matrix system is detailed as follows:

One-Dimensional Matrix

1. List a series of desirable features or performance goals in a column to the left side of the page.

2. Evaluate each individual feature or goal in terms of its contribution to the overall objective. Place a number in each row reflecting the relative percentage contribution

of each feature to the objectives as a whole, such that the sum of all these numbers adds up to 100 percent.

3. List the names of each candidate under consideration for acquisition as column headings across the top of the page.

4. Draw a matrix of boxes joining the columns and rows. Then draw a diagonal across each box from upper left to lower right.

5. In the top right triangle of each box, evaluate the performance of each candidate with respect to the goal. Use a 10 for "fulfills goals perfectly," and 0 for "not qualified."

6. In the bottom left triangle, place the product of the evaluation, determined in (5) above, multiplied by the percentage number found in (2) above.

7. Sum the figures in each column from the bottom left triangles only. This sum represents the utility value of each candidate.

8. Divide the sums in each candidate's column by the cost of that candidate and determine a utility/cost ratio. The higher values are preferred, reflecting a higher number of utility value points per dollar spent.

9. Plot the utility value (*y*-axis, or *ordinate*) versus the cost (*x*-axis, or *abscissa*) of each candidate. Begin at the origin and draw an envelope through all points closest to the *y*-axis. Normally the selection is made from one of the points on the envelope.

EXAMPLE 14.9

A contractor needs to purchase a supply of hard hats for all job personnel on a large construction site. Four different brands are available, each with its own special features and price. Determine the preferred selection.

Solution.   The contractor decides on the performance goals for the hard hat and lists them in the left-hand column of Figure 14.2. Each goal is weighted for relative importance to the overall objective. Next each candidate brand of available hard hat is listed at the column head and the evaluation process begun. Each brand of hard hat is rated from 1 to 10 according to its ability to fulfill the individual performance goals, and the rating number is entered in the upper right triangle of each matrix rectangle. Then the ratings are multiplied by the relative importance, and the product is entered into the lower left triangle of each matrix rectangle.

The products are summed vertically, and the resulting sum is the utility value of each candidate brand of hard hat. The utility value is divided by the cost to find the U/C ratio, and a graphical plot is made, as shown in Figure 14.3.

It is evident from the U/C ratio that brand D is the better buy in terms of utility rating points per dollar expended. However, for top-quality rating, an

Performance goals	Relative importance of this goal to overall objectives	Brand A — Rating × importance	Brand A — Rating 1 to 10	Brand B — Rating × importance	Brand B — Rating 1 to 10	Brand C — Rating × importance	Brand C — Rating 1 to 10	Brand D — Rating × importance	Brand D — Rating 1 to 10
Safety	50%	450	9	300	6	350	7	350	7
Appearance	15%	30	2	135	9	45	3	75	5
Weight	15%	45	3	30	2	105	7	75	5
Comfort	20%	60	3	80	4	140	7	100	5
Total utility value	100%	$\Sigma = 585$		545		640		600	
($)Cost/each		$30.00		$22.00		$27.50		$20.00	
Utility/Cost		$\frac{585}{30} = 19.5$		$\frac{545}{22} = 24.77$		$\frac{640}{27.50} = 23.27$		$\frac{600}{20} = 30.00$	

Figure 14.2 One-dimensional U/C matrix for Example 14.9.

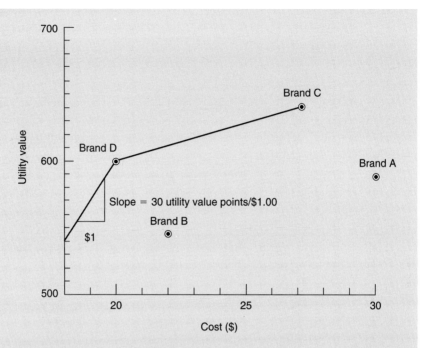

Figure 14.3   One-dimensional U/C graph for Example 14.9.

extra 40 utility points may be obtained by purchasing brand C for an additional $7.50 over brand D. The extra points are purchased at a rate of 40/$7.50 = 5.33 utility points per dollar, which is a far more expensive rate than the 1.5 utility points per dollar paid for brand D. Brands B and A are obviously out of the competition, since they provide fewer utility points at a higher price. Therefore, the evident choice is brand D.

Two-Dimensional Utility/Cost Ratio

The two-dimensional matrix can be employed where a series of desirable system features contribute to achieving the performance goals, and each candidate may contain from one to all of the system features to some degree.

EXAMPLE 14.10

Assume a county engineer needs a good, dependable earthmoving excavator that can operate with reasonable economy as a front end loader and as a trencher. The following steps are used to set up the two-dimensional utility/cost matrix.

1. Establish the primary goals and objectives of the system and enter these system goals as column headings at the top of the matrix. In this case the system goals are the following:
   a. Front end loading
   b. Trenching
   c. Economy of operation
   d. Durability and dependability

2. Weight the system goals according to their importance in terms of their percent contribution to the overall objectives.
   In this example assume the county engineer's priorities are expressed as follows:

a. Front end loading	35%
b. Trenching	20%
c. Economy of operation	20%
d. Durability and dependability	25%
Total	100%

3. Each candidate machine has certain system features that in some way contribute to the attainment of the goals listed above. These system features are listed in the left-hand column of the matrix. For the example, they are listed as follows:
   a. Front bucket size and design
   b. Rear backhoe size and design
   c. Engine type and horsepower
   d. Quality of construction

4. Consider in turn each system goal listed at the top of each column, and in each matrix rectangle below the column head evaluate (in terms of percent) how much each system feature contributes to each system goal. Thus, for the system goal of "front end loading" the front bucket size and design will contribute heavily to achieving the goal, whereas rear backhoe size and design will have very little bearing on front end loading. The matrix now appears as shown in Figure 14.4.

5. Each candidate machine is rated (evaluated) on a basis of 1 to 10 on each of the system features using one full matrix for each candidate. Thus, if candidate machine A has an exceptionally good front end bucket size and design, it rates a 10 in the upper left triangle of the appropriate matrix rectangles, as shown in Figure 14.5.

6. The ratings found in (5) above are multiplied times the percentage found in (4) and entered into the lower left trapezoid of each matrix rectangle.

Candidate Machine A

System features \ System goods	Front end loading 35%	Trenching 20%	Economy of operations 20%	Durability and dependability 25%	Total 100%
Front end bucket size and design	40	0	5	20	
Rear backhoe size and design	0	45	5	25	
Engine type and horsepower	30	25	50	20	
Quality of construction	30	30	40	35	
Total	100	100	100	100	

Figure 14.4 Matrix features versus matrix goals for machine A.

Candidate Machine A, cost = $27,000

System features \ System goods	Front end loading 35%	Trenching 20%	Economy of operations 20%	Durability and dependability 25%	Total 100%
Front end bucket size and design	10 40 / 400	0 / 0	7 5 / 35	6 20 / 120	
Rear backhoe size and design	0 / 0	6 45 / 270	6 5 / 30	5 25 / 125	
Engine type and horsepower	3 30 / 90	6 25 / 150	6 50 / 300	7 20 / 140	
Quality of construction	5 30 / 150	6 30 / 180	7 40 / 280	6 35 / 210	
Total	100 / 640 × 35 = 22,400	100 / 600 × 20 = 12,000	100 / 645 × 20 = 12,900	100 / 595 × 25 = 14,875	$\Sigma = 62{,}175$

$$U/C = \frac{62{,}175}{\$27{,}000} = 2.30$$

Figure 14.5 Two-dimensional U/C matrix for Example 14.10.

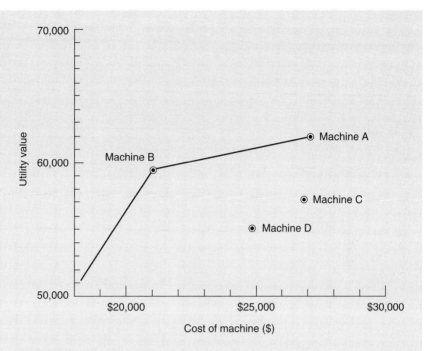

Figure 14.6 Two-dimensional U/C graph for Example 14.10.

7. In each column, the products entered into the trapezoids are summed verti-cally and entered into the trapezoid at the bottom of the column.

8. The sums found in (7) above are multiplied by the percentage amounts found at the top of the column; the product is entered in the lower right tri-angle at the bottom of each column, and summed to the right-hand column for the total utility value score for this candidate machine.

9. The utility value score is divided by the cost of the machine to determine utility/cost ratio, and the utility value versus the cost are plotted on a graph with other competing candidate machines for greater visual clarity, as shown in Figure 14.6.

For Example 14.10, candidate machine B is the preferred selection based on the U/C ratio.

SPREADSHEET APPLICATIONS

Spreadsheets can be of great value in solving benefit/cost problems. Templates can be de-veloped in which values can be changed and quick comparisons made of different oppor-tunities. The procedure for comparing cost alternatives using B/C ratios is the same as that developed in Chapter 11. First determine if each individual project is attractive based on its initial B/C ratio. Projects that meet this minimum requirement, similar to comparing to the MARR, will then be compared. Then make incremental comparisons of the differences between project choices. Reject the higher cost alternative if the B/C ratio is less than 1.0 and accept the lower cost alternative.

	A	B	C	D	E	F
1	**EXAMPLE 14.8**					
2	Solving the previous Example 14.8 using the EXCEL spreadsheet software					
3						
4	Interest Rate = 12%					
5						
6		Project G	Project H	Project F	Project I	Project J
7						
8	First cost ($) (P)	−18,000	−22500	−25000	−28400	−32000
9	Annual net income ($)	3800	4800	5000	6000	6600
10	Useful life (years)	8	8	8	8	8
11						
12	Present worth benefit ($)	$18,877	$23,845	$24,838	$29,806	$32,786
13	$P_B = A(P/A, 12\%, 8)$					
14	B/C for each alternative	1.05	1.06	0.99	1.05	1.02
15	Decision	Accept G	Accept H	Reject F	Accept I	Accept J
16	Reject Project F with a B/C ratio of less than 1.0 and make incremental comparisons of the					
17	remaining project choices					
18						
19	**INCREMENTAL ANALYSIS OF PROJECT CHOICES**					
20		H - G	I - H	F not include	J - I	
21	Delta benefit	$4,968	$5,961		$2,981	
22	Delta cost	4,500	5,900		3,600	
23						
24	Incremental B/C	1.10	1.01		0.83	
25	Decision	Reject G	Reject H	Accept I and Reject J		
26	Because the B/C ratio of the incremental analysis of J-I is less than 1.0, choose the lower					
27	cost alternative I and accept the higher cost alternative J.					
28						
29	CELL CONTENTS: The following cell contents will allow us to evaluate the problem using a spreadsheet.					
30	Row 12 contents are typical of cell B12:	NPV(B4,B9,B9,B9,B9,B9,B9,B9,B9)				
31	Row 14 contents are typical of cell B14:	ABS(B12/B8)				
32	Row 15 contents are typical of cell B15:	IF(B14>1, Accept G , Reject G)				
33	Row 21 contents are typical of cell B21:	(C12-B12)				
34	Row 22 contents are typical of cell B22:	ABS(C8-B8)				
35	Row 24 contents are typical of cell B24:	B21/B22				
36	Row 25 contents are typical of cell B25:	IF(B24<=1, Accept G , Reject G)				
37						

SUMMARY

This chapter presents a useful method of analysis of public works projects, called the benefit/cost analysis, in which the ratio of benefits to costs are determined on an equivalent basis. If the ratio is greater than 1.0, benefits exceed costs, and the project deserves further consideration.

In determining the ratio, problems can arise in defining benefits and costs. While benefits are widely understood to mean cost savings, accident reductions, intangible service, and so forth to the users of the facility, and costs are widely understood to mean the cost to construct the facility, a problem arises in considering the costs of operating and maintaining the facility. Such costs can be considered as either increases in the costs (denominator) or as reductions in benefits (disbenefits) in the numerator. Thus, there are two forms of the B/C ratio.

$$\text{conventional } B/C = \frac{B_n}{C_n + M_n} \tag{14.1}$$

$$\text{modified } B/C = \frac{B_n - M_n}{C_n} \tag{14.2}$$

Sign convention: In the *numerator,* any increase in benefits is positive, while any reduction in costs is also positive. In the *denominator,* any increase in costs is positive, while any reduction in costs is negative. Use of these two formulas can result in different conclusions, and the reader is cautioned to closely examine the arguments presented in the chapter concerning which formula to use. Limitations of this method are also presented, which are very important to its judicious use.

PROBLEMS FOR CHAPTER 14, BENEFIT/COST ANALYSIS

Problem Group A

Given data for alternatively continuing an existing facility or constructing a new one, formulate the conventional and/or modified B/C ratios.

A.1. A group of citizens is petitioning to have a certain gravel road paved. The county engineer works up the following estimates of cost data: $i = 6$ percent, $n = 15$ years.

	Existing gravel	Proposed pavement
Annual maintenance	$2,500/mi	$1,000/mi
Annual gradient increase in maintenance	$200/mi/yr	$10/mi/yr
Road-user costs, annual	$10,000/mi	$4,000/mi
Annual gradient increase in road-user costs	$2,000/mi/yr	$1,000/mi/yr
New capital investment required	0	$20,000/mi

In addition, there will be some safety benefits that have not been quantified.

Find B/C by both conventional and modified methods and comment briefly, in one or two sentences, on your recommendation for which to use. Assume there is another project competing for these funds, and it shows $B/C = 2.5$. Comment on which project should have priority and why. Would you recommend further study? Why or why not?

A.2. Same as Problem A.1, except $i = 8$ percent, and $n = 25$ years.

A.3. Compare these two new alternate routes at an interest rate of 6 percent.

	Existing		Alternate A		Alternate B	
	Life	Cost	Life	Cost	Life	Cost
R/W	∞	0	∞	$400,000	∞	$1,000,000
Pavement	5 yr	$40,000	10 yr	100,000	20 yr	200,000
Structure	10 yr	50,000	20 yr	100,000	40 yr	500,000
Maintenance		20,000/yr		10,000/yr		50,000/yr
User costs		1,000,000/yr		970,000/yr		850,000/yr

Note: The annual cost of a capital asset with an infinite life is simply the cost of the interest payments.

(*Ans.* Choose alternative B.)

A.4. Same as Problem A.3, except that $i = 8$ percent.

(*Ans.* Choose alternative A.)

A.5. The university is considering building 12 additional racquetball courts at a cost of $50,000 each. Assuming their life is 25 years, annual maintenance for each court will be $150 the first year, increasing by $20 per year per year, and there will be no salvage value. An average of two students use each court 5 hours per day for 220 days each year, and a "benefit" of $1.50 is assigned to each student-hour of use. Use the modified benefit/cost ratio analysis and a rate of return of 6 percent.

(a) Is this project justified?

(b) What $/hour benefit would result in a modified B/C of 1.25?

[*Ans.* (a) Since the B/C ratio is 0.76, the project is not justified.]

A.6. A state can install a new traffic light system at an installed cost of $295,000. The system will cost $5,000 per year to maintain. This system is estimated to save police traffic-officer time worth $20,000 yearly. The motorist will save time valued at $35,000 yearly, but will incur extra gasoline and car operation costs of $8,000 yearly. For an i of 8 percent and an economic life of 20 years for the system with no salvage value, what are the conventional and modified benefit/cost ratios for this project?

A.7. A bottleneck exists on University Avenue west of 34th Street where a two-lane bridge constricts the flow of traffic. Using $i = 8$ percent for road bond funds, assume the following data and find the modified B/C for a proposed widening project.

traffic count, year around average = 20,000 vehicles/day
annual gradient in traffic count = 1,000 vehicles/day/yr (second year 21,000 vehicles/day, third year 22,000 vehicles/day, etc.)
traffic mix and costs = commercial traffic, 20% at $10.00/hr
noncommercial, 80% at $2.00/hr

	Existing	Proposed
Cost to DOT for maintenance	$2,000/yr	$4,000/yr
Annual maintenance gradient	200/yr/yr	400/yr/yr
Travel time to pass bottleneck area	6 min	4 min
Capital cost	0	$6,000,000
Estimated life	25 yr	25 yr

(*Ans.* modified $B/C = 2.19$)

A.8. A harbor dredging and enlarging project is proposed for a coastal city. If the harbor is dredged and enlarged, larger container-type ships will be able to dock at the city, resulting in savings of approximately $0.50 per ton on shipping charges. Current tonnage through the port is 1,000,000 tons per year, increasing at the rate of 50,000 tons per year. The cost of the dredging is estimated at $5,000,000. Maintenance of the current harbor now runs about $600,000 per year, increasing about $50,000 per year each year. The life of the project is estimated at 20 years. If the harbor is dredged, the maintenance will cost $800,000 per year and increase by $40,000 per year each year.
    (a) Find the conventional and modified B/C ratios, assuming that bond funds are available for the project at 6 percent.
    (b) Find the conventional and modified B/C ratios, assuming bond funds are available at 9 percent.

A.9. Same as Problem A.8(a), except that current tonnage (at time zero) is 1,000,000 tons and is increasing at the rate of 4 percent per year.

A.10. A new regional park facility is planned, and you are asked to do (a) a benefit/cost study, and (b) a rate of return analysis. The data follow:

Land: Purchase price, $1,000,000. Life is perpetual. This land was bringing in a tax return of $10,000/yr, which will now be lost to the community.

Improvements: Swimming pool, tennis courts, ball fields, picnic area, pavilion. Construction cost $5,000,000. Life, 40 yr. O & M costs are $400,000/yr, increasing $10,000/yr each year.

Use: It is estimated that an average of 1,000 people/day, 365 days/yr will use this facility at the outset. This figure should grow by about 60 people/day/yr (1,060/day in the second year, 1,120/day in the third year, etc.). If this proposed facility were not constructed, these people would have to travel an average of 10 extra miles round trip to find a comparable park. The extra time and travel costs are estimated at $2,000/person. Therefore, the new park is assumed to save each user $2.00/visit.

(a) Find the modified benefit/cost ratio if funds are available at 8 percent.

(*Ans.* $B/C = 1.35$)

(b) Find the rate of return on the capital investment of the taxpayers' funds.

(*Ans.* $i = 10.2\%$)

A.11. Same as Problem A.10(a), except the average of 1,000 people per day (at time zero) should grow by about 6 percent per year.

A.12. A new road is under consideration for connecting two municipalities in the Smoky Mountains by a more direct route. The existing route is 25 mi long and has an annual maintenance cost of $4,800 per mile. Every 10 years major maintenance and resurfacing costs an additional $1,000,000, and these services are due at the present time. The new direct route will be just 15 mi long and will cost $15,000,000. Annual maintenance is estimated at $3,200 per mile, and major maintenance and resurfacing every 10 years is expected to cost $520,000.

   The average speed on the existing route is 45 mph, whereas the new route should allow a 55 mph average. The average annual traffic is estimated at 800,000 vehicles per year, of which 20 percent is commercial. Traffic is expected to increase at about 40,000 vehicles per year until reaching capacity at 2,000,000 vehicles per year. The cost of time for commercial traffic is estimated at $18.00 per hour and for noncommercial traffic at $4.50 per hour. Time charges do not include vehicle operating costs, which are estimated on the existing route at $0.45 per vehicle mile for commercial traffic and $0.18 per vehicle mile for noncommercial. Operating costs on the new route should be 10 percent less per mile. The anticipated life of both proposals is 40 years, with zero salvage value for both. Using $i = 8$ percent, compare the B/C ratio using the modified and the conventional methods.

(*Ans.* conventional $B/C = 3.53$)

A.13. The federal government is planning a hydroelectric project for a river basin. The project will provide electric power, flood control, irrigation, and recreation benefits in the amounts estimated as follows:

Initial cost	−$27,500,000
Power sales	+$ 900,000/yr increasing by $50,000/yr
Flood control savings	+$ 350,000/yr increasing by $20,000/yr
Irrigation benefits	+$ 250,000/yr increasing by $12,000/yr
Recreation benefits	+$ 100,000/yr increasing by $5,000/yr
Operating and maintenance costs	−$ 200,000/yr increasing by $10,000/yr
Loss of production from lost land	−$ 100,000/yr increasing by $5,000/yr

The dam would be built with money borrowed at 6 percent, and the life of the project is estimated as 50 years. The salvage value at the end of 50 years is estimated to be zero. Find the B/C ratio by the (a) conventional and (b) modified methods.

A.14. A university is considering building 12 additional tennis courts at a cost of $65,000 *each*. Assuming their life is 25 years, annual maintenance for each court will be $100 the first year, increasing by $20 each year thereafter, and there will be no salvage value, is this project justifiable if an average of two students use each court five hours per day for 210 days each year and a "benefit" of $3.00 is assigned to each student-hour of use? Use the *modified benefit/cost ratio analysis* and a rate of return of 6 percent.

A.15. Same as Problem A.14, except the life is 15 years, and $i = 8$ percent.

A.16. Same as Problem A.14, except the students are expected to use the facility 150 days each year.

(*Ans.* Since the modified $B/C = 0.90$, the project is not justified.)

Problem Group B

The existing facility may no longer be adequate, so two or more proposed facilities are compared.

B.1. The benefits and costs for four alternative public works projects are given below. Which alternative, if any, should be selected, assuming a minimum attractive rate of return of 7 percent? The "do nothing" alternative is available.

Alternative	Equivalent annual benefits ($)	Equivalent annual costs ($)
A	230,000	191,000
B	142,000	151,000
C	210,000	164,000
D	112,000	76,000

(*Ans.* Choose alternative C.)

B.2. Assume a new direct route from Podunk to Cut-n-Shoot is proposed with two alternate alignments. The data on each alternative are given in the following table. The anticipated life of the road is 25 years, with a negligible salvage value. Anticipated traffic on each of the three alternatives is estimated to be 1,200,000 vehicles this year, and increase each year thereafter by 100,000 vehicles per year. Using the modified benefit/cost ratio method, determine which alternative, if any, should be selected if the minimum attractive rate of return is 7 percent.

	Existing situation	Site A	Site B
Construction cost ($)	0	30,000,000	80,000,000
Annual maintenance ($)	2,000,000	8,000,000	5,000,000
User costs per vehicle trip ($)	58.52	42.37	39.16

B.3. A dam is under construction. The initial investment, based on dam height, is shown below. Also shown are estimated annual savings from reduced flood damage. Based on benefit/cost ratio analyses, which dam height should be selected? Use $i = 8$ percent and assume an infinite life for the dam.

Dam height (ft)	Initial cost ($)	Annual savings ($)
22	21,000,000	1,700,000
24	24,600,000	2,000,000
26	28,400,000	2,300,000

(*Ans.* Choose the 24-ft height.)

B.4. Determine the preferred alternative, using the modified benefit/cost ratio method. The interest rate is 5 percent, and the economic life of each alternative is 25 years.

	Existing	A	B
Initial cost ($)	0	800,000	1,300,000
Annual maintenance ($)	35,000	30,000	20,000
Gradual increase in maintenance/yr ($)	2,000	0	2,000
Annual road user cost ($)	350,000	400,000	250,000

B.5. The benefits and costs for four public works alternatives are given below. Which alternative should be selected?

Alternative	Equivalent annual benefit ($)	Equivalent annual cost ($)
A	182,000	91,500
B	167,000	79,500
C	115,000	88,500
D	95,000	50,000

(*Ans.* Choose alternative A.)

B.6. A small city is experiencing rapid growth and has taken on a number of new city employees over the past several years in new service areas such as traffic, planning, recreation, and so on.

    The one operator at the city hall manual switchboard can no longer handle all the calls, and the telephone company offers the following alternatives for service for the next 10 years (use $i = 7\%$ and, for simplicity, use $n = 10$ yr rather than 120 months).

1. Add a second operator. Each operator gets $880 per month ($10,560/yr), including fringe benefits. This salary is expected to increase by $50 per month each year for each operator ($600/yr). A second manual switchboard would cost $12,000 plus $100 per month ($1,200/yr) service charges for each switchboard. The existing switchboard service could be sold for $1,000 if not continued in service.

2. Install an automatic system for $65,000 plus $600 per month ($7,200/yr) on a 10-year service contract. Assume zero salvage value.

(a) Find the modified B/C ratio. (*Hint:* Consider alternative 1 as the existing system and alternative 2 as the proposed. All operating expenses are either M_p or M_f.)
(b) Does the conventional B/C equation give valid results? Why?
(c) Can the rate of return method be used to compare these alternatives? Why?

Problem Group C

Formulate a B/C study of your own.

C.1. Make up a hypothetical B/C study of one of the proposed new installations listed below or an equivalent of your own selection.

Assume you are a consultant to the city council. They want to know what the benefits and costs to the community will be if the proposed installation is constructed in or near your city.

Set up a hypothetical situation of your own devising. Assume any reasonable values. The main value of this assignment will come from applying your own creative imagination to determining methods of solving this very common but difficult problem. Assuming a reasonable amount of time and financing, spell out in detail how you propose to determine the value of each item in the B/C study. Find the resulting B/C.

Try to keep your solution process as simple as possible. The more complex the process gets, the less it will be understood and used.

shopping center	civic auditorium
water works	university activities center
sewage treatment plant	bond issue to build new schools
steel mill	municipal electric generating plant

Problem Group D

B/C method for company investments.

D.1. A firm is considering three alternatives as part of a production program. The data follow:

	A	B	C
Installed cost ($)	10,000	20,000	15,000
Uniform annual benefit ($)	1,765	3,850	1,530
Useful life (yr)	10	20	30

Assuming a minimum attractive rate of return of 15 percent, which alternative, if any, would you select, based on the B/C ratio method?

(*Ans.* Choose B.)

D.2. Two possible pollution-control devices are under consideration. If a firm's minimum attractive rate of return is 20 percent, which alternative should be selected, using the B/C method?

	A	B
Purchase price ($)	720,000	1,250,000
Annual maintenance cost ($)	82,000	23,000
Useful life (yr)	10	10
Salvage value ($)	0	0

D.3. Estimates for alternate plans in the design of a waste-water facility follow:

	Plan R	Plan S
First cost ($)	50,000	90,000
Life (yr)	20	30
Salvage value ($)	10,000	0
Annual costs ($)	11,000	5,600

For $i = 12$ percent, which plan should be selected, using the B/C method?

(*Ans.* Choose S.)

D.4. Three alternative proposals are available for investment. Using the incremental B/C method, determine which alternative should be selected, assuming $i = 20$ percent.

	T	U	V
Initial investment ($)	12,000	24,000	16,000
Annual net income ($)	2,300	5,500	3,600
Salvage value ($)	0	0	0
Life (yr)	14	15	16

PART THREE

Evaluating Alternative Investments

CHAPTER 15

Taxes

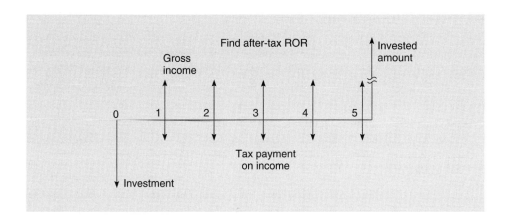

KEY EXPRESSION IN THIS CHAPTER

IRS = The Internal Revenue Service, the agency of the federal government responsible for collecting the income tax.

The citizens of our country require our government to provide a wide variety of public services, such as schools and colleges, national defense, law enforcement, fire protection, construction and operation of roads, parks, and so on. The government supports these services through various kinds of taxes, including property, sales, and income taxes. For the investor, these taxes are a necessary part of the cost of doing business and must be accounted for when estimating costs versus income on any investment.

SOURCES OF TAX REVENUE

Traditionally the greater part of the revenues needed to support the various levels of government has come from three major sources.

Level of government	Major source of revenue
Local (county, city, schools)	Property ad valorem tax
State	Sales tax
Federal	Income tax

421

Over the years, however, as the need for additional sources of revenue became more urgent, the various government levels sought tax revenues wherever they could be found, and some mixing of tax revenue sources resulted. Currently, the largest single source of operating revenue for the federal government is still the income tax. Only about 4 percent of federal revenue comes from excise (sales) taxes. State governments still obtain about one-third of their revenue from sales taxes, while income taxes now account for more than one-third of state revenue. On the local level, while property taxes still account for the majority of revenues, local sales taxes are common, and over 4,000 local jurisdictions also assess income taxes against their citizen taxpayers. These taxing jurisdictions include counties, cities, and school boards. The main criteria for evaluating a potential tax source are generally the following:

1. The cost of collection should be low. (For example, the cost of collecting income tax is less than 2 percent of revenue received, whereas the cost of collecting a $1.00 toll at a bridge could easily exceed 20% of the revenues.)

2. The amount of tax collected should relate to the ability to pay. (For example, the ad valorem property tax relates to the value of the property owned, sales tax relates to purchasing power, and the graduated income tax is proportional to income.)

Millage

Property is usually taxed on the basis of millage. One mill equals one-tenth of one cent (1 mill = 0.1 cent), and the tax rate expressed in millage signifies the number of mills the tax collector expects to collect on each dollar of assessed property value. For instance, if the property is valued by the tax assessor at $30,000 (or "the assessed valuation is $30,000") and the millage rate is 25 mills, then the tax owed will be

$$\$30,000 \times (25 \text{ mills}/\$1 \text{ valuation}) = 750,000 \text{ mills}$$

Since 1,000 mills = $1, the tax bill is $750.

Another method of calculating the same amount of property tax is to recognize that each mill of tax represents a one-dollar tax on a thousand dollars worth of property (1 mill = $1 tax per $1,000 property valuation). Therefore 25 mills of tax would represent $25 tax per thousand dollars of property. Since the property in this case is assessed at $30,000, then the tax is calculated at $25 per thousand times 30 to yield a tax of $25 × 30 = $750.

The county property appraiser (tax assessor) sets a value on each piece of property in the county, while the local governing boards set the millage rates annually. The county tax collector then sends out bills and collects taxes.

Property Taxes

Most local governments (county, municipality, school board) depend on property taxes as a primary source of revenue. Property taxes are usually assessed in proportion to the value of the property and thus are termed *ad valorem* taxes. Basically there are two types of property for tax purposes.

1. *Real property* consists of relatively unmovable property such as land, buildings, and fixed attachments thereto. It is subject to annual ad valorem tax usually by local government only (city, county, schools).

2. *Personal property* consists of any kind of readily movable property. It is subdivided into (a) tangible, and (b) intangible.
   a. *Tangible personal property* is any readily movable property that has inherent value (rather than being a piece of paper representing value), for example, a bulldozer of other movable construction equipment, a car, clothing, jewelry, and appliances that are not fixed to a building. A kitchen range that is freestanding would usually be classified as personal property, whereas a built-in range is real property. Personal property held for income-producing purposes is usually taxed by local government, but personal property held for private use only often is exempted or ignored by taxing authorities.
   b. *Intangible property* consists largely of paper that represents a right or title to the value inherent in some other object, property, or enterprise, for example, money, bonds, stocks, mortgages, notes accounts receivable, and other certificates of indebtedness or promises to pay. Intangible property is often subject to a one-time or annual ad valorem state or local tax, but usually at a very low rate. (For example, one typical state's tax rate varies from 0.1 mill to 1 mill on annual levies, and is 2 mills on nonrecurring assessments such as mortgages.)

Sales Taxes

The largest single source of income for state government in about two-thirds of the states is the sales tax. This usually consists of a fixed percentage of tax on retail sales (services may also be taxed), which is collected by the retailer at the time of sale to the customer and forwarded periodically to the state treasury. The amount typically ranges from 3 to 7 percent. Some cities and counties are empowered by their respective states to levy an additional sales tax and have done so.

Value-Added Tax (VAT)

Several western European countries have been generating significant revenues through a tax called a value-added tax (VAT, first introduced in France in 1954). As a material moves from the raw material stage to the finished product, whenever it changes ownership, the product is taxed at a fixed rate on the value of the product after all costs of manufacture (materials and other expenses) have been deducted. For instance, assume a pipe manufacturer buys a batch of PVC plastic for $100,000 and spends an additional $125,000 in manufacturing costs in turning it into pipe that sells for $250,000. For this manufacturer, the total cost of the product is $225,000, but the market value is $250,000. Therefore, the value added is judged to be the net $25,000 difference. If the VAT rate is

10 percent, then the manufacturer pays a tax calculated as follows:

$$
\begin{array}{rr}
\text{cost of material} = & -\$100,000 \\
\text{cost of manufacture} = & -\ \ 125,000 \\
\text{market value} = & +\ \ 250,000 \\
\hline
\text{value added} = & \$\ \ 25,000 \\
\text{VAT rate} = & 10\% \\
\hline
\text{VAT paid by pipe manufacturer} = & \$\ \ \ 2,500
\end{array}
$$

Then assume that the manufacturer sells the pipe in bulk to a retailer for $250,000. The retailer incurs additional retailing costs of $50,000 in retailing the pipe to individual customers, and is able to sell the pipe for a total market price of $340,000. The retailer pays an additional VAT calculated as:

$$
\begin{array}{rr}
\text{cost of pipe} = & -\$250,000 \\
\text{retailing costs} = & -\ \ \ 50,000 \\
\text{market sales value} = & +\ \ 340,000 \\
\hline
\text{value added} = & \$\ \ 40,000 \\
\text{VAT rate} = \times & 10\% \\
\hline
\text{VAT paid by retailer} = & \$\ \ \ 4,000
\end{array}
$$

Of course in the end the consumer pays through the increased price of the product, so that the impact is similar to a sales tax (both are a tax on consumption rather than a tax on income, or a tax on the value of property owned). There is some interest and support in legislative circles for introducing a VAT in the United States at the federal level.

FEDERAL INCOME TAX LAW

Federal tax law originates in the U.S. House of Representatives, which is a body of 435 individuals each with his or her own ideas of what fair and proper tax law should be. The law begins as a proposal or bill in a committee where members with somewhat different viewpoints reach as equitable a compromise as possible. The bill then may go through other committees, and finally to the floor. At each point some additional input, revisions, and compromises may occur until finally the bill is in a form satisfactory to a majority of the members of both houses (including the Senate with another 100 members), at which point it is voted upon, passed, and, upon approval by the president, becomes law. Obviously it cannot be expected that the resulting document will be an epitome of clarity, nor be without some ambiguity. After the bill becomes law, it is the responsibility of the Internal Revenue Service (IRS) to interpret the law and administer it. In doing this, they have additional input from the justice department and the courts. Where taxpayers disagree with the interpretation given by the IRS, the matter can be taken to the courts, where a binding decision is made either upholding or overturning the IRS interpretation in whole or in part. The IRS of course takes note of the findings of the courts in any future interpretations of the tax law. Since tax law does become rather involved (there are over 10,000 pages of current federal regulations regarding income taxes), it is recommended that specialists be consulted when important investment decisions with tax ramifications are involved. The elements of tax law presented in this text are intended to acquaint the student with the general effect of taxes on investment decisions. In order to clarify and illustrate, problem situations are typically simplified and tax law generalized.

Partners with the Federal Government

In a real sense the federal government is a partner in every taxpaying business. When the business prospers, the government shares in that prosperity, since a large percentage of the profit is paid to the government in taxes. Conversely, when business loses a dollar, the government also loses the part of this dollar that would come to the government as tax. After income taxes and all other expenses for the year are met, the remaining business profit may be invested in new plant and equipment or distributed to the stockholders as dividends. Many firms invest about one-half the profit in expansion and distribute the remaining one-half in dividends. When the dividends are received by the stockholders, they are normally taxed a second time as personal income. If the individual stockholder is in the 28 percent tax bracket, the corporate net income dollar, already reduced to $0.66 by the corporate tax of 34 percent, is further reduced by a personal income tax of $28\% \times \$0.66 = \$0.18$. Thus the stockholder-owner of the corporation gets to keep about $\$1 - \$0.34 - \$0.18 = \$0.48$ out of every corporate dollar of profit. The federal government by comparison receives $0.52. Thus it is obviously in the interest of government itself, as well as the national interest, for Congress to pass laws aimed at creating a climate of prosperity for American business and industry.

Investment Tax Credit

Tax law can be a powerful tool that many governments have learned to use to help stimulate their national economy. Experience teaches that gains in productivity are necessary for increases in standards of living. One key to increasing productivity is to update and modernize productive plant and equipment. Many governments of nations around the world provide tax incentives to business to encourage modernization of productive capacity. With this concept in mind, the U.S. Congress passed its first investment tax credit (ITC) provision in 1962, allowing a percentage of the first cost of certain qualifying investments in plant and equipment as a direct deduction from the amount of federal income tax owed. For instance an investment tax credit allowance of 10 percent on a qualifying investment costing $100,000 results in a *reduction* of $10,000 in the amount of taxes owed. The percentage amount allowed for tax credit changes from time to time, depending upon whether the government feels the economy needs stimulation or restraint. For instance, in 1966 the investment credit was suspended entirely, and then restored in 1967, changed in 1981, and virtually eliminated again in 1986 all according to the Congress' interpretation of national needs.

Congress tends to treat the ITC as a stimulus to the economy, to be applied when the economy sags, and removed as the economy becomes more vigorous. Although the bulk of the ITC provisions were swept out of the 1986 Tax Reform Act, they are likely to be reintroduced whenever the Congress again feels the economy needs such stimulation. Some ITC provisions that endured for extended periods of time include (1) a 20 percent ITC for spending on research at an increased level over the average expenditures of the three preceding years, (2) a 20 percent ITC for rehabilitation of certified historic buildings, and a 10 percent ITC for rehabilitation of other buildings placed in service before 1936.

Tax *Credit* Compared to Tax *Deduction*

Even though the terms *tax credit* and *tax deduction* sound very much alike, there is a significant difference between the two. A tax credit is subtracted from the amount of tax itself, while a tax deduction is deducted from income, before calculating the tax. Therefore a dollar of tax credit saves the taxpayer more than a dollar of tax deduction. For example, assume a business has $100,000 worth of taxable income and is in the 35% tax bracket, so it is about to pay an income tax of

$$35\% \times \$100,000 = \$35,000 \text{ tax payment due}$$

The business now finds it can spend $20,000 on a useful item that is eligible as a tax credit. The tax credit amount is deducted directly from the $35,000 tax due, resulting in a reduction in the tax payment as follows:

$$\$35,000 - \$20,000 \text{ (tax credit)} = \$15,000 \text{ tax payment due}$$

If the $20,000 expenditure does not qualify as a tax credit item but is eligible as a tax deductible item, then the $20,000 is deducted from the $100,000 taxable income before the tax is calculated:

$$(\$100,000 - \$20,000) \times 35\% = \$28,000 \text{ tax payment due}$$

In this example, a $20,000 tax credit is worth $13,000 more to the taxpayer than a tax deduction.

$$\$28,000 - \$15,000 = \$13,000 \text{ tax savings from reclassification}$$

The government may define and set the rates for tax credits and tax deductions in order to influence the economy. In recessionary times the allowances can be increased to stimulate the economy. If the economy becomes overheated and inflationary, the allowances can be reduced or eliminated to slow the economy.

Capital Gains

Capital gains are profits made from the sale of capital assets. Capital assets are properties held for investment or other qualifying purposes. For instance, if an acre of land is purchased for $10,000 and, after a specified minimum holding period, is sold for $15,000, the resulting profit represents a capital gain of $5,000.

A preferential reduced tax rate on long-term capital gains has been a popular feature of the tax code since 1921. For several brief time periods capital gains income was treated as ordinary income. More typically, however, capital gains are accorded preferential treatment. For instance, if shares of stock are sold for more than the purchase price, part of the money received represents the original purchase price, plus any other capital costs involved in the purchase and ownership, while the balance represents a profit. Profits on qualified assets held longer than a stipulated time period (frequently 12 or 18 months) may be considered as long-term capital gains, and a portion of the gain (60 percent, for example) may be excluded from taxable income. For example, assume an investor in the 28 percent tax bracket purchases stock for $10,000. Two years later he sells the stock for $12,000, for a capital gain of $2,000. If the current tax law allowed 60 percent of the capital gain to be excluded from taxable income, then only 40 percent of the gain would be taxed at the investor's ordinary income tax level of 28 percent.

Thus,

$$
\begin{aligned}
\text{selling price} &= \$12,000 \\
\text{purchase price} &= \underline{10,000} \\
\text{capital gain} &= \$\ 2,000
\end{aligned}
$$

40% of capital gain reported as ordinary income

$$
\begin{aligned}
0.4 \times \$2,000 &= \$800 \\
\text{income tax due at 28\%, } 0.28 \times \$800 &= \$\underline{\underline{224}} \text{ tax}
\end{aligned}
$$

If the profit results from the sale of an asset held for less than the stipulated time period, the profit may be considered to be a short-term capital gain and is usually taxed at the same rate as ordinary income.

Tax law sometimes stipulates specific exceptions to the capital gains treatment, such as assets held as stock for regular business. Under this provision, investors in real estate sometimes found that profit from sale of real estate held more than one year was ruled taxable as ordinary income, if the IRS had reason to regard them as dealers, and the land as their stock in trade. This occurred where an investor made a regular habit of buying and selling real estate, or bought large tracts and subdivided for resale.

Personal Income Taxes

Generally tax laws, as envisioned by our legislators, start out as fairly simple straightforward concepts. However, as the members of Congress try to deal with a wide variety of taxpayer situations and different kinds of income and expenses, the provisions of the law become increasingly complex. In addition, as time passes on, legislators may receive new enlightenment on the subject, or the economy changes its pattern, and so the legislators feel obliged to change the tax laws accordingly. Since tax law is prone to change, by necessity this discussion will deal only with the basics of tax law, and the reader is advised to consult with a tax accountant or tax attorney for current regulations concerning any particular investment situation.

Taxable income is defined as income minus allowable deductions. Income includes both *earned* income (wages, salaries, payments for services) as well as *passive* income (investment income, interest, dividends, etc.). Allowable deductions include most of the costs incurred in earning the income. In addition the law typically allows some deductions for personal exemptions, mortgage interest payments, charitable contributions, some state and local taxes, and other similar allowances. The income tax payable is calculated by subtracting the allowable deductions from the income to find the taxable income. With a graduated income tax, the taxable income is then subject to a stepped tax rate. All taxable income above a minimum threshold of exemption and up to a specified step level is taxed at the low beginning rate. Then the remaining taxable income above the specified step level is taxed at one or more higher rates, as illustrated in Example 15.1.

EXAMPLE 15.1

Assume the following tax rates are in effect:

> Tax rate for joint filers (based on table of adjusted income)
> Personal exemptions   = $2,550 for each dependent
> Standard deduction   = $6,700 for joint filer

Then assume a certain married couple filing jointly have income and deductible figures as follows:

$$
\begin{aligned}
\text{Gross income before taxes} &= \quad \$61,280 \\
\text{Allowable deductions for mortgage interest, etc.} &= -\$15,240 \\
\text{2 personal exemptions at } \$2,550/\text{ea} &= -\$\ 5,100 \\
\text{Standard deduction for joint filer} &= \underline{-\$\ 6,700} \\
\text{Net taxable income} &\quad \$34,240
\end{aligned}
$$

The tax is then calculated by using the table provided with the IRS 1040 form. For taxable income between $34,200 and $34,250, the tax owed for persons filing jointly is

$$\text{Tax owed} = \$5,134$$

The average tax rate is determined simply be dividing the amount of tax owed, $5,134, by the taxable income, $34,240. Therefore, the tax rate paid by this joint filer is

$$\$5,134/\$34,240 = \underline{15\%}$$

Corporate Income Taxes

Corporations, like individuals, must pay income tax on profits they earn, usually on a graduated scale. Their taxable income (TI) is determined by taking total earned income and reducing it by allowable deductions:

$$\text{TI} = \text{earned income minus allowable deductions}$$

Allowable deductions include all normal costs incurred in doing business, such as the cost of labor, materials, utilities, depreciation on capital items (Chapter 16), depletion allowance on consumption of natural resources, research and development expenditures, advertisement costs, and interest costs on debt.

The Internal Revenue Service (IRS) tax rate schedule for corporations in one recent year is given in the following table:

Tax rate	Taxable income
15%	First $50,000
25%	Next $25,000
34%	All over $75,000

The tax rate percentages and taxable income thresholds change according to changes passed by Congress and approved by the president, but the general method of calculation generally endures.

Other Income Taxes

In addition to the income tax paid to the federal government, about 44 out of 50 states and over 40 cities levy income taxes also, although the rate is usually much lower than the federal income tax rate.

Effective Tax Rate

Since the tax rate of both individuals and corporations increases on a graduated scale, the question arises as to what tax rate should be used when analyzing a proposed investment. There are two basic approaches to this question.

Marginal or Incremental Tax Rate   One approach is to treat the return from any proposed investment as new income added on top of present income and thus subject to tax at the highest marginal or incremental rate applicable. For example, if a taxpaying firm already has a taxable income (TI) of $140,000, then using the example schedule of rates, every $1 of added income from the new investment will be taxed at the rate for a TI of $140,000, 34 percent rate.

Average Tax Rate   The other approach is to use an income tax rate that is equal to the average tax rate for the total TI, including the added new income. For example, the taxpaying corporation earning $140,000 (TI) anticipates an additional TI of $10,000 from the proposed investment. The *average* tax rate is calculated as follows:

$$
\begin{aligned}
\text{present TI} &= \quad \$140,000 \\
\text{proposed additional TI} &= + \quad 10,000 \\
\text{new total TI} &= \quad \$150,000
\end{aligned}
$$

Tax on the new total TI

$$
\begin{aligned}
0.15 \times \$50,000 &= \$\ 7,500 \\
0.25 \times 25,000 &= \quad 6,250 \\
0.34 \times 75,000 &= \underline{\quad 25,500} \\
\text{total tax} &\quad \$39,250
\end{aligned}
$$

$$
\text{average tax rate} = \frac{\$39,250}{\$150,000} = \underline{\underline{26.17\%}}
$$

While the tax on the *increment* of income is 34 percent, the *average* tax is only 26.17 percent. While many analysts use the average tax in calculating feasibility studies, the marginal tax rate usually is considered more appropriate for incremental investments.

For the convenience of the student, the federal income tax is usually shown in textbooks as a given percentage of taxable income, where taxable income is the income left over after subtraction of allowable expenses and deductions. Thus the percentage used in textbook examples and problems usually represents the marginal tax bracket, or the percent tax on each *additional* dollar earned by the taxpayer in question. For instance, using the simplified textbook approach, a taxpayer in the 28 percent tax bracket with a gross income of $50,000 and allowable expenses and deductions of $30,000 would pay a tax of

$$
(\$50,000 - \$30,000) \times 0.28 = \$5,600 \text{ income tax}
$$

EFFECT OF TAXES ON CASH FLOW

When evaluating the feasibility of any proposal subject to taxes, the taxes are treated as just another expense of doing business.

EXAMPLE 15.2

Assume an investor has an opportunity to invest $10,000 for five years in a proposal estimated to yield the net cash flow *before taxes* illustrated in Figure 15.1.

Since the $10,000 principal is returned intact at EOY 5, the *before*-tax rate of return is obviously

$$
\$1,000/\$10,000 = 10\% \text{ ROR}
$$

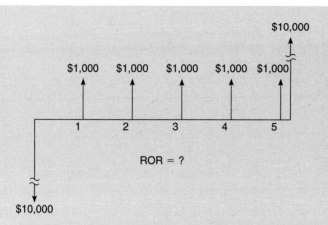

Figure 15.1 Before-tax cash flow diagram for Example 15.2.

However, if the investor is required to pay taxes on the income and is in the 28 percent tax bracket, the tax will amount to $0.28 of each $1 of income or interest earned by this investment. The original $10,000 invested at EOY 0 is *not* taxed upon withdrawal from the investment at EOY 5, since this is old investment capital returned and not new earned income or interest. (If the $10,000 invested at EOY 0 were claimed as a tax deduction, then the $10,000 income *would be* taxed, normally as capital gains.) The after-tax cash flow diagram now appears as in Figure 15.2.

The after-tax rate of return now is

(annual income − income tax)/(intact principal) = ROR

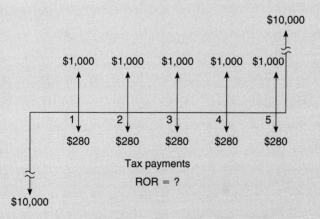

Figure 15.2 After-tax cash flow diagram for Example 15.2.

or,

$$(\$1,000 - 0.28 \times \$1,000)/\$10,000 = \underline{7.2\% \text{ ROR after taxes}}$$

Notice that the numerator (annual income − income tax) can be expressed more simply as [income × (1 − tax rate)], or in this case [$1,000 × (1 − 0.28)], or $1,000 × 0.72.

For the more common cases where the original investment is *not* returned intact, the after-tax rate of return is calculated as in previous rate of return problems, with one simple innovation; that is, the taxes are now calculated and added on as a new cost item. An example involving taxes follows.

EXAMPLE 15.3

A client proposes a project with estimated net cash flows as illustrated in Figure 15.3, and asks you to calculate the after-tax rate of return. The client is in the 28 percent tax bracket (of every $1 of net income, $0.28 is paid in taxes). For this investment, assume that the $2,000 profit from resale (capital gain) is taxable as ordinary income and is *not* eligible for any reduced rate for capital gain.

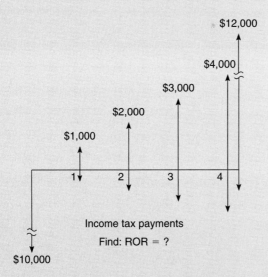

Figure 15.3   Cash flow diagram for Example 15.3.

Solution.   All the annual net incomes are taxable at 28 percent, so the client retains $0.72 out of every $1 of income after taxes. At EOY 4, $12,000 of principal is returned where only $10,000 was invested. Therefore only the extra $2,000 (from $12,000 − $10,000 = $2,000) is taxable at 28 percent, and the net after-tax amount received at EOY 4 is $10,000 + $2,000 × 0.72 = $11,440. Thus the after-tax net cash flow diagram will appear as in Figure 15.4, with after-tax incomes equal to 0.72 times before-tax incomes.

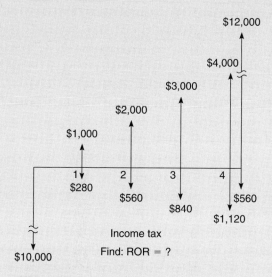

Figure 15.4   After-tax cash flow diagram for Example 15.3.

The after-tax rate of return is determined as in previous chapters by formulating present worth equations and substituting trial values of i until the value of i is found at which the total present worth equals zero.

$$
\begin{array}{lrr}
 & 15\% & 20\% \\
P_1 = -\$10,000 = & -\$10,000 & -\$10,000 \\
P_2 = +(1{,}000 - 280)(P/G, i, 5)(F/P, i, 1) = & +\quad 4{,}782 & +\quad 4{,}239 \\
P_3 = +(12{,}000 - 560)(P/F, i, 4) = & +\quad 6{,}541 & +\quad 5{,}517 \\
\hline
 & +\$\ 1{,}323 & -\$\quad 244
\end{array}
$$

$$
i = 15\% + 5\% \times \frac{\$1{,}323}{\$1{,}323 + \$244} = 19.22 = \underline{\underline{19.2\% \text{ after-tax ROR}}}
$$

After-Tax Revenues

For every $1 of taxable income an individual or business receives, the government wants to share a certain percentage by means of taxes. The taxpayer in the 28 percent tax bracket gets to keep $0.72 out of every $1 of income taxable at that rate and pays the other $0.28

in taxes. Thus, when entering such revenues into time-value calculations, simply multiply the pretax revenue by $(1 - \text{tax rate})$ in order to obtain the after-tax revenue.

EXAMPLE 15.4

Find the after-tax present worth of taxable revenues of $10,000 per year for ten years with $i = 10$ percent. The taxpayer is in the 34 percent tax bracket.

Solution. Find the after-tax cash flow as follows:

$$\text{after-tax cash flow} = \text{taxable income} \times (1 - \text{tax rate})$$
$$\text{after-tax cash flow} = \$10,000/\text{yr} \times (1 - 0.34)$$
$$\text{after-tax cash flow} = \$10,000/\text{yr} \times 0.66$$
$$\text{then } P = \$10,000 \times 0.66 \underbrace{(P/A, 10\%, 10)}_{6.1445} = \underline{\underline{\$40,554}}$$

The After-Tax Cost of Business Expenses

The government recognizes legitimate business expenses and does not tax revenue dollars that are used to pay these expenses. For instance, if a consulting engineer receives gross revenues of $140,000 per year but has business expenses of $40,000 per year, then the government only taxes the $100,000 net income after expenses.

For every $1 spent on expenses, the engineer saves the tax rate times that $1. Thus every dollar of expense only costs the consultant $1 \times (1 - \text{tax rate})$. If the consultant is in the 28 percent tax bracket and takes a short business trip which costs $100, it really only costs the consultant $100 \times 0.72 = \$72$. Here is how it works.

	Before trip	After $100 trip
Gross income	$140,000	$140,000
Expenses	−40,000	−40,100
Net income before taxes	$100,000	$ 99,900
Tax bracket	×0.28	× 0.28
Income tax	$ 28,000	$ 27,972
After-tax income		
Before trip	$ 72,000	72,000
After trip		71,928
After-tax cost of $100 trip		72

Thus the $100 trip actually reduced the consultant's after-tax income by only $72, and it is evident that the government shares both in our income as well as our business expenses required to earn that income.

SUMMARY

Income tax and other tax effects on investments are significant and should be considered carefully during feasibility studies. Thus, the engineer should keep abreast of current trends in federal, state, and local tax laws and their effect on economic analysis of engineering projects. In a real sense the federal government is a silent partner, sharing in a company's profits and easing the burden of some expenses. Taxes are assessed as a percentage of taxable income, usually on a graduated basis: the higher the taxable income, the higher the tax rate. Taxes reduce the income received from investments and thus significantly affect cash flow. On a profitable investment, the after-tax rate of return will be less than the before-tax rate of return.

Sometimes there are special provisions enacted into tax law which significantly influence the cash flow. They are (1) depreciation (covered in the next chapter), (2) investment tax credit, and (3) special provisions for capital gain and recapture.

The specific tax rates and examples contained in this text should be used as guidelines *only,* because the laws as well as the interpretations thereof are subject to change.

PROBLEMS FOR CHAPTER 15, TAXES

Problem Group A

Find taxes or tax rate.

A.1. John and Mary have one child and earned $24,762 during the year. They had allowable deductions of $5,680 and use the tax schedule shown in Example 15.1 for joint filers.

   (a) How much tax do they owe?

   (*Ans.* $1,212)

   (b) What marginal tax bracket are they in?

   (*Ans.* 15%)

   (c) What is their average tax rate?

   (*Ans.* 15%)

A.2. A corporation estimates its total income this year will be $1,750,000, of which $1,432,566 will be allowable deductions that are not subject to the corporate tax schedule shown in this chapter.

   (a) How much tax will they owe?

   (b) What is their average tax rate?

Problem Group B

Find the before-tax and after-tax rates of return. Capital gains are treated as ordinary income.

B.1. An investor buys a corporate bond for $1,000. The bond pays $100 cash interest per year for 10 years. At EOY 10 the bond matures and the $1,000 investment is refunded. The

investor pays an annual income tax of 25 percent on the interest income plus an annual intangible tax of 0.1 percent on the $1,000 face value of the bond. What is the after-tax rate
of return?

(*Ans.* 7.4%)

B.2. Assume you purchased a lot five years ago for $2,000 and now have an offer to sell for
$10,000. You have been paying annual real estate taxes at 30 mills on an assessed value of
$1,500. Assume real estate taxes are deductible from ordinary income at the end of the year
in which paid. Assume you are in the 28 percent tax bracket. If you sell, the capital gain (sale
price less cost) is treated as ordinary income and taxed at the same 28 percent rate. What is
the rate of return after taxes?

B.3. Your firm is in the 34 percent tax bracket and has $10,000 to invest. (a) Tax-free municipal
bonds maturing in 20 years are available with 6 percent coupons (coupons paying 6 percent
of the face value annually). One such bond with a face value of $10,000 can be purchased for
$9,200.

   An alternative (b) $10,000 corporate bond (income taxable) matures in 20 years with
7 percent coupons. It can be purchased for $9,000. The annual coupon income is taxable at
ordinary income rates, and the difference between the $9,000 purchase price and $10,000
redemption value (capital gain) is taxable as ordinary income at the time of redemption.
Which is the better investment if both bonds are of equal quality or risk?

(*Ans.* (a) 6.8%; (b) 5.4%)

B.4. Your client is in the 35 percent tax bracket and is considering an investment in a proposed
project whose anticipated cash flow is tabulated below.

EOY	Expenditure ($)	Income ($)
0 purchase land	100,000	
1 construction	1,000,000	
2 operating expenses and income	200,000	350,000
3 operating expenses and income	220,000	375,000
4 operating expenses and income	240,000	400,000
5 operating expenses and income	260,000	425,000
6 operating expenses and income	280,000	450,000
6 resale		1,500,000

Assume there is no allowable depreciation taken on this project. The capital investment
consists of the expenditures for "purchase land" and "construction," and the difference
between these expenditures and the "resale" income (capital gain) is taxed as ordinary
income. The client asks you to determine if the proposed project meets his MARR objective
of 15 percent after taxes.

B.5. You are offered the opportunity of investing in a new engineering office-supply business. For
an investment of $10,000, you are told you can expect to receive $15,000 at the end of four
years.

(a) What before-tax rate of return will you receive on this investment?

(*Ans.* 10.7%)

(b) Assume the entire capital gains profit on the investment will be taxed as ordinary income, and you expect to be in the 28 percent tax bracket. What after-tax rate of return will you receive on the investment?

(*Ans.* 8.0%)

B.6. As an alternative to receiving $15,000 in one lump sum payment, as stated in Problem B.5, you are offered $3,500 for each of four years (for a total of $14,000 over the four years). Each year, you count $3,500 as payment on the principal and $1,000 as earned income, and your tax bracket is 28 percent. What after-tax rate of return will you receive on this investment?

B.7. Compare the two alternative proposals in Problems B.5 and B.6 and discuss intangible factors that could affect your decision as to which alternative you would prefer.

(*Ans.* The return is better in Problem B.6, and the cash flow occurs earlier.)

Problem Group C

Problems illustrating preferential tax rate on capital gains and investment tax credit (ITC).

C.1. A client of yours who is in the 42 percent tax bracket purchased 80 acres for $80,000 five year ago. Expenses have averaged $2,000 per year and have been listed as tax-deductible items in the year in which they occurred. Therefore, they have not added to the book value (tax basis) of the land, and the actual cost to the client was [expense × (1 − tax rate)]. Your client now has two offers from people who would like to buy the land. Offer A is for the entire 80 acres for $150,000. The $70,000 profit would be taxable as capital gains; therefore, she pays a 42 percent tax on 40 percent of the capital gain.

Offer B involves subdividing the 80 acres into 5-acre lots and selling for $3,000 per acre, but this will also cause your client to be classified as a dealer, and the profit is taxable as ordinary income. Find the after-tax rate of return for each offer.

(*Ans.* Offer A, 10.4%; Offer B, 15.6%)

C.2. The corporation in Problem A.2 is considering a major capital investment of $175,000, which qualifies for an ITC of 10 percent. If they make this investment, how much will their taxes be reduced?

C.3. A major corporation (more than $100,000 per year in profits) purchased some common stock in another company four years ago for $2,758,500. They now have an opportunity to sell the stock for $5,752,400. Assume they are in the 46 percent tax bracket and only 40 percent of the capital gain is subject to income tax. If they sell, how much income tax will they pay on the capital gain?

(*Ans.* $550,878)

CHAPTER **16**

Depreciation

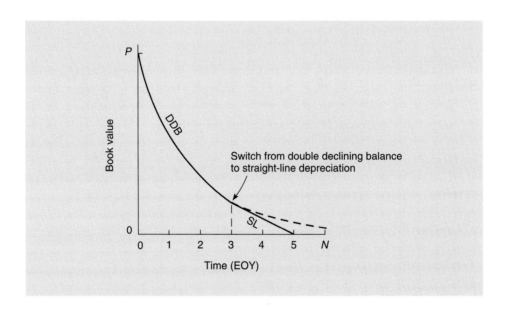

KEY EXPRESSIONS IN THIS CHAPTER

P = Purchase price of depreciable asset = book value at time zero.

N = Depreciable life of the asset.

F = Estimated salvage value of asset at time N (often shown as S).

m = Age of asset at time of calculation.

D_m = Annual depreciation amount taken for year m.

R_m = Depreciation rate for year m, expressed as a decimal value.

R = Depreciation rate for declining balance depreciation method.

BV_m = Book Value of asset at a particular age, m, $(BV_0 = P)$

SL = Straight-Line method of depreciation.

SOY = Sum of Years. Sum of the ordinal digits for years 1 through N.

$SOYD$ = Sum-of-Years Digits method of depreciation accounting.

DB = Declining Balance method of depreciation accounting.

DDB = Double Declining Balance method of depreciation accounting.

DBV = Depreciated Book Value (usually used with reference to some point in time, such as DBV at EOY 5).

BOY = Beginning of year; used to avoid confusion with events that occur at the end of the year (EOY).

DEPRECIATION: A LOSS IN VALUE

Depreciation represents the loss in value of a property such as a machine, building, vehicle, or other investment over a period of time, caused by one or more of the following:

1. Unrepaired wear
2. Deterioration
3. Obsolescence
4. Reduced demand

    1. Wear accumulates as a function of hours of use, severity of use, and level of preventive maintenance. For instance, each hour of wear on moving parts of an engine brings them closer to the point of requiring repair or salvage. From time to time, decisions are made whether or not to invest in repairs, and to what extent. The investment in repair is economically justifiable only if the value added by the repair exceeds the cost of the repair. When the engine reaches the point in its life where the cost of a needed repair exceeds the value added to the engine, it is typically sold for whatever it will bring, or junked.

    2. Deterioration, the gradual decay, corrosion, or erosion of property, occurs as a function of time and severity of exposure conditions. It is similar to wear depreciation in many ways, except it occurs whether or not the property has moving parts in actual use. Deterioration can usually be controlled by good maintenance. Many structures that are hundreds of years old (in a few instances, one or two thousand years old) are still in usable condition, usually due to a consistent level of good maintenance.

    3. Obsolescence depreciation is the reduction in value and marketability of a product due to competition from newer and/or more productive models. Obsolescence can be subdivided into two types: (a) technological and (b) style and taste.

        a. Technological obsolescence usually can be measured in terms of productivity. In some areas technological obsolescence typically has occurred at a reasonably constant rate. For instance, manufacturers of earthmoving equipment indicate that they have succeeded in increasing the productivity rate of their products by about 5 percent per year over the past 20 years. In other areas, relatively abrupt technological breakthroughs occur. For instance, the invention of the transistor and computer chip revolutionized the electronics, communications, and information industries in a relatively few short years. When a newer, more productive product (computer, earthmover, aircraft, or whatever) appears on the market, the existing competitive products lose market resale value by comparison and ultimately are junked or sold at a reduced price, having lost value due to technological obsolescence.

        b. Style obsolescence depreciation occurs as a function of customers' taste. This is much less predictable, although just as real in terms of lost value. As styles change, both men and women typically refuse to purchase clothing, cars, houses, and so forth, that do not incorporate features considered "up-to-date."

    4. Reduced demand normally results in a decreased value (increased depreciation) of an asset in accordance with the marketplace "law" of supply and demand. Reduced

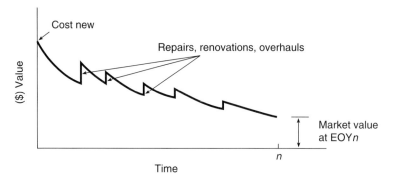

Figure 16.1   Schematic graph of depreciation.

demand results from a number of factors, such as a general business recession, change in development patterns, consumer habits, or even changes in weather patterns. The onset of a prolonged dry weather cycle, for example, may reduce the demand for drainage facilities and enhance the demand for irrigation, resulting in demand-induced changes in depreciation.

The effect of the combined types of depreciation is shown graphically in Figure 16.1. The typical pattern is a gradually decreasing value, interrupted by jogs indicating decisions to invest in repairs, overhaul, or renovations. For each such investment, the value added is presumed to exceed the cost. Finally, the time comes when the cost exceeds the value added, and the item is junked, sold, or abandoned.

MEASURING DEPRECIATION

Since depreciation represents the loss in value over a period of time, the depreciation of the actual value of an item may be traced by simply graphing the market value as a function of time. In determining market value, several variables are usually encountered. For instance, in the automobile market there is one market price (sometimes more than one, depending upon terms, trade-in, relative bargaining power, etc.) for retail and one for wholesale, with other variables affected by mileage, condition, location, and so forth.

As an example, a graph of values of a typical automobile as it depreciates with time appears in Figure 16.2, showing various depreciation methods.

At times a vehicle or structure may *appreciate* rather than depreciate. When the price of gasoline rises, some types of fuel-efficient small cars may actually increase in value. Buildings typically appreciate in value over the prime years of their lives due to the rising cost of construction of their newer competitors and the growing scarcity of good locations. Therefore, the actual depreciation (or appreciation) is the actual change in market value. This can and frequently does follow an erratic course. To simplify the bookkeeping involved, several standard methods of depreciation have been devised based upon the

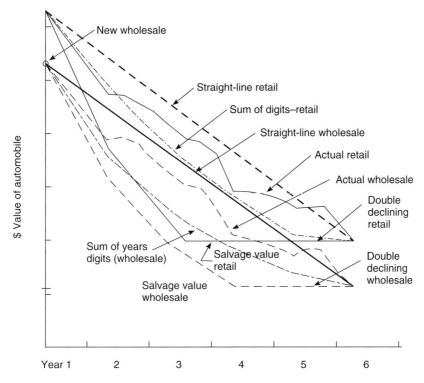

Figure 16.2 Graph of actual depreciation of an automobile.

premise that values of all depreciable properties eventually will decline from the new purchase price value to zero value.

TWO APPROACHES TO DEPRECIATION ACCOUNTING: ACTUAL DOLLAR VALUE AND TAX ACCOUNTING

Depreciation accounting is the systematic division of the depreciable value of a capital investment into annual allocations over a period of years. There are two basic approaches to depreciation accounting, each with a separate and distinct goal.

1. Depreciation for actual dollar value (nontax) accounting for management needs: The aim of this approach to depreciation is to determine a depreciated book value (DBV) for each future year that will approximate the actual declining value of a firm's depreciable assets. Therefore, this depreciation approach attempts to track the actual market value of the capital assets being depreciated. To enable managers to know their own cost of delivering their product, the actual depreciation costs must be reasonably approximated. These depreciation calculations are based upon a realistic depreciation life over which the book value is depreciated annually, from the cost new down to an estimated salvage or resale value.

2. Depreciation for tax accounting: Depreciation values for tax purposes are periodically rewritten by Congress, and are frequently subject to interpretation by the Internal Revenue Service and the tax courts. Therefore, a competent tax advisor should be consulted before applying the guidelines presented herein to any specific situation. Since depreciation of property used in the production of income typically is a tax-deductible expense, the objective here is to claim the legal maximum of depreciation as soon as allowed in order to reduce taxes as much as is possible while not exceeding the strict requirements of the tax laws. The tax laws do *not* require depreciation to track market values. These depreciation calculations normally employ the minimum depreciation life span allowed by law, over which the book value is depreciated to the lowest allowable value, usually zero. (Occasionally a taxpayer will benefit more by saving depreciation for use in some future year, and therefore will forgo the opportunity to take the maximum depreciation as early as possible.)

Since the purposes of tracking actual depreciated value versus tax value are so divergent, two different sets of depreciation accounting figures are employed simultaneously for the same property. This double accounting is quite legal and ethical and is done openly. While the government requires strict adherence to its tax regulations regarding depreciation accounting, it does not expect that the methods allowed or required by tax law are going to provide the type of depreciation information required to efficiently manage a business.

THE THREE KEY VARIABLES AND THREE STANDARD METHODS FOR CALCULATING DEPRECIATION

Values for three important variables are required in order to calculate depreciation by any depreciation method:

1. The value of the *purchase price* or cost when new.
2. The length of the *economic life* (the estimated time between purchase new and disposal at resale or salvage value) or recovery period for tax purposes.
3. The estimated resale *salvage value* (usually zero for tax purposes).

Based on these three variables, accountants have devised a number of methods of accounting for depreciation at a regular, predictable rate over the expected economic life. Of these methods, three have found wide acceptance in professional practice.

1. Straight-line method.
2. Sum-of-years digits method.
3. Declining balance method.

Straight-Line Method of Depreciation Accounting

Straight-line (SL) depreciation is the simplest method to apply and is the most widely used method of depreciation. The annual depreciation, D_m, is constant, and thus the book value, BV_m, decreases by a uniform amount each year. The equation for SL depreciation follows:

$$\text{depreciation rate, } R_m = \frac{1}{N} \tag{16.1}$$

$$\text{annual depreciation, } D_m = R_m(P - F) = \frac{P - F}{N} \tag{16.2}$$

$$\text{book value, } BV_m = P - mD_m \tag{16.3}$$

Straight-line depreciation is illustrated in Example 16.1 and Figure 16.3.

Four Examples Illustrating Depreciation Used for Asset-Value Accounting

In Examples 16.1 through 16.4, assume that even though the tax law may permit a shorter depreciation period down to zero salvage value for this asset, this particular owner is a government organization not concerned with depreciation for tax purposes. They estimate that a truck, for example, purchased new for $80,000, actually will be kept for seven years and then sold for $24,000.

EXAMPLE 16.1 **(SL depreciation to $24,000 salvage value at EOY 7)**

Given the data below, find the annual depreciation and plot a graph (Figure 16.3) showing the depreciation each year, using the straight-line method and a seven-year depreciation life.

$$\text{truck, purchase price, } P = \$80,000$$
$$\text{resale value after 7 yr, } F = \$24,000$$

Solution

$$\text{purchase price, } P = \$80,000$$
$$\text{future resale or salvage value, } F = \underline{\quad 24,000}$$
$$\text{value to depreciate in } N = 7 \text{ equal installments, } P - F = \overline{\$56,000}$$
$$\text{depreciation taken each year, } D_m = (P - F)/N = \$56,000/7 = \$\ 8,000$$

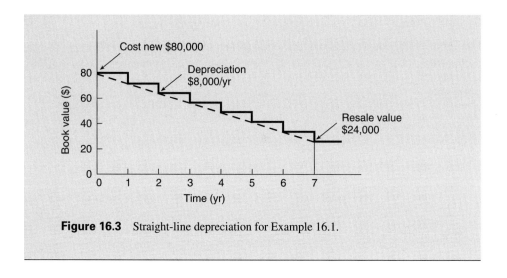

Figure 16.3   Straight-line depreciation for Example 16.1.

The Sum-of-Years Digits Method

The sum-of-years digits method is an *accelerated depreciation* (fast write-off) method, which is a term applied to any method that permits depreciation at a rate faster than that of the straight-line method. This method calculates depreciation for each year as the total original depreciable value times a certain fraction (see Equation 16.5). The fraction is found as follows: the denominator of the fraction is the sum of the digits, where each digit equals the age of the item at the end of each year of life as in $1 + 2 + 3$, and so on. The numerator is the number that represents the years of life remaining for the investment.

Equations for Sum-of-Years Digits Method (SOYD)

The sum of the ordinal digits for each of the years 1 through N is

1. $\text{SOY} = \dfrac{N(N+1)}{2}$ (16.4)

Example: $N = 7$

$\therefore \text{SOY} = \dfrac{7(7+1)}{2} = 28$

or, without using the equation,

$\text{SOY} = 1 + 2 + 3 + 4 + 5 + 6 + 7 = 28$

2. The annual depreciation D_m, for the mth year (at any age, m) is

$$D_m = (P - F)\dfrac{N - m + 1}{\text{SOY}}$$ (16.5)

Example: (using the figures from Example 16.1)
The depreciation allowed for the third year ($m = 3$) is found as follows:

$$D_m = (80,000 - 24,000)\frac{7 - 3 + 1}{28} = 10,000$$

3. Using SOY depreciation, the book value at the end of year m is

$$BV_m = P - (P - F)\left[\frac{m(N - m/2 + 0.5)}{SOY}\right] \tag{16.6}$$

An example of sum-of-years digits depreciation is given below.

EXAMPLE 16.2 (SOYD depreciation to $24,000 salvage value at EOY 7)

Using the same $80,000 truck as in Example 16.1, find and plot the allowable depreciation, using the SOYD method and a seven-year depreciation life.

Solution. A machine with a seven-year life uses a denominator of $1 + 2 + 3 + 4 + 5 + 6 + 7 = 28$, or $SOY = N(N + 1)/2 = 7(7 + 1)/2 = 28$. The depreciation allowed for the first year is $\frac{7}{28}$ of the total depreciation; the allowance for the second year is $\frac{6}{28}$ of the total, and so on to $\frac{1}{28}$ for the seventh year. Using the sum-of-years digits method,

$$\text{truck, purchase price, } P = \$80,000$$
$$\text{resale value after } N = 7 \text{ yr, } F = \underline{\quad 24,000}$$
$$\text{depreciable value, } (P - F) = \$56,000$$

Year, m	Depreciation allowed for this year, D_m	Book value, BV_m at EOY m
1	$\frac{7}{28} \times \$56,000 = \$14,000$	$\$80,000 - \$14,000 = \$66,000$
2	$\frac{6}{28} \times 56,000 = 12,000$	$66,000 - 12,000 = 54,000$
3	$\frac{5}{28} \times 56,000 = 10,000$	$54,000 - 10,000 = 44,000$
4	$\frac{4}{28} \times 56,000 = 8,000$	$44,000 - 8,000 = 36,000$
5	$\frac{3}{28} \times 56,000 = 6,000$	$36,000 - 6,000 = 30,000$
6	$\frac{2}{28} \times 56,000 = 4,000$	$30,000 - 4,000 = 26,000$
7	$\frac{1}{28} \times 56,000 = \underline{2,000}$	$26,000 - 2,000 = 24,000$
	$\$56,000$	

In this example the book value at EOY is calculated by subtracting the depreciation for a given year from its book value at the beginning of the year (BOY).

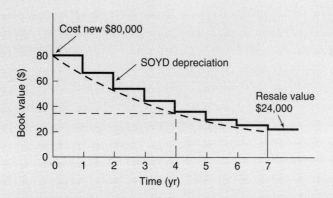

Figure 16.4 SOYD depreciation for Examples 16.2 and 16.3.

A graph of values from Example 16.2 is shown in Figure 16.4. To find the book value in any year, either use the equation

$$BV_m = P - (P - F)\left[\frac{m(N - m/2 + 0.5)}{SOY}\right]$$

or do the following:

1. Find the unused depreciation by
   a. summing the fractions for the remaining years and
   b. multiplying this sum times the total depreciable value ($56,000 in this case).

2. Add this unused depreciation to the resale (salvage) value ($24,000 in this example).

EXAMPLE 16.3 **(SOYD)**

Find the book value at the end of the fourth year for Example 16.2.

Solution. The fractions for the remaining three years of the seven-year life total $\frac{3}{28} + \frac{2}{28} + \frac{1}{28} = \frac{6}{28}$. Multiplying times the depreciable value yields $\frac{6}{28} \times$ $56,000 = $12,000. The book value at the end of the fourth year then equals the unused depreciation plus the salvage value, or $12,000 + $24,000 = $36,000 book value at end of fourth year.

Equation 16.6, of course, yields the same value.

$$BV_4 = 80,000 - (56,000)\frac{4(7 - \frac{4}{2} + 0.5)}{28} = \$36,000, \text{ book value at EOY 4}$$

Declining Balance Methods

The declining balance methods are accelerated depreciation methods that provide for a larger share of the cost of depreciation to be written off in the early years and less in the later years. This system more nearly approximates the actual decline in the market value for many types of property, especially mechanical equipment. Using this system the annual depreciation is taken as 2.0 (for the 200 percent rate, called *double declining balance method*, or DDB), or 1.5 (for the 150 percent rate), or 1.25 (for the 125 percent rate), times the current book value of the property divided by the total years of economic life. For example, if the economic life is seven years, then if it were not for accelerated depreciation, $\frac{1}{7}$ of the value could logically be deducted each year. The DDB method doubles this deduction to $\frac{2}{7}$ of the *current value each year.* The machine is depreciated down to the estimated future resale value, F, and then no more depreciation is taken. Theoretically the book value, BV_m, is not supposed to fall below salvage value, F. However, as discussed later, tax law typically allows depreciated values to fall to zero. Then resale income is treated as a separate event.

Equations for Declining Balance Methods (DB)

The abbreviation R represents the depreciation rate for declining balance depreciation.

1. The depreciation rate, R, is the depreciation multiple divided by the estimated life, N

$$\text{for double declining balance depreciation, } R = 2/N$$
$$1.75 \text{ declining balance depreciation, } R = 1.75/N$$
$$1.5 \text{ declining balance depreciation, } R = 1.5/N$$

2. The depreciation, D_m, for any given year, m, and any given depreciation rate, R, is

$$D_m = RP(1 - R)^{m-1} \quad \text{or} \quad D_m = (BV_{m-1})R \qquad (16.7)$$

3. The book value for any year, BV_m, is:

$$BV_m = P(1 - R)^m \quad \text{provided} \quad BV_m \geq F \qquad (16.8)$$

4. The age, m, at which book value, BV_m, will decline to any future value F, is

$$m = \frac{\ln(F/P)}{\ln(1 - R)} \qquad (16.9)$$

Note that the depreciation amount D_m is determined by the book value only and is *not* influenced by the salvage value F (excepting that $BV_m - D_{m-1} \geq F$). Therefore, the book value during later years may exhibit any of the following three tendencies:

Case 1: If F is zero or very low, then BV_m may never reach F.

Case 2: BV may intersect F before N (BV is not permitted to be less than F).

Case 3: BV may intersect F at N (very rare occurrence).

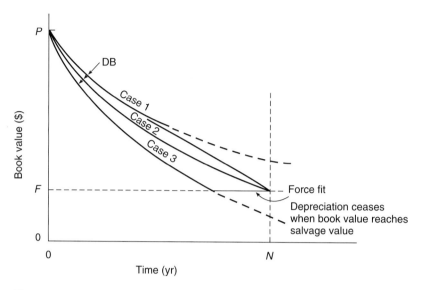

Figure 16.5　DB depreciation.

It is economically undesirable for the book value at the end of economic life, N, to be greater than the salvage value (Case 1), since all allowable depreciation should be used as early as possible.

Graphically these three situations are given in Figure 16.5. Since the book value must never be less than salvage value in the declining balance method, the latter values of BV_m are usually force fit to equal F, as depicted in Figure 16.5.

An example of double declining balance depreciation is shown below, using the same $80,000 truck as in Example 16.1.

EXAMPLE 16.4　(DDB depreciation to $24,000 salvage value at EOY 7)

Find the depreciation allowed each year, using double (200 percent) declining balance method, (DDB) and a seven-year depreciation life.

$$\text{depreciation rate, } R = \tfrac{2}{7}$$
$$\text{truck purchase price, } P = \$80,000$$
$$\text{resale value, } F, \text{ after } N = 7 \text{ yr} = \underline{24,000}$$
$$\text{depreciable value, } P - F = \$56,000$$

During the fourth, year, the calculated depreciation of $\tfrac{2}{7} \times \$29,154 = \$8,330$ would, if used, reduce the book value to $20,824, which is less than the salvage value of $24,000. In this case the fourth year's depreciation is reduced to $5,154 in order to force the book value to equal the salvage value, F.

Solution

Year, *m*	Book value × DDB factor $(BV_m \times R)$	Depreciation, D_m allowed for this year
1	$\$80,000 \times \frac{2}{7}$	= $\$22,857$
2	$- 22,857$	
	$\overline{57,143} \times \frac{2}{7}$	= $16,327$
3	$- 16,327$	
	$\overline{40,816} \times \frac{2}{7}$	= $11,662$
4	$- 11,662$	
	$\overline{29,154} \times \frac{2}{7}$	= $(8,330)$ N/A, use $\$5,154$
	$- 8,330$	This is less than the resale value $BV_m - D_{m-1} \geq F$ of $\$24,000$, so only $\$29,154$
	$(20,284)$	$-\$24,000 = \$5,154$ is allowed for depreciation this year
5	$\$24,000$	0
6	24,000	0 } Truck fully depreciated
7	24,000	0
		$\$56,000$ total depreciation using double (200%) declining balance method of depreciation

A graph of the values obtained by the double declining balance depreciation method is shown in Figure 16.6.

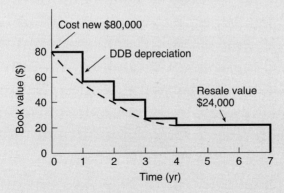

Figure 16.6 DDB depreciation for Example 16.4.

COMPARISON OF DEPRECIATION ACCOUNTING METHODS

A comparison of the three methods for the truck example is shown in Figure 16.7. Examination of the graph reveals the following:

1. The SL method takes an *equal fraction* of the *depreciable value* each year. Thus, when trying to set up a rental rate for a dragline or other depreciable equipment, the SL method is the simplest since a constant rate for depreciation charges can easily be established.
2. The SOYD method takes a *changing fraction* of the *depreciable value*.
3. The DB methods take an *equal fraction* of the *current book value*.
4. Both the SOYD and DB methods accelerate the depreciation during the early years at the expense of later years. This can have a significant influence on the income tax consequences of an investment.

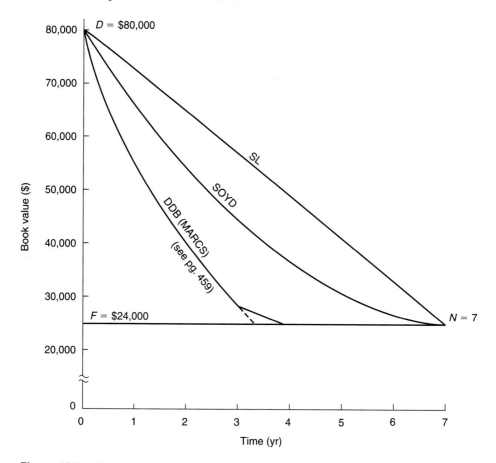

Figure 16.7   Comparison of depreciation methods.

5. The DB method does not allow book values to fall below the salvage value. A one-time change to the SL method (see Example 16.8) in later years is permitted to make the transition to F at EOY N.

Depreciation for Tax Accounting

When a machine (say a truck) is purchased for business use, one might naively assume that the purchase price is deductible from taxable income in the year of purchase. For instance, if a firm with $100,000 taxable income buys a truck for $80,000, would it not be logical to assume that the $80,000 purchase price is deductible from taxable income, leaving only $20,000 of taxable income for that year? The IRS holds that when purchasing any capital item with a life of more than one year, the purchase price money is not really "spent," but only "traded" for the capital item of equal value. Therefore, at the time of purchase the buyer has not lost any dollar value. The buyer's list of assets simply shows $80,000 *less* under cash, but an equivalent $80,000 amount *more* under value of property owned. The loss in value *does* occur, *not* when the truck is purchased, but as the truck begins to age and wear as it is used in service. This loss in value is accounted for by an annual depreciation allowance. Therefore, depreciation is simply an allocation of a portion of the purchase cost of the truck to each year of the depreciation life of the truck. Tax laws usually permit the item to be depreciated all the way from cost new to zero, so the depreciation for tax purposes is literally an allocation of the purchase price to the years permitted for tax depreciation. Contrast this tax treatment to the non-tax treatment of depreciation accounting, which does not permit the depreciated book value to drop below the actual salvage value or resale price. Using depreciation for tax accounting, if a truck is purchased for $80,000, and if the owner uses straight-line depreciation over a five-year life, he is allowed to deduct $80,000/5 years = $16,000 per year from taxable income as depreciation expense. While this depreciation is not itself a cash flow item, it does *cause* a positive cash flow by reducing taxes, since depreciation is tax deductible. (In this text it is always assumed that the firm is profitable enough to owe more taxes than the deductions will shelter.) The actual cash value of the annual depreciation allowance may be calculated as follows. Assume the firm that bought the truck is in the financial situation shown in the table below:

	Without depreciation	With depreciation
Taxable income before depreciation,	$100,000	$100,000
Amount of depreciation for 1 yr	0	− 16,000
Taxable income after depreciation	$100,000	$ 84,000
Tax bracket	× 34%	× 34%
Tax owed	$ 34,000	$ 28,560
After-tax income with depreciation	$100,000 − 28,560 =	$ 71,440
After-tax income without depreciation	$100,000 − 34,000 =	$ 66,000
Amount of income added due to depreciation	=	$ 5,440

This savings of $5,440 may be calculated more quickly and easily by simply multiplying the amount of the depreciation by the tax bracket:

$$\text{tax saved} = \$16,000 \times 0.34 = \$5,440$$

Conclusion.   Depreciation, while not a cash flow item itself, *results* in a positive cash flow (savings) of income tax payments. The amount of savings in a given year is calculated as the amount of depreciation allowed in that year times the tax bracket. This assumes that the taxable income after subtracting depreciation is greater than, or at least equal to, zero.

EXAMPLE 16.5   **(After-tax effect of SL depreciation to zero salvage value at EOY 5)**

Given a truck, with the following data,

1. Use the straight-line method to find the annual depreciation for tax purposes.

2. Plot a graph showing the depreciation each year.

3. Find the after-tax PW of the purchase, depreciation, and resale value.

$$\text{purchase price, } P = \$80,000$$
$$\text{depreciation life, minimum allowed by IRS, } N = 5 \text{ yr}$$
$$\text{salvage value, minimum allowed by IRS at EOY 5, } F_1 = \$0$$
$$\text{actual resale value at EOY 7, } F_2 = \$10,000$$

The firm is in the 34 percent tax bracket, and $i = 12$ percent.

Solution.   The truck is depreciated from $80,000 to zero book value over the five-year period, as allowed by the IRS. At the EOY 7, the truck is sold for $10,000. Since the truck is depreciated to zero book value, the entire $10,000 income from resale is treated as taxable income for the year in which it is received (EOY 7).

1. SL depreciation taken each year, $D_m = (P - F_1)/N = \$80,000/5 = \$16,000/\text{yr}$.

2. A graph of the values is shown in Figure 16.8.

3. Depreciation is the annual allocation of a portion of the purchase price.

Therefore the owner can deduct the depreciation from taxable income before calculating the income tax owed. The cash value of depreciation to the owner is simply the amount of annual depreciation times the tax rate over the depreciation life. The PW of depreciation is found as the present worth of these annual benefits over the five-year period, as follows:

$$\text{depreciation, } P_2 = \$16,000 \times 0.34(P/A, 12\%, 5) = \$19,610$$

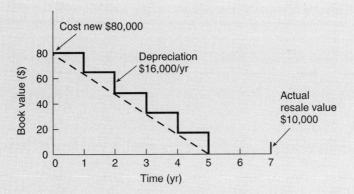

Figure 16.8 Straight-line depreciation for Example 16.5.

When the truck is sold for any amount higher than the depreciated book value, the difference between the book value and the resale value represents taxable income for the year of sale. In this case the book value is zero, so the entire amount of the resale is taxable. (*Note:* If the truck is sold *before* the end of the depreciable life, any gain over the DBV is taxable as ordinary income. If the truck is sold for less than the DBV, then the loss is normally deductible.) Since the firm is in the 34 percent tax bracket, it pays 34 percent of the resale value to the IRS, and keeps 66 percent at EOY 7. The after-tax PW of the resale is calculated as

$$\text{resale, } P_3 = \$10,000 \times 0.66(P/F, 12\%, 7) = \$2,986$$

In summary, the after-tax PW of the purchase cost, depreciation, and resale is

$$
\begin{aligned}
\text{purchase, } P_1 &= -\$80,000 \\
\text{depreciation, } P_2 &= +\ 19,610 \\
\text{resale, } P_3 &= +\quad 2,986 \\
\hline
\text{total, } P_{\text{total}} &= -57,404
\end{aligned}
$$

Conclusion. The PW of the after-tax cost of purchase, depreciation, and resale of the $80,000 truck using the SL method of depreciation is $57,404. A later example will compare the PW using SL depreciation to the PW using the SOYD and the DDB methods. Note in passing that in order to pay for this portion of the truck's cost, the firm must charge its clients an equivalent amount of *after-tax* dollars (where the amount charged the customer for these costs is found as the after-tax cost divided by 1 minus the tax rate). For this example, the NPW of the amount charged to the customer is

$$P = \$57,404/(1 - 0.34) = \underline{\underline{\$86,976}}$$

When this amount is collected, it is divided as follows:

$$\text{NPW of pretax amount collected from customer} = \$86{,}976$$
$$\text{NPW of tax paid to IRS, } 0.34 \times \$86{,}976 = -29{,}572$$
$$\text{NPW of after-tax income to owner} = \underline{\$57{,}404}$$

EXAMPLE 16.6   (After-tax effect of SOYD depreciation to zero salvage value at EOY 5)

Given the same $80,000 truck as in Example 16.1, with the data below,

1. Use the SOYD method to find the annual depreciation for tax purposes.
2. Plot a graph showing the depreciation each year.
3. Find the after-tax PW of the purchase, depreciation, and resale value.

$$\text{purchase price } P = \$80{,}000$$

depreciation life (assume Current Tax Law restricts depreciation life on this class of items to no less than $N = 5$ years)

$$\text{resale value, minimum allowed by IRS, } F_1 = \$0$$
$$\text{actual resale value at EOY 7, } F_2 = \$10{,}000$$

This firm is in the 34 percent tax bracket, and $i = 12$ percent.

Solution

1. The truck is depreciated from $80,000 to zero book value over the five-year period. With a five-year depreciation life, the denomination becomes

$$1 + 2 + 3 + 4 + 5 = 15$$

or, by the equation, $\text{SOY} = N(N + 1)/2 = 5(5 + 1)/2 = 15$.

The depreciable value $(P - F)$ is still $80,000, and the SOYD depreciation taken each year is calculated as shown in the table following.

Year, m	Depreciation allowed this year, D_m	Book value BV_m at EOY m
1	$5/15 \times \$80{,}000 = \$26{,}667$	$\$80{,}000 - \$26{,}667 = \$53{,}333$
2	$4/15 \times\ 80{,}000 =\ 21{,}333$	$53{,}333 -\ 21{,}333 =\ 32{,}000$
3	$3/15 \times\ 80{,}000 =\ 16{,}000$	$32{,}000 -\ 16{,}000 =\ 16{,}000$
4	$2/15 \times\ 80{,}000 =\ 10{,}667$	$16{,}000 -\ 10{,}667 =\ 5{,}333$
5	$1/15 \times\ 80{,}000 =\ \underline{\ 5{,}333}$	$5{,}333 -\ 5{,}333 =\ 0$
	$\$80{,}000$	

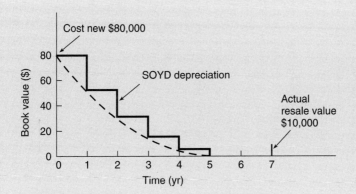

Figure 16.9 SOYD depreciation for Example 16.6.

2. A graph of the values is shown in Figure 16.9.

3. The PW of the cash value of depreciation to the owner is found in a manner similar to Example 16.1. Since the SOYD method results in a different amount of depreciation for each year, the depreciation benefit for each year is multiplied by an appropriate P/F factor to find the PW, as follows:

$$
\begin{aligned}
\text{depreciation, EOY 1, } P_2 &= \$26{,}667 \times 0.34(P/F, 12\%, 1) = \$\ 8{,}095 \\
\text{EOY 2, } P_3 &= \$21{,}333 \times 0.34(P/F, 12\%, 2) = \$\ 5{,}782 \\
\text{EOY 3, } P_4 &= \$16{,}000 \times 0.34(P/F, 12\%, 3) = \$\ 3{,}872 \\
\text{EOY 4, } P_5 &= \$10{,}667 \times 0.34(P/F, 12\%, 4) = \$\ 2{,}305 \\
\text{EOY 5, } P_6 &= \$\ 5{,}333 \times 0.34(P/F, 12\%, 5) = \underline{\$\ 1{,}029} \\
\text{total PW of depreciation} &= \$21{,}083
\end{aligned}
$$

The PW of the value of the resale at EOY 7 is calculated as in Example 16.1, as follows:

$$\text{resale, } P_3 = \$10{,}000 \times 0.66(P/F, 12\%, 7) = \$2{,}986$$

In summary, the after-tax PW of the purchase cost, depreciation, and resale is

$$
\begin{aligned}
\text{purchase, } P_1 &= -\$80{,}000 \\
\text{depreciation, } P_d &= +\ 21{,}083 \\
\text{resale, } P_r &= \underline{+\ \ \ 2{,}986} \\
\text{total, } P_{\text{total}} &= -\$55{,}931
\end{aligned}
$$

Conclusion. The after-tax cost of purchase, depreciation, and resale using the SOYD method of depreciation is \$55,931. Compare this to the NPW = \$57,404 found using the SL method. The saving results from the earlier recovery of depreciation allowed by the SOYD.

Declining Balance Methods Used for Tax Accounting

Note that the declining balance methods mathematically will not permit the depreciated book value to reach zero, since only a fraction of the current DBV is subtracted at the end of every year. Therefore, the IRS typically permits a one-time switch to straight-line depreciation in whatever year of age that the depreciation by SL will yield greater depreciation than the DB method.

Example 16.7 illustrates the effect of using double declining balance depreciation on the same $80,000 truck as used in Example 16.1.

EXAMPLE 16.7   (After-tax effect of DDB depreciation to zero at EOY 5)

Given a truck with the data below,

1. Find the annual depreciation for tax purposes, using the double (200 percent) declining balance method (DDB).
2. Plot a graph showing the depreciation each year.
3. Find the after-tax PW of the purchase, depreciation, and resale value.

$$\text{purchase price, } P = \$80,000$$
$$\text{depreciation life, minimum allowed by IRS, } N = 5 \text{ yr}$$
$$\text{resale value, minimum allowed by IRS, } F_1 = \$0$$
$$\text{actual resale value at EOY 7, } F_2 = \$10,000$$

The firm is in the 34 percent tax bracket, and $i = 12$ percent.

Solution

1. The truck is depreciated from $80,000 to zero book value over the five-year period, as in the previous examples. To find the DDB depreciation allowance each year, the following table can be constructed:

Year, m	Depreciated book value × DDB factor $(BV_m \times R)$	DDB depreciation D_m allowed for this year	SL depreciation from EOY m to end depreciation life
1	$80,000 × 2/5 = − 32,000	$32,000	$80,000/5 = $16,000
2	48,000 × 2/5 = − 19,200	19,200	48,000/4 = 12,000
3	28,800 × 2/5 = − 11,520	11,520	28,800/3 = 9,600
4	17,280 × 2/5 =	6,912 not used	17,280/2 = 8,640[a]
4	Switched to SL − 8,640	8,640	
5	8,640	8,640	
		$80,000 Total depreciation used	

[a] Note that at EOY 4 the allowable SL depreciation ($8,640) is greater than the allowable DDB depreciation ($6,912). Therefore, the one-time switch is made to SL depreciation, and the SL value of $8,640 per year is used for the depreciation for EOY 4 and 5.

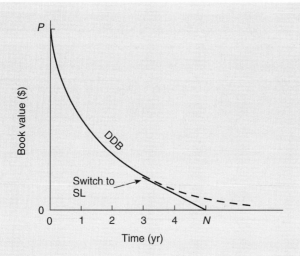

Figure 16.10 DB depreciation for Example 16.7.

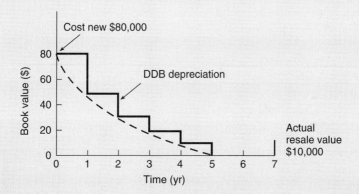

Figure 16.11 DDB depreciation for Example 16.7.

2. Graphs of the values appear in Figures 16.10 and 16.11.

3. The cash value of depreciation to the owner is calculated in a manner similar to that of Example 16.7. Since each of the yearly DDB amounts is different, a separate P/F factor for each year must be used to find the PW.

depreciation, EOY 1, $P_2 = \$32,000 \times 0.34(P/F, 12\%, 1) = \$\ 9,714$
EOY 2, $P_3 = \$19,200 \times 0.34(P/F, 12\%, 2) = \$\ 5,204$
EOY 3, $P_4 = \$11,520 \times 0.34(P/F, 12\%, 3) = \$\ 2,788$
EOY 4, $P_5 = \$\ 8,640 \times 0.34(P/F, 12\%, 4) = \$\ 1,867$
EOY 5, $P_6 = \$\ 8,640 \times 0.34(P/F, 12\%, 5) = \underline{\$\ 1,667}$
total PW of depreciation $= \$21,240$

Resale: The PW of the resale value, as explained in Example 16.5, is calculated as follows.

$$\text{resale, } P_7 = \$10,000 \times 0.66(P/F, 12\%, 7) = \$2,986$$

The after-tax PW of the purchase cost, depreciation, and resale is summarized as:

$$\begin{aligned}
\text{purchase, } P_1 &= -\$80,000 \\
\text{depreciation, } P_2 &= +\ \ 21,240 \\
\text{resale, } P_3 &= +\ \ \underline{\ \ 2,986} \\
\text{total, } P_{\text{total}} &= -\$55,774
\end{aligned}$$

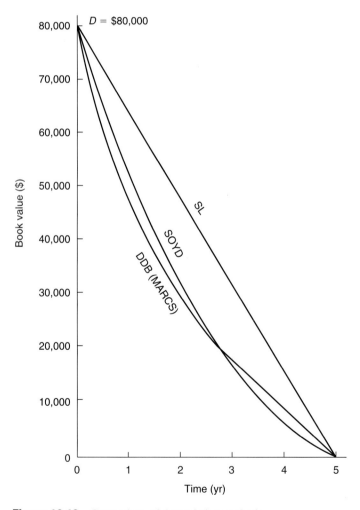

Figure 16.12   Comparison of depreciation methods.

Comparison of the After-Tax PW of Three Depreciation Methods

The PWs of the after-tax cost of purchase, depreciation, and resale by the three depreciation methods are compared as follows.

Method	After-tax PW
SL	−$57,404
SOYD	− 55,931
DDB	− 55,554

In this case, the DDB method yields the lowest after-tax cost. Compared to the cost found by using the SL method of −$57,404, there is a PW saving of $1,850, or 2.31 percent of the truck's $80,000 price. This illustrates the value of earlier recovery of the truck's depreciation cost by use of the DDB method (see Figure 16.12).

ACCELERATED COST RECOVERY SYSTEM (ACRS AND MACRS)

From time to time Congress makes earnest new efforts to clarify and simplify the tax law regarding depreciation. In the recent past, Congress has adopted two successive systems of depreciation: the first was known as the Accelerated Cost Recovery system (ACRS), and a later modification was appropriately named the Modified Accelerated Cost Recovery System (MACRS). These both involve application of the declining balance methods of depreciation. The ACRS became effective in 1981 and was superseded in 1986 by MACRS. Both systems feature tables relating allowable recovery periods for selected classes of depreciable property.

Nomenclature

Recovery property is any depreciable, tangible property (personal or real) used in business or held for the production of income.

Recovery period is the depreciation life of recovery property for tax purposes. The recovery period is the time period over which capital cost is recovered (depreciated). The recovery period is usually shorter than the useful life of the capital investment in order to provide a quicker write-off and thus encourage capital investment, stimulate the economy, and improve productivity. The recovery period for representative classes of recoverable capital items are listed in Table 16.1.

Class life is the median service life (midpoint of all service lives in a record sample used in the same activity) for a class of assets. For example, if the experience of a certain contractor indicates that the median life for their air compressors is 8.6 years (under their particular service conditions), then the recovery period is 5 years, as determined from Table 16.1 (the "asset depreciation range" of 8.6 years falls between 4 and 10 years, and is therefore eligible for a 5-year depreciation life).

TABLE 16.1 MACRS, recovery period for representative classes of recoverable capital assets (subject to change by action of Congress and the President)

Property with a median class life (years) of	Including assets such as	Allowable cost-recovery period (depreciation life)	Depreciation method
≤4 yr	Short-lived assets	3 yr	200% declining balance
>4 to <10	Autos, light trucks, computers, typewriters, research and experimentation equipment	5	200% declining balance
10 to <16	Most types of manufacturing equipment, single-purpose barns, greenhouses, office furniture, commercial airplanes	7	200% declining balance
16 to <20	Railroad tracks, ships	10	200% declining balance
20 to <25	Municipal waste-water treatment plants, land improvements, billboards	15	150% declining balance
>25	Farm buildings	20	150% declining balance
	Residential rental property	27.5	Straight-line
	Nonresidential rental property	31.5	Straight-line

Note: The tax law includes numerous qualifications and exceptions and is subject to change. This table illustrates the general provisions of the MACRS but is not intended as an official guide. Competent tax professionals should be consulted before investing.

MACRS General Rules for Capital Cost Recovery for Both Personal and Real Property MACRS

1. The recovery property may be depreciated down to zero value. Salvage (resale) value need not be accounted for in depreciation, but any gain over the depreciated value will be taxed as ordinary income, as shown in Figures 16.13 and 16.14.

2. Both new and used property are treated the same.

3. For properties with allowable cost-recovery periods of 20 years or less, either the accelerated capital recovery system (ACRS) or straight-line recovery (SL) may be used to calculate depreciation, at the discretion of the taxpayer.

All assets are depreciated to zero salvage value over the depreciation life listed. All properties with allowable cost-recovery periods of 10 years or less may use the 200 percent

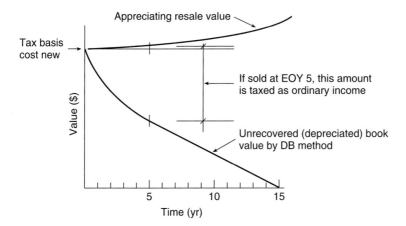

Figure 16.13   Personal property recovered by declining balance method.

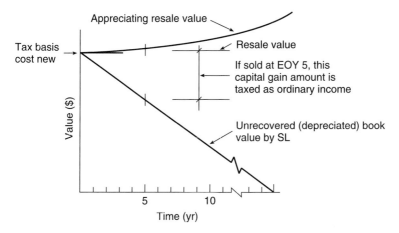

Figure 16.14   All improved real property acquired after 1986 is recovered by SL.

DB method, while those with 15- and 20-year lives may use the 150 percent declining balance method. Where the declining balance methods are specified, the taxpayer is allowed to elect to use the straight-line method, but once the election is made it may not be rescinded. When the declining balance methods are used, a switch to the straight-line (SL) method is permitted in the first year in which the SL method will yield a greater depreciation allowance. The SL method depreciates the *remaining* depreciated book value down to zero over the *remaining* depreciation life.

Half-Year Convention for Personal Property

Personal property placed into service or disposed of during a taxable year is treated as placed into service or disposed of at the midpoint of that year. This allows one-half the depreciation allowance for the first year and one-half for the last year of service. This system is illustrated in Example 16.8. (Exception: If more than 40 percent of the personal property is put into service during the last 3 months of the year, then the *midquarter* convention is used, which treats all personal property as if it were placed in service at midquarter.)

EXAMPLE 16.8 **(Half-year convention [same principles as the midquarter convention])**

A pickup truck is purchased for $10,000 and placed in service in September during year 1. The truck is eligible for 200 percent declining balance depreciation over a five-year depreciation life. The owner's tax year corresponds with the calendar year, and this property qualifies for the half-year rule. Find the depreciation for each of the six tax years into which the depreciation life extends.

Solution. The pickup is placed in service in September, the third quarter of the year, and the owner is entitled to one-half of one year's depreciation for this first year. The last one-half year of depreciation may be credited to year 6. The table below illustrates the calculations necessary to determine the depreciation deductions that the owner would receive beginning in taxable year 1 and ending in taxable year 6.

Year	Book value BOY	Years credited this year	Depreciation factor (2DB for yr 1, 2, 3) (SL for yr 4, 5, 6)	Depreciation this year
1	$10,000	0.5	$0.5 \times (2/5) \times \$10,000 =$	$2,000
2	8,000	1	$(2/5) \times 8,000 =$	3,200
3	4,800	1	$(2/5) \times 4,800 =$	1,920
4	2,880			

There are now 2 remaining years of depreciation in calendar tax years 4 and 5, plus 0.5 of a year of credit remaining at the beginning of tax year 6, for a total of 2.5 remaining years. The depreciation by 2DB for year 4 would be $(2/5) \times 2,880 = \$1,152$. However, if the switch is made to straight-line, the allowable depreciation remains the same at $(1/2.5) \times 2,880 = \$1,152$. Therefore the switch is made to SL, as follows.

Year	Book value BOY	Years credited this year	Depreciation factor	Depreciation this year
4	$2,880	1	$1/2.5 \times \$2,880 =$	$1,152
5	1,728	1	$1/2.5 \times 2,880 =$	1,152
6	576	0.5	$0.5/2.5 \times 2,880 =$	576
			Total depreciation =	$10,000

DEPRECIATION VERSUS AMORTIZATION

Depreciation should not be confused with amortization. A common example of amortization occurs when a debt P is amortized (paid out) by n end-of-period payments, each of amount A, where i is the rate at which interest periodically accrues on the unpaid balance, and $A = P(A/P, i, n)$. The differences between depreciation and amortization are outlined as follows.

1. Depreciation does *not* involve cash flow (except at times of purchase and resale), whereas amortization *does* involve periodic cash flow payments of principal and interest.

2. Depreciation does *not* involve interest and simply provides a timetable for orderly accounting for the loss in property value from purchase price to resale value. Amortization does involve periodic payments of interest accruing at rate i on the unpaid balance, in addition to payments on principal.

3. The depreciation allocated for each year is a non–cash flow expense which is usually tax deductible for the allocated year. Thus, while depreciation is not a cash flow item, it often does affect after-tax cash flow. For amortization, typically only the interest portion (not the principal) of the periodic payment is tax deductible.

SPREADSHEET APPLICATIONS

Many spreadsheet programs have predefined macros to perform calculations for depreciation. The EXCEL spreadsheet program is used here to illustrate the power of preprogrammed computer software for solving engineering economy problems involving depreciation. Students should learn to develop tables or templates, as shown in the following example problem, that can accommodate or be easily adapted to a variety of problem types. Students thus learn to appreciate the power of these programs for applications within an engineering firm, where time becomes a critical factor in developing client recommendations based on financial considerations.

	A	B	C	D	E	F
1	EXAMPLE 16.8					
2	Solving our previous Example 16.8 using the EXCEL spreadsheet software					
3						
4	Interest Rate = 12%					
5	Tax Bracket = 34%					
6	P = $ 80,000.00					
7	N (years) = 5					
8	SALVAGE = $0.00					
9						
10	Determining at the end of which year to switch from DDB to SL depreciation					
11	EOY	Beginning of Year BV	DDB Depreciation	SL Depreciation	Depreciation Amount Used	DECISION
12						
13	1	$ 80,000	$32,000	$16,000	$32,000	Don't Switch
14	2	$ 48,000	$19,200	$12,000	$19,200	Don't Switch
15	3	$ 28,800	$11,520	$ 9,600	$11,520	Switch to SL*
16	4	$ 17,280	$ 6,912	$ 8,640	$ 8,640	
17	5	$ 8,640	$ 4,147	$ 8,640	$ 8,640	
18						
19	*Switch to SL after the end of this year					
20						

(Spreadsheet continued)

	A	B	C	D	E	F
21	Determine the cash equivalent value of the depreciation to the owner					
22	EOY	Cash Value of Depreciation				
23			In order to determine the value of the depreciation, we			
24	1	$ 9,714	must also consider the income tax (34%) consequences,			
25	2	$ 5,204	which can be seen in our cell formulas below.			
26	3	$ 2,788				
27	4	$ 1,867				
28	5	$ 1,667				
29						
30	Total	$21,240	This is the cash equivalent value realized by depreciating the truck.			
31						
32	CELL CONTENTS: The following cell contents were used to evaluate the problem with a spreadsheet.					
33	Cell B15:	(B14-E14)				
34	Cell C14:	DDB(B6,B8,B7,A14)		Double declining balance function		
35	Cell C15:	DDB(B6,B8,B7,A15)				
36	Cell D14:	SLN(B14,B8,B7)		Straight-line depreciation function		
37	Cell D15:	IF(D14<C14,SLN(B15,B8,$B$7-A14),D14)		Combined SLN and logical IF		
38	Cell E14:	IF(C14>D14, C14, D14)				
39	Cell E15:	IF(C15>D15, C15, D15)				
40	Cell F14:	IF(D15>C15, "Switch to SL*", "Don't Switch") Decision logic with text printout				
41						
42	There are many other functions in spreadsheets that can be used to formulate very creative reports					
43	and output for problems such as this one. When using spreadsheets, the presentation should be					
44	made in a series of logical steps. The primary function of an engineering economy study is to					
45	provide understandable information that can be used to make strategic business decisions. The					
46	spreadsheet provides an excellent means of presentation that can easily be followed and under-					
47	stood by those who need to make decisions based upon such presentations.					
48						

SUMMARY

Depreciation accounting is a systematic allocation of the loss in value of a capital asset over its depreciable life. Depreciation as reflected by the depreciated *book* value does not necessarily track *market* value. There are several methods of depreciation accounting currently being used. They include straight-line, sum-of-years digits, and declining balance. The straight-line method is the simplest and most often used and allocates the depreciable cost of the asset uniformly over its life. Sum-of-years digits and declining balance methods accelerate the rate of depreciation during the early life of the asset. These methods usually yield the most advantageous income tax effects, but they are slightly more complicated than the straight-line method.

Depreciation can have a significant effect on after-tax rates of return for alternatives involving investment in depreciable assets and is an important aspect of engineering feasibility studies.

SUMMARY OF DEPRECIATION EQUATIONS USED IN THIS CHAPTER

Equations for Straight-Line Method

The equation for SL depreciation is

$$\text{depreciation rate, } R_m = \frac{1}{N} \tag{16.1}$$

$$\text{annual depreciation, } D_m = R_m(P - F) = \frac{P - F}{N} \tag{16.2}$$

$$\text{book value, } BV_m = P - mD_m \tag{16.3}$$

Equations for Sum-of-Years Digits Method

The sum of the ordinal digits for each of the years 1 through N is

1. $\text{SOY} = \dfrac{N(N + 1)}{2}$ (16.4)

2. The annual depreciation D_m, for the mth year (at any age, m) is

$$D_m = (P - F)\frac{N - m + 1}{\text{SOY}} \tag{16.5}$$

3. The book value at end of year m is

$$BV_m = P - (P - F)\left[\frac{m(N - m/2 + 0.5)}{\text{SOY}}\right] \tag{16.6}$$

Equations for Declining Balance Methods

The abbreviation R represents the depreciation rate for declining balance depreciation.

1. The depreciation rate, R, is the depreciation multiple divided by the estimated life, n.

$$\text{for double declining balance depreciation, } R = 2/n$$
$$1.75 \text{ declining balance depreciation, } R = 1.75/n$$
$$1.5 \text{ declining balance depreciation, } R = 1.5/n$$

2. The depreciation, D_m, for any given year, m, and any given depreciation rate, R, is

$$D_m = RP(1 - R)^{m-1} \quad \text{or} \quad D_m = (BV_{m-1})R \tag{16.7}$$

3. The book value for any year BV_m is

$$BV_m = P(1 - R)^m \quad \text{provided} \quad BV_m \geq F \tag{16.8}$$

4. The age, m, at which book value, BV_m, will decline to any future value, F, is

$$m = \frac{\ln(F/P)}{\ln(1 - R)} \tag{16.9}$$

PROBLEMS FOR CHAPTER 16, DEPRECIATION

Note: Do *not* use the half-year convention unless specifically requested to do so.

Problem Group A

Before-income-tax application of depreciation methods.

A.1.   A friend purchased an asphalt plant 6 years ago for $250,000 and expects to sell it when it is 12 years old for $70,000. An offer to sell now has been received. Find what the current book value is by SL, SOYD, and DDB methods. Assume a 10-year depreciation life and zero DBV at EOY 10.

(*Ans.* SL $100,000; SOYD $45,455; DDB $65,536)

A.2.   Assume you are looking for a building in which to locate your professional office, and one is available for $500,000. The appraisal is divided into $100,000 for the land and $400,000 for the building. The building can be depreciated over a 31.5-year period by the SL method. Resale value after 20 years is $800,000 for the building.

   (a) If you buy the building in September, how much can you deduct for depreciation for the first calendar year (use the half-month convention)?

   (b) How much in the second year?

A.3.   The local property tax assessor (not the IRS) is willing to accept the depreciated book value calculated by double declining balance method as the basis for his personal property assessments if you will provide the calculations. His property tax is collected at the *end* of each year. The tax rate is 30 mills (3 percent tax) on the depreciated book value of the property at the *beginning* of the year.

   (a) Compared to the DBV by the straight-line method, how much cash flow can be saved on property purchased for $100,000 with a DBV at EOY 5 of $40,000?

   (*Ans.* $3,000)

   (b) What is the annual equivalent of this amount if $i = 8$ percent? (*Hint:* First find the PW of each year's savings.)

   (*Ans.* $589/yr)

Problem Group B

Find after-tax PW of savings or ROR for DDB versus SL.

B.1.   A friend of yours has just purchased a pickup for use in his business and expects to keep it for four years and then sell it. The friend is in the 28 percent tax bracket and asks for suggestions on which depreciation method to use for tax purposes. Assume that $i = 12$ percent and show how much (in terms of present worth of tax payments over the four-year period) can be saved by using the double declining balance method as compared to the straight-line method. Assume the pickup cost $18,000 new and sells at the end of four years for $8,000. It can be depreciated by DDB or SL to zero salvage over a five-year life.

   (*Ans.* SL, $P = $2,279; DDB, $P = $2,551)

B.2.   An automatic block-making machine is available for $50,000. Your best estimates indicate that it will be worth $10,000 when you expect to dispose of it at the end of five years. It is capable of producing 100,000 blocks per year at a net profit (not including cost of the machine) before taxes of $0.10 per block. Use a 28 percent tax rate. It can be depreciated to zero DBV in seven years.

(a) Find the cash flow after taxes for the machine using (i) straight-line depreciation and (ii) double declining balance depreciation.

(b) Find the rate of return on the investment in the machine using (i) SL depreciation and (ii) DDB depreciation.

B.3. You are consultant to a contracting firm with an annual before-tax income averaging $500,000 per year subject to income tax of 36 percent on all income over $365,000. During the course of your consulting, you find the concrete plant can be made to yield extra revenue over the next 10 years by any one of the following mutually exclusive alternatives. Funds to finance these alternative proposals are available out of working capital. Working capital is currently earning the company 10 percent after taxes. Find which of the following alternatives has the highest after-tax present worth at $i = 10$ percent. Assume all taxes are paid at the end of the year.

(a) An extra $22,000 of before-tax revenue may be earned by investing $100,000 in new equipment. The $100,000 may be depreciated for tax purposes over a period of 10 years at $10,000 per year. At the end of 10 years, the equipment should have a $20,000 resale value.

(*Ans.* $P = \$16,346$)

(b) Increase before-tax revenue by $20,000 per year for the 10-year period by rearranging the flow of materials. This requires disassembly of the several large steel structures, but no capital investment. The $100,000 cost of doing the work will be spent and charged off as a tax-deductible expense in one lump sum at the end of the first year.

(*Ans.* $P = \$20,468$)

(c) An extra $18,000 per year of taxable revenue for the 10-year period may be added by hiring an additional plant technician at a cost of $10,000 (tax deductible) the first year, increasing $1,000 per year thereafter.

(*Ans.* $P = \$16,810$)

B.4. You client is interested in building and operating a small shopping center of 100,000 ft^2 and asks you to calculate the rental rate needed per square foot in order to make a 15 percent rate of return after taxes on all funds invested. Your client is in the 50 percent tax bracket for ordinary income. The future costs and income for the project are estimated as follows.

EOY	
0	Purchase land. $100,000
1	Finish construction of $2,000,000 worth of buildings and other improvements depreciable over a 31.5-yr life by the straight-line depreciation method. Begin rental and operation of the center, but rental income and operating costs are not credited or debited until EOY 2.
2	First year's operating and maintenance costs charged at EOY 2 = 100,000/yr. First year's income deposited after income taxes are paid at a tax rate of 50% × net income (where net income = gross income − O & M − depreciation).
3–6	Same as EOY 2. (O & M costs continue at $100,000/yr.)
6	Sell the shopping center for original cost plus appreciation estimated at 4%/yr compounded. The difference between the depreciated book value and the sale price is taxable as ordinary income.

(a) Draw a cash flow diagram indicating the time and amount of each cash flow occurrence.

(b) Find the rental rate in terms of $/ft^2/yr to obtain 15 percent return after taxes on all funds invested.

Comment: For convenience, make equation time zero coincide with EOY 1, and do *not* use the half-month convention.

B.5. Your engineering office is considering the purchase of a new computer for $100,000. You expect to sell it at EOY 5 for $40,000. The computer may be depreciated to a DBV of zero at EOY 5.

(a) Fill in the depreciation schedule below for the years indicated.

Straight-line depreciation

Year	Book value at beginning of the year	$ Value × multiplier	=	Depreciation taken this year	Ans.
1	_____	_____ × _____	=	_____	$20,000
2	_____	_____ × _____	=	_____	20,000

Double declining balance depreciation

| 1 | _____ | _____ × _____ | = | _____ | 40,000 |
| 2 | _____ | _____ × _____ | = | _____ | 24,000 |

(b) The computer would be 100 percent financed and will be paid off in five equal annual installment payments with interest at 10 percent on the unpaid balance. The first payment is due at EOY 1. Fill in the amortization schedule for the years indicated.

Amortization schedule

Year	Amount of equal annual payment	Interest payment for this year	Principal payment for this year	Balance owed after EOY payment	Ans.
0				_____	$100,000
1	_____	_____	_____	_____	83,620
2	_____	_____	_____	_____	65,603

(c) Assume your engineering firm now earns a net cash flow income of $60,000 per year after all salaries and expenses, but before considering payments of principal and interest on the computer, and before accounting for computer depreciation and before paying income tax. Your firm is in the 28 percent tax bracket and uses DDB depreciation. (i) How much income tax would your firm owe for the second year's operation at EOY 2? (ii) What is the net cash flow income after taxes and P & I payments for the second year?

(*Ans.* (i) $7,739; (ii) $25,881)

Problem Group C

Find the after-tax incremental rate of return.

C.1. Your firm is replacing its fleet of executive sedans and has narrowed the selection down to two models with the life-cycle costs listed below. Recommend a selection based on *after-tax incremental rate of return.*

The firm is in the 38 percent tax bracket.

Assume the following items are tax deductible: depreciation, O & M costs.

The cars will be driven 15,000 miles per year and will be sold at EOY 3.

Depreciation method: straight-line to zero salvage at EOY 5.

Capital gain (the difference between depreciated book value and resale value) is taxed as ordinary income at the time of resale.

Model	Cost new ($)	Resale value at EOY 3($)	O & M costs ($/mi)	Gradient O & M costs annual increase ($/mi/yr)
A	17,000	8,500	0.370	0.030
B	19,500	9,600	0.310	0.022

(a) Find the incremental rate of return *after taxes*. (*Hint:* Try 15%.)

(*Ans.* 16.8% or 16.3% interpolated)

(b) Assume the incremental rates of return for two other alternatives are as listed below. Draw an arrow diagram showing the four alternatives and the direction of each arrow between the alternatives.

	B	C	D
A	Find	17.3	11.1
B		20.6	9.3
C			4.5

(c) On the arrow diagram, show by dashed arrows the method of reaching the better alternative of the four if the MARR is 15 percent.

C.2. A contractor needs to buy a new excavator and currently has two models under consideration. The life-cycle costs of each model are tabulated below. In addition, the following data are supplied concerning income taxes for the contractor.

ordinary income tax bracket 36%

depreciation method straight-line depreciated to zero at EOY 7

Capital gain (the resale value gain over DBV) is taxable at ordinary rates.

Type	A	B
Initial cost	$120,000	$160,000
Resale value at EOY 10	$ 18,000	$ 40,000
Productivity (yd³/hr)	200	290
Productivity decline gradient (productivity drop in yd³/hr each year)	12	8
O & M	$ 80,000/yr	$ 90,000/yr
O & M	$ 8,400/yr/yr	$ 9,000/yr/yr

The income for each cubic yard of earth moved is $0.10/yd^3$, and the excavator will work 2,000 hours per year.

(a) Find the incremental rate of return *after taxes*. (*Hint:* Try 15%.)

(b) Assume the incremental rates of return for the other alternatives are as listed below. Draw an arrow diagram showing the four alternatives and the direction of each arrow between the alternatives.

	B	C	D
A	Find	13.6	22.4
B		10.5	23.0
C			29.7

(c) Show by dashed arrows the method of reaching the better alternative of the four if the MARR is 15 percent.

CHAPTER 17

Bond Financing for Public Works and Corporate Investment

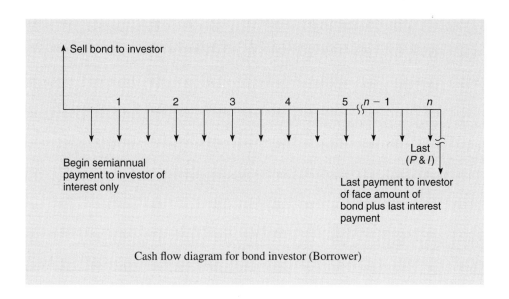

Cash flow diagram for bond investor (Borrower)

FINANCING PUBLIC WORKS

Like the governments of many other nations around the world, our government consists of a three-level structure: (1) federal, concerned with national and international affairs, (2) state, concerned with improving the conditions within state boundaries, and (3) local, concerned with the problems on the county, city and other areas of interest. According to the U.S. Constitution, residual power not specifically delegated to the federal or local governments remains with the governments of the individual states.

The budgets for each of these levels of government usually are subdivided into an operating budget and a capital budget. The operating budget includes all expenditures necessary for maintaining the day-to-day operating and maintenance functions of each level. Salaries and wages are usually the largest single item in an operating budget, which also includes supplies and other consumables with a short usable life expectancy (usually less than one year).

Expenditures for items with lives of greater than one year are usually defined as capital expenditures and belong in the capital budget. (As a matter of practice operating budgets often contain a few capital items such as typewriters, desks, etc.) Thus funding for new buildings, lands, and other large projects with a long life expectancy normally is done through the capital budget.

On the state and local level, often the income from regular tax revenues is totally consumed by the day-to-day operating expenses of government, leaving little for needed capital expenditures. Thus, when large capital expenditures are needed, funds to finance the project usually must be borrowed. Long-term borrowing for state and local government is usually done through bond issues. (In one recent year, approximately 60 percent of state and local construction is financed by borrowed money secured by bond issues.) These bonds are simply long-term public debts of authorized governmental agencies that have been marketed in convenient denominations backed with a dependable source of income, bearing interest at an agreed rate, with the principal (face value or redemption value) to be repaid on stipulated maturity dates. At all levels, the bond market for public works is large. In one recent year alone, more than 200 billion dollars was borrowed by issuing bonds from all levels of government is this country, as compared to only 40 billion dollars borrowed by privately owned corporations issuing bonds. Purchasers of bonds include a wide variety of individual and corporate entities, with varying investment objectives. Among the purchasers of a large percentage of bonds (lenders) are organizations and firms with long-term capital to invest such as insurance companies, pension funds, mutual funds, and some banks and savings institutions.

BOND DENOMINATIONS AND PAYMENT PERIODS

While corporate bonds are normally issued with face value of $1,000, many municipal bonds are issued in $5,000 denominations. Both types of bonds have a stated nominal interest rate, payments period, and maturity date (or life). An example might be a $5,000 municipal bond bearing 8 percent interest payable semiannually and maturing in 20 years. This means the bond issuer (governmental agency) agrees to pay the bond purchaser (investor) $0.04 \times \$5,000 = \$200$ every six months for 20 years (40 payments). Thus, the actual interest payment is 4 percent/semiannually while the nominal annual interest is 8 percent, and the effective interest rate is 8.16 percent. With the last payment, the investor receives a repayment of the $5,000 face value of the bond. The cash flow to the purchaser is shown in Figure 17.1.

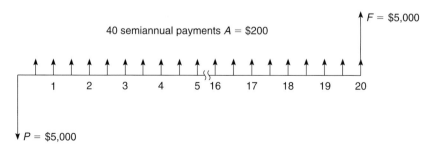

Figure 17.1   Cash flow for bond investment.

Notice that although the bond is termed a $5,000 bond, and in 20 years will be re-deemed for $5,000, the actual market purchase price of the bond, P, will vary with the current market interest rate. The investor who pays $5,000 for the bond will earn a nom-inal annual interest rate of 8 percent and an effective interest rate of 8.16 percent [$i_e = (1 + 0.08/2)^2 - 1 = 0.0816$]. But if interest rates are higher on competitive in-vestments, the investor may not be willing to pay $5,000 for the bond; instead, he may offer to purchase the bond for, say, $4,500. If the seller agrees, what interest rate will be earned? *Note that the payments are not changed.* The bond, issued as a $5,000 bond, pays interest of $200 every six months. At maturity, the purchaser surrenders the last coupon together with the bond, and in turn receives $5,200. To determine the rate of return, sim-ply set up an equivalent net worth equation and solve for the unknown i. (The number of compounding periods must relate correctly to the interest per period, of course).

$$\text{NPW} = -\$4,500 + 200(P/A, i, 40) + 5,000(P/F, i, 40)$$
$$\text{by trial and error: } i = \underline{\underline{4.55\%}}$$
$$i_n = 2 \times 4.55 = \underline{\underline{9.10\%}}$$
$$i_e = (1.0455^2 - 1)/100 = \underline{\underline{9.30\%}}$$

Since the purchase price is lower than the $5,000 face amount, the interest earned by the purchaser is greater than the 8 percent interest rate shown on the bond itself. If the bond sells for more than $5,000, the resulting earned interest would be less than the rate shown on the bond.

While bonds can be bought and sold between individuals if desired, most transactions go through a broker. Some brokers specialize in bonds only, while other brokers handle stocks, bonds, and other financial paper. Because most bonds are backed by the good re-liable sources of income of the governmental entity-issuer, they usually are readily mar-ketable and are bought and sold in large quantities every business day. Thus, during the life of a bond (time until it matures or is called*), it may have numerous owners. Prices fluctuate each day, depending upon a number of variables such as demand, money supply, rating of the bond, public confidence in the governmental entity-issuer, and the antici-pated inflation rate.

TAX-FREE MUNICIPAL BONDS

To encourage investment in these bonds, the federal law provides that the interest earned from most state and local bonds is not subject to federal income taxes. Therefore, tax-paying investors will lend money at a lower interest rate for this tax-exempt type of gov-ernment bond and still receive competitive after-tax rates of return. Tax-exempt bonds are

---

* Some bond agreements (or "covenants") provide that the bonds may be "called," or redeemed after a cer-tain date at the option of the issuer, sometimes at par (100 percent of face value) and sometimes at a premium over par.

frequently referred to as "municipal bonds" or "municipals," even though they may be issued by a state, county, school board, or special government authority (e.g., Florida Turnpike Authority) as well as by a municipality. Interestingly the income from most federal bonds (such as Savings Bonds) is not tax-exempt.

There are several different types of bonds currently available depending upon the assets backing the bonds:

1. *General obligation bonds (GO bonds).* These bonds are backed by the full faith and credit of the city, county, or state issuing them. Before issuing GO bonds, a public referendum normally is required so that the citizens may vote on whether or not they wish to pledge all of the community real estate assets to repay the bonds if called upon. Usually they are also asked to pledge a small increase in real estate taxes to repay the bonds over a set number of years. Since GO bonds do have such low risk due to excellent collateral, investors will frequently buy them at $\frac{1}{2}$ percent to 1 percent lower interest rates than other types of municipal bonds.

2. *Revenue bonds (sometimes called revenue certificates).* These bonds are backed by some special source of revenue, such as cigarette tax, sales tax, or receipts from the facility being financed (college dormitories, stadiums, parking garage, electric power plant, etc.). Normally these do not require a public referendum, but simply action by the governing authority, such as the city commission, county commission, school board, and so on. These bonds are frequently used to borrow funds to construct a facility whose receipts will then be used to pay off the bonds. Examples are Florida Turnpike Authority, Chesapeake Bay Bridge and Tunnel, Gainesville Electric Water and Sewer and many others. If the revenue from the pledged source is inadequate, the bondholders usually have lien rights on the revenues (but seldom on the properties themselves). From time to time an isolated few bonds do default temporarily. The Chesapeake Bay Bridge and Tunnel, and the West Virginia Turnpike projects are well-known examples of bonds that were in default on payments for some time. The bondholders will most likely be paid off some day, but meanwhile the bonds resell for considerably less than their face value.

3. *Special assessment bonds.* These bonds, as the name implies, represent money borrowed for a project of special benefit to a few property owners (such as a sewer line or a residential street-paving project that benefits the fronting property owners). The principal and interest is paid from special annual assessments against the improved property for a specified number of years.

4. *Industrial revenue bonds.* Local governments are permitted to finance and/or build facilities to lure new industries or businesses into their areas. The objective of course is to provide more jobs, growth, and so on, and generally stimulate the local economy. The money to finance these projects can be obtained by selling industrial revenue bonds. (The project does not need to be "industrial" in order to qualify for "industrial revenue" bonds.) Typically the income from rental or lease of the facility is the only backing for the bonds. If a large, local industrial tenant shuts down, or moves out, or fails to pay rent, the bonds could be in trouble. Also the tax exemptions on these bonds have been limited by current tax law. For these reasons, industrial revenue bonds typically are required to yield higher interest rates before investors will purchase them.

5. *Housing authority bonds.* Many local governments have housing authorities to plan, construct, and manage housing for low-income families. For example, the Gainesville Housing Authority is assisted by the Housing Assistance Administration, an agency of the federal government, which is authorized to guarantee annual contributions toward repayments of bonds issued to finance the housing. Thus, these bonds indirectly have federal government backing.

BOND RATINGS

Investors who buy the bonds are very concerned about the quality of the bonds and the risk involved. Therefore, evaluating and rating of all types of bonds is a necessary part of the bond market. Many fixed-income securities, such as bonds, are regularly evaluated and rated in terms of their overall quality, that is, the perceived ability of the bond issuer to pay the obligation involved in issuing the bond. Three major private rating organizations provide this service; Standard & Poor's, Moody's Investor Service, and Fitch Investor Service. They each use a combination of letters to indicate *relative* quality of bonds. The highest rate is AAA (Standard & Poor's) or Aaa (Moody's), which represents issuers with a strong capacity to pay principal and interest. A slightly lower quality rating would be AA (Aa), and the ratings descend to D (Standard & Poor's) for issues in default with principal and/or interest payments in arrears. While differences do occasionally appear, the rating services generally report similar ratings and are heavily relied upon by investors.

MECHANICS OF FUNDING PUBLIC WORKS
THROUGH A MUNICIPAL BOND ISSUE

The steps through which a municipal bond issue for public works or other investment by a local government must go before the money for the project is in hand will vary somewhat with the agency, the type of bond issue, and the state within which they are issued, but will follow generally the steps listed below for a typical municipality. As two typical cases, consider (1) a revenue bond issue for a parking facility, and (2) a general obligation bond issue for a swimming pool, recreation center, and park facility for a city.

1. *Revenue bond.*
   a. Under the direction of the mayor and city council, the city manager and staff study the city's needs. They draw up a list of needed facilities of top priority. They recommend financing these facilities by borrowing money on revenue bonds.
   b. A bond consultant is brought in. The consultant studies the city's financial position and renders an opinion and recommendations on how much can be borrowed, a schedule of repayments, how well the bonds should sell, and what bond rating and interest rate may be expected.

c. The city council passes a motion authorizing the city attorney to draw up a bond covenant (often in consultation with a bond counsel) and seek validation by the method required in that particular state.

d. The bond covenant is drawn and approved in resolution form by the city council. The bond covenant, as the name implies, is a list of promises made to the potential bond buyers. The promises include safeguards of the investment, such as pledges of escrow, ratio of income to principal and interest (P & I) payments, retaining competent consultants, and so on.

e. The bond issue is then submitted for validation. This procedure varies from state to state. In North Carolina and New Jersey there are state commissions to which municipal bond issues must be submitted for approval. In Texas, bond issues must be submitted to the attorney general's office for approval. Other states have other procedures. In Florida bond issues must be validated by the circuit court after a properly advertised hearing. Appeals by aggrieved parties may be taken to the Florida Supreme Court. In accordance with these requirements, for the example at hand, the city attorney requests the circuit court to hold a hearing on the proposed bond issue, and if no reasonable arguments to the contrary are presented, the court approves (validates) the bond issue and rules that it is a legal exercise of the city's powers. The validation proceeding protects the bond buyer by preventing subsequent suits by irate taxpayers or others seeking to challenge the city's action. City officials are legally barred from acting outside of their constituted authority. Should a plaintiff prove in a successful suit that the bond money was obtained for purposes or by procedures not specifically authorized by law, the bonds could subsequently become worthless. After successful validation proceedings, any would-be plaintiff is informed that the question of the legality of the bonds has already been tried and upheld by the court.

f. The bond consultant tries to aim for a lull or dip in interest rates on the bond market to maximize the amount of money the city will receive and announces the proposed sale of the bonds. Prospectuses are sent out, and presentation dates and a sale date are arranged.

g. Prospective buyers and bond raters are invited to interview top city officials. An informal presentation and sales presentation are given in support of the bonds. The presentation and discussions normally include glowing comments on the city's growth rate, stability, progressive outlook, and the rosy future for the area.

h. Upon request, one or more of the rating services (Moody's, Standard & Poor's) rate the bonds (for a fee). A city may expend considerable effort to convince the rating services that a high rating is warranted, since a high rating results in a lower interest rate.

i. The sale date arrives, and with it come wholesale buyers (bond brokers) with sealed bids in terms of a net interest rate. These are opened at the specified time, and the bids are read aloud. A few days later, after checking on qualifications, the qualified bidder with *lowest net interest rate* pays the city the present worth price and in return gets the bonds. The buyer usually is a broker or brokerage group who then retails them at the going market rate (hopefully at a reasonable markup from what they paid) to individual investors, banks, or other institutions.

j. As the investors hold the bonds and the interest payments come due, the bonds are serviced by a designated fiscal agent, usually a bank that contracts this service for a fee. Servicing includes accepting regularly scheduled deposits from the city for payment of interest coupons and redemption of the bonds as they mature or are called.

2. *General obligation* (GO) *bonds* differ only in that a referendum is held between (d) and (e). A referendum is not just a routine step in which the approval of the voters automatically occurs. Ordinarily, between one-half to one-quarter of all GO bond referenda fail. A number of good articles have been written on planning successful campaigns to promote voter approval of GO bonds, and some commercial firms offer services to plan and carry out election campaigns aimed at passing important bond issues. A typical flowchart for bond financing is shown in Figure 17.2.

FEASIBILITY OF PROJECTS FINANCED BY REVENUE BONDS

Since a bond is essentially an IOU promising to pay interest and principal on the agreed-upon dates, potential buyers of the bonds are quite interested in learning as much as possible about the anticipated costs and income (and consequent ability to make P & I payments when due) of the project that the bond funds are to finance. For this reason, the sponsors of the project will usually have available a feasibility study for the project. These feasibility studies are typically performed by a consulting engineer during the initial planning stages of the project and are often the basis upon which the sponsors make a firm decision on whether or not to proceed with the project. A part of the study usually details how much money must be borrowed by means of bonds and provides a schedule of repayments of the bond funds. This repayment schedule shows how much revenue is needed for payment of interest as well as principal needed to repay the bonds that mature each year. In order to adequately protect their investment, bond buyers prefer that either or both of two standard provisions be included in the bond covenant. The first provision is that a reserve fund be set aside, usually equal to one year's payment of principal and interest. The second provision is that the project's user rates be set so that the anticipated income from the project designated for P & I payments is equal to between 1.25 and 2.0 times the actual amount needed for P & I payments. The actual feasibility study and accompanying background information frequently exceeds 50 or more pages in length, including color pictures and attractive binding. A simplified example of the project cost analysis involved in a typical report follows.

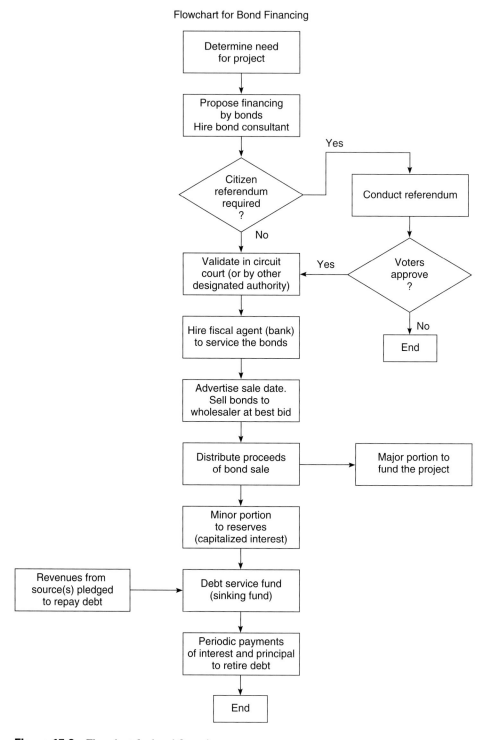

Figure 17.2 Flowchart for bond financing.

EXAMPLE 17.1

This is an outline of a simplified feasibility study for an authority-owned parking garage financed by revenue bonds.

Assume that a downtown parking authority (DPA), which is an agency of the city authorized to issue revenue bonds, proposes to construct a parking garage in a specified downtown location. They engage a consulting engineer (CE) to perform a feasibility study, which includes a determination of hourly parking charges. The CE presents the following information based on calculations and current field surveys.

Interest: The project is to be financed by revenue bonds. The principal and interest (P & I) on the bonds will be repaid by the income revenue collected from parking fees. The interest rate on the bonds of this quality is currently estimated to be 7 percent. (The actual interest rate will not be known until the bonds are let out for bid.)

Construction cost: The construction cost for the project is estimated to be $600,000.

Cost of land: Estimated cost of the land required for the structure is $200,000.

Life of project: The project is expected to be in service for 30 years after the date of completion.

Salvage value: After 30 years of service the parking structure is estimated to have a market value of $400,000, either for resale, refinancing, or salvage. However, any estimate of actual market value that far into the future is questionable at best, and not considered as high-quality security for the bondholders. Therefore the anticipated income from salvage is not considered as income available for repayment of the bonds. Whatever value remains at that time will simply be an additional asset of the DPA, accruing over the life of the structure.

Time of construction: Two full years are expected to pass between the time the bonds are sold and the time the parking garage is open for business. The bond funds not needed for current progress payments to the contractor or other similar costs will be in interest-bearing accounts during this time and will earn income for the project. However, this income will be allocated for contingencies and will not be counted in these calculations.

First income: One full year's income is expected to accrue during the first year of operation. That portion not allocated to O & M will be credited at the end of year (EOY) 3.

Expected patronage: Since the project is located in a downtown area that is expected to suffer a continuing shortage of parking spaces, the level of occupancy should continue to be relatively high and constant over the life of the structure. Current surveys indicate that similar parking garages in this area are experiencing 90 percent occupancy for nine hours per day, five days per week, 52 weeks per year.

Operating and maintenance costs (O & M): These costs are expected to average $80,000 per year during the first year of operation and increase at a rate of 5 percent per year thereafter. O & M costs will commence when the project opens at EOY 2, and will be paid out of working capital, which in turn is constantly replenished from current income.

Working capital: To finance the day-to-day operation of the parking structure, the DPA needs working capital of $15,000 (approximately two months of O & M costs). This is actually a revolving fund from which current expenses are paid and to which income is deposited. It will be needed for the entire life of the structure. The DPA decides to amortize this fund as part of capital costs, so that when the bonds are paid off, the working capital fund will still be available for the continued operation of the structure if needed. When no longer needed, it becomes an asset of the DPA accrued over the life of the project.

Parking fee rate increases: Parking fees may be increased each year by an amount sufficient to cover the increasing O & M costs.

Bond P & I: The bonds are scheduled for sale at EOY 0, as funds are required at this time to buy land and begin construction. During the two-year construction period, interest only will be paid, with one interest payment due at EOY 1, and another at EOY 2. Extra funds must be borrowed to finance these interest payments, since the $800,000 previously designated is needed to buy land and construct the building. The first payment of both interest and principal is scheduled for EOY 3. The payment of principal in this case means that a certain number of bonds will be scheduled to mature at that date. In succeeding years from EOY 3 through EOY 32, bonds will be scheduled to mature in accordance with the amount of income available for payment of principal as determined by the modified amortization schedule.

Bond covenants to protect the bondholders: To enhance the marketability to the bonds (and hopefully induce investors to buy at a lower interest rate), the DPA covenants to do the following:

1. Set up a reserve fund equal to the first year's payment of P & I.

2. Set the fees for parking at a level sufficient to cover P & I payments by a factor of 1.25.

How much to borrow? The project needs the funds listed below on the dates noted. Since bonds are usually sold in $1,000 denominations, the total amount borrowed and the maturity schedule for the bonds will be rounded to reflect this.

1. $800,000 for land and construction of the parking structure, needed at time zero.

2. Working capital fund of $15,000 needed to begin parking operations at EOY 2.

3. Interest payments during the two-year construction period at 7 percent on whatever funds are borrowed. Payments of interest are due at EOY 1 and EOY 2. The amount needed is as follows. Let B represent the amount borrowed, including the $800,000 for land and building, and the $15,000 working capital fund, as well as the two annual payments of 7 percent interest. Then,

$$B = \$815,000 + 2 \times 0.07B$$

Solving for B yields

$$B = \$947,674$$

This is the total amount needed to cover the $815,000 plus the two interest payments.

4. Reserve fund equal to one year's payment of P & I. The reserve fund may be treated as a self-sustaining trust fund whose assets are reinvested elsewhere at 7 percent if possible. The interest income from the investment is used to set up the reserve fund. These bonds would be scheduled to mature at the end of the 30-year service life of the parking structure, at which time the reserve fund is liquidated and the bonds paid off. If the reinvestment rate is more or less than 7 percent, an annual charge or credit is added to account for the difference.

A cash flow diagram is shown in Figure 17.3 to illustrate the incomes and expenditures involved.

Repayments: The bonds may be repaid in equal annual installments or by any other reasonable means mutually convenient to the lenders and borrowers. If the DPA were anticipating higher income in future years, they could schedule more bonds for maturity in later years. For this example, equal annual

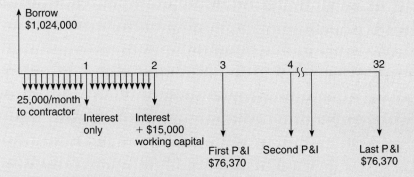

Figure 17.3 Cash flow diagram for Example 17.1.

payments are used. Annual repayments will be made on the $947,674 amount calculated above. In addition a future lump sum payment will be scheduled at the end of 30 years to liquidate the reserve fund, using the proceeds from that fund. Thus, the equal annual payments required are

$$A = \$947,674(A/P, 7\%, 30) = \$76,370$$

The amount required in the reserve fund is equal to one annual payment of P & I, so the total amount to be borrowed is

$$
\begin{array}{r}
\$\ 947,674 \\
\underline{76,370}
\end{array}
$$

$1,024,044, rounded to the nearest $1,000 to $1,024,000 by reducing the reserve fund by $44 to $76,326

These funds will be spent as follows.

EOY		
0	$ 76,326	establish reserve fund trust, and reinvest at 7%
	200,000	purchase land
0-2	600,000	progress payments to contractor during construction
1	66,337	interest on $947,674 at 7% during construction
2	66,337	interest on $947,674 at 7% during construction
2	15,000	working capital revolving fund for operations
	$1,024,000	

The repayment schedule for the bonds may be determined by using an amortization schedule, with minor modifications. The bonds are usually sold in denominations of $1,000, and therefore must be scheduled to mature in whole multiples of $1,000. A normal amortization schedule may be modified to take this into account, as shown in Table 17.1

Income, charges for parking: The amount charged for parking must meet two criteria:

1. The amount charged must be sufficient to cover O & M costs, as well as P & I costs to repay the bonds.

2. The amount charged must be competitive with that of other parking structures in the area.

The amount required to cover costs is as follows. The amount of space-hours

TABLE 17.1 Modified 30-yr Amortization Schedule: EOY[a]

EOY[a]	P, Bonds outstanding[b]	Interest payment I at 7%	Calculated annual payment, A, P & I	Income available for payment of P, $(A - I)$	Used for payment of P (value of bonds maturing this year)	Residue of P & I left over this year	Cumulative residue not used for P
3	$947,674	$66,337	$76,370	$10,033	$10,000	+$ 33	$ 33
4	937,674	65,637	76,370	10,733	10,000	+ 733	769
5	927,674	64,937	76,370	11,433	12,000	− 567	202
6	915,674	64,097	76,370	12,273	12,000	+ 273	475
7	903,674						

[a]End-of-year counting from time bonds were issued at beginning of construction period.
[b]Bonds outstanding except self-sustaining reserve fund, to be liquidated at the end of 30 years with proceeds from the fund plus income that year. ($1,024,000 − 76,326 = $947,647.)

483

available per year is (assuming 100 spaces and 90 percent occupancy):

100 spaces × 9 hr/day × 5 days/week
    × 52 weeks/yr × 0.9 occupancy = 210,600 space-hr/yr
hourly charge for P & I $76,370/210,600 = $0.363/hr
25% for bond covenant coverage of 1.25 =   0.091/hr
hourly charge for O & M $80,000/210,600 =   <u>0.380/hr</u>

        total hourly charge required $0.834/hr

During the second year the O & M costs are expected to rise by 5 percent, so the charges would be raised by $0.05 \times 0.380 = \$0.019$/hr. Other increases are expected to follow in like manner throughout the 30-year life of the project.

Allocating the hourly charges: According to the above results, the hourly charges during the first year should average $0.834 per hour. In order to reach this average, it may be desirable (or more equitable) to charge different amounts to different classes of customers. For instance higher occupancy is encouraged by charging all-day parkers less per hour than transient cars parked for a shorter period of time. If studies indicate that 30 percent of the cars are parked all day, and 70 percent are transient, then a tentative schedule of rates could be determined using the following equation, where H represents the hourly rate for all-day parkers, and T represents the hourly rate for transients.

$$0.3 \times H\$/hr + 0.7 \times T\$/hr = \$0.834/hr$$

Since there are two unknowns in the equation, the value of one will have to be assigned. For this case, suppose that a reasonable daily rate for all-day parkers (H) is $4.00 per day. The hourly rate for transient parkers may then be derived as

$$0.3 \times \$4.00/9 \text{ hr} + 0.7T = \$0.834$$
$$T = (-0.3 \times \tfrac{4}{9} + 0.834)/0.7 = \$1.00/hr$$

If desired, this could be broken down further into different charges per hour for one hour, two hours, one-half day, and so forth, provided that the volume of each type of customer could be estimated with reasonable accuracy. The resulting charges are customarily rounded upward to the nearest convenient nickel, dime, or other convenient module of our monetary system. Thus, for the above example, if all the projections, prove accurate, the following is one of a number of alternative rate structures sufficient to amortize the structure. In addition, adequate safeguards are provided for the bondholders, and a significant equity accrues for the DPA.

all-day parkers—$4.00/day
hourly parkers—$1.00/hr

The consultant must evaluate whether or not these rates are competitive and comment on his findings.

CORPORATE BONDS FOR CAPITAL PROJECTS

Private corporations normally do a major part of their capital financing by issuing corporate bonds. These are not tax-exempt and frequently not as secure as many municipals, and therefore require higher interest rates to attract investors' money. There are two principal types:

1. Mortgage bonds, backed by some guaranteed right, usually a mortgage to the property being financed.
2. Debenture, backed by the full faith and credit of the corporation, but not by a specific mortgage.

INTEREST RATE

The interest rate on bonds is influenced not only by risk and by supply and demand for loan funds, but also by inflation. Conservative investors seem to traditionally anticipate about 3 to 4 percent return on a good liquid investment (can be easily sold or liquidated, "turned into cash," should the desire or need occur) *after* inflation and taxes. On municipal bonds, since there are usually no taxes and the security is typically sound, inflation is a prime consideration. Bonds are typically long-term investments, and the investors who buy them are concerned about long-term inflation.

Even though short peaks in inflation in excess of 10 percent do occur from time to time, in the United States the average long-term inflation rate historically has averaged less than 6 percent. In this environment investors are typically satisfied with yields of the current inflation rate plus about 3 to 4 percent on tax-exempt bonds of high quality. In the late 1970s and early 1980s inflation reached unprecedented levels (peaks of up to 20 percent), and investors were reluctant to purchase bonds with traditionally lower yields. Many government agencies were forced to either cancel prospective bond issues (and their associated construction projects) until more favorable rates cold be obtained, or pay uncharacteristically high interest rates for their borrowed money. Such high interest bonds usually contain clauses allowing the issuer to "call" them (buy them back) when interest rates go down. The issuer then reissues the bond at a lower interest rate.

Buyers of lower quality bonds expect a higher rate of return (interest) to compensate for the additional risk. (If a bond has a risk probability of 2 chances in 100 of defaulting, then approximately 2 percent additional return is needed to compensate for the risk.)

RESALE VALUE

As discussed earlier, there is an active market in trading bonds of all types, which is separate from the issuing agency or corporation. While there are "registered" issues, the more common type of bond is the "bearer bond": the issuer must redeem the periodic coupons, for the value stated, to the bearer, and must pay the cash value of the surrendered bond at maturity or upon call. Once the issuer makes the initial sale and receives its funds, the issuer no longer enters into any of the subsequent sales transactions, nor records ownership changes of bearer bonds.

When a bond is resold before maturity, the buyer may offer more or less than the face value, as presumed earlier. The following example illustrates the calculations involved.

EXAMPLE 17.2

A $1,000 bond paying 6 percent semiannually ($30 semiannually) has five years to go until maturity. The current market for these bonds is 8 percent. That is, investors are willing to buy the bond if they can receive a nominal interest rate of 8 percent on their investment. How much will the bond be worth?

Solution: The cash flow diagram for this problem is shown in Figure 17.4. Find the present worth of the remaining ten coupons representing $30 income every six months plus redemption value of $1,000 after five years, at 8 percent. Since the payments are semiannual, $n = 5 \times 2 = 10$, and $i = 8/2 = 4.0$ percent.

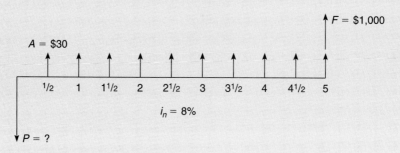

Figure 17.4　Cash flow diagram for Example 17.2.

PW of 10 semiannual $= P_1 = A(P/A, i, n) = \$30\underbrace{(P/A, 4\%, 10)}_{8.1109} = \$243$
    payments

PW of the \$1,000 $= P_2 = F(P/F, i, n) = \$1,000\underbrace{(P/F, 4\%, 10)}_{0.6756} = \underline{\quad 676}$
redemption value
    in 5 yr

$$\text{total } P_1 + P_2 = \underline{\underline{\$919}}$$

The bond should sell for \$919 for an 8 percent nominal interest rate yield to maturity ($i_e = 8.16\%$).

TWO YIELDS

Bonds with long-term maturities are frequently listed with two different figures for yields. The first is "current yield," and the second is "yield to maturity." The current yield is simply the amount of coupon interest received each year divided by the *current* price of the bond. To determine the current yield for Example 17.2, assume a 6 percent bond has two semiannual coupons for \$30 payable each year, and the bond is currently selling for \$919; the current yield is determined simply as \$60/\$919 $= 6.53$ percent.

The "yield to maturity" is simply the rate of return, including the redemption value (face value) of the bond upon maturity, which is shown in Example 17.2 as $i = 8\%$ or $i_e = 8.16\%$. For bonds with long lives (time to maturity), the current yield closely approximates the yield to maturity.

THE BOND MARKET

Daily reports on activities of selected portions of the bond market are available in most daily newspapers. The reports are confined largely to trading on the New York Stock Exchange market and to corporate issues rather than municipals. Interpretation of the data in a typical bond exchange market report listing is not difficult, as illustrated by the following example. Step 1 of course is to find a particular bond issue of interest. An example illustrating the meaning of each item in a typical bond listing follows.

EXAMPLE 17.3

Explain the American Telephone and Telegraph Company bond listed as:

ATT 8.80 s 25 14 114 $64\frac{5}{8}$ $-\frac{1}{8}$.

Solution

Item 1. Abbreviation of the company name (American Telephone & Telegraph).

Item 2. The next number and letter following the name is 8.80 s. This tells the interest coupon rate carried by this particular bond and the payment period. That is, assuming the face amount of the bond is $1,000, the bond coupons can be cashed in for 8.80 percent of $1,000 or $88.00 each year. The *s* means serial payments. If the bond has semiannual coupons, there would be two coupons due each year, dated six months apart, at $44.00 each.

Item 3. The next number, 25, gives the last two digits of the maturity date for the bond, the year 2025. In that year the bond will be redeemed by AT&T for the face amount, usually $1,000.

Item 4. The next numbers indicate the current yield is 14 percent instead of the coupon rate of 8.80 percent. This results from the bond selling at below the $1,000 face value.

Item 5. The number of bonds sold on the day reported is shown in the next column; in this case 114 bonds were sold.

Item 6. The market price of the last bond sold for the day (closing market price) is listed in the next column as a percent face value. In this case the listed price is $64\frac{5}{8}$. If the bond has a face value of $1,000 (as many bonds do) then the last bond sold before market closing sold for $64\frac{5}{8}$ percent of $1,000 or $646.25.

Item 7. The change from the last closing price is listed in the next column as $-\frac{1}{8}$. This means the bonds are selling for $\frac{1}{8}$ of 1 percent lower, or in this case $1.25 less for a $1,000 bond than at the close of the previous market day.

BUYING AND SELLING BONDS

Bonds may be purchased through almost any broker listed in the yellow pages of the local telephone directory under "Stock and Bond Brokers." A typical commission is $10 for the purchase or sale of a bond with a face value of $1,000. While these commissions are comparatively small, they should be taken into account when calculating the yield or value of a bond.

Investment Decisions

Since the market value of a bond fluctuates with the market until maturity, care must be taken not to confuse the value or yield of a bond at the present time with the value or yield at the time of purchase, or any other time. For example, should a $1,000 bond with a 7.9 percent coupon be sold in order to invest in some alternative of equal quality but yielding a 10 percent return? The answer depends not on the fact that the bond originally cost $1,000 but on the current market price and resulting yield to maturity. Whatever has been spent previously is gone, and is classified as "sunk cost." Sunk cost has little or no bearing on current yield or yield to maturity (an exception may occur where credit for tax losses is available). The following example illustrates the principal of sunk costs when comparing a bond currently owned with an alternative investment.

EXAMPLE 17.4

Assume that a friend of yours had $1,000 to invest five years ago this month, so she purchased a Consolidated Edison Company (Con Ed) bond selling at 99 ($990 + $10 commission). The bond paid 7.9 percent annually by means of the usual coupons redeemable for $79 per year on this anniversary date. The maturity date on the bond is 14 years from today. The friend now has an opportunity to invest any amount in a mutual fund that is currently paying 10 percent on whatever is invested. The friend asks your advice on whether to sell the bond and invest the proceeds in the mutual fund at 10 percent. You look in the paper and find the following listing for the current selling price of the bond. An additional expense of $10 commission is involved if the bond is sold. Assume the mutual fund is the same quality (risk) investment as the Con Ed bond.

bond listing: Con Ed 7.90 xx* 12.5 25 $63\frac{3}{8}$.

- **a.** What is the current yield for this bond?
- **b.** What is the yield to maturity for an investor buying this bond at today's price as listed above plus $10 commission?
- **c.** Would you advise the friend that the yield on the bond is about 7.9 percent and the bond should be sold and the proceeds reinvested in the mutual fund at 10 percent?
- **d.** Calculate the yield to maturity of the bond from the friend's point of view as a current *owner* of the bond, and advise her on the basis of comparative yields whether or not to keep the bond or reinvest in the mutual fund.

Solution

- **a.** The current yield may be read directly from the market quote as 12.5 percent or calculated as 7.9/63.375 = <u>12.47 percent.</u>

---

* The maturity date is 14 years from now.

b. The yield to maturity for a *new investor* is calculated by trial and error, as in the rate of return problems in Chapter 10.

	$i = 13\%$	$i = 14\%$
cost of purchase $P_1 = (-\$63\frac{3}{8} \times 10) - \$10 =$	$-\$643.75$	$-\$643.75$
annual coupons $P_2 = +79(P/A, i, 14) =$	$+ 497.90$	$+ 474.16$
maturity value $P_3 = +1,000(P/F, i, 14) =$	$+ 180.68$	$+ 159.71$
	$+\$ 34.83$	$-\$ 9.88$

Interpolation yields

$$i = \frac{34.83}{34.83 + 9.88} \times 1\% + 13\% = 13.78\% = \underline{13.8\%} \text{ yield to maturity}$$

c. No. The yield on the bond at the time of purchase *was* about 7.9 percent, but it no longer is. It currently is higher than the alternative 10 percent available, so the bond should be retained. The difference in the $1,000 price paid originally for the bond and the currently available price of $623.75 (net after payment of commission) is lost. If interest rates go down, part of the loss could be recovered. But if the decision is made on the basis of the current market price, then the bond should be retained for the best yield.

d. The friend's point of view should see a bond with a present value of $623.75 (net after paying $10 sales commission), yielding $79 annually for another 14 years, and $1,000 lump sum at the end of 14 years. All other income and expense has occurred in the past and should be disregarded (except where tax losses or gains may be involved). The rate of return for *the friend* may be calculated in a way similar to that in part (b) above, as follows:

	$i = 14\%$	$i = 15\%$
cash value (net) $P_1 = (-\$63\frac{3}{8} \times 10) + \$10 =$	$-\$623.75$	$-\$623.75$
annual coupons $P_2 = +79(P/A, i, 14) =$	$+ 474.16$	$+ 452.23$
maturity value $P_3 = +1,000(P/F, i, 14) =$	$+ 159.71$	$+ 141.33$
	$+\$ 10.12$	$-\$ 30.19$

Extrapolation yields

$$i = \frac{10.12}{10.12 + 30.19} \times 1\% + 14\% = 14.25\% = \underline{14.3\%} \text{ yield to maturity}$$

In other words, the friend, if she sells, will lose $1,000 − 623.75 = $376.25 and will be able to invest only $623.75 in the mutual fund earning 10 percent. But, her $623.75 *value* in the bond will earn 14.3% percent at maturity, so she is better off by keeping the bond.

BEFORE- AND AFTER-TAX COMPARISONS

When comparing investment opportunities, investors are most concerned about the after-tax rate of return rather than income before taxes. Tax payments are just another normal expense involved in gaining income, and the gross income is not nearly as important as how much is left over after all expenses (including taxes) are paid. When considering investment opportunities in bonds, yields of tax-exempt municipals routinely compete with taxable corporates on the basis of yield after taxes. The following example illustrates such a comparison after the cost of tax payments is taken into consideration.

EXAMPLE 17.5   (After-tax rate of return)

Assume your firm has some idle cash that should be kept fairly liquid but at an acceptable after-tax interest rate. Your boss is particularly interested in two alternative bonds and asks you to calculate the after-tax yield to maturity on each of them.

The first is an American Airlines $10\frac{7}{8}$ bond that matures 11 years from now, rated BBB, currently selling at $95\frac{1}{4}$. The second is a tax-exempt Florida Turnpike 7.10 bond maturing 24 years from now, rated BBB, and selling for 97.

Your firm is in the 38 percent tax bracket (capital gains are taxed as ordinary income) and expects to pay a $10 commission on each bond purchased. The bonds are assumed to have a $1,000 redemption value with no commission paid upon redemption.

Solution.   The after-tax yield is found by the customary trial, error, and interpolation method. Both bonds have coupons that come due semiannually, but a reasonably close result may be obtained by treating the income more simply on an annual basis. For a rough estimate helpful in approximating the first trial, *i*, find the current yield and adjust it for appreciation or depreciation upon redemption. In this case the current market price is only slightly under the redemption value, so the yield to maturity should be only slightly under the current yield. For convenience, the problem may be worked in terms of the percentage values given in the market quotes, rather than in terms of actual cost and income (which are simply the percentages multiplied by 10).

Florida Turnpike Bond (tax-exempt). Current yield $7.1/(97 + 1) = 7.2$ percent. Since the yield to maturity is probably less than 8 percent but greater than 7 percent, use 7 percent for a first trial i.

Because the positive sign on the result indicates that the actual return is greater than 7 percent, the next trial is made at 8 percent.

Since this is a tax-free bond (the interest payments are not taxable), the before-tax yield to maturity is almost the same as the after-tax yield. Only the 2-point ($20) capital gain is taxed (at ordinary rates), and that tax is payable in the year of redemption of the bond, as shown in the solution below.

$$
\begin{array}{rcrr}
\text{trial } i = & & 7\% & 8\% \\
P_1 = & & -98.00 & -98.00 \\
P_2 = 7.1(P/A, i\%, 24) = & & +81.43 & +74.75 \\
P_3 = (100 - 0.38 \times (100 - 98))\underbrace{(P/F, 1\%, 24)} & = & +19.57 & +15.65 \\
& & \overline{} & \overline{} \\
0.19715 & & +3.00 & -7.60 \\
0.15770 & & &
\end{array}
$$

Interpolation then yields:

$$
\frac{3.00}{3.00 + 7.60} \times 1\% + 7\% = 7.28 = \underline{7.3\%} \text{ yield to maturity}
$$

American Airlines Bond. Current yield $10.875/(95\frac{1}{4} + 1) = 11.30$ percent. This is a considerably greater before-tax current yield than the Turnpike bond. The first trial i is estimated by finding the current yield after taxes and adjusting as necessary. The tax rate is 38 percent, so the after-tax income is 62 percent of the coupon amount. Thus, current after-tax yield is $0.62 \times 10.875/96.25 = 7.01$ percent. The yield to maturity will be slightly more, since the redemption value less the ordinary tax on capital gains is still slightly more than the purchase price plus commission. Calculating the yield using the first trial i of 7 percent leads to a positive sum, so the second trial is at 8 percent.

$$
\begin{array}{rcrr}
\text{trial } i = & & 7\% & 8\% \\
\text{cost} = P_1 = & & -96.25 & -96.25 \\
\text{interest income} = P_2 = 0.62 \times 10.875(P/A, i\%, 11) = & & +50.56 & +48.14 \\
\text{redemption income} = P_3 & & & \\
= [100 - 0.38(100 - 96.25)]\underbrace{(P/F, i\%, 11)} & = & +46.83 & +42.28 \\
& & \overline{} & \overline{} \\
.47509 & & +1.14 & -5.83 \\
.42888 & & &
\end{array}
$$

Interpolating yields

$$
1\% \times \frac{1.14}{1.14 + 5.83} + 7\% = 7.16 = \underline{7.2\%} \text{ after-tax yield to maturity (ytm)}
$$

> **Discussion:** The comparative after-tax yields of the two bonds are
>
> Florida Turnpike = 7.3%
> American Airlines = 7.2%
>
> While the difference in current before-tax yields is significant (11.4 versus 7.3%), the difference in after-tax ytm is not large. The ratings on the two bonds are the same, so their relative risks are approximately the same. The difference of 0.1 percent on the yield makes a difference of $1 per year on the interest earned for a $1,000 bond. The bonds are essentially equal except for maturity dates. If interest rates are expected to rise in the future, the shorter-term American Airlines bond has an advantage. If interest rates are expected to decline, the longer-term bond is the better selection.

SUNK COSTS AND BUY-SELL DECISIONS

Investors in bonds are dealing with a constantly fluctuating market, although the changes in the bond market are not usually as rapid and large as changes in stock or commodity markets. Sometimes the interest rates decline, with consequent rises in the average price of bonds, and other times the reverse happens. Astute investors frequently try to upgrade their portfolios by selling current holdings and buying more attractive issues. The decision to sell or buy is usually based on a number of factors, such as the quality of the bond (fiscal soundness, marketability of product, amount and type of bond collateral pledged, etc.) as well as the income and yield. When comparing the yields on current versus prospective holdings, a question sometimes arises concerning the true value of the current holding. If the investor paid 96 for a corporate bond five years ago (95 market + 1 commission), and the current market value of the bond is 84, which figure should be used for current value? Actually the current value is really the amount of cash that can be realized in exchange for the bond (see Example 17.4d). In most cases this is the current market value, less commission, plus or minus any tax considerations. Where a capital gain is involved, income taxes must be paid (or credited). In the example at hand the loss in value normally will result in a tax-credit benefit. Thus, assuming a taxable $1,000 corporate bond with a market value of 84, upon sale the investor would receive $840 less a $10 commission. Since this is somewhat less than the $960 paid for the bond, a capital-loss credit normally is available. The original purchase price of the bond does not enter into the calculation of current value except in considering the tax loss or gain. The loss in value is considered as a "sunk cost" and should not be confused with the value of the bond to the investor. When comparing rates of return between this bond and a prospective replacement, the current cash value calculated above should always be employed.

SPREADSHEET APPLICATIONS

	A	B	C	D	E
1	EXAMPLE 17.1				
2	Solving example 17.1 using the EXCEL spreadsheet software				
3					
4	Given:	Land & construction parking	$ 800,000		
5		Working capital at EOY 2	$ 15,000		
6		Interest (%)	0.07		
7		N (year)	30		
8		2 annual pmt due at EOY 1&2			
9					
10	Step 1:	Total amt to borrow (B) is:		$ 947,674	
11		Cell D10 : B=(800,000 + 15,000) + 2* 0.07B			
12					
13	Step 2:	Annual Pmt required is:		$ 76,370	
14		Cell D13 = PMT(C6,C7,-D10,0)			
15					
16	Step 3:	Total Amt required in reserve fund is:		$ 1,024,044	
17		Cell D16 = D10 + D13			
18					
19	Step 4:	The interest pmt is:		$ 66,337	
20		Cell D19 =D10 * C6			
21					
22	Step 5:	Total hourly charge :			
23		Based on spaces	100		
24		Occupancy (%)	0.9		
25		Hours/day	9		
26		Days/week	5		
27		Week/year	52		
28		Bond Covenant Coverage	0.25		
29		Operaing and Maintenance	$ 80,000		
30					
31		Total spaces	$ 210,600	space-hr/yr	
32		Cell C31 = C23 * C25 * C26 * C27 * C24			
33		Hourly Charge for P & I	$ 0.363	/hr	
34		Cell C33 = D13 / C31			
35		25% for bond covenant	$ 0.091	/hr	
36		Cell C35 = C28 * C33			
37		Hourly Charge for O & M	$ 0.380	/hr	
38		Cell C37 = C29 */C 31			
39		Total Hourly Charge is:	$ 0.833	/hr	
40					
41	Step 6:	Allocating the Hourly Charge:			
42		Car all-day park (H) (%)	0.30		
43		Transient (T) (%)	0.70		
44		All day Parker	$ 4.00	per day	
45					
46		Transit Hourly Park:	$ 1.00		
47		0.3 * $4.00/9 hr + 0.7 T = $0.834			
48		Cell C46 = (C39-(C42 * C44 / C25)) / C43			
49					
50	Procedure:				
51	Step 1 through Step 3 are used to determine the amount required in the reserve				
52	fund which is equal to one annual payment of P& I. The total amout is $1,024,044.				
53	Step 4 is used to determine the interest payment which is calculated to be $66,337.				
54	Step 5 is used to find the total hourly charge required which is calculated to be $0.833/hr.				
55	Step 6 is used to find the rate for all-day parking and for hourly parking. The all-day rate is				
56	calculated to be $4.00 per day. The transit hourly parking is calculated to be $1.00/hr.				
57	Table 17.1 shows partial the modified 30 year amortization schedule.				

SUMMARY

Bonds are commonly accepted as a means of acquiring needed funds for capital improvements for both government and private industry. Bonds are a vital source of funding for large numbers of engineering projects. This chapter presents the cost analysis of bonds, both municipal (tax-exempt) and corporate. A bond is simply a long-term public debt of an issuer that has been marketed in a convenient denomination, usually through a broker. The bond has a face-redemption value, together with a series of interest payments usually designated by a nominal interest rate of the face value, and a maturity date upon which the bond may be redeemed for its face value.

There is an active bond market in which bonds are bought and sold each business day. Depending upon a number of variables, including the demand, money supply, bond rating, and yield on competing investments, the bonds are traded for more or less than their face value. Thus, the actual interest earned on an investment in bonds may be less than or greater than the interest paid on the face amount.

PROBLEMS FOR CHAPTER 17, BOND FINANCING FOR PUBLIC WORKS AND CORPORATE INVESTMENT

Problem Group A

No tax considerations. Given values for four of the five variables P, A, F, i, n, find the fifth.

A.1.   A $5,000 bond with 8 percent annual coupons for $400 will mature in 20 years. It can be purchased for $4,300.

  (a) Find the current yield.

  (*Ans.* 9.30%)

  (b) Find the yield to maturity

  (*Ans.* 9.60%)

A.2.   Same as Problem A.1, except the 8 percent coupons are paid semiannually ($200 every six months).

A.3.   A coupon bond is purchased for $5,500. Coupons for $350 are clipped from the bond and cashed in at the bank once a year. (The bond has a face value of $5,000 with 7 percent coupons.) At the end of 20 years the last coupon for $350 is turned in at the bank, and the bond is cashed in for $5,000. What is the before-tax rate of return?

  (*Ans.* 6.13%)

A.4.   An investor desires a municipal bond yielding 8 percent to maturity. A friend has one with a $5,000 face value, maturing in ten years and with 20 more semiannual coupons for $125 each coupon. What price should be offered to obtain an 8 percent yield? (Use $n = 20$.)

A.5.   A $10,000 bond with 7 percent coupons payable annually was purchased four years ago for $9,800 and sold today for $9,500.

  (a) What before-tax rate of return was realized on this investment by the original purchaser?

  (*Ans.* 6.45%)

(b) If the new purchaser holds the bond to maturity (eight more years) what before-tax rate of return will be realized?

(*Ans.* 7.87%)

A.6. A port authority toll bridge is currently yielding net revenues of $100,000 per year, increasing by $10,000 per year. The bridge is old and needs replacement, and will be out of service for two years during reconstruction. Revenue bonds will be sold to finance the replacement. If the bonds pay an interest rate $i = 7$ percent, what is the maximum amount of bonds that can be sold now and paid off within 25 years from now, assuming the full amount of net revenues is used to pay off the bonds? Assume the existing bridge is taken out of service now, and the first income for repayment of bonds occurs at the end of year 3 in the amount of $130,000, increasing by $10,000 per year thereafter.

A.7. Same as Problem A.6, except that while the bonds have a coupon interest rate of 7 percent, investors want to receive 9 percent on their investment.

(*Ans.* $1,595,000)

A.8. A transportation authority wants to finance the construction of a bridge by borrowing money on bonds. How much can they borrow if the estimated toll receipts occur as listed below? The bonds are expected to have a 7 percent interest rate and be paid off by the end of the twenty-fifth year. O & M costs will begin at EOY 3 at $40,000 per year and increase by $2,000 per year each year through EOY 25.

EOY	Toll receipts
1	None (bridge under construction)
2	None
3	$15,000
4	$35,000
5–7	(Increase $20,000/yr)
8	$115,000
9	$120,000
10–24	(Increase $5,000/yr)
25	$200,000

Problem Group B

No taxes. Compare alternatives. Sunk costs.

B.1. Three years ago your client paid $1,000 for a 5 percent bond (annual coupons of $50 each) that will mature 10 years from now. The client now wants to sell the bond and finds it has a current market value of $620. In addition there is a brokerage fee of $10 per bond on each sale or purchase. The client is thinking of purchasing a 9.5 percent bond (coupons of $95 per year) priced at $920 plus commission that matures 15 years from now. Find the yield to maturity on each and compare.

(*Ans.* old bond 11.9%; new bond 10.3%)

B.2. Your firm intends to invest some of its funds in bonds and is particularly interested in two bonds issued by the same organization. The after-tax yield to maturity on both is 7.96 percent. The only difference is that bond A has a maturity date 3 years from now, and bond B matures in 25 years.

(a) What advantage(s) does bond A have over B?

(b) What advantage(s) does B have over A?

B.3. Four years ago you purchased a computer science (Comp Sci) bond for $987.00 including commission. The bond has coupons redeemable for $60 per year, and it is expected to be worth $1,000 upon maturity eight years from now. Assume that the anniversary date on the latest coupon just happened to coincide with yesterday's date, so it will be a full year before the next coupon comes due. Assume that the maturity date on the bond will also coincide with yesterday's date but in the year of maturity. You now have an opportunity to invest in a mutual fund that is currently paying 10 percent. The current selling price of the bond is shown in the listing below. An additional expense of $10 commission is involved if the bond is sold. Assume the mutual fund is the same quality (risk) investment as the bond. Find the (yield to maturity) for the bond and compare with the mutual fund.

bond listing: Comp Si 6s xx 10.2 59 . . .   (xx = a date 8 yr from now)

(*Ans.* 15.60% better than mutual fund)

B.4. From the following daily bond market report for a Holiday corporation, make the appropriate identification matchings for a $1,000.00 bond.

Bond quotations

Holiday (1)	7¾ s 20 (2)(3)(4) Item description	9.5 (5)	88 (6)	97 (7)	+½ (8) Identify (1) to (8)
(a) Maturity date—last two digits				_____	_____
(b) Market price of last bond sold				_____	_____
(c) Company name				_____	_____
(d) Semiannual coupons				_____	_____
(e) Current yield %				_____	_____
(f) Interest coupon rate				_____	_____
(g) Number of bonds sold				_____	_____
(h) Change from previous closing price		_____			_____

(i) How much will you receive for each coupon cashed? $ _____

(j) In what year will the bond mature? _____

(k) If you paid a $10 commission along with the purchase price of the bond, how much total would you pay for buying one bond? $ _____

(l) Normally, what is the bond's redemption value at maturity? $ _____

(m) T or F For a bond selling below its face value, the yield to maturity % is expected to be lower than the current yield %.

B.5. An investor has a choice of investing $9,500 in an eight-year certificate of deposit yielding 9 percent compounded monthly (0.75 percent per month), or investing in an unrated $10,000 bond yielding 8 percent payable semiannually ($400 each six months) and maturing in eight more years. The bond can be purchased for $7,800 today.

(a) Compare the rates of return for the two investments.

(*Ans.* CD = 9.38%; bond = 12.38%)

(b) Discuss the relative merits of the two investments, including (1) the fact that the certificate of deposit cannot be cashed in early without paying a substantial penalty; (2) if the bond is purchased, what happens to the difference in money between the two investments ($1,700 in this case)? (3) consider the unknown risk associated with purchasing an unrated bond.

Problem Group C

Involves taxes.

C.1. An investor is seeking a corporate bond with at least a 7 percent yield to maturity after taxes. A corporate bond with a face value of $5,000 is offered for sale at 80 (for $4,000). The bond will mature in 10 years and has 8 percent coupons ($400/yr). The investor expects to pay an annual income tax of 30 percent on the interest income and an annual intangible tax of 0.1 percent on the face value of the bond. In addition, the capital gain difference between the $4,000 purchase price of the bond and the $5,000 redemption amount is taxable at the time of redemption at the ordinary rate. Does the after-tax yield exceed 7 percent? Show proof.

(*Ans.* 7.88%)

C.2. In an effort to put idle cash to work, your client purchased ten Gen Inst bonds four years ago today, at 89 + 1 for commission. The bonds pay interest of 5 percent once a year on an anniversary date coinciding with yesterday's date and mature 20 years from now. The bonds are currently selling at the following market price.

Gen Inst 5s xx 6.9 $72\frac{3}{4}$.

If the bonds are sold, an additional cost of $10 per bond will be incurred.

   Your client now has an opportunity to invest in tax-free municipal bonds that have a yield to maturity of 6 percent.

   Your client is in the 40 percent tax bracket on all income. Losses are eligible for a tax credit against other income.

   (a) What is the current yield for this bond before taxes?
   (b) Find whether or not your client should sell the current holdings (10 bonds) and reinvest the funds in the tax-free municipals. Do this by determining whether a higher rate of return is obtained by leaving the cash invested in the bonds or selling the bonds and investing in municipals at 6 percent (calculate net return after paying commission costs). Assume either bond would be held until maturity.

C.3. Calculate the after-tax rate of return for the original investor of Problem A.5 if the tax rate is 27.2 percent.

(*Ans.* 4.29%)

CHAPTER 18

Home Ownership and Mortgage Financing: Owning Versus Renting

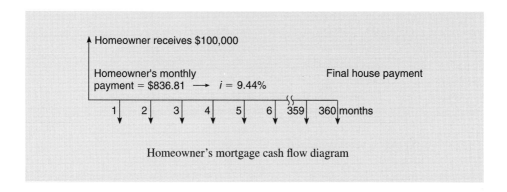

Homeowner's mortgage cash flow diagram

MORTGAGE FINANCING FOR HOME OWNERSHIP

Currently about 2 million housing units are constructed annually at a cost of well over $100 billion (comprising over one-third of the total construction market and about 3 percent of the nation's entire gross national product).

In the early days of our country's history, when the business of mortgage lending was in its horse-and-buggy stages, mortgages were difficult to obtain, and prospective home buyers were expected to make down payments of about 50 percent or more of the purchase price. In addition the life of the mortgage was typically shorter, say five years, and the amount borrowed was not amortized but was repayable on the due date in one lump sum. Consequently, as few as one-third of American homes were owner-occupied (compared to about two-thirds now), and mortgage foreclosures were not uncommon.

Federal Housing Administration (FHA) and Veterans Administration (VA)

The mortgage lending industry has progressed and enlarged considerably since those early days. A major turning point occurred in 1934 when the Federal Housing Administration (FHA) was conceived by Congress to insure mortgages against default. Since 1934 the FHA has written mortgage loan insurance aggregating over $200 billion, covering over 10 million homes, more than a million living units in multifamily projects, and over 35 million property improvement loans. The FHA is one of the very few government agencies that is entirely self-supporting. It derives its income from fees, insurance premiums, and investments, and its insurance reserves are well over a billion dollars. The FHA does not build houses or lend money. It acts only as an *insurer* of privately made loans from approved lenders. Any FHA-approved mortgage lender, such as a bank, savings and loan association, or other financial institution can make insured mortgage or home improvement and rehabilitation loans.

Only a low minimum down payment normally is required by the FHA in order to qualify for an insured mortgage loan. The amount of down payment is revised and updated from time to time, and is ordinarily about 3 percent of the purchase price for houses in the lower price brackets.

Several years after the establishment of the FHA, the Veterans Administration (VA) was given a similar charge to guarantee loans for the nation's 30,000,000 veterans who served in the Armed Forces. The VA does not require any down payment on home purchases by qualified veterans.

The creation of the FHA and VA had a very stabilizing influence on the industry and gave lenders more confidence in mortgages as an investment. Over the years this brought a greatly increased inflow of mortgage loan funds into this financial market area, since the FHA insured against loss through foreclosure. (An extra $\frac{1}{2}$ percent of the unpaid balance is charged the borrowers to finance the insurance fund. If a mortgage is seriously in default, the mortgage may be purchased from the mortgage holder with money from this fund. The FHA then forecloses if necessary and resells the property.) One effect of the FHA and VA on the housing market has been to make housing available to many who could previously not afford the high down payments. The average down payment recently has dropped to about 7 percent, while over 15 percent of homes are sold with no down payment at all.

By far the largest number of home mortgage loans are termed *conventional loans,* under the terms of which the borrower usually pays 20 to 25 percent of the purchase price as a down payment and borrows the remainder from major lending institutions such as savings and loan associations. Since the mortgage on the property is the only protection the lender has, borrowers must demonstrate some ability to repay the loan.

Often the original lender does not keep the mortgage but sells it to an investor. Then the original lender may use the money from the sale to lend on more new mortgages, keeping the discount points charged as profit on the turnover. Among the larger institutions, the Government National Mortgage Association (GNMA) purchases mortgages insured by the FHA and VA so that mortgage lenders may have additional cash to loan on more mortgages. The Federal National Mortgage Association (FNMA), a private-public corporation, also buys mortgages at auction from private mortgage lending institutions in periods of short money supply. These organizations were created by Congress to insure a healthy and efficient mortgage market since so much of our nation's new housing and commercial building is dependent upon mortgage financing.

MORTGAGES IN THE INVESTMENT MARKET

There is an active market for buying and selling mortgages. The professionals who do this as a business usually call themselves mortgage bankers or mortgage brokers. Mortgage brokers simply act as middlemen between borrower and lender. Mortgage bankers, on the other hand, offer a complete line of services to the investor. They will find and screen the property owner or potential buyer in need of a mortgage, lend the money, sell the mortgage to the investor, and service the mortgage. Servicing includes collection, bookkeeping, following up on delinquent accounts, carrying escrow accounts for insurance and taxes, and periodically checking on the condition of the property. Typically about one-quarter of the mortgage banker's gross income comes from origination fees (front-end charges or discount points paid by the borrower), and one-half comes from service fees to the investor. All of these services greatly assist in bringing money into the mortgage money market. The greater the supply of money available, the lower the interest rate, and the easier it is for the prospective home buyer to obtain a mortgage loan.

WHAT IS A MORTGAGE?

A real estate mortgage and note is simply a promise to repay borrowed funds using real estate as collateral. They are normally long-term notes (terms of 20 to 35 years are common). In the event of default the mortgage lender has a claim against the property and typically has recourse to court action requiring sale of the property at public auction to satisfy that claim.

A mortgage involves two basic items, (1) the note, and (2) the mortgage itself.

1. The note provides evidence of the debt. It describes the dollar amount, the interest rate, and terms and dates of repayment.
2. The mortgage gives the lender an interest in the property until the note is paid off. In 30 states (including Florida), the mortgage constitutes a lien on the property and goes with the property until paid off regardless of transfers of ownership. In most of the other 20 states the mortgage transfers title to the lender until the note is paid off.

Both the note and the mortgage must be properly executed in order to constitute a complete legal document.

Unless otherwise stipulated, mortgages are inherently freely transferable by either party. Mortgages are *not* transferable only if so stipulated in the agreement between the lender and borrower. Currently most lenders insert what is called a "paragraph 17" in the mortgage agreement, which in effect says that the mortgage is assumable only upon approval of the lender, and frequently requires an additional payment or increase in interest rate. In the absence of such a stipulation the mortgage lender is free to sell the mortgage, and the borrower is free to sell the property and let the new owner take over the mortgage with the property without permission of the other party. However, in case of default, each person who has previously assumed the mortgage may be liable for any loss on the mortgage. The holder of the mortgage may look to each person who has sold the property subject to the mortgage as an individual guarantor of the mortgage amount.

SECOND AND THIRD MORTGAGES

A property owner may owe to more than one lender simultaneously, using the same property as collateral for all the loans. A mortgage securing a second loan is a second mortgage. Third and more mortgages can be taken out; the number is only limited by each subsequent lender's faith in the ability of the borrower to repay. Normally, each successive loan is subsidiary to the one before it. In case of default the last lender may offer to take over all obligations and payments to the prior lenders in exchange for a quit-claim deed from the owner. For instance, if the value of the property warrants it, the third mortgage holder can offer to take over payments to mortgage holders numbers 1 and 2, in hopes of salvaging something from his interest. The owner clears up delinquent debts on the property and salvages a sagging credit rating. In case of foreclosure the first priority usually goes to payment of delinquent taxes and any other government claims. Then (provided there are no prior liens) the first mortgage holder gets the balance due (together with interest on delinquent payments) before the second mortgage holder gets anything. The third is in the same position with respect to the second. Consequently, at a sheriff's foreclosure auction sale the first mortgage holder is usually interested in bidding the price up high enough to cover the amount due on the first mortgage. The second mortgage holder may take over the bidding at this point if the value of the property exceeds the amount of the first mortgage and he is able to raise the cash to pay off the first mortgage balance (or assume the first mortgage if the lender permits).

Assume, for example, the first mortgage balance due is $100,000, and the second mortgage is $50,000. The second mortgage holder feels the property is worth only $140,000 due to some recent deterioration. If the highest bid during the course of the auction is $120,000, the second mortgage holder has the option of letting it go to this bidder, who would pay the auctioneer $120,000. Out of this $120,000 the first mortgage holder receives $100,000 (assuming that taxes and other priority liens have been paid), and the second mortgage holder would net $20,000 after the first mortgage was paid in full. Instead of netting only $20,000, the second mortgage holder would logically bid the price up as far as $140,000 in expectation of either getting the $40,000 or more in cash if someone else is successful bidder, or at least getting the property worth $140,000 in exchange for $100,000 to pay off the first mortgage if he is the highest bidder. If no one else bids higher, his only obligation is to pay off the first-mortgage $100,000 balance (the first mortgage holder might let the new buyer simply resume monthly payments on the first mortgage, after paying up any delinquencies). Any auction sales price over the $150,000, of course, results in the overage going to the owner, providing there are no other valid liens on the property.

Foreclosure does not free the borrower of any unpaid obligation. If the sale brings less than the mortgages owed, the borrower may still be found personally liable for the difference.

MORTGAGE TERMS, I AND N

Residential first mortgages typically have a life of from 15 to 35 years, with the majority being 25 to 30 years. Second mortgages typically run about 10 years, although any mutually agreeable life is possible. The interest rate depends upon supply and demand of mortgage money and the risk. For many years first mortgage interest rates hovered typically between 4 and 6 percent, with periodic dips or peaks on either side. Then in the mid-1960s mortgage interest rates began a gradual rise. In the early 1980s mortgage interest rates peaked at about 15 percent when inflation reached double digits, and then declined to more modest levels. When interest rates rose, many lenders found themselves with a lot of money loaned out on long-term mortgages at low fixed rates. Some of these lenders, such as the savings and loan banks (S & Ls), had loaned money from the savings accounts of their depositors (which they are chartered to do), and the depositors began to withdraw their savings in order to invest them at higher interest elsewhere. Many S & Ls were caught in the crunch between depositors' withdrawals and the S & Ls inability to pay these withdrawals because the money was loaned out on fixed-rate, long-term mortgages that declined in resale value as interest rates rose. Those that survived began lending some of their mortgage money at variable rates of interest. Typically, variable-rate mortgage agreements stipulate that the interest rate will be adjusted periodically (every year, two years, or other mutually agreeable period) and that the adjusted interest rate will be some recognized standard (such as the prime rate, or the U.S. Treasury T-bill rate, or other) plus several percentage points added on. In addition for the borrower's benefit there is often a floor and a ceiling on the percentage interest, and a limit on the percentage change per year. With variable-rate mortgages, the borrower assumes most of the risk of inflation, at least between the floor and the ceiling rates. The analysis of variable-rate mortgages is similar in kind, but slightly more complex in degree, to that of fixed-rate mortgages. Fixed interest rates are assumed throughout this chapter. Most lending institutions now offer a choice between fixed-rate or variable-rate mortgages. Variable-rate mortgages are offered at lower interest rates because there is less risk to the lender. However, fixed-rate mortgages are still popular with borrowers because of the comforting assurance that the monthly payments are fixed for the life of the mortgage.

High Interest Rates Exert a Depressing Effect on the Economy

Whenever interest rates increase, fewer borrowers apply and qualify for mortgage loans due to the high payments required. As a result, real estate construction and resales sag. From the developer's viewpoint, higher interest rates are an added cost of doing business, and must be added to the other costs of the project. Whenever interest rates rise, the costs of proposed projects all over the country also rise, and an increasing number of them become financially unfeasible due to the higher interest costs. This causes a slowdown in the entire construction industry, and tends to affect the economy as a whole in the same way. By the same reasoning, falling interest rates have a stimulating effect on both the construction industry and the economy as a whole.

INCREASING THE YIELD

Sometimes lending institutions participate in various government-sponsored lending programs that involve ceilings on interest rates for mortgages. If the current mortgage market interest rate is higher than the government ceiling, the lending institution may either stop lending under that program or utilize some other method for reaching the market rate. Methods currently in use to increase the yield above the ceiling include (1) mortgage discounts, (2) other front-end charges, and (3) nonassignment provisions.

1. *Mortgage discounts* are called "points" if they are expressed in terms of a percentage of the amount of the loan. For instance, if a lender charges "3 points" as a condition of making a $100,000 mortgage loan, he wants 3 percent or $0.03 \times \$100,000 = \$3,000$, in *addition* to the down payment. The points may be added to the balance due or deducted from the amount loaned, depending on the policy adopted by the lender. Therefore, the lender expects the borrower to repay either $103,000 plus interest in exchange for a loan of $100,000, or to repay $100,000 plus interest for a loan amount of $97,000.

The rule-of-thumb equivalency between points and interest rates is: 8 points charged at the time the loan is made costs the equivalent of approximately 1 percent interest added to the interest rate over the term of the loan. (This is fairly accurate for mortgages of 25 to 30 years but loses accuracy rapidly with shorter mortgage lives.) For instance, assume a homebuyer needs a $100,000 mortgage, and finds there is a choice between two alternative mortgage plans:

1. $100,000 mortgage plus 8 points, 9 percent, 25 years.

2. $100,000 mortgage plus zero points, 10 percent, 25 years.

The homebuyer must determine whether it is better to pay 8 points for a mortgage at a lower 9 percent interest rate, or pay no points but pay a higher 10 percent interest rate. Since the 25-year terms are the same, the mortgages may be evaluated by comparing the monthly payments, as follows.

1. For the first alternative mortgage plan, the 8 points indicates that 8 percent of the $100,000 amount borrowed will be added to the amount owed, and the homebuyer's monthly repayments will be based on a principal balance owed of $108,000. The monthly repayments are calculated as

$$A = \$108,000(A/P, 9/12\%, 300) = \$906.33/\text{month}$$

2. For the second alternative mortgage plan, the monthly repayments are based upon a $100,000 principal balance, but a higher 10 percent interest rate, as follows.

$$A = \$100,000(A/P, 10/12\%, 300) = \$908.70/\text{month}$$

By comparison of the monthly payments (both are for 300 months), the homebuyer can see that the 8 points extra paid on plan 1 is closely equivalent to the 1 percent difference in interest rates on plan 2. Thus, on a long-term mortgage, each 1 point (paid in advance at the time of closing) is approximately equivalent to an increase of $\frac{1}{8}$ percent interest rate.

2. *Other front-end charges.* Sometimes a borrower will be charged a loan initiation fee, credit check fee, loan discount fee, lender's inspection fee, assumption fee, settlement or closing fee, document preparation fee, and so on. These have the same effect of increasing the effective interest to the borrower. The amount of increased effective interest can be calculated as in the case of points.

3. *Nonassignment provisions* resulting in early payoff of mortgage balance. Nonassignment provisions require the mortgage to be paid in full if the property is sold. Since the average person moves about once every 5 years, most 25- to 30-year mortgages will be paid off early if they contain a nonassignment clause.

While the 8-point discount is approximately equivalent to 1 percent difference in interest rate for a long-term mortgage, the equivalencies change dramatically for a short-term mortgage or a mortgage that is paid off before maturity. For example, in the extreme case, if the $100,000 mortgage were paid off after just one year, under plan 1 (9 percent mortgage plus 8 points) the borrower would repay approximately 17 percent interest, whereas after one year under plan 2 (10 percent mortgage and no points) only 10 percent interest would be repaid. Thus over a short life, the points can be much more costly than the otherwise equivalent increase in interest rate. Another more typical example of the effect of an early payoff on the actual interest rate is shown in the following.

EXAMPLE 18.1

Two years ago a homebuyer took out a $100,000 mortgage at 9 percent for 360 months (30 years), with monthly repayments. The mortgage also required 4 points added on, so the initial principal balance was $104,000. Now two years later the homebuyer has a lucrative job offer in a distant city and has decided to sell the house. Due to a strict paragraph 17, she must pay off the balance on the mortgage.

a. Find the actual interest rate if the loan were amortized with the normal monthly installments over the 360-month period.

b. Find the actual interest rate if the unpaid principal balance of the loan were paid off at EOY 2 (EOM 24).

Solution

a. The monthly payments are found as follows:

$$A = \$104,000(A/P, 9/12\%, 360) = \$836.81$$

The lender's cash flow diagram is shown in Figure 18.1. Given A, P, and n, solve for i to find the actual mortgage loan rate. Since the homebuyer actually received only \$100,000 (to give to the seller), the actual interest is calculated on the basis of $P = \$100,000$:

$$A/P = 836.81/100,000.00 = 0.0083681 = (A/P, i, 360)$$

By interpolation, the actual $i = 0.7870$ percent per month, and the nominal $i = 12 \times 0.7870 = \underline{9.44 \text{ percent per year.}}$

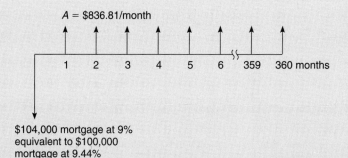

$A = \$836.81/\text{month}$

1   2   3   4   5   6   359   360 months

\$104,000 mortgage at 9%
equivalent to \$100,000
mortgage at 9.44%

Figure 18.1   Lender's cash flow diagram for Example 18.1a.

Conclusion.   The extra 4 points charged on the mortgage raised the nominal interest rate for the mortgage from 9 to 9.44 percent over the 30-year repayment period. This increase could have been approximated by applying the rule of thumb that indicates an increase of about $\frac{1}{8}$ percent of interest for each 1 point charged when taking out the loan.

b. To determine the interest rate if the loan is paid off at EOY 2 (Figure 18.2):

1. Find the payoff balance at EOY 2.

2. Sketch the cash flow diagram.

3. Solve the ROR problem and find i.

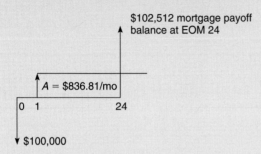

Figure 18.2 Lender's cash flow diagram for Example 18.1b.

1. The payoff balance at EOM 24 is the PW of the remaining 336 payments after 24 payments have been made:

$$P = A(P/A, i, n) = \$836.81(P/A, 9/12\%, 336) = \$102{,}512$$

2. The cash flow diagram (for the lender) shows

$$\text{EOM } 0 = \$100{,}000 \text{ cost}$$
$$\text{EOM 1 through EOM } 24 = \$836.81/\text{month income}$$
$$\text{EOM } 24 = \$102{,}512 \text{ payoff balance income}$$

3. The ROR problem is solved as follows:

	0.75%	1%
$P_1 = -\$100{,}000 =$	$-\$100{,}000$	$-\$100{,}000$
$P_2 = 836.81(P/A, i, 24) = +$	$18{,}317$	$+\quad 17{,}776$
$P_3 = 102{,}512(P/F, i, 24) = +$	$85{,}683$	$+\quad 80{,}736$
	$+\quad 4{,}000$	$-\quad 1{,}488$

By interpolation, the ROR is found as

$$i = 0.75\% + 0.25 \times 4{,}000(1{,}488 + 4{,}000) = 0.930\%/\text{month}$$

The nominal annual rate is found as $12 \times 0.930 = 11.16$ percent per year $= \underline{11.2 \text{ percent per year}}$ (compared to 9.44 percent per year if amortized over 25 years).

Conclusion.   Where points are involved, paying off the mortgage balance at an early date raises the actual interest compared to amortizing the loan over its normal life.

Wrap-Around Mortgage (WAM)

When a property owner decides to sell a property, and that property happens to have an existing mortgage with a low interest rate, frequently there is a reluctance to give up the low-interest mortgage. This is especially true in times of rising interest rates, if selling involves paying off the balance owed on the low-interest mortgage, and refinancing a new larger mortgage at a higher interest rate. One method by which the advantages of the low-interest mortgage can be retained by the seller is by use of a WAM. An example will show how a WAM works:

1. The property owner bought a house five years ago for $75,000.
2. The down payment at the time was 20 percent, or $15,000.
3. The balance was financed under a mortgage, at 7.5 percent, for 30 years, for $60,000.
4. The payments under the mortgage are

$$A = \$60,000(A/P, 7.5/12\%, 360) = \$419.53/\text{month}$$

5. Now, five years later, the owner decides to sell the house at its current market value of $100,000.
6. The mortgage balance is paid down to

$$P = \$419.53(P/A, 7.5/12\%, 300) = \$56,770.46$$

7. The seller therefore has an equity in the property of ($100,000 − $56,770.46 =) $43,229.54.
8. The seller would like to find a buyer with a down payment of $43,229.54, but most of the buyers with that amount of cash are looking for bigger, higher priced homes. So the seller gets a bid from a prospective buyer who offers a 20 percent down payment on the $100,000 house amounting to $20,000.
9. To finance the sale, the seller begins by offering to let the buyer take over the first mortgage of $56,770.46.

10. Seller will also write a new, second mortgage for the balance of ($43,229.54 − $20,000 =) $23,229.54 at 12 percent and 10 years,

with payments of $A_2 = \$23,229.54(A/P, 1\%, 120) = \$333.28/month$
together with the first mortgage payments of   419.53/month
to make a total monthly payment for the buyer of   \$752.81/month

11. The bank that holds the first mortgage examines the prospective buyer's financial statement and finds they cannot qualify the buyer for payments of over $700.00 per month. They suggest lowering the monthly payments. The seller proposes a WAM. Under the WAM, the buyer makes the same down payment of $20,000 and agrees to make mortgage payments to the seller on the WAM mortgage balance of $80,000 at 9.5 percent for 25 years. The payments on the WAM are

$$A_3 = \$80,000(A/P, 9.5/12\%, 300) = \$698.96/month$$

a saving of $53.85 per month over the previous plan, and just within the bank's guidelines.

12. The seller continues to pay the bank $419.53 under the old mortgage, while collecting $698.96 under the new WAM mortgage.

What are the advantages for the buyer?

1. The buyer gets financing for a purchase for which the buyer otherwise might not qualify.
2. The buyer gets lower monthly payments for the first 10 years.

What are the advantages for the seller?

1. The seller sells the house, which otherwise might not sell this soon or at this high a price.
2. The seller earns interest on the full $80,000 amount of the buyer's mortgage, while only $43,230 of this mortgage is seller's actual equity. Thus the seller earns 9.5 percent on his own equity as well as on the $56,770 that is borrowed from the bank at 7.5 percent.

What rate of return does the seller earn on the seller's actual equity? Solve as a rate of return problem.

$P = \$23,229.54$, seller's equity
$A = \$698.96/month − 419.53 = \$279.43/month$, net to seller
$n = 300$ months

$$P/A = (P/A, i\%, n)$$
$23,229.54/279.43 = 83.1319$
by interpolation, $i = 1.166\%/month = 13.99\% \approx \underline{\underline{14.0\%/yr}}$ nominal

If the buyer were permitted to assume the first mortgage, then the seller's investment would consist of a second mortgage for $23,229.54 earning 12 percent interest for 10 years.

With the buyer taking a WAM, the seller's investment consists of a WAM with a seller's equity of $23,229.54 earning 14.0 percent interest for 25 years.

The IROR can be calculated as follows: Comparing the second mortgage with the WAM, the second mortgage provides $333.28 per month income for 120 months, while the WAM provides $279.43 per month for 300 months.

Since the two are mutually exclusive (take one or the other but not both), then accepting the WAM means giving up ($333.28 − $279.43 =) $53.85 per month for 120 months in order to gain $279.43 per month for the last 180 months of the 300-month mortgage life. The equation may be solved for present value as

	trial $i =$	1.25%	1.50%
first 120 months, $P_1 = -\$ 53.85(P/A, i\%, 120) =$		−$3,338	−$2,989
second 180 months, $P_2 = +\$279.43(P/A, i\%, 180)$			
$\times (P/F, i\%, 120) =$		4,496	2,907
		+$1,158	−$ 82

The interest rate at which $P_1 + P_2 = 0$ is found by interpolation

$$i = 1.25\% + 0.25\% \times 1{,}158/(1{,}158 + 82) = 1.48\%/\text{month}$$
$$\text{IROR}, i_n = 1.48 \times 12 = 17.76 = 17.8\%/\text{yr}$$

Conclusion.   The incremental investment of $53.85 per month for the first 120 months earns a ROR of 17.8 percent per year. Therefore, assuming that the seller has a choice between a WAM and a second mortgage (and assuming that the risks and cash flows are acceptable), the seller may find either of two circumstances regarding the provision of financing for the sale of the house.

1. If financing (by either the WAM or the second mortgage) *must* be provided in order to sell the house, and if the house must be sold, then the concern is only the IROR earned by the increment of investment:
   a. If the seller's MARR < 17.8 percent, take the WAM, since the increment of investment in the WAM earns 17.8 percent.
   b. But, if the seller's MARR > 17.8 percent, take the second mortgage and invest the extra income during the first 120 months at the MARR.

2. If it is *optional* whether the financing is provided, then the concern is to invest the seller's equity of $23,230 wherever it can earn the higher return:
   a. If the MARR < 14.0 percent, take the WAM.
   b. But, if the MARR > 14.0 percent, invest in the MARR (the second mortgage alone at 12 percent loses out to the WAM at 14 percent).

INCOME TAX BENEFITS FROM HOME OWNERSHIP

In order to encourage home ownership the United States Congress has permitted certain expenses to be deducted on income tax returns. The most important of these to the average home owner are the deductions for interest payments and real estate taxes. The largest part of each payment on a new mortgage consists of interest (note Example 18.1b). Therefore this income tax deduction is significant to the average home owner. In addition, should the homeowner sell the home for more than he paid for it (a usual situation), additional tax concessions are available. Instead of paying income tax at regular rates, the profit usually can be tax-deferred. The tax deferral provision in a recent tax law permits within certain limits (up to $125,000) the amount of the resale price to be reinvested in another, higher priced home. After age 54 a one-time tax exemption of $125,000 generally was available.

AD VALOREM TAX BENEFITS AND COSTS ASSOCIATED WITH HOME OWNERSHIP

In addition, some states permit a homestead exemption to local real estate taxes for owner-occupied homes. If the tax millage rate is 30 mills, a qualified home in Florida with a $25,000 homestead exemption saves $0.03 \times \$25,000 = \$750$ per year in local property taxes. To qualify for homestead exemption in most states, the owner must live in the home as a permanent residence.

Ad valorem taxes may be levied by any authorized level of government, including cities, counties, school boards, and special tax authorities. In Florida there is a state-administered ceiling of 10 mills on the county *and* city, and 8 mills on the school board. The voters acting in referenda can vote themselves additional ad valorem taxes for special projects (bond issues) if they so desire. Thus for the city dweller in Florida the maximum tax without referenda is 28 mills plus special tax authorities. As a practical matter the average level is frequently just above 30 mills, since worthy bond projects frequently meet with voter approval despite the increased tax required. A mill is one-tenth of a cent and refers to the tax on a dollar's worth of assessed value. Thus, if the ad valorem tax rate is at 30 mills, a property assessed at $100,000 pays $0.03 \times \$100,000 = \$3,000$ taxes.

COMPARING MORTGAGES AT DIFFERENT INTEREST RATES

Homebuyers sometimes encounter situations where either (a) the existing first and second mortgages may be assumed, or (b) the property may be refinanced with one new mortgage replacing the two existing mortgages. The better of the two alternatives may be determined by comparing (a) the combined interest rate of the two mortgages to (b) the interest rate of the proposed replacement. For example, suppose a house has a present first mortgage at 6 percent plus a second mortgage at 10 percent. The mortgages could be assumed by the buyer, or he could refinance at 8 percent. Which is the most economical alternative? The following example illustrates a solution to this type of problem.

EXAMPLE 18.2

The data on the two mortgages are provided in the table below.

Mortgage	Balance remaining	Interest rate	Monthly payments	Number of payments remaining
First	$21,720	6%	$241.14	120
Second	50,944	10%	489.72	243

Should the new owner refinance at 8 percent?

Solution. The equivalent combined interest rate may be determined by finding the common i at which the present worth of the monthly payments equals the sum of the balance remaining on the two mortgages.

The sum of the balances is

$$\begin{aligned} \text{first mortgage} &= \$21,720 \\ \text{second mortgage} &= \underline{50,944} \\ \text{sum of first and second mortgages} &= \$72,664 \end{aligned}$$

Since the 10 percent mortgage is the larger of the two, the resulting interest rate will be greater than midway between 6 percent and 10 percent, or greater than 8 percent. Therefore, try 9 percent for a first trial.

The PW of the first and second mortgages at 9 percent is the PW of the remaining payments, and is compared to the sum of both mortgages as follows:

	trial i =	9/12%	10/12%
first mortgage $P_1 = \$241.14(P/A, i, 120) =$		$19,036	$18,246
second mortgage $P_2 = \$489.72(P/A, i, 243) =$		54,671	50,944
sum of first and second mortgages $=$		$-$ 72,664	$-$ 72,664
net		$+\$$ 1,043	$-\$$ 3,474

The equivalent combined interest rate is found by interpolation as follows:

$$i = 9/12\% + \frac{1,043}{1,043 + 3,474} \times 1/12\% = \underline{\underline{0.7692\%/\text{month}}}$$

The nominal rate is then

$$0.7692 \times 12 \text{ months/yr} = 9.23\% = \underline{9.2\%}$$

The combined interest rate of the two mortgages is $\underline{9.2 \text{ percent}}$. Since this is greater than the 8 percent rate available by refinancing, the new owner appears to pay less interest under alternative (b), *refinancing*. (Before a final decision is made, the new buyer should check to see if points and closing costs are required for the refinancing and how much they will raise the effective interest rate). By inspection it is evident that if the two existing mortgages are retained, the combined interest rate will increase slightly with each payment, until at EOM 120 the 6 percent note will be paid out. At this point the interest rate levels off at 10 percent since the entire debt remaining at the time is all under the 10 percent mortgage.

Reviewing the preceding steps in the example, the combined interest rate may be found by

Step 1. Find the monthly payments, A_1 and A_2.

Step 2. Find the sum P_1 and P_2 of the unpaid balance remaining on the mortgages.

Step 3. Select a logical trial i and find the present worths P_3 and P_4 of each of the series A_1 and A_2 using the same trial i. Find $P_3 = A_1 (P/A, i_{\text{trial}}, n_1)$ and $P_4 = A_2(P/A, i_{\text{trial}}, n_2)$.

Step 4. By trial and error and interpolation, find the i at which

$$P_3 + P_4 = P_1 + P_2.$$

This is the combined effective interest rate of the two mortgages.

EFFECT OF DIFFERENT MORTGAGE LIVES

When the lives of the two mortgages are different ($n_1 \neq n_2$), the combined rate changes with time. It gradually approaches the dominant i as the shorter mortgage is paid off. In the example case, after 120 months the first mortgage is paid off, and the effective i obviously becomes 10 percent, the i of the second mortgage. If the two mortgages have the same life ($n_1 = n_2$), then the effective rate remains the same throughout the life of both mortgages.

AMORTIZATION SCHEDULES

An amortization schedule is simply a tabulation of the periodic repayments A, required to pay off an amount P, over n periods with interest at i percent. The amortization schedule usually includes a listing of the due dates, amount of each payment due, amount of interest paid, amount of principal paid, and the total amount of principal remaining after each payment is made. For instance, the amortization schedule for a $100,000 mortgage payable monthly over 30 years (360 months) with $i = 9$ percent (9/12%/month) appears as:

(1) Date due	(2) Monthly payment due	(3) Interest payment	(4) Payment on principal	(5) Unpaid balance remaining
				$100,000.00
July 1	$804.62	$750.00	$54.62	99,945.38
Aug. 1	804.62	749.59	55.03	99,890.35
Sept. 1	804.62	749.18	55.44	99,834.91
Oct. 1	804.62	748.76	55.86	99,779.05

The significance of the figures in each column is discussed in turn.

Column (1): date due. This typically begins about 30 days after the loan is made, or "closed," but is usually arranged for mutual convenience. The due date need not be on the first of the month, but often is a date agreeable and convenient for both parties. Interest normally is paid in arrears, at the end of each time period, rather than in advance at the beginning of each period. Thus the interest paid on July 1 is owed for use of the money during the month of June. If the closing were, say, May 20, then the interest for the last 10 days of May would probably be prepaid at the time of closing. Then the interest paid on July 1 would be calculated the same as for every other month as (9%/12) × (unpaid principal). For the July 1 payment, that amounts to $(0.09/12) \times$ $100,000 = $750.

Column (2), *monthly payment due*, is calculated as $A = P(A/P, i, n)$, or in this case

$$A = 100,000(A/P, 9/12\%, 360) = \$804.62$$

Column (3), *interest payment,* is calculated as noted above in (1).

Column (4), *payment on principal*, is simply the portion of the monthly payment left after the interest payment is deducted, or $804.62 − $750.00 = $54.62 for the July 1 payment.

Column (5): *The unpaid balance remaining* begins as the full amount of the mortgage and is reduced each month by the payments of the principal from column (4). It eventually reduces to zero at the end of 360 months.

A full amortization schedule usually is too tedious to do by hand; it is typically run on a computer. The amortization schedule contains information needed when comparing alternatives, calculating tax deductions for interest, and buying or selling properties under mortgage.

OUTLINE OF COMPARISONS: RENT AN APARTMENT OR BUY A HOME

One of the important choices a person periodically has to make over the years is the selection of a place to live. The selection frequently involves a choice between renting and buying. In addition, friends or clients sometimes will ask advice regarding the relative merits of each alternative. The following may be used as a checklist when calculating numerically the present worth or annual cost comparisons.

1. Renting a residence—advantages
    a. No down payment required (but first and last months' rent plus damage deposit may be required).
    b. Maintenance is usually taken care of by the landlord. Often there is no yard to mow. In an apartment complex any major facilities such as pool, sauna, tennis courts, game room, laundry facilities are all maintained by the management.
    c. No direct real estate taxes and insurance charges (although the landlord naturally must charge enough extra rent to cover these items).
    d. Housing cost is a known amount predetermined for the life of the lease (1 to 3 years). It is x dollars per month, with no worries about unexpected bills for leaky roofs or faucets.
    e. In place of an equity in a house, a renter may wisely invest the amount of the down payment saved and have a large sum at the end of 30 years. For instance, a $3,000 lump sum invested at 8 percent for 30 years = $30,188. In addition, if there is, say, a $25 per month saving in the total rent costs over ownership costs, and if this could be invested at 8 percent, it would result in a balance of about $37,259 after 30 years. However, if housing values continue to appreciate at their historic rate, the probabilities are remote of renters having lower monthly costs than owners over more than the first few years.
    f. For multi-unit apartments, the economies of scale realized by the landlord in building and maintaining a large number of units are, one would hope, passed on to the tenant.
    g. Mobility is high. If the need or desire to move occurs, the penalties involved in breaking a lease usually are not prohibitive.
2. Renting a residence—disadvantages
    a. No buildup of equity.
    b. No right to alter the premises without landlord approval.
    c. Inflation brings increases in the rent charged. (Although it also should bring increases in salary sufficient to pay it. The down payment saved and invested in some inflation hedge should also increase in value.)
3. Owning your own home—advantages
    a. Inflation and other market forces normally tend to raise the resale value of the property (with some exceptions).
    b. Mortgage payments remain unchanged (for fixed-rate mortgages only), while rents habitually escalate.

   c. Ownership equity increases to 100 percent over the life of the mortgage.

   d. Income tax deductions normally are allowed for expenditures for interest and property taxes. During the first few years, most of the mortgage payment is attributable to interest, so a large proportion of the housing cost is tax deductible.

   e. Pays no profit to the landlord (although the landlord's profit could well consist of the difference between retail prices of the building and maintenance paid by the homeowner, and the wholesale price of building and maintenance paid by the landlord).

4. Owning your own home—disadvantages

   a. Requires a down payment (none required for renter, but he may have to deposit first and last months' rent plus damage deposit). Down payment involves loss of interest that would result if this money had been invested.

   b. Monthly housing costs are not definitely known since all maintenance and repairs (including unexpected repairs) are the responsibility of the homeowner.

   c. If deflation should occur, the homeowner could be burdened with high mortgage payments.

SUMMARY

Millions of Americans who own their own homes are familiar with mortgage financing with monthly repayments. Through such federal agencies as the FHA and VA, loans are guaranteed, and down payments are reduced to levels that make home ownership an attainable goal for almost everyone.

A mortgage has two parts: (a) a note that describes the terms of the debt, and (b) the mortgage itself, which gives the lender an interest in the property until the note is paid off.

Mortgages vary in length of time for payout; most first mortgages run from 25 to 35 years. Mortgages may carry either fixed interest rates or varying rates. Money-lending institutions usually include front-end charges to increase their yield. One such charge is termed *points*, expressed in terms of a percentage of the amount of the loan. For example, a lender who requires 3 points on a $100,000 loan is requiring $3,000 in addition to the down payment. The points may be added to the principal balance due or deducted from the amount loaned. Some mortgage terms (such as paragraph 17) may restrict the homeowner from selling the mortgage along with the house to another party (through an assumption agreement). If the property changes hands, refinancing may be required if stipulated in the agreement.

One of the main economic advantages to home ownership is the reduction in income taxes through deductions for local property taxes and interest on the mortgage. In deciding between selling or renting a home, all of the advantages and disadvantages should be carefully evaluated.

PROBLEMS FOR CHAPTER 18, HOME OWNERSHIP AND MORTGAGE FINANCING: OWNING VERSUS RENTING

Problem Group A

Find the monthly payments, the balance due on mortgages, effective i, and combined i. No taxes.

A.1. A developer advertises a house for sale at $100,000 with 5 percent down. A 25-year conventional mortgage is available at 8 percent if the buyer is willing to pay 4 points. (Add 4 percent to mortgage balance.)

(a) What are the monthly payments? (Use the equation, with $n = 300$.)

(*Ans.* $762.55/month)

(b) What is the approximate actual interest rate, using end-of-year payments?

(*Ans.* 8.33%)

(c) What is the actual interest rate, using $n = 300$? Compare with the results obtained in (b).

(*Ans.* $0.705 \times 12 = 8.46\%/\text{yr}$)

A.2. Assume you purchased a home five years ago at a price of $75,000. You paid 5 percent down and obtained a 30-year, 7 percent mortgage. You paid no points. You now have decided to sell it, and the market value is $100,000. You offer to sell to a prospective buyer for 5 percent down, taking back a second mortgage for the difference between the first mortgage balance and the selling price less the down payment. For the second mortgage, $n = 10$ years, and $i = 10$ percent. (To simplify the calculations, assume end-of-year payments on both mortgages.)

(a) What are the payments on the first mortgage?
(b) What is the balance due on the first mortgage?
(c) What is the amount of the second mortgage?
(d) What are the payments on the second mortgage?

A.3. Assume that five years ago you purchased your present residence under the following terms:

$$
\begin{aligned}
\text{mortgage terms} &= \quad 8\%, \text{ plus 5 points on the mortgage, 30 yr.} \\
\text{price} &= \quad \$61,500 \\
\text{less down payment} &= \quad -\,6,000 \\
\text{amount loaned by mortgage} &= \quad \$55,500 \\
\text{plus 5 points} &= \times \quad \underline{1.05} \\
\text{amount of mortgage balance owed} &= \quad \$58,275
\end{aligned}
$$

Now (five years after purchase) you are ready to move. You put the house up for sale for $84,000, asking $6,000 down (plus your closing costs), and a second mortgage at 10 percent for 10 years. Find the following:

(a) The amount of annual payments on the first mortgage.

(*Ans.* $5,176.42)

(b) The balance due on the first mortgage, five years after your purchase of the house.

(*Ans.* $55,257)

(c) The total amount required to be financed by the second mortgage.

(*Ans.* $28,743)

(d) The *annual* payments on the second mortgage.

(*Ans.* $4,678/yr)

(e) If you would like to sell the second mortgage for cash, and the expected rate of return is 15 percent for investors in this type of paper at present, how much cash can you get for it?

(*Ans.* $23,477)

A.4. A house is for sale for $80,000 with no down payment and a mortgage at 8.5 percent, with $n = 30$ years. In order to obtain this attractive financing for the buyer, the seller is required to pay a 6-point discount on the mortgage, which means the seller only gets $(0.94 \times \$80,000 =)$ $75,200 for the house. Consequently the seller offers to sell you the house for $75,200 if you buy the house for cash (or obtain your own financing so that the seller gets cash). Find the actual interest rate, i, that the buyer pays if the house is purchased for $80,000 with the mortgage, assuming that the actual cash value of the house is $75,200.

A.5. A house is for sale with the following mortgages.

	%	Unpaid balance	Time remaining
First mortgage	7%	50,160	12 yr
Second mortgage	11%	34,600	20 yr

(a) What is the equivalent combined interest rate?

(*Ans.* 9.03%)

(b) If the two mortgages can be consolidated into one mortgage for $84,760 at 9 percent for 20 years,

(i) Will the annual payments rise or fall?

(*Ans.* fall)

(ii) Will the interest rate rise or fall from the level found in (a)?

(*Ans.* fall slightly)

Note: Use end-of-year annual payment for a close approximation.

Problem Group B

Find the equivalent annual or monthly costs. Compare alternatives. No taxes.

B.1. You need to buy a house and have narrowed the available selection to two that are both equally desirable. The down payment is the same for both. Which has the lower yearly mortgage payments?

(a) Costs $80,000 with a $20,000 down payment. Balance of $60,000 payable over 20 years at 9 percent.

(*Ans.* $6,573/yr)

(b) Costs $90,000 with $20,000 down and mortgage at 8 percent, 20 years.

(*Ans.* $7,130/yr)

B.2. A young married couple asks your advice on selecting a home. They have narrowed the choices down to two: a mobile home or a custom-built home. They already own a lot zoned for either. (For estimating purposes, use end-of-year annual payments instead of monthly.) ($i = 12\%$.)

1. Buy a large mobile home for $42,000. The down payment is $2,000, and the mortgage is 10 years at 10 percent. The resale value in 10 years is expected to be about $16,000. Annual taxes, $120 for a license plate.

2. Buy a custom-built home, same square footage, for $84,000. The down payment is $8,000, and the mortgage is 30 years at 9 percent. The resale value of the house over the near term is expected to increase by 3 percent per year. Annual taxes are 30 mills on a $50,000 assessed valuation.

   (a) Over the 10-year period, which has the lowest equivalent annual cost?
   (b) Which has the lower equivalent annual cost if the custom-built house appreciates at a net rate of 6 percent compounded?

B.3. You are being assigned to a distant city for a five-year period. You find you can either rent a suitable house to live in on a five-year lease for $1,050 per month or you can buy the same house for $105,000. The average ownership costs over the five-year period are estimated as follows:

Real estate taxes are 3 percent of purchase price per year.

Maintenance, $150 per month.

Insurance, $300 per year.

Rebate on income tax, $1,200 per year.

(a) If $i = 9$ percent and the house is purchased for cash (no mortgage), what would the minimum sales price have to be in five years to reach the break-even point with renting?

(*Ans.* $110,400)

(b) If the house could be bought with no down payment and an 8 percent, 30-year mortgage loan ($i = 8\%$ for the mortgage, 9% for everything else), what would the break-even selling price be in five years?

(*Ans.* $104,200)

Note: Assume all series payments are made at EOY, and there are no taxes except as listed in the problem.

B.4. A solar hot-water heating system costs about $2,000 new with about $100 per year maintenance costs. An electric hot-water heater of the same capacity costs about $400 new with about $60 worth of electricity per month (use $720 per year). A competitive gas hot-water heater costs $600 new with about $40 worth of gas per month (use $480 per year). Any of the three can be selected for a new residence now being designed. The mortgage will be at 8 percent. The anticipated life of each system is 15 years. Which is the most economical?
(a) For the homeowner.
(b) For the developer.

B.5. A prospective house buyer has a choice between gas and electricity for heat, A/C, hot water, and kitchen. He gets estimates that the gas installation will cost about $6,000 new plus $0.06/1,000 Btu, while electricity costs about $4,000 plus $0.22/1,000 Btu. The estimated life of each is 15 years, and each would be financed under the mortgage at 9 percent. How many Btu's per year must be used at the break-even point between gas and electric?

(*Ans.* 13,720K Btu/yr)

Problem Group C

Arithmetic and geometric gradients, with tax considerations.

C.1. Your client is about to start a three-year assignment in a distant city and has located a suitable house whose owner is willing either to sell the house for $90,000 or rent it for $10,800 per year. Find which is less expensive for your client. Assume end-of-year payments and your client's $i = 12$ percent.

Alternate 1.   Purchase a home.

   Purchase expenses

$$\text{purchase price} = \$90,000$$
$$\text{mortgage loan/value ratio} = 0.80$$
$$\text{mortgage, } 0.80 \times \$90,000 = \$72,000 \text{ amount received}$$
$$\text{points required to obtain mortgage} = 3 \text{ points}$$
add on the points to find the beginning balance
$$\text{of mortgage for repayment} = 1.03 \times \$72,000 = \$74,160$$
$$\text{mortgage } i = 9.5\%$$
$$\text{mortgage term} = 25 \text{ yr}$$
$$\text{down payment} = \$18,000$$
purchaser's closing costs in addition
$$\text{to down payment} = \$700$$

Expenses of owning and maintaining the home:
1. Ad valorem property taxes, $1,200 per year, increasing 4 percent every year.
2. Maintenance and insurance, $1,800 per year, increasing 4 percent every year.
Income and expenses of reselling the house at EOY 3:
Resale value of the house increases at 4 percent per year compounded. Upon resale, the net proceeds to the owner are reduced by the following
1. A 6 percent sales commission to the real estate broker.
2. An additional $500 for the seller's share of closing costs.
3. A tax, at ordinary tax rates, on the net capital gain from the sale (the net capital gain is the purchase price plus mortgage points and purchaser's closing costs, subtracted from resale price less seller's closing costs and commission).
Taxes:
The client is in the 28 percent income tax bracket. The tax-deductible items are
1. Ad valorem property taxes.
2. Interest-payment portion of the mortgage payment.
Alternate 2.   Rent at $10,800 per year ($900 per month), increasing 4 percent per year at the end of every year.
(a) Find the total cost of purchase (the sum of the beginning balance of mortgage for repayment, plus down payment, plus the client-purchaser's closing costs).

   (*Ans.* $92,860)

(b) Construct an amortization schedule for the mortgage over the four-year period, using the "beginning balance of mortgage for repayment" as given above, and end-of-year payments.

Payment no.	Payment amount	Interest	Principal	Balance
				$74,160
1	$7,858	$7,045	$813	73,347
etc.				

(c) For the expenses of owning and maintaining the home, construct a table showing the amount of after-tax cash flow occurring for each of the three years. Then find the present worth for each year. Then sum to find the after-tax PW of all the owning and maintenance expenses.

(*Ans.* PW of owning and maintenance expenses for EOY $1 + 2 + 3$, after-tax $P = \$-20,823$)

(d) Find the sale price at EOY 3. Subtract the resale sales commission and the client-seller's closing cost to find the net to client-seller *before* capital gain tax, and payment of the mortgage balance at EOY 3.

(*Ans.* net resale $=$ resale price $-$ commission $-$ seller's closing costs $= \$94,663$)

(e) Calculate the capital gain tax. To find the amount of the capital gain, subtract the answer found in (a) from the answer found in (d). The capital gain tax is the ordinary tax rate times the capital gain.

(*Ans.* capital gains tax $= \$-505$)

(f) Find the after-tax net to client-seller at EOY 3 by subtracting the balance due on the mortgage at EOY 3, and subtracting the capital gain tax found in (e) from the net to client-seller before tax found in (d). Find the PW of the after-tax net by multiplying by the appropriate P/F factor.

(*Ans.* PW of after-tax net proceeds to seller from resale, $P = \$14,411$)

(g) Find the PW of all after-tax costs of owning a house for three years by summing the PW of all the cash flows. This includes the actual cash paid out at purchase (down payment plus buyer's closing costs), plus the PW of after-tax owning and maintenance costs found in (c), plus the PW of net income from selling found in (f).

(*Ans.* PW of all after-tax costs and incomes from owning the house for 3 yr, $P = -\$23,383$)

(h) Find the PW of renting for three years. Since the rent increases every year, use the geometric gradient equation.

(*Ans.* PW of renting for 3 yr, $P = -\$26,911$)

(i) Compare the PW of purchase to the PW for renting.

(*Ans.* Purchase is $3,528 less expensive; however, renting involves less risk, fewer ownership and management problems, and more mobility.)

(j) Find the beginning rental rate at EOY 1 that makes the economic comparison equal, assuming the rental increases continue at the same percentage as previously.

[*Ans.* break-even rent $= \$9,384$/yr (versus \$10,800 available)]

C.2. Same as Problem C.1, with the following changes:

$$\begin{aligned}
\text{client's interest rate} &= 10\% \\
\text{purchase price} &= \$100{,}000 \\
\text{mortgage interest rate} &= 8.5\% \\
\text{mortgage term} &= 25 \text{ yr} \\
L/V \text{ ratio} &= 0.85 \\
\text{mortgage points, add on} &= 5 \text{ points} \\
\text{beginning mortgage balance} &= \$89{,}250 \\
\text{down payment} &= \$15{,}000 \\
\text{buyer's closing costs in addition to down payment} &= \$600 \\
\text{maintenance and insurance} &= \$1{,}500/\text{yr} \\
\text{annual increase in maintenance and insurance} &= 6\%/\text{yr} \\
\text{property taxes} &= \$1{,}050/\text{yr} \\
\text{annual increase in property tax} &= 6\%/\text{yr} \\
\text{resale value increase in purchase price} &= 6\%/\text{yr} \\
\text{income tax bracket} &= 34\% \\
\text{resale sales commission} &= 6\% \\
\text{seller's closing costs upon resale} &= \$700 \\
\text{rental rate of alternative rental} &= \$11{,}400/\text{yr} \\
\text{increase in rental rate} &= 5\%/\text{yr}
\end{aligned}$$

Answer questions (a) through (i) for Problem C.1.

C.3. As a consultant you are asked to determine which is the lower cost alternative for acquiring housing over a four-year period, assuming the following conditions. The choice is between purchase or rental. Assume end-of-year payments and $i = 11$ percent.

Alternative 1.   Purchase a home.

   Purchase expenses:

$$\begin{aligned}
\text{purchase price} &= \$120{,}000 \\
\text{mortgage loan/value ratio} &= 0.80 \\
\text{mortgage, } 0.80 \times \$120{,}000 &= \$96{,}000 \text{ amount received} \\
\text{points required to obtain mortgage} &= 4 \text{ points} \\
\text{add on the points to find the beginning balance} & \\
\text{of mortgage for repayment} = 1.04 \times \$96{,}000 &= \$99{,}840 \\
\text{mortgage } i &= 9\% \\
\text{mortgage term} &= 30 \text{ yr} \\
\text{down payment} &= \$24{,}000 \\
\text{purchaser's closing costs in addition} & \\
\text{to down payment} &= \$1{,}000
\end{aligned}$$

Expenses of owning and maintaining the home:

1. Ad valorem property taxes, $3,000 per year, increasing 5 percent every year.
2. Maintenance and insurance, $2,800 per year, increasing 5 percent every year.

Income and expenses of reselling the house at EOY 4:

Resale value of the house increases at 5 percent per year compounded. Upon resale, the net proceeds to the owner are reduced by the following:

1. A 6 percent sales commission to the real estate broker.
2. An additional $800 for the seller's share of closing costs.
3. A tax, at ordinary tax rates, on the net capital gain from the sale (the net capital gain is the purchase price plus mortgage points and closing costs, subtracted from resale price less closing costs and commission).

Taxes:

The owner is in the 30 percent income tax bracket. The tax deductible items are

1. Ad valorem property taxes.
2. Interest-payment portion of the mortgage payment.

Alternative 2. Rent at $14,400 per year ($1,200 per month), increasing 5 percent per year at the end of every year.

(a) Find the total cost of purchase (the sum of the beginning balance of mortgage for repayment, plus down payment, plus the purchaser's closing costs).

(b) Construct an amortization schedule for the mortgage over the four-year period, using the "beginning balance of mortgage for repayment" as given above, and end-of-year payments.

(c) For the expenses of owning and maintaining the home, construct a table showing the amount of after-tax cash flow occurring for each of the four years. Then find the present worth for each year. Then sum to find the after-tax PW of all the owning and maintenance expenses.

(d) Find the sale price at EOY 4. Subtract the resale sales commission and the seller's closing cost, to find the net to seller *before* capital gain tax and payment of the mortgage balance at EOY 4.

(e) Calculate the capital gain tax. To find the amount of the capital gain, subtract the answer found in (a) from the answer found in (d). The capital gain tax is the ordinary tax rate times the capital gain.

(f) Find the after-tax net to seller at EOY 4 by subtracting the balance due on the mortgage at EOY 4 and subtracting the capital gain tax found in (e) from the net to seller before-tax found in (d). Find the PW of the after-tax net by multiplying by the appropriate P/F factor.

(g) Find the PW of all after-tax costs of owning a house for four years by summing the PW of all the cash flows. This includes the actual cash paid out at purchase (down payment plus buyer's closing costs), plus the PW of after-tax owning and maintenance costs found in (c), plus the PW of net income from selling found in (f).

(h) Find the PW of renting for four years. Since the rent increases every year, use the geometric gradient equation.

(i) Compare the PW of purchase to the PW for renting.

(j) Find the beginning rental rate at EOY 1 that makes the economic comparison equal, assuming the rental increases continue at the same percentage as previously.

C.4. Same as Problem C.3, with the following changes:

$$\text{interest rate} = 11\%$$
$$\text{purchase price} = \$75,000$$
$$\text{mortgage interest rate} = 8\%$$

$$\text{mortgage term} = 25 \text{ yr}$$
$$L/V \text{ ratio} = 0.90$$
$$\text{mortgage points, add on} = 4 \text{ points}$$
$$\text{beginning mortgage balance} = \$70,200$$
$$\text{down payment} = \$7,500$$
$$\text{buyer's closing costs in addition to down payment} = \$500$$
$$\text{maintenance and insurance} = \$1,200/\text{yr}$$
$$\text{annual increase in maintenance and insurance} = 5\%/\text{yr}$$
$$\text{property taxes} = \$720/\text{yr}$$
$$\text{annual increase in property tax} = 5\%/\text{yr}$$
$$\text{resale value increase in purchase price} = 5\%/\text{yr}$$
$$\text{income tax bracket} = 28\%$$
$$\text{resale sales commission} = 6\%$$
$$\text{seller's closing costs upon resale} = \$600$$
$$\text{rental rate of alternative rental} = \$8,400/\text{yr}$$
$$\text{increase in rental rate} = 5\%/\text{yr}$$

Answer questions (a) through (i) for Problem C.3.

Problem Group D

Mortgages. No taxes.

D.1. WAM mortgage. Assume that five years ago you purchased a house for $100,000, with a 20 percent down payment and a 9 percent, 30-year mortgage, payable monthly. Now you are ready to sell the house for $140,000. You have a prospective buyer with a $20,000 down payment. You offer either of the following:

1. A WAM at 10.5 percent, 25 years, with monthly payments.

2. A second mortgage at 12 percent, 10 years, monthly payments.

    (a) Compare the buyer's monthly payments under each alternative.

    (b) Find the ROR to the seller under each alternative.

    (c) Draw the cash flow diagram illustrating the increment of investment and find the IROR to the seller for the increment of investment.

    (d) What values of MARR lead to what choices under circumstances of **(i)** must sell, must finance; **(ii)** optional to sell and finance?

CHAPTER 19

Investment Property

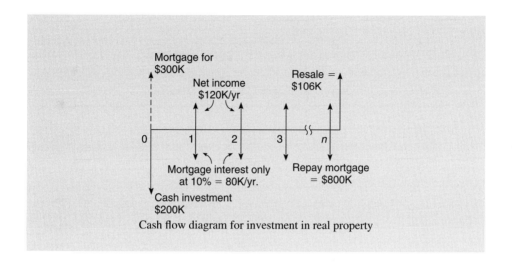

Cash flow diagram for investment in real property

LAND INVENTORY

For purposes of discussion, real estate investment property may be subdivided into (a) undeveloped and agricultural land, and (b) developed property. The United States contains about 2.26 billion acres of land area. As yet less than 5 percent of this area has been developed for urban and transportation uses. About 80 percent of the total land area is used for crop land, pasture, range, and forest land. Another 6 percent is used as recreational, wildlife, and other special uses. About 12 percent is essentially unused land, including desert, mountains, tundra, and swamps. About three-fifths of the land area of the United States is in private ownership, and two-fifths is owned by federal, state, and local governments.

Much of the privately held undeveloped and agricultural land will eventually be developed in response to the increasing public demand for more residences, shopping centers, increased industrial production, and so on. Currently about 1.2 million additional acres of land each year are being developed. Meanwhile, many of the owners of the remaining undeveloped land are hoping that prospects for future development will cause the value of their land to increase rapidly enough to justify their cost of investment in the land.

FEATURES OF INVESTMENT REAL ESTATE

Undeveloped Land as an Investment

Investors in undeveloped land (except agricultural land) typically incur continuing costs and little, if any, income while waiting for a profitable opportunity to sell or develop the property. These costs include the following:

1. Loss of use of whatever net equity is invested in the land. The amount of this equity is calculated as the net present worth of the market value less mortgage balance owed, selling costs, and taxes on capital gain, if any.

2. Annual mortgage payments, including interest on the current mortgage balance.

3. Annual real estate taxes.

4. Annual maintenance costs, if any. The property may require construction and maintenance of access trails, fences, culverts, and so on.

   In most cases the income consists of the anticipated profit at the time of resale, but may also include small annual amounts for timber, hunting rights, agricultural revenues, and so on. A typical example of a cost analysis for an investment in undeveloped land follows. For tax purposes, the term *mill* is often used. One mill of taxes is equal to $1.00 per $1,000 of value, or 1/10 of a cent per dollar.

EXAMPLE 19.1

An investor group holds 3,000 acres of undeveloped land, which they purchased at $1,000 per acre five years ago. The land is assessed at $500 per acre, and the tax rate is 20 mills, so they have paid $10 per acre per year in taxes. At the end of the fourth year they put in some access trails at a cost of $50,000. How much must the property sell for at the end of the fifth year in order to earn a before-tax rate of return of 15 percent on their investment?

Solution.   Use future values, referred to the time of resale of the property.

Future value of cost

$$
\begin{aligned}
\text{purchase price, } F_1 &= 3{,}000 \text{ acres} \\
&\times \$1{,}000/\text{acre}(F/P, 15\%, 5) = -\$6{,}034{,}100 \\
\text{taxes, } F_2 = 3{,}000 \text{ acres} \times \$10/\text{acre/yr}(F/A, 15\%, 5) &= -\$\ \ 202{,}300 \\
\text{access road, } F_3 = \$50{,}000(F/P, 15\%, 1) &= -\$\ \ \ \ 57{,}500 \\
\hline
\text{selling price required after 5 yr for 15\% return} &= \ \ \$6{,}293{,}900
\end{aligned}
$$

If this group had been able to acquire the property with a small down payment, their selling price could be much lower (or their rate of return much higher), as illustrated by the next example.

EXAMPLE 19.2

Assume, for instance, the group of investors had been able to acquire the property in Example 19.1 for 20 percent down, paying interest on the balance at 10 percent, with no payments on the principal due until the end of five years.

$$\text{purchase price of land} = \$1,000/\text{acre} \times 3,000 \text{ acres} = \$3,000,000$$
$$\text{down payment} = 0.20 \times \$3,000,000 = \$\ \ 600,000$$
$$\text{mortgage} = \$3,000,000 - \$600,000 = \$2,400,000$$
$$\text{annual interest payments} = 0.10 \times \$2,400,000 = \$\ \ \ 240,000/\text{yr}$$
$$\text{annual taxes} = 3,000 \text{ acres} \times \$10/\text{acre} = \$\ \ \ \ 30,000/\text{yr}$$

Find the selling price to realize 15 percent compounded on the $600,000 investment.

Solution

$$\text{down payment } F_1 = \$600,000(F/P, 15\%, 5) = -\$1,206,800$$
$$\text{taxes and interest } F_2 = \$270,000/\text{yr}(F/A, 15\%, 5) = -\$1,820,400$$
$$\text{access road } F_3 = \$50,000(F/P, 15\%, 1) = -\$\ \ \ 57,500$$
$$\text{future value of investment at 15\%, } excluding \text{ mortgage}$$
$$F_{1+2+3} = \quad \$3,084,700$$
$$+\text{balance owed on mortgage (interest has been paid)} = \quad \$2,400,000$$
$$\text{selling price required after 5 yr for 15\% return} = \quad \$5,484,700$$

Note this selling price is $809,200 *less* than is required without the financing. This is because the $2,400,000 mortgage only has to earn 10 percent instead of 15 percent.

Developed Property

Developed investment property can be further subdivided into the following kinds:

1. Residential
2. Commercial
3. Industrial
4. Governmental-institutional-miscellaneous

The largest dollar volume is in residential investment property, mostly apartment houses.

Residential rental property (apartments and single-family homes) provides housing for about 100,000,000 Americans, as well as providing investment income for many millions of rental property owners.

As an investment, rental real estate has several features that set it apart from other types of investment.

CHANGES IN VALUE

Appreciation

Real estate typically (but certainly not always) increases in value with time. In general, resale values for real estate vary with location and time, as well as other factors. For developed real estate, an important influence is the quality of maintenance on the property in question, as well as maintenance of nearby property in the surrounding neighborhood. Nationally, resale prices for residential property have risen, on average, over 4 percent per year for the past 20 years.

Depreciation

All structures have finite lives. At one extreme, it is not uncommon for buildings as young as 25 years of age or younger to be torn down to make room for newer more productive structures. At the other extreme, there are many buildings well over 100 years of age that have been properly maintained, or restored or renovated, and have continued in service almost indefinitely. Therefore, it is difficult to predict the length of useful life for any particular building. However, it is logically inevitable that some day every structure will be demolished either by design, accident, or forces of nature. Recognizing this, the Internal Revenue Service (IRS) permits a depreciation allowance for real estate improvements. This allowable depreciation life changes from time to time at the direction of Congress. For instance, if Congress wants to encourage investment in additional housing or other types of real estate development, one of the measures it can take is to shorten the depreciation period. Conversely, lengthening the permitted depreciation life makes real estate investments less attractive. At one time the specified depreciation life for buildings was as long as 40 years, and at another time, as short as 15 years (5 years for rehabilitated historic structures). Depreciation is a recognized cost of doing business, so an annual depreciation allowance is permitted for the theoretical cost of depreciation, without regard to the actual depreciation or appreciation of the particular property in question. However, at the time the property is sold, a compensating tax is levied on any difference between depreciated book value and actual sales price. The depreciation allowance then amounts only to a temporary tax shelter, rather than an avoidance of taxes. See Chapters 15 and 16 for more detailed treatment of taxes and depreciation.

TAXES ON REAL ESTATE INVESTMENTS

The various levels of government obtain tax revenues from real estate investments in several ways.

1. Real estate taxes are assessed annually by local government (county, school board, municipality, or special taxing authority) with a tax on the *value* of all nonexempt property within their jurisdiction (thus they are referred to as *ad valorem* taxes). In order to assure equity in taxing, the tax assessor tries to determine the value of the property as fairly as possible. He may use a formula that includes multipliers for

square footage, number of bathrooms, general condition, and so on. Fair market value often averages out to about 80 to 90 percent of the actual sales price. The tax assessors are allowed to adjust the assessed values downward from the sales price to take into account sales commissions, financing costs, and so on. On the one hand the local tax assessors usually are subject to state audit to be sure the assessments are high enough, and on the other hand, they are subject to sharp criticism from the taxpayers if the assessments are too high.

2. Income taxes are paid on the taxable annual income from real estate investments, just as on any other investment. Depreciation allowances frequently shelter some share of this income until the time the property is sold. Trading the property instead of selling it can sometimes extend the life of the tax shelter, but some depreciation benefits may be lost.

3. Capital gains are taxed at the taxpayer's ordinary tax level, as discussed in Chapter 15.

See Chapter 16 for further details on taxes for real estate properties.

OWNER'S VIEWPOINT: INVESTMENT RETURN

The three variables that most influence a rental investment are (1) leverage, (2) mortgage interest rate and mortgage term (payout period), and (3) depreciation schedule. Each is discussed in turn.

Leverage

The amount of borrowed money available to help finance the investment as compared to the market price of the investment is referred to as the amount of leverage, or loan-to-value ratio (L/V). Leverage is common to many other types of investments as well. (A familiar example is margin trading on the stock or commodity exchange.) Rental properties can typically be mortgaged to 75 or 80 percent of their purchase price, and it is not uncommon to mortgage 90 percent or more. Thus, it is normal for an investor to purchase a $1,000,000 apartment complex by paying $200,000 or less in cash and arranging for a mortgage to finance the remainder. Most investors try for as much leverage as they can get. Examples 19.2 and 19.3 both illustrate how an increase in leverage can increase the rate of return.

EXAMPLE 19.3

An apartment house can be constructed that is estimated to earn a profit of $120,000 per year before income taxes. The total project cost is estimated as $1,000,000 which may be paid either (a) all in cash, or (b) by a cash investment of $200,000 and a mortgage for the $800,000 balance at 10 percent interest on the unpaid balance, annual payments of interest only, with lump sum principal due at some future date. The resale value remains at $1,000,000.

Find which alternative is the more profitable in terms of before-tax rate of return.

Solution

a. Without leverage, the rate of return on the cash transaction is

$$\frac{\$120,000}{\$1,000,000} = 12\%$$

b. With leverage, there is a 10 percent mortgage. The annual interest payment is

$$0.10 \times \$800,000 = \$80,000/\text{yr}$$

The net income before tax then is

$$\$120,000 - \$80,000 = \$40,000/\text{yr}$$

The before-tax rate of return is

$$\$40,000/\$200,000 = \underline{20\% \text{ return}} \text{ on the } \$200,000 \text{ investment.}$$

Conclusion.   The leveraged investment returning 20 percent ROR is preferable to the nonleveraged all-cash investment yielding 12 percent. Thus, the leveraging results in a significant increase in the rate of return. It should be recognized that high leverage usually involves more risk, since the income may decline during economic recessions, but the mortgage payments keep right on coming due. Of course the first simple key to profit through leverage is to borrow at a lower interest rate than is earned on the investment, or to paraphrase an old saying, "Borrow low, invest high." The second key is to be sure the cash flow from the investment is adequate to meet the debt service requirements. Many a millionaire has gone broke because impatient creditors demanded cash now while the millionaire's assets were tied up in investments that would take months to liquidate at anywhere near fair market value.

Mortgage Interest Rate and Payout Period

Small changes in interest rates can have a large influence on profitability. If the interest rate in Example 19.3 were raised to 12 percent, the effect would be

$$
\begin{aligned}
\text{mortgage} &= \$800,000 \\
\text{net income before interest} &= \$120,000 \\
\text{interest payment on } \$800,000 \text{ at 12 percent} &= \underline{\$\ 96,000} \\
\text{net income after interest} &= \$\ 24,000
\end{aligned}
$$

percent return on $200,000 = 24,000/200,000 = 12\%$, instead of the previous 20%

Depreciation Schedule

Depreciation is a tax-deductible cost of doing business. The IRS has recently permitted only straight-line depreciation on income-producing buildings purchased or placed in service after 1986. Tax requirements change by acts of Congress based on economic necessity; therefore, the application of these principles to actual investment situations should be made according to the current tax code. (Buildings acquired in years past whose depreciation began under any one of the previously allowed depreciation systems are allowed to continue under those systems until the ownership changes. Other depreciation systems discussed in Chapter 16 may be encountered for a number of years to come.) The depreciation allowance may be deducted from income during years of ownership. However, upon sale of the property, any difference between depreciated book value and sale price is taxed. At worst, the depreciation deduction is equivalent to an interest-free loan from the government until the property is sold.

DETERMINATION OF PROJECT FEASIBILITY, AFTER TAXES

Taxes are a normal cost of doing business for most investors, and must be taken into account when calculating net rates of return. Example 19.4 illustrates a normal study of a proposed rental project over a three-year period. This example finds the year-to-year after-tax rate of return (YYATROR) as well as the overall after-tax rate of return (OATROR). The same techniques may be applied to solve for most other variables related to the problem should they happen to be unknowns for any given problem situation. Projects may be examined over longer periods of time by repetitions of the same procedures for as many years as desired. These calculations are readily adaptable to a computer program, and if several projects are to be examined, the computer becomes a real time-saver. The computer can range all of the relevant variables and put out a large amount of significant data concerning the feasibility of each project in a relatively short time.

EXAMPLE 19.4   **(Find the YYATROR [year-to-year after-tax rate of return] and the OATROR [overall after-tax rate of return] for a proposed apartment project, and compare with the MARR)**

A developer client asks your opinion on a proposal to construct an apartment complex with estimated data as shown. The developer expects to sell the property either at EOY 2 or at EOY 3, and asks you to calculate the YYATROR and the OATROR and advise on the results. For convenience, use EOY payments. (*Note:* K = $1,000.)

Points are a cost of borrowing, in percent, which are added to the loan value by the lending institution. In this case, the value of the loan is $1,440K, but with the points from the lender added, the loan amount that must be repaid to the lender becomes $(1,440K \times 1.03) = \$1,490.4K$.

(Points normally are depreciable and are to be added to the depreciable basis. Points are not added to the initial market value when calculating appreciation.)

$$
\begin{aligned}
\text{land cost (20 acres at \$15K/acre)} &= \$\ \ 300K \\
\text{construction cost (depreciable improvements)} &= \$1,500K \\
\text{total value of land} + \text{construction} &= \$1,800K \\
\text{developer's mortgage at 80\% } L/V, \text{ receives} & \\
(0.8 \times \$1,800K = \$1,440K) \text{ at 10\%, 25 yr} &= \$1,440K \\
\text{points} &= 3.5 \\
\text{amount developer must repay, (add points to} & \\
\text{mortgage), } 1.035 \times \$1,440K &= \$1,490.4K \\
\text{down payment (\$1,800K} - \$1,440K = \$360K) &= \$360K \\
\text{rent, gross annual income at 100\% occupancy} &= \$280K \text{ at EOY 1} \\
\text{geometric gradient increase in rental income} &= 4\%/\text{yr compounded} \\
\text{vacancy rate} &= 5\% \\
\text{O \& M costs at EOY 1} &= \$100K \text{ at EOY 1} \\
\text{increase in O \& M} &= 4\%/\text{yr compounded} \\
\text{increase in resale value of this property} &= 4\%/\text{yr compounded} \\
\text{depreciation method} &= \text{straight-line} \\
\text{depreciate only the improvements plus points,} & \\
\text{to zero salvage at EOY 27.5} & \\
\text{tax bracket for ordinary income} = t &= 0.28
\end{aligned}
$$

Capital gains from resale of the project are taxed as ordinary income at the 28 percent tax rate. Allowable deductions from taxable income include depreciation, O & M, and interest payments.

Solution.   The problem can be solved by following a logical series of steps:

Step 1.   Using the initial mortgage balance (including the 3.5 points added on) of $1,490.4K at EOY 0, set up an amortization schedule through EOY 3. The amount of the payment is calculated as

$$A = \$1,490.4K(A/P, 10\%, 25) = \$164.194K/\text{yr}$$

The amortization schedule is filled in as follows:

EOY	Payment	Interest	Principal	Balance
0				$1,490.400K
1	$164.194K	$149.040K	$15.154K	1,475.246
2	164.194	147.525	16.670	1,458.576
3	164.194	145.858	18.337	1,440.239

Step 2. Calculate the straight-line depreciation schedule. The depreciable basis is found at BOY 1 as

$$depreciable\ improvements = \$1,500.000K$$
$$points\ on\ the\ mortgage = \$\ \ \ \ 50.400K$$
$$total\ depreciable\ basis = \$1,550.400K$$

The allowable depreciation life is 27.5 years, so the annual depreciation allowance is calculated as

$$annual\ depreciation = \$1,550.400K/27.5\ yr = \$56.378K/yr$$

Then the depreciation schedule is filled in.

Year	Straight-line depreciation taken this year	Depreciated BV of depreciable basis at EOY	Depreciated BV including land at EOY
0		$1,550.400K	$1,850.400K
1	$56.378K	1,494.022	1,794.022
2	56.378	1,437.644	1,737.644
3	56.378	1,381.265	1,681.265

Step 3. Calculate the cash flow from operations (including the tax benefits of depreciation) for the years shown. Do **not** include cash flow from purchase or sale in this step.

Name of CF Item	EOY 1 after-tax cash flow	EOY 2 after-tax cash flow	EOY 3 after-tax cash flow
Cash flow in from operations			
(Sample calculations for EOY 1)			
rent × occupancy × $(1 - t)$, where t = the tax rate			
Rent $280K × 0.95 × 0.72 =$	$191.520K	$199.181K	$207.148K
Depreciation $56.378K × 0.28 =$	15.786	15.786	15.786
Cash flow out from operations			
O & M costs $100K × 0.72 =$	− 72.000	− 74.880	− 77.875
Interest payment 149.040 × 0.72 =	− 107.309	− 106.218	− 105.017
Principal payment	− 15.154	− 16.670	− 18.337
Sum net ATCF, operations, at EOY	$ 12.843K	$ 17.199K	$ 21.704K

Step 4.   Calculate the owner's equity at the times indicated.

Name of item	EOY 2 AT amount	EOY 3 AT amount
Resale value $1,800K	$\times\ 1.04^2 = \$1{,}946.880K$	$\times\ 1.04^3 = \$2{,}024.755K$
Mortgage balance	$-\ 1{,}458.576$	$-\ 1{,}440.239$
Capital gains tax[a]	$-\quad 58.586$	$-\quad 96.177$
Owner's equity $=$	$\$\ \ 429.718K$	$\$\ \ 488.339K$

[a]*Note:* The capital gains tax (CGT) is calculated as CGT $=$ [(resale value) $-$ (depreciated BV)] $\times$ (ordinary tax rate); thus for EOY 2, CGT $=$ ($1,946.880 $-$ $1,737.644) $\times$ 0.28 $=$ $58.586K.

Step 5.   (YYAT profit in dollars). Calculate the owner's year-to-year after-tax profit (or loss) in dollars (the operating cash flow plus the increase in equity) from EOY 2 to EOY 3.

Answer: The YYAT profit $=$ (equity EOY 3 $-$ equity EOY 2) $+$ AT cash flow EOY 3; thus, YYAT profit $=$ ($488.339K $-$ $429.718K) $+$ $21.704K $=$ $80.326K.

Step 6.   (YYATROR). Calculate the owner's year-to-year after-tax rate of return from EOY 2 to EOY 3.

Answer: YYATROR $=$ (YYAT profit)/(equity at BOY) $=$ $80.326K/$429.718K $=$ 18.69%.

Step 7.   (OATROR). Calculate the owner's overall after-tax rate of return for the period EOY 0 to EOY 3.

Answer: This is solved as a rate of return problem, using all the cash flows from the initial investment until EOY 3.

	$i=$	14%	15%
Down payment, $P_1 = -\$360K$		$-\$360.K$	$-\$360.K$
ATCF at EOY 1, $P_2 = 12.843K(P/F, i, 1) =$		11.266K	11.168K
ATCF at EOY 2, $P_3 = 17.199K(P/F, i, 2) =$		13.234K	13.005K
ATCF at EOY 3, $P_4 = (488.339K + 21.704K(P/F, i, 3) =$		344.264K	335.363K
		$\$\ \ 8.764K$	$-\$\ \ 0.464K$

The OATROR is found by interpolation:

$$\text{OATROR} = 14\% + 1\% \times 8.764/(8.764 + 0.464) = 14.95\%$$

Step 8. Assume the time is now EOY 0 (before a commitment has been made to construct or buy the project). For what values of MARR would you recommend this investment, assuming that all payments occur as anticipated?

Answer: At the end of three years, the anticipated OATROR is 14.9 percent. Therefore, if the client's MARR is less than 14.9 percent, this project should be of interest. If the client's MARR is higher than 14.9 percent, the funds are better placed in some other investment at the MARR. (Further computation shows on OATROR of over 16 percent at EOY 5, if the predictions of input data prove valid.)

Step 9. Assume the time is now EOY 2, and your client already owns the project. For what values of MARR would you recommend retention until at least EOY 3?

Answer: The client can earn 18.7 percent YYATROR by holding on to the project for one more year (if all assumptions prove to be correct). Therefore, for a MARR of less than 18.7 percent, the project should be retained.

Step 10. What caveats (qualifications) would you issue along with your recommendation?

Answer: If the project appears feasible thus far, a more detailed analysis should be undertaken using essentially the same approach but involving a careful study of the particular market in which this project will compete, in order to verify and refine the input data. The client's tax situation should be examined carefully, and a tax accountant should review the data. In addition the input variables should be ranged to determine sensitivity, and a risk analysis undertaken to find the best, worst, and most probable outcomes. The client should understand that any attempt to forecast future trends of the input variables involves normal risks associated with investments of this type, and that while this procedure provides a good method of analysis, the actual outcome is heavily dependent upon how close the estimated input data approximate the actual costs and incomes from the market over the coming years.

Computer solution. A computer program, a computer spreadsheet solution, and a graph (Figure 19.1) of the resulting YYATROR and the OATROR for this problem follow.

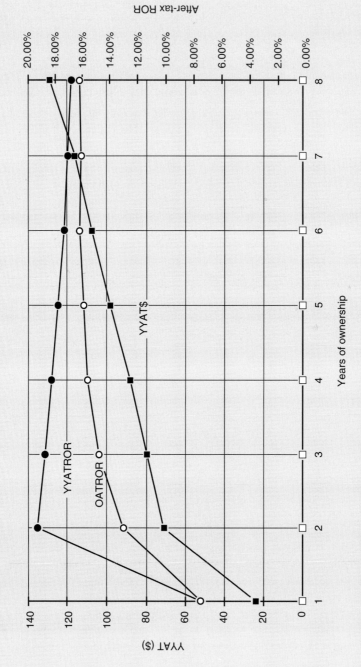

Figure 19.1 Graph for Example 19.4.

SPREADSHEET SOLUTIONS FOR EXAMPLE 19.3

	A	B	C	D	E	F	G
1	EXAMPLE 19.4						
2	The following spreadsheet example illustrates the use of spreadsheets to solve problems						
3	such as example 19.4						
4							
5	$300.000	Land cost					
6	$1,5000.000	Cost of Depreciable Improvements					
7	$1,800.000	Total cost A1 + A2					
8	10.00%	Mortgage i/year					
9	25	Mortgage term, years					
10	0.80	L/V ratio					
11	3.50	Mortgage points, add on					
12	$1,490.400	Begin Mortgage Balance					
13	$360.000	Down Payment					
14	$280.000	Rent Income EOY 1					
15	4%	Increase in rent per year					
16	5%	Vacancy rate					
17	$100.000	O&M costs at EOY 1					
18	4%	Increase in O&M per year					
19	4%	Increase in resale value per year					
20	1	SL depreciation method					
21	27.5	SL depreciation life, years					
22	0.28	t, tax bracket					
23							
24		1) Amortization Schedule					
25	Payment #	Paymt Amount	Interest	Principal	Balance		
26	0				1,490.400		
27	1	164.194	149.040	15.154	1,475.246		
28	2	164.194	147.525	16.670	1,458.576		
29	3	164.194	145.858	18.337	1,440.239		
30	4	164.194	144.024	20.171	1,420.068		
31	5	164.194	142.007	22.188	1,397.880		
32	6	164.194	139.788	24.406	1,373.474		
33	7	164.194	137.347	26.847	1,346.627		
34	8	164.194	134.663	29.532	1,317.095		
35							
36	2) Straight-line Deprec. Sched.						
37	EOY	SLBV at BOY	Depr this year	SLBV at EOY	SLBV + land		
38	1	1,550.400	56.378	1,494.022	1,794.022		
39	2	1,494.022	56.378	1,437.644	1,737.644		
40	3	1,437.644	56.378	1,381.265	1,681.265		
41	4	1,381.265	56.378	1,324.887	1,624.887		
42	5	1,324.887	56.378	1,268.509	1,568.509		
43	6	1,268.509	56.378	1,212.131	1,512.131		
44	7	1,212.131	56.378	1,155.753	1,455.753		
45	8	1,155.753	56.378	1,099.375	1,399.375		
46							

Spreadsheet (continued)

	A	B	C	D	E	F	G
47		3) AT Cash Flow, Operations					
48	EOY	AT Rent	AT SL Deprec	AT O&M	AT Interest	Principal	AT Cash Flow
49	0						−360.000
50	1	191.520	15.786	−72.000	−107.309	−15.154	12.843
51	2	199.181	15.786	−74.880	−106.218	−16.670	17.199
52	3	207.148	15.786	−77.875	−105.017	−18.337	21.704
53	4	215.434	15.786	−80.990	−103.697	−20.171	26.362
54	5	224.051	15.786	−84.230	−102.245	−22.188	31.175
55	6	233.013	15.786	−87.599	−100.647	−24.406	36.146
56	7	242.334	15.786	−91.103	−98.890	−26.847	41.280
57	8	252.027	15.786	−94.747	−96.957	−29.532	46.577
58							
59		4) Owner's AT Equity					
60	EOY	Resale Price	Mortgage Bal.	Cap Gain Tax	Owner's Equity		
61	0	1,800.000	−1,490.400		309.600		
62	1	1,872.000	−1,475.246	−21.834	374.921		
63	2	1,946.880	−1,458.576	−58.586	429.718		
64	3	2,024.755	−1,440.239	−96.177	488.339		
65	4	2,105.745	−1,420.068	−134.640	551.037		
66	5	2,189.975	−1,397.880	−174.011	618.084		
67	6	2,277.574	−1,373.474	−214.324	689.776		
68	7	2,368.677	−1,346.627	−255.619	766.432		
69	8	2,463.424	−1,317.095	−297.934	848.395		
70							
71		5) YYAT Profit in $.			6) YYATROR	7) OATROR	
72	EOY	ATCF+ ΔEquity					
73	1	27.763			7.71%	7.71%	
74	2	71.997			19.20%	13.22%	
75	3	80.326			18.69%	14.95%	
76	4	89.060			18.24%	15.71%	
77	5	98.222			17.82%	16.10%	
78	6	107.838			17.45%	16.30%	
79	7	117.935			17.10%	16.40%	
80	8	128.541			16.77%	16.44%	
81							
82	8) EOY Cash Flow of Project Sold at EOY (n).						
83	EOY1	EOY2	EOY3	EOY4	EOY5	EOY6	
84	−360.000	−360.000	−360.000	−360.000	−360.000	−360.000	
85	387.763	12.843	12.843	12.843	12.843	12.843	
86		446.917	17.199	17.199	17.199	17.199	
87			510.044	21.704	21.704	21.704	
88	EOY 7	EOY 8		577.399	26.362	26.362	
89	−360.000	−360.000	Purchase EOY	0	649.259	31.175	
90	12.843	12.843	At cash flow EOY	1		725.923	
91	17.199	17.199		2			
92	21.704	21.704		3			
93	26.362	26.362		4			
94	31.175	31.175		5			
95	36.146	36.146		6			
96	807.711	41.280	At cash flow EOY	7			
97		894.973	At cash flow resale EOY	8			

SPREADSHEET FORMULAS FOR EXAMPLE 19.3

	A	B	C	D	E	F	G
1	300,000	Land cost					
2	1,500,000	Cost of Depreciable Improvements					
3	=A1 + A2	Total cost A1 + A2					
4	0.1	Mortgage i/year					
5	25	Mortgage term, years					
6	0.8	L/V ratio					
7	3.5	Mortgage points, add on					
8	=A3*A6*(1 + A7/100)	Begin Mortgage Balance					
9	=A3*(1 − A6)	Down Payment					
10	280,000	Rent Income EOY 1					
11	0.04	Increase in rent per year					
12	0.05	Vacancy rate					
13	100,000	O&M costs at EOY 1					
14	0.04	Increase in O&M per year					
15	0.04	Increase in resale value per year					
16	1	SL depreciation method					
17	27.5	SL depreciation life, years					
18	0.28	t, tax bracket					
19							
20		1) Amortization Schedule					
21	Payment #	Paymt Amount	Interest	Principal	Balance		
22	0				=A8		
23	1	=−PMT(A4,A5,A8)	=E22*($A$4)	=B23 − C23	=E22 − D23		
24	2	=B23	=E23*($A$4)	=B24 − C24	=E23 − D24		
25	3	=B24	=E24*($A$4)	=B25 − C25	=E24 − D25		
26	4	=B25	=E25*($A$4)	=B26 − C26	=E25 − D26		
27	5	=B26	=E26*($A$4)	=B27 − C27	=E26 − D27		
28	6	=B27	=E27*($A$4)	=B28 − C28	=E27 − D28		
29	7	=B28	=E28*($A$4)	=B29 − C29	=E28 − D29		
30	8	=B29	=E29*($A$4)	=B30 − C30	=E29 − D30		
31							

Spreadsheet (continued)

	A	B	C	D	E	F	G
32		2) Straight-Line Deprec. Sched.					
33	EOY	SLBV at BOY	Depr this year	SLBV at EOY	SLVB + land		
34	1	= A2 + (A1+A2)*A6*A7/100	= $B$34/$A$17	= B34 – C34	= D34 + $A$1		
35	2		= C34	= B35 – C35	= D35 + $A$1		
36	3		= C35	= B36 – C36	= D36 + $A$1		
37	4		= C36	= B37 – C37	= D37 + $A$1		
38	5		= C37	= B38 – C38	= D38 + $A$1		
39	6		= C38	= B39 – C39	= D39 + $A$1		
40	7		= C39	= B40 – C40	= D40 + $A$1		
41	8		= C40	= B41 – C41	= D41 + $A$1		
42							
43		3) AT Cash Flow, Operations					
44	EOY	AT Rent	AT SL Deprec	AT O&M	AT Interest	Principal	AT Cash Flow
45	0						= – $A$9
46	1	= $A$10*(1 – $A$18)*(1 – $A$12)	= C34*$A$18	=–$A$13*(1 – $A$18)	= – C23*(1 – $A$18)	= – D23	= sum(B46:F46)
47	2	= B46*(1 + $A$11)	= C35*$A$18	= D46*(1 + $A$14)	= – C24*(1 – $A$18)	= – D24	= sum(B47:F47)
48	3	= B47*(1 + $A$11)	= C36*$A$18	= D47*(1 + $A$14)	= – C25*(1 – $A$18)	= – D25	= sum(B48:F48)
49	4	= B48*(1 + $A$11)	= C37*$A$18	= D48*(1 + $A$14)	= – C26*(1 – $A$18)	= – D26	= sum(B49:F49)
50	5	= B49*(1 + $A$11)	= C38*$A$18	= D49*(1 + $A$14)	= – C27*(1 – $A$18)	= – D27	= sum(B50:F50)
51	6	= B50*(1 + $A$11)	= C39*$A$18	= D50*(1 + $A$14)	= – C28*(1 – $A$18)	= – D28	= sum(B51:F51)
52	7	= B51*(1 + $A$11)	= C40*$A$18	= D51*(1 + $A$14)	= – C29*(1 – $A$18)	= – D29	= sum(B52:F52)
53	8	= B52*(1 + $A$11)	= C41*$A$18	= D52*(1 + $A$14)	= – C30*(1 – $A$18)	= – D30	= sum(B53:F53)
54							
55		4) Owner's At Equity					
56	EOY	Resale Price	Mortgage Bal.	Cap Gain Tax	Owner's Equity		
57	0	= $A$3	= – E22		= B57 + C57		
58	1	= B57*(1 + $A$15)	= – E23	= – (B58 – E34)*$A$18	= B58 + C58 + D58		
59	2	= B58*(1 + $A$15)	= – E24	= – (B59 – E35)*$A$18	= B59 + C59 + D59		
60	3	= B59*(1 + $A$15)	= – E25	= – (B60 – E36)*$A$18	= B60 + C60 + D60		
61	4	= B60*(1 + $A$15)	= – E26	= – (B61 – E37)*$A$18	= B61 + C61 + D61		
62	5	= B61*(1 + $A$15)	= – E27	= – (B62 – E38)*$A$18	= B62 + C62 + D62		
63	6	= B62*(1 + $A$15)	= – E28	= – (B63 – E39)*$A$18	= B63 + C63 + D63		

Spreadsheet (continued)

	A	B	C	D	E	F	G
64	7	=B63*(1 + $A$15)	=-E29	=-(B64 - E40)*$A$18	= B64 + C64 + D64		
65	8	=B64*(1 + $A$15)	=-E30	=-(B65 - E41)*$A$18	= B65 + C65 + D65		
66							
67		5) YYAT Profit in $.		6) YYATROR	7) OATROR		
68	EOY	ATCF + Δ Equity					
69	1	=G46 + E58 - A9		=B69/A9	=IRR(A80:A81,D69)		
70	2	=G47 + E59 - E58		=B70/E58	=IRR(B80:B82,E69)		
71	3	=G48 + E60 - E59		=B71/E59	=IRR(C80:C83,E70)		
72	4	=G49 + E61 - E60		=B72/E60	=IRR(D80:D84,E71)		
73	5	=G50 + E62 - E61		=B73/E61	=IRR(E80:E85,E72)		
74	6	=G51 + E63 - E62		=B74/E62	=IRR(F80:F86,E73)		
75	7	=G52 + E64 - E63		=B75/E63	=IRR(A85:A92,E74)		
76	8	=G53 + E65 - E64		=B76/E64	=IRR(B85:B93,E75)		
77							
78		8) EOY Cash Flow of Pr					
79	EOY1	EOY2	EOY3	EOY4	EOY5	EOY6	
80	=-$A$9	=-$A$9	=-$A$9	=-$A$9	=-$A$9	=-$A$9	
81	=G46+E58	=$G$46	=$G$46	=$G$46	=$G$46	=$G$46	
82	=$G$47	=G47+E59	=$G$47	=$G$47	=$G$47	=$G$47	
83	=$G$48		=G48 + E60	=$G$48	=$G$48	=$G$48	
84	EOY 7	EOY 8		=G49 + E61	=$G$49	=$G$49	
85	=-$A$9	=-$A$9			=G50 + E62	=$G$50	
86	=$G$46	=$G$46				=G51 + E63	
87	=$G$47	=$G$47					
88	=$G$48	=$G$48					
89	=$G$49	=$G$49					
90	=$G$50	=$G$50					
91	=$G$51	=$G$51					
92	=$G$52+E64	=G52					
93	=$G$53 + E65	=G53 + E65					

SUMMARY

Like most other problems, normally real estate investment problems can be solved by dividing them up into simple components, solving each component, and summing the total. This chapter addresses a representative group of the many factors influencing such investments. These factors include credit terms for mortgage financing, depreciation, capital gains, recapture, and after-tax rate of return. If other variables are encountered, they usually can be handled by one of the standard approaches discussed so far.

PROBLEMS FOR CHAPTER 19, INVESTMENT PROPERTY

Problem Group A

Find the required rental income or sales price to fulfill stated objectives. No income taxes.

A.1. Your client is interested in purchasing a small, 10-unit office complex. The purchase price is $500,000 with 10 percent down payment required. In addition to the down payment, closing costs and front-end charges require an additional cash outlay of $2,000. The mortgage is for 30 years at 10 percent. The operating costs, including property taxes, are estimated at 40 percent of the gross rental income. The resale value in 30 years is estimated at $700,000. Approximate the following answers by using end-of-year annual payments and income. Your institutional client pays property taxes but no income taxes.

(a) How much monthly rent must be collected from each of the 10 tenants to develop a 15 percent rate of return before taxes over the 30-year life of the building if it stays 100 percent rented?

(b) More realistically, occupancy should average 85 percent, and rents should increase by 5 percent per year. What should the beginning rent be?

(c) What is the first-year cash flow under the conditions of (b)?

A.2. You have just completed a preliminary estimate for a proposed warehouse. The total estimated first cost for the project is $1,000,000, with a resale value after 30 years of $200,000. An investor is interested in having you design and build it if the investment will earn a 10 percent before-tax return. How much annual rent must be charged (in excess of property taxes and operating expenses) in order to obtain a 10 percent return on the investment?

A.3. Your client has located a 2,000 acre tract of land available for $1,000 per acre. The tract is suitable for development into a country-club-type residential community around an 18-hole championship golf course. The tract would contain 2,000 lots averaging $\frac{1}{2}$ acre each, with the remaining land used for golf course, lakes, open space, and right-of-way. Your client feels that 200 lots will sell the first year and that sales will continue at that same rate of 200 lots per year for ten years. The sales strategy is to start the prices as low as possible the first year and increase them 20 percent per year. (For example, if the first year's price is 1.0, then the price the second year is 1.20, the third year is 1.44, and the fourth year is 1.728, etc.) The following development costs are anticipated as of the *end* of the year listed.

Year	Cost	Item
0	$2,000,000	Purchase of land (begin land sales)
1	2,000,000	Site grading, roads, utilities, taxes, etc.
2	2,000,000	Construct golf course, tennis etc., maintenance
3	200,000	Maintenance and taxes, additional development
4	220,000	Same
⋮	⋮	Increase 20,000/yr/yr
10	340,000	Same

Your client wants to make 18 percent before income taxes on invested capital, and so will charge the project at 18 percent interest on all funds invested therein. How much should each residential lot sell for (**a**) the first year; (**b**) the tenth year? Assume that all income received from the sale of lots is credited to the project as of the end of the year in which the lot is sold.

Problem Group B

Find the rate of return, i, *before* taxes.

B.1. An 80-acre tract of land is available for $100,000 cash. The owner will accept as an alternative a down payment of $20,000 and six additional end-of-year annual payments of $20,000. To what interest rate would the alternative payment plan be equivalent (before taxes)? (*Hint:* Try 10 percent.)

B.2. A client asks your opinion on a proposal to construct a warehouse rental property in a local industrial park. The estimated terms are shown below. Your client specifically wants to know the anticipated rate of return before taxes if the property is sold after 10 years.

$$\begin{aligned}
\text{total first cost of project} &= \$2,000,000 \\
\text{first-year gross income} &= 240,000/\text{yr} \\
\text{annual increase in income each year} &= 12,000/\text{yr/yr} \\
\text{first year O \& M costs} &= 100,000/\text{yr} \\
\text{annual increase in costs each year} &= 5,000/\text{yr/yr} \\
\text{annual increase in resale value} &= 5\%/\text{yr compounded}
\end{aligned}$$

B.3. A friend bought a 40-acre farm eight years ago and, having previously sold part of it, at EOY 5 wants to sell the remainder of it at the present time. Using the figures shown below, find the before-tax rate of return from the entire transaction.

EOY			
0	Purchase price	−	$65,000
1–5	Annual taxes and O & M	−	8,000/yr
1–5	Annual income	+	12,000/yr
5	Sell 25 acres at $3,000/acre	+	75,000
6–8	Annual taxes and O & M	−	6,000/yr
6–8	Annual income	+	8,200/yr
8	Sell remaining 15 acres at $4,000/acre	+	60,000

B.4.   A client needs a preliminary estimate of before-tax rate of return on a proposed high-rise apartment named Plush Manor (balconies overlooking the sea for every apartment) with 100 apartment units costing $60,000 per unit. Rental rate per unit is set at $1,200 per month, with 90 percent occupancy expected. Taxes, maintenance, and other operating costs (excluding depreciation) are estimated at $4,200 per year for each unit, regardless of whether occupied or not. Life expectancy is 20 years, with 50 percent salvage value at the end of that time. What is the rate of return before income taxes?

Problem Group C

Find the rental rate, selling price, purchase price, or net profit for a given after-tax rate of return or after-tax profit.

C.1.   Your client is interested in building and operating a small shopping center of 100,000 ft^2 and asks you to calculate the rental rate needed per square foot in order to make a 15 percent rate of return after taxes on all funds invested. Your client is in the 30 percent tax bracket for ordinary income. The future costs and income for the project are estimated as follows.

EOY	
0	Purchase land for $100,000.
1	Finish construction of $2,000,000 worth of buildings and other improvements depreciable to zero over a 31.5-yr life by straight-line depreciation. Begin rental and operation of the center, but rental income and operating costs are not credited or debited until EOY 2.
2	First year's operating and maintenance costs charged at EOY 2 = 100,000/yr. First year's income deposited after income taxes are paid at a tax rate of 30% net income (where net income = gross income − O & M − depreciation).
3–6	Same as EOY 2. (O & M costs continue at $100,000/yr.)
6	Sell the shopping center for original cost plus appreciation estimated at 4%/yr compounded. The difference between the depreciated book value and the sales price is treated as a capital gain taxed at ordinary rate).

   (a) Draw a cash flow diagram indicating the time and amount of each cash flow occurrence.
   (b) Find the rental rate in terms of dollars per square foot each year to obtain 15 percent return after taxes on all funds invested.

C.2.   A client purchased some undeveloped land ten years ago for $10,000, and has paid an average of $300 per year in taxes and other expenses on the property. Now a customer is interested in purchasing the land.
   (a) What should be the minimum selling price in order to make a 20 percent profit *before* income taxes on all funds invested in the land?
   (b) The client is in the 30 percent tax bracket (and has been for the last 10 years) and pays ordinary rate taxes on the capital gain. What is the selling price in order to make 20 percent profit *after* income taxes?

C.3.   The developer of a 360-unit seafront condominium asks you, as consultant, to calculate how much the condominium units should sell for. You work up the following time and cost schedule. All costs and income are credited at the end of the month in which incurred.

Time	
0	Land purchase, $1,000,000
Months 1 through 6	Plan and design the project. Designer is paid $600,000 in equal monthly installments of $100,000 at the end of each month.
Months 7 through 18	Construct the project. Contractor is paid $500,000 at the end of each month for 12 months, plus an additional $1,000,000 retainage paid to the contractor at the end of the twelfth month of construction (total of $7,000,000).
Months 19 through 36	Sales of the 360 units are expected to progress at the rate of 20 units per month. To aid sales promotion, the developer wants to start prices low and raise them 1% per month over the 18-month sales period. Sales promotions and commission = 10% of sales price.

EOM

12	$50,000 for	Ad valorem taxes, insurance, and other tax-deductible items.
24	$200,000 for	Same
36	$200,000 for	Same

Main office overhead chargeable to this project = $10,000 per month for 36 months.

The developer pays 1.5 percent per month interest on all funds used to finance the project, desires to make $3,000 per unit profit after taxes on each unit sold, and is in the 42 percent tax bracket. Assume taxes are paid in monthly installments when due.

(a) What must the selling price be for the first 20 condominiums sold in the nineteenth month of the project life (credited at the end of the nineteenth month)?

(b) What must the selling price be for the last 20 condominiums in the thirty-sixth month?

C.4. You are interested in investing in a warehouse that is currently for sale. The following data are available (assume EOY payments):

1. Current rental income is $100,000 per year, with an estimated increase in rental income of 5 percent per year.
2. Estimated sales price for warehouse in six years is $1,400,000.
3. Estimated property tax, maintenance, and insurance is $40,000 per year, with an estimated increase of 5 percent per year.
4. A 15-year mortgage for $750,000 is available at 10 percent with equal annual payments.

Find the maximum price you can pay to net a 20 percent rate of return before income taxes.

C.5. Same as Problem C.4, but add the following assumptions.

5. Assume the following tax deductions are available as a part of the investment and that you are in the 34 percent tax bracket. (a) $50,000 per year for depreciation of warehouse; (b) $40,000 per year property tax, maintenance, insurance, and interest on mortgage, with an estimated increase of 5 percent per year.
6. Assume also you will pay 34 percent tax on the capital gains of $500,000 upon selling the property in six years. Find the maximum price you can pay now to net an after-tax rate of return of 15 percent.

C.6. A land subdivider advertises 5-acre parcels at $2,000 per acre (total = $10,000/parcel) on some rural acreage he is subdividing. He advertises 6 percent, 10-year mortgages with no down payment. In order to raise cash, he is having to resell these mortgages to investors, and due to the risk involved, the investors are requiring a 20 percent interest return.

(a) How much is the subdivider actually getting for the land in dollars per parcel after discounting the mortgages (before taxes)?

(b) Assume the subdivider bought the land for $1,000 per acre and paid $100 per acre in legal and surveying fees. He is in the 28 percent tax bracket and pays ordinary income tax on all net profit. How much is the subdivider netting per parcel *after* taxes?

C.7. A landowner wants to sell 100 acres of land. He has two offers, and asks your advice on which is the better offer. He paid $100 per acre for the land 15 years ago and has been paying property taxes of $300 per year ever since. He has been in the 30 percent tax bracket the entire time. The terms of each offer to purchase are

Offer A. $1,000 per acre cash now. The capital gain on the sale is taxed at the ordinary rate.

Offer B. A real estate salesperson offers to sell the land for him in 5-acre parcels for $2,000 per acre ($10,000/parcel) for a 10 percent commission "off the top" ($200/acre commission taken out of the down payment). The salesperson suggests terms of $1,000 down and a mortgage for nine additional end-of-year annual payments of $1,000, no interest charged. The owner finds he could sell these mortgages for cash if he will discount them to yield a 15 percent rate of return. Assume that with these terms the land would sell quickly and he could get his cash reasonably soon. The gain on the sale is taxed as ordinary income.

C.8. Your client has an opportunity to purchase 670 acres along the Suwannee River for $1,200 per acre. You investigate the site and find there are a total of 500 good home sites outside the flood plain with the remainder used for recreation areas and green space. After consultation with realtors in the area you estimate lot sales at the rate of 100 lots per year for five years. As a sales strategy, you intend to start the lots at a relatively low price and raise the price by 20 percent every year. (For example, if the first year's price is 1.0, then the price during the second year is 1.2, the third year it rises to 1.44, etc.) The following development costs are anticipated as of the *end* of the year listed. Since the land is considered stock in trade, all costs are tax deductible, and all income is taxable at $t = 30$ percent.

Year	Cost	Item
0	$804,000	Purchase the land
1	932,000	Install roads, water, and sewer. *Begin sale of lots,* but income is credited at EOY 2.
2	120,000	Maintenance, taxes, additional facilities (boat ramp, tennis, etc.)
3	140,000	Same Increase by $20,000/yr/yr
6	200,000	Same. Sell last lots. Turn common areas over to Homeowners Association or the county.

Your client wants to make a 15 percent after-tax rate of return on all funds invested in the project.

(a) How much should the first 100 lots sell for in terms of dollars per lot?

(b) How much should the last 100 lots sell for in terms of dollars per lot? Assume that all income from the lots is credited to the project at the end of the year in which the lots are sold.

Problem Group D

Find the YYATROR, OATROR, ATROR, and other significant values.

D.1. Your client is interested in developing a small, private, medical clinic with characteristics as follows:

$$
\begin{aligned}
\text{value of land only} &= \$200K \\
\text{value of depreciable improvements} &= \$800K \\
\text{mortgage} &= \$900K \text{ at } 12\%, 25 \text{ yr} \\
\text{points on the mortgage} &= 4 \text{ points, add on to initial} \\
&\quad \text{mortgage balance due} \\
\text{down payment} &= \$100K \\
\text{gross annual rental income} &= \$220K \\
\text{geometric gradient increase in rental income} &= 5\%/\text{yr compounded} \\
\text{vacancy rate} &= 5\% \\
\text{O \& M costs at EOY } 1 &= \$120K \\
\text{geometric gradient increase in O \& M} &= 5\%/\text{yr compounded} \\
\text{market value of property increases by} &= 5\%/\text{yr compounded} \\
\text{depreciation period} &= 31.5 \text{ yr} \\
\text{depreciation method} &= \text{straight-line to zero salvage} \\
\text{amount to depreciate (basis)} &= \text{improvements and points} \\
\text{client's tax bracket for ordinary income} &= 35\%
\end{aligned}
$$

Following is the standard question format for Problems D.1, D.2, D.3, and D.4.

Capital gain is taxed at the ordinary tax rate. *The answers shown below are for Problem D.1.*

1. How much is the initial mortgage balance at EOY 0, including points added on?

(*Ans.* $936,000)

2. For the mortgage of this problem, set up a neat amortization table with properly labeled columns for EOY 0 through EOY 2.
Amortization table headings:

Answer:

Year	Payment	Interest	Principal	Balance
0				$936,000
1	$119,340	$112,320	$7,020	928,980
2	119,340	111,478	7,862	921,118

3. Using straight-line depreciation, calculate the depreciation schedule for this property through the first two years. The depreciable basis is the improvements and points only.

Answer:

Year	Depreciation this year	Depreciated book value at EOY	DBV plus land cost at EOY
0		$836,000 (Depreciable basis)	
1	$26,540	809,460	$1,009,460
2	26,540	782,921	982,921

4. Calculate the owner's cash flow from operations (including the effects of depreciation) for EOY 1 and EOY 2.

Answer:

	EOY 1		EOY 2	
Cash flow *in* from operations	Before-tax amount	After-tax amount	Before-tax amount	After-tax amount
Rent	220K × 0.95 × 0.65 =	$135,850	×1.05 =	$142,643
Depreciation	26,540 × 0.35 =	9,289		9,289
Cash flow *out* from operations				
O & M	− 120K × 0.65 =	− 78,000	× 1.05 =	− 81,900
Interest	−112,320 × 0.65 =	− 73,008	114,478 × 0.72 =	− 72,460
Principal		− 7,020		− 7,862
Σ AT cash flow from operations		−$12,889		−$ 10,291

5. Calculate **(a)** the owner's cash investment for EOY 0, and **(b)** the owner's after-tax equity for EOY 1 and EOY 2.

Answer: Owner's cash investment and owner's equity

Item name (list names of items needed to calculate the owner's equity)	EOY 1 after-tax amount	EOY 2 after-tax amount
Resale price	+$1,050,000	+1,102,500
Mortgage balance	− 928,980	− 921,118
Capital gain tax	− 14,189	− 41,853
Σ Owner's equity	+$ 106,831	+ 139,530

6. (YYAT profit). Calculate the owner's year-to-year after-tax profit in dollars at EOY 1 and EOY 2.
Answer: EOY 1 −6,058 and EOY 2 +22,407 .

7. (YYATROR). Calculate the owner's year-to-year after-tax rate of return at EOY 1 and EOY 2.
Answer: EOY 1 −6.06% and EOY 2 +20.97%.

8. (OATROR). Calculate the owner's overall after-tax rate of return for the period EOY 1 and EOY 2.
Answer: EOY 0–1 −6.06% and EOY 0–2 +7.42% .

9. For what values of MARR would you recommend purchase at EOY 0 (assume resale at EOY 2, and that all input data will prove valid).
Answer: Purchase if MARR < 7.4 percent.

10. For what value of MARR would you recommend retention if your client already owns the property and the time is already EOY 1.
Answer: Retain if MARR < 21.0 percent.

D.2. A residential rental property can be acquired on the following terms:

value of land only = $100K
value of depreciable improvements = $500K
initial mortgage = $540K at 11%, 25 yr
points on the mortgage = 4 points, add on to initial mortgage
down payment = $60K
gross annual rental income at 100% occupancy = $150K
geometric gradient increase in rental income = 5%/yr compounded
vacancy rate = 5%
O & M costs at EOY 1 = $95K
geometric gradient increase in O & M = 5%/yr compounded
market value of property increases by = 5%/yr compounded
depreciation period = 27.5 yr
depreciation method = straight-line to zero salvage
amount to depreciate (basis) = improvements + points
client's tax bracket for ordinary income = 28%

Answer the standard question format shown for Problem D.1.

D.3. Your client has an opportunity to build a small private airport in a developing area of the state, with the data described below.

value of land only at EOY 0 = $400K
value of improvements at EOY 0 = $3,600K
initial mortgage = $3,800K at 10%, 20 yr
points on the mortgage = 6 points, add on to initial mortgage
down payment = $200K
gross annual income at EOY 1 = $480K
geometric gradient increase in rental income = 5%/yr compounded
vacancy rate = 5%
O & M costs at EOY 1 = $210K
geometric gradient increase in O & M = 6%/yr compounded
market value of property increases by = 7%/yr compounded
depreciation period = 31.5 yr
depreciation method = straight-line to zero salvage
amount to depreciate (basis) = improvements + points
client's tax bracket for ordinary income = 28%

Answer the standard question format shown for Problem D.1.

D.4. A retirement village is proposed in an ideal sunbelt site, with the data described below.

value of land only at EOY 0 = $500K
value of improvements at EOY 0 = $5,500K
initial mortgage = $5,400K at 9%, 30 yr
points on the mortgage = 4 points, add on to initial
mortgage balance
down payment = $600K
gross annual rental income at 100% occupancy = $700K
geometric gradient increase in income = 4%/yr compounded
vacancy rate = 5%
O & M costs at EOY 1 = $240K

geometric gradient increase in O & M = 3%/yr compounded
market value of property increases by = 3%/yr compounded
depreciation period = 27.5 yr
depreciation method = straight-line to zero salvage
amount to depreciate (basis) = improvements + points
client's tax bracket for ordinary income = 34%

Answer the standard question format shown for Problem D.1.

D.5. Your client has just located some land for sale that might make a good location for a new shopping center. Calculate the rate of return using the following estimates of cost and income. (*Hint:* Find i between 10 and 20 percent.)

EOY		Cost	Income
		(All figures are × 1,000)	
0	Purchase the land	$– 100	0
1	Complete the design	– 80	0
2	Complete construction	–1,000	0
3	Operating expense and income	– 200	+ 100
4	Operating expense and income	– 210	+ 150
5	Operating expense and income	– 220	+ 200
6	Operating expense and income	– 230	+ 250
7	Operating expense and income	– 240	+ 300
8	Operating expense and income	– 250	+ 350
9	Operating expense and income	– 260	+ 400
10	Operating expense and income	– 270	+ 450
10	Resell shopping center and land		+3,600

Assume the cost of design and construction both may be depreciated by SL over a 31.5-year period (both first credited at EOY 3). $t = 34$ percent.

Problem Group E

Find the arithmetic gradient, G.

E.1. A proposed shopping center is under consideration with the following costs and income estimates.

Original cost of land is $500,000, and the cost of the buildings is $3,500,000, for a total of $4,000,000. The buildings can be depreciated by SL over a 31.5-year period. This project will be financed by $1,000,000 of owner's capital and $3,000,000 from a mortgage at 8 percent and 30 years with payments of interest only made annually. At the end of 12 years the owner expects to pay off the balance of the mortgage in one lump sum and sell the center for $5,000,000. The difference between the SL depreciated book value and resale price is capital gain taxed at ordinary tax rates. Before-tax rental income, after deducting O & M costs, should net $300,000 in the first year and should increase by a certain gradient amount per year after that. What arithmetic gradient increase in rents is necessary for the owner to make a 15 percent compounded after-tax return on the cash investment? $t = 30$ percent.

Assume all yearly payments are made at EOY.

CHAPTER 20

Equipment Replacement Analysis

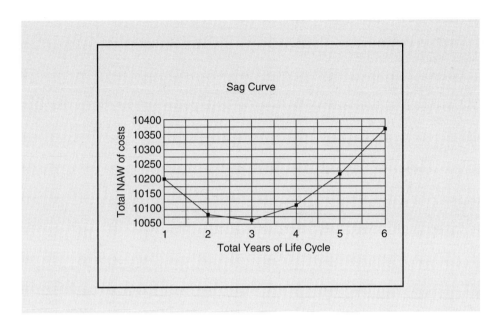

KEY EXPRESSIONS IN THIS CHAPTER

$ACRS$ = Accelerated Capital Recovery System (faster depreciation than SL).

Challenger = A candidate attempting to replace the defender.

Defender = The machine now in service, being compared to a potential replacement challenger.

DBV = Depreciated Book Value.

EOY = End of Year.

ETZ = Equation Time Zero, the beginning point for the time-value equation under consideration.

L = The life-cycle time of the challenger.

N = The remaining life of the defender.

NAW = Net Annual Worth.

NPW = Net Present Worth.

$P\Sigma D$ = The sum of all of the challenger's present worth.

$$PTZ = \text{Problem Time Zero (the present time, now).}$$
$$SL\ depreciation = \text{Straight-Line depreciation.}$$
$$SLBV = \text{Straight-Line Book Value.}$$
$$t = \text{income tax bracket, tax rate.}$$

ECONOMIC LIFE

With the exception of land, most economic investments have at least four distinct "lives"; (1) the actual physical life, (2) the depreciation life, (3) the service life, and (4) the economic life. The following paragraphs and diagram explain these concepts in further detail.

Actual physical life, as the name implies, is the period of time over which the investment is actually used. If a car is used over a 10-year period of time before it is junked (even if owned by several different owners) then the car has an "actual physical life" of 10 years.

Depreciation life is the period of time over which the investment is depreciated on the accounting books. Each new owner may depreciate the asset from its value at the time of purchase, according to an appropriate IRS depreciation schedule. Two different types of depreciation are practiced and for two different reasons: (1) tax-deductible depreciation, which must be done in strict accordance with current tax regulations (which from time to time are changed by majority vote of the U.S. Congress), and (2) actual depreciation, which reflects the loss in actual market value of the car. In general, tax-deductible depreciation is allowed on most capital investments (land is a notable exception) that are acquired for use in a business or are held to produce income. The depreciation period starts each time the property is placed in service by a new owner and ends when (a) the property is taken out of service, or (b) all of the allowable depreciation is deducted, or (c) the property is no longer used for business purposes (even though it may still be in nonbusiness service).

The *service life* is the period of time over which a property is held for one particular purpose or level of service.

Figure 20.1 graphically illustrates the first three different types of lives. Referring to Figure 20.1, assume the car was initially purchased as the company president's car to be used for company business. Assume the tax law allowed a three-year depreciation period, either by straight-line depreciation or by the accelerated capital recovery system. In this example the president only drove the car for two years, but since the car was still owned by the same firm, the firm could continue to take tax-deductible depreciation for the full three-year allowable term. When the three-year term for allowable depreciation expired, no more depreciation could be deducted, even though the car continued in service. When the president passed the car on to the department manager, the first *service life* ended, and the car entered a second service life at a different level of service. The second service life came to an end when the firm sold the car. Whenever depreciated property is resold, under

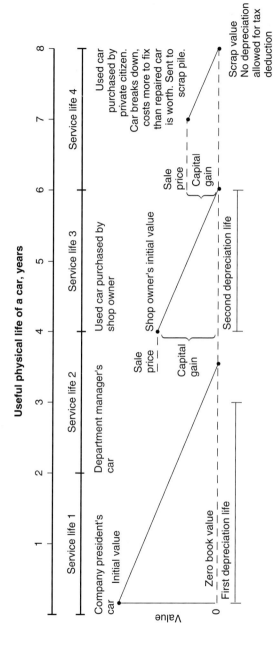

Figure 20.1 The physical life, the actual economic lives, and the depreciation lives of a car.

553

current Tax Law the gain in resale price over depreciated book value is taxed as ordinary income. In this case since the car is depreciated to zero, the firm pays a tax on the entire resale price of the car. The next owner of the car is the shop owner, who buys the car to use in the business. Therefore, the shop owner is entitled to deduct depreciation for three more years, using the price paid for the used car as the basis from which to calculate depreciation deductions. The third service life ends when the car is sold to a private citizen. The citizen cannot deduct depreciation, since the car was not used for business purposes during this service life. When the car breaks down for the last time, and the citizen finds that the cost of the repairs exceeds the value of the repaired car, then finally the car is scrapped, and the fourth service life comes to an end.

The fourth type of life listed above is the *economic life,* sometimes described as the time period for which "the equivalent net annual worth (NAW) is minimum." In actual practice this time period is difficult to determine exactly since full and accurate economic data often are not available [owners of expensive equipment often keep inadequate cost records on their individual equipment units and terminate the ownership life (sell, trade, or scrap) based on intuition, or rules of thumb, without benefit of any economic study]. Of course there are different methods of economic analysis, and these, when compared, frequently show some variation in the resulting estimate of economic life span. In addition, even when considering only one method, differing opinions frequently arise over the numerical values for the input data. Under such circumstances it is sometimes difficult to determine with confidence the point in time at which the NAW is minimum. Therefore, the derived estimates of future economic life spans should be treated as estimates only— helpful guidelines to be used with caution.

WHAT TO DO *NOW?*

While the estimate of life span is a necessary and useful value needed for estimating costs and budgeting for the future, another important product of an economic analysis is the information it provides on what action to take now, at the *present time.* Should the defender be kept or replaced now? Based upon projections of the costs and incomes of the defender versus those of the challenger, the relative merits of the competing (and frequently mutually exclusive) courses of action can be compared numerically. The decision to keep or replace the defender *now* can be based upon an examination of these relative merits. In addition, at appropriate intervals of time the data can be updated and the analysis rerun. Usually the analysis will provide some indication of how soon the break-even point will occur, and about how much it will cost to advance or delay the replacement date. This in turn provides an indication of how frequently the analysis should be rerun.

The following example illustrates a simplified economic analysis to determine the theoretical end point of the economic life. The *theoretical* designation indicates that the end point found in this example is only valid if the replacement has *identical* costs. The *actual* end point occurs when a better challenger becomes available. More on that later.

EXAMPLE 20.1

Your client is considering the purchase of an automobile and asks for your help in determining the theoretical optimum economic life.

$$\text{client's interest rate } i = 10\%$$
$$\text{purchase price of automobile} = \$24,000$$
$$\text{decline in resale value with each year of age} = 20\% \text{ of previous year's value}$$
$$\text{operating and maintenance costs} = \$3,000/\text{yr}$$
$$\text{arithmetic gradient in O \& M costs} = \$1,200/\text{yr/yr}$$

Find the theoretical optimum economic life.

Solution. The decline in resale value of 20 percent per year indicates that the resale value in each subsequent year will be $(1 - 0.2 =)$ 0.8 times the value for each prior year, or

$$\text{the resale value at EOY } 1 = 0.8 \times \$24,000 = \$19,200$$
$$\text{the resale value at EOY } 2 = 0.8 \times \$19,200 = \$15,360$$
etc.

The NAW is found for the auto for values of economic life, n, ranging from 1 to 6. Example calculations for finding the NAW for an economic life of one year ($n = 1$) follow:

$$\text{cost new } A_1 = -24,000(A/P, 10\%, 1) = -\$26,400$$
$$\text{resale at EOY } 1 \; A_2 = +19,200(A/F, 10\%, 1) = \quad 19,200$$
$$\text{O \& M costs } A_3 = -3,000 = - \quad 3,000$$
$$\text{gradient O \& M } A_4 = -1,200(A/G, 10\%, 1) = \underline{\qquad 0}$$
$$\text{total NAW} = -\$10,200/\text{yr}$$

The NAWs for life cycles of a larger number of years are calculated in a similar manner, yielding the table below.

Total years of life-cycle	(1) Purchase price	(2) Annual worth of purchase $(A/P) \times$ (1)	(3) Resale value (80% of previous)	(4) Annual worth of resale $(A/F) \times$ (3)	(5) Annual O & M	(6) Annual O & M gradient $(A/G) \times$ 1200	(7) Total net annual worth	
1	−24,000	−26,400	19,200	19,200	−3,000	0	−10,200	
2		−13,829	15,360	7,314	−3,000	−571	−10,086	
3		−9,650	12,288	3,712	−3,000	−1,124	−10,063	Low NAW
4		−7,572	9,830	2,118	−3,000	−1,657	−10,111	
5		−6,331	7,864	1,288	−3,000	−2,172	−10,215	
6		−5,510	6,291	815	−3,000	−2,669	−10,364	

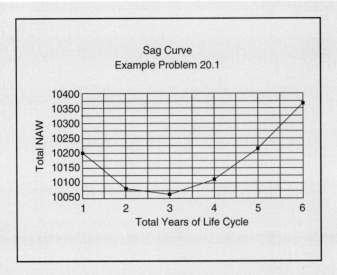

Sag Curve
Example Problem 20.1

The net annual worths tabulated in the right-hand column show that the net annual worth is minimum for a *three*-year life cycle, and therefore the theoretical *optimum economic* life is three years. (Comparing with the other types of lives, the tax-*depreciation* life is determined by the current tax regulations, whereas both the *physical* life and the *service* lives will be terminated at the discretion of the owner.)

In Example 20.1 the annual worth of *owning* the auto (annual cost of purchase less the annual worth of resale) *decreases* as the life increases. However, the annual worth of *operating and maintaining* the auto *increases* each year. Summing these factors leads to an optimum economic life in which the annual cost is minimum.

This finding of a three-year life in Example 20.1 requires some qualification. It is valid only if at the end of every three years another auto could be purchased for the same price ($24,000), and if all other costs of the replacement auto were the same. Under these circumstances the lowest cost alternative is to replace with a new auto every three years. A common rationale supporting this type of solution is to assume that the costs of purchasing and operating will rise at about the same inflationary rate as incomes and thus are counterbalancing. It also must be assumed that all costs and incomes approximate a series of payments, rather than some costs occurring as lump sums at wider intervals.

Time Intervals

Most of the examples used in this text assume end-of-year payments. Other intervals of time may be employed if greater detail is desired. With appropriate adjustments of *i* values, intervals such as quarters, months, and so on may be used.

The Better Challenger

Perhaps the most important task when determining the realistic actual economic life of an existing defender is to separately analyze and compare the costs of the replacement challenger. Even though the defender already may be older than the sag-point age in the cost curve and may incur increasingly higher costs with each successive year, it is possible that the only available challengers may be so high-priced that it is still cheaper to retain the defender for one or more years of additional service. Therefore, the best time to retire the defender may *not* be at the theoretical optimum economic life for the defender (sag point on the defender's NAW cost curve), but rather at the optimum economic life of the *combination* of defender plus a series of replacement challengers (sag point of a matrix of combined NAW values for defender's lives of length *N*, and challenger's lives of length *L*).

Since the costs of purchase, as well as O & M costs, for new replacement autos usually rise over time, then the new auto should not be purchased until the projected annual cost of the existing auto (the lowest NAW for one year or more of defender life) exceeds the lowest NAW of any of the new replacement challengers.

Intangibles

In the case of items such as an auto, certain intangible factors may enter the analysis. As the defender ages, a *prestige* or *status* cost may need consideration if relevant to the actual problem situation. For example, some firms might find that it is detrimental to the company image if its executives or salespersons drive "older" cars. Depending on just how detrimental, a cost penalty could be added for age. This can be included in the solution in the form of one additional equation with, for instance, a cost gradient of $1,000 or so per year of age as a prestige cost.

DEFENDER'S CURRENT VALUE

With the passing of time, many factors change, including current costs, technologies, and markets. As a result, the economic life of most investments should be reevaluated periodically. When reevaluations are undertaken, two questions that frequently arise are, "Has the defender's value changed?" and "what is the defender's real value?" Should the defender be valued at the original cost new, the replacement value now, the depreciated book value, or the current net market value? *The only relevant value is the current net market value*. The current net market value is the amount of cash (or cash equivalent if the owner receives an IOU or trade-in credit) received by the owner after all expenses of selling are paid, including advertisements, commissions, and income taxes. The original cost, depreciated book value, or replacement cost for comparable equipment have no direct bearing on the real value to be used in an analysis of the economic life of the defender.

EXAMPLE 20.2

A city public works department (no taxes) has a concrete mixer that was purchased 10 years ago for $25,000. Lately the mixer is requiring a great deal of maintenance to keep it running. Due to the improved technology of available replacements, the market value of the presently owned mixer has dropped to zero. Anticipated annual maintenance costs are estimated at $5,000 for the coming year (payable at EOY), and will increase by $2,000 per year for each subsequent year. Find the optimum economic life of the mixer for $i = 9$ percent.

Solution. Since this machine has a market value of zero, no cash is foregone if the machine is retained in service, nor is there a future income from resale value in the future. The only costs involved are the maintenance costs shown in Figure 20.2. The net annual worths of these costs are shown in the following table.

Remaining years defender is kept	Calculations for defender's NAW
1	$5,000/yr for 1 yr, if kept until EOY 1
2	$5,000 + 2,000(A/G, 9\%, 2) = $5,960/yr until EOY 2
3	$5,000 + 2,000(A/G, 9\%, 3) = $6,890/yr until EOY 3

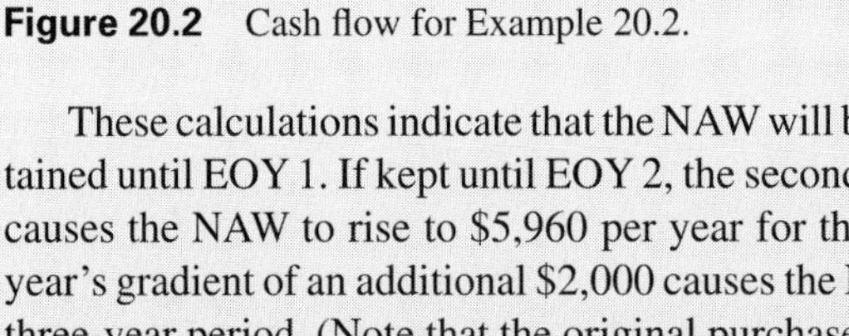

Figure 20.2   Cash flow for Example 20.2.

These calculations indicate that the NAW will be $5,000 if the machine is retained until EOY 1. If kept until EOY 2, the second year's cost is $7,000, which causes the NAW to rise to $5,960 per year for the two-year period. The third year's gradient of an additional $2,000 causes the NAW to rise to $6,890 for the three-year period. (Note that the original purchase price of the machine has no effect upon the problem at this point. As soon as the machine was purchased, the actual purchase price became a sunk cost, and the only relevant value is the current net resale value.) Clearly the NAW of the costs for this machine is rising with each year that it is retained in service. Therefore, if this machine were competing only against other identical replacements, then the economic life would be one year, and the optimum trade-in time would be EOY 1. (That is, if there were several of these used machines, worth zero market value, available in identical condition stored nearby, then the existing machine should be replaced at EOY 1, after giving one year of service for $5,000. Then another identical used machine is put into service for one more year at $5,000. This continues until all the available identical machines are used up. Then the second year in the life of the machine would have to be employed at a cost of $7,000, or the department would have to find a lower cost alternative challenger.)

Usually there are no identical used machines available ready to go on line, and a comparison must be made among whatever challengers are available.

EXAMPLE 20.3

Assume that the defender in Example 20.2 is carried over into Example 20.3 and a challenger is added as follows:

Challenger cost data

cost new (if replacement is postponed a few years into the
     future, there still is no escalation in cost new) = $10,000
                            O & M costs = $4,000/yr
no gradients in O & M costs optimum cycle time = 5 yr
                    resale value at EOY 5 = $3,500

The NAW for the challenger is calculated as

$$\text{cost new, } A_1 = \$10,000(A/P, 9\%, 5) = -\$2,571/\text{yr}$$
$$\text{O \& M, } A_2 = 4,000/\text{yr} = -\ 4,000$$
$$\text{resale at fifth year, } A_3 = 3,500(A/F, 9\%, 5) = +\ \ \ \ 585$$
$$\text{NAW} = -\$5,986/\text{yr}$$

Conveniently there are no gradients for this challenger, and it can be purchased at any time for this price (the present time, or at EOY 1, or any other future time). Once purchased, it will be kept for the predetermined optimum time period of five years, and consequently the equivalent annual cost remains constant. Finding the optimum replacement time is simplified, because it is evident that if the defender is kept for one more year, the annual cost for that one year will be $5,000, whereas if the challenger is accepted (and kept for five years), the NAW will rise to $5,986 per year. However, if at the EOY 1 the defender is kept for a second year, the cost goes to $7,000 for that year, whereas the challenger is still available at an equivalent $5,986 per year. Therefore, the optimum replacement time for the defender is at EOY 1, and the optimum economic life for the defender is one more year, as shown in Figure 20.3.

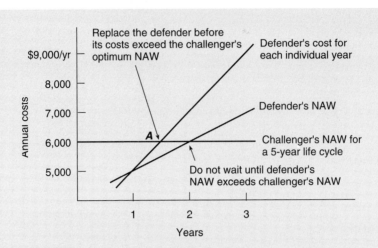

Figure 20.3   Defender versus challenger for Example 20.3.

Several elements of Example 20.3 are worth further examination. Comparing only the NAWs, the defender's NAW is lower than the challenger's NAW until EOY 2, giving the erroneous impression that EOY 2 is the proper trade time. To avoid this error, the cost of each successive individual year of the defender's life must be compared separately against the challenger's optimum NAW. In this example the defender's first year cost of $5,000 wins out over the challenger's NAW of $5,986. The defender's second year cost of $7,000 loses to the challenger's NAW of the same $5,986, and the trade-in is made at EOY 1.

EFFECT OF RISING DEFENDER COST CURVES

In cases where the defender's cost curve is rising steeply, an additional check should be made to verify or correct the defender's N-value, as illustrated in Example 20.3. Since the defender's NAW will be reduced (compared to the actual dollar cost of the defender for the Nth year) by a steeply rising cost curve, the actual cost of the defender for the Nth year of service should be found and compared to the best available challenger NAW. The results from a rising defender cost curve tend to favor smaller (shorter) N-value, as in Example 20.3.

OPTIMUM LIFE BY MINIMUM LIFE-CYCLE COSTS

One of the more common objectives of equipment-cost studies is to find a good estimate of the optimum life-cycle time. Should the presently owned equipment (e.g., an automobile, pump, computer, dragline, etc.) be replaced now or held for another year or more?

The second and equally important part of the problem is to find the estimated life span and the optimum life-cycle time for the challenger once it is acquired? Other facets of the same problem are, What is the penalty paid for each year short of, or beyond, the optimum economic life? Other practical considerations include the options of purchasing used equipment as replacement challengers, as well as the various alternative methods of financing, leasing, or other options.

Comparing Life-Cycle Times by Infinite Series

The use of infinite series is an important method for determining optimum life-cycle times. The optimum defender replacement time, and subsequent challenger life-cycle time, could be found simply by finding the NPW of each reasonable combination of defender remaining life plus challenger life cycle and then selecting the lower NPW of costs (or the higher NPW of net incomes). However, one qualification for using the NPW approach is that only investments with equal lives can be compared by NPW. Therefore, the first major problem stems from the different combined total of years that results from comparing one combination of defender life plus challenger life cycle to another combination. For instance, a defender life of one year plus a challenger life of four years yields a total combination life of five years. Comparing this combination to another combination of three-year defender life plus a five-year challenger life requires an NPW comparison between a five-year total and an eight-year total. These two could be compared over a total of 40 years, but then the next combination would require a different total. This complexity can be overcome by assuming an *infinite* series of challenger replacements in order to ensure an equal total of combined lives for all combinations of defender and challenger lives. For example, the NPW is found of a one-year life for the defender plus an infinite series of four-year challenger replacement life cycles. This can be compared to any other combination, say a three-year defender life plus an infinite series of five-year challenger replacement life cycles. Each combination has the same infinite life span, and thus satisfies the requirement of equal lives in order to be eligible for comparison by NPW. For an added advantage, the equations for an infinite series are characteristically much simpler in form and require less effort to solve.

The mechanics of setting up the equations involve the following steps.

Step 1.　Find the NPW of all of the defender's payments (all costs and incomes). Use a series of trial N values, where N represents the remaining life of the defender. For example let N vary from zero to four years.

Step 2.　Translate all challenger payments (all costs and incomes) into an equivalent lump sum located at the *end* of the life of each replacement challenger. Use a series of trial L values, where L is the life cycle of the replacement challenger. For example, let L vary from one to five years.

Step 3.　This series of equally spaced lump sums (representing challenger payments) now forms an infinite series of periodic payments spaced L years apart. These periodic payments may remain constant, or may increase (or decrease) arithmetically or geometrically. The NPW of each infinite series is first found at the ETZ located at the *beginning* of the first replacement challenger's life. This lump sum is then transferred back in time over N years (the trial length of the defender's life) to PTZ.

Step 4.   Using trial-value combinations of N and L, add the NPW of the defender to the NPW of the challenger for each combination. Using these sums, create a matrix of the sums of the paired NPW values. The defender's life, N, goes from zero (replace defender now), to some reasonable value, say four years, descending down each column of the matrix. The challenger's life, L, proceeds from one year to some reasonable value, say five years, along each row of the matrix. Thus the value in the matrix cell at the top row, left column represents the NPW for a zero life for the defender (this cost is zero) added to the NPW of an infinite series of one-year replacement cycles for the challenger.

Step 5.   From the matrix, select the least NPW of net cost or the greatest NPW of net income. Then look to the row and column headings to find the N and L values that correspond to that optimum NPW value. The N value is the optimum defender trade-in time, and the corresponding L value is the optimum challenger life-cycle time.

The next example (Example 20.4, similar to those in Problem Group B at the end of the chapter) illustrates a case in which there are no taxes, and the challenger's series payments within each cycle increase with an arithmetic gradient, but the equivalent periodic payments at the end of each life cycle are constant. (That is, there *is* an increase in O & M costs with age within a life cycle, but there *is no* increase in purchase price, or other payments from life cycle to life cycle). Also, productivity remains constant, and is not considered in the first example. In the second example (Example 20.5, similar to those in Problem Group C at the end of the chapter) some payments (incomes and costs) increase geometrically and some arithmetically from life cycle to life cycle, and there *are* taxes.

EXAMPLE 20.4   **(Find the optimum retirement time for the defender and the optimum life-cycle time for the replacement challenger by finding the NPW of all reasonable combinations of defender life plus an infinite series of challenger lives)**

A state department of transportation (DOT) maintenance division (no taxes) owns an excavator (defender) which is a candidate for replacement. Using the data below find the following:

a. When should the DOT owner replace the defender now?

b. If not now, find the optimum remaining life for the defender (best time to replace the defender).

c. Find the optimum life-cycle time for the replacement challenger.

$i = 15$ percent.

	Defender	Challenger
Current market value	$80,000	
Cost new		$130,000
Decline in value of excavator/yr	− 20%/yr	− 20%/yr
O & M costs	$36,000/yr	$ 30,000/yr
Gradient O & M	$8,000/yr/yr	$ 6,500/yr/yr

Solution. As outlined above, the defender's NPW is calculated for a sequence of trial life spans, N (for this example trial values of N range from 0 to 4). Then the NPW of the challenger is calculated for an infinite series of trial life spans, L (trial values of L range from 1 to 5). The challenger's payments are all translated to equivalent lump sum amounts at the end of each challenger life cycle of trial duration L, and then the lump sum amounts are treated as an infinite periodic series. The resulting matrix solution is shown below. The complete calculations and equations are shown in the spreadsheet example presented on the following pages. $P\Sigma D$ signifies the sum (Σ) of Present worths (P) of all Defender's (D) incomes and costs. $P\Sigma D$ is the sum of all of the Challenger's present worths.

NPW of combined	Challenger, L =		1	2	3	4	5
	Defender, N						
Defender + Challenger life							
$P\Sigma D$ (N = 0) + $P\Sigma C$,							
where L ranges from 1 to 5 =		0	−503,333	−495,271	−491,104	−489,943	−491,075
$P\Sigma D$ (N = 1) + $P\Sigma C(P/F, 15\%,1)$,							
where L ranges from 1 to 5		1	−493,333	−486,323	−482,699	−481,689	−482,674
$P\Sigma D$ (N = 2) + $P\Sigma C(P/F, 15\%,2)$,							
where L ranges from 1 to 5		2	−486,452	−480,356	−477,205	−476,327	−477,184
$P\Sigma D$ (N = 3) + $P\Sigma C(P/F, 15\%,3)$,							
where L ranges from 1 to 5		3	−482,783	−477,483	−474,742	−473,979	−474,724
$P\Sigma D$ (N = 4) + $P\Sigma C(P/F, 15\%,4)$,							
where L ranges from 1 to 5		4	−482,118	−477,508	−475,126	−474,462	−475,109

Inspection of the following matrix, on page 567, reveals that the least negative NPW of −$473,979 occurs at

$N = 3$ yr, defender's estimated remaining life,
$L = 4$ yr, challenger's estimated future life cycle

In response to the DOT, owner's basic question, "What do you advise us to do now? Should we keep the defender in service or replace it with the challenger?" these data indicate that the defender should be kept in service. The data also suggest that the defender should be the least costly machine to use for about three more years, and then should be replaced with a challenger whose life-cycle time is estimated at about four years. The penalty for delaying one year on the retirement of the defender (replacing at EOY 4 instead of EOY 3) is found by subtracting the NPW of −$473,979 (found above) from the lowest NPW in the $N = 4$ row. This is found in the matrix as −$474,462 at $N = 4$, $L = 4$, and subtraction produces the following results.

$$N = 4, \qquad\qquad L = 4, = \quad \$474,462$$
$$N = 3, \qquad\qquad L = 4, = - \quad 473,979$$
$$\text{cost of 1-yr delay} = \quad \$ \quad 483$$

In this case the cost of delay is negligible compared to the NPW totals. It can be concluded that replacement in either year 3 or year 4 is acceptable. Further examination of the matrix reveals that the cost curves have relatively shallow slopes for a year or two in each direction, allowing the DOT owner a window of opportunity of about three years within which to make the best possible arrangements for disposing of the defender and acquiring the replacement challenger. Thus the matrix provides valuable information on the cost (or lack thereof) of delays or advances in the schedule. Of course the prudent owner will update the analysis at frequent intervals. For expensive or sensitive equipment, the studies could be done as frequently as the owner desires. Once the computer program is set up, reruns can be produced quickly and easily at the convenience of the owner.

Increasing Costs for Future Challenger Replacements

In Example 20.4, when considering replacement costs for the infinite series of challenger replacements, the simplistic assumption was made that prices would not change, and that each subsequent replacement would have the same purchase price as the first one. More realistically, however, the prices of replacements do increase with time, due to technological advances (usually accompanied by increases in productivity) and to inflation. One simple method of accounting for this normal increase in costs of future replacements is to assume that all costs, including the purchase price, will increase by some estimated percentage value, r_1, each year. Then for replacements that occur L years apart, the periodic rate of increase can be estimated as a factor of $(1 + r_1)^L$. For instance, if the escalation rate is 7 percent per year and the replacement cycle is four years, then each subsequent purchase price will be higher than the previous one by a factor of $1.07^4 = 1.3108$. If the first replacement costs $10,000, then four years later the second one is expected to cost $10,000 × 1.3108 = $13,108. Since all future replacement costs are subject to the same escalation, the NPW of the series of increasing lump sums at the end of each challenger cycle can be found by use of the infinite series geometric gradient equation (see Chapter 6, "Geometric Gradient").

Other Challenger Costs and Incomes

Each category of challenger cost or income can be transferred to an equivalent lump sum payment at the end of the first challenger's trial life at EOY L. Then this lump sum is considered as the first payment in a periodic geometric infinite series. The following steps are involved:

1. Translate each payment or series of payments into an equivalent lump sum at the end of each life cycle.

2. Assume an infinite series of subsequent replacement challengers, each with the same trial life span, L, and each with all payments translated into one end-of-life lump sum. The PW of this infinite series of lump sum payments is found at ETZ (the ETZ is at the beginning of the first challenger's life) by using the infinite series equation. If payments are assumed to increase as a periodic geometric gradient over a period of L years, then the infinite series geometric gradient equation is used.

3. The NPW of the infinite series of replacement challengers is found at PTZ by translating the PW back over the trial life N of the defender.

Either arithmetic or geometric gradients may be used, or a combination of both. Double gradients may be used if needed (see Chapter 21, "Double Gradients").

Developing the Equations Representing the Defender-Challenger Payments

For a real-life study, the actual payment history of the defender (as well as the history of equipment expected to act like the challenger would) should be examined, and realistic estimates made of the anticipated future payments (all costs and incomes) for both the defender and the challenger. If these estimated cost and income payments are plotted on graph paper, some will approximate a straight line with a slope that is either positive, negative, or zero (level). These straight-line series of payments may be represented by a uniform annual series, plus an arithmetic gradient. Other payment series that approximate a curved line usually can be approximated by a geometric gradient. Still other payments may appear as one-time lump sum payments (such as a major overhaul), and can be added to the NPW by use of a P/F factor. After careful study and consideration of factors influencing future payments, a series of time-value equations is devised to represent each of the defender's payment components anticipated for the remaining years of defender life. Then the same type of estimates are made for each future life cycle of a series of replacement challenger machines, out to an infinite horizon. The sum of these two series represents the total NPW of payments for the remaining life of the defender plus an infinite series of replacement challengers. Thus all payments are accounted for, and the model is complete. These equations contain two unknown (dependent) variables:

$N =$ the remaining life of the defender
$L =$ the expected life cycle of each of the sequential series of replacement challengers

A sequence of trial values for these unknown life spans are substituted in the equation series, until the sum of the NPWs of the infinite series attains an identifiable minimum for costs, or maximum for net income.

Spreadsheet Solution for Example 20.4

#	A	B	C	D	E	F	G	H	I
1	Example 20.4: Defender/Challenger, no tax, no inflation								
2	Find the optimum retirement time for the Defender, and the optimum life-cycle time for the replacement Challenger.								
3	Note: All Defender costs are identified with a "D", and all Challenger costs are identified with a "C".								
4									
5	INPUT DATA, GENERAL								
6		$i =$	15%						
7									
8			Defender			Challenger			
9	Defender's Present Resale Value	P1D =	$80,000						
10	Challenger's Cost New				F1C =	$130,000			
11	Decline in Value per year of age	=	-20%			-20%			
12	O&M Costs, $/year	A1D =	$36,000		A1C =	$30,000			
13	Gradient O&M, $/yr/yr	G1D =	$8,000		G1C =	$6,500			
14									
15	SOLUTION: Step 1. Find the NPW ($P\Sigma D$) of the Defender for a remaining life, ranging								
16	from N = 0 (immediate replacement) to N = 4 years (keep for 4 more years).								
17									
18	DEFENDER'S EQUATIONS (Subscripted "D")								
19	Defender's Trial Life	N =			0	1	2	3	4
20	Present Resale Value, at PTZ =	P1D =			-80,000	-80,000	-80,000	-80,000	-80,000
21	Defender's Market Value at EOY N,	F1D =	P1D(F/P,-20%,N) =		80,000	64,000	51,200	40,960	32,768
22	PW of Future Resale at EOY N =	P2D =	F1D(P/F,15%,N) =		80,000	55,652	38,715	26,932	18,735
23	PW of Annual O&M Costs =	P3D =	A1D(P/A,15%,N) =		0	-31,304	-58,526	-82,196	-102,779
24	PW of Gradient O&M Costs =	P4D =	G1D(P/G,15%,N) =		0	0	-6,049	-16,569	-30,291
25									
26	Defender's NPW =	PΣD =	P1D+P2D+P3D+P4D =		0	-55,652	-105,860	-151,834	-194,335
27									
28	Step 2. Temporarily assume N = 0. Then find the NPW ($P\Sigma C$) of the replacement Challenger plus								
29	an infinite series of future replacement Challengers. Assume trial lives for the Challengers from L = 1 to L = 5.								
30									
31	CHALLENGER'S EQUATIONS, (subscripted "C") (Assume Defender's life N = 0).								
32	Trial Life	L =			1	2	3	4	5
33	Effective i,	ie =	(1+i)^L-1 =		0.15	0.3225	0.520875	0.749006	1.011357

	A	B	C	D	E	F	G	H	I
34									
35	Cost New+ ∞ Series of Replacmnts =	$P1C =$	$F1C+F1C(P/A,ie, ∞) =$		−996,667	−533,101	−379,580	−303,563	−258,540
36	Challenger's Resale Value, EOY $L =$	$F2C =$	$F1C(F/P, -20\%, L) =$		104,000	83,200	66,560	53,248	42,598
37	Resale at EOY $L+ ∞$ Series Of Resales	$P2C =$	$F2C(P/A,ie, ∞) =$		693,333	257,984	127,785	71,092	42,120
38	PW of Annual O&M Costs =	$P3C =$	$A1C(F/A,15\%,L)(P/A,ie,∞) =$		−200,000	−200,000	−200,000	−200,000	−200,000
39	PW of Gradient O&M =	$P4C =$	$G1C(F/G,15\%,L)(P/A,ie,∞) =$		0	−20,155	−39,309	−57,471	−74,655
40									
41	Challenger's NPW at EOY $N =$	$PΣC =$	$P1C+P2C+P3C+P4C =$		−503,333	−495,271	−491,104	−489,943	−491,075
42									
43	Step 3. Develop a matrix of values representing combinations of lives of Defender plus replacement Challengers. The top row								
44	represents immediate replacement by Challengers with life cycles ranging from 1 year to 5 years. The second row								
45	represents a Defender with 1 more year of life, then at EOYI it is subsequently replaced with Challengers with life cycles ranging from 1 to 5 years.								
46									
47									
48			Challenger, $L =$		1	2	3	4	5
49			Defender , $N = /$						
50	$PΣD (N = 0) + PΣC$ where L ranges from 1 to 5 =			0	−503,333	−495,271	−491,104	−489,943	−491,075
51	$PΣD (N = 1) + PΣC (P/F, 15\%,1)$, where L ranges from 1 to 5			1	−493,333	−486,323	−482,699	−481,689	−482,674
52	$PΣD (N = 2) + PΣC (P/F, 15\%,2)$, where L ranges from 1 to 5			2	−486,452	−480,356	−477,205	−476,327	−477,184
53	$PΣD (N = 3) + PΣC (P/F, 15\%,3)$, where L ranges from 1 to 5			3	−482,783	−477,483	−474,742	−473,979	−474,724
54	$PΣD (N = 4) + PΣC (P/F, 15\%,4)$, where L ranges from 1 to 5			4	−482,118	−477,508	−475,126	−474,462	−475,109
55									
56									
57	Step 4. The lowest cost for the combined lives is found at $N = 3$, and $L = 4$. Therefore the conclusion is that the								
58	Defender should be retained for about 3 more years, and the replacement Challenger at that time is expected to have								
59	a life-cycle time of about 4 years. In addition, the penalties for early or late termination can be estimated								
60	by subtracting the lowest cost found above from the appropriate surrounding cost figures. For instance, the								
61	cost penalty for terminating the Defender now rather than at EOY 3 is found as the difference between $489,943,								
62	(found at $N = 0$, $L = 4$) and the lowest cost figure found above. This totals $489,943 − $473,979 = $15,964 in terms of NPW.								

Spreadsheet Formulas for Example 20.4

Columns B–I, to the right of column A, continue on subsequent pages. Row numbers and stub heads apply across all columns B–I.

	A
	A
1	**Example 20.4:** Defender/Challenger, no tax, no inflation
2	Find the optimum retirement time for the Defender, and the optimum life-cycle time for the replacement Challenger.
3	Note: All Defender costs are identified with a "D", and all Challenger costs are identified with a "C".
4	
5	INPUT DATA, GENERAL
6	$i =$
7	
8	
9	Defender's Present Resale Value
10	Challenger's Cost New
11	Decline in Value per year of age
12	O&M Costs, $/year
13	Gradient O&M, $/yr/yr
14	
15	SOLUTION: Step 1. Find the NPW ($P\Sigma D$) of the Defender for a remaining life, ranging
16	from $N = 0$ (immediate replacement) to $N = 4$ years (keep for 4 more years).
17	
18	DEFENDER'S EQUATIONS (Subscripted "D")
19	Defender's Trial Life
20	Present Resale Value, at PTZ =
21	Defender's Market Value at EOY N,
22	PW of Future Resale at EOY $N =$
23	PW of Annual O&M Costs =
24	PW of Gradient O&M Costs =
25	
26	Defender's NPW =
27	
28	Step 2. Temporarily assume $N = 0$. Then find the NPW ($P\Sigma C$) of the replacement Challenger plus
29	an infinite series of future replacement Challengers. Assume trial lives for the Challengers from $L = 1$ to $L = 5$.
30	
31	CHALLENGER'S EQUATIONS, (subscripted "C") (Assume Defender's life $N = 0$).
32	Trial Life
33	Effective i,
34	
35	Cost New+ ∞ Series of Replacmnts =
36	Challenger's Resale Value, EOY $L =$
37	Resale at EOY $L + \infty$ SeriesOfResales =
38	PW of Annual O&M Costs =
39	PW of Gradient O&M =
40	
41	Challenger's NPW at EOY $N =$
42	
43	Step 3. Develop a matrix of values representing combinations of lives of Defender plus replacement
44	Challengers. The top row represents immediate replacement by Challengers with life cycles ranging from
45	1 year to 5 years. The second row represents a Defender with 1 more year of life, subsequently replaced
46	with Challengers with life cycles ranging from 1 to 5 years.
47	
48	
49	NPW of combined Defender + Challenger lives.
50	$P\Sigma D$ ($N = 0$) + $P\Sigma C$ where L ranges from 1 to 5 =
51	$P\Sigma D$ ($N = 1$) + $P\Sigma C$ (P/F, 15%,1), where L ranges from 1 to 5
52	$P\Sigma D$ ($N = 2$) + $P\Sigma C$ (P/F, 15%,2), where L ranges from 1 to 5
53	$P\Sigma D$ ($N = 3$) + $P\Sigma C$ (P/F, 15%,3), where L ranges from 1 to 5
54	$P\Sigma D$ ($N = 4$) + $P\Sigma C$ (P/F, 15%,4), where L ranges from 1 to 5
55	
56	
57	Step 4. The lowest cost for the combined lives is found at $N = 3$, and $L = 4$. Therefore the conclusion is that the
58	Defender should be retained for about 3 more years, and the replacement Challenger at that time is expected to have
59	a life cycle time of about 4 years. In addition, the penalties for early or late termination can be estimated
60	by subtracting the lowest cost found above from the appropriate surrounding cost figures. For instance the
61	cost penalty for terminating the Defender now rather than at EOY 3 is found as the difference in $489,943,
62	(found at $N = 0$, $L = 4$) and the lowest cost figure found above. This totals $489,943 – $473,979 = $15,964 in terms of NPW.

B	C	D	E
0.15			
	Defender		
P1D =	80000		
			F1C =
=	−0.2		
A1D =	36000		A1C =
G1D =	8000		G1C =
N =			0
P1D =			= −$C$9
F1D =	P1D(F/P, −20%,N) =		= C9
P2D =	F1D(P/F,15%,N) =		= PV($B$6,E19,,−E21)
P3D =	A1D(P/A, 15%,N) =		= PV($B$6,E19,$C$12)
P4D =	G1D(P/G,15%,N) =		= $C$13*(PV($B$6,E19,1)−E19/(1+$B$6)^E19)/$B$6
PΣD =	P1D + P2D+P3D+P4D =		=SUM(E20,E22:E24)
L=			1
ie =	(1+i)^L−1 =		=(1+$B$6)^E32−1
P1C =	F1C+F1C(P/A,ie,∞) =		=FV($B$6,E32,,$F$10)/E33
F2C =	F1C(F/P,−20%,L) =		=$F$10*(1+$F$11)
P2C =	F2C(P/A,ie,∞) =		=E36/E33
P3C =	A1C(F/A,15%,L)(P/A,ie, ∞) =		=FV($B$6,E32,$F$12)/E33
P4C =	G1C(F/G,15%,L)(P/A,ie, ∞) =		=−$F$13*((FV($B$6,E32,−1)−E32)/$B$6)/E33
PΣC =	P1C + P2C + P3C+P4C =		=SUM(E35,E37:E39)
	Challenger, L =		= 1
	Defender, N = /		
		= 0	=E41
		1	=F$26+PV($B$6,$D51,,−$E$41)
		2	=$G$26+PV($B$6,$D52,,−$E$41)
		3	=$H$26+PV($B$6,$D$53,,−E41)
		4	=$1$26+PV($B$6,$D$54,,−E41)

Spreadsheet Formulas for Example 20.4 (continued)

	F	G
1		
2		
3		
4		
5		
6		
7		
8	Challenger	
9		
10	130000	
11	−0.2	
12	30000	
13	6500	
14		
15		
16		
17		
18		
19	1	2
20	=−$C$9	=−$C$9
21	=E21*(1+$C$11)	=F21*(1+$C$11)
22	=PV($B$6,F19,,−F21)	=PV($B$6,G19,,−G21)
23	=PV($B$6,F19,$C$12)	=PV($B$6,G19,$C$12)
24	= −$C$13*(−PV($B$6,F19,1)−F19/(1+$B$6)^F19)/$B$6	=−$C$13*(−PV($B$6,G19,1)−G19/(1+$B$6)^G19)/$B$6
25		
26	=SUM(F20,F22:F24)	=SUM(G20,G22:G24)
27		
28		
29		
30		
31		
32	2	3
33	=(1+$B$6)^F32−1	=(1+$B$6)^G32−1
34		
35	=FV($B$6,F32,,$F$10)/F33	=FV($B$6,G32,,$F$10)/G33
36	=E36*(1+$F$11)	=F36*(1+$F$11)
37	=F36/F33	=G36/G33
38	=FV($B$6,F32,$F$12)/F33	=FV($B$6,G32,$F$12)/G33
39	=−$F$13*((FV($B$6,F32,−1)−F32)/$B$6)/F33	=−$F$13*((FV($B$6,G32,−1)−G32)/$B$6)/G33
40		
41	=SUM(F35,F37:F39)	=SUM(G35,G37:G39)
42		
43		
44		
45		
46		
47		
48	2	3
49		
50	=F41	=G41
51	=F$26+PV($B$6,$D$51,,−$F$41)	=$F$26+PV($B$6,$D$51,,−G41)
52	=$G$26+PV($B$6,$D$52,,−F$41)	=$G$26+PV($B$6,$D$52,,−G$41)
53	=H$26+PV($B$6,$D$53,,−F$41)	=H$26+PV($B$6,$D$53,,−G$41)
54	=I$26+PV($B$6,$D$54,,−F$41)	=I$26+PV($B$6,$D$54,,−G$41)
55		
56		
57		
58		
59		
60		
61		
62		

H	I
3	4
=−$C$9	=−$C$9
=G21*(1+$C$11)	=H21*(1+$C$11)
=PV($B$6,H19,,−H21)	=PV($B$6,I19,,−I21)
=PV($B$6,H19,$C$12)	=PV($B$6,I19,$C$12)
= −$C$13*(−PV($B$6,H19,1)−H19/(1+$B$6)^H19)/$B$6	=−$C$13*(−PV($B$6,I19,1)−I19/(1+$B$6)^I19)/$B$6
=SUM(H20,H22:H24)	=SUM(I20,I22:I24)
4	5
=(1+$B$6)^H32−1	=(1+$B$6)^I32−1
=FV($B$6,H32,,$F$10)/H33	=FV($B$6,I32,,$F$10)/I33
=G36*(1+$F$11)	=H36*(1+$F$11)
=H36/H33	=I36/I33
=FV($B$6,H32,$F$12)/H33	=FV($B$6,I32,$F$12)/I33
=−$F$13*((FV($B$6,H32,−1)−H32)/$B$6)/H33	=−$F$13*((FV($B$6,I32,−1)−I32)/$B$6)/I33
=SUM(H35,H37:H39)	=SUM(I35,I37:I39)
4	5
=H41	=I41
=$F$26+PV($B$6,$D$51,,−H41)	=$F$26+PV($B$6,$D$51,,−I41)
=$G$26+PV($B$6,$D$52,,−H$41)	=$G$26+PV($B$6,$D$52,,−I$41)
=H$26+PV($B$6,$D$53,,−H$41)	=H$26+PV($B$6,$D$53,,−I$41)
=I$26+PV($B$6,$D$54,,−H$41)	=I$26+PV($B$6,$D$54,,−I$41)

EXAMPLE 20.5

A contractor owns a three-year-old excavator and is considering the purchase of a replacement. Given the data shown (refer to Chapter 15, "Taxes" and Chapter 16, "Depreciation"),

a. Find the optimum trade-in time for the presently owned defender excavator.

b. Find the optimum life-cycle time for the replacement challenger.

Tax rules

1. Depreciation is by the straight-line (SL) method, and the depreciated book value at the EOY is the straight-line book value (SLBV).

2. All annual income is taxable, and all annual expenses, and depreciation are deductible from taxable income. All payments are assumed paid at EOY.

3. Upon resale, any gain above the SLBV is subject to recapture tax at the ordinary income tax rate, t.

	Item name used in computer program	
General		
1. Owner's investment rate, i	i	15%
2. Inflation rate,	r_1	7%
3. Owner's income tax bracket, t	t	36%
Defender		
1. Original purchase price, cost new	P_1D	$180,000
2. Present resale value at PTZ	P_2D	$ 90,000
3. Rate of decline of resale value	r_2D	— 20%/yr/yr
4. Annual repair cost, paid at EOY	A_1D	$ 10,000/yr
5. Repair cost (arithmetic gradient)	G_1D	$ 2,500/yr
6. O & M costs at EOY 1 (geometric gradient)	C_1D	$100,000/yr
7. O & M cost growth rate	r_3D	12%/yr
8. Depreciation life		5 yr
9. Present age		3 yr
10. Income	A_2D	$155,000
11. Income gradient (arithmetic)	G_2D	—$ 1,000

Defender equation series. For the sake of simplicity, only one representative equation of each type is used in this example.

Defender's depreciation. Depreciation is an accounting over time for the loss in value of property. Depreciation is not a cash flow item, but does result in a cash flow due to the reduction of the income tax payment. So the annual after-tax cash flow resulting from depreciation (the tax payments saved) is calculated simply as

$$A_3D = \$/\text{yr depreciation} \times \text{tax bracket}$$
$$= (P_1D/\text{depreciation life of defender}) \times t$$

Since the defender is already three years old and the depreciation life allowed for this machine is five years, the PW of the depreciation for up to two years remaining is calculated as follows:

$$P_3D = A_3D \times (P/A, i, N_2)$$

where if N + age of defender $\leq$ depreciation life of defender, then $N_2 = N$, but if N + age of defender > depreciation life of defender, then N_2 = depreciation life of defender − age of defender (e.g., if the defender's trial remaining life is $N = 1$ yr, then N + age of defender = $1 + 3 = 4$, and since $4 < 5$, then $N_2 = N = 1$. There is yet one more year of allowable depreciation left after this year.)

Defender's current market value (lump sum). All previous costs and income associated with the defender before now (before the PTZ) have become sunk costs. The first cost of keeping the defender is only the current net cash value after taxes. This net cash value is the amount of cash foregone if the owner keeps the defender (or the net cash the owner could have if the defender were sold now). The owner would receive the current market resale value of the defender, less the selling expenses (advertising, commissions, etc.) and the capital gains tax (if any). The capital gains tax is paid on any gain above the present depreciated book value. Therefore the after-tax NPW of the current market value of the defender is calculated as follows:

$$P_4D = P_2D - (P_2D - \text{SLBV}) \times t$$

where P_4D = net after-tax present worth of the defender's current market value

P_2D = defender's present resale value (market value after all expenses of selling have been paid, except capital gains tax)

SLBV = present depreciated book value of the defender at the present time, PTZ. This machine is depreciated by SL depreciation; therefore,

SLBV = $P_1D \times$ (depreciation life of defender − age of defender)/depreciation life of defender, where SLBV is never less than zero

If the defender is depreciated by the double declining balance method permitted under the accelerated capital recovery system (ACRS), then the

ACRSBV = $P_1D \times (1 - 2/\text{depreciation life of defender})$ (age of defender)

in which ACRSBV is the depreciated book value using the double declining balance method. (*Note:* In the rare case where P_2D < depreciated book value, the P_4D equation is still valid, since the owner typically is entitled to a tax deduction for the additional loss of value beyond the depreciated book value.)

Defender's future after-tax resale value. At the end of the defender's remaining life (at EOY N), it will be sold at current market value (resale value) for that year, and an income to the owner will result. For the example problem, the rate of decline in market value, r_2D, is estimated at $r_2D = -20$ percent per year. That is, the resale price next year is expected to be (100% − 20% =) 80% of this year's resale price, and so on for each succeeding year. Beginning from the present resale price, the sales price in each future year will be

$$\text{future resale value} = \text{present resale value} \times (1 - r_2D)^N$$
$$F_1D = P_2D \times (1 - r_2D)^N$$

If the defender's resale price is higher than the depreciated book value, a capital gains tax will be levied at ordinary tax rates on any gain over the depreciated book value. For straight-line depreciation the future depreciated book value over a five-year period is found in the usual manner:

$$\text{future depreciated book value} = \text{cost new} \times (\text{depreciation life} \\ - \text{defender's age})/\text{depreciation life}$$
$$\text{SLBV} = P_1D \times (\text{depreciation life of defender} \\ - \text{age of defender})/\text{depreciation life of} \\ \text{defender, if age of defender} \\ \leq \text{depreciation life of defender}$$

Of course the depreciated book value cannot be negative, so this equation is valid only for values where age of defender ≤ depreciation life of defender:

if age of defender > depreciation life of defender, then $\text{SLBV} = 0$

To find the net return to the owner after taxes,

1. Calculate the amount of capital gains tax due as capital gains tax $= (F_1D - \text{SLBV}) \times t$

2. Then subtract the capital gains tax from the resale price:

$$\text{future after-tax net resale value} = \text{future resale value} - (\text{future} \\ \text{resale value} - \text{future} \\ \text{depreciated book value}) \times \text{tax} \\ \text{rate}$$

or

$$\text{future net after taxes, } F_2D = F_1D - (F_1D - \text{SLBV}) \times t$$

Normally the future resale value is higher than the future depreciated book value, since the annual allowable depreciation for tax purposes is usually more than the annual decline in actual market value.

3. Next, the present worth is found as the future net after taxes times the present worth factor, or

$$P_5D = F_2D \times (P/F, i, N)$$

in which P_5D = present worth of defender's net resale value after taxes.

If the ACRS depreciation is employed, a different equation is required to provide the correct depreciated book value for the defender in each year of trial life.

Defender's repair costs. This group of costs is represented by an annual uniform series plus an arithmetic gradient, and can be calculated as follows:

$$P_6D = A_1D \times (P/A, i, N)$$
$$P_7D = G_1D \times (P/G, i, N)$$

in which

P_6D = the present worth of the defender's annual repair costs
P_7D = the present worth of the defender's arithmetic gradient in repair costs

Defender's operating and maintenance (O & M) costs. This group of costs is represented by a geometric gradient with a rate of increase of $r_3D = 12$ percent per year. The present worth is found as

$$P_8D = C_1D \times (P/A, w, N)/(1 + r_3D) \qquad \text{geometric gradient equation}$$

Since

$$r_3D < i, \quad \text{then} \quad w = [(1 + i)/(1 + r_3D)] - 1$$

in which

P_8D = the present worth of the defender's operating and maintenance costs
C_1D = the first year's operating and maintenance cost for the defender, chargeable at EOY 1 from now
w = a derived quantity used in the geometric gradient equation

Defender's income. Income may occur with gradients (either arithmetic or geometric), in either single or double variety. For instance, income could be a function both of production and unit price; for example, production at EOY

$1 = 100,000$ yd^3/yr at \$1.00/yd^3. A double gradient could then be employed to represent an anticipated decrease in future production as the machine ages, with a concurrent increase in market price. For instance, production could *decrease* by 1,000 yd^3/yr as the machine ages, but the price might *increase* by \$0.06/yd^3 each year. These double gradients may be either arithmetic or geometric. See Chapter 21 for further information on double gradients. For simplicity, in this example assume that income occurs by means of an annual payment series, together with a single arithmetic gradient. Then the income is represented by

$$\text{income, } P_9 D = A_2 D \times (P/A, i, N)$$
$$\text{gradient income, } P_{10} D = G_2 D \times (P/G, i, N)$$

where $A_2 D$ is the defender's annual income, and $G_2 D$ is the defender's gradient income.

Challenger data.   The data for the challenger follows:

Challenger

1. Cost new (if purchased at PTZ)	$P_1 C = \$190,000$
2. Rate of decline of resale value	$r_2 C = -20\%/\text{yr}$
3. Repair costs at EOY 1	$A_1 C = \$8,000/\text{yr}$
4. Repair costs (arithmetic gradient)	$G_1 C = \$2,000/\text{yr/yr}$
5. O & M at EOY 1	$C_1 C = \$79,000$
6. O & M geometric gradient growth rate	$r_3 C = 12\%/\text{yr}$
7. Depreciation life	$= 5$ yr
8. Income at EOY 1	$A_2 C = \$161,000/\text{yr}$
9. Gradient income (arithmetic)	$G_2 C = \$3,500/\text{yr/yr}$

Challenger equation series.   This step is divided into four parts:

1. Develop equations for the first challenger replacement, assuming temporarily that $N = 0$ and the challenger goes into service immediately. Then at the end of the first challenger's life (at EOY L), it is replaced with an infinite series of subsequent challengers. Let the challenger's life-cycle time, L, vary from 1 to some reasonable number of years, say 6.

2. Total all the NPWs for the infinite series of challengers with the PTZ still located at the time the first challenger is put into service. This is the NPW for a combination of a defender with a zero life together with an infinite series of challengers with lives varying from 1 to 6 years. When the matrix is arrayed, the top line represents the defender being replaced immediately (defender life $N = 0$), and an infinite series of replacement challengers with six trial life cycles, $L = 1$ through 6.

3. Now assume for a moment that the defender has 1 additional year of life, $N = 1$. If the defender has a life of 1 more year, the starting point for the challenger's costs do not begin until EOY 1. This means that all costs will increase by the amount of price escalation (12 percent in this example) due to the 1-year delay, and then will be discounted back to present time at the investment rate (15 percent). Therefore for a defender's life of 1 year ($N = 1$), the challenger's values found in (2) above are multiplied by $(F/P, r_1, N)$ to represent the effect of 1 year's price escalation, and then reduced by a multiplier of $(P/F, i, N)$ to represent the interest discount (due to the 1-year postponement before the payments are due). Then add the NPW of the defender's 1 year of remaining life to the NPW for the challenger. Five additional values are generated in this step for the combination of defender $N = 1$, and challenger $L = 1$ through 6 years.

4. In sequence try additional increments of trial life, N, for the defender, using $N = 2, 3$, up to some reasonable value, say 5 years. For each additional year of defender's life, six more values of defender-challenger combination lives are generated. The result is a matrix of values, showing the sum of defender's lives from zero to N, added to challenger's lives from 1 to L. These steps are illustrated below.

The equations for the first challenger replacement are developed in a manner similar to those for the defender, except that the payments made during the challenger's life are all transferred to an equivalent lump sum at the *end* of the challenger's life (at EOY L). For this example, assume that subsequent challenger replacements will act similarly to the first challenger replacement but with values for the variables modified to suit advances in technology and other predicted trends in market prices and other factors. Since the payments increase at the end of each life cycle by an equivalent escalation factor, r_e, the series of end-of-life lump sum equivalents constitute a series of payments that increase as a geometric gradient with the purchase of each successive challenger replacement. Since the time span, L, between these lump sum expenditures usually is longer than the one-year compounding period for the interest rate, the effective interest rate, i_e, must be calculated and used over each trial life. These effective rates are found as follows:

$$\text{effective interest rate, } i_e = (1 + i)^L - 1$$
$$\text{effective price escalation rate, } r_e = (1 + r_1)^L - 1$$

where L is the trial life span of the challenger. For the geometric gradient with $i_e > r_e$, the derived value, w_e, is

$$w_e = \frac{1 + i_e}{1 + r_e} - 1$$

The challenger equations are listed as follows:

$$\text{purchase cost, } P_1C = P_1C(F/P, i, L)(P/C, w_e, \infty)$$

where P_1C is the challenger's cost new, a lump sum expenditure at the beginning of the first challenger life cycle.

$(F/P, i, L)$ translates this lump sum expenditure (first in a series) to the end of the first challenger life cycle, so that the ETZ for the geometric gradient will be at the beginning of the first challenger's life cycle.

$(P/C, w_e, \infty)$ finds the present worth of an infinite series of lump sum replacement expenditures, with each expenditure increasing by a factor of $(1 + r_e)$ times the previous challenger's cost new.

$$\text{depreciation, } P_3C = (P_1C \times t/\text{depreciation life of challenger})$$
$$\times (F/A, i, L)(P/C, w_e, \infty)$$

$$\text{resale minus recap tax, } P_4C = P_1C \times (1 - r_2C)^L - \{P_1C \times (1 - r_2C)^L$$
$$- (P_1C \times \text{age of challenger/depreciation}$$
$$\text{life of challenger})\} \times (P/C, w_e, \infty)$$

where the SLBV cannot be $<$ zero.

$$\text{Repairs, } P_5C = A_1C \times (F/A, i, L)(P/C, w_e, \infty)$$
$$P_6C = G_1C \times (F/A, i, L)(P/C, w_e, \infty)$$
$$\text{O \& M, } P_7C = C_1C \times (P/C, w, L)(F/P, i, L)(P/C, w_e, \infty)$$

Sum all the equations to obtain the six values representing the present worths of each trial span of the challenger's life-cycle time (for this example $L = 1$ through 6). These values temporarily assume that ETZ = PTZ, so they are actually valid only for defender $N = 0$ and challenger $L = 1$ through 6.

Those categories that are tax-deductible expenses or taxable incomes are multiplied by $(1 - t)$ to obtain their effective after-tax values. In our sample problem the equations are summed as follows.

For the defender:

$$P_{\text{total}}D = P_3D + P_4D + P_5D + (1 - t)(P_6D + P_7D + P_8D + P_9D + 10D)$$

in which

$P_{\text{total}}D =$ the after-tax total of the present worths of the defender's capital and operating costs

For the challenger:

$$P_{total}C = P_2C + P_3C + P_4C + (1 - t)(P_5C + P_6C + P_7C + P_8C)$$

in which

$P_{total}C$ = the after-tax total of the present worths of an infinite series of replacement challengers

Now assume the defender is given one more year of life ($N = 1$), and therefore the challenger's starting date is moved back to EOY 1. This requires multiplying all challenger values by

$$(F/P, r_1, N)(P/F, i, N)$$

This multiplication increases the payments to reflect price escalation, and then reduces them by the interest rate discount due to postponement for one year.

A matrix is then created which represents

$$P_{combined\ total} = P_{total}D + P_{total}C$$

in which P combined total = after-tax total of the NPW of the remaining life of the defender plus the NPW of the life cycle costs of an infinite series of replacement challengers.

The solution matrix for this is shown on page 583. [For studies for tax-exempt owners (such as federal, state or local Government owners), the tax multipliers and the depreciation equations are omitted.]

Examining the resulting matrix shows the highest NPW located at

$$N = 2 \quad \text{and} \quad L = 5$$

These data indicate that the most profitable strategy is to keep the defender for about two more years, and then replace it with a challenger on a five-year replacement cycle. If the defender is kept until EOY 3, the penalty in terms of NPW will amount to the difference between the value just found and the value located in the next lower box of the matrix, or

$$\begin{array}{r} \$\ 7{,}039 \\ -3{,}682 \\ \hline \$\ \ 3{,}357 \end{array}$$

The penalty for keeping the defender for one extra year is $3,357 in terms of NPW. Using the same approach, the penalty for any other delay or advance in the schedule can be determined.

Spreadsheet Formulas for Example 20.5

	A	B	C	D	E	F	G
1	Example 20.5: Defender/Challenger, w/ taxes, income & inflation.						
2	Input Data, General						
3	Owner's After-Tax Investment rate, i_r = I1 =	15%					
4	Inflation Rate =	= R1 =	7%				
5	Owner's Income Tax rate, t	= TX =	36%				
6		Defender	Challenger				
7	Cost New, $	DCN =	$180,000	$190,000	CON		
8	Resale Value, $ now at (PTZ)	DPRV =	$90,000				
9	Decline in Resale Value, $ per year	DR2 =	–20%	–20%	CR1		
10	Repair Cost $/yr at EOY 1	DREPA =	$10,000	$8,000	CREPA		
11	Arith Grad Repair Cost, $/yr/yr	DREPG=	$2,500	$2,000	CREPG		
12	O&M Cost, $/yr at EOY 1 =	DO&MC =	$100,000	$79,000	CO&MC		
13	O&M Costs Rate of Increase, %/yr	DO&MR5 =	12%	12%	CO&MR5		
14	Depreciation Life, years =	DDL =	5	5	CDL		
15	Defender's Present Age, yrs	DPA =	3				
16	Annual Income, $/yr =	DINCA =	$155,000	$161,000	CINCA		
17	Arith. Grad. income, $/yr/yr	DINCG=	($1,000)	$3,500	CINCG		
18							
19	Defender's Equations						
20	StrtLineBkValu at PTZ, PDBV=DCN(DDL–DPA)DDL	$72,000					
21	P1, After Tax Resale Value, at PTZ, P1=	($83,520)	($83,520)	($83,520)	($83,520)	($83,520)	($83,520)
22							
23	Defender's Trial Life, N = DCN (DP2–DPA–N)/DDL	0	1	2	3	4	5
24							
25	Future Deprec Bk Valu, FDBV=CN(DL–DA–N)/DL	$72,000	$36,000	$0	($36,000)	($72,000)	($108,000)
26	If FDBV < 0, then FDBV = 0	$72,000	$36,000	$0	$0	$0	$0
27	Future Resale Value, FRV=PRV*(1+R2)^N	$90,000	$72,000	$57,600	$46,080	$36,864	$29,491

	A	B	C	D	E	F	G
28	FutAftrTaxResaleValu at EOY N, F1=FRV −(FRV −FDBV)*TX	$83,520	$59,040	$36,864	$29,491	$23,593	$18,874
29							
30	Depreciation AftrTx Credit, $/yr, A2 = (CN/DL)*TX	$12,960	$12,960	$12,960	$12,960	$12,960	$12,960
31	where if N+DA ≤ DDL, then N2 = N	0	1	2	2	2	2
32	but if N+DA > DDL, then N2 = DDL−DA	12,960	12,960	12,960	0	0	0
33		0	11,270	21,069	21,069	21,069	21,069
34	P2 = A2*(P/A, i, N2) = P2, PW of Depreciation,	0	11,270	21,069	21,069	21,069	21,069
35							
36	P3, PW of Future Resale, P3 = F1(P/F, i, N) =	$83,520	$51,339	$27,874	$19,391	$13,489	$9,384
37	P4A, PW of Repair Costs, P4A = DREPA(P/A, i, N) =	$0	($8,696)	($16,257)	($22,832)	($28,550)	($33,522)
38	P4B, PW of Repair Gradient, P4B = DREPG(P/G, i, N) =	$0	$0	($1,890)	($5,178)	($9,466)	($14,438)
39	Maintenance Geom. Gradient, W5=[(1+i)/(1+R5)]−1	0.02678571	0.026785714	0.026785714	0.026785714	0.02678571	0.026785714
40	P5, PW MaintGeoGrad, P5=MANC*(P/C,W5,N)/(1+R5)	$0	($86,957)	($171,645)	($254,123)	($334,451)	($412,682)
41							
42	P6, PW of Income, P6=Income(P/A, i, N) =	$0	$134,783	$251,985	$353,900	$442,522	$519,584
43	P7, PW of Gradient Income, P7=Grad Inc(P/G, i, N) =	$0	$0	($756)	($2,071)	($3,786)	($5,775)
44	Total,							
45	DfndrsNPW, PTD=P1−P2−P3+(1−TX)(P4+P5+P6+P7) =	$0	$4,132	$4,743	$1,545	($6,550)	($19,040)
46							
47	Challenger's Trial Life Span, L	1	2	3	4	5	6
48							
49	Challenger's Equations						
50	Effective i, IE = (1+i)^L − 1	0.15	0.3225	0.520875	0.74900625	1.01135719	1.313060766
51	Effective r, RE = (1+R1)^L − 1	0.07	0.1449	0.225043	0.31079601	0.40255173	0.500730352
52	Derived, we = (1+IE)/(1+RE) − 1	0.07476636	0.155122718	0.241487034	0.334308494	0.43406988	0.541290054
53	(P/C, we, ∞) = 1/[(1+RE]we]	12.5	5.630630631	3.380296925	2.282009658	1.64256084	1.231026172
54	(F/P, i, L) =	1.15	1.3225	1.520875	1.74900625	2.01135719	2.313060766

Spreadsheet Formulas for Example 20.5 (continued)

	A	B	C	D	E	F	G
55							
56	In this section, all Challenger equations assume $N = 0$. (Challenger goes into service immediately).						
57	PW of Challenger Purchase plus an ∞ Series of Challenger Replacements						
58	Cost New+Replacements, PC1=CCN(F/P, i, L)(P/C, we, ∞)=	($2,731,250)	($1,414,837)	($976,792)	($758,337)	($627,718)	($541,013)
59							
60	PW of After-Tax Depreciation,						
61	(F/A, i, L) =	1	2.15	3.4725	4.993375	6.74238125	8.753738438
62	Cutoff Depreciation at End of Challenger's Deprec Life, if ChTrialLf>ChDprLf, then (F/A,i,ChDprLf), if ChTrlLf=<ChDprLf, then (F/A,i,ChTrlLf)						
63	(F/A, i, trial L or deprec L)	1	2.15	3.4725	4.993375	6.74238125	6.74238125
64	PW of Deprec, PC2=(CCN*TX/DL)(F/A,i,L)(P/C,we,∞)=	$171,000	$165,608	$160,577	$155,883	$151,503	$113,545
65							
66	Resale minus Recap Tax, PC3=						
67	Resale Value at EOY L = CCN*(1+CR1)^L =	152,000	121,600	97,280	77,824	62,259	49,807
68	Deprec Book Value = CCN(DL−L)/DL=	152,000	114,000	76,000	38,000	0	−38,000
69	If ChDpBkVlu < 0, then ChDpBkVlu = 0	152,000	114,000	76,000	38,000	0	0
70	PC3=[RsIVlu−(RsIVlu−DpBkVlu)*TX]*(P/C,we,∞)=	$1,900,000	$669,279	$302,940	$144,879	$65,449	$39,241
71							
72	Repairs, PC4 = [CRPA*(F/A,i,L)+CRPG(F/G,i,L)](P/C,we,∞)						
73	ChFWRepairsAnnual, = CRPA(F/A,i,L)=	−8,000	−17,200	−27,780	−39,947	−53,939	−70,030
74	ChFWRepairsGrad, = CRPG(F/G,i,L)=	0	−2,000	−6,300	−13,245	−23,232	−36,717
75	PC4=(FWRepAnnual+FWRepGrad)(P/C,we,∞)	($100,000)	($108,108)	($115,201)	($121,385)	($126,758)	($131,408)
76							
77	Maintenance Geom Grad, PC5 = ChO&MC(P/C,CW5,L)(F/P,i,L)(P/C,we,∞)						
78	CW5 = (1+i1I)/(1+ChO&MR5)−1	0.02678571					
79	F5 = ChO&MC(P/C,CW5,L)(F/P,i,L)	−79,000	−179,330	−305,327	−462,115	−655,741	−893,327
80	PC5 = F5(P/C,we,∞)	($987,500)	($1,009,741)	($1,032,096)	($1,054,552)	($1,077,094)	($1,099,709)

	A	B	C	D	E	F	G
81							
82	Income, $PC6 = [CINCA*(F/A,i,L)+CINCG(F/G,i,L)](P/C,we,\infty)=$						
83	CFIncA, = CINCA(F/A,i,L) =	161,000	346,150	559,073	803,933	1,085,523	1,409,352
84	CFIncG, = CINCG(F/G,i,L)=	0	3,500	11,025	23,179	40,656	64,254
85	$PC6=(CFIncAn+CFIncGrad)(P/C,we,\infty)=$	$2,012,500	$1,968,750	$1,927,099	$1,887,478	$1,849,817	$1,814,047
86	In the matrix below the Challenger is moved from PTZ to EOY N by $(F/P, R1, N)$, where $R1$ is the inflation rate, and then back to PTZ by $(P/F,I1,N)$,						
87	where $I1$ is the investment rate. Then the Defender's NPW is added to the Challenger's NPW for a total infinite cycle NPW.						
88	Total, $PTC=PC1+PC2+PC3+(1-TX)(PC4+PC5+PC6)$						
89		($68,250)	($35,373)	($14,202)	($2,190)	$2,653	($15,152)
90	Matrix, ChallengerLife, L --–>	1	2	3	4	5	6
91	Defender Life, $N/$ 0	−68,250	−35,373	−14,202	−2,190	2,653	−15,152
92	1	−59,370	−28,780	−9,082	2,095	6,600	9,966
93	2	−54,341	−25,879	−7,552	2,848	7,039	−8,374
94	3	−53,429	−26,947	−9,894	−219	3,682	−10,660
95	4	−57,700	−33,060	−17,193	−8,190	−4,562	−17,905
96	5	−66,632	−43,706	−28,943	−20,567	−17,190	−29,606

Spreadsheet Formulas for Example 20.5

Columns B and C, to right of A, are on subsequent pages. Line numbers and stub copy apply across all columns.

	A
1	**Example 20.5:** Defender/Challenger, w/ taxes, income & inflation.
2	Input Data, General
3	Owner's After-Tax Investment rate, i, = I1 =
4	Inflation Rate = = R1 =
5	Owner's Income Tax rate, t = TX =
6	
7	Cost New , $ DCN =
8	Resale Value, $ now at (PTZ) DPRV =
9	Decline in Resale Value, $ per year DR1 =
10	Repair Cost $/yr at EOY 1 DREPA =
11	Arith Grad Repair Cost, $/yr/yr DREPG =
12	O&M Cost, $/yr at EOY 1 = DO&MC =
13	O&M Costs Rate of Increase, %/yr DO&MR5 =
14	Depreciation Life, years = DDL =
15	Defender's Present Age, yrs DPA =
16	Annual Income, $/yr = DINCA =
17	Arith. Grad. income, $/yr/yr DINCG =
18	
19	Defender's Equations
20	StrtLineBkValu at PTZ, PDBV=DCN(DDL−DPA)/DDL
21	$P1$, After Tax Resale Value, at PTZ, $P1 =$
22	
23	Defender's Trial Life, $N =$
24	
25	Future Deprec Bk Valu, FDBV=CN(DL−DA−N)/DL
26	If FDBV < 0, then FDBV = 0
27	Future Resale Value, FRV=PRV*(1+R1)^N
28	FutAftrTaxResaleValu at EOY N, F1=FRV − (FRV−FDBV)*TX
29	
30	Depreciation AftrTx Credit, $/yr, A2 = (CN/DL)*TX
31	where if N+DA ≤ DDL, then $N2 = N$
32	but if N+DA > DDL, then $N2 =$ DDL −DA
33	
34	$P2$, PW of Depreciation, $P2 = A2*(P/A, i, N2) =$
35	
36	$P3$, PW of Future Resale, $P3 = F1(P/F, i, N) =$
37	$P4A$, PW of Repair Costs, $P4A =$ DREPA$(P/A, i, N) =$
38	$P4B$, PW of Repair Gradient, $P4B =$ DREPG$(P/G, i, N) =$
39	Maintenance Geom, Gradient, $W5=[(1+i)/(1+R5)]−1$
40	$P5$, PW MaintGeoGrad, $P5=$MANC$*(P/C,$W5$,N,)/(1+$R5$) =$
41	
42	P6, PW of Income, $P6=$Income$(P/A, i, N) =$
43	P7, PW of Gradient Income, $P7=$Grad Inc$(P/G, i, N) =$
44	Total,
45	DfndrsNPW, PTD=$P1−P2−P3+(1−$TX$)(P4+P5+P6+P7) =$
46	
47	Challenger's Trial Life Span, L
48	
49	Challenger's Equations
50	Effective i, IE = $(1+i)$^L −1
51	Effective r, RE = $(1+$R1$)$^L −1

B	C
0.15	
0.07	
0.36	
Defender	Challenger
180000	190000
90000	
−0.2	−0.2
10000	8000
2500	2000
100000	79000
0.12	0.12
5	5
3	
155000	161000
−1000	3500
=$B$7*($B$14−$B$15)/$B$14	
=−$B$28	=−$B$28
0	=B23+1
=$B$7*($B$14−$B$15−B23)/$B$14	=$B$7*($B$14−$B$15−C23)/$B$14
=IF(B25<0,0,B25)	=IF(C25<0,0,C25)
=$B$8*(1+$B$9)^B23	=$B$8*(1+$B$9)^C23
=B27−(B27−B26)*$B$5	=C27−(C27−C26)*$B$5
=($B$7/$B$14)*$B$5	=($B$7/$B$14)*$B$5
=IF(B23+$B$15<=$B$14,B23,$B$14−$B$15)	=IF(C23+$B$15<=$B$14,C23,$B$14−$B$15)
=IF(B23+$B$15<=$B$14,B30,$B$31)	=IF(C23+$B$15<=$B$14,C30,$B$31)
=−PV($B$3,B23,B32)	=−PV($B$3,C31,C32)
=B33	=IF(C33>0,C33,B34)
=−PV($B$3,B23,,B28)	=−PV($B$3,C23,,C28)
=PV($B$3,B23,$B$10)	=PV($B$3,C23,$B$10)
=$B$11*(PV($B$3,B23,1)+B23/(1+$B$3)^B23)/$B$3	=$B$11*(PV($B$3,C23,1)+C23/(1+$B$3)^C23)/$B$3
=((1+$B$3)/(1+$B$13))−1	=((1+$B$3)/(1+$B$13))−1
=(1/(1+$B$13))*PV(B39,B23,$B$12)	=(1/(1+$B$13))*PV(C39,C23,$B$12)
0	=−PV($B$3,C23,$B$16)
0	=−$B$17*((PV($B$3,C23,1)+C23/(1+$B$3)^C23)/$B$3)
=B21+B34+B36+(1−$B$5)*(B37+B38+B40+B42+B43)	=C21+C34+C36+(1−$B$5)*(C37+C38+C40+C42+C43)
=1	=B47+1
=(1+$B$3)^B47−1	=(1+$B$3)^C47−1
=(1+$B$4)^B47−1	=(1+$B$4)^C47−1

Spreadsheet Formulas for Example 20.5 (continued)

	A
52	Derived, $we = (1+IE)/(1+RE)-1$
53	$(P/C, we, \infty) = 1/[(1+RE)we]$
54	$(F/P, i, L,) =$
55	
56	In this section, all Challenger equations assume $N = 0$, (Challenger goes into service immediately).
57	PW of Challenger Purchase plus an ∞ Series of Challenger Replacements
58	Cost New+Replacements, $PC1=CCN(F/P,i, L)(P/C, we, \infty)=$
59	
60	PW of After-tax Depreciation,
61	$(F/A, i, L) =$
62	Cutoff Depreciation at End of Challenger Deprec Life, if ChTrialLf>ChDprLf, then $(F/A, i, ChDprLf)$, if ChTrlLf=
63	$(F/A, i,$ trial L or deprec $L)$
64	PW of Deprec, $PC2=(CCN*TX/DL)(F/A, i, L)(P/C, we, \infty)=$
65	
66	Resale minus Recap Tax, $PC3 =$
67	Resale Value at EOY $L =CCN*(1+CR1)^\wedge L=$
68	Deprec Book Value $= CCN(DL-L)/DL=$
69	If ChDpBkVlu < 0, then ChDpBkVlu = 0
70	$PC3=[RslVlu-(RslVlu-DpBkVlu)*TX]*(P/C,we,\infty)=$
71	
72	Repairs, $PC4=[CRPA*(F/A,i,L)+CRPG(F/G,I,L)](P/C,we,\infty)$
73	ChFWRepairsAnnual, $= CRPA(F/A,i,L) =$
74	ChFWRepairsGrad, $= CRPG(F/G,i,L)$
75	$PC4=(FWRepAnnual+FWRepGrad)(P/C,we,\infty)$
76	
77	Maintenance Geom Grad, $PC5= ChO\&MC(P/C,CW5,L)(F/P,i,L)(P/C,we,\infty)$
78	$CW5 = (1+I1)/(1+ChO\&MR5)-1$
79	$F5 = ChO\&MC(P/C, CW5,L)(F/P,i,L)$
80	$PC5= F5(P/C,we,\infty)$
81	
82	Income, $PC6 = [CINCA*(F/A,i,L)+CINCG(F/G,i,L)](P/C,we,\infty)=$
83	CFIncA, $= CINCA(F/A,i,L) =$
84	CFIncG, $= CINCG(F/G,i,L) =$
85	$PC6=(CFIncAn+CFIncGrad)(P/C,we,\infty) =$
86	In the matrix below the Challenger is moved from PTZ to EOY N by $(F/P,R1,N)$, where $R1$ is the inflation rate,
87	where I1 is the investment rate. Then the Defender's NPW is added to the Challenger's NPW for a total infinite
88	Total, $PTC=PC1+PC2+PC3+(1-TX)(PC4+PC5+PC6)$
89	
90	Matrix, ChallengerLife, L --->
91	Defender Life, $N/$ 0
92	=1
93	=2
94	=3
95	=4
96	=5

B	C
=(1+B50)/(1+B51)−1	=(1+C50)/(1+C51)−1
=1/((1+B51)*B52)	=1/((1+C51)*C52)
=−FV($B$3,B47,,1)	=−FV($B$3,C47,,1)
=−$C$7*B54*B53	=−$C$7*C54*C53
=−FV($B$3,B47,1)	=−FV($B$3,C47,1)
<ChDprLf, then (*F/A,i*,ChTrlLf)	
=B61	=IF(C47>$C$14,B63,C61)
=($C$7*$B$5/$C$14)*B63*B53	=($C$7*$B$5/$C$14)*C63*C53
=$C$7*(1+$C$9)^B47	=$C$7*(1+$C$9)^C47
=$C$7*($C$14−B47)/$C$14	=$C$7*($C$14−C47)/$C$14
=IF(B68<0,0,B68)	=IF(C68<0,0,C68)
=(B67−(B67−B69)*$B$5)*B53	=(C67−(C67−C69)*$B$5)*C53
=FV($B$3,B47,$C$10)	=FV($B$3,C47,$C$10)
=$C$11*(FV($B$3,B47,1)+B47)/$B$3	=$C$11*(FV)$B$3,C47,1)+C47)/$B$3
=(B73+B74)*B53	=(C73+C74)*C53
=(1+$B$3)/(1+$C$13)−1	
=−FV($B$3,B47,,PV($B$78,B47,$C$12)/(1+$C$13))	=−FV($B$3,C47,,PV($B$78,C47$C$12)/(1+$C$13))
=B79*B53	=C79*C53
=−FV($B$3,B47,$C$16)	=−FV($B$3,C47,$C$16)
=−$C$17*(FV($B$3,B47,1)+B47)/$B$3	=−$C$17*(FV($B$3,C47,1)+C47)/$B$3
=(B83+B84)*B53	=(C83+C84)*C53
=(B58+B64+B70+(1−$B$5)*(B75+B80+B85))	(C58+C64+C70+(1−$B$5)*(C75+C80+C85))
1	2
=$B$45+B89	=$B$45+C89
=$C$45+PV($B$3,$B$90,,B89)*FV($B$4,$A92,,1)	=$C$45+PV($B$3,$B$90,,C89)*FV($B$4,$A92,,1)
=$D$45+PV($B$3,$C$90,,B89)*FV($B$4,$A93,,1)	=$D$45+PV($B$3,$C$90,,C89)*FV($B$4,$A93,,1)
=$E$45+PV($B$3,$D$90,,B89)*FV($B$4,$A94,,1)	=$E$45+PV($B$3,$D$90,,C89)*FV($B$4,$A94,,1)
=$F$45+PV($B$3,$E$90,,B89)*FV($B$4,$A95,,1)	=$F$45+PV($B$3,$E$90,,C89)*FV($B$4,$A95,,1)
=$G$45+PV($B$3,$F$90,,B89)*FV($B$4,$A96,,1)	=$G$45+PV($B$3,$F$90,,C89)*FV($B$4,$A96,,1)

SENSITIVITY ANALYSIS

Engineering estimators, planners, and analysts are constantly called upon to predict future costs and incomes. Obviously this occupation is hazardous at best. Many variables must be considered, each of which can influence the results. For instance, future operating costs will be influenced by both labor and material costs. These are difficult enough to predict over the short term, but many estimates require long-term predictions. To increase the accuracy of the results, many estimators perform a sensitivity analysis. In outline form this consists of the following steps:

1. List the variables most likely to affect the future cost figures being estimated.
2. Determine a probable range over which these variables may fluctuate.
3. Determine the effect on the estimated future cost figures of the variables fluctuating over their probable range.

If the bottom-line cost is significantly affected by a particular variable, then the cost estimate is sensitive to that variable. The variables that show the greatest effect on the bottom-line total merit special attention and analysis, whereas those variables to which the totals are relatively insensitive deserve less expenditure of time and effort.

Sensitivity analysis may be further refined by introducing a probability basis for ranging the variables. For instance, if labor costs next year for a given group of workers have the probability of rising by the percentages shown below, then an expected value may be determined as follows:

Increase in labor costs	Probability of this increase	Increase × probability
5%	20%	1%
6%	50%	3%
7%	30%	2.1%
Expected increase in labor costs		6.1%

The computer greatly facilitates sensitivity studies, permitting many repetitious calculations to be made with a minimum of effort.

MAINTENANCE LEVEL

In a similar manner to economic life, there often is a maintenance level that either maximizes profit or minimizes cost. Obviously wear and tear accumulate when maintenance programs are at a level insufficient to prevent this accumulation. Some maintenance programs are the result of considerable thought, calculation, and planning. Other maintenance programs just happen on the spur of the moment whenever someone becomes

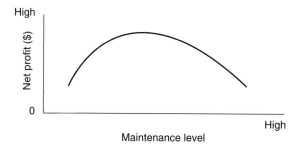

Figure 20.4   Net profit versus maintenance level.

concerned enough to do the maintenance. Considering maintenance as the independent variable and profit (or rate of return or similar measure) as the dependent variable, a characteristic graph of the relationship between the two is shown in Figure 20.4.

Ordinarily the curve is nearly flat on top with decline in net profit with either too high or too low a level of maintenance. To determine the optimum level of maintenance for a particular capital investment, estimates of income and costs using different types of maintenance programs should be made and a measure of profitability derived. Gross profit and intangibles should also be evaluated. A simplified example follows.

EXAMPLE 20.6

A van is needed for a public dial-a-ride service. At this point two maintenance programs are being considered, with costs and income as shown.

	Maintenance level A	Maintenance level B
First cost,	$100,000	$100,000
O & M costs,	$50,000/yr	$13,000/yr
	(High level of maintenance)	(Low level of maintenance)
Expected life	20 yr	5 yr
	(Long life)	(Life shortened by poor level of maintenance)
Income	$60,000/yr	$60,000/yr decreasing $10,000/yr

Which level of maintenance should be selected if $i = 8$ percent?

Solution

Plan A

$$\text{annual first cost, } A_1 = -100{,}000(A/P, 8\%, 20) = -\$10{,}190$$
$$\text{annual income} -\text{O \& M, } A_3 - A_2 = 60{,}000 - 50{,}000 = + 10{,}000$$
$$\underline{ - \quad 190}$$

Plan B

$$\text{annual first cost, } A_1 = -100,000(A/P, 8\%, 5) = -\$25,050$$
$$\text{annual income} - \text{gradient, } A_3 = 60,000 - 10,000(A/G, 8\%, 5) = \quad 41,535$$
$$\text{annual O \& M costs, } A_2 = -13,000 = - \quad 13,000$$
$$+\$ \ 3,485$$

In this case the lower maintenance level and consequent shorter life and declining income of Plan B actually yield a greater profit on an annualized basis.

However, in evaluating the gross profit and intangibles, note that the life of the investment is only five years, and *no* consideration has been given to the possible increases in first costs as new replacements are purchased.

Also, in considering intangibles, the lack of maintenance, while in this case more profitable over the short term, could lead to long-term erosion of public confidence and good will, and decrease in value of related investments and other costs to owner, investor, and community.

Maintenance costs may also depend upon the quality of the product. An engineer purchasing equipment is frequently confronted with a choice between the alternatives of (A) high first cost with resulting low maintenance requirements or (B) low first cost and high maintenance. The variables to be evaluated include the following:

1. Estimate of maintenance costs for each level of first-cost quality. Will wage rates and prices of maintenance supplies rise in the future? (Yes favors A, with low maintenance requirements.) If yes, how rapidly? (Anticipated near-term rises favor alternative A, with low maintenance requirements.)

2. Estimate of expected life span for each level of quality.
   a. Is this life adequate to accomplish the objective?
   b. If future replacements are necessary, will the replacement cost rise? If so, how much?

3. Source of funds.
   a. Capital funds to purchase new equipment may be difficult to acquire, while funds for maintenance may be approved routinely.
   b. If privately owned, tax considerations may favor maintenance (favors alternative B, causing less drain on limited capital, higher tax-deductible maintenance costs).

4. Tax considerations. Engineering investments involving a trade-off of capital costs versus O & M costs are frequently influenced by tax considerations. The basic reason is that for tax purposes, maintenance costs are usually deductible in the year in which they occur, while capital costs usually are deducted as they are depreciated over the entire life of the investment. Thus, when faced with a choice between alternative of the same equivalent before-tax cost, tax considerations may favor the maintenance alternative.

SUMMARY

Replacement analysis involves two separate, yet interrelated, economic analyses. First the costs for the existing unit (termed *defender*) must be determined, and then the costs for the most eligible *challengers* are found. Then a decision is needed on whether the defender must be replaced now or continued in service for at least one more time period (usually a year, but shorter analysis periods could be used; e.g., a month, three months, etc.). The optimum economic life is the life for which the sum of the defender's NPW plus the replacement challenger's NPW is optimum (least costs, or greatest net income). It is convenient to use an infinite series of replacements, since it allows the normal increases in replacement purchase price and other costs to be taken into account, and the equations are simpler.

PROBLEMS FOR CHAPTER 20, EQUIPMENT REPLACEMENT ANALYSIS

(*Note:* Where an annual decline in value of the property is given, the decline factor is always multiplied by the previous year's value. For instance in Problem A.1, the cost new = $200K, and the decline in value is 22 percent per year. Therefore,

$$\text{the value at EOY } 1 = (1 - 0.22) \times \$200K = \$156K.$$
$$\text{the value at EOY } 2 = (1 - 0.22) \times \$156K = \$121.7K, \text{ etc.}$$

These EOY values are assumed to be the net resale market values before taxes.)
All annual payments occur at EOY unless otherwise specified.

Problem Group A

Find the optimum NAW for defender only, no taxes. For Group A problem solutions to be valid, it must be assumed that all future replacements will be identical to the original purchase, and that all costs for future replacements will be the same as for the original purchase. The same effect may be roughly approximated if funding levels increase at the same rate as the increase in replacement costs (if inflation affects income and cost to the same degree, then the effects are mutually offsetting, somewhat).

A.1. A city public works department is considering the purchase of an asphalt-mixing plant with cost data tabulated below.

$$\text{interest rate, } i = 12\%$$
$$\text{cost new} = \$200,000$$
$$\text{decline in plant's value per year} = 22\%$$
$$\text{O \& M costs} = \$80,000/\text{yr}$$
$$\text{gradient in O \& M costs} = \$12,000/\text{yr/yr}$$

(a) Find the optimum economic life by finding the life cycle with the lowest NAW.

(*Ans.* 3 yr NAW = $146,200/yr)

(b) If at the end of the optimum economic life, funds are not available to replace the plant for an additional two years, find the *extra* cost incurred in terms of NAW.

(*Ans.* $1,429/yr)

A.2. A state road maintenance department is considering the purchase of a rock ripper for use in a rock quarry.

$$\text{interest rate, } i = 10\%$$
$$\text{cost new} = \$20,000$$
$$\text{decline in ripper value per year} = 20\%$$
$$\text{O \& M costs} = \$2,500/yr$$
$$\text{gradient O \& M costs} = \$1,000/yr/yr$$

(a) Find the optimum economic life by finding the life cycle with the lowest NAW.

(b) If at the end of the optimum economic life, funds are not available to replace the ripper for one additional year, find the *extra* cost incurred in terms of NAW.

A.3. A city has a consulting services department (CSD) that derives its income from charges for services to other city departments. The CSD is considering the purchase of an expensive computer software package that the CSD can choose to resell to the vendor at any time in the future, but at a discount of 23 percent for every year it has been used.

$$\text{interest rate, } i = 15\%$$
$$\text{cost new} = -\$120,000$$
$$\text{decline in resale price} = 23\%/yr$$
$$\text{net annual income} = +\$43,000/yr$$
$$\text{gradient } decrease \text{ in annual income} = -\$7,000/yr/yr$$

(a) Find the optimum economic life by comparing the NAW for each life cycle.

(*Ans.* 4 yr, NAW $= +\$132/yr$)

(b) If at the end of the optimum economic life, funds are not available to replace the software package for one additional year, find the *extra* loss in income incurred in terms of NAW.

(*Ans.* $\$136/yr$)

Problem Group B

Find the optimum life for the defender and challenger, no taxes. For all Group B problems, assume that the defender will be replaced with an infinite series of replacement challengers. All costs for future replacement challengers will be the same as for the first replacement challenger (zero inflation, or offsetting effects of rising income and rising costs).

B.1. A public works department now owns a three-year-old excavator, and is considering the purchase of a replacement. Use $i = 12$ percent.

	Defender	Challenger
Present resale value	$150,000	
Cost new		$250,000
Decline in resale value	20%/yr	20%/yr
O & M	$90,000/yr	$65,000/yr
Gradient in O & M	$ 8,000/yr/yr	$12,000/yr/yr

(a) Find the optimum retirement time for the defender. Assume the optimum economic life-cycle time for the infinite series of challenger replacements has been found as four years.

Determine the optimum defender life by finding the lowest total NPW of the defender life cycle plus an infinite series of challenger replacements with four-year life cycles.

(*Ans.* $L = 5$ yr, NPW $= \$1,170K$)

(b) If at the defender's optimum retirement time funds are not available to replace the defender for one additional year, find the *extra* cost incurred in terms of NPW.

(*Ans.* $\$1,793$)

(c) (Omit unless specifically assigned by the instructor.) Using a computer, find the optimum retirement time for the defender, and the optimum life-cycle time for an infinite series of replacement challengers. Derive a matrix of total NPW values, with L varying from at least 1 to 6, and N varying from 0 to 5.

(*Ans.* $N = 4$ yr, $L = 5$ yr, NPW $= \$1,170K$)

B.2. A city water and waste-water department has a four-year-old sludge pump and is considering the purchase of a replacement. Use $i = 10$ percent.

	Defender	Challenger
Present resale value	$30,000	
Cost new		$50,000
Decline in resale value	20%/yr	20%/yr
O & M	$18,000/yr	$13,000/yr
Gradient in O & M	$ 1,600/yr/yr	$ 2,400/yr/yr

(a) Find the optimum retirement time for the defender. Assume that the optimum economic life-cycle time for the infinite series of challenger replacements has been found as three years. Determine the optimum defender life by finding the lowest total NPW of the defender life cycle plus an infinite series of challenger replacements with three-year life cycles.

(b) If at the defender's optimum retirement time funds are not available to replace the defender for one additional year, find the *extra* cost incurred in terms of NPW.

(c) (Omit unless specifically assigned by the instructor.) Using a computer, find the optimum retirement time for the defender, and the optimum life-cycle time for an infinite series of replacement challengers. Derive a matrix of total NPW values, with L varying from at least 1 to 6, and N varying from 0 to 5.

B.3. A city building inspection department derives revenue for the city by charging inspection fees. At the heart of the system is a computer currently worth $150,000. The department head is considering the replacement of this computer by a newer more efficient model. Assume the interest rate, $i = 12$ percent.

	Defender	Challenger
Present resale value	+$150,000	
Cost new		+$250,000
Decline in resale value	20%/yr	20%/yr
Net Income	+$ 48,000/yr	+$ 85,000/yr
Gradient decline in income	+$ 5,000/yr/yr	+$ 10,000/yr/yr

(a) Find the optimum retirement time for the defender. Assume the optimum economic life-cycle time for the infinite series of challenger replacements has been found as five years. Determine the optimum defender life by comparing the total NPW for each defender life cycle plus an infinite series of challenger five-year life cycle replacements.

   (*Ans.* $N = 0$, NPW $= +\$89{,}972$)

(b) If at the defender's optimum retirement time funds are not available to replace the defender for one additional year, find the loss incurred in terms of NPW.

   (*Ans.* $\$2{,}065$)

(c) (Omit unless specifically assigned by the instructor.) Using a computer, find the optimum retirement time for the defender, and the optimum life-cycle time for an infinite series of replacement challengers. Derive a matrix of total NPW values, with L varying from at least 1 to 6, and N varying from 0 to 5.

   (*Ans.* $N = 0$, $L = 5$, NPW $= +\$89{,}972$)

Problem Group C

Defender versus infinite series of replacement challengers, after taxes. Assume all future costs of replacement challengers increase at inflation rate, R_1, per year.

Tax rules

1. Depreciation is by the SL method, and the depreciated book value at EOY is the SLBV.
2. All annual expenses, and depreciation are deductible from taxable income. All annual income is taxable.
3. Upon resale, any gain above SLBV is subject to tax at the ordinary income tax rate, t.

C.1. A contractor owns a concrete pump and is considering the purchase of a replacement.

$$\text{owner's after-tax investment rate, } i = 14\%$$
$$\text{inflation rate} = 6\%$$
$$\text{owner's income tax bracket} = 33\%$$

	Defender	Challenger
Defender's cost new	$90,000	
Present resale value	$45,000	
Cost new		$95,000
Decline in resale value	20%/yr	20%/yr
Repair costs	$22,000/yr	$17,000/yr
Arithmetic gradient in repair costs	$ 4,000/yr/yr	$ 5,000/yr/yr
O & M	$31,000 at EOY 1	$22,000/yr
Geometric gradient in O & M	12%	12%
Depreciation life	5 yr	5 yr
Defender's present age	3 yr	

(a) Find the optimum retirement time for the defender. Assume the optimum economic life-cycle time for the infinite series of challenger replacements has been found as five years. Determine the optimum defender life by finding the lowest total NPW of the defender life cycle plus infinite series of challenger replacements with five-year life cycles.

(b) If at the defender's optimum retirement time funds are not available to replace the defender for one additional year, find the *extra* cost incurred in terms of NPW.

(c) (Omit unless specifically assigned by the instructor.) Using a computer, find the optimum retirement time for the defender, and the optimum life-cycle time for an infinite series of replacement challengers. Derive a matrix of total NPW values, with L varying from at least 1 to 6, and N varying at least from 0 to 5.

C.2. A contractor owns a tunneling machine and is thinking of replacing it.

owner's after-tax investment rate, $i = 14\%$

inflation rate $= 6\%$

owner's income tax bracket $= 30\%$

	Defender	Challenger
Defender's cost new	$350,000	
Present resale value	$190,000	
Cost new		$400,000
Decline in resale value	20%/yr	20%/yr
Repair costs	$ 30,000/yr	$ 42,000/yr
Arithmetic gradient in repair costs	$ 6,000/yr/yr	$ 12,000/yr/yr
O & M	$ 82,000 at EOY 1	$ 92,000/yr
Geometric gradient in O & M	12%	12%
Depreciation life	5 yr	5 yr
Defender's present age	3 yr	
Income from tunneling machine	$225,000	$275,000
Arithmetic gradient in income	−$ 10,000/yr/yr	+$ 3,500/yr/yr

(a) Find the optimum retirement time for the defender. Assume that the optimum economic life-cycle time for the infinite series of challenger replacements has been found as five years. Determine the optimum defender life by finding the lowest total NPW of defender life cycle plus infinite series of challenger replacements with five-year life cycles.

(b) If at the defender's optimum retirement time funds are not available to replace the defender for one additional year, find the *extra* cost incurred in terms of NPW.

(c) (Omit unless specifically assigned by the instructor.) Using a computer, find the optimum retirement time for the defender, and the optimum life-cycle time for an infinite series of replacement challengers. Derive a matrix of total NPW values, with L varying from at least 1 to 6, and N varying at least from 0 to 5.

C.3. Your client owns a dump truck and asks your advice on when to replace it.

owner's after-tax investment rate, $i = 20\%$

inflation rate $= 8\%$

owner's income tax bracket $= 35\%$

	Defender	Challenger
Defender's cost new	$100,000	
Present resale value	$ 40,000	
Cost new		$120,000
Decline in resale value	20%/yr	20%/yr
Repair costs	$ 10,000/yr	$ 8,000/yr
Arithmetic gradient in repair costs	$ 2,000/yr/yr	$ 3,000/yr/yr
O & M	$ 52,000 at EOY 1	$ 45,000/yr
Geometric gradient in O & M	12%	12%
Depreciation life	5 yr	5 yr
Defender's present age	4 yr	
Income from dump truck	$ 86,000	$104,000
Arithmetic gradient in income	−$ 2,000/yr/yr	+$ 4,000/yr/yr

(a) Find the optimum retirement time for the defender. Assume that the optimum economic life-cycle time for the infinite series of challenger replacements has been found as five years. Determine the optimum defender life by finding the lowest total NPW of the defender life cycle plus infinite series of challenger replacements with five-year life cycles.

(b) If at the defender's optimum retirement time funds are not available to replace the defender for one additional year, find the *extra* cost incurred in terms of NPW.

(c) (Omit unless specifically assigned by the instructor.) Using a computer, find the optimum retirement time for the defender, and the optimum life-cycle time for an infinite series of replacement challengers. Derive a matrix of total NPW values, with L varying from at least 1 to 6, and N varying at least from 0 to 5.

CHAPTER 21

Double Gradients, Arithmetic and Geometric

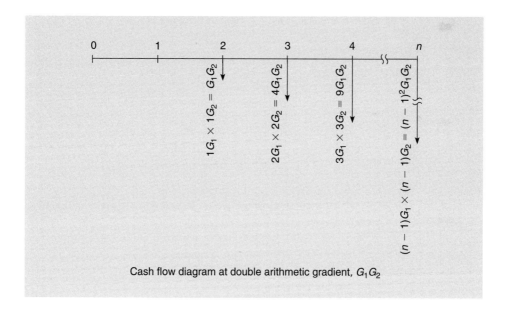

Cash flow diagram at double arithmetic gradient, $G_1 G_2$

INTRODUCTION

Change is accepted as a normal condition of life, and calculations of equivalent worths of proposed projects or equipment purchases must reflect these realities. In previous chapters regular changes in a series of payments were accounted for by either arithmetic gradients or geometric gradients. Where the first gradient payment itself is subject to a second gradient, the solution requires the use of a double gradient. For instance,

1. Utilities may experience a first gradient in *volume* (number of units sold per given time period) together with a second gradient in *unit price* per unit sold for water, waste water, electricity, gas, and so on.

2. Transportation facilities may experience a first gradient in the *volume* of users (persons, vehicles, passengers, etc.), as well as a second gradient in *unit cost* or savings per user for their facilities (roads, intersections, terminals, public carriers, etc.)

3. Private business managers may experience a first gradient in the *number of units* produced or consumed as well as a second gradient in the *cost or income per unit*.

For these situations, a double gradient usually can be designed to represent the activity. Furthermore a significant error can occur if the double gradient is ignored. Both the double arithmetic gradient as well as the double geometric gradient are presented in this chapter, in that order.

DERIVATION OF THE DOUBLE ARITHMETIC GRADIENT

The general form of an equation for the double arithmetic gradient may be developed using the following notations and assumptions.

G_1 represents the first of the two gradients, say, an increasing number of vehicles per year using a new toll facility.

G_2 represents the second of two gradients, with, say, an increase in toll charges per vehicle.

Beginning at the EOY 2, the increment in toll receipts (represented by the increment in vehicles times the increment in charges) is deposited at the end of each year (or other compounding time period) into an account earning interest compounded at rate i. (Since this is an arithmetic gradient, the first deposit does not occur until the EOY 2. If a review of gradient fundamentals is needed, refer back to Chapters 5 and 6.) The increment in vehicles is represented initially by G_1 at EOY 2 and increases to $2G_1$ at EOY 3, to $3G_1$ at EOY 4, and so on until the number of vehicles reaches $(n - 1)G_1$ at EOY n.

By the same reasoning, the incremental toll per vehicle is represented by G_2 at EOY 2, increasing to $2G_2$ at EOY 3 and so on. Therefore, the incremental toll per vehicle deposited at the end of the nth year is $(n - 1)G_2$. The total amount deposited the first year is zero. The total amount deposited the second year is the product of G_1 vehicles times G_2 incremental toll per vehicle, or simply G_1G_2. At the end of the third year the total amount deposited is $2G_1$ vehicles times $2G_2$ tolls per vehicle, or $4G_1G_2$. The deposit at the end of the nth year will amount to the product of $(n - 1)^2G_1G_2$, as shown in Figure 21.1. If each of these deposits bears interest compounded at the end of each year, the accumulation of principal and interest at the end of the nth year may be summed as the future value, F, of a series. This series can be examined one year at a time, as shown below, where F_n is the future value of the series at the end of year n.

$$n = 1, \quad F_1 = 0$$

In the arithmetic gradient series there is no deposit at the EOY 1.

$$n = 2, \quad F_2 = G_1G_2$$

The first deposit occurs at the EOY 2, with no time for interest to accumulate.

$$n = 3, \quad F_3 = G_1G_2(1 + i) + 2^2G_1G_2$$

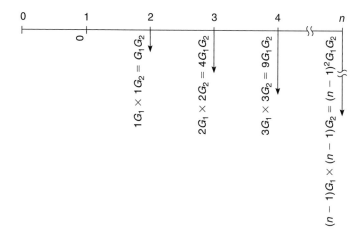

Figure 21.1 Cash flow diagram representing a double arithmetic gradient, G_1G_2.

At the EOY 3, the second deposit occurs, and interest for one period accumulates on the first deposit.

The general term for the future value of the series in the nth year is

$$n = n, \quad F_n = G_1G_2(1 + i)^{n-2} + 2^2G_1G_2(1 + i)^{n-3} + \cdots + (n - 1)^2G_1G_2$$

The equation is made slightly less cumbersome by dividing both sides by G_1G_2, as follows.

$$F/G_1G_2 = (1 + i)^{n-2} + 2^2(1 + i)^{n-3} + 3^2(1 + i)^{n-4} + \cdots + (n - 1)^2 \tag{21.1}$$

or in notation form

$$F = G_1G_2(F/GG, i\%, n) \tag{21.2}$$

To find the present worth of a double arithmetic gradient, simply multiply the future worth by the P/F factor of $1/(1 + i)^n$, to yield

$$P/G_1G_2 = 1/(1 + i)^2 + 2^2/(1 + i)^3 + 3^2/(1 + i)^4 + \cdots + (n - 1)^2/(1 + i)^n \tag{21.3}$$

In notation form this is recognized as

$$P = G_1G_2(P/GG, i\%, n) \tag{21.4}$$

In a similar manner the equivalent annual worth may be found by multiplying by either the appropriate A/F or A/P factors. These equations were programmed into a computer to produce the tables of GG factors at the end of this chapter (Table 21.1). These tables greatly reduce the time and effort required to solve problems involving double gradients. An application of the GG factor in solving a typical problem is shown in the following example.

EXAMPLE 21.1   (Given, G_1, G_2, i, n, find P)

A proposed new signalized intersection is under study to serve a rapidly developing area, and a comparison of benefits versus costs is needed. All savings are credited at EOY, and the intersection is designed to serve the traffic needs until EOY 10. The data follow (vpd = vehicles per day, vpy = vehicles per year):

The road will be opened at EOY 1.

Savings per vehicle during year 2, credited at EOY 2 = $0.05/vehicle.

Savings during each subsequent year, increase = $0.05/vehicle/yr.

Number of vehicles using the new road = 1,000 vpd.

Increase in vpd for each subsequent year = 1,000 vpd/yr. (i.e., during the second year there will be 1,000 vpd, each saving $0.05/vehicle credited at EOY 2; during the third year there will be 2,0000 vpd, each saving $0.10/vehicle credited at EOY 3, etc.)

The intersection construction will cost $1,000,000, but will not cause any increase in the department's operating and maintenance budget.

Compare the PW of the benefits (savings to motorists) to the cost of the project, using $i = 8$ percent.

Solution.   The two gradients are found as

$$\text{traffic gradient, } G_1 = 1,000 \text{ vpd/yr} \times 365 \text{ days/yr}$$
$$= 365,000 \text{ vpy/yr}$$
$$\text{savings/vehicle gradient, } G_2 = \$0.05/\text{vpy}$$

Then n = the number of gradient payments (9) plus one, so $n = 10$. (There is no gradient payment at EOY 1; see Chapter 5.) The present worth of the savings until EOY 10 is found as

$$P = G_1 G_2 (P/GG, 8\%, 10) = 365,000 \times 0.05 \times 154.305 \text{ (from Table 21.1)}$$
$$P = \underline{\$2,816,000}$$

The PW of the $2,816,000 benefits compare quite favorably with the $1,000,000 cost of the project. Note that if a solution to the problem had been attempted by using the P/G factor (instead of the P/GG factor) the PW of savings benefit would have shown as only $474,000, far below the actual PW of the benefits. The results would have been off by a factor of almost 6 $(P/GG \div P/G = 154.305 \div 25.976 = 5.94)$.

DOUBLE ARITHMETIC GRADIENT PLUS UNIFORM SERIES

Many common problem situations involve both the double gradient and the uniform series. For instance, a city water department is already serving a number of customers, and expects to add more customers as well as raise the rate. They need to know the equivalent present worth of the expected income.

The general solution for finding the equivalent worth of a combined series is found through application of the following general equations.

For the present worth:

$$P = A_1A_2(P/A, i, n) + (G_1A_2 + G_2A_1)(P/G, i, n) + G_1G_2(P/GG, i, n) \qquad (21.5)$$

For the annual worth:

$$A = A_1A_2 + (G_1A_2 + G_2A_1)(A/G, i, n) + G_1G_2(A/GG, i, n) \qquad (21.6)$$

For the future worth:

$$F = A_1A_2(F/A, i, n) + (G_1A_2 + G_2A_1)(F/G, i, n) + G_1G_2(F/GG, i, n) \qquad (21.7)$$

The following example illustrates a problem of combined uniform series and double gradient.

EXAMPLE 21.2   (Given A_1, A_2, G_1, G_2, i, n; find P)*

A city-owned water department has a request to serve a new growth area consisting of 400 dwelling units (du) initially, with 100 more added during each subsequent year for 12 more years. The expansion will cost the water department an estimated $650,000, and to pay for the expansion the public service commission will allow them a surcharge not to exceed $0.60/kgal the first year, plus an increment of $0.10/kgal/yr for each subsequent year. The surcharge will be removed at EOY 12. The city asks you to determine if this is going to

* *Note:* kgal = 1,000 gallons.

be sufficient to pay for the new water facilities by the EOY 12. Each dwelling unit uses an average of 100 kgal/yr. Assume all surcharge payments are accounted for at EOY, and $i = 10$ percent.

Solution

    Step 1. Sort out the data and find

initial water use, $A_1 = 400\ \text{du} \times 100\ \text{kgal/du/yr} = 40,000\ \text{kgal/yr}$
gradient use, $G_1 = 100\ \text{du/yr} \times 100\ \text{kgal/du/yr} = 10,000\ \text{kgal/yr/yr}$
initial revenue, $A_2 = \$0.60/\text{kgal}$
gradient revenue, $G_2 = \$0.10/\text{kgal/yr}$
                    $n = 12\ \text{yr}$
                    $i = 10\%$

    Step 2. Find the PW of the surcharge revenue. An examination of the problem shows that the income the first year is the product of the consumption in kgal/yr (A_1) times the rate charged per kgal (A_2), or

$$A_1A_2 = 40,000\ \text{kgal/yr} \times \$0.60/\text{kgal} = \$24,000/\text{yr}$$

Beginning with EOY 2, two single-gradient increases provide additional income, calculated as follows:

1. The gradient increase in consumption, (G_1), of 10,000 kgal/yr each year at the base rate of \$0.60/kgal yields an arithmetic gradient of

$$G_1A_2 = 10,000\ \text{kgal/yr/yr} \times \$0.60/\text{kgal} = \$6,000/\text{yr/yr}$$

2. The gradient increase in rate (G_2) of \$0.10/kgal/yr on the base amount of consumption (A_1) of 40,000 kgal/yr yields a gradient of

$$G_2A_1 = \$0.10/\text{kgal/yr} \times 40,000\ \text{kgal/yr} = \$4,000/\text{yr/yr}$$

Finally there is the double gradient increase resulting from the increase in consumption (G_1) of 10,000 kgal/yr/yr, multiplied by the increase in rate (G_2) of \$0.10/kgal/yr. This double gradient amounts to

$$G_1G_2 = 10,000\ \text{kgal/yr} \times \$0.10/\text{kgal/yr} = \$1,000/\text{yr/yr}$$

The present worth of all of these components can be summed in equation form as follows:

$$P = A_1A_2(P/A, i, n) + (G_1A_2 + G_2A_1)(P/G, i, n) + G_1G_2(P/GG, i, n)$$
$$P = 40,000 \times 0.60(P/A, 10\%, 12) + (10,000 \times 0.60$$
$$+\ 0.10 \times 40,000)(P/G, 10\%, 12) + 10,000 \times 0.10(P/GG, 10\%, 12)$$
$$P = 163,500 + 299,000 + 207,300 = \underline{\underline{\$669,800\ \text{PW of surcharge}}}$$

Step 3.   Compare the PW of the surcharge revenue ($669,800) with the estimated cost of the expansion ($650,000), and find that there is adequate revenue to meet the cost, provided that all of the assumptions are fulfilled. Note that the double gradient (*GG* factor) accounts for $207,300 of the total $669,900 of revenue. Omitting this *GG* portion would result in serious error.

Unit Costs

Where units of production or consumption are concerned, the unit costs often involve double gradients, such as a gradient in production coupled with a gradient in price. The resulting costs can be found as outlined in Example 21.3.

EXAMPLE 21.3   (Given P, A_1, G_1, G_2, i, n, for an excavator; find the equivalent annual payment, A_2, in terms of $/yd^3)

The county (no taxes) has a "borrow pit" operation and asks your help in finding the break-even cost to charge for excavation, using the equipment whose data are listed below, and assuming the charge will be increased by $0.05/yd^3 at the end of each year.

Cost of excavator	P_1	$100,000
Production	A_1	222.5 yd^3/hr
Gradient in production	G_1	−10 yd^3/hr/hr
Break-even charge for excavated material in terms of $/yd^3	A_2	Unknown (find)
Gradient in charge for excavated material in $/yd^3/yr	G_2	$0.05 yd^3/yr
O & M costs	A_3	$38.00/hr
Gradient increase in O & M costs	G_3	$3.20/hr/yr
Major maintenance every 2 yr; first at EOY 2, last at EOY 10	A_4	$10,000/each 2 yr
Gradient increase in major maintenance (cost is $12,200 at EOY 4, $14,400 at EOY 6, etc.)	G_4	$2,200/each/each
Resale value at EOY 10	F_1	$10,000
Remaining life		10 yr

Assume the following:
All payments are made at EOY.
The excavator works 2,000 hours per year.
$i = 15$ percent.
Find the cost of production in terms of A_2/yd^3.

Solution.   The problem can be solved in terms of equivalent annual worth, as follows.

Cost

Cost now

$$A_5 = \$ -100,000 \underbrace{(A/P, 15\%, 10)}_{0.19925} = \hspace{4cm} -19,925$$

O & M

$$A_6 = \$ -38.00/\text{hr} \times 2,000 \text{ hr/yr} = \hspace{3cm} -76,000$$

Gradient O & M

$$A_7 = \$ -3.20/\text{hr/yr} \times 2,000 \text{ hr/yr} \underbrace{(A/G, 15\%, 10)}_{3.38320} = \hspace{0.5cm} -21,652$$

Major maintenance every 2 yr

$$A_8 = \$ -10,000 \underbrace{(A/F, 15\%, 2)}_{0.46512} = \hspace{4cm} -4,651$$

Gradient in major maintenance

$$i_e = 1.15^2 - 1,$$
$$(A/G, i_e, 5) = (1/i_e) - n/[(1 + i_e)^n - 1] = 1.4590$$
$$A_9 = \$ -2,200 \underbrace{(A/G, i_e, 5)}_{1,4590} \underbrace{(A/F, 15\%, 2)}_{0.46512} = \hspace{1cm} -1,493$$

Income

Income from charges for excavated material:

$$A_{10} = A_1 \times A_2 = 222.5 \text{ yd}^3/\text{hr} \times A_2\$/\text{hr} \times 2,000 \text{ hr/yr} = 445,000 A_2/\text{hr}$$
$$A_{11} = (A_1 \times G_2 + A_2 \times G_1) \times 2,000 \times \underbrace{(A/G, 15\%, 10)}_{3.3832}$$

$$A_{11} = [222.5 \times 0.05 + A_2 \times (-10)] \times 2,000 \times 3.3832 = 75,276 - 67,664 A_2$$
$$A_{12} = G_1 G_2 (A/GG, 15\%, 10) \times 2,000 \text{ hr/yr}$$
$$A_{12} = -10 \times 0.05 \times (18.9412) \times 2,000 \text{ hr/yr} = \hspace{1cm} -18,940$$

Salvage

$$A_{13} = 10K \underbrace{(A/F, 15\%, 10)}_{0.04925} = \hspace{4cm} +493$$

Total

$$A_{5-13} = \hspace{4cm} -66,892 + 377,366 A_2$$

Solving for A_2 yields,

$$A_2 = 66,892/377,366 = \$0.177/\text{yd}^3$$

In this case if the double gradient, *GG*, had been ignored, the *calculated* charge for excavation would have been more than 10 percent below the *actual break-even* cost.

COMBINING MORE THAN TWO GRADIENTS

What happens when *more* than two gradients combine? When all other methods fail, we can simply calculate the cash flow amounts due to the combined gradients and use the basic *P/F* equation to find the equivalent values at the PTZ. The following example illustrates this fall-back method as applied to a triple gradient situation.

EXAMPLE 21.4

Assume the income for an excavator is controlled by *three* uniform series plus their respective gradients, as follows:

Production (first uniform series)	$A_1 =$	26 yd^3/hr
Δ Production (first gradient)	$G_1 =$	−0.5 yd^3/hr/yr
Hours of operation per year		
(second uniform series)	$A_2 =$	2,000 hr/yr
Downtime factor the first year	$=$	0.06
Δ Downtime factor (second gradient)	$G_2 =$	+0.03/yr
Income from production (third uniform series)	$A_3 =$	$1.00/yd^3
Δ Income (third gradient)		
$i = 15\%$	$G_3 =$	$+0.08/yd^3/yr

Find the NPW of the production for the first three years.

Solution. The production for each individual year may be calculated and then brought back to PTZ by use of the *P/F* factor, as follows:

Year						PW of product
1	26 yd^3/hr × 2,000 hr/yr × 0.94 × $1.00/yd^3 × $(P/F, 15\%, 1) =$					$ 42,504
Δ	−0.5		− 0.03 + 0.08			
2	25.5 yd^3/hr × 2,000 hr/yr × 0.91 × 1.08		× $(P/F, 15\%, 2) =$			$ 37,900
Δ	−0.5		− 0.03 + 0.08			
3	25.0	× 2,000	× 0.88 × 1.16	× $(P/F, 15\%, 3) =$		$ 33,560
PW of production for years 1 through 3 $=$						$113,964

DOUBLE GEOMETRIC GRADIENT

Double geometric gradients can occur where changes occur as a given percent of each of two interactive base amounts. The following method can be used to solve problems of this type. The cash flow diagram for $C \times D$ is shown in Figure 21.2.

Given a series of periodic payments defined as follows:

1. The first payment has a value $\$C \times D$.

2. Each subsequent value of C is changing (increasing or decreasing) by a proportionate value, r.

3. Each subsequent value of D is changing by a proportionate value, s.

4. All amounts involved are receiving or paying interest at i rate, on the balance compounding each period for n periods.

5. q is a derived value from $q = (1 + r)(1 + s) - 1$.

Then,

$$(1 + r)(1 + s) = (1 + q)$$

If $q > i$, then

$$w = \frac{1 + q}{1 + i} - 1 \quad \text{and} \quad P = \frac{CD}{(1 + i)}(F/A, w, n)$$

If $q < i$, then

$$w = \frac{1 + i}{1 + q} - 1 \quad \text{and} \quad P = \frac{CD}{(1 + q)}(P/A, w, n)$$

If $n \to \infty$, then

$$P = \frac{CD}{(1 + q)w}$$

If $q = i$, then

$$P = CDn/(1 + q) = CDn/(1 + i)$$

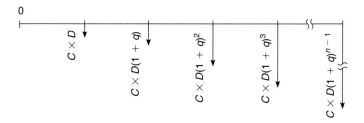

Figure 21.2   Cash flow diagram representing a double geometric gradient, $C \times D$.

Example Problem Illustrating the Use of the
Double Geometric Gradient

EXAMPLE 21.5

Find the equivalent net present worth of the income from an excavator with production values and sales prices listed below. Assume the following:

$$i = 15\%$$
$$n = 10$$

Production:

$$C = 222.5 \text{ yd}^3/\text{hr}$$
$$r = -4.5\%/\text{yr}$$
$$D = \$0.10/\text{yd}^3$$
$$s = +10\%/\text{yr}$$

Solution.   Find q from the relationship $(1 + q) = (1 + r)(1 + s)$.

$$q = (1 - 0.045)(1 + 0.10) - 1 = 0.0505$$

Since $q < i$, then

$$w = (1 + i)/(1 + q) - 1 = 1.15/1.0505 - 1$$
$$w = 1.09472 - 1 = 0.09472$$
$$P = \{CD/(1 + q)\}(P/A, w, n)$$

Then

$$P = \{CD/(1 + q)\}\{[(1 + w)^n - 1]/(1 + w)^n w\}$$
$$= \{CD/1.0505\}\{[1.09472^{10} - 1]/(1.09472^{10} \times 0.09472)\}$$
$$P = CD \times 5.98425$$

The present value of the production equals $CD \times 5.98425$. Since the value of CD is given as

$$CD = 222.5 \text{ yd}^3/\text{hr} \times \$0.10/\text{yd}^3 \times 2\text{K hr/yr} = \$44.5\text{K/yr}$$

then

$$P = \$44.5\text{K/yr} \times 5.98425 = \underline{\underline{\$266,300}}$$

If desired, the annual worth may be found as follows:

$$A = \$266.3\text{K} \times \underbrace{(A/P, 15\%, 10)}_{0.19925} = \$53,060/\text{yr}$$

COMPARE THE GEOMETRIC GRADIENT TO THE ARITHMETIC GRADIENT

If the -4.5 percent geometric gradient in production is multiplied times the initial production figure of 222.5 yd^3/hr, the initial gradient amounts to ($-0.045 \times 222.5 =$) -10 yd^3/hr the first year. If production were assumed to change by an *arithmetic* gradient of -10 yd^3/hr fixed amount each year (instead of the actual geometric gradient), then the result would have been as follows.

$$\text{base production, } P_1 = A_1A_2 \times 2K \times (P/A, 15\%, 10)$$
$$= 222.5 \times 0.10 \times 2K \times 5.0187 = \$223,332$$
$$\text{gradient production, } P_2 = (A_1G_2 + A_2G_1)(P/G, 15\%, 10)$$
$$= (222.5 \times 0.01 - 0.10 \times 10) \times 2K \times 16.979 = \quad 41,599$$
$$GG \text{ production, } P_3 = G_1G_2(P/GG, 15\%, 10) =$$
$$= 0.01 \times 10 \times 2K \times 95.0617 = \quad \underline{19,012}$$
$$\text{calculated as an } \textit{arithmetic} \text{ gradient } P_{\text{total}} = \underline{\underline{\$245,919}}$$

Compare to the NPW of \$266,300 if calculated as a *geometric* gradient. The differences increase an *n* increases.

SUMMARY

Double gradients are needed to solve a wide variety of problems involving gradient changes in two interdependent variables.

In general, providers of capital facilities as well as goods and services may experience a gradient in volume of units of production or consumption (customers, passengers, vehicles, or other units used, sold, purchased, etc.), as well as a gradient in the unit payment (cost, income, savings, etc.).

For example, in future years a municipal water department expects both a gradient increase in numbers of customers as well as increased rates. The gradients could be expressed as either arithmetic or geometric gradients. If the increments are represented by *arithmetic gradients,* then

$$A_1 = \text{the present number of customers}$$
$$G_1 = \text{the gradient change in customers}$$
$$A_2 = \text{the present rate charged per customer}$$
$$G_2 = \text{the gradient rate charged per customer}$$

Then, for instance, to find the NPW, the following equation may be used:

$$P = A_1 A_2 (P/A, i, n) + (A_1 G_2 + G_1 A_2)(P/G, i, n) + G_1 G_2 (P/GG, i, n)$$

If the increments are represented by *geometric gradients,* then

$C =$ the present number of customers

$r =$ the % rate of increase in the present number of customers

$D =$ the present water rate charged per customer

$s =$ the % rate of increase in water rates charged per customer

$i =$ the investment interest rate

Then, $q = (1 + r)(1 + s) - 1$, and

if $q > i$, then $w = (1 + q)/(1 + i) - 1$ and $P = \{CD/(1 + i)\}(F/A, w, n)$

if $q < i$, then $w = (1 + i)/(1 + q) - 1$ and $P = \{CD/(1 + q)\}(P/A, w, n)$

if $q = i$, then $P = Cn/(1 + q) = Cn/(1 + i)$

These problems could involve, for example:

Transportation

Number of vehicles increases	Cost or savings per vehicle increases
Number of passengers increases	Cost per passenger changes
Individual miles per year declines with vehicle age	Maintenance cost per mile increases

Construction

Equipment output units per hour declines with age	Price per unit of output increases

Probability problems

Change in frequency of occurrence	Change in cost of each occurrence

General

Increase in customers	Decreased cost to customer
Change in occupancy rate	Change in rental rate
Change in units produced	Change in cost per unit

When more than two interdependent gradients occur, the problem can be solved by individually calculating each item of cash flow, treating them as an irregular sequence of lump sums, and finding the equivalent present worth at PTZ by using the P/F relationships.

TABLE 21.1 Interest tables for the double arithmetic gradient

n	i = 1% F/GG	P/GG	A/GG	i = 2% F/GG	P/GG	A/GG	i = 3% F/GG	P/GG	A/GG	n
1	0.00000	0.00000	0.00000	0.00000	0.00000	0.00000	0.00000	0.00000	0.00000	1
2	1.00000	0.98030	0.49751	1.00000	0.96117	0.49505	1.00000	0.94260	0.49251	2
3	5.01000	4.86266	1.65341	5.02000	4.73046	1.64031	5.03000	4.60316	1.62736	3
4	14.0601	13.5115	3.46274	14.1204	13.0451	3.42594	14.1809	12.5995	3.38962	4
5	30.2007	28.7349	5.92054	30.4028	27.5368	5.84215	30.6063	26.4013	5.76484	5
6	55.5027	52.2861	9.02187	56.0109	49.7360	8.87917	56.5245	47.3384	8.73855	6
7	92.0577	85.8639	12.7618	93.1311	81.0762	12.5272	94.2203	76.6097	12.2963	7
8	141.978	131.115	17.1354	143.994	122.897	16.7767	146.047	115.291	16.4239	8
9	207.398	189.632	22.1377	210.874	176.450	21.6178	214.428	164.341	21.1070	9
10	290.472	262.961	27.7639	296.091	242.898	27.0410	301.861	224.613	26.3315	10
11	393.377	352.593	34.0090	402.013	323.324	33.0366	410.917	296.855	32.0833	11
12	518.311	459.974	40.8682	531.053	418.732	39.5951	544.244	381.722	38.3486	12
13	667.494	586.502	48.3364	685.674	530.048	46.7070	704.572	479.779	45.1134	13
14	843.169	733.525	56.4090	868.388	658.129	54.3628	894.709	591.508	52.3640	14
15	1047.60	902.350	65.0809	1081.76	803.760	62.5530	1117.55	717.313	60.0869	15
16	1283.08	1094.23	74.3473	1328.39	967.661	71.2683	1376.08	857.526	68.2683	16
17	1551.91	1310.40	84.2035	1610.96	1150.49	80.4993	1673.36	1012.41	76.8951	17
18	1856.43	1552.00	94.6444	1932.18	1352.83	90.2368	2012.56	1182.17	85.9538	18
19	2198.99	1820.19	105.665	2294.82	1575.24	100.471	2396.94	1366.94	95.4313	19
20	2581.98	2116.05	117.261	2701.72	1818.18	111.194	2829.84	1566.82	105.315	20
21	3007.80	2440.62	129.428	3155.75	2082.09	122.395	3314.74	1781.84	115.591	21
22	3478.88	2794.92	142.160	3659.87	2367.34	134.066	3855.18	2011.99	126.247	22
23	3997.67	3179.91	155.453	4217.06	2674.28	146.198	4454.84	2257.23	137.271	23
24	4566.64	3596.53	169.301	4830.41	3003.17	158.781	5117.48	2517.46	148.650	24
25	5188.31	4045.68	183.701	5503.01	3354.26	171.806	5847.01	2792.56	160.371	25
30	9176.32	6808.12	263.802	9868.80	5448.28	243.265	10642.3	4384.50	223.694	30
36	16245.3	11354.2	377.122	17771.6	8712.03	341.798	19522.4	6735.86	308.528	36
40	22616.1	15190.2	462.626	25030.0	11335.9	414.391	27849.5	8537.46	369.351	40
48	40174.4	24918.6	656.202	45535.3	17601.1	573.828	52032.3	12591.7	498.353	48
60	81560.4	44894.9	998.662	95960.3	29247.0	841.376	114443.	19424.7	701.873	60
120	783778.	237481.	3407.16	>10**6	106427.	2346.51	>10**6	51508.2	1591.08	120
180	>10**6	533967.	6408.50	>10**6	174005.	3581.50	>10**6	67580.5	2037.38	180
240	>10**6	856001.	9425.30	>10**6	215115.	4339.75	>10**6	73086.6	2194.42	240
300	>10**6	>10**6	12116.4	>10**6	236056.	4733.57	>10**6	74658.1	2240.06	300
360	>10**6	>10**6	14331.7	>10**6	245661.	4917.15	>10**6	75061.0	2251.88	360
INF	INF	2010000	20100.0	INF	252500.	5050.00	INF	75185.2	2255.56	INF

TABLE 21.1 *(continued)*

n	i = 5%			i = 6%			i = 8%			n
	F/GG	P/GG	A/GG	F/GG	P/GG	A/GG	F/GG	P/GG	A/GG	
1	0.00000	0.00000	0.00000	0.00000	0.00000	0.00000	0.00000	0.00000	0.00000	1
2	1.00000	0.90703	0.48780	1.00000	0.89000	0.48544	1.00000	0.85734	0.48077	2
3	5.05000	4.36238	1.60190	5.06000	4.24847	1.58940	5.08000	4.03267	1.56481	3
4	14.3025	11.7667	3.31835	14.3636	11.3773	3.28340	14.4864	10.6479	3.21483	4
5	31.0176	24.3031	5.61341	31.2254	23.3334	5.53928	31.6453	21.5373	5.39415	5
6	57.5685	42.9585	8.46358	58.0989	40.9575	8.32922	59.1769	37.2915	8.06673	6
7	96.4469	68.5430	11.8456	97.5849	64.8995	11.6258	99.9111	58.2972	11.1973	7
8	150.269	101.708	15.7365	152.440	95.6427	15.4019	156.904	84.7703	14.7513	8
9	221.783	142.963	20.1135	225.586	133.524	19.6310	233.456	116.786	18.6951	9
10	313.872	192.690	24.9543	320.122	178.754	24.2870	333.133	154.305	22.9960	10
11	429.565	251.158	30.2366	439.329	231.433	29.3441	459.783	197.193	27.6221	11
12	572.044	318.535	35.9389	586.689	291.566	34.7772	617.566	245.244	32.5427	12
13	744.646	394.902	42.0396	765.890	359.079	40.5616	810.971	298.192	37.7279	13
14	950.878	480.258	48.5176	980.843	433.828	46.6733	1044.85	355.730	43.1490	14
15	1194.42	574.537	55.3523	1235.69	515.612	53.0888	1324.44	417.518	48.7784	15
16	1479.14	677.613	62.5232	1534.84	604.182	59.7851	1655.39	483.193	54.5896	16
17	1809.10	789.305	70.0106	1882.93	699.252	66.7399	2043.82	552.382	60.5573	17
18	2188.56	909.390	77.7949	2284.90	800.501	73.9315	2496.33	624.704	66.6572	18
19	2621.98	1037.61	85.8569	2746.00	907.587	81.3387	3020.04	699.779	72.8663	19
20	3114.08	1173.66	94.1779	3271.75	1020.15	88.9412	3622.64	777.231	79.1627	20
21	3669.79	1317.24	102.740	3868.06	1137.81	96.7191	4312.45	856.693	85.5256	21
22	4294.28	1468.00	111.525	4541.14	1260.19	104.653	5098.45	937.811	91.9355	22
23	4992.99	1625.57	120.515	5297.61	1386.90	112.725	5990.32	1020.24	98.3741	23
24	5771.64	1789.60	129.694	6144.47	1517.55	120.917	6998.55	1103.67	104.824	24
25	6636.22	1959.69	139.045	7089.14	1651.76	129.212	8134.43	1187.77	111.269	25
30	12479.9	2887.56	187.840	13572.2	2363.05	171.673	16192.0	1609.12	142.934	30
36	23865.8	4120.60	249.027	26563.6	3260.44	222.997	33358.2	2089.04	178.289	36
40	35055.8	4979.53	290.198	39669.3	3856.74	256.325	51693.4	2379.49	199.545	40
48	69700.8	6701.18	370.699	81745.3	4986.33	318.615	115493.	2872.20	235.636	48
60	169939.	9097.75	480.618	211734.	6418.59	397.155	343544.	3392.79	274.131	60
120	>10**6	15252.4	764.814	>10**6	9246.46	555.298	>10**6	4040.89	323.303	120
180	>10**6	16276.0	813.923	>10**6	9518.94	571.152	>10**6	4062.05	324.964	180
240	>10**6	16388.8	819.448	>10**6	9536.11	572.167	>10**6	4062.49	324.999	240
300	>10**6	16399.1	819.955	>10**6	9536.99	572.220	>10**6	4062.50	325.000	300
360	>10**6	16399.9	819.997	>10**6	9537.04	572.222	>10**6	4062.50	325.000	360
INF	INF	16400.0	820.000	INF	9537.04	572.222	INF	4062.50	325.000	INF

TABLE 21.1 (continued)

	i = 10%			i = 15%			i = 20%			
n	F/GG	P/GG	A/GG	F/GG	P/GG	A/GG	F/GG	P/GG	A/GG	n
1	0.00000	0.00000	0.00000	0.00000	0.00000	0.00000	0.00000	0.00000	0.00000	1
2	1.00000	0.82645	0.47619	1.00000	0.75614	0.46512	1.00000	0.69444	0.45455	2
3	5.10000	3.83171	1.54079	5.15000	3.38621	1.48308	5.20000	3.00926	1.42857	3
4	14.6100	9.97883	3.14803	14.9225	8.53199	2.98846	15.2400	7.34954	2.83905	4
5	32.0710	19.9136	5.25315	33.1609	16.4868	4.91267	34.2880	13.7796	4.60761	5
6	60.2781	34.0254	7.81249	63.1350	27.2950	7.21235	66.1456	22.1520	6.66124	6
7	102.306	52.4991	10.7836	108.605	40.8287	9.81361	115.375	32.1990	8.93276	7
8	161.537	75.3580	14.1254	173.896	56.8469	12.6683	187.450	43.5948	11.3612	8
9	241.690	102.500	17.7982	263.980	75.0397	15.7264	288.940	55.9984	13.8921	9
10	346.859	133.729	21.7638	384.578	95.0617	18.9412	427.728	69.0804	16.4772	10
11	481.545	168.779	25.9857	542.264	116.556	22.2702	613.273	82.5392	19.0751	11
12	650.700	207.333	30.4289	744.604	139.172	25.6745	856.928	96.1101	21.6502	12
13	859.770	249.045	35.0601	1000.29	162.576	29.1190	1172.31	109.569	24.1731	13
14	1114.75	293.548	39.8480	1319.34	186.460	32.5725	1575.78	122.732	26.6197	14
15	1422.22	340.468	44.7627	1713.24	210.548	36.0072	2086.93	135.453	28.9710	15
16	1789.44	389.435	49.7763	2195.23	234.592	39.3992	2729.32	147.623	31.2128	16
17	2224.39	440.083	54.8626	2780.61	258.381	42.7277	3531.18	159.162	33.3349	17
18	2735.83	492.062	59.9973	3486.59	281.734	45.9751	4526.42	170.017	35.3304	18
19	3333.41	545.039	65.1577	4333.57	304.500	49.1269	5755.70	180.158	37.1959	19
20	4027.75	598.699	70.3230	5344.61	326.557	52.1712	7267.84	189.575	38.9304	20
21	4830.52	652.752	75.4740	6546.30	347.809	55.0988	9121.41	198.269	40.5350	21
22	5754.58	706.927	80.5932	7969.25	368.184	57.9027	11386.7	206.258	42.0125	22
23	6814.04	760.979	85.6648	9648.63	387.628	60.5779	14148.0	213.564	43.3673	23
24	8024.44	814.686	90.6744	11624.9	406.108	63.1214	17506.6	220.218	44.6047	24
25	9402.88	867.848	95.6092	13944.7	423.606	65.5316	21584.0	226.256	45.7306	25
30	19543.7	1120.02	118.811	32875.6	496.523	75.6205	59003.5	248.565	49.9233	30
36	42656.6	1379.91	142.604	85086.4	555.569	83.8830	186366.	262.930	52.2604	36
40	68944.4	1523.32	155.774	155780.	581.564	87.5615	393912.	268.009	53.6383	40
48	168996.	1741.92	176.005	501725.	612.307	91.9583	>10**6	272.754	54.5594	48
60	589311.	1935.46	194.184	>10**6	630.201	94.5517	>10**6	274.622	54.9255	60
120	>10**6	2098.17	209.819	>10**6	637.031	95.5547	>10**6	275.000	55.0000	120
180	>10**6	2099.99	209.999	>10**6	637.037	95.5556	>10**6	275.000	55.0000	180
240	>10**6	2100.00	210.000	>10**6	637.037	95.5556	>10**6	275.000	55.0000	240
300	>10**6	2100.00	210.000	>10**6	637.037	95.5556	>10**6	275.000	55.0000	300
360	>10**6	2100.00	210.000	>10**6	637.037	95.5556	>10**6	275.000	55.0000	360
INF	INF	2100.00	210.000	INF	637.037	95.5556	INF	275.000	55.0000	INF

TABLE 21.1 (continued)

	i = 30%			i = 50%			
n	F/GG	P/GG	A/GG	F/GG	P/GG	A/GG	n
1	0.00000	0.00000	0.00000	0.00000	0.00000	0.00000	1
2	1.00000	0.59172	0.43478	1.00000	0.44444	0.40000	2
3	5.30000	2.41238	1.32832	5.50000	1.62963	1.15789	3
4	15.8900	5.56353	2.56829	17.2500	3.40741	2.12308	4
5	36.6570	9.87280	4.05359	41.8750	5.51440	3.17536	5
6	72.6541	15.0522	5.69567	87.8125	7.70919	4.22556	6
7	130.450	20.7894	7.41918	167.719	9.81619	5.21321	7
8	218.585	26.7963	9.16205	300.578	11.7281	6.10214	8
9	348.161	32.8314	10.8749	514.867	13.3929	6.87528	9
10	533.609	38.7070	12.5203	853.301	14.7975	7.52934	10
11	793.692	44.2869	14.0712	1379.95	15.9536	8.07012	11
12	1152.80	49.4805	15.5098	2190.93	16.8862	8.50870	12
13	1642.64	54.2349	16.8260	3430.39	17.6261	8.85859	13
14	2304.43	58.5271	18.0157	5314.59	18.2050	9.13381	14
15	3191.76	62.3563	19.0796	8167.88	18.6526	9.34767	15
16	4374.29	65.7376	20.0222	12476.8	18.9952	9.51208	16
17	5942.58	68.6970	20.8501	18971.2	19.2550	9.63729	17
18	8014.35	71.2669	21.5719	28745.8	19.4506	9.73187	18
19	10742.7	73.4832	22.1968	43442.8	19.5967	9.80278	19
20	14326.5	75.3827	22.7344	65525.1	19.7053	9.85561	20
21	19024.4	77.0017	23.1944	98687.7	19.7855	9.89472	21
22	25172.7	78.3747	23.5859	148473.	19.8444	9.92354	22
23	33208.5	79.5339	23.9175	223193.	19.8875	9.94466	23
24	43700.1	80.5085	24.1971	335318.	19.9190	9.96008	24
25	57386.1	81.3248	24.4321	503553.	19.9418	9.97129	25
30	219433.	83.7532	25.1355	>10**6	19.9893	9.99468	30
36	>10**6	84.7736	25.4341	>10**6	19.9987	9.99934	36
40	>10**6	85.0106	25.5039	>10**6	19.9997	9.99984	40
48	>10**6	85.1552	25.5466	>10**6	20.0000	9.99999	48
60	>10**6	85.1832	25.5550	>10**6	20.0000	10.0000	60
120	>10**6	85.1852	25.5556	>10**6	20.0000	10.0000	120
180	>10**6	85.1852	25.5556	>10**6	20.0000	10.0000	180
240	>10**6	85.1852	25.5556	>10**6	20.0000	10.0000	240
300	>10**6	85.1852	25.5556	>10**6	20.0000	10.0000	300
360	>10**6	85.1852	25.5556	>10**6	20.0000	10.0000	360
INF	INF	85.1852	25.5556	INF	20.0000	10.0000	INF

PROBLEMS FOR CHAPTER 21, DOUBLE GRADIENTS, ARITHMETIC AND GEOMETRIC

Problem Group A

Compare alternatives involving double arithmetic gradients. No taxes.

A.1. Two alternate alignments are under consideration for the location of a proposed highway. The characteristics of each are listed in the table. The motoring public is assumed to borrow and lend at an average $i = 10$ percent. Both routes will cost the same to construct and maintain. You are asked to find the route with the least user cost over the 10-year period.

	A	B
Traffic count the first year	10,000 vehicles/day	15,000 vehicles/day
Gradient increase in each subsequent year	1,000 vehicles/day/yr	0
Operating cost per vehicle traversing the entire route the first year	$1.00/vehicle	$1.25/vehicle
Gradient increase in each subsequent year	$0.10/vehicle/yr	$0.10/vehicle/yr

(*Ans.* alternate A, $A = \$7,164$/yr; alternate B, $A = \$8,883$/yr)

A.2. A contractor buys a new scraper for $250,000. He expects to keep it for five years and then trade it in on a new one. He asks you to determine if the extra amount he is charging per cubic yard of excavation is sufficient to fund the purchase of a new scraper at the end of the next five-year period. The following data are provided as a basis for your calculations.

Heavy equipment prices for new equipment are rising at the rate of 5.5 percent compounded annually, so the replacement five years from now will cost more than the present model. During the first year, the scraper moves dirt at an average rate of 300 yd^3/hr, but with age and loss of efficiency, the rate declines by 30 yd^3/hr each year. The contractor's equipment works an average of 1,800 hr/yr.

The contractor is charging $0.10/yd^3 this year to cover the replacement cost, but expects to raise the amount by $0.02/yd^3 each year starting one year from now. He deposits this extra $0.10/yd^3 charge into an account bearing 10 percent interest compounded annually. The existing machine can be traded in on the new machine for an estimated $20,000 trade-in allowance at the end of the five-year period. Will the balance in the account plus the trade-in value be sufficient to purchase the new machine?

A.3. Find the equivalent annual cost of a bulldozer over a five-year life span, using the following data. Use $i = 10$ percent.

$$\text{cost new} = \$80,000$$
$$\text{resale price after 5 yr} = \$35.000$$
$$\text{cost per hour to operate and maintain the first year} = \$16.00/\text{hr}$$
$$\text{gradient in hourly costs each year} = \$1.10/\text{hr/yr}$$
$$\text{number of hours used the first year} = 2,000 \text{ hr/yr}$$
$$\text{gradient in hours of use per year} = -100 \text{ hr/yr}$$

(Machine is used 1,900 hr the second year, 1,800 hr the third year, and so on.)

(*Ans. A* = $47,879/yr)

A.4. At the present time a certain downstream area floods when it is subject to a 10-year storm. (The area has a 1 in 10 probability, or $p = 0.10$, of flooding this year.) With increasing development upstream, this probability will increase by $p = 0.01$ each year for the next 20 years (next year $p = 0.11$, the third year $p = 0.12$ etc.). If the area floods this year, the cost of flood damage is estimated at $100,000. Due to the increasing value of the property in the area, if flooding occurs in any future year, the damage is expected to increase by $10,000 per year each year after this year. A flood-prevention structure can be built to protect the area for the next 20 years at an estimated cost of $200,000. Is the present worth of the probable flood damage greater or less than the cost of the flood-prevention structure? Which alternative do you recommend and why? Use $i = 8$ percent.

A.5. Assume a friend purchased a car four years ago for $9,600. Expenses have been as listed in the table below. At what minimum price can she sell the car now if her goal is to keep mileage costs under $0.28/mi. Use $i = 10$ percent.

	Year			
	1	2	3	4
Gas (cents/gal)	120	130	140	150
Tires (cents/mi)	1.6	1.7	1.8	1.9
Repairs and maintenance (cents/mi)	2.2	2.6	3.0	3.4
License and insurance (cents/mi)	4.0	4.2	4.4	4.6
mi/yr driven	20,000	18,000	16,000	14,000

The car gets 20 miles to the gallon. Make the simplifying assumption that all car costs are paid at the end of the year in which incurred.

(*Ans. F* = $3,882)

A.6. A city police department wants to acquire a helicopter to patrol in high crime areas. They can either lease or buy one. You are asked to compare the costs of the two alternative plans.

The helicopter will fly 1,000 hours per year, and be sold at EOY 5. Use $i = 10$ percent.

Plan A. Lease. A helicopter can be obtained on a long-term lease for $150/hr the first year (payable at EOY), with the hourly rate escalating by 3.5 percent per year for each year thereafter (geometric gradient).

Plan B. Buy

$$\text{cost new, purchased at EOY } 0 = \$243,000$$
$$\text{resale value at EOY } 5 = \$125,000$$

The operating and maintenance (O & M) costs in dollars per hour increase as the helicopter ages, as shown below.

O & M costs	Charged at EOY
$100/hr	1
110	2
120	3
130	4
140	5 Sell helicopter

(a) Find the NPW of the lease alternative.
(b) Find the NPW of the buy alternative, and compare with the lease cost.
(c) Find the NPW of the O & M costs (only) if the hours flown per year *de*crease by *50* hours per year (e.g., 1,000 hours flown the first year of life, 950 the second year, 900 the third, etc.).

A.7. A highway department has a long-term need for fill dirt, and thinks it can save money by purchasing its own borrow pit together with a dragline to excavate the material from the borrow pit. They now pay $1/yd^3 for borrow covered by the costs listed below.

Excavation. Excavation rate begins at 100 yd^3/hr when the machine is new, but production declines as the machine ages.

Excavation	At EOY
100 yd^3/hr	1
96 yd^3/hr	2
92 yd^3/hr	3
88 yd^3/hr	4
84 yd^3/hr	5
80 yd^3/hr	6
76 yd^3/hr	7 Sell at EOY 7

Equipment costs

$$\text{cost new, purchased EOY } 0 = \$280,000$$
$$\text{resale value at EOY } 7 = \$ \ 50,000$$

The machine works 2,000 hours per year. Use $i = 10$ percent.

Borrow-pit cost

$$\text{cost to purchase borrow pit now} = \$200,000$$
$$\text{resale value at EOY } 7 = \$150,000$$

(a) Find the NPW of the cost of owning and operating the borrow pit and equipment, plus the savings from not purchasing borrow.
(b) Find the NPW if the hours of use per year decrease by 100 hours per year (2,000 hr of work during the first year. 1,900 for second year, 1,800 for the third, etc.)

A.8.   Regarding their fleet of automobiles for the next four years, a state agency asks your assistance in determining which is more economical: (1) lease or (2) purchase. The agency's $i = 10$ percent, and the data below are for each auto. All mileage costs are debited to the EOY.

   1. *Lease alternative:* Cost = $10,000 per year for four years paid in advance at the beginning of each year. This price covers all ownership and O & M costs.

   2. *Purchase Alternative:*

$$\text{cost new, purchased at EOY } 0 = \$15,000$$
$$\text{resale value at EOY } 4 = \$4,000$$

Each auto is driven 20,000 miles per year. The operating and maintenance (O & M) costs in dollars per mile increase as the auto ages, as shown below.

O & M costs	Charged at EOY
$0.30/mi	1
0.32	2
0.34	3
0.36	4 Sell at EOY

   (a) Find the NPW of each alternative, and compare.
   (b) Find the NPW of the purchase alternative if the miles driven per year *de*crease by 2,000 mi/yr (e.g., 20,000 miles driven the first year of the auto's life, 18,000 the second year, 16,000 the third, etc.).

A.9.   A city public works department needs to acquire an additional new pickup truck. They can either lease or buy one. You are asked to compare the costs of the two alternative plans. All mileage costs are charged at EOY. The pickup will be driven 15,000 mi/yr. Use $i = 10$ percent.

   Plan A. Lease. A pickup can be obtained on a lease for $0.50/mi the first year (payable at EOY), with the rate escalating by 4 percent per year for each year thereafter (geometric gradient).

   Plan B. Buy

$$\text{cost new, purchased at EOY } 0 = \$16,000$$
$$\text{resale value at EOY } 3 = \$\ 9,000$$

The operating and maintenance (O & M) costs in dollars per mile increase as the pickup gets older, as shown below.

O & M costs	Charged at EOY
$0.30	1
0.35	2
0.40	3 Sell pickup

(a) Find the NPW of the lease alternative.
(b) Find the NPW of the buy alternative, and compare with the lease cost.
(c) Find the NPW of the O & M costs if the miles driven per year *decrease* by *3,000* mi/yr within each three-year cycle (e.g., 15,000 miles driven the first year of life, 12,000 the second year, 9,000 the third).

Problem Group B

After-tax problems.

B.1. A firm of consulting engineers needs to buy an airplane for transportation to remote project sites. The plane will fly *1,000* hours per year the first year, *decreasing* by *100* hours per year (e.g., 1,000 hours flown the first year of life, 900 the second year, 800 the third, etc.). The firm's after-tax MARR is 15 percent.

cost new, purchased at EOY 0 = $243,000
resale value at EOY 6 = $125,000

Taxes. The firm is in the 34 percent tax bracket for ordinary income. Interest, depreciation, major overhauls, and O & M payments are tax deductible.

Resale. Upon resale, the difference between resale value and SLBV is taxed as ordinary income.

Financing. Assume 75 percent of the purchase price is financed at 10 percent simple interest, paid annually at EOY, with the principal balance due and payable at EOY 4.

Depreciation. Use straight-line depreciation to zero salvage value, with a five-year recovery (depreciation) life.

Operating and maintenance (O & M). O & M costs in dollars per hour increase as the plane ages, as shown below.

O & M costs	Charged at EOY
$100/hr	1
110	2
120	3
130	4
140	5
150	6 Sell plane

Major overhaul (MO). At EOY 3 the plane needs a major overhaul at a cost of $20,000.
(a) Find the NPW after taxes for:
  1. Cost new, excluding the loan, interest, and repayment of principal.

   (*Ans.* −$243,000)

  2. Financing, including the loan, interest, and repayment of principal.

   (*Ans.* +$43,710)

  3. After-tax cash flows resulting from depreciation.

   (*Ans.* +$55,390)

4. Resale value including recapture tax.

(*Ans.* +$35,670)

5. O & M costs.

(*Ans.* −$351,100)

6. Major overhaul.

(*Ans.* −$13,150)

7. Total NPW after taxes.

(*Ans.* −$472,500)

(b) Whenever the plane is used on a client's project, the cost is billed against that project. What should be the billing rate (in $/hr) the first year in order to recover all costs including taxes, and assuming the billing rate will be raised by $25 per hour at the end of each year?

(*Ans.* $195.90/hr)

B.2. A surveying firm with an expanding business needs to add a new four-wheel-drive off-road vehicle to their fleet. The vehicle will be driven 15,000 mi/yr for the first year, decreasing by 1,000 mi/yr (e.g., 15,000 miles the first year of the cycle, 14,000 miles for the second year, 13,000 miles the third year, etc.). They plan to sell it at EOY 4. The firm's after-tax MARR is *15* percent.

$$\text{cost new, purchased at EOY } 0 = \$16,000$$
$$\text{resale value at EOY } 4 = \$\ 5,000$$

All mileage costs are credited at the EOY.

Taxes. The ordinary tax rate is $t = 34$ percent. Interest, depreciation, major overhauls, and O & M payments are tax deductible.

Resale. Upon resale, the difference between resale value and SLBV is taxed as ordinary income.

Financing. Assume 75 percent of the purchase price is financed at 10 percent simple interest, paid annually at EOY, with the principal balance due and payable at EOY 3.

Depreciation. Use straight-line depreciation to zero salvage value, with a five-year recovery (depreciation) life. (*Note:* Depreciation life > economic life.)

O & M costs. The operating and maintenance costs in dollars per mile increase as the vehicle gets older, as shown below.

O & M costs	Charged at EOY
$0.30	1
0.35	2
0.40	3
0.45	4 Sell vehicle

Major overhauls (MO). At EOY 2 the vehicle needs a major overhaul at a cost of $4,000.

(a) Find the NPW after taxes for the following:
  1. Cost new, excluding the loan, interest, and repayment of principal.
  2. Financing, including the loan, interest, and repayment of principal.
  3. After-tax cash flows resulting from depreciation.
  4. Resale value after tax.
  5. O & M costs.
  6. Major overhauls.
  7. Total NPW after taxes.

(b) Find the *billing rate*. Whenever the vehicle is used on a client's project, the mileage cost is billed against that project. What should be the mileage billing rate (in $/mi) the first year in order to recover all costs including taxes, and assuming the billing rate will be raised by $0.05/mi at the end of each year.

Problem Group C

Double geometric gradient, no tax.

C.1. A county engineer's budget receives $0.04 a gallon from a tax on fuel sold in the county. He asks your help in predicting what that budget will be in future years. Assume discrete EOM compounding.

$$\text{vehicles now registered in the county} = 40{,}000$$
$$\text{expected increase in vehicle registration} = 0.3\%/\text{month}$$
$$\text{average fuel consumption now} = 17 \text{ mi/gal}$$
$$\text{expected decrease in fuel consumption} = 0.5\%/\text{month}$$
$$\text{average annual mileage per vehicle} = 12{,}000 \text{ mi/yr}$$

Average mileage is expected to remain constant in future year, and assume the interest rate for the county is $i = 1$ percent per month.

(a) Find the PW of all fuel-tax receipts expected in the county for the next 10 years.

(*Ans.* $5,970,000)

(b) Assume the county wants to pledge one-half of these funds to a road-improvement program. Money will be borrowed on bonds to be repaid at 9 percent over a 10-year time period (interest compounded monthly). How much bond money for the road program can be raised now by pledging one-half the fuel-tax revenues for the next 10 years?

(*Ans.* $3,381,000)

C.2. The estimated data for a proposed highway intersection improvement are shown.

$$\text{present traffic count} = 4{,}000{,}000 \text{ vehicles/yr}$$
$$\text{anticipated increase in traffic} = 5\%/\text{yr compounded}$$
$$\text{savings in vehicle operating costs attributable to the}$$
$$\text{intersection improvement} = \$0.10/\text{vehicle}$$
$$\text{anticipated increase in savings} = 7\%/\text{yr compounded}$$
$$n = 20 \text{ yr}$$
$$i = 10\%$$

Find the PW of the savings in vehicle operating costs over the life of the project.

CHAPTER 22

Utility Rate Studies

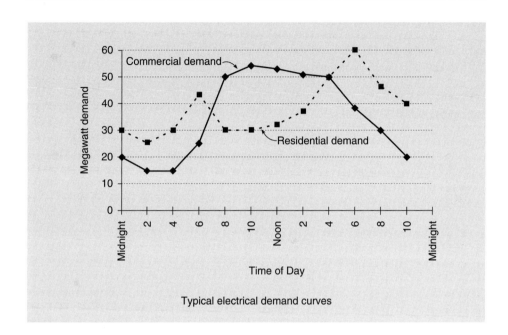

Typical electrical demand curves

RATE STUDIES

A utility rate study is simply an analysis of the utility's finances in which the cost of service to each class of customer is determined and compared to the rates paid by each class. Utilities usually try to set rates so that each class of customer pays for its own fair share of costs. Most of these rate studies are carried out by engineers who are familiar with both the utility system and with the techniques of cost analysis.

Rate studies will vary in level of sophistication and detail depending upon the complexity of the system, the requirements of the system's policy board, and the performance and capability of the consultant. A method that is adequate for a small utility having only one class of customer, and therefore a very simple rate schedule, will serve as the first example. Later, a second example will deal with the problems of more complex utility systems.

Costs fall basically into two categories:

1. Ownership costs, also called fixed costs, capital costs, plant costs, installation costs, and so forth.

2. Operating costs, including maintenance, repairs, customer accounts, and so forth; also called variable costs.

These costs may be further subdivided for more complex studies as desired. The allocation of these costs into rates is typically done by charging a flat base rate per customer to cover the cost of constructing and owning the utility and connecting it to the customer's property. This is sometimes called a "ready to serve" charge. In the simplest case this base rate amounts to the total monthly ownership costs divided by the total number of customers. If a water system has ownership costs equivalent to $128,000 per month with 32,000 customers, then the flat rate per customer is $128,000/32,000 = $4.00 per customer.

The operating costs are covered by charging an additional unit cost per unit of product used. If the operating costs are $80,000 per month to operate a water system selling 320,000 kgal (1 kgal = 1,000 gallons*) of water per month, then the variable rate is $80,000/320,000 = $0.25/kgal.

Therefore, the rate for this utility system, with only one class of customer is

1. $4.00 per month base charge to each customer regardless of consumption to cover ownership costs plus

2. $0.25/kgal for each 1,000 gallons consumed by the customer per month. Thus, a customer using 6 kgal in one month receives a bill for $5.50, and a customer using 10 kgal is billed for $6.50.

MORE THAN ONE CLASS OF CUSTOMER

As utility systems increase in size, they frequently are required to service more than one type of customer. The question then arises, should the large customer be charged the same rate as the small customer, or should customers be divided into classifications with different rates? Following the original premise that each customer should pay its own fair share, an analysis can be made to determine the actual costs attributable to each customer class. The system's costs are subdivided into appropriate categories, such as ownership costs and operating costs, and each class of customer pays a fair share of each.

ELECTRIC UTILITY SYSTEMS RATE STUDIES

An electric utility is similar to other rate-making organizations in that there are ownership costs and operating costs, but they are called by different names: demand costs and energy costs.

Demand Costs

The cost of ownership usually is called a *demand* cost, since it relates to the amount of generating capacity in kilowatts (kW) which may be required (demanded) by the system

---

* The symbol k is the first letter of the world *kilo,* meaning 1,000. The custom of using M (the Roman letter for 1,000) is not followed here because of the confusion that arises when M is incorrectly assumed to be one million.

or a class of customer. The electric utility cannot store its product and therefore installs generating capacity that exceeds system demand by a safe margin. The margin should be sufficient to account for generator outages during peak load, and for growth until a new generating unit is ready to come on line. In effect, each customer requires a certain number of kilowatts reserved from the system's total kilowatts of generating capacity to be ready to serve the customer on demand, at the flick of a switch. These kilowatts of capacity can be shared, of course, with off-peak customers. But for all practical purposes, a certain number of kilowatts of generating capacity is standing by, reserved and ready on demand for the personal use of each customer at the time of peak load. Part of the reserved kilowatts of demand capacity is used daily, while another significant part is used only a few hours each year, either at the summer peak (late afternoon of a hot summer day) or the winter peak (usually a cold winter morning). The remaining part (about 10–15 percent) of the reserved kilowatts is needed for emergency reserve to be used only during breakdown or maintenance downtime for the rest of the system. Thus, reasonably, each customer should expect to pay a fair share of the demand costs, which include all of the capital costs for purchase and installation of all plant equipment, in addition to the costs of maintaining the plant and equipment in ready-to-serve, standby condition. Demand costs usually are calculated in terms of dollars per kilowatt. The nomenclature used to describe system and customer-class demand follows.

1. System peak demand is the maximum demand (kW) imposed on the system, sometimes called *system coincident peak demand,* or simply coincident peak.

2. *Class noncoincident peak demand* is the maximum demand of a customer class (in kW), which does not occur simultaneously with the system peak.

3. *Class coincident peak demand* refers to the maximum coincident demand (kW) of any one class of customers.

Examples of the variation and types of demand are given in Figure 22.1.

Energy Costs

The costs of operating the system and producing kilowatt-hours (kWh) of electrical energy are called energy costs. These are the costs of turning the generators, including labor, repairs, supplies, maintenance, and a base charge for fuel. Since fuel costs have risen so rapidly, a separate fuel escalation charge is added later to compensate for the incremental fuel costs above the base fuel cost. Energy production costs are normally treated as a constant unit cost (in terms of $/kWh), although any given generator can usually produce energy for slightly less cost at optimum load than at other loads. Energy production costs usually are assumed to be the same in dollars per kilowatt-hour for each class of customer.

Customer Costs

In addition to the familiar owning (demand) and operating (energy) costs, electric utility costs usually involve one or more additional cost categories not directly related to the kilowatts of demand nor to the kilowatt-hours of energy consumed. For instance,

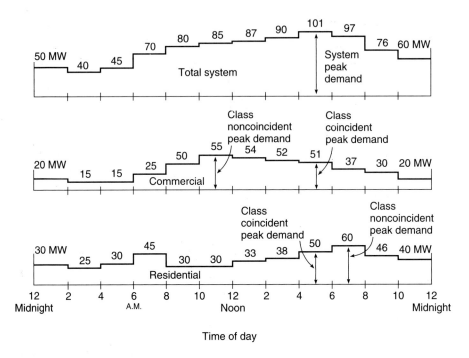

Figure 22.1   Example of power demand for two customer classes.

"customer" costs include billing costs, meter-reading costs, and any other costs related directly to servicing the customer. Since there is negligible difference in the expense of billing a large customer or a small customer, each customer can equitably share a prorated portion of the cost of customer service. Customer costs usually are calculated first in terms of dollars per customer per year, but they may be billed to the customer either as a flat monthly charge per customer, or prorated over the first few kilowatt-hours each month and charged in terms of dollars per kilowatt-hour.

Revenue-Related Costs

Another cost category sometimes encountered is costs related to revenue, such as the gross receipts tax passed on to state government. Usually this charge is calculated first as a percent increase, and then applied evenly to all charges, including both flat monthly charges as well as unit dollar per kilowatt-hour charges.

BASIC APPROACH TO ELECTRIC RATE MAKING

An electric rate schedule should pass two important tests:

   1. It must return adequate revenue to cover all costs of the system.

2. It should treat each class of customer on the same equitable basis, fairly apportioning the cost of serving to the customer for whose benefit the cost is incurred.

To accomplish this, all the costs of the electric utility system are first allocated to an appropriate cost category, such as demand costs, energy costs, customer costs, and revenue related costs. Then each cost category is totaled and apportioned out appropriately to each class of customer, as illustrated in the following example.

EXAMPLE 22.1   (Electric utility problem)

A municipally owned (no income taxes) utility system has three classes of customers (residential, commercial, and bulk power). The bulk power customer purchases directly from the transmission line, and does not use the distribution system. The following data on the system are provided:

1. Customers. There are 20,000 residential customers, 2,000 commercial customers, and 1 bulk power customer.

2. Expenditures are $14,084,000 per year consisting of the cost items listed in Table 22.1.

TABLE 22.1   Summary statement of expenditures for fiscal year for Example 22.1

Power production expense: electric generation		Cost category
a. Operations	$ 3,198,000	Energy
b. Maintenance	351,000	Demand
c. Transmission expense	160,000	Demand
c. Distribution expense	511,000	Demand
c. Engineering expense	148,000	Demand
d. Customer accounts	281,000	Customer
e. Sales promotion expense	56,000	Demand
f. System security	21,000	Demand
g. Administration and general		
Employee pensions and benefits	508,000	50% demand
		50% energy
State utility tax	224,000	Revenue
Remaining A & G	940,000	Demand and revenue
h. Transfer to bond, PIF, general government	6,187,000	Demand
Subtotal	$12,585,000	
Fuel-escalation charges collected separately on a flat cost-of-energy per kWH basis ($/kWH)	1,499,000	
Total expenditure for the fiscal year	$14,084,000	

3. Class peak demand. The residential and commercial customers peak simultaneously with the system peak, while the bulk power customer is at 80 percent of its customer class peak when the system peaks. The customer class peaks and the system coincident peaks are listed below.

	Class noncoincident peak demand	Class coincident peak demand
Residential	84,000 kW	84,000 kW
Commercial	40,500 kW	40,500 kW
Bulk power	50,500 kW × 0.80 =	40,400 kW
System coincident peak demand		164,900 kW

4. Energy sales.

	Annual totals	Average monthly/customer
20,000 residential customers	354.08×10^6 kWH	1,475 kWH
2,000 commercial customers	206.02×10^6 kWH	8,584 kWH
1 bulk power customer	159.00×10^6 kWH	13,250 kWH
Total sales	719.10×10^6 kWH	

5. Book value (BV) of the system.

		Percent of value
Generating and transmission	$45,750,000	79.8
Distribution system	11,580,000	20.2
Total BV of electric utility	$57,330,000	100.0

6. Total expenditures for the system are $14,084,000 per year, summed from the individual cost items listed in Table 22.1.

Determine equitable rates to charge each customer for electric service.

Solution.   The problem is to determine what fair share of the system's cost should be borne by each of the three customer classes. A rate schedule can then be constructed that will adequately apportion the costs to the appropriate customer classes.

The first step is to examine each of the cost items in Table 22.1 and allocate each item to one or more of the four cost categories: demand, energy, customer, and revenue. Systematic examination of each of these items in the summary statement of expenditures leads to the allocations shown in Table 22.2 and discussed below.

TABLE 22.2 Summary of cost allocations for Example 22.1

	Demand-related costs based on kW of CP demand	Energy-related costs based on kWh of energy consumed	Customer-related costs based on the no. of customers	Revenue-related costs based on $ revenue received
Power production expense, electric generation, operations, maintenance	$ 351,000	$3,198,000[a]		
Transmission expense	160,000			
Distribution expense	511,000			
Engineering expense	148,000			
Customer accounts			$281,000	
Sales promotion expense	56,000			
System security	21,000			
Administrative and general employee pensions and benefits	254,000	254,000		
State utility tax				$224,000
Remaining A & G	554,000			386,000
Transfer to bond, PIF, general government	6,187,000			
	$8,242,000	$3,452,000	$281,000	$610,000

[a]Does not include $1,499,000 in fuel escalation charges, which are billed and collected separately.

a. The entire expenditure entitled "Operations" under "Power production expense: electric generation" is related to operating rather than owning the system and so is allocated to the *energy* category. All costs incurred in actually turning the generators are considered energy-related costs.

b. The "maintenance" expenditure under "Power production expense electric generation" is necessary to keep the plant in standby ready-to-serve condition, and so is a cost allocated to ownership, or *demand*.

c. Transmission expense, distribution expense, and engineering expense are all incurred due to the installation rather than operation of the utility and so are logically considered ownership costs and charged as *demand-related costs*.

d. Customer accounts are charged to the third category of costs, the *customer-related costs*.

e. The sales promotion expense of $56,000 is usually justified by its proponents as necessary to increase the load factor for the whole system, which reduces the ownership cost per customer. Therefore, it is attributed to ownership and charged to the *demand* category.

f. System security is required whether the generator turns or not, so is attributable to *demand*.

g. Administrative and general. The largest single item here is $508,000 for employee pensions and benefits. This is logically divided between *demand* and *energy* on the basis of the approximate proportion of personnel servicing each category, assumed here as 50–50. Another large item, $224,000 for state utility tax is charged as a percent of the customer's bill and passed on to the state, and so is *revenue-related*.

h. The annual equivalent costs for depreciation, financing, and other capital investment costs are reasonably approximated by the current annual transfer item in the summary. The transfer includes the annual payment for principal and interest on electric revenue bonds, plant improvement fund, disaster fund, and general government (in lieu of taxes). Some of these transfers are required under the bond covenants. The transfer to general government in lieu of taxes compensates city government for property that would generate tax revenue if privately held, since it doubtless generates as much general government costs (police and fire protection, road and traffic) as equivalent privately held property. These costs constitute justifiable charges against all electric customers on an equitable basis and properly are ownership costs allocated to *demand*.

Summary of cost allocations. Based on the foregoing discussions, a summary of cost allocations may be derived from the summary statement of expenditures (Table 22.2).

This table plus the fuel escalation charges add up to the actual $14,084,000 spent on the system, as shown below.

$$
\begin{aligned}
\text{demand (fixed plant costs)} &= \$\ \ 8,242,000 \\
\text{energy (operating costs)} &= \ \ \ 3,452,000 \\
\text{customer accounts} &= \ \ \ \ \ \ 281,000 \\
\text{revenue related} &= \ \ \ \ \ \ 610,000 \\
\text{fuel escalation (billed and collected separately)} &= \ \ \underline{1,499,000} \\
&= \$14,084,000
\end{aligned}
$$

Assumptions for allocation of costs to customer's class rates. After allocating the actual system costs to the four cost categories, the costs must still be allocated from the four cost categories to the customer's rates by classes.

In arriving at an allocation of costs to rates, the following are usually assumed unless the utility's policy-making board has adopted policies to the contrary.

1. Each new class of customer is entitled to be charged the pooled costs of the system rather than the incremental costs of serving the new customer class. Under this system the new capacity and any new fuel source cost (for instance, the incremental costs of high-priced oil versus limited supply of lower priced gas) of serving the new customer class raise the average cost of all old customers. (While the pooled-cost concept is widely accepted, the contrary concept of charging impact fees to large new customers who cause large incremental cost increases to the system certainly is worth consideration and discussion.)

2. There is a reasonably uniform load factor among customers within a given class, and therefore all customers within a class equally share the demand costs of the class.

3. Fixed costs for the systems are allocated on the basis of demand, which in turn is derived by any one of several methods in current use.

Three common demand-allocation methods. Demand-related costs are usually allocated to classes of service in an average cost of service study using one of the three methods described below.

1. *Peak responsibility method.* Demand costs are allocated to classes of service based on the proportion of class contribution to peak at the time of the system coincident peak (CP). This method charges demand costs to customer classes that create peak demand and requires peak demand-meeting facilities to be installed.

2. *Noncoincident demand method.* Demand costs are allocated to classes of service based on the proportion of the summation of the class noncoincident peak (NCP) demand. This method charges demand costs to customer classes based on facility requirements necessary to serve the class maximum demand and gives credit for diversity benefits.

3. *Average and excess demand method.* Demand costs are allocated to classes of service on the basis of the extent of facilities as well as contribution to system peak demands. The maximum demand portion of demand-related costs is allocated to customer classes on the basis of peak responsibility or noncoincidental group demands. The balance of costs is allocated to customer classes on the basis of peak responsibility or noncoincidental group demands. This balance is allocated on the basis of average demands utilizing system load factors.

Only the first method, called the peak responsibility method, will be illustrated here. The load curves for the system are shown in Figure 22.2, subdivided into the three customer classes described. Under this method the three customer classes will share the costs of peak demand in proportion to the ratio of class coincident peak demand to the system coincident peak demand.

$$\text{residential} = \frac{84,000 \text{ kW}}{164,900 \text{ kW}} = 0.509$$

$$\text{commercial} = \frac{40,500 \text{ kW}}{164,900 \text{ kW}} = 0.246$$

$$\text{bulk power} = \frac{40,400 \text{ kW}}{164,900 \text{ kW}} = \underline{0.245}$$

$$\text{total} = 1.000$$

The annual cost for demand totals $8,242,000 for the generation, transmission, and distribution parts of the system. However, the bulk power customer uses only the generation and transmission parts of the system and should not be charged for distribution costs. Therefore, the calculations are separated.

Generation and transmission account for 79.8 percent of the book value of the system (distribution accounts for the remaining 20.2 percent), and thus all customers using generation and transmission should pay for 79.8 percent of the demand cost, or $0.798 \times \$8,242,000 = \$6,577,116$. Each customer class is charged according to its proportion of peak load.

$$
\begin{array}{rcl}
\text{residential} = \$6,577,116 \times 0.509 & = & \$3,347,752 \\
\text{commercial} = \$6,577,116 \times 0.246 & = & 1,617,971 \\
\text{bulk power} = \$6,577,116 \times 0.245 & = & \underline{1,611,393} \\
& & \overline{\$6,577,116}
\end{array}
$$

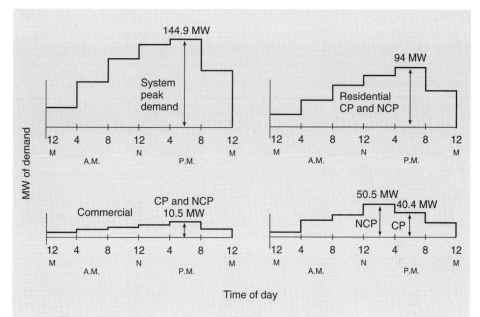

Figure 22.2 Graph of customer demand versus time of day.

The demand costs of the distribution system are not used by the bulk power customer, so those costs should be charged only to residential and commercial customers. Their cost allocations in proportion to the remaining peak demand (after deducting bulk power peak) is

$$\text{demand cost attributable to the distribution}$$
$$\text{system} = 0.202 \times \$8,242,000 = \$1,664,884$$

$$\text{residential} = \$1,644,884 \times \frac{84,000 \text{ kW}}{84,000 \text{ kW} + 40,500 \text{ kW}} = 1,123,295$$

$$\text{commercial} = \$1,664,884 \times \frac{40,500 \text{ kW}}{84,000 \text{ kW} + 40,500 \text{ kW}} = \underline{541,589}$$

$$\$1,664,884$$

There are two basic alternative methods of billing these demand charges to the customers, plus any number of combinations of the two methods.

Alternative Method 1: Lump sum monthly demand-charge billing. Bill each customer a fixed lump sum demand charge monthly. Each class of customer has a different fixed sum, but all customers within a class are treated as if they had the same demand requirements on the system. Using this system, the monthly bills would be calculated as follows.

	Demand charge
Residential	
Generation and transmission	$3,347,752
Distribution	1,123,295
Total demand charge to residential ÷ (Number of customers × months/yr)	$4,471,147 ÷ (20,000 customers × 12 months) = $18.63/customer/month
Commercial	
Generation and transmission	$1,617,971
Distribution	541,589
Total demand charge to commercial ÷ (Number of customers × months/yr)	$2,159,560 ÷ (2,000 customers × 12 months) = $89.98/customer/month
Bulk power	
Generation and transmission	$1,611,393/12 months − $134,284/month

With this method, the large, all-electric residence pays the same demand charges as the small cottage using gas for cooking, heat, and hot water. Obviously this system favors the higher-than-average-demand customers within each customer class.

Alternative Method 2: demand-proportional-to-energy charge allocated to kilowatt-hours.   The second method assumes that demand is proportional to energy consumption. Advocates of this method reason that the residence consuming twice the average energy in kilowatt-hours must have twice the peak load demand. This of course assumes that the load curves all have the same profile shape and vary only in magnitude. The annual demand charge is obtained by dividing demand cost by the energy consumption in kilowatt-hours. The resulting charge in terms of dollar per kilowatt-hour is billed to the customer on the basis of kilowatt-hours consumed. Electric rates calculated by this method would be

residential   (demand charges for generation, transmission and distribution)/(energy consumed in kWh/yr)
$4,471,047/354.08 × 10^6 kWh = $0.01263/kWh

commercial   (demand charges for generation, transmission and distribution)/(energy consumed in kWh/yr)
$2,159,560/206.02 × 10^6 kWh = $0.010482/kWh

bulk power   (demand charges for generation and transmission)/(energy consumed in kWh/yr)
$1,611,393/159.00 × 10^6 kWh = $0.010135/kWh

This method favors those customers who have higher-than-normal peak demand compared to their energy consumption. This type of customer is described as having a low "load factor."

The load factor for each customer is simply defined as the "amount of power purchased by each customer relative to the peak load requirement."

Thus, for example, a 100 percent load factor indicates a customer is using peak load 100 percent of the time. A 40 percent load factor may indicate the full kilowatt-hour peak load is used an average of 40 percent of the time, or 40 percent of the kilowatt-hour peak load is used all the time (except for a short peak), or any other combination.

The load factor for the whole class of residential customers in this example problem is

$$\frac{354.08 \times 10^6 \text{ kWh}}{8,760 \text{ hr/yr} \times 84,000 \text{ kW}} = 0.481 = 48\%$$

Customers with a lower-than-average load factor would be favored by alternative method 2, while those with higher load factors would pay a little more than their fair share.

Energy cost allocation to customers' rates. After allocating the demand costs to customers' monthly billing rates, the allocation of energy costs is relatively simple. The total cost of producing energy ($3,452,000) is divided by the total energy sold (719.10 $\times$ 10^6 kWh), and a cost in dollars per kilowatt-hour is determined. For the example problem, this yields

$$\$3,452,000/719.10 \times 10^6 \text{ kWh} = \$0.0048/\text{kWh}$$

This charge is the same for all customers in all classes, since it costs the same in dollars per kilowatt-hour to turn the generator for the small customer as for the large. There are energy losses in the system, so the denominator is the net energy sold at the meters rather than the gross energy generated.

Customer costs allocated to customer rates. The customer costs are assumed to be about the same for one customer as for another. So each customer regardless of customer class is billed the same monthly fixed charge to cover customer costs. The amount of that charge per customer per month is calculated as the (total customer costs)/(total number of customers $\times$ 12 months/yr), or

$$\frac{\$281,000/\text{yr}}{(20,000 \text{ residential} + 2,000 \text{ commercial} + 1 \text{ bulk customer})} \times 12 \text{ months/yr}$$

$$= \$1.07/\text{month/customer}$$

Revenue costs allocated to customer rates. The revenue costs are based on the amount of gross revenue (less fuel-adjustment charges) collected by the system and are simply passed through to the state tax collector. The amount is calculated as a percentage and is the same for all customer classes. The percentage is calculated as follows:

$$\frac{\text{amount of tax (\$610,000)}}{\text{gross revenue less fuel adjustment (\$14,084,000} - \$1,499,000)}$$
$$= 0.0485, \quad \text{or} \quad 4.85\%$$

Rate schedule.   All of these charges (demand, energy, customer, revenue) are added together to make up a rate schedule, shown as follows.
Using demand method 1:

Residential

demand	$18.63/month
energy	$0.048/kWh
customer	$1.07/month
revenue	multiply × 1.0485

To obtain a usable billing rate for residential customers, the two monthly rates are combined ($18.63/month + $1.07/month = $19.70/month) and multiplied by the revenue multiplier. Also the energy charge is multiplied by the revenue multiplier to yield the following billing schedule of rates.

monthly base total = $19.70/month × 1.0485 = $20.66/month
energy total = 0.0048 × 1.0485 = $0.00503/kWh

A typical 1,000 kWh monthly bill amounts to

$20.66/month + $0.00503/kWh × 1,000 kWh = $25.69/month

Commercial

demand	$89.98/month
energy	$0.0048/kWh
customer	$1.07/month
revenue	multiply × 1.0485

Using the same process as for residential rates, the commercial rates are established as follows.

monthly base total = $91.05/month × 1.0485 = $95.47/month
energy total = $0.00503/kWh

A typical 10,000 kWh bill amounts to

$$\$95.47/\text{month} + 10,000 \text{ kWh} \times \$0.00503/\text{kWH} = \underline{\underline{\$145.77/\text{month}}}$$

Bulk Power

demand	$134,284/month
energy	$0.0048/kWh
customer	$1.07/month
revenue	multiply × 1.0485

monthly base total = $134,284/month × 1.0485 = $140,797
energy total = $0.00503/kWh

A typical 13.25×10^6 kWh bill amounts to

$$\$140,797 + 0.00503/\text{kWh} \times 13.25 \times 10^6 = \underline{\underline{\$207,444/\text{month}}}$$

If the alternate method 2 for demand allocation is used, the resulting monthly rates are as follows.

Residential

demand	$0.01263/kWh
energy	$0.0048/kWh
customer	$1.07/month
revenue	multiply × 1.0485

monthly base = $1.07/month × 1.0485 = $1.12/month
energy total = ($0.01263 + $0.0048) × 1.0485 = $0.01828/kWh

A typical 1,000 kWh monthly bill amounts to

$$\$1.12/\text{month} + 0.01828 \times 1,000 = \underline{\underline{\$19.40/\text{month}}}$$

The 1,000 kWh monthly bill is lower than the average 1,475 kWh, so the resulting monthly bill is lower using demand method 2.

Commercial

demand	$0.010482/kWh
energy	$0.0048/kWh
customer	$1.07/month
revenue	multiply × 1.0485

monthly base = \$1.07/month × 1.0485 = \$1.12/month
energy total = (\$0.010482 + \$0.0048) × 1.0485 = \$0.016023/kWh

A typical 10,000 kWh monthly bill amounts to

$$\$1.12/month + 0.016023 \times 10,000 = \underline{\$161.35/month}$$

The 10,000 kWh monthly bill is higher than the 8,584 kWh average, so the monthly charge is higher using demand method 2.

Bulk power

demand	\$0.010135/kWh
energy	\$0.0048/kWh
customer	\$1.07/month
revenue	multiply × 1.0485

monthly base = \$1.07/month × 1.0485 = \$1.12/month
energy total = (\$0.010135 + \$0.0048) × 1.0485 = \$0.015659/kWh

A typical 13.25×10^6 kWh monthly bill amounts to

$$\$1.07/month + \$0.015659 \times 13.50 \times 10^6 \text{ kWh} = \underline{\$207,483/month}$$

The 13.25×10^6 kWh bill is the average for this bulk power customer. Using method 2, the monthly bill is higher in the fifth significant figure because of round-off in the rates.

Additional Fuel-Adjustment Charges

Before the late 1960s when fuel prices began their rapid escalation in cost, fuel oil cost about \$2 per barrel with little fluctuation over long periods of time. With productivity increases and greater efficiencies in generating equipment, periodic rate reductions were common. Whenever a rate revision was in order, public hearings were held, and in due course the rates were appropriately changed. After fuel prices began their rapid rise, it became impractical to have a rate hearing every time the price of fuel went up. In addition the utilities received unpredictable allocations of cheaper, price-regulated gas interspersed with higher foreign oil, so the fuel cost varied from month to month. As a result, an automatic fuel-adjustment pass-through charge is allowed, so that all fuel costs over some benchmark price for oil are billed to the customer separately. Thus, if the excess cost of fuel (over the base cost) amounts to \$875,000 for a given month and the amount of energy sold that month to all customers amounts to 60,000,000 kWh, then the fuel-adjustment charge to every customer is

$$\frac{\$870,000/\text{month}}{60,000,000 \text{ kWh/month}} = \$0.0145/\text{kWh}$$

Usually there is no revenue percentage added to this fuel-adjustment charge.

SUMMARY

Utility rate studies are often carried out by engineers who are familiar with both the utility system and with techniques of cost analysis. Rates are generally calculated in two parts—a base rate, which is usually a fixed cost per month for the service, regardless of the amount of product used and a cost per unit of product consumed. Added to these costs are costs for fuel adjustment.

In formulating electric rates several critical points require sound engineering judgment and experience.

1. First, sound judgment and familiarity with the system are required to properly allocate the electric utility system costs to the four cost categories of demand, energy, customer, and revenue. High allocations to demand favor the off peak customer if the peak responsibility method is used, and favor low-peak-compared-to-load-factor customers when most other methods are used. A high allocation to energy raises rates to the bulk power customer and lowers them to residential and commercial. A change of $1,000,000 from demand to energy (8% change) raises bulk rates about 4 percent and lowers residential about 1 percent. A change in coincident peak of 17 percent from 100 percent of noncoincident peak to 83 percent of noncoincident peak lowers the bulk power rate 11 percent.

2. Second, good local data should be used when determining the load curves for each customer class. These load curves are needed for determining class demand peaks and the resulting allocations of system demand costs. Local data can be obtained by means of special recording meters located on representative sample customer locations, or by installing recording meters on designated substations that serve representative customer classes. If no other means are available, data from neighboring utilities systems may be used, but obviously constant monitoring of local data is preferred.

PROBLEMS FOR CHAPTER 22, UTILITY RATE STUDIES

Problem Group A

Determine the rate.

A.1. A small city requests that you make up a rate schedule for them adequate to just cover their costs for a water utility system now being constructed. The system will have 8,000 customers at the beginning of year one. The whole system is financed with 20-year special assessment bonds with $i = 8$ percent. Assume the life of the system is 20 years, with $500,000 resale value at the end of that time. They estimate costs and consumption as follows.

Capital costs	$3,000,000
Operating costs	
Maintenance	$120,000/yr escalating $5,000/yr
Operations	40,000/yr escalating $4,000/yr
Consumption	1,000,000 gal/day

(a) Find a rate that will pay for the system *without* rate increases. The rate should include: **(i)** a flat monthly charge, plus **(ii)** a unit charge in dollars per 1,000 gal consumed.

(*Ans.* $3.07/month and $0.612/kgal)

(b) Find a rate that will pay for the system *with* rate increases. Begin with **(i)** a flat rate of $2.00 per month, plus **(ii)** a unit charge of $0.25 per 1,000 gal.

(*Ans.* Flat rate $2.00/month and increases of $0.152/month at the end of each year. Unit charge of $0.25/kgal, plus increases of $0.051/kgal at the end of each year.)

A.2. A county is going to provide solid waste collection service (garbage pickup) for a new community within its boundaries. The estimated costs are shown:

$$i = 12\%/\text{yr}$$
$$\text{purchase price of garbage truck} = \$200,000$$
$$\text{expected life of truck} = 12 \text{ yr}$$
$$\text{resale value of truck at EOY } 12 = \$20,000$$
$$\text{first year O \& M costs payable at EOY } 1 = \$90,000$$
$$\text{annual increase in O \& M costs} = 4\%/\text{yr}$$
$$\text{major overhaul (MOH) at EOY } 4 = \$25,000$$
$$\text{subsequent MOHs every 2 yr (EOY 6, 8, 10)}$$
$$\text{increase by 5\%/yr number of customers} = 1,000$$

(a) If the users' fee is paid annually at EOY and increases by 4 percent per year, how much should each user pay at EOY 1?

(b) If the bill is paid monthly instead of annually, how much would the monthly bills be during the first year? All costs remain the same as above, and the bills remain constant throughout each 12-month period. The county deposits the monthly receipts into a fund earning interest at 1 percent per month.

A.3. A new sewage treatment plant has an anticipated life of 20 years, after which it should be replaced. The local government prudently decided to charge enough extra in their rate structure to finance the new plant 20 years from now. Taking into account inflation and other factors the best guess is the new plant will cost $20,000,000 at a time 20 years from now. The current plant will process 150,000,000 gal per month at the start. This will increase by 1,000,000 gal per month for the next 20 years.

Assume that the sewage billing will be on the basis of water used, since the sewage is not metered but the water is. It is estimated that for every gallon of water metered, 0.7 gal goes into the sewer. How much should be charged per 1,000 gal of water billed in order to pay for the new sewage treatment plant 20 years from now? Nominal annual $i = 7$ percent, $n = 240$.

(*Ans.* $0.111/kgal)

A.4. The county commission requests your help in setting fares for a proposed bus transportation system currently under consideration. The following data are provided:

$$\text{cost per bus, new} = \$60,000/\text{each}$$
$$\text{expected life} = 20 \text{ yr}$$
$$\text{interest} = 7\%$$
$$\text{salvage value at the end of 20 yr} = \$10,000$$
$$\text{O \& M costs (including overhead)} = \$1.00/\text{mi the first year, increasing by}$$
$$\$0.10/\text{mi each year thereafter}$$

The buses will run 12 hours per day, 6 days per week, 52 weeks per year at an average speed, including stops, of 10 mph. Each bus is expected to carry an average of 80,000 passenger trips per year, and the annual revenue is calculated as the fare per passenger times the 80,000 passenger trips per year.

As a simplifying assumption, all O & M costs and fare receipts are accounted for at the *end* of the year in which they occur.

(a) If the fare ($/passenger) is constant over the 20-year period, what fare should be charged each passenger in order for the bus system to break even at the end of 20 years?

(b) If the average passenger rides 2 mi per trip, what is the equivalent cost per passenger mile? (*Hint:* Divide $/passenger by 2 mi/passenger.)

(c) If travel by private auto costs $0.15/mi now and is expected to increase by $0.015/mi every year for the next 20 years, compare the costs per passenger-mile to determine if it would be less expensive to move people by car or by bus for this particular bus system. (Compare the annual equivalent $/passenger-mile for each alternative.)

(d) For the bus system, if fares are $0.35 the first year, what increase in fares per year (arithmetic gradient) is required to let the system break even at the end of 20 years? (Find answers in $/each fare/yr.)

(e) If fares are $0.35 each year and increases are limited to $0.05 per fare per year, find the present worth of the subsidy required per bus.

(f) Assume the federal government offers to pay for the cost of the new buses, so that the county receives them free of charge, if the county will pay for O & M costs. How much subsidy is the federal government providing in terms of equivalent dollars per fare for each passenger? (Assume the fare and equivalent subsidy are constant over the 20-yr period.) In other words, how much does the government subsidy amount to in terms of dollars per passenger, assuming that the subsidy per passenger does not change over the 20-year life of the system?

A.5. A small municipality asks your help in setting water rates. The estimated costs are shown:

$$i = 12\%/\text{yr}$$
$$\text{cost of water works} = \$2,000,000$$
$$\text{expected life} = 30 \text{ yr}$$
$$\text{salvage value of water works at EOY 30} = \$100,000$$
$$\text{first year O \& M costs payable at EOY 1} = \$400,000$$
$$\text{annual increase in O \& M costs} = 4\%/\text{yr}$$
$$\text{major overhaul (MOH) at EOY 3} = \$250,000$$
$$\text{subsequent MOHs every 5 yr (EOY 8, 13, 18, etc.)}$$
$$\text{increase by 5\%/yr number of customers} = 1,000$$

(a) If the water bill is paid annually at EOY and increases by 4 percent per year, how much should each user pay at EOY 1?

(*Ans.* $628.66)

(b) If the bill is paid monthly instead of annually, how much would the monthly bills be for the first year? All costs remain the same as above. The monthly billing rate is adjusted upward once a year at the beginning of each subsequent 12-month period and then remains constant throughout that 12-month period. The municipality deposits the monthly receipts into a fund bearing interest at 1 percent per month.

(*Ans.* $49.57/month)

Problem Group B

Is the rate adequate?

B.1.   As consultant to a small municipal utility (income tax exempt), you are requested to report on the feasibility of a proposed water main extension. The following data have been supplied:

$$i = 7\%$$
$$\text{expected life} = 25 \text{ yr}$$
$$\text{extension length} = 10,000 \text{ ft water line}$$
$$\text{cost of materials and installation} = \$8.10/\text{ft}$$
current monthly billing rate for general
$$\text{residential customers} = \$0.75/1,000 \text{ gal}$$
$$\text{average current consumption} = 10,000 \text{ gals/month/customer}$$
$$\text{current cost to utility} = \$0.45/1,000 \text{ gal for O \& M and}$$
$$\$0.30/1,000 \text{ gal for P \& I}$$
(amortize capital costs)

The proposed water main will serve a subdivision with 160 lots. Full development of the subdivision is expected to take five years and to occur at the rate of about 20 percent per year. Thus, the first year only 20 percent of 160, or 32 customers will need water, the second year there will be 40 percent, and so on. (For simplification, use the end-of-year convention, with 20 percent of full receipts occurring at the end of the first year, 40 percent at the end of the second, and so on. Assume the utility installs the full 10,000 ft of line at the beginning of the first year.)

(a) Will the current schedule of amortization charges be adequate to pay for capital costs of this installation?

(*Ans.* No)

(b) If not, how much is the shortage or surplus?

(*Ans.* Shortage of $23,960 NPW)

(c) If the income from the amortization charge is not enough to make the P & I payments on this installation, and the difference is to be made up as a one-time hook-on capital facilities charge, how much should this charge be per meter if the meters were installed at the *beginning* of the first year? (*Hint:* Find the PW of the shortage and divide by the total number of customers.)

(*Ans.* $150)

(d) How much should be the capital facilities charge for meters installed at the *end* of the first year? [*Hint:* Each new hook-on should pay accrued interest from the EOY 0, or $F = (F/P, i, n)$.]

(*Ans.* $160.50)

B.2. A small municipality owns its own electric utility and sells electricity to customers within the urban area. Since the generator is subject to downtime (both scheduled and unscheduled), the city commission has decided to intertie with a larger neighboring system. They ask you to determine whether or not the intertie will result in a net profit or loss to their system. They provide you with the following data. Assume $i = 10$ percent.

Existing system

$$\text{generator size} = 20 \text{ MW rated capacity}$$
$$\text{load factor} = 50\% \text{ (actually generates only 50\%}$$
$$\text{of the product of the rated}$$
$$\text{MW} \times \text{total h/yr)}$$
$$\text{average sales price to customer} = \$0.05/\text{kWh}$$
$$\text{average fuel, operations \& maintenance cost} = \$0.03/\text{kWh}$$
$$\text{overhead and administration cost} = \$500,000/\text{yr}$$
$$\text{annual payments on } \$5,000,000 \text{ first cost of}$$
$$\text{20-MW plant } (A/P, 10\%, 40) = \$511,300/\text{yr}$$
$$\text{average downtime} = 3\%$$

Proposed intertie

$$\text{original cost} = \$300,000$$
$$\text{estimated life of intertie} = 40 \text{ yr (no salvage value)}$$
$$\text{maintenance costs of intertie} = \$15,000/\text{yr}$$
$$\text{cost of electricity purchased through the intertie} = \$0.04/\text{kWh}$$

Assume the intertie plus the existing generator will provide electricity 100 percent of the time.

(a) Find the equivalent annual cost of the intertie (including the additional costs and income) and tell whether it is a net income or a net cost to the system.

(b) Same as (a) except assume the following costs and incomes escalate at the rate of 5 percent per year compounded (geometric gradient) over the 40-year life of the intertie: (i) average sales price to customer; (ii) average fuel, operations, and maintenance costs; (iii) overhead and administrative costs; (iv) maintenance costs of intertie; (v) cost of electricity purchased through the intertie.

Note: More information is provided in the problem than is needed for the solution.

B.3. A small utility company is being formed for the purpose of supplying water and waste-water service to a new community. Construction of the water and waste-water system will begin now and take 24 months to complete. The construction contract calls for payments of $100,000 per month to the contractor, for a total contract amount of $2,400,000. The first payment is due at EOM 1. A local bank has agreed to lend money to the utility at 1 percent per month. At EOM 1 the utility borrows the $100,000 from the bank and pays the contractor. Each subsequent month the same procedure is followed. After construction is completed, at EOM 25 the utility starts collecting from its customers so that it can begin making repayments to the bank. These repayments will be made annually, with the first repayment at EOY 3 amounting to $200,000. Each subsequent payment thereafter will be 8 percent larger than the previous one. Interest continues on the unpaid balance at the original rate.

(a) How much does the utility company owe the bank at EOM 24?

(b) Given the repayment schedule described above, the final repayment will be made at which EOY?

Problem Group C

Find the rate of return.

C.1.   A private utility company is considering the feasibility of servicing a large, new subdivision. You are provided with the following income and cost projections and are requested to find the before-tax rate of return, i. (*Hint:* Try 10%.)

EOY		EOY capital cost	EOY net operating income
0	Begin design	−$ 50,000.00	
1	Begin construction	− 300,000.00	
2	Complete construction commence		
	operations	− 500,000.00	
3	Operations underway, subdivision		
	in growth stages		−$100,000.00
4			− 60,000.00
5			− 20,000.00
6			+ 20,000.00
7			+ 60,000.00
8			+ 100,000.00
9	Subdivision fully developed (change		
	of gradient)		+ 140,000.00
10			+ 145,000.00
11			+ 150,000.00
	(Increase $5,000.00/yr through		
	thirty-fifth year)		
35			+ 270,000.00
35	Salvage value		+ 500,000.00

(*Ans.* 10.13%)

Problem Group D

Find the errors in a proposal for a rate increase.

D.1.   A small, new firm is supplying utilities for a new development. Their proposed rates are based on the following statement of costs. Point out the errors in the statement and correct them.

$$
\begin{aligned}
\text{capital recovery of investment of \$500,000 at 6\%, } n = 40 &= \$\ 33,230\\
\text{operating costs for maintenance supplies, labor} &= \ \ 75,000\\
\text{bond interest at 6\%} &= \ \ 30,000\\
\text{overhead costs} &= \ \ 10,000\\
\text{depreciation} &= \ \ 12,500\\
\text{major overhaul costs every 10 yr} = \$50,000/10 &= \ \ \ \ 5,000\\
\text{amortization of bond issue over 20 yr} = \$500,000/20 &= \ \ 25,000\\
\text{federal income taxes, state income taxes, local property taxes} &= \ \ \underline{\ \ 50,000}\\
&\ \ \ \ \$245,730
\end{aligned}
$$

Problem Group E

Electric utility rate making.

E.1.   Use the data below to develop an electric utility rate schedule that will balance all the estimated costs with revenue. Assume the large power customers do not use the distribution system. Omit the revenue cost category.

Number of customers	Customer classes	Annual consumption (kWh)	Noncoincident class (kWp)	Coincident peak ÷ noncoincident peak
32,900	Residential	340,354,000	83,950	1.00
3,500	Commercial	262,508,000	72,000	0.90
8	Large power	56,148,000	17,325	0.80
	Total	659,010,000		

System peak demand: 162,610 kWP

Allocated annual system costs

Demand	$21,297,000/yr
Energy	3,294,000
Customer	891,000
Total	$25,482,000/yr

Book value of the system

Generating	$53,022,000
Transmission	12,835,000
Distribution	23,171,000
Total value	$89,028,000

Find the schedule of rates for each class of customer. Use method 2 for allocating demand costs to the customer's electric rates.

E.2.   Same as Problem E.1, but use demand method 1.

Problem Group F

Double arithmetic gradient, *GG*.

F.1.   Same as Problem A.1, except at the beginning of year 1 the system will have, initially, 6,000 customers, each consuming 125 gal per day and the number of customers will increase by 500 customers per year.

   (a)

   (*Ans.* Flat rate $2.49/month plus $0.514/kgal)

   (b)

   (*Ans.* Flat rate $2.00/month plus increases of $0.0691/month at the end of each year. Unit charge of $0.25/kgal plus increases of $0.0594/kgal at the end of each year.)

CHAPTER 23

Inflation

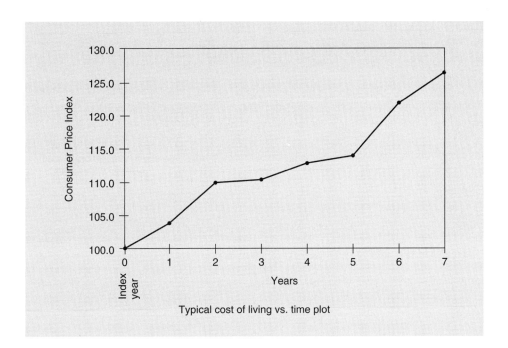

Typical cost of living vs. time plot

KEY EXPRESSIONS IN THIS CHAPTER

i_i = Inflation rate

i_c = Combined interest-inflation rate. Rates calculated for loans or investments, including the combined effects of interest charges and inflation. This rate is not discounted for the effects of inflation on buying power. It is also known as the nominal market interest rate, apparent rate, or published rate (see Equation 23.1).

i_r = Real interest rate after inflation, or real increase in buying power, periodically compounded (see Equation 23.2).

NATURE OF INFLATION

Throughout this text, money (expressed in dollars) serves as a convenient, easily understood, readily accepted measure of the value, or worth of goods and services. The concept of comparing alternatives on the basis of *equivalent* worth is valid as long as the basic value of the dollar remains constant. Unfortunately, the basic value of the dollar, expressed in terms of the goods and services it can purchase, often changes with time. Inflation occurs when the price of goods and services increases (inflates) as the purchasing power of money declines. For example, if a given loaf of bread costs $1.00 now and $1.12 a year later, the price of the bread has been *inflated* by $0.12 in one year. If there has been no corresponding increase in the size of the loaf or quality of the product, then money is simply not worth as much (in terms of bread), and the 12 percent price increase is attributed to inflation.

Inflation is an important factor in cost analysis and has been observed throughout much of the commercial history of the civilized world. (For instance, in 301 A.D. the Emperor Diocletian claimed that inflation threatened to destroy the economy of the Roman Empire.) Inflation is often treated as affecting all goods and services equally, but actually the rate of price increase often varies greatly from one item to another. For instance, gasoline may double in price during the same time period that the price of bread increases by a much smaller amount.

INFLATION AND PRICE INDEXES

Price trends for groups of items of special interest are often condensed into an index format. Essentially, the index is the ratio of the price of a specific group of goods and services at one date to the price of this same group at a later date. By taking the total price of the group at a specified "starting date" as 100, the subsequent prices can be reported as percentages of this base price (although the percent sign is usually not shown). There are a number of indexes prepared in the United States. One of the most widely known is the Consumer Price Index (CPI) prepared by the Bureau of Labor Statistics, U.S. Department of Commerce. The CPI is based on the price of about 400 different goods and services, and is used by many agencies as one measure of the inflation rate in the United States. It was started in 1913 and has been altered somewhat through the ensuing years. Base years are updated periodically.

These cycles in our economy are influenced through adjustments to interest rates by the Federal Reserve Board (FRB) as a measure to influence the flow of money through the economy. Raising interest rates slows the flow, and lowering the rates increases the flow, thus stimulating the economy. In recent years these measures have been fairly successful in keeping inflation at reasonable levels.

Over the past several decades the cost of goods and services has fluctuated widely, with the majority of the fluctuations resulting in a growth in the prices for such goods and services. Using the CPI, for example, the growth in the CPI averaged between 1 and 2 percent per year for the period between 1955 and 1965. Between 1965 and 1975 the growth of the CPI steadily increased from around 1 percent to about 7 percent per year. After 1975, the CPI first grew at an alarming rate, reaching double-digit proportions by the early 1980s. Then, responding to high interest rates and a supply of goods and services that exceeded demand, the CPI slowed down to less than 4 percent for several years. In the future the CPI growth rate will undoubtedly fluctuate with the economy, the supply-demand ratio of labor, raw materials, finished products, and other factors.

Perhaps more germane to the engineering community are the indexes that reflect the increases in the cost of construction goods and services in the United States. There are a number of these indexes, including the widely used Construction Cost Index and the Building Cost Index, published by the *Engineering News Record.* The Construction Cost Index was established in 1921, following the tracking of selected costs for the preceding 12 years. Designed as a general-purpose index for construction costs, it is a weighted-aggregate index of the prices of constant quantities of structural steel, portland cement, lumber, and common labor, and was given a relative price of 100 in 1913. The Building Cost Index was introduced in 1938 to weigh the impact of skilled labor in construction cost trends. As such, the labor component of the index is the average union wage rate, plus fringes, for carpenters, bricklayers, and structural iron workers. The material component is the same as for the Construction Cost Index. The Building Cost Index also uses a base year of 1913. These indexes are reported weekly in the *Engineering News Record,* partially summarized quarterly, and completely summarized yearly (for a complete summary of these indexes see the issue for the third week in March each year).

Following is a partial summary of these two index values at five-year intervals, plus the average annual percentage rate of change.

Year	Construction cost index		Building cost index	
	Index value	Average % rate/yr	Index value	Average % rate/yr
1950	510	—	375	—
1955	660	5.3	469	4.6
1960	824	4.5	559	3.6
1965	971	2.3	627	2.3
1970	1,385	7.4	836	5.6
1975	2,212	9.8	1,306	9.3
1980	3,237	7.9	1,943	8.3
1985	4,182	5.3	2,425	4.5
1990	4,732	2.5	2,702	2.2
1995	5,471	2.9	3,111	2.8
1955-1995 (40 yr)		5.3		4.8

Examining these indexes reveals that the rate of increase has fluctuated during the last 30 years. Also note that although both indexes track similar goods and services, they are somewhat different. It is important to note that any of the indexes in use today are only *indicators* of inflationary trends and not necessarily representative of specific situations, goods, or services.

HOW INFLATION WORKS

An example of how inflation works is shown in the following. Assume that current wages and prices for a given worker, time, and place are

$$\text{pay for work} = \$10/\text{hr}$$
$$\text{price for bread} = \$1/\text{loaf}$$

Therefore, this worker can buy 10 loaves of bread with one hour's pay.

Assume inflation occurs at a rate of 10 percent per year ($i_i = 10\%$). Then one year later, if both of these items have kept pace with inflation,

$$\text{pay for work} = \$11/\text{hr}$$
$$\text{price for bread} = \$1.10/\text{loaf}$$

Therefore, despite the pay increase the worker can still buy only 10 loaves with one hour's pay. A dollar is simply worth about 10 percent less than one year previously. If all costs and incomes keep pace exactly with inflation, then no one's purchasing power is decreased (or increased). However, many people are on fixed incomes and are not able to raise their income to compensate for inflation. Even though the nominal amount of their income or wealth is unchanged, their buying power declines together with their standard of living in many instances. Others find themselves in higher tax brackets with less buying power, while some manage to profit from inflation as property prices increase while mortgage payments remain constant.

Investors Seek a Gain in Purchasing Power

Normally investors are not satisfied just to keep pace with inflation, as this nets no real gain on an investment. To be successful, an investment must result in a net gain in buying power over and above the increase required to keep up with inflation.

For example, if an investor desires a real return of 4 percent increase in buying power after inflation ($i_r = 4\%$) and inflation is at 10 percent ($i_i = 10\%$), then for an investment of $1,000 for a duration of one year, the cash flow diagram is as shown in Figure 23.1. To regain the same buying power that the $1,000 had at the beginning of the year (with no real increase in purchasing power), F must equal $P(1 + i_i)^n$ or,

$$F = \$1,000 \times 1.10 = \$1,100$$

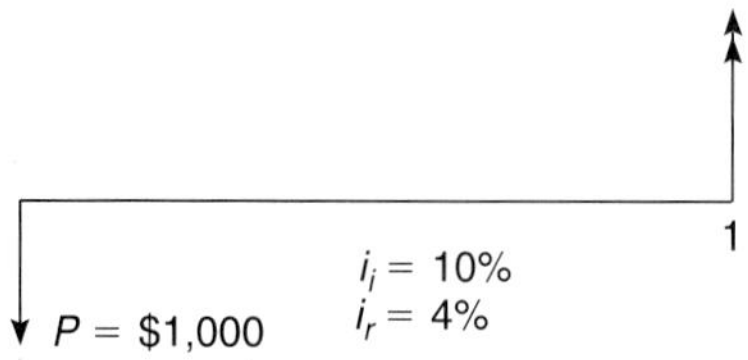

$F = ?$

1

$i_i = 10\%$
$i_r = 4\%$

$P = \$1,000$

Figure 23.1 Cash flow diagram.

This $1,100 at EOY 1 will buy only the same 1,000 loaves of bread or 100 hours of work that $1,000 would buy at EOY 0. An investor who wants a 4 percent increase in buying power must invest at a rate that will yield $1,100(1 + i_r)^n or $1,100(1.04) = $1,144. Expressed as an equation, the investment must yield an FW of

$$F = P(1 + i_i)^n (1 + i_r)^n$$

so that

$$F = \$1,000(1.10)(1.04) = \$1,144$$

Expressed in terms of interest, the combined interest-inflation rate to account for real return as well as inflation must be

$$\text{combined interest-inflation } i_c = (1 + i_i)(1 + i_r) - 1$$

so that

$$i_c = (1.10)(1.04) - 1 = 0.144 \quad \text{or} \quad 14.4\%$$

If the investment is for five years, the same procedure is used to find F.

$$
\begin{aligned}
F_1 &= P(1 + i_i)^n \times (1 + i_r)^n \\
F_1 &= \$1,000(1.10^5 \times 1.04^5) \\
F_1 &= \$1,000(1.10 \times 1.04)^5 \\
F_1 &= \$1,000 \times 1.144^5 = \$1,959
\end{aligned}
$$

Thus, a return of $1,959 on every $1,000 invested would provide a real increase in buying power of 4 percent per year (before taxes) compounded annually over the five-year period, with inflation at 10 percent.

REAL INTEREST RATE

It is evident from the foregoing that in order to net a real increase in buying power at a rate i_r, the combined interest-inflation rate of return, i_c, must exceed the inflation rate i_r, and can be calculated as combined interest-inflation, i_c.

$$i_c = (1 + i_i)(1 + i_r) - 1 \tag{23.1}$$

Conversely, the rate of return after inflation (ROR i_r) can be found as real gain in buying power,

$$i_r = \frac{1 + i_c}{1 + i_i} - 1 \tag{23.2}$$

The next example illustrates the use of the equations in determining the real after-inflation rate of return, i_r, compared to the combined interest-inflation rate of return, i_c.

EXAMPLE 23.1   (Given P, F, i_i, n, find i_c, i_r)

Given

$$P = \$1,000$$
$$F = \$2,000$$
$$n = 4 \text{ yr}$$
$$i_i = 10\%$$

Find (a) ROR i_c, the combined interest-inflation rate of return; (b) ROR i_r, the real after-inflation rate of return.

Solution

a. The ROR i_c is found in the same way as the ROR found in earlier chapters, as follows:

$$F/P = (1 + i_c)^n$$

Thus,

$$2,000/1,000 = (1 + i_c)^4$$

and

$$i_c = 2^{0.25} - 1 = 0.1892 \quad \text{or} \quad \underline{18.9\%}$$

b. The real ROR i_r is found from Equation 23.2 as

$$i_r = \frac{(1 + i_c)}{(1 + i_i)} - 1$$

Thus,

$$i_r = \frac{1.1892}{1.1} - 1 = 0.0811 \quad \text{or} \quad \underline{\underline{8.1\%}} \qquad \begin{array}{l}\text{real interest gain}\\ \text{in buying power}\end{array}$$

Thus, the real gain in buying power from this investment equals 8.1 percent compounded per year over the four-year period. Using the $1 loaf of bread and $10-per-hour labor wage to illustrate, for each $100 invested (equal to 100 loaves or 10 hours of labor), the purchasing power of the balance in the account increases enough at the end of the first year to buy an additional 8.1 loaves of bread or 0.81 hours of labor, even though the price of bread and labor rose 10 percent during the year. These increases in purchasing power are compounded annually at 8.1 percent throughout the four-year investment period.

INFLATION AND SERIES PAYMENTS

Many series payments in an inflationary environment can be classified in one of two general categories.

1. *Equal payment series.* The amounts remain constant, but the buying power of each payment shrinks due to inflation; for example, bonds yield a series of future fixed-interest payments, but with inflation each payment will buy less.

2. *Series of increasing payments with constant buying power.* Each payment in the series is larger than the previous one by an amount sufficient to maintain constant buying power (for example, an investment in a rental building with rents geared to rise with inflation). Another type of example is any project incurring a commitment for a future series of maintenance person-hours per year relatively fixed in number, such as a highway, utility, or similar project. The costs per person-hour increase as wages rise each year with inflation, but the number of person-hours remains relatively constant.

Series of Fixed Payments with Declining Buying Power

An example follows illustrating how to find the real rate of return (rate of compound increase in buying power) for a given P, A, F, and i_i.

EXAMPLE 23.2   (Given $P, A, F, n, i_i,$ find i_c, i_r)

Given

$$P = -\$1,000$$
$$A = +\$150/\text{yr}$$
$$F = +\$1,000$$
$$n = 3$$
$$i_i = 10\%$$

Find i_c, the combined interest-inflation rate of return, and i_r, the real rate of return. See Figure 23.2.

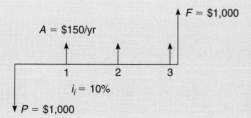

Figure 23.2   Cash flow diagram for Example 23.2.

Solution.   Solve for ROR i_c using the present worth method (or any other valid method).

$$PW = 0 = -\$1,000 + 150(P/A, i_c, 3) + 1,000(P/F, i_c, 3)$$

Solving the combined interest-inflation rate, $i_c = 15$ percent. To determine the real interest rate, i_r, assume as before that

$$\text{pay for work} = \$10/\text{hr at PTZ}$$
$$\text{price for bread} = \$1/\text{loaf at PTZ}$$

If the 10 percent inflation raises the wages for work as well as the cost of bread, then at EOY 1 work earns $11 per hour, which still only buys 10 loaves of bread, since bread is $1.10 per loaf.

The interest payment of $150 at EOY 1 is worth only $150/($11/hr) = 13.64 hours of work instead of 15 hours. Thus from the initial investment of $1,000 or 100 hours worth of work, the return the first year is only 13.64 hours worth. In subsequent years the table below shows the value of the interest payment in terms of hours of work. Assume inflation at 10 percent per year.

Hourly wage	EOY	Investment	Return
$10/hr	0	$1,000 or 100 hr	
$11/hr	1		$150 or 13.64 hr
$12.10/hr	2		$150 or 12.40 hr
$13.31/hr	3		$150 or 11.27 hr
$13.31/hr	3		$1,000 or 75.13 hr

The cash flow diagram for this real investment and return is shown in Figure 23.3. Based on the buying power of the investment and return, the real rate of return i_r, may be calculated as follows.

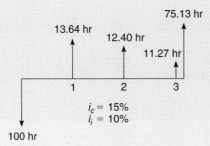

Figure 23.3   Cash flow diagram in terms of hours of work.

EOY		4%	5%
0	$P_1 =$	−100.00 hr	−100.00 hr
1	$P_2 = 13.64\,(P/F, i, 1)$		
	4% 0.96154	+ 13.12	+ 12.99
	5% 0.95239		
2	$P_3 = 12.40\,(P/F, i, 2)$		
	4% 0.92456	+ 11.46	+ 11.25
	5% 0.80703		
3	$P_4 = 11.27\,(P/F, i, 3)$		
	4% 0.88900	+ 10.02	+ 9.74
	5% 0.86384		
3	$P_5 = 75.13\,(P/F, i, 3)$	+ 66.79	+ 64.90
		+ 1.39 hr	− 1.12 hr

$$i_r = 4\% + \frac{1.39}{1.39 + 1.12} \times 1\% = 4.55\%$$

Thus the real interest rate of return, $i_r = 4.55$ percent, or the investment earned additional purchasing power at the rate of 4.55 percent compounded annually over and above the inflation rate of 10 percent.

A faster route to the same conclusion involves use of the i_r equation (23.2), as follows. Since $i_c = 15$ percent and $i_i = 10$ percent then

$$i_r = \frac{1 + i_c}{1 + i_i} - 1 = \frac{1.15}{1.10} - 1 = 0.0455 \quad \text{or} \quad \underline{4.55\%}$$

Thus the same resulting ROR $i_r = 4.55\%$ is determined more easily.

INFLATION IN ENGINEERING COST ESTIMATING

Engineers are often required to estimate the cost of building an engineered project which may take several years to complete. If inflation rates are high, these rates must be considered if accurate estimates are to be made. The following example illustrates the problem.

EXAMPLE 23.3

A construction contractor is preparing a fixed-price bid for the construction of an industrial plant, which is expected to take two years to construct. Detailed estimates (assuming zero inflation) indicate the costs listed below will be incurred for the project.

$$\begin{array}{r} \text{labor, equipment, materials, etc.} = \$14{,}652{,}285 \\ \text{supervision, overhead, etc.} = \underline{4{,}463{,}854} \\ \text{total direct cost} = \$19{,}116{,}169 \end{array}$$

These direct costs are expected to occur at a constant monthly rate calculated at \$19,116,169/24 months = \$796,507/month. In addition the contractor adds a profit and contingencies factor of 4 percent as a reasonable return on investment and to help offset some of the risks involved in this long-term 24-month contract. The owner will pay the contractor in one lump sum upon completion of the project at EOM 24, requiring the contractor to obtain funds to construct the project. The contractor plans to finance the construction of the project by borrowing 24 payments of \$796,507 per month at the *beginning* of each month. The bank agrees to open a line of credit with the contractor that permits borrowing whatever funds are needed at 1 percent per month interest on the current balance outstanding. The contractor calculates the interest charges to add to the contract bid price as follows (see Figure 23.4):

$$\begin{aligned} \text{interest charges} = F - An &= A(F/A, 1\%, 24)(F/P, 1\%, 1) - 24A \\ &= \$796{,}507 \times 26.973 \times 1.01 - 24 \times \$796{,}507 \\ &= \$21{,}699{,}025 - 19{,}116{,}169 = \$2{,}582{,}856 \end{aligned}$$

Interest charges for financing the construction = \$2,582,856. The total bid price follows:

$$\begin{array}{r} \text{total direct cost} = \$19{,}116{,}169 \\ \text{4\% add-on for profit and contingencies} \\ 0.04 \times \$19{,}116{,}169 = 764{,}647 \\ \text{interest cost due the bank for financing} = \underline{2{,}582{,}856} \\ \text{total bid price} = \$22{,}463{,}672 \end{array}$$

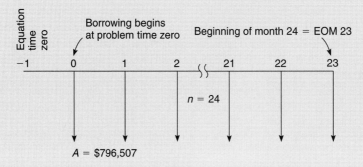

Figure 23.4 Cash flow diagram for borrowing.

The contractor did *not* take inflation into account, and let us assume that inflation actually occurs at the rate of 4 percent per year (or that inflation *was* taken into account but turned out to be 4 percent higher than anticipated). The *monthly* rate of inflation is 4/12 percent per month, and results in the contractor's total direct costs rising by 4/12 percent per month. Therefore, to cover total direct costs the contractor borrows $796,507 at the beginning of the first month (to cover the payments for labor, material, overhead, etc., that month), but at the beginning of the second month the contractor must borrow 4/12 percent more (4/12% = 0.003333), figured as

$$\text{amount needed at BOM 2} = (1.003333) \times \$796,507$$
$$= \$799,162$$

Each subsequent month the amount borrowed is increased again by a factor of 1.003333, creating a geometric series of borrowings from the bank. The amount owed the bank at the end of the twenty-fourth month is calculated as a geometric gradient with ETZ for the PW at EOM minus 1. Then the lump sum PW is translated to future worth at EOM 24, as follows:

$$F = C(P/C, w, 24)(F/P, 1\%, 25), \text{ (see Chapter 6)}$$

where

$$w = [(1 + i)/(1 + r)] - 1 = (1.01/1.0033) - 1 = 0.0066445$$
$$F = (\$796,507/1.003333)(P/A, 0.66445\%, 24)(F/P, 1\%, 25)$$
$$F = \underline{\$22,515,608}$$

The contractor had estimated that the amount owed the bank at EOM 24 would be $21,699,025. Due to 4/12 percent per month inflation in construction costs, the amount borrowed plus interest accumulated to $22,515,608 at EOM 24.

at EOM 24, the contractor owes the bank, $F = -\$22,515,608$.
at EOM 24 the owner owes the contractor $= + \underline{\text{ 22,463,672}}$ bid price
contractor's *loss* on this project $= -\$51,936$

Since the contractor had anticipated making a profit on the job and ended up with a loss, the total loss due to the 4 percent inflation amounted to

$$\text{estimated 4\% profit and contingencies} = \$764,647$$
$$\text{actual loss on this project} = \underline{\quad 51,936}$$
$$\text{net decrease in income due to inflation} = \$816,583$$

This analysis assumes that no unforeseen additional costs occur during construction and that the owner pays on time at the close of the project.

In order to account for the effects of inflation, engineers typically build into their estimates provisions for increases in the cost of goods and services they are providing. If possible, they obtain firm quotes for materials to be delivered at some specified time in the future and, using appropriate indexes, include escalation factors in their cost estimates. Finally, they hope that their assumed inflation rates are realistic.

TAXES AND INFLATION

In an environment of inflation, taxes are treated as just another expense, as in previous chapters. For example, assume an investor is in the 27 percent tax bracket and therefore is allowed to keep 73 percent of each $150 interest payment (or $0.73 \times \$150 = \$109.50$ after-tax income), as shown in the cash flow diagram (Figure 23.5). The after-tax real interest rate of return after inflation (at $i_i = 10\%$) can then be calculated by the usual method for finding the ROR as follows.

$$
\begin{array}{rrr}
 & 1\% & 0\% \\
\hline
P_1 = \$1,000 = & -1,000.00 & -1,000.00 \\
P_2 = (0.73 \times \$150)\,\underbrace{(P/F, i, 1)}_{1\%\ 0.9901}\,/\,\underbrace{(F/P, i_i, 1)}_{1.100} = & +\quad 98.56 & +\quad 99.55 \\
P_3 = (0.73 \times \$150)\,\underbrace{(P/F, i, 2)}_{1\%\ 0.9803}\,/\,\underbrace{(F/P, i_i, 2)}_{1.210} = & +\quad 88.71 & +\quad 90.50 \\
P_4 = (0.73 \times \$150 + \$1,000)\,\underbrace{(P/F, i, 3)}_{1\%\ 0.9706}\,/\,\underbrace{(F/P, i_i, 3)}_{1.331} = & +\quad 809.08 & +\quad 833.58 \\
\hline
P_{\text{total}} = & -\quad 3.65 & +\quad 23.63 \\
\end{array}
$$

$$
i_r = \frac{23.63}{23.63 + 3.65} \times 1\% = 0.86 = \underline{\underline{0.9\%}} \text{ real rate of return}
$$

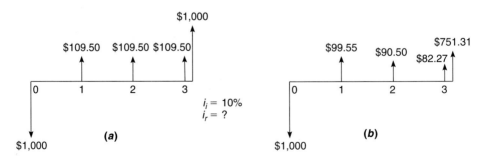

(a)

(b)

Figure 23.5   (a) Actual cash flow for 27% tax bracket. (b) Real after-tax purchasing power in terms of PTZ dollars. Purchasing power of interest payments $= (150 - 40.5)/(1 + i_i)^n$.

Thus the real *after-tax* purchasing power of the investment increases at only 0.9 percent compounded annually.

Alternate Method

The after-tax real rate of return can be obtained a little more expeditiously by first finding the after-tax ROR i_c and then deriving the real after-tax ROR i_r.

Since the $1,000 investment in this example is returned nominally intact at EOY 3, then the after-tax combined interest-inflation rate of return can be calculated simply as

$$i_c = \frac{150 \times 0.73}{1,000} = 10.95\%$$

The after-tax real rate, i_r, is determined from Equation 23.2 as

$$i_r = \frac{1 + i_c}{1 + i_i} - 1$$

and

$$i_r = \frac{1.1095}{1.100} - 1 = 0.0086 \quad \text{or} \quad 0.9\%$$

Conclusion. Use of the i_r equation produces the results more rapidly once a clear mental image of the process is obtained.

Example of a Series of Payments with Constant Buying Power and Nominal Increases in Payment Amounts with Inflation

Many investors in engineering projects desire a fixed level of income in terms of buying power; consequently the nominal income must rise with inflation. Examples of such projects include rental buildings, production facilities such as electric generating plants, and many others.

EXAMPLE 23.4

Consider again the previous example. With the inflation rate equal to zero, the anticipated income is shown in the cash flow diagram (Figure 23.6) yielding ROR $i_c = 15$ percent. Now, assume that inflation occurs at a rate of 10 percent ($i_i = 10\%$) and that the cash flows are going to increase proportionately to inflation much as an investor in a rental building might plan to increase the rental income with inflation.

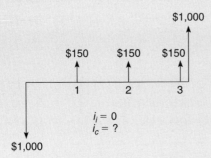

Figure 23.6   Cash flow diagram A, with zero inflation ($i_i = 0$). Buying power remains constant.

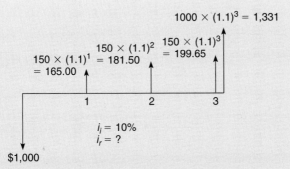

Figure 23.7   Cash flow diagram B, with inflation at rate $i = 10\%$. Rents and resale value increase with inflation.

Now the cash flow diagram A (Figure 23.6) represents the buying power value of the nominal cash flows in the cash flow diagram B (Figure 23.7).

Given that the cash flow diagram B is the actual cash flow, the combined interest-inflation rate of return, i_c, can be found together with the real rate of return, i_r.

Solution.   Since the third i value (of the three i's: i_c, i_i, i_r) can be found if any two are already known, and one is already known ($i_i = 10\%$), then the easiest-to-find of the other pair should be sought. This is obviously i_r, since i_r involves the simplest series.

The i_r value is obtained directly from cash flow diagram A (Figure 23.6) showing the buying power of the investment and return. (If the resale value were different from the initial cost, the usual trial-and-error system for determining ROR i_r would be used.) Since the resale value is the same as the initial cost, then

$$i_r = \frac{150}{1,000} = 15\%$$

Then the combined interest-inflation rate of return, i_c, can be derived as

$$i_c = (1 + i_r)(1 + i_i) - 1 = (1.15 \times 1.10) - 1 = 0.265 = 26.5\%$$

Conclusion. When inflation is at $i_i = 10$ percent, in order to realize a real rate of increase in buying power of $i_r = 15$ percent, a combined interest-inflation rate of $i_c = 26.5$ percent is required.

COMPARISON OF ALTERNATIVES IN AN ENVIRONMENT OF INFLATION

Often when selecting among alternatives, one alternative features a high cost and low maintenance, while a competing alternative has the opposite feature. The following example illustrates the effect that inflation has on this choice. Both alternatives in this example have the same total payment cash flow of $1,200,000 over the 20-year period, in order to better illustrate the effects of the difference in timing of payments and interest rates.

EXAMPLE 23.5 **(Comparison of two alternative proposals, with the same total cash flow but different timing)**

Assume a proposed road construction project can be paved either with material A or material B. Material A features a low first cost and high maintenance, while material B has a high first cost but low maintenance. If there were zero inflation, the cash flow diagrams for the two materials would appear as in Figures 23.8 and 23.9. In other words, these are the anticipated payments in terms of dollars with a *constant* purchasing power. ($1,000 is abbreviated as $K below.)

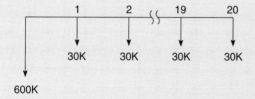

Figure 23.8 Material A cash flow.

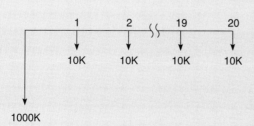

Figure 23.9 Material B cash flow.

Material A

$$\begin{aligned}
\text{construction cost} &= \$\ 600K \\
\text{O \& M 20 yr} \times \$30K/yr &= \underline{\ \ \ \ 600K} \\
\text{total cash flow} &= \$1,200K
\end{aligned}$$

Material B

$$\begin{aligned}
\text{construction cost} &= \$1,000K \\
\text{O \& M 20 yr} \times \$10K/yr &= \underline{\ \ \ \ 200K} \\
\text{total cash flow} &= \$1,200K
\end{aligned}$$

Solution.   To find the more economical alternative under these conditions, an IROR can be calculated on the incremental investment of $400K required if material B is selected rather than the lower cost material A. The value of i_c at which the NPW $= 0$ is found as follows:

$$\Delta P = -\text{material A} + \text{material B } i_c = 0\%$$

$$\text{construction, } \Delta P_1 = +\$600K - \$1,000K = -\$400K$$

$$\text{O \& M, } \Delta P_2 = (+30K - 10K)(P/A, i, 20) = \underline{+\$400K}$$
$$0$$

Only one trial rate, i_c, is necessary, since NPW $= 0$ when $i_c = 0$.

Conclusion.   If inflation is zero ($i_i = 0$) and the cash flows of two alternatives total the same, then the alternative with the lowest first cost has the lowest present worth for all positive interest rates ($i_c > 0$). (The lowest first cost alternative will incur the least amount borrowed at interest for the shortest time.) Thus for MARR > 0, select material A in this case.

Now assume inflation is anticipated at a sustained rate of 5 percent ($i_i = 5\%$). Then the actual cash flow diagram in terms of current (decreasing purchasing power) dollars is shown in Figures 23.10 and 23.11. The increasing dollar amounts shown are needed to buy a constant level of service for operating and maintaining the road.

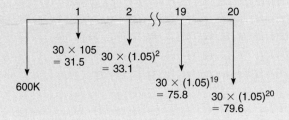

Figure 23.10   Inflated expenditures required to maintain material A.

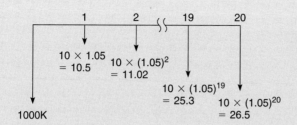

Figure 23.11 Inflated expenditures required to maintain material B.

To find the IROR for this case, the geometric gradient is used, since the dollar amounts increase at the inflation rate of 5 percent per year ($i_i = 5\%$).

The combined interest-inflation rate, i_c, at which $\Delta P_{\text{total}} = 0$, is found as follows.

$$i_c = 5\%$$

$$\text{construction cost, } \Delta P_1 = (+\$600K - \$1,000K) = \$ - 400K$$

$$c_A = 1.05 \times 30 = 31.5$$

$$c_B = 1.05 \times 10 = 10.5$$

$$\text{O \& M costs, } \Delta P_2 = \frac{Cn}{1+r} = \frac{(+31.5 - 10.5) \times 20}{1.05} = \underline{+400K}$$
$$0$$

Conclusion. When inflation is at 5 percent ($i_i = 5\%$), then for equal, total, real cash flows (equal buying power of the total cash flow for each alternative) the IROR $i_c = 5$ percent. Therefore, if the MARR $i_c > 5$ percent for the combined interest-inflation rate value (infers that $i_r > 0$), then use material A. And if the MARR $i_c < 5$ percent for the combined interest-inflation rate (infers that $i_r < 0$), then use material B.

The general conclusion is that whenever the total real buying power cash flows of the two alternatives are equal, then

1. For MARR $i_c > i_i$, the alternative with the lowest initial cost is more economical.

2. Conversely, for MARR $i_c < i_i$, the alternative with the higher initial cost is more economical.

PROJECTS WITH FINANCING

Some roads are financed with borrowed money (often money is obtained by selling road bonds), while others are paid out of current tax revenue. The current tax revenues should also be considered as borrowed funds, since most of the taxpayers owe debts of their own, and had the taxes not been levied, the taxpayers could use these funds to repay debts. Therefore, interest should be charged against the funds used in construction projects even when channeled through current tax revenues. The next example illustrates the point.

EXAMPLE 23.6   **(Shows effect of financing)**

Find the more economical selection if the project outlined in Example 23.5 is now financed by road bonds at 5 percent.

Solution.   The cash flow diagrams now show a receipt of the borrowed construction funds in addition to annual amortization payments of principal plus interest over the 20-year period. With road bonds at $i = 5\%$, the annual repayment amount, A, is

$$A = P(A/P, 5\%, 20) = P \times 0.08024$$
$$A_A = \$600K \times 0.08024 = \$48.14K$$
$$A_B = \$1,000K \times 0.08024 = \$80.24K$$

These repayments are combined with the O & M payments of Example 23.4 corrected for inflation and shown in Figures 23.12 and 23.13.

Even though the owner's net investment in the project is zero for both alternatives (so that there is no increment of investment and thus no real IROR), the interest rate at which the present worths equal each other can be calculated.

Assume

$$i_i = 5\%$$
$$i_{bonds} = 5\%$$
$$n = 20 \text{ yr}$$

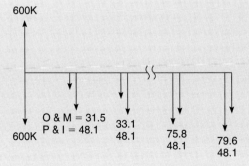

Figure 23.12   Constant P & I versus inflating O & M.

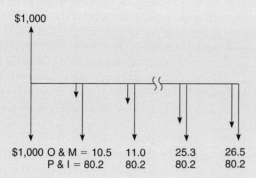

Figure 23.13 Constant P & I versus inflating O & M.

Then for the geometric gradient (for O & M costs)

$$r = i_i = 0.5$$

Since i_c is not yet known, a trial value of $i_c = 0.10$ is assumed. Then

$$P \& I = \Delta P - A+B = (48.14 - 80.24) \underbrace{P/A, i, 20)}_{12.462} = -\$400$$

$$O \& M = \Delta P - A+B = \frac{cn}{1+r} = \frac{(31.5 - 10.5) \times 20}{1.05} = \$400$$

$$\Delta P_{total} = \quad 0$$

Conclusion

1. If bond money can be borrowed at the same rate as the inflation rate (if $i_{bonds} = i_i$), and if the combined interest-inflation rate is greater than the inflation rate (if $i_c > i_i$), then the alternative with the lower first cost is more economical (material A is preferred). (Money earns more if left with the taxpayer longer, and this A series total begins with a smaller amount and gradually increases to exceed the totals for the B series.) But conversely if the combined interest-inflation rate is lower than the inflation rate (if $i_c < i_i$) then the higher first cost is preferred (material B is preferred). (Money earns more if the taxpayer invests it through taxes to pay for roads.)

2. If the real rate of return is positive ($i_r > 0$, which implies that $i_c > i_i$), and bond funds are available at less than the market rate ($i_{bonds} < i_c$) (since bonds are usually low-risk investments, the interest rate on bonds is often lower than average), then the lower first cost alternative (material A) is the better selection (less money paying less interest for less time, and purchasing power is rising).

SUMMARY

Inflation plays an important role in today's economic planning. To select the most economical from among alternatives, the inflationary loss in purchasing power of future income and costs must be accounted for and real, incremental rates of return calculated to determine the true anticipated benefits of proposed projects.

The real rate of return, i_r, can be calculated using the following equation:

$$i_r = \frac{(1 + i_c)}{(1 + i_i)} - 1$$

PROBLEMS FOR CHAPTER 23, INFLATION

Problem Group A

Find the amounts to invest in order to obtain a given objective.

A.1. A young engineer on her twenty-third birthday decides to start saving (investing) toward building up a retirement fund. She feels that $500,000 worth of buying power in terms of today's dollars will be adequate to see her through her sunset years starting at her sixty-third birthday.

    (a) If the inflation rate is estimated at 5 percent for the intervening period, how much must the investment be worth on her sixty-third birthday?

    (*Ans.* $3,520,000)

    (b) If her rate of return from her savings (investment) is estimated at 15 percent and she expects to make annual end-of-year deposits (with the first deposit on her twenty-third birthday and the last on her sixty-third) and each subsequent deposit will be 5 percent more than the previous one, how much should her first deposit amount to?

    (c) What will be the amount of her last deposit?

    (*Ans.* $8,240)

A.2. Same as Problem A. 1, except that the inflation rate is 6 percent, and her annual end-of-year deposits are expected to increase by $100 per year per year.

Problem Group B

Find the ROR.

B.1. A client expects to deposit $1,000 per year for 10 years (first deposit at EOY 1, last deposit at EOY 10) into an account that pays interest compounded at 8 percent per year. If inflation is estimated at 5 percent per year, find the real interest rate, i_r, in terms of real compound increase in buying power.

B.2. A home buyer purchases a house for $70,000. He pays $10,000 down and gets a $60,000 mortgage at 10 percent and 30 years payout period. If inflation occurs at 6 percent, what is the real interest rate, i_r?

B.3. Find the real rate of return, i_r, on the investments illustrated by each of the following cash flow diagrams (see Figure B.3 on next page).

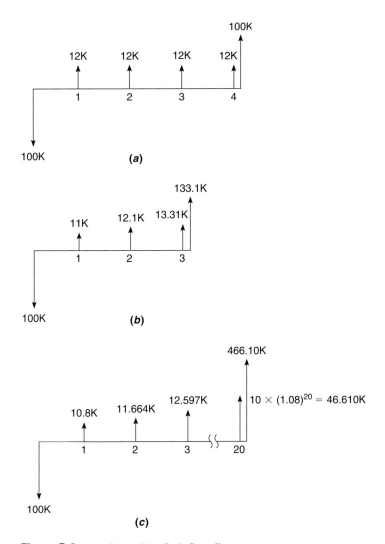

Figure B.3 (a), (b), and (c) Cash flow diagrams.

(a) Given: $i_i = 8$ percent.
(b) Given: $i_i = 8$ percent.
(c) Given: $i_i = 10$ percent.

B.4. An engineering graduate receives a starting salary of $35,000 per year. If the engineer's goal is to have an increase in real income of about 4 percent per year and inflation is estimated to average 5.3 percent per year:
(a) What average total rate of annual salary increase must be obtained?
(b) What should be the engineer's annual salary five years after graduating? Ten years after graduating? Twenty years after graduating?

B.5. An engineer received a starting salary of $16,200 per year upon graduation in 1975. In 1985, the annual salary was $36,000.

  (a) What combined annual interest rate has the salary increased by during the 10-year period?

  (*Ans.* 8.3%)

  (b) Assuming the Construction Cost Index accurately reflects the inflation rate during this 10-year period, what real rate of annual increase in salary did the engineer receive?

  (*Ans.* 1.6%)

B.6. A contractor bids on a construction project with data as shown. The duration of the project $= 30$ months. The owner will pay one lump sum when the project is completed at EOM 30. The contractor can obtain financing for all project direct costs from the bank at 1 percent per month interest on the balance outstanding. The contractor borrows at the beginning of the month for all costs incurred during that month.

$$\begin{aligned}
\text{cost of labor, materials, and equipment} &= \$24{,}346{,}986 \\
\text{cost of supervision, overhead, etc.} &= \underline{\$11{,}457{,}954} \\
\text{total direct costs} &= \$35{,}804{,}940 \\
\text{profit and contingencies} &= \$\ 2{,}666{,}426
\end{aligned}$$

  The contractor expects monthly costs to be $\$35{,}804{,}940/30$ months $= \$1{,}193{,}940$/month, and anticipates borrowing this amount from the bank on the first of the month for 30 months. The first borrowing is at the beginning of the first month and the last at the beginning of the thirtieth month, with repayment to the bank scheduled at EOM 30, when the owner pays the contractor.

  (a) How much should be added to the above sums just to pay the interest charges to the bank? (After calculating the correct amount, add this amount to the figures above to find the total bid price. This is the total amount the owner owes the contractor at EOM 30.)

  (b) Assume inflation increases the amount the contractor must borrow by 3/12 percent per month after the first borrowing. How much does the contractor owe the bank at EOM 30?

  (c) Assuming the owner pays on time at EOM 30, did the contractor include enough in profit and contingencies to cover the extra cost of inflation and still leave a viable profit? Assume 75 percent of the profit and contingencies amount was for profit and that inflation was the only contingency cost that occurred. How much profit (or loss) does the contractor have compared to how much was originally estimated **(i)** in terms of dollars, **(ii)** in terms of percent of total direct costs?

Problem Group C

Find the after-tax rate of return i_c.

C.1. An investor is considering the purchase of a bond with a maturity amount of $10,000 and date 10 years from now. The coupons on the bond provide for $900 interest paid annually at the end of each year until maturity. Assume $i_i = 6$ percent and the investor is in the 28 percent tax bracket for ordinary income. If the bond can be purchased now for $8,000 including commission, find the real rate of return, i_r.

  (*Ans.* $i_r = 3.4\%$)

C.2. Find the real after-tax ROR for the before-tax income illustrated in the cash flow diagram below. Given: 28 percent tax bracket and $i_i = 8.5$ percent.

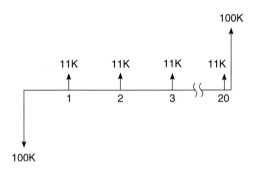

Figure C.2 Cash flow diagram.

C.3. Find the real after-tax IROR for the before-tax costs and income listed below. Given: 34 percent tax bracket; $i_i = 3$ percent; depreciation by SL to zero at EOY 5.

	A	B
Cost new ($)	100K	150K
Resale value at EOY 10 ($)	20K	30K
O & M costs ($)	55K/yr	45K/yr
Grad. O & M costs—both increases 5K/yr/yr		

Interest Tables

INTEREST TABLES

	i = 0.5% Present sum, P			i = 0.5% Uniform series, A			i = 0.5% Future sum, F			
n	P/F	P/A	P/G	A/F	A/P	A/G	F/P	F/A	F/G	n
1	0.99502	0.9950	0.0000	1.00000	1.00500	0.0000	1.0050	1.0000	0.0000	1
2	0.99007	1.9851	0.9803	0.49875	0.50375	0.4987	1.0100	2.0050	1.0000	2
3	0.98515	2.9702	2.9603	0.33167	0.33667	0.9966	1.0150	3.0150	3.0050	3
4	0.98025	3.9505	5.9011	0.24813	0.25313	1.4937	1.0201	4.0301	6.0200	4
5	0.97537	4.9258	9.8026	0.19801	0.20301	1.9900	1.0252	5.0502	10.050	5
6	0.97052	5.8963	14.655	0.16460	0.16960	2.4854	1.0303	6.0755	15.100	6
7	0.96569	6.8620	20.175	0.14073	0.14573	2.9800	1.0355	7.1058	21.175	7
8	0.96089	7.8229	27.175	0.12283	0.12783	3.4738	1.0407	8.1414	28.281	8
9	0.95610	8.7790	34.826	0.10891	0.11391	3.9667	1.0459	9.1821	36.423	9
10	0.95135	9.7304	43.386	0.09777	0.10277	4.4588	1.0511	10.228	45.505	10
11	0.94661	10.677	52.852	0.08866	0.09366	4.9501	1.0564	11.279	55.833	11
12	0.94191	11.618	63.213	0.08107	0.08607	5.4405	1.0616	12.335	67.112	12
13	0.93722	12.556	74.460	0.07464	0.07964	5.9301	1.0669	13.397	79.448	13
14	0.93256	13.488	86.583	0.06914	0.07414	6.4189	1.0723	14.464	92.845	14
15	0.92792	14.416	99.574	0.06436	0.06936	6.9069	1.0776	15.536	107.31	15
16	0.92330	15.339	113.42	0.06019	0.06519	7.3940	1.0830	16.614	122.84	16
17	0.91871	16.258	128.12	0.05651	0.06151	7.8803	1.0884	17.697	139.46	17
18	0.91414	17.172	143.66	0.05323	0.05823	8.3657	1.0939	18.785	157.15	18
19	0.90959	18.082	160.03	0.05030	0.05530	8.8504	1.0994	19.879	175.94	19
20	0.90506	18.987	177.23	0.04767	0.05267	9.3341	1.1049	20.979	195.82	20
21	0.90056	19.888	195.24	0.04528	0.05028	9.8171	1.1104	22.084	216.80	21
22	0.89608	20.784	214.06	0.04311	0.04811	10.299	1.1159	23.194	238.88	22
23	0.89162	21.675	233.67	0.04113	0.04613	10.780	1.1215	24.310	262.08	23
24	0.88719	22.562	254.08	0.03932	0.04432	11.261	1.1271	25.432	286.39	24
25	0.88277	23.445	275.26	0.03765	0.04265	11.740	1.1328	26.559	311.82	25
26	0.87838	24.324	297.22	0.03611	0.04111	12.219	1.1384	27.691	338.38	26
27	0.87401	25.198	319.95	0.03469	0.03969	12.697	1.1441	28.830	366.07	27
28	0.86966	26.067	343.43	0.03336	0.03836	13.174	1.1498	29.974	394.90	28
29	0.86533	26.933	367.66	0.03213	0.03713	13.651	1.1556	31.124	424.87	29
30	0.86103	27.794	392.63	0.03098	0.03598	14.126	1.1614	32.280	456.00	30
32	0.85248	29.503	444.76	0.02889	0.03389	15.075	1.1730	34.608	521.72	32
34	0.84402	31.195	499.75	0.02706	0.03206	16.020	1.1848	36.960	592.11	34
36	0.83564	32.871	557.56	0.02542	0.03042	16.962	1.1966	39.336	667.22	36
48	0.78710	42.580	959.91	0.01849	0.02349	22.543	1.2704	54.097	1219.5	48
60	0.74137	51.725	1448.6	0.01433	0.01933	28.006	1.3488	69.770	1954.0	60
120	0.54963	90.073	4823.5	0.00610	0.01110	53.550	1.8194	163.87	8775.8	120
180	0.40748	118.50	9031.3	0.00344	0.00844	76.211	2.4540	290.81	22163.	180
240	0.30210	139.58	13415.	0.00216	0.00716	96.113	3.3102	462.04	44408.	240
300	0.22397	155.20	17603.	0.00144	0.00644	113.41	4.4649	692.99	78598.	300
360	0.16604	166.79	21403.	0.00100	0.00600	128.32	6.0225	1004.5	128903.	360
INF	0.00000	200.00	40000.	0.0000	0.00500	200.00	INF	INF	INF	INF

Present sum, P | Uniform series, A | Future sum, F

n	P/F	P/A	P/G	A/F	A/P	A/G	F/P	F/A	F/G	n
1	0.99256	0.9925	0.0000	1.00000	1.00750	0.0000	1.0075	1.0000	0.0000	1
2	0.98517	1.9777	0.9851	0.49813	0.50563	0.4981	1.0150	2.0075	1.0000	2
3	0.97783	2.9555	2.9408	0.33085	0.33835	0.9950	1.0226	3.0225	3.0075	3
4	0.97055	3.9261	5.8525	0.24721	0.25471	1.4906	1.0303	4.0452	6.0300	4
5	0.96333	4.8894	9.7058	0.19702	0.20452	1.9850	1.0380	5.0755	10.075	5
6	0.95616	5.8456	14.486	0.16357	0.17107	2.4782	1.0458	6.1136	15.150	6
7	0.94904	6.7946	20.180	0.13967	0.14717	2.9701	1.0537	7.1594	21.264	7
8	0.94198	7.7366	26.774	0.12176	0.12926	3.4607	1.0616	8.2131	28.424	8
9	0.93496	8.6715	34.254	0.10782	0.11532	3.9501	1.0695	9.2747	36.637	9
10	0.92800	9.5995	42.606	0.09657	0.10417	4.4383	1.0775	10.344	45.911	10
11	0.92109	10.520	51.817	0.08755	0.09505	4.9252	1.0856	11.421	56.256	11
12	0.91424	11.434	61.874	0.07995	0.08745	5.4109	1.0938	12.507	67.678	12
13	0.90743	12.342	72.763	0.07352	0.08102	5.8954	1.1020	13.601	80.185	13
14	0.90068	13.243	84.472	0.06801	0.07551	6.3786	1.1102	14.703	93.787	14
15	0.89397	14.137	96.987	0.06324	0.07074	6.8605	1.1186	15.813	108.49	15
16	0.88732	15.024	110.29	0.05906	0.06656	7.3412	1.1269	16.932	124.30	16
17	0.88071	15.905	124.38	0.05537	0.06287	7.8207	1.1354	18.059	141.23	17
18	0.87416	16.779	139.24	0.05210	0.05960	8.2989	1.1439	19.194	159.29	18
19	0.86765	17.646	154.86	0.04917	0.05667	8.7759	1.1525	20.338	178.49	19
20	0.86119	18.508	171.23	0.04653	0.05403	9.2516	1.1611	21.491	198.82	20
21	0.85478	19.362	188.32	0.04415	0.05165	9.7261	1.1698	22.652	220.32	21
22	0.84842	20.211	206.14	0.04198	0.04948	10.199	1.1786	23.822	242.97	22
23	0.84210	21.053	224.66	0.04000	0.04750	10.671	1.1875	25.001	266.79	23
24	0.83583	21.889	243.89	0.03818	0.04568	11.142	1.1964	26.188	291.79	24
25	0.82961	22.718	263.80	0.03652	0.04402	11.611	1.2053	27.384	317.98	25
26	0.82343	23.542	284.38	0.03498	0.04248	12.080	1.2144	28.590	345.36	26
27	0.81730	24.359	305.63	0.03355	0.04105	12.547	1.2235	29.804	373.96	27
28	0.81122	25.170	327.54	0.03223	0.03973	13.012	1.2327	31.028	403.76	28
29	0.80518	25.975	350.08	0.03100	0.03850	13.477	1.2419	32.260	434.79	29
30	0.79919	26.775	373.26	0.02985	0.03735	13.940	1.2512	33.502	467.05	30
32	0.78733	28.355	421.46	0.02777	0.03527	14.863	1.2701	36.014	535.31	32
34	0.77565	29.912	472.07	0.02593	0.03343	15.781	1.2892	38.564	608.61	34
36	0.76446	31.446	524.99	0.02430	0.03180	16.694	1.3086	41.152	687.02	36
48	0.69861	40.184	836.84	0.01739	0.02489	22.069	1.4314	57.520	1269.4	48
60	0.63870	48.173	1313.5	0.01326	0.02076	27.266	1.5656	75.424	2056.5	60
120	0.40794	78.941	3998.5	0.00517	0.01267	50.652	2.4513	193.51	9801.9	120
180	0.26055	98.593	6892.6	0.00264	0.01014	69.900	3.8380	378.40	26454.	180
240	0.16641	111.14	9494.1	0.00150	0.00900	85.421	6.0091	667.88	57051.	240
300	0.10629	119.16	11636.	0.00089	0.00839	97.654	9.4084	1121.1	109483.	300
360	0.06789	124.28	13312.	0.00055	0.00805	107.11	14.730	1830.7	196099.	360
INF	0.00000	133.33	17777.	0.0000	0.00750	133.33	INF	INF	INF	INF

i = 1%

	Present sum, P			Uniform series, A			Future sum, F			
n	P/F	P/A	P/G	A/F	A/P	A/G	F/P	F/A	F/G	n
1	0.99010	0.9901	0.0000	1.00000	1.01000	0.0000	1.0100	1.0000	0.0000	1
2	0.98030	1.9704	0.9803	0.49751	0.50751	0.4975	1.0201	2.0100	1.0000	2
3	0.97059	2.9410	2.9214	0.33002	0.34002	0.9933	1.0303	3.0301	3.0100	3
4	0.96098	3.9019	5.8044	0.24628	0.25628	1.4875	1.0406	4.0604	6.0401	4
5	0.95147	4.8534	9.6102	0.19604	0.20604	1.9801	1.0510	5.1010	10.100	5
6	0.94205	5.7954	14.320	0.16255	0.17255	2.4709	1.0615	6.1520	15.201	6
7	0.93272	6.7281	19.916	0.13863	0.14863	2.9602	1.0721	7.2135	21.353	7
8	0.92348	7.6516	25.981	0.12069	0.13069	3.4477	1.0828	8.2856	28.567	8
9	0.91434	8.5660	33.695	0.10674	0.11674	3.9336	1.0936	9.3685	36.852	9
10	0.90529	9.4713	41.843	0.09558	0.10558	4.4179	1.1046	10.462	46.221	10
11	0.89632	10.367	50.806	0.08645	0.09645	4.9005	1.1156	11.566	56.683	11
12	0.88745	11.255	60.568	0.07885	0.08885	5.3814	1.1268	12.682	68.250	12
13	0.87866	12.133	71.112	0.07241	0.08241	5.8607	1.1380	13.809	80.932	13
14	0.86996	13.003	82.422	0.06690	0.07690	6.3383	1.1494	14.947	94.742	14
15	0.86135	13.865	94.481	0.06212	0.07212	6.8143	1.1609	16.096	109.69	15
16	0.85282	14.717	107.27	0.05794	0.06794	7.2886	1.1725	17.257	125.78	16
17	0.84438	15.562	120.78	0.05426	0.06426	7.7613	1.1843	18.430	143.04	17
18	0.83602	16.398	134.99	0.05098	0.06098	8.2323	1.1961	19.614	161.47	18
19	0.82774	17.226	149.89	0.04805	0.05805	8.7016	1.2081	20.810	181.09	19
20	0.81954	18.045	165.46	0.04542	0.05542	9.1693	1.2201	22.019	201.90	20
21	0.81143	18.857	181.69	0.04303	0.05303	9.6354	1.2323	23.239	223.91	21
22	0.80340	19.660	198.56	0.04086	0.05086	10.099	1.2447	24.471	247.15	22
23	0.79544	20.455	216.06	0.03889	0.04889	10.562	1.2571	25.716	271.63	23
24	0.78757	21.243	234.18	0.03707	0.04707	11.023	1.2697	26.973	297.34	24
25	0.77977	22.023	252.89	0.03541	0.04541	11.483	1.2824	28.243	324.32	25
26	0.77205	22.795	272.19	0.03387	0.04387	11.940	1.2952	29.525	352.56	26
27	0.76440	23.559	292.07	0.03245	0.04245	12.397	1.3082	30.820	382.08	27
28	0.75684	24.316	312.50	0.03112	0.04112	12.851	1.3212	32.129	412.91	28
29	0.74934	25.065	333.48	0.02990	0.03990	13.304	1.3345	33.450	445.03	29
30	0.74192	25.807	355.03	0.02875	0.03875	13.755	1.3478	34.784	478.48	30
32	0.72730	27.269	399.58	0.02667	0.03667	14.653	1.3749	37.494	549.40	32
34	0.71297	28.702	446.15	0.02484	0.03484	15.544	1.4025	40.257	625.77	34
36	0.69892	30.107	494.62	0.02321	0.03321	16.428	1.4307	43.076	707.68	36
48	0.62026	37.974	820.14	0.01533	0.02633	21.597	1.6122	61.222	1322.2	48
60	0.55045	44.955	1192.8	0.01224	0.02224	26.533	1.8167	81.669	2166.9	60
120	0.30299	69.700	3334.1	0.00435	0.01435	47.834	3.3003	230.03	11003.	120
180	0.16678	83.321	5330.0	0.00201	0.01201	63.969	5.9958	499.58	31958.	180
240	0.09181	90.819	6878.6	0.00101	0.01101	75.739	10.892	989.25	74925.	240
300	0.05053	94.946	7978.8	0.00053	0.01053	84.032	19.788	1878.8	157885.	300
360	0.02782	97.218	8720.4	0.00029	0.01029	89.699	35.949	3494.9	313496.	360
INF	0.00000	100.00	10000.	0.0000	0.01000	100.00	INF	INF	INF	INF

	i = 1.5%			i = 1.5%			i = 1.5%			
	Present sum, P			Uniform series, A			Future sum, F			
n	P/F	P/A	P/G	A/F	A/P	A/G	F/P	F/A	F/G	n
1	0.98522	0.9852	0.0000	1.00000	1.01500	0.0000	1.0150	1.0000	0.0000	1
2	0.97066	1.9558	0.9706	0.49628	0.51128	0.4962	1.0302	2.0150	1.0000	2
3	0.95632	2.9122	2.8833	0.32838	0.34338	0.9900	1.0456	3.0452	3.0150	3
4	0.94218	3.8543	5.7098	0.24444	0.25944	1.4813	1.0613	4.0909	6.0602	4
5	0.92826	4.7826	9.4228	0.19409	0.20909	1.9702	1.0772	5.1522	10.151	5
6	0.91454	5.6971	13.995	0.16053	0.17553	2.4565	1.0934	6.2295	15.303	6
7	0.90103	6.5982	19.401	0.13656	0.15156	2.9404	1.1098	7.3229	21.532	7
8	0.88771	7.4859	25.615	0.11858	0.13358	3.4218	1.1264	8.4328	28.855	8
9	0.87459	8.3605	32.612	0.10461	0.11961	3.9007	1.1433	9.5593	37.288	9
10	0.86167	9.2221	40.367	0.09343	0.10843	4.3772	1.1605	10.702	46.848	10
11	0.84893	10.071	48.856	0.08429	0.09929	4.8511	1.1779	11.863	57.550	11
12	0.83639	10.907	58.057	0.07668	0.09168	5.3226	1.1956	13.041	69.414	12
13	0.82403	11.731	67.945	0.07024	0.08524	5.7916	1.2136	14.236	82.455	13
14	0.81185	12.543	78.499	0.06472	0.07972	6.2582	1.2317	15.450	96.692	14
15	0.79985	13.343	89.697	0.05994	0.07494	6.7223	1.2502	16.682	112.14	15
16	0.78803	14.131	101.51	0.05577	0.07077	7.1839	1.2689	17.932	128.82	16
17	0.77639	14.907	113.94	0.05208	0.06708	7.6430	1.2880	19.201	146.75	17
18	0.76491	15.672	126.94	0.04881	0.06381	8.0997	1.3073	20.489	165.95	18
19	0.75361	16.426	140.50	0.04588	0.06088	8.5539	1.3269	21.796	186.44	19
20	0.74247	17.168	154.61	0.04325	0.05825	9.0056	1.3468	23.123	208.24	20
21	0.73150	17.900	169.24	0.04087	0.05587	9.4549	1.3670	24.470	231.36	21
22	0.72069	18.620	184.38	0.03870	0.05370	9.9018	1.3875	25.837	255.83	22
23	0.71004	19.330	200.00	0.03673	0.05173	10.346	1.4083	27.225	281.67	23
24	0.69954	20.030	216.09	0.03492	0.04992	10.788	1.4295	28.633	308.90	24
25	0.68921	20.719	232.63	0.03326	0.04826	11.227	1.4509	30.063	337.53	25
26	0.67902	21.398	249.60	0.03173	0.04673	11.664	1.4727	31.514	367.59	26
27	0.66899	22.067	267.00	0.03032	0.04532	12.099	1.4948	32.986	399.11	27
28	0.65910	22.726	284.79	0.02900	0.04400	12.531	1.5172	34.481	432.09	28
29	0.64936	23.376	302.97	0.02778	0.04278	12.961	1.5399	35.998	466.58	29
30	0.63976	24.015	321.53	0.02664	0.04164	13.388	1.5630	37.538	502.57	30
32	0.62099	25.267	359.69	0.02458	0.03958	14.235	1.6103	40.688	579.21	32
34	0.60277	26.481	399.16	0.02276	0.03776	15.073	1.6590	43.933	662.20	34
36	0.58509	27.660	439.83	0.02115	0.03615	15.900	1.7091	47.276	751.73	36
48	0.48936	34.042	703.54	0.01437	0.02937	20.666	2.0434	69.565	1437.6	48
60	0.40930	39.380	988.16	0.01039	0.02539	25.093	2.4432	96.214	2414.3	60
120	0.16752	55.498	2359.7	0.00302	0.01802	42.518	5.9693	331.28	14085.	120
180	0.06857	62.095	3316.7	0.00110	0.01610	53.416	14.584	905.62	48375.	180
240	0.02806	64.795	3870.6	0.00043	0.01543	59.736	35.632	2308.8	137924.	240
300	0.01149	65.900	4163.6	0.00017	0.01517	63.180	87.058	5737.2	362484.	300
360	0.00470	66.353	4310.7	0.00007	0.01507	64.966	212.70	14113.	916906.	360
INF	0.00000	66.666	4444.4	0.0000	0.01500	66.666	INF	INF	INF	INF

i = 2%

n	P/F	P/A	P/G	A/F	A/P	A/G	F/P	F/A	F/G	n
	Present sum, P			Uniform series, A			Future sum, F			
1	0.98039	0.9803	0.0000	1.00000	1.02000	0.0000	1.0200	1.0000	0.0000	1
2	0.96117	1.9415	0.9611	0.49505	0.51505	0.4950	1.0404	2.0200	1.0000	2
3	0.94232	2.8838	2.8458	0.32675	0.34675	0.9868	1.0612	3.0604	3.0604	3
4	0.92385	3.8077	5.6173	0.24262	0.26262	1.4752	1.0824	4.1216	6.0804	4
5	0.90573	4.7134	9.2402	0.19216	0.21216	1.9604	1.1040	5.2040	10.202	5
6	0.88797	5.6014	13.680	0.15853	0.17853	2.4422	1.1261	6.3081	15.406	6
7	0.87056	6.4719	18.903	0.13451	0.15451	2.9208	1.1486	7.4342	21.714	7
8	0.85349	7.3254	24.877	0.11651	0.13651	3.3960	1.1716	8.5829	29.148	8
9	0.83676	8.1622	31.572	0.10252	0.12252	3.8680	1.1950	9.7546	37.731	9
10	0.82035	8.9825	38.955	0.09133	0.11133	4.3367	1.2189	10.949	47.486	10
11	0.80426	9.7868	46.997	0.08218	0.10218	4.8021	1.2433	12.168	58.435	11
12	0.78849	10.575	55.671	0.07456	0.09456	5.2642	1.2682	13.412	70.604	12
13	0.77303	11.348	64.947	0.06812	0.08812	5.7230	1.2936	14.680	84.016	13
14	0.75788	12.106	74.799	0.06260	0.08260	6.1786	1.3194	15.973	98.696	14
15	0.74301	12.849	85.202	0.05783	0.07783	6.6309	1.3458	17.293	114.67	15
16	0.72845	13.577	96.128	0.05365	0.07365	7.0799	1.3727	18.639	131.96	16
17	0.71416	14.291	107.55	0.04997	0.06997	7.5256	1.4002	20.012	150.60	17
18	0.70016	14.992	119.45	0.04670	0.06670	7.9681	1.4282	21.412	170.61	18
19	0.68643	15.678	131.81	0.04378	0.06378	8.4073	1.4568	22.840	192.02	19
20	0.67297	16.351	144.60	0.04116	0.06116	8.8432	1.4859	24.297	214.86	20
21	0.65978	17.011	157.79	0.03878	0.05878	9.2759	1.5156	25.783	239.16	21
22	0.64684	17.658	171.37	0.03663	0.05663	9.7054	1.5459	27.299	264.94	22
23	0.63416	18.292	185.33	0.03467	0.05467	10.131	1.5769	28.845	292.24	23
24	0.62172	18.913	199.63	0.03287	0.05287	10.554	1.6084	30.421	321.09	24
25	0.60953	19.523	214.25	0.03122	0.05122	10.974	1.6406	32.030	351.51	25
26	0.59758	20.121	229.19	0.02970	0.04970	11.391	1.6734	33.670	383.54	26
27	0.58586	20.706	244.43	0.02829	0.04829	11.804	1.7068	35.344	417.21	27
28	0.57437	21.281	259.93	0.02699	0.04699	12.214	1.7410	37.051	452.56	28
29	0.56311	21.844	275.70	0.02578	0.04578	12.621	1.7758	38.792	489.61	29
30	0.55207	22.396	291.71	0.02465	0.04465	13.025	1.8113	40.568	528.40	30
32	0.53063	23.468	324.40	0.02261	0.04261	13.823	1.8845	44.227	611.35	32
34	0.51003	24.498	357.88	0.02082	0.04082	14.608	1.9606	48.033	701.69	34
36	0.49022	25.488	392.04	0.01923	0.03923	15.380	2.0398	51.994	799.71	36
48	0.38654	30.673	605.96	0.01260	0.03260	19.755	2.5870	79.353	1567.6	48
60	0.30478	34.760	823.69	0.00877	0.02877	23.696	3.2810	114.05	2702.5	60
120	0.09289	45.355	1710.4	0.00205	0.02205	37.711	10.765	488.25	18412.	120
180	0.02831	48.568	2174.2	0.00058	0.02058	44.755	35.320	1716.0	76802.	180
240	0.00863	49.568	2374.9	0.00017	0.02017	47.911	115.88	5744.4	275222.	240
300	0.00263	49.868	2453.9	0.00005	0.02005	49.208	380.23	18961.	933086.	300
360	0.00080	49.959	2483.5	0.00002	0.02002	49.711	1247.5	62328.	>10**6	360
INF	0.00000	50.000	2500.0	0.0000	0.02000	50.000	INF	INF	INF	INF

| | i = 2.5% | | | i = 2.5% | | | i = 2.5% | | | |
| | Present sum, P | | | Uniform series, A | | | Future sum, F | | | |
n	P/F	P/A	P/G	A/F	A/P	A/G	F/P	F/A	F/G	n
1	0.97561	0.9756	0.0000	1.00000	1.02500	0.0000	1.0250	1.0000	0.0000	1
2	0.95181	1.9274	0.9518	0.49383	0.51883	0.4938	1.0506	2.0250	1.0000	2
3	0.92860	2.8560	2.8090	0.32514	0.35014	0.9835	1.0768	3.0756	3.0250	3
4	0.90595	3.7619	5.5263	0.24082	0.26582	1.4691	1.1038	4.1525	6.1006	4
5	0.88385	4.6458	9.0622	0.19025	0.21525	1.9506	1.1314	5.2563	10.253	5
6	0.86230	5.5081	13.373	0.15655	0.18155	2.4280	1.1596	6.3877	15.509	6
7	0.84127	6.3493	18.421	0.13250	0.15750	2.9012	1.1886	7.5474	21.897	7
8	0.82075	7.1701	24.166	0.11447	0.13947	3.3704	1.2184	8.7361	29.444	8
9	0.80073	7.9708	30.572	0.10046	0.12546	3.8355	1.2488	9.9545	38.180	9
10	0.78120	8.7520	37.603	0.08926	0.11426	4.2964	1.2800	11.203	48.135	10
11	0.76214	9.5142	45.224	0.08011	0.10511	4.7533	1.3120	12.483	59.338	11
12	0.74356	10.257	53.403	0.07249	0.09749	5.2061	1.3448	13.795	71.822	12
13	0.72542	10.983	62.108	0.06605	0.09105	5.6549	1.3785	15.140	85.617	13
14	0.70773	11.690	71.309	0.06054	0.08554	6.0995	1.4129	16.519	100.75	14
15	0.69047	12.381	80.975	0.05577	0.08077	6.5401	1.4483	17.931	117.27	15
16	0.67362	13.055	91.080	0.05150	0.07660	6.9766	1.4845	19.380	135.20	16
17	0.65720	13.712	101.59	0.04793	0.07293	7.4091	1.5216	20.864	154.58	17
18	0.64117	14.353	112.49	0.04467	0.06967	7.8375	1.5596	22.386	175.45	18
19	0.62553	14.978	123.75	0.04176	0.06676	8.2619	1.5986	23.946	197.84	19
20	0.61027	15.589	135.35	0.03915	0.06415	8.6823	1.6386	25.544	221.78	20
21	0.59539	16.184	147.25	0.03679	0.06179	9.0990	1.6796	27.183	247.33	21
22	0.58086	16.765	159.45	0.03465	0.05965	9.5109	1.7216	28.862	274.51	22
23	0.56670	17.332	171.92	0.03270	0.05770	9.9193	1.7646	30.584	303.37	23
24	0.55288	17.885	184.63	0.03091	0.05591	10.324	1.8087	32.349	333.96	24
25	0.53939	18.424	197.58	0.02928	0.05428	10.724	1.8539	34.157	366.31	25
30	0.47674	20.930	265.12	0.02278	0.04778	12.666	2.0976	43.902	556.10	30
35	0.42137	23.145	335.88	0.01821	0.04321	14.512	2.3732	54.928	797.12	35
36	0.41109	23.556	350.27	0.01745	0.04245	14.869	2.4325	57.301	852.05	36
40	0.37243	25.102	408.22	0.01484	0.03984	16.262	2.6850	67.402	1096.1	40
48	0.30567	27.773	524.03	0.01101	0.03601	18.868	3.2714	90.859	1714.3	48
50	0.29094	28.362	552.60	0.01026	0.03526	19.483	3.4371	97.484	1899.3	50
60	0.22728	30.909	693.86	0.00735	0.03235	22.352	4.3997	135.99	3039.6	60
100	0.08465	36.614	1125.9	0.00231	0.02731	30.752	11.813	432.54	13301.	100
120	0.05166	37.933	1269.3	0.00136	0.02636	33.463	19.358	734.32	24573.	120
180	0.01174	39.530	1496.6	0.00030	0.02530	37.861	85.171	3366.8	127475.	180
240	0.00267	39.893	1570.1	0.00007	0.02507	39.357	374.73	14949.	588381.	240
300	0.00061	39.975	1591.7	0.00002	0.02502	39.817	1648.7	65910.	>10**6	300
360	0.00014	39.994	1597.7	0.00000	0.02500	39.950	7254.2	290129	>10**6	360
INF	0.00000	40.000	1600.0	0.0000	0.02500	40.000	INF	INF	INF	INF

	i = 3% Present sum, P			i = 3% Uniform series, A			i = 3% Future sum, F			
n	P/F	P/A	P/G	A/F	A/P	A/G	F/P	F/A	F/G	n
1	0.97087	0.9708	0.0000	1.00000	1.03000	0.0000	1.0300	1.0000	0.0000	1
2	0.94260	1.9134	0.9426	0.49261	0.52261	0.4926	1.0609	2.0300	1.0000	2
3	0.91514	2.8286	2.7728	0.32353	0.35353	0.9803	1.0927	3.0909	3.0300	3
4	0.88849	3.7171	5.4383	0.23903	0.26903	1.4630	1.1255	4.1836	6.1209	4
5	0.86261	4.5797	8.8887	0.18835	0.21835	1.9409	1.1592	5.3091	10.304	5
6	0.83748	5.4171	13.076	0.15460	0.18460	2.4138	1.1940	6.4684	15.613	6
7	0.81309	6.2302	17.954	0.13051	0.16051	2.8818	1.2298	7.6624	22.082	7
8	0.78941	7.0196	23.480	0.11246	0.14246	3.3441	1.2667	8.8923	29.744	8
9	0.76642	7.7861	29.611	0.09843	0.12843	3.8031	1.3047	10.159	38.636	9
10	0.74409	8.5302	36.308	0.08723	0.11723	4.2565	1.3439	11.463	48.796	10
11	0.72242	9.2526	43.533	0.07808	0.10808	4.7049	1.3842	12.807	60.259	11
12	0.70138	9.9540	51.248	0.07046	0.10046	5.1485	1.4257	14.192	73.067	12
13	0.68095	10.635	59.419	0.06403	0.09403	5.5872	1.4685	15.617	87.259	13
14	0.66112	11.296	68.014	0.05853	0.08853	6.0210	1.5125	17.086	102.87	14
15	0.64186	11.937	77.000	0.05377	0.08377	6.4500	1.5579	18.598	119.96	15
16	0.62317	12.561	86.347	0.04961	0.07961	6.8742	1.6047	20.156	138.56	16
17	0.60502	13.166	96.028	0.04595	0.07595	7.2935	1.6528	21.761	158.72	17
18	0.58739	13.753	106.01	0.04271	0.07271	7.7081	1.7024	23.414	180.48	18
19	0.57029	14.323	116.27	0.03981	0.06981	8.1178	1.7535	25.116	203.89	19
20	0.55368	14.877	126.79	0.03722	0.06722	8.5228	1.8061	26.870	229.01	20
21	0.53755	15.415	137.55	0.03487	0.06487	8.9230	1.8602	28.676	255.88	21
22	0.52189	15.936	148.50	0.03275	0.06275	9.3185	1.9161	30.536	284.55	22
23	0.50669	16.443	159.65	0.03081	0.06081	9.7093	1.9735	32.452	315.09	23
24	0.49193	16.935	170.97	0.02905	0.05905	10.095	2.0327	34.426	347.54	24
25	0.47761	17.413	182.43	0.02743	0.05743	10.476	2.0937	36.459	381.97	25
26	0.46369	17.876	194.02	0.02594	0.05594	10.853	2.1565	38.553	418.43	26
28	0.43708	18.764	217.53	0.02329	0.05329	11.593	2.2879	42.930	497.69	28
30	0.41199	19.600	241.36	0.02102	0.05102	12.314	2.4272	47.575	585.84	30
35	0.35538	21.487	301.62	0.01654	0.04654	14.037	2.8138	60.462	848.73	35
36	0.34503	21.832	313.70	0.01580	0.04580	14.368	2.8982	63.275	909.19	36
40	0.30656	23.114	351.75	0.01326	0.04326	15.650	3.2620	75.401	1180.0	40
45	0.26444	24.518	420.63	0.01079	0.04079	17.155	3.7816	92.719	1590.6	45
48	0.24200	25.266	457.48	0.00958	0.03958	18.008	4.1322	104.40	1880.2	48
50	0.22811	25.729	477.48	0.00887	0.03887	18.557	4.3839	112.79	2093.2	50
60	0.16973	27.675	583.05	0.00613	0.03613	21.067	5.8916	163.05	3435.1	60
70	0.12630	29.123	676.08	0.00434	0.03434	23.214	7.9178	230.59	5353.1	70
80	0.09398	30.202	756.08	0.00311	0.03311	25.035	10.640	321.36	8045.4	80
90	0.06993	31.002	823.63	0.00226	0.03226	26.566	14.300	443.34	11778.	90
100	0.05203	31.598	879.85	0.00165	0.03165	27.844	19.218	607.28	16909.	100
120	0.02881	32.373	963.86	0.00089	0.03089	29.773	34.711	1123.7	33456.	120
180	0.00489	33.170	1076.3	0.00015	0.03015	32.448	204.50	6783.4	220115.	180
240	0.00083	33.305	1103.5	0.00002	0.03002	33.134	1204.8	40128.	>10**6	240
INF	0.00000	33.333	1111.1	0.0000	0.03000	33.333	INF	INF	INF	INF

| | i = 4% | | | i = 4% | | | i = 4% | | | |
| | Present sum, P | | | Uniform series, A | | | Future sum, F | | | |
n	P/F	P/A	P/G	A/F	A/P	A/G	F/P	F/A	F/G	n
1	0.96154	0.9615	0.0000	1.00000	1.04000	0.0000	1.0400	1.0000	0.0000	1
2	0.92456	1.8860	0.9245	0.49020	0.53020	0.4902	1.0816	2.0400	1.0000	2
3	0.88900	2.7750	2.7025	0.32035	0.36035	0.9738	1.1248	3.1216	3.0400	3
4	0.85480	3.6299	5.2669	0.23549	0.27549	1.4510	1.1698	4.2464	6.1616	4
5	0.82193	4.4518	8.5546	0.18463	0.22463	1.9216	1.2166	5.4163	10.408	5
6	0.79031	5.2421	12.506	0.15076	0.19076	2.3857	1.2653	6.6329	15.824	6
7	0.75992	6.0020	17.065	0.12661	0.16661	2.8433	1.3159	7.8982	22.457	7
8	0.73069	6.7327	22.180	0.10853	0.14853	3.2944	1.3685	9.2142	30.355	8
9	0.70259	7.4353	27.801	0.09449	0.13449	3.7390	1.4233	10.582	39.569	9
10	0.67556	8.1109	33.881	0.08329	0.12329	4.1772	1.4802	12.006	50.152	10
11	0.64958	8.7604	40.377	0.07415	0.11415	4.6090	1.5394	13.486	62.158	11
12	0.62460	9.3850	47.247	0.06655	0.10655	5.0343	1.6010	15.025	75.645	12
13	0.60057	9.9856	54.454	0.06014	0.10014	5.4532	1.6650	16.626	90.670	13
14	0.57748	10.563	61.961	0.05467	0.09467	5.8658	1.7316	18.291	107.29	14
15	0.55526	11.118	69.735	0.04994	0.08994	6.2720	1.8009	20.023	125.59	15
16	0.53391	11.652	77.744	0.04582	0.08582	6.6720	1.8729	21.824	145.61	16
17	0.51337	12.165	85.958	0.04220	0.08220	7.0656	1.9479	23.697	167.43	17
18	0.49363	12.659	94.349	0.03899	0.07899	7.4530	2.0258	25.645	191.13	18
19	0.47464	13.133	102.89	0.03614	0.07614	7.8341	2.1068	27.671	216.78	19
20	0.45639	13.590	111.56	0.03358	0.07358	8.2091	2.1911	29.778	244.45	20
21	0.43883	14.029	120.34	0.03128	0.07128	8.5779	2.2787	31.969	274.23	21
22	0.42196	14.451	129.20	0.02920	0.06920	8.9406	2.3699	34.248	306.19	22
23	0.40573	14.856	138.12	0.02731	0.06731	9.2972	2.4647	36.617	340.44	23
24	0.39012	15.247	147.10	0.02559	0.06559	9.6479	2.5633	39.082	377.06	24
25	0.37512	15.622	156.10	0.02401	0.06401	9.9925	2.6658	41.645	416.14	25
26	0.36069	15.982	165.12	0.02257	0.06257	10.331	2.7724	44.311	457.79	26
28	0.33348	16.663	183.14	0.02001	0.06001	10.990	2.9987	49.967	549.19	28
30	0.30832	17.292	201.06	0.01783	0.05783	11.627	3.2434	56.084	652.12	30
35	0.25342	18.664	244.87	0.01358	0.05358	13.119	3.9460	73.652	966.30	35
36	0.24367	18.908	253.40	0.01289	0.05289	13.401	4.1039	77.598	1039.9	36
40	0.20829	19.792	286.53	0.01052	0.05052	14.476	4.8010	95.025	1375.6	40
45	0.17120	20.720	325.40	0.00826	0.04826	15.704	5.8411	121.02	1900.7	45
48	0.15219	21.195	347.24	0.00718	0.04718	16.383	6.5705	139.26	2281.5	48
50	0.14071	21.482	361.16	0.00655	0.04655	16.812	7.1066	152.66	2556.6	50
55	0.11566	22.108	393.68	0.00523	0.04523	17.807	8.6463	191.15	3403.9	55
60	0.09506	22.623	422.99	0.00420	0.04420	18.697	10.519	237.99	4449.7	60
70	0.06422	23.394	472.47	0.00275	0.04275	20.196	15.571	364.29	7357.2	70
80	0.04338	23.915	511.11	0.00181	0.04181	21.371	23.049	551.24	11781.	80
90	0.02931	24.267	540.72	0.00121	0.04121	22.282	34.119	827.98	18449.	90
100	0.01980	24.505	563.12	0.00081	0.04081	22.980	50.504	1237.6	28440.	100
120	0.00904	24.774	592.24	0.00036	0.04036	23.905	110.66	2741.5	65539.	120
180	0.00086	24.978	620.59	0.00003	0.04003	24.845	1164.1	29078.	724256.	180
INF	0.00000	25.000	625.00	0.0000	0.04000	25.000	INF	INF	INF	INF

	i = 5%			i = 5%			i = 5%			
	Present sum, P			Uniform series, A			Future sum, F			
n	P/F	P/A	P/G	A/F	A/P	A/G	F/P	F/A	F/G	n
1	0.95238	0.9523	0.0000	1.00000	1.05000	0.0000	1.0500	1.0000	0.0000	1
2	0.90703	1.8594	0.9070	0.48780	0.53780	0.4878	1.1025	2.0500	1.0000	2
3	0.86384	2.7232	2.6347	0.31721	0.36721	0.9674	1.1576	3.1525	3.0500	3
4	0.82270	3.5459	5.1028	0.23201	0.28201	1.4390	1.2155	4.3101	6.2025	4
5	0.78353	4.3294	8.2369	0.18097	0.23097	1.9025	1.2762	5.5256	10.512	5
6	0.74622	5.0756	11.968	0.14702	0.19702	2.3579	1.3401	6.8019	16.038	6
7	0.71068	5.7863	16.232	0.12282	0.17282	2.8052	1.4071	8.1420	22.840	7
8	0.67684	6.4632	20.970	0.10472	0.15472	3.2445	1.4774	9.5491	30.982	8
9	0.64461	7.1078	26.126	0.09069	0.14069	3.6757	1.5513	11.026	40.531	9
10	0.61391	7.7217	31.652	0.07950	0.12950	4.0990	1.6288	12.577	51.557	10
11	0.58468	8.3064	37.498	0.07039	0.12039	4.5144	1.7103	14.206	64.135	11
12	0.55684	8.8632	43.624	0.06283	0.11283	4.9219	1.7958	15.917	78.342	12
13	0.53032	9.3935	49.987	0.05646	0.10646	5.3215	1.8856	17.713	94.259	13
14	0.50507	9.8986	56.553	0.05102	0.10102	5.7132	1.9799	19.598	111.97	14
15	0.48102	10.379	63.288	0.04634	0.09634	6.0973	2.0789	21.578	131.57	15
16	0.45811	10.837	70.159	0.04227	0.09227	6.4736	2.1828	23.657	153.15	16
17	0.43630	11.274	77.140	0.03870	0.08870	6.8422	2.2920	25.840	176.80	17
18	0.41552	11.689	84.204	0.03555	0.08555	7.2033	2.4066	28.132	202.64	18
19	0.39573	12.085	91.327	0.03275	0.08275	7.5569	2.5269	30.539	230.78	19
20	0.37689	12.462	98.488	0.03024	0.08024	7.9029	2.6533	33.066	261.31	20
21	0.35894	12.821	105.66	0.02800	0.07800	8.2416	2.7859	35.719	294.38	21
22	0.34185	13.163	112.84	0.02597	0.07597	8.5729	2.9252	38.505	330.10	22
23	0.32557	13.488	120.00	0.02414	0.07414	8.8970	3.0715	41.430	368.61	23
24	0.31007	13.798	127.14	0.02247	0.07247	9.2139	3.2251	44.502	410.04	24
25	0.29530	14.093	134.22	0.02095	0.07095	9.5237	3.3863	47.727	454.54	25
26	0.28124	14.375	141.25	0.01956	0.06956	9.8265	3.5556	51.113	502.26	26
28	0.25509	14.898	155.11	0.01712	0.06712	10.411	3.9201	58.402	608.05	28
30	0.23138	15.372	168.62	0.01505	0.06505	10.969	4.3219	66.438	728.77	30
35	0.18129	16.374	200.58	0.01107	0.06107	12.249	5.5160	90.320	1106.4	35
36	0.17266	16.546	206.62	0.01043	0.06043	12.487	5.7918	95.836	1196.7	36
40	0.14205	17.159	229.54	0.00828	0.05828	13.377	7.0399	120.80	1616.0	40
45	0.11130	17.774	255.31	0.00626	0.05626	14.364	8.9850	159.70	2294.0	45
48	0.09614	18.077	265.25	0.00532	0.05532	14.894	10.401	188.02	2800.5	48
50	0.08720	18.255	277.91	0.00478	0.05478	15.223	11.467	209.34	3186.9	50
55	0.06833	18.633	297.51	0.00367	0.05367	15.966	14.635	272.71	4354.2	55
60	0.05354	18.929	314.34	0.00283	0.05283	16.606	18.679	353.58	5871.6	60
70	0.03287	19.342	340.84	0.00170	0.05170	17.621	30.426	588.52	10370.	70
80	0.02018	19.596	359.64	0.00103	0.05103	18.352	49.561	971.22	17824.	80
90	0.01239	19.752	372.74	0.00063	0.05063	18.871	80.730	1594.6	30092.	90
100	0.00760	19.847	381.74	0.00038	0.05038	19.233	131.50	2610.0	50200.	100
120	0.00287	19.942	391.97	0.00014	0.05014	19.655	348.91	6958.2	136765.	120
INF	0.00000	20.000	400.00	0.0000	0.05000	20.000	INF	INF	INF	INF

	i = 6%			i = 6%			i = 6%			
	Present sum, P			Uniform series, A			Future sum, F			
n	P/F	P/A	P/G	A/F	A/P	A/G	F/P	F/A	F/G	n
1	0.94340	0.9434	0.0000	1.00000	1.06000	0.0000	1.0600	1.0000	0.0000	1
2	0.89000	1.8333	0.8900	0.48544	0.54544	0.4854	1.1236	2.0600	1.0000	2
3	0.83962	2.6730	2.5692	0.31411	0.37411	0.9611	1.1910	3.1836	3.0600	3
4	0.79209	3.4651	4.9455	0.22859	0.28859	1.4272	1.2624	4.3746	6.2436	4
5	0.74726	4.2123	7.9345	0.17740	0.23740	1.8836	1.3382	5.6370	10.618	5
6	0.70496	4.9173	11.459	0.14336	0.20336	2.3304	1.4185	6.9753	16.255	6
7	0.66506	5.5823	15.449	0.11914	0.17914	2.7675	1.5036	8.3938	23.230	7
8	0.62741	6.2097	19.841	0.10104	0.16104	3.1952	1.5938	9.8974	31.624	8
9	0.59190	6.8016	24.576	0.08702	0.14702	3.6133	1.6894	11.491	41.521	9
10	0.55839	7.3600	29.602	0.07587	0.13587	4.0220	1.7908	13.180	53.013	10
11	0.52679	7.8868	34.870	0.06679	0.12679	4.4212	1.8983	14.971	66.194	11
12	0.49697	8.3838	40.336	0.05928	0.11928	4.8112	2.0122	16.869	81.165	12
13	0.46884	8.8526	45.962	0.05296	0.11296	5.1919	2.1329	18.882	98.035	13
14	0.44230	9.2949	51.712	0.04758	0.10758	5.5635	2.2609	21.015	116.91	14
15	0.41727	9.7122	57.554	0.04296	0.10296	5.9259	2.3965	23.276	137.93	15
16	0.39365	10.105	63.459	0.03895	0.09895	6.2794	2.5403	25.672	161.20	16
17	0.37136	10.477	69.401	0.03544	0.09544	6.6239	2.6927	28.212	186.88	17
18	0.35034	10.827	75.356	0.03236	0.09236	6.9597	2.8543	30.905	215.09	18
19	0.33051	11.158	81.306	0.02962	0.08962	7.2867	3.0256	33.760	246.00	19
20	0.31180	11.469	87.230	0.02718	0.08718	7.6051	3.2071	36.785	279.76	20
21	0.29416	11.764	93.113	0.02500	0.08500	7.9150	3.3995	39.992	316.54	21
22	0.27751	12.041	98.941	0.02305	0.08305	8.2166	3.6035	43.392	356.53	22
23	0.26180	12.303	104.70	0.02128	0.08128	8.5099	3.8197	46.995	399.93	23
24	0.24698	12.550	110.38	0.01968	0.07968	8.7950	4.0489	50.815	446.92	24
25	0.23300	12.783	115.97	0.01823	0.07823	9.0722	4.2918	54.864	497.74	25
26	0.21981	13.003	121.46	0.01690	0.07690	9.3414	4.5493	59.156	552.60	26
27	0.20737	13.210	126.86	0.01570	0.07570	9.6029	4.8223	63.705	611.76	27
28	0.19563	13.406	132.14	0.01459	0.07459	9.8568	5.1116	68.528	675.46	28
29	0.18456	13.590	137.31	0.01358	0.07358	10.103	5.4183	73.639	743.99	29
30	0.17411	13.764	142.35	0.01265	0.07265	10.342	5.7434	79.058	817.63	30
35	0.13011	14.498	155.74	0.00897	0.06897	11.431	7.6860	111.43	1273.9	35
40	0.09722	15.046	185.95	0.00646	0.06646	12.359	10.285	154.76	1912.7	40
45	0.07265	15.455	203.11	0.00470	0.06470	13.141	13.764	212.74	2795.4	45
50	0.05429	15.761	217.45	0.00344	0.06344	13.796	18.420	290.33	4005.6	50
55	0.04057	15.990	229.32	0.00254	0.06254	14.341	24.650	394.17	5652.8	55
60	0.03031	16.161	239.04	0.00188	0.06188	14.790	32.987	533.12	7885.4	60
65	0.02265	16.289	246.94	0.00139	0.06139	15.160	44.145	719.08	10901.	65
70	0.01693	16.384	253.32	0.00103	0.06103	15.461	59.075	967.93	14965.	70
80	0.00945	16.509	262.54	0.00057	0.06057	15.903	105.79	1746.6	27776.	80
90	0.00528	16.578	268.39	0.00032	0.06032	16.189	189.46	3141.0	50851.	90
100	0.00295	16.617	272.04	0.00018	0.06018	16.371	339.30	5638.3	92306.	100
120	0.00092	16.651	275.68	0.00005	0.06006	16.556	1088.1	18119.	299997.	120
INF	0.00000	16.666	277.77	0.0000	0.06000	16.666	INF	INF	INF	INF

	Present sum, P			Uniform series, A			Future sum, F			
n	P/F	P/A	P/G	A/F	A/P	A/G	F/P	F/A	F/G	n
1	0.93458	0.9345	0.0000	1.00000	1.07000	0.0000	1.0700	1.0000	0.0000	1
2	0.87344	1.8080	0.8734	0.48309	0.55309	0.4830	1.1449	2.0700	1.0000	2
3	0.81630	2.6243	2.5060	0.31105	0.38105	0.9549	1.2250	3.2149	3.0700	3
4	0.76290	3.3872	4.7947	0.22523	0.29523	1.4155	1.3108	4.4399	6.2849	4
5	0.71299	4.1002	7.6466	0.17389	0.24389	1.8649	1.4025	5.7507	10.724	5
6	0.66634	4.7665	10.978	0.13980	0.20980	2.3032	1.5007	7.1532	16.475	6
7	0.62275	5.3892	14.714	0.11555	0.18555	2.7303	1.6057	8.6540	23.628	7
8	0.58201	5.9713	18.788	0.09747	0.16747	3.1465	1.7181	10.259	32.282	8
9	0.54393	6.5152	23.140	0.08349	0.15349	3.5517	1.8384	11.978	42.542	9
10	0.50835	7.0235	27.715	0.07238	0.14238	3.9460	1.9671	13.816	54.520	10
11	0.47509	7.4986	32.466	0.06336	0.13336	4.3296	2.1048	15.783	68.337	11
12	0.44401	7.9426	37.350	0.05590	0.12590	4.7025	2.2521	17.888	84.120	12
13	0.41496	8.3576	42.330	0.04965	0.11965	5.0648	2.4098	20.140	102.00	13
14	0.38782	8.7454	47.371	0.04434	0.11434	5.4167	2.5785	22.550	122.15	14
15	0.36245	9.1079	52.446	0.03979	0.10979	5.7582	2.7590	25.129	144.70	15
16	0.33873	9.4466	57.527	0.03586	0.10586	6.0896	2.9521	27.888	169.82	16
17	0.31657	9.7632	62.592	0.03243	0.10243	6.4110	3.1588	30.840	197.71	17
18	0.29586	10.059	67.621	0.02941	0.09941	6.7224	3.3799	33.999	228.55	18
19	0.27651	10.335	72.599	0.02675	0.09675	7.0241	3.6165	37.379	262.55	19
20	0.25842	10.594	77.509	0.02439	0.09439	7.3163	3.8696	40.995	299.93	20
21	0.24151	10.835	82.339	0.02229	0.09229	7.5990	4.1405	44.865	340.93	21
22	0.22571	11.061	87.079	0.02041	0.09041	7.8724	4.4304	49.005	385.79	22
23	0.21095	11.272	91.720	0.01871	0.08871	8.1088	4.7405	53.436	434.80	23
24	0.19715	11.469	96.254	0.01719	0.08719	8.3933	5.0723	58.176	488.23	24
25	0.18425	11.653	100.67	0.01581	0.08581	8.6391	5.4274	63.249	546.41	25
26	0.17220	11.825	104.98	0.01456	0.08456	8.8773	5.8073	68.676	609.66	26
27	0.16093	11.986	109.16	0.01343	0.08343	9.1072	6.2138	74.483	678.34	27
28	0.15040	12.137	113.22	0.01239	0.08239	9.3289	6.6488	80.697	752.82	28
29	0.14056	12.277	117.16	0.01145	0.08145	9.5427	7.1142	87.346	833.52	29
30	0.13137	12.409	120.97	0.01059	0.08059	9.7486	7.6122	94.460	920.86	30
35	0.09366	12.947	138.13	0.00723	0.07723	10.668	10.676	138.23	1474.8	35
40	0.06678	13.331	152.29	0.00501	0.07501	11.423	14.974	199.63	2280.5	40
45	0.04761	13.605	163.75	0.00350	0.07350	12.036	21.002	285.74	3439.2	45
50	0.03395	13.800	172.90	0.00246	0.07246	12.528	29.457	406.52	5093.2	50
55	0.02420	13.939	180.12	0.00174	0.07174	12.921	41.315	575.92	7441.8	55
60	0.01726	14.039	185.76	0.00123	0.07123	13.232	57.946	813.52	10764.	60
65	0.01230	14.109	190.14	0.00087	0.07087	13.476	81.272	1146.7	15453.	65
70	0.00877	14.160	193.51	0.00062	0.07062	13.666	113.99	1614.1	22455.	70
80	0.00446	14.222	198.07	0.00031	0.07031	13.927	224.23	3189.0	44155.	80
90	0.00227	14.253	200.70	0.00016	0.07016	14.081	441.10	6287.1	88531.	90
100	0.00115	14.269	202.20	0.00008	0.07008	14.170	867.71	12381.	175452.	100
120	0.00033	14.281	203.51	0.00002	0.07002	14.250	3357.7	47954.	683345.	120
INF	0.00000	14.285	204.08	0.0000	0.07000	14.285	INF	INF	INF	INF

	i = 8%			i = 8%			i = 8%			
	Present sum, P			Uniform series, A			Future sum, F			
n	P/F	P/A	P/G	A/F	A/P	A/G	F/P	F/A	F/G	n
1	0.92593	0.9259	0.0000	1.00000	1.08000	0.0000	1.0800	1.0000	0.0000	1
2	0.85734	1.7832	0.8573	0.48077	0.56077	0.4807	1.1664	2.0800	1.0000	2
3	0.79383	2.5771	2.4450	0.30803	0.38803	0.9487	1.2597	3.2464	3.0800	3
4	0.73503	3.3121	4.6500	0.22192	0.30192	1.4039	1.3604	4.5061	6.3264	4
5	0.68058	3.9927	7.3724	0.17046	0.25046	1.8464	1.4693	5.8666	10.832	5
6	0.63017	4.6228	10.523	0.13632	0.21632	2.2763	1.5868	7.3359	16.699	6
7	0.58349	5.2063	14.024	0.11207	0.19207	2.6936	1.7138	8.9228	24.035	7
8	0.54027	5.7466	17.806	0.09401	0.17401	3.0985	1.8509	10.636	32.957	8
9	0.50025	6.2468	21.808	0.08008	0.16008	3.4910	1.9990	12.487	43.594	9
10	0.46319	6.7100	25.976	0.06903	0.14903	3.8713	2.1589	14.486	56.082	10
11	0.42888	7.1389	30.265	0.06008	0.14008	4.2395	2.3316	16.645	70.568	11
12	0.39711	7.5360	34.633	0.05270	0.13270	4.5957	2.5181	18.977	87.214	12
13	0.36770	7.9037	39.046	0.04652	0.12652	4.9402	2.7196	21.495	106.19	13
14	0.34046	8.2442	43.472	0.04130	0.12130	5.2730	2.9372	24.214	127.68	14
15	0.31524	8.5594	47.885	0.03683	0.11683	5.5944	3.1722	27.152	151.90	15
16	0.29189	8.8513	52.264	0.03298	0.11298	5.9046	3.4259	30.324	179.05	16
17	0.27027	9.1216	56.588	0.02963	0.10963	6.2037	3.7000	33.750	209.37	17
18	0.25025	9.3718	60.842	0.02670	0.10670	6.4920	3.9960	37.450	243.12	18
19	0.23171	9.6036	65.013	0.02413	0.10413	6.7696	4.3157	41.446	280.57	19
20	0.21455	9.8181	69.089	0.02185	0.10185	7.0369	4.6609	45.762	322.02	20
21	0.19866	10.016	73.062	0.01983	0.09983	7.2940	5.0338	50.422	367.78	21
22	0.18394	10.200	76.925	0.01803	0.09803	7.5411	5.4365	55.456	418.20	22
23	0.17032	10.371	80.672	0.01642	0.09642	7.7786	5.8714	60.893	473.66	23
24	0.15770	10.528	84.299	0.01498	0.09498	8.0066	6.3411	66.764	534.55	24
25	0.14602	10.674	87.804	0.01368	0.09368	8.2253	6.8484	73.105	601.32	25
26	0.13520	10.810	91.184	0.01251	0.09251	8.4351	7.3963	79.954	674.43	26
27	0.12519	10.935	94.439	0.01145	0.09145	8.6362	7.9880	87.350	754.38	27
28	0.11591	11.051	97.568	0.01049	0.09049	8.8288	8.6271	95.338	841.73	28
29	0.10733	11.158	100.57	0.00962	0.08962	9.0132	9.3172	103.96	937.07	29
30	0.09938	11.257	103.45	0.00883	0.08883	9.1897	10.062	113.28	1041.0	30
35	0.06763	11.654	116.09	0.00580	0.08580	9.9610	14.785	172.31	1716.4	35
40	0.04603	11.924	126.04	0.00386	0.08386	10.569	21.724	259.05	2738.2	40
45	0.03133	12.108	133.73	0.00259	0.08259	11.044	31.920	386.50	4268.8	45
50	0.02132	12.233	139.59	0.00174	0.08174	11.410	46.901	573.77	6547.1	50
55	0.01451	12.318	144.00	0.00118	0.08118	11.690	68.913	848.92	9924.0	55
60	0.00988	12.376	147.30	0.00080	0.08080	11.901	101.25	1253.2	14915.	60
65	0.00672	12.416	149.73	0.00054	0.08054	12.060	148.78	1847.2	22278.	65
70	0.00457	12.442	151.53	0.00037	0.08037	12.178	218.60	2720.0	33126.	70
80	0.00212	12.473	153.80	0.00017	0.08017	12.330	471.95	5886.9	72586.	80
90	0.00098	12.487	154.99	0.00008	0.08008	12.411	1018.9	12723.	157924.	90
100	0.00045	12.494	155.61	0.00004	0.08004	12.454	2199.7	27484.	342306.	100
120	0.00010	12.498	156.03	0.00001	0.08001	12.488	10253.	128150.	>10**6	120
INF	0.00000	12.500	156.25	0.0000	0.08000	12.500	INF	INF	INF	INF

| | $i = 9\%$ | | | | $i = 9\%$ | | | | $i = 9\%$ | | | |
| | Present sum, P | | | | Uniform series, A | | | | Future sum, F | | | |
n	P/F	P/A	P/G		A/F	A/P	A/G		F/P	F/A	F/G	n
1	0.91743	0.9174	0.0000		1.00000	1.09000	0.0000		1.0900	1.0000	0.0000	1
2	0.84168	1.7591	0.8416		0.47847	0.56847	0.4784		1.1881	2.0900	1.0000	2
3	0.77218	2.5312	2.3860		0.30505	0.39505	0.9426		1.2950	3.2781	3.0900	3
4	0.70843	3.2397	4.5113		0.21867	0.30867	1.3925		1.4115	4.5731	6.3681	4
5	0.64993	3.8896	7.1110		0.16709	0.25709	1.8282		1.5386	5.9847	10.941	5
6	0.59627	4.4859	10.092		0.13292	0.22292	2.2497		1.6771	7.5233	16.925	6
7	0.54703	5.0329	13.374		0.10869	0.19869	2.6571		1.8280	9.2004	24.449	7
8	0.50187	5.5348	16.887		0.09067	0.18067	3.0511		1.9925	11.028	33.649	8
9	0.46043	5.9952	20.571		0.07680	0.16680	3.4312		2.1718	13.021	44.678	9
10	0.42241	6.4176	24.372		0.06582	0.15582	3.7977		2.3673	15.192	57.699	10
11	0.38753	6.8051	28.248		0.05695	0.14695	4.1509		2.5804	17.560	72.892	11
12	0.35553	7.1607	32.159		0.04965	0.13965	4.4910		2.8126	20.140	90.452	12
13	0.32618	7.4869	36.073		0.04357	0.13357	4.8181		3.0658	22.953	110.59	13
14	0.29925	7.7861	39.963		0.03843	0.12843	5.1326		3.3417	26.019	133.54	14
15	0.27454	8.0606	43.806		0.03406	0.12406	5.4346		3.6424	29.360	159.56	15
16	0.25187	8.3125	47.584		0.03030	0.12030	5.7244		3.9703	33.003	188.92	16
17	0.23107	8.5436	51.282		0.02705	0.11705	6.0023		4.3276	36.973	221.93	17
18	0.21199	8.7556	54.886		0.02421	0.11421	6.2686		4.7171	41.301	258.90	18
19	0.19449	8.9501	58.386		0.02173	0.11173	6.5235		5.1416	46.018	300.20	19
20	0.17843	9.1285	61.777		0.01955	0.10955	6.7674		5.6044	51.160	346.22	20
21	0.16370	9.2922	65.050		0.01762	0.10762	7.0005		6.1088	56.764	397.38	21
22	0.15018	9.4424	68.204		0.01590	0.10590	7.2232		6.6586	62.873	454.14	22
23	0.13778	9.5802	71.235		0.01438	0.10438	7.4357		7.2578	69.531	517.02	23
24	0.12640	9.7066	74.143		0.01302	0.10302	7.6384		7.9110	76.789	586.55	24
25	0.11597	9.8225	76.926		0.01181	0.10181	7.8316		8.6230	84.700	663.34	25
26	0.10639	9.9289	79.586		0.01072	0.10072	8.0155		9.3991	93.324	748.04	26
27	0.09761	10.026	82.124		0.00973	0.09973	8.1906		10.245	102.72	841.36	27
28	0.08955	10.116	84.545		0.00885	0.09885	8.3571		11.167	112.96	944.09	28
29	0.08215	10.198	86.842		0.00806	0.09806	8.5153		12.172	124.13	1057.0	29
30	0.07537	10.273	89.028		0.00734	0.09734	8.6656		13.267	136.30	1181.1	30
35	0.04899	10.566	98.359		0.00464	0.09464	9.3082		20.414	215.71	2007.9	35
40	0.03184	10.757	105.37		0.00296	0.09296	9.7957		31.409	337.88	3309.8	40
45	0.02069	10.881	110.56		0.00190	0.09190	10.160		48.327	525.85	5342.8	45
50	0.01345	10.961	114.32		0.00123	0.09123	10.426		74.357	815.08	8500.9	50
55	0.00874	11.014	117.03		0.00079	0.09079	10.626		114.40	1260.0	13389.	55
60	0.00568	11.048	118.96		0.00051	0.09051	10.768		176.03	1944.7	20942.	60
65	0.00369	11.070	120.33		0.00033	0.09033	10.870		270.84	2998.2	32592.	65
70	0.00240	11.084	121.29		0.00022	0.09022	10.942		416.73	4619.2	50546.	70
80	0.00101	11.099	122.43		0.00009	0.09009	11.029		986.55	10950.	120784.	80
90	0.00043	11.106	122.97		0.00004	0.09004	11.072		2335.5	25939.	287213.	90
100	0.00018	11.109	123.23		0.00002	0.09002	11.093		5529.0	61422.	681363.	100
120	0.00003	11.110	123.41		0.00000	0.09000	11.107		30987.	344289.	>10**6	120
INF	0.00000	11.111	123.45		0.0000	0.09000	11.111		INF	INF	INF	INF

| | i = 10% | | | i = 10% | | | i = 10% | | | |
| | Present sum, P | | | Uniform series, A | | | Future sum, F | | | |
n	P/F	P/A	P/G	A/F	A/P	A/G	F/P	F/A	F/G	n
1	0.90909	0.9090	0.0000	1.00000	1.00000	0.0000	1.1000	1.0000	0.0000	1
2	0.82645	1.7355	0.8264	0.47619	0.57619	0.4761	1.2100	2.1000	1.0000	2
3	0.75131	2.4868	2.3290	0.30211	0.40211	0.9365	1.3310	3.3100	3.1000	3
4	0.68301	3.1698	4.3781	0.21547	0.31547	1.3811	1.4641	4.6410	6.4100	4
5	0.62092	3.7907	6.8618	0.16380	0.26380	1.8101	1.6105	6.1051	11.051	5
6	0.56447	4.3552	9.6841	0.12961	0.22961	2.2235	1.7715	7.7156	17.156	6
7	0.51316	4.8684	12.763	0.10541	0.20541	2.6216	1.9487	9.4871	24.871	7
8	0.46651	5.3349	16.028	0.08744	0.18744	3.0044	2.1435	11.435	34.358	8
9	0.42410	5.7590	19.421	0.07364	0.17364	3.3723	2.3579	13.579	45.794	9
10	0.38554	6.1445	22.891	0.06275	0.16275	3.7254	2.5937	15.937	59.374	10
11	0.35049	6.4950	26.396	0.05396	0.15396	4.0640	2.8531	18.531	75.311	11
12	0.31863	6.8136	29.901	0.04676	0.14676	4.3884	3.1384	21.384	93.842	12
13	0.28966	7.1033	33.377	0.04078	0.14078	4.6987	3.4522	24.522	115.22	13
14	0.26333	7.3666	36.800	0.03575	0.13575	4.9955	3.7975	27.975	139.75	14
15	0.23939	7.6060	40.152	0.03147	0.13147	5.2789	4.1772	31.772	167.72	15
16	0.21763	7.8237	43.416	0.02782	0.12782	5.5493	4.5949	35.949	199.49	16
17	0.19784	8.0215	46.581	0.02466	0.12466	5.8071	5.0544	40.544	235.44	17
18	0.17986	8.2014	49.639	0.02193	0.12193	6.0525	5.5599	45.599	275.99	18
19	0.16351	8.3649	52.582	0.01955	0.11955	6.2861	6.1159	51.159	321.59	19
20	0.14864	8.5135	55.406	0.01746	0.11746	6.5080	6.7275	57.275	372.75	20
21	0.13513	8.6486	58.109	0.01562	0.11562	6.7188	7.4002	64.002	430.02	21
22	0.12285	8.7715	60.689	0.01401	0.11401	6.9188	8.1402	71.402	494.02	22
23	0.11168	8.8832	63.146	0.01257	0.11257	7.1084	8.9543	79.543	565.43	23
24	0.10153	8.9847	65.481	0.01130	0.11130	7.2880	9.8497	88.497	644.97	24
25	0.09230	9.0770	67.696	0.01017	0.11017	7.4579	10.834	98.347	733.47	25
26	0.08391	9.1609	69.794	0.00916	0.10916	7.6186	11.918	109.18	831.81	26
27	0.07628	9.2372	71.777	0.00826	0.10826	7.7704	13.110	121.10	940.99	27
28	0.06934	9.3065	73.649	0.00745	0.10745	7.9137	14.421	134.21	1062.1	28
29	0.06304	9.3696	75.414	0.00673	0.10673	8.0488	15.863	148.63	1196.3	29
30	0.05731	9.4269	77.076	0.00608	0.10608	8.1762	17.449	164.49	1344.9	30
35	0.03558	9.6441	83.987	0.00369	0.10369	8.7086	28.102	271.02	2360.2	35
40	0.02209	9.7790	88.952	0.00226	0.10226	9.0962	45.259	442.59	4025.9	40
45	0.01372	9.8628	92.454	0.00139	0.10139	9.3740	72.890	718.90	6739.0	45
50	0.00852	9.9148	94.888	0.00086	0.10086	9.5704	117.39	1163.9	11139.	50
55	0.00529	9.9471	96.561	0.00053	0.10053	9.7075	189.05	1880.5	18255.	55
60	0.00328	9.9671	97.701	0.00033	0.10033	9.8022	304.48	3034.8	29748.	60
65	0.00204	9.9796	98.470	0.00020	0.10020	9.8671	490.37	4893.7	48287.	65
70	0.00127	9.9873	98.987	0.00013	0.10013	9.9112	789.74	7887.4	78174.	70
80	0.00049	9.9951	99.560	0.00005	0.10005	9.9609	2048.4	20474.	203940.	80
90	0.00019	9.9981	99.811	0.00002	0.10002	9.9830	5313.0	53120.	530302.	90
100	0.00007	9.9992	99.920	0.00001	0.10001	9.9927	13780.	137796.	>10**6	100
120	0.00001	9.9998	99.986	0.00001	0.10000	9.9987	92709.	927081.	>10**6	120
INF	0.00000	10.000	130.00	0.0000	0.10000	10.000	INF	INF	INF	INF

	i = 11% Present sum, P			i = 11% Uniform series, A			i = 11% Future sum, F			
n	P/F	P/A	P/G	A/F	A/P	A/G	F/P	F/A	F/G	n
1	0.90090	0.9009	0.0000	1.00000	1.11000	0.0000	1.1100	1.0000	0.0000	1
2	0.81162	1.7125	0.8116	0.47393	0.58393	0.4739	1.2321	2.1100	1.0000	2
3	0.73119	2.4437	2.2740	0.29921	0.40921	0.9305	1.3676	3.3421	3.1100	3
4	0.65873	3.1024	4.2502	0.21233	0.32233	1.3699	1.5180	4.7097	6.4521	4
5	0.59345	3.6959	6.6240	0.16057	0.27057	1.7922	1.6850	6.2278	11.161	5
6	0.53464	4.2305	9.2972	0.12638	0.23638	2.1976	1.8704	7.9128	17.389	6
7	0.48166	4.7122	12.187	0.10222	0.21222	2.5863	2.0761	9.7832	25.302	7
8	0.43393	5.1461	15.224	0.08432	0.19432	2.9584	2.3045	11.859	35.085	8
9	0.39092	5.5370	18.352	0.07060	0.18060	3.3144	2.5580	14.164	46.945	9
10	0.35218	5.8892	21.521	0.05980	0.16980	3.6544	2.8394	16.722	61.109	10
11	0.31728	6.2065	24.694	0.05112	0.16112	3.9788	3.1517	19.561	77.831	11
12	0.28584	6.4923	27.838	0.04403	0.15403	4.2879	3.4984	22.713	97.392	12
13	0.25751	6.7498	30.929	0.03815	0.14815	4.5821	3.8832	26.211	120.10	13
14	0.23199	6.9818	33.944	0.03323	0.14323	4.8618	4.3104	30.094	146.31	14
15	0.20900	7.1908	36.870	0.02907	0.13907	5.1274	4.7845	34.405	176.41	15
16	0.18829	7.3791	39.695	0.02552	0.13552	5.3793	5.3108	39.189	210.81	16
17	0.16963	7.5487	42.409	0.02247	0.13247	5.6180	5.8950	44.500	250.00	17
18	0.15282	7.7016	45.007	0.01984	0.12984	5.8438	6.5435	50.395	294.50	18
19	0.13768	7.8392	47.485	0.01756	0.12756	6.0573	7.2633	56.939	344.90	19
20	0.12403	7.9633	49.842	0.01558	0.12558	6.2589	8.0623	64.202	401.84	20
21	0.11174	8.0750	52.077	0.01384	0.12384	6.4491	8.9491	72.265	466.04	21
22	0.10067	8.1757	54.191	0.01231	0.12231	6.6282	9.9335	81.214	538.31	22
23	0.09069	8.2664	56.186	0.01097	0.12097	6.7569	11.026	91.147	619.52	23
24	0.08170	8.3481	58.065	0.00979	0.11979	6.9555	12.239	102.17	710.67	24
25	0.07361	8.4217	59.832	0.00874	0.11874	7.1044	13.585	114.41	812.84	25
26	0.06631	8.4880	61.490	0.00781	0.11781	7.2443	15.079	127.99	927.26	26
27	0.05974	8.5478	63.043	0.00699	0.11699	7.3753	16.738	143.07	1055.2	27
28	0.05382	8.6016	64.496	0.00626	0.11626	7.4981	18.579	159.81	1198.3	28
29	0.04849	8.6501	65.854	0.00561	0.11561	7.6131	20.623	178.39	1358.1	29
30	0.04368	8.6937	67.121	0.00502	0.11502	7.7205	22.892	199.02	1536.5	30
35	0.02592	8.8552	72.253	0.00293	0.11293	8.1594	38.574	341.59	2787.1	35
40	0.01538	8.9510	75.778	0.00172	0.11172	8.4659	65.000	581.82	4925.6	40
45	0.00913	9.0079	78.155	0.00101	0.11101	8.6762	109.53	986.63	8560.3	45
50	0.00542	9.0416	79.734	0.00060	0.11060	8.8185	184.56	1668.7	14716.	50
55	0.00322	9.0616	80.771	0.00035	0.11035	8.9134	311.00	2818.2	25120.	55
60	0.00191	9.0735	81.446	0.00021	0.11021	8.9762	524.05	4755.0	42682.	60
65	0.00113	9.0806	81.881	0.00012	0.11012	9.0172	883.06	8018.7	72307.	65
70	0.00067	9.0848	82.161	0.00007	0.11007	9.0438	1488.0	13518.	122258.	70
INF	0.00000	9.0909	82.644	0.0000	0.11000	9.0909	INF	INF	INF	INF

	$i=12\%$			$i=12\%$			$i=12\%$			
	Present sum, P			Uniform series, A			Future sum, F			
n	P/F	P/A	P/G	A/F	A/P	A/G	F/P	F/A	F/G	n
1	0.89286	0.8928	0.0000	1.00000	1.12000	0.0000	1.1200	1.0000	0.0000	1
2	0.79719	1.6900	0.7971	0.47170	0.59170	0.4717	1.2544	2.1200	1.0000	2
3	0.71178	2.4018	2.2207	0.29635	0.41635	0.9246	1.4049	3.3744	3.1200	3
4	0.63552	3.0373	4.1273	0.20923	0.32923	1.3588	1.5735	4.7793	6.4944	4
5	0.56743	3.6047	6.3970	0.15741	0.27741	1.7745	1.7623	6.3528	11.273	5
6	0.50663	4.1114	8.9301	0.12323	0.24323	2.1720	1.9738	8.1151	17.626	6
7	0.45235	4.5637	11.644	0.09912	0.21912	2.5514	2.2106	10.089	25.741	7
8	0.40388	4.9676	14.471	0.08130	0.20130	2.9131	2.4759	12.299	35.830	8
9	0.36061	5.3282	17.356	0.06768	0.18768	3.2574	2.7730	14.775	48.130	9
10	0.32197	5.6502	20.254	0.05698	0.17698	3.5846	3.1058	17.548	62.906	10
11	0.28748	5.9377	23.128	0.04842	0.16842	3.8952	3.4785	20.654	80.454	11
12	0.25668	6.1943	25.952	0.04144	0.16144	4.1896	3.8959	24.133	101.10	12
13	0.22917	6.4235	28.702	0.03568	0.15568	4.4683	4.3634	28.029	125.24	13
14	0.20462	6.6281	31.362	0.03087	0.15087	4.7316	4.8871	32.392	153.27	14
15	0.18270	6.8108	33.920	0.02682	0.14682	4.9803	5.4735	37.279	185.66	15
16	0.16312	6.9739	36.367	0.02339	0.14339	5.2146	6.1303	42.753	222.94	16
17	0.14564	7.1196	38.697	0.02046	0.14046	5.4353	6.8660	48.883	265.69	17
18	0.13004	7.2496	40.908	0.01794	0.13794	5.6427	7.6899	55.749	314.58	18
19	0.11611	7.3657	42.997	0.01576	0.13576	5.8375	8.6127	63.439	370.33	19
20	0.10367	7.4694	44.967	0.01388	0.13388	6.0202	9.6462	72.052	433.77	20
21	0.09256	7.5620	46.818	0.01224	0.13224	6.1913	10.803	81.698	505.82	21
22	0.08264	7.6446	48.554	0.01081	0.13081	6.3514	12.100	92.502	587.52	22
23	0.07379	7.7184	50.178	0.00956	0.12956	6.5010	13.552	104.60	680.02	23
24	0.06588	7.7843	51.692	0.00846	0.12846	6.6406	15.178	118.15	784.62	24
25	0.05882	7.8431	53.104	0.00750	0.12750	6.7708	17.000	133.33	902.78	25
26	0.05252	7.8956	54.417	0.00665	0.12665	6.8921	19.040	150.33	1036.1	26
27	0.04689	7.9425	55.636	0.00590	0.12590	7.0049	21.324	169.37	1186.4	27
28	0.04187	7.9844	56.767	0.00524	0.12524	7.1097	23.883	190.69	1355.3	28
29	0.03738	8.0218	57.814	0.00466	0.12466	7.2071	26.749	214.58	1546.5	29
30	0.03338	8.0551	58.782	0.00414	0.12414	7.2974	29.959	241.33	1761.1	30
35	0.01894	8.1755	62.605	0.00232	0.12232	7.6576	52.799	431.66	3305.5	35
40	0.01075	8.2437	65.115	0.00130	0.12130	7.8987	93.051	767.09	6059.1	40
45	0.00610	8.2825	66.734	0.00074	0.12074	8.0572	163.98	1358.2	10943.	45
50	0.00346	8.3045	67.762	0.00042	0.12042	8.1597	289.00	2400.0	19563.	50
55	0.00196	8.3169	68.408	0.00024	0.12024	8.2251	509.32	4236.0	34841.	55
60	0.00111	8.3240	68.810	0.00013	0.12013	8.2664	897.59	7471.6	61763.	60
65	0.00063	8.3280	69.058	0.00008	0.12008	8.2922	1581.8	13173.	109241.	65
70	0.00036	8.3303	69.210	0.00004	0.12004	8.3082	2787.8	23223.	192944.	70
INF	0.00000	8.3333	69.444	0.0000	0.12000	8.3333	INF	INF	INF	INF

	i = 13%			i = 13%			i = 13%			
	Present sum, P			Uniform series, A			Future sum, F			
n	P/F	P/A	P/G	A/F	A/P	A/G	F/P	F/A	F/G	n
1	0.88496	0.8849	0.0000	1.00000	1.13000	0.0000	1.1300	1.0000	0.0000	1
2	0.78315	1.6681	0.7831	0.46948	0.59948	0.4694	1.2769	2.1300	1.0000	2
3	0.69305	2.3611	2.1692	0.29352	0.42352	0.9187	1.4429	3.4069	3.1300	3
4	0.61332	2.9744	4.0092	0.20619	0.33619	1.3478	1.6304	4.8498	6.5369	4
5	0.54276	3.5172	6.1802	0.15431	0.28431	1.7571	1.8424	6.4802	11.386	5
6	0.48032	3.9975	8.5818	0.12015	0.25015	2.1467	2.0819	8.3227	17.867	6
7	0.42506	4.4226	11.132	0.09611	0.22611	2.5171	2.3526	10.404	26.189	7
8	0.37616	4.7987	13.765	0.07839	0.20839	2.8685	2.6584	12.757	36.594	8
9	0.33288	5.1316	16.428	0.06487	0.19487	3.2013	3.0040	15.415	49.351	9
10	0.29459	5.4262	19.079	0.05429	0.18429	3.5161	3.3945	18.419	64.767	10
11	0.26070	5.6869	21.686	0.04584	0.17584	3.8134	3.8358	21.814	83.187	11
12	0.23071	5.9176	24.224	0.03899	0.16899	4.0935	4.3345	25.650	105.00	12
13	0.20416	6.1218	26.674	0.03335	0.16335	4.3572	4.8980	29.984	130.65	13
14	0.18068	6.3024	29.023	0.02867	0.15867	4.6050	5.5347	34.882	160.63	14
15	0.15989	6.4623	31.261	0.02474	0.15474	4.8374	6.2542	40.417	195.51	15
16	0.14150	6.6038	33.387	0.02143	0.15143	5.0552	7.0673	46.671	235.93	16
17	0.12522	6.7290	35.387	0.01861	0.14861	5.2589	7.9860	53.739	282.60	17
18	0.11081	6.8399	37.271	0.01520	0.14620	5.4491	9.0242	61.725	336.34	18
19	0.09806	6.9379	39.036	0.01413	0.14413	5.6265	10.197	70.749	398.07	19
20	0.08678	7.0247	40.685	0.01235	0.14235	5.7917	11.523	80.946	468.82	20
21	0.07680	7.1015	42.221	0.01081	0.14081	5.9453	13.021	92.469	549.76	21
22	0.06796	7.1695	43.648	0.00948	0.13948	6.0880	14.713	105.49	642.23	22
23	0.06014	7.2296	44.971	0.00832	0.13832	6.2204	16.626	120.83	747.73	23
24	0.05323	7.2828	46.196	0.00731	0.13731	6.3430	18.788	136.83	867.93	24
25	0.04710	7.3299	47.326	0.00643	0.13643	6.4565	21.230	155.62	1004.7	25
26	0.04168	7.3716	48.368	0.00565	0.13565	6.5614	23.990	176.85	1160.3	26
27	0.03689	7.4085	49.327	0.00498	0.13498	6.6581	27.109	200.84	1337.2	27
28	0.03264	7.4412	50.209	0.00438	0.13438	6.7474	30.633	227.95	1538.0	28
29	0.02889	7.4700	51.017	0.00387	0.13387	6.8296	34.615	258.58	1766.0	29
30	0.02557	7.4956	51.759	0.00341	0.13341	6.9052	39.115	293.19	2024.6	30
35	0.01388	7.5855	54.614	0.00183	0.13183	7.1998	72.068	546.68	3936.0	35
40	0.00753	7.6343	56.408	0.00099	0.13099	7.3887	132.78	1013.7	7490.0	40
45	0.00409	7.6608	57.514	0.00053	0.13053	7.5076	244.64	1874.1	14070.	45
50	0.00222	7.6752	58.187	0.00029	0.13029	7.5811	450.73	3459.5	26227.	50
55	0.00120	7.6830	58.590	0.00016	0.13016	7.6260	830.45	6380.4	48656.	55
60	0.00065	7.6872	58.831	0.00009	0.13009	7.6530	1530.0	11761.	90015.	60
65	0.00035	7.6895	58.973	0.00005	0.13005	7.6692	2819.0	21677.	166247.	65
70	0.00019	7.6908	59.056	0.00003	0.13003	7.6788	5193.8	39945.	306732.	70
INF	0.00000	7.6923	59.171	0.0000	0.13000	7.6923	INF	INF	INF	INF

| | $i = 14\%$ | | | $i = 14\%$ | | | $i = 14\%$ | | | |
| | Present sum, P | | | Uniform series, A | | | Future sum, F | | | |
n	P/F	P/A	P/G	A/F	A/P	A/G	F/P	F/A	F/G	n
1	0.87719	0.8771	0.0000	1.00000	1.14000	0.0000	1.1400	1.0000	0.0000	1
2	0.76947	1.6466	0.7694	0.46729	0.60729	0.4672	1.2996	2.1400	1.0000	2
3	0.67497	2.3216	2.1194	0.29073	0.43073	0.9129	1.4815	3.4396	3.1400	3
4	0.59208	2.9137	3.8956	0.20320	0.34320	1.3370	1.6889	4.9211	6.5796	4
5	0.51937	3.4330	5.9731	0.15128	0.29128	1.7398	1.9254	6.6101	11.500	5
6	0.45559	3.8886	8.2510	0.11716	0.25716	2.1218	2.1949	8.5355	18.110	6
7	0.39964	4.2883	10.648	0.09319	0.23319	2.4832	2.5022	10.730	26.646	7
8	0.35056	4.6388	13.102	0.07557	0.21557	2.8245	2.8525	13.232	37.376	8
9	0.30751	4.9463	15.562	0.06217	0.20217	3.1463	3.2519	16.085	50.609	9
10	0.26974	5.2161	17.990	0.05171	0.19171	3.4490	3.7072	19.337	66.695	10
11	0.23662	5.4527	20.356	0.04339	0.18339	3.7333	4.2262	23.044	86.032	11
12	0.20756	5.6602	22.639	0.03667	0.17667	3.9997	4.8179	27.270	109.07	12
13	0.18207	5.8423	24.824	0.03116	0.17116	4.2490	5.4924	32.088	136.34	13
14	0.15971	6.0020	26.900	0.02661	0.16661	4.4819	6.2613	37.581	168.43	14
15	0.14010	6.1421	28.862	0.02281	0.16281	4.6990	7.1379	43.842	206.01	15
16	0.12289	6.2650	30.705	0.01962	0.15962	4.9011	8.1372	50.980	249.86	16
17	0.10780	6.3728	32.430	0.01692	0.15692	5.0888	9.2764	59.117	300.84	17
18	0.09456	6.4674	34.038	0.01462	0.15462	5.2629	10.575	68.394	359.95	18
19	0.08295	6.5503	35.531	0.01266	0.15266	5.4242	12.055	78.969	428.35	19
20	0.07276	6.6231	36.913	0.01099	0.15099	5.5734	13.743	91.024	507.32	20
21	0.06383	6.6869	38.190	0.00954	0.14954	5.7111	15.667	104.76	598.34	21
22	0.05599	6.7429	39.365	0.00830	0.14830	5.8380	17.861	120.43	703.11	22
23	0.04911	6.7920	40.446	0.00723	0.14723	5.9549	20.361	138.29	823.55	23
24	0.04308	6.8351	41.437	0.00630	0.14630	6.0623	23.212	158.65	961.84	24
25	0.03779	6.8729	42.344	0.00550	0.14550	6.1610	26.461	181.87	1120.5	25
26	0.03315	6.9060	43.172	0.00480	0.14480	6.2514	30.166	208.33	1302.3	26
27	0.02908	6.9351	43.928	0.00419	0.14419	6.3342	34.389	238.49	1510.7	27
28	0.02551	6.9606	44.617	0.00366	0.14366	6.4099	39.204	272.88	1749.2	28
29	0.02237	6.9830	45.244	0.00320	0.14320	6.4791	44.693	312.09	2022.1	29
30	0.01963	7.0026	45.813	0.00280	0.14280	6.5422	50.950	356.78	2334.1	30
35	0.01019	7.0700	47.951	0.00144	0.14144	6.7824	98.100	693.57	4704.0	35
40	0.00529	7.1050	49.237	0.00075	0.14075	6.9299	188.88	1342.0	9300.1	40
45	0.00275	7.1232	49.996	0.00039	0.14039	7.0187	363.67	2590.5	18182.	45
50	0.00143	7.1326	50.437	0.00020	0.14020	7.0713	700.23	4994.5	35318.	50
55	0.00074	7.1375	50.691	0.00010	0.14010	7.1020	1348.2	9623.1	68343.	55
60	0.00039	7.1401	50.835	0.00005	0.14005	7.1197	2595.9	18535.	131965.	60
65	0.00020	7.1414	50.917	0.00003	0.14003	7.1298	4998.2	35694.	254496.	65
70	0.00010	7.1421	50.961	0.00001	0.14001	7.1355	9623.6	68733.	490451.	70
INF	0.00000	7.1428	51.020	0.0000	0.14000	7.1428	INF	INF	INF	INF

	Present sum, P			Uniform series, A			Future sum, F			
n	P/F	P/A	P/G	A/F	A/P	A/G	F/P	F/A	F/G	n
1	0.86957	0.8695	0.0000	1.00000	1.15000	0.0000	1.1500	1.0000	0.0000	1
2	0.75614	1.6257	0.7561	0.46512	0.61512	0.4651	1.3225	2.1500	1.0000	2
3	0.65752	2.2832	2.0711	0.28798	0.43798	0.9071	1.5208	3.4725	3.1500	3
4	0.57175	2.8549	3.7864	0.20027	0.35027	1.3262	1.7490	4.9933	6.6225	4
5	0.49718	3.3521	5.7751	0.14832	0.29832	1.7228	2.0113	6.7423	11.615	5
6	0.43233	3.7844	7.9367	0.11424	0.26424	2.0971	2.3130	8.7537	18.358	6
7	0.37594	4.1604	10.192	0.09036	0.24036	2.4498	2.6600	11.066	27.112	7
8	0.32690	4.4873	12.480	0.07285	0.22285	2.7813	3.0590	13.726	38.178	8
9	0.28426	4.7715	14.754	0.05957	0.20957	3.0922	3.5178	16.785	51.905	9
10	0.24718	5.0187	16.979	0.04925	0.19925	3.3832	4.0455	20.303	68.691	10
11	0.21494	5.2337	19.128	0.04107	0.19107	3.6549	4.6523	24.349	88.995	11
12	0.18691	5.4206	21.184	0.03448	0.18448	3.9082	5.3502	29.001	113.34	12
13	0.16253	5.5831	23.135	0.02911	0.17911	4.1437	6.1527	34.351	142.34	13
14	0.14133	5.7244	24.972	0.02469	0.17469	4.3624	7.0757	40.504	176.69	14
15	0.12289	5.8473	26.693	0.02102	0.17102	4.5649	8.1370	47.580	217.20	15
16	0.10686	5.9542	28.296	0.01795	0.16795	4.7522	9.3576	55.717	264.78	16
17	0.09293	6.0471	29.782	0.01537	0.16537	4.9250	10.761	65.075	320.50	17
18	0.08081	6.1279	31.156	0.01319	0.16319	5.0843	12.375	75.836	385.57	18
19	0.07027	6.1982	32.421	0.01134	0.16134	5.2307	14.231	88.211	461.41	19
20	0.06110	6.2593	33.582	0.00976	0.15976	5.3651	16.366	102.44	549.62	20
21	0.05313	6.3124	34.644	0.00842	0.15842	5.4883	18.821	118.81	652.06	21
22	0.04620	6.3586	35.615	0.00727	0.15727	5.6010	21.644	137.63	770.87	22
23	0.04017	6.3988	36.498	0.00628	0.15628	5.7039	24.891	159.27	908.50	23
24	0.03493	6.4337	37.302	0.00543	0.15543	5.7978	28.625	184.16	1067.7	24
25	0.03038	6.4641	38.031	0.00470	0.15470	5.8834	32.919	212.79	1251.9	25
26	0.02642	6.4905	38.691	0.00407	0.15407	5.9612	37.856	245.71	1464.7	26
27	0.02297	6.5135	39.289	0.00353	0.15353	6.0319	43.535	283.56	1710.4	27
28	0.01997	6.5335	39.828	0.00306	0.15306	6.0960	50.065	327.10	1994.0	28
29	0.01737	6.5508	40.314	0.00255	0.15265	6.1540	57.575	377.17	2321.1	29
30	0.01510	6.5659	40.753	0.00230	0.15230	6.2066	66.211	434.74	2698.3	30
35	0.00751	6.6166	42.358	0.00113	0.15113	6.4018	133.17	881.17	5641.1	35
40	0.00373	6.6417	43.283	0.00056	0.15056	6.5167	267.86	1779.0	11593.	40
45	0.00186	6.6542	43.805	0.00028	0.15028	6.5829	538.76	3585.1	23600.	45
50	0.00092	6.6605	44.095	0.00014	0.15014	6.6204	1083.6	7217.7	47784.	50
55	0.00046	6.6636	44.255	0.00007	0.15007	6.6414	2179.6	14524.	96461.	55
60	0.00023	6.6651	44.343	0.00003	0.15003	6.6529	4384.0	29220.	194400.	60
65	0.00011	6.6659	44.390	0.00002	0.15002	6.6592	8817.7	58778.	391424.	65
INF	0.00000	6.6666	44.444	0.0000	0.15000	6.6666	INF	INF	INF	INF

	i = 20%			i = 20%			i = 20%			
	Present sum, P			Uniform series, A			Future sum, F			
n	P/F	P/A	P/G	A/F	A/P	A/G	F/P	F/A	F/G	n
1	0.83333	0.8333	0.0000	1.00000	1.20000	0.0000	1.2000	1.0000	0.0000	1
2	0.69444	1.5277	0.6944	0.45455	0.65455	0.4545	1.4400	2.2000	1.0000	2
3	0.57870	2.1064	1.8518	0.27473	0.47473	0.8791	1.7280	3.6400	3.2000	3
4	0.48225	2.5887	3.2986	0.18629	0.38629	1.2742	2.0736	5.3680	6.8400	4
5	0.40188	2.9906	4.9061	0.13438	0.33438	1.6405	2.4883	7.4416	12.208	5
6	0.33490	3.3255	6.5806	0.10071	0.30071	1.9788	2.9859	9.9299	19.649	6
7	0.27908	3.6045	8.2551	0.07742	0.27742	2.2901	3.5831	12.915	29.579	7
8	0.23257	3.8371	9.8830	0.06061	0.26061	2.5756	4.2998	16.499	42.495	8
9	0.19381	4.0309	11.433	0.04808	0.24808	2.8364	5.1597	20.798	58.994	9
10	0.16151	4.1924	12.887	0.03852	0.23852	3.0738	6.1917	25.958	79.793	10
11	0.13459	4.3270	14.233	0.03110	0.23110	3.2892	7.4300	32.150	105.75	11
12	0.11216	4.4392	15.466	0.02526	0.22526	3.4841	8.9161	39.580	137.90	12
13	0.09346	4.5326	16.588	0.02062	0.22062	3.6597	10.699	48.496	177.48	13
14	0.07789	4.6105	17.600	0.01689	0.21689	3.8174	12.839	59.195	225.98	14
15	0.06491	4.6754	18.509	0.01388	0.21388	3.9588	15.407	72.035	285.17	15
16	0.05409	4.7295	19.320	0.01144	0.21144	4.0851	18.488	87.442	357.21	16
17	0.04507	4.7746	20.041	0.00944	0.20944	4.1975	22.186	105.93	444.65	17
18	0.03756	4.8121	20.680	0.00781	0.20781	4.2975	26.623	128.11	550.58	18
19	0.03130	4.8435	21.243	0.00646	0.20646	4.3860	31.948	154.74	678.70	19
20	0.02608	4.8695	21.739	0.00536	0.20536	4.4643	38.337	186.68	833.44	20
21	0.02174	4.8913	22.174	0.00444	0.20444	4.5333	46.005	225.02	1020.1	21
22	0.01811	4.9094	22.554	0.00369	0.20369	4.5941	55.206	271.03	1245.1	22
23	0.01509	4.9245	22.886	0.00307	0.20307	4.6475	66.247	326.23	1516.1	23
24	0.01258	4.9371	23.176	0.00255	0.20255	4.6942	79.496	392.48	1842.4	24
25	0.01048	4.9475	23.427	0.00212	0.20212	4.7351	95.396	471.98	2234.9	25
26	0.00874	4.9563	23.646	0.00176	0.20176	4.7708	114.47	567.37	2706.8	26
27	0.00728	4.9636	23.835	0.00147	0.20147	4.8020	137.37	681.85	3274.2	27
28	0.00607	4.9696	23.999	0.00122	0.20122	4.8291	164.84	819.22	3956.1	28
29	0.00506	4.9747	24.140	0.00102	0.20102	4.8526	197.81	984.06	4775.3	29
30	0.00421	4.9789	24.262	0.00085	0.20085	4.8730	237.37	1181.8	5759.4	30
35	0.00169	4.9915	24.661	0.00034	0.20034	4.9406	590.66	2948.3	14566.	35
40	0.00068	4.9966	24.846	0.00014	0.20014	4.9727	1469.7	7343.8	36519.	40
45	0.00027	4.9986	24.931	0.00005	0.20005	4.9876	3657.2	18281.	91181.	45
50	0.00011	4.9994	24.969	0.00002	0.20002	4.9945	9100.4	45497.	227236.	50
55	0.00004	4.9997	24.986	0.00001	0.20001	4.9975	22644.	113219.	565820.	55
60	0.00002	4.9999	24.994	0.00000	0.20000	4.9989	56347.	281733	>10**6	60
INF	0.00000	5.0000	25.000	0.0000	0.20000	5.0000	INF	INF	INF	INF

| | i = 25% | | | i = 25% | | | i = 25% | | | |
| | PRESENT SUM, P | | | Uniform series, A | | | Future sum, F | | | |
n	P/F	P/A	P/G	A/F	A/P	A/G	F/P	F/A	F/G	n
1	0.80000	0.8000	0.0000	1.00000	1.25000	0.0000	1.2500	1.0000	0.0000	1
2	0.64000	1.4400	0.6400	0.44444	0.69444	0.4444	1.5625	2.2500	1.0000	2
3	0.51200	1.9520	1.6640	0.26230	0.51230	0.8524	1.9531	3.8125	3.2500	3
4	0.40960	2.3616	2.8928	0.17344	0.42344	1.2249	2.4414	5.7656	7.0625	4
5	0.32768	2.6892	4.2035	0.12185	0.37185	1.5630	3.0517	8.2070	12.828	5
6	0.26214	2.9514	5.5142	0.08882	0.33882	1.8683	3.8147	11.258	21.035	6
7	0.20972	3.1611	6.7725	0.06634	0.31634	2.1424	4.7683	15.073	32.293	7
8	0.16777	3.3289	7.9469	0.05040	0.30040	2.3872	5.9604	19.841	47.367	8
9	0.13422	3.4631	9.0206	0.03876	0.28876	2.6048	7.4505	25.802	67.209	9
10	0.10737	3.5705	9.9870	0.03007	0.28007	2.7971	9.3132	33.252	93.011	10
11	0.08590	3.6564	10.846	0.02349	0.27349	2.9663	11.641	42.566	126.26	11
12	0.06872	3.7251	11.602	0.01845	0.26845	3.1145	14.551	54.207	168.83	12
13	0.05498	3.7801	12.261	0.01454	0.26454	3.2437	18.189	68.759	223.03	13
14	0.04398	3.8240	12.833	0.01150	0.26150	3.3559	22.737	86.949	291.79	14
15	0.03518	3.8592	13.326	0.00912	0.25912	3.4529	28.421	109.68	378.74	15
16	0.02815	3.8874	13.748	0.00724	0.25724	3.5366	35.527	138.10	488.43	16
17	0.02252	3.9099	14.108	0.00576	0.25576	3.6083	44.408	173.63	626.54	17
18	0.01801	3.9279	14.414	0.00459	0.25459	3.6697	55.511	218.04	800.17	18
19	0.01441	3.9423	14.674	0.00366	0.25366	3.7221	69.388	273.55	1018.2	19
20	0.01153	3.9538	14.893	0.00292	0.25292	3.7667	86.736	342.94	1291.7	20
21	0.00922	3.9631	15.077	0.00233	0.25233	3.8045	108.42	429.68	1634.7	21
22	0.00738	3.9704	15.232	0.00186	0.25186	3.8364	135.52	538.10	2064.4	22
23	0.00590	3.9763	15.362	0.00148	0.25148	3.8634	169.40	673.62	2602.5	23
24	0.00472	3.9811	15.471	0.00119	0.25119	3.8861	211.75	843.03	3276.1	24
25	0.00378	3.9848	15.561	0.00095	0.25095	3.9051	264.69	1054.7	4119.1	25
26	0.00302	3.9879	15.637	0.00076	0.25076	3.9211	330.87	1319.4	5173.9	26
27	0.00242	3.9903	15.700	0.00061	0.25061	3.9345	413.59	1650.3	6493.4	27
28	0.00193	3.9922	15.752	0.00048	0.25048	3.9457	516.98	2063.9	8143.8	28
29	0.00155	3.9938	15.795	0.00039	0.25039	3.9550	646.23	2580.9	10207.	29
30	0.00124	3.9950	15.831	0.00031	0.25031	3.9628	807.79	3227.1	12788.	30
35	0.00041	3.9983	15.936	0.00010	0.25010	3.9858	2465.1	9856.7	39287.	35
40	0.00013	3.9994	15.976	0.00003	0.25003	3.9946	7523.1	30088.	120195.	40
45	0.00004	3.9998	15.991	0.00001	0.25001	3.9980	22958.	91831.	367146.	45
INF	0.00000	4.0000	16.000	0.0000	0.25000	4.0000	INF	INF	INF	INF

| | $i = 30\%$ | | | $i = 30\%$ | | | $i = 30\%$ | | | |
| | Present sum, P | | | Uniform series, A | | | Future sum, F | | | |
n	P/F	P/A	P/G	A/F	A/P	A/G	F/P	F/A	F/G	n
1	0.76923	0.7692	0.0000	1.00000	1.30000	0.0000	1.3000	1.0000	0.0000	1
2	0.59172	1.3609	0.5917	0.43478	0.73478	0.4347	1.6900	2.3000	1.0000	2
3	0.45517	1.8161	1.5020	0.25063	0.55063	0.8270	2.1970	3.9900	3.3000	3
4	0.35013	2.1662	2.5524	0.16163	0.46163	1.1782	2.8561	6.1870	7.2900	4
5	0.26933	2.4355	3.6297	0.11058	0.41058	1.4903	3.7129	9.0431	13.477	5
6	0.20718	2.6427	4.6656	0.07839	0.37839	1.7654	4.8268	12.756	22.520	6
7	0.15937	2.8021	5.6218	0.05687	0.35687	2.0062	6.2748	17.582	35.276	7
8	0.12259	2.9247	6.4799	0.04192	0.34192	2.2155	8.1573	23.857	52.859	8
9	0.09430	3.0190	7.2343	0.03124	0.33124	2.3962	10.604	32.015	76.716	9
10	0.07254	3.0915	7.8871	0.02346	0.32346	2.5512	13.785	42.619	108.73	10
11	0.05580	3.1473	8.4451	0.01773	0.31773	2.6832	17.921	56.405	151.35	11
12	0.04292	3.1902	8.9173	0.01345	0.31345	2.7951	23.298	74.327	207.75	12
13	0.03302	3.2232	9.3135	0.01024	0.31024	2.8894	30.287	97.625	282.08	13
14	0.02540	3.2486	9.6436	0.00782	0.30782	2.9685	39.373	127.91	379.70	14
15	0.01954	3.2682	9.9172	0.00598	0.30598	3.0344	51.185	167.28	507.62	15
16	0.01503	3.2832	10.142	0.00458	0.30458	3.0892	66.541	218.47	674.90	16
17	0.01156	3.2948	10.327	0.00351	0.30351	3.1345	86.504	285.01	893.38	17
18	0.00889	3.3036	10.478	0.00269	0.30269	3.1718	112.45	371.51	1178.3	18
19	0.00684	3.3105	10.601	0.00207	0.30207	3.2024	146.19	483.97	1549.9	19
20	0.00526	3.3157	10.701	0.00159	0.30159	3.2275	190.05	630.16	2033.8	20
21	0.00405	3.3198	10.782	0.00122	0.30122	3.2479	247.06	820.21	2664.0	21
22	0.00311	3.3229	10.848	0.00094	0.30094	3.2646	321.18	1067.2	3484.2	22
23	0.00239	3.3253	10.900	0.00072	0.30072	3.2781	417.53	1388.4	4551.5	23
24	0.00184	3.3271	10.943	0.00055	0.30055	3.2890	542.80	1806.0	5940.0	24
25	0.00142	3.3286	10.977	0.00043	0.30043	3.2978	705.64	2348.8	7746.0	25
26	0.00109	3.3297	11.004	0.00033	0.30033	3.3049	917.33	3054.4	10094.	26
27	0.00084	3.3305	11.026	0.00025	0.30025	3.3106	1192.5	3971.7	13149.	27
28	0.00065	3.3311	11.043	0.00019	0.30019	3.3152	1550.2	5164.3	17121.	28
29	0.00050	3.3316	11.057	0.00015	0.30015	3.3189	2015.3	6714.6	22285.	29
30	0.00038	3.3320	11.068	0.00011	0.30011	3.3218	2620.0	8729.9	29000.	30
35	0.00010	3.3329	11.098	0.00003	0.30003	3.3297	9727.8	32422.	107960.	35
INF	0.00000	3.3333	11.111	0.0000	0.30000	3.3333	INF	INF	INF	INF

| | $i = 40\%$ | | | | $i = 40\%$ | | | | $i = 40\%$ | | |
| | Present sum, P | | | | Uniform series, A | | | | Future sum, F | | |
n	P/F	P/A	P/G	A/F	A/P	A/G	F/P	F/A	F/G	n
1	0.71429	0.7142	0.0000	1.00000	1.40000	0.0000	1.4000	1.0000	0.0000	1
2	0.51020	1.2244	0.5102	0.41667	0.81667	0.4166	1.9600	2.4000	1.0000	2
3	0.36443	1.5889	1.2390	0.22936	0.62936	0.7798	2.7440	4.3600	3.4000	3
4	0.26031	1.8492	2.0199	0.14077	0.54077	1.0923	3.8416	7.1040	7.7600	4
5	0.18593	2.0351	2.7637	0.09136	0.49136	1.3579	5.3782	10.945	14.864	5
6	0.13281	2.1679	3.4277	0.06126	0.46126	1.5811	7.5295	16.323	25.809	6
7	0.09486	2.2628	3.9969	0.04192	0.44192	1.7663	10.541	23.853	42.133	7
8	0.06776	2.3306	4.4712	0.02907	0.42907	1.9185	14.757	34.394	65.986	8
9	0.04840	2.3790	4.8584	0.02034	0.42034	2.0422	20.661	49.152	100.38	9
10	0.03457	2.4135	5.1696	0.01432	0.41432	2.1419	28.925	69.813	149.53	10
11	0.02469	2.4382	5.4165	0.01013	0.41013	2.2214	40.495	98.739	219.34	11
12	0.01764	2.4559	5.6106	0.00718	0.40718	2.2845	56.693	139.23	318.08	12
13	0.01260	2.4685	5.7617	0.00510	0.40510	2.3341	79.371	195.92	457.32	13
14	0.00900	2.4775	5.8787	0.00363	0.40363	2.3728	111.12	275.30	653.25	14
15	0.00643	2.4839	5.9687	0.00259	0.40259	2.4029	155.56	386.42	928.55	15
16	0.00459	2.4885	6.0376	0.00185	0.40185	2.4262	217.79	541.98	1314.9	16
17	0.00328	2.4918	6.0901	0.00132	0.40132	2.4440	304.91	759.78	1856.9	17
18	0.00234	2.4941	6.1299	0.00094	0.40094	2.4577	426.87	1064.7	2616.7	18
19	0.00167	2.4958	6.1600	0.00067	0.40067	2.4681	597.63	1491.5	3681.4	19
20	0.00120	2.4970	6.1827	0.00048	0.40048	2.4760	836.68	2089.2	5173.0	20
21	0.00085	2.4978	6.1998	0.00034	0.40034	2.4820	1171.3	2925.8	7262.2	21
22	0.00061	2.4984	6.2126	0.00024	0.40024	2.4865	1639.9	4097.2	10188.	22
23	0.00044	2.4989	6.2222	0.00017	0.40017	2.4899	2295.8	5737.1	14285.	23
24	0.00031	2.4992	6.2293	0.00012	0.40012	2.4925	3214.2	8033.0	20022.	24
25	0.00022	2.4994	6.2347	0.00009	0.40009	2.4944	4499.8	11247.	28055.	25
INF	0.00000	2.5000	6.2500	0.0000	0.40000	2.5000	INF	INF	INF	INF

	i = 50%			i = 50%			i = 50%		
	Present sum, P			Uniform series, A			Future sum, F			
n	P/F	P/A	P/G	A/F	A/P	A/G	F/P	F/A	F/G	n
1	0.66667	0.6666	0.0000	1.00000	1.50000	0.0000	1.5000	1.0000	0.0000	1
2	0.44444	1.1111	0.4444	0.40000	0.90000	0.4000	2.2500	2.5000	1.0000	2
3	0.29630	1.4074	1.0370	0.21053	0.71053	0.7368	3.3750	4.7500	3.5000	3
4	0.19753	1.6049	1.6296	0.12308	0.62308	1.0153	5.0625	8.1250	8.2500	4
5	0.13169	1.7366	2.1563	0.07583	0.57583	1.2417	7.5937	13.187	16.375	5
6	0.08779	1.8244	2.5953	0.04812	0.54812	1.4225	11.390	20.781	29.562	6
7	0.05853	1.8829	2.9465	0.03108	0.53108	1.5648	17.085	32.171	50.343	7
8	0.03902	1.9219	3.2196	0.02030	0.52030	1.6751	25.628	49.257	82.515	8
9	0.02601	1.9479	3.4277	0.01335	0.51335	1.7596	38.443	74.886	131.77	9
10	0.01734	1.9653	3.5838	0.00882	0.50882	1.8235	57.665	113.33	206.66	10
11	0.01156	1.9768	3.6994	0.00585	0.50585	1.8713	86.497	170.99	319.99	11
12	0.00771	1.9845	3.7841	0.00388	0.50388	1.9067	129.74	257.49	490.98	12
13	0.00514	1.9897	3.8458	0.00258	0.50258	1.9328	194.62	387.23	748.47	13
14	0.00343	1.9931	3.8903	0.00172	0.50172	1.9518	291.92	581.85	1135.7	14
15	0.00228	1.9954	3.9223	0.00114	0.50114	1.9656	437.89	873.78	1717.5	15
16	0.00152	1.9969	3.9451	0.00076	0.50076	1.9756	656.84	1311.6	2591.3	16
17	0.00101	1.9979	3.9614	0.00051	0.50051	1.9827	985.26	1968.5	3903.0	17
18	0.00068	1.9986	3.9729	0.00034	0.50034	1.9878	1477.8	2953.7	5871.5	18
19	0.00045	1.9991	3.9810	0.00023	0.50023	1.9914	2216.8	4431.6	8825.3	19
20	0.00030	1.9994	3.9867	0.00015	0.50015	1.9939	3325.2	6648.5	13257.	20
21	0.00020	1.9996	3.9907	0.00010	0.50010	1.9957	4987.8	9973.7	19905.	21
22	0.00013	1.9997	3.9935	0.00007	0.50007	1.9970	7481.8	14961.	29879.	22
23	0.00009	1.9998	3.9955	0.00004	0.50004	1.9979	11222.	22443.	44841.	23
24	0.00006	1.9998	3.9969	0.00003	0.50003	1.9985	16834.	33666.	67284.	24
25	0.00004	1.9999	3.9978	0.00002	0.50002	1.9990	25251.	50500.	100951.	25
INF	0.00000	2.0000	4.0000	0.0000	0.50000	2.0000	INF	INF	INF	INF

Summary of Depreciation Equations

EQUATIONS FOR STRAIGHT LINE METHOD

The equation for SL depreciation is

$$\text{depreciation rate, } R_m = \frac{1}{N}$$

$$\text{annual depreciation, } D_m = R_m(P - F) = \frac{P - F}{N}$$

$$\text{book value, } BV_m = P - mD_m$$

EQUATIONS FOR SUM-OF-YEARS DIGITS METHOD

The sum of the ordinal digits for each of the years 1 through N is

1. $\text{SOY} = \dfrac{N(N + 1)}{2}$

2. The annual depreciation, D_m, for the mth year (at any age, m) is

$$D_m = (P - F)\frac{N - m + 1}{\text{SOY}}$$

3. The book value at end of year m is

$$BV_m = P - (P - F)\left[\frac{m(N - m/2 + 0.5)}{\text{SOY}}\right]$$

EQUATIONS FOR DECLINING BALANCE METHODS

The abbreviation $R = $ depreciation rate for declining balance depreciation.

1. The depreciation rate, R, is the depreciation multiple divided by the estimated life, n.

    for double declining balance depreciation, $R = 2/n$
            1.75 declining balance depreciation, $R = 1.75/n$
            1.5 declining balance depreciation, $R = 1.5/n$

2. The depreciation, D_m, for any given year, m, and any given depreciation rate, R, is:

    $$D_m = RP(1 - R)^{m-1} \quad \text{or} \quad D_m = (BV_{m-1})R$$

3. The book value for any year BV_m is:

    $$BV_m = P(1 - R)^m \quad \text{provided} \quad BV_m \geq F$$

4. The age, m, at which book value, BV_m, will decline to any future value, F, is:

    $$m = \frac{\ln(F/P)}{\ln(1 - R)}$$

EQUATION FOR SINKING FUND METHOD

The book value at the end of year m is

$$BV_m = P - (P - F)(A/F, i, n)(F/A, i, m)$$

The amount of depreciation allowed in any one year can be determined by either of two methods.

1. Simply subtract the accumulated depreciation at the end of the prior year from the depreciation at the end of the current year.

2. The depreciation allowed for any year m may be determined as the future value of one lump sum deposit of A made $(m - 1)$ years previously. The depreciation allowed for each individual year may be totaled to equal the full allowable depreciation for the full n years.

SALT LAKE CITY

Farmington

°Bountiful

Sandy o

Kaysville o

OGDEN

Jordan

Bingham

•Saltair

Antelope Island

•Black Rock

•Garfield

Mud Isl.

TOOELE◎

←Fremont Isl.

S A L T   L A K E

Stansbury Isl.

Carrington Isl.

...rong's Knob

Alkali Flats

LAKESIDE MTS

E

N ←   → S

W

...reat Salt Lake Desert

41°

Winter Solstice
1991

REFUGE

For Pat —
This book comes to
you through the love
of Mick in the name
of family and landscape.
Refuge
In the birds —
In each other —

Fondly,
Terry Tempest Williams

The Secret Language of Snow (with Ted Major)

Pieces of White Shell: A Journey to Navajoland

Between Cattails

Coyote's Canyon

Earthly Messengers

REFUGE

*An Unnatural
History of Family and Place*

Terry Tempest Williams

Pantheon Books   New York

"The Clan of One-Breasted Women" by Terry Tempest Williams was originally
published in Northern Lights, January 1990, Volume VI, No. 1.

Grateful acknowledgment is made to the following for permission to reprint previ-
ously published material:
Atlantic Monthly Press: "Wild Geese" from Dream Work by Mary Oliver.
Copyright © 1986 by Mary Oliver. Reprinted by permission of Atlantic Monthly
Press.
Harcourt Brace Jovanovich, Inc.: "The Peace of Wild Things" from Openings
by Wendell Berry. Reprinted by permission of Harcourt Brace Jovanovich, Inc.

Library of Congress Cataloging-in-Publication Data

Williams, Terry Tempest.
Refuge: an unnatural history of family and place/Terry Tempest Williams.
p. cm.
ISBN 0-679-40516-X
1. Natural history—Utah—Great Salt Lake Region. 2. Williams, Terry
Tempest—Health. 3. Breast—Cancer—Patients—Utah—Biography.
4. Birds—Utah—Bear River Migratory Bird Refuge—Ecology. 5. Bird
populations—Utah—Bear River Migratory Bird Refuge. I. Title.
QH105.U8W55. 1991
362.1'969949'0092—dc20   91-8104
[B]
CIP

Book Design by Anne Scatto

Manufactured in the United States of America

9 8 7 6 5 4 3 2

For
Diane Dixon Tempest
who understood landscape as refuge

WILD GEESE

You do not have to be good.
You do not have to walk on your knees
for a hundred miles through the desert, repenting.
You only have to let the soft animal of your body love
*   what it loves.*
Tell me about despair, yours, and I will tell you mine.
Meanwhile the world goes on.
Meanwhile the sun and the clear pebbles of the rain
are moving across the landscapes,
over the prairies and deep trees,
the mountains and the rivers.
Meanwhile the wild geese, high in the clean blue air
are heading home again.
Whoever you are, no matter how lonely,
the world offers itself to your imagination,
calls to you like the wild geese, harsh and exciting—
over and over announcing your place
in the family of things.

—MARY OLIVER,
Dream Work

CONTENTS

ix

Contents

Contents

CONTENTS

xiii

REFUGE

PROLOGUE

Everything about Great Salt Lake is exaggerated—the heat,
the cold, the salt, and the brine. It is a landscape so surreal
one can never know what it is for certain.

In the past seven years, Great Salt Lake has advanced and
retreated. The Bear River Migratory Bird Refuge, devastated by the flood, now begins to heal. Volunteers are
beginning to reconstruct the marshes just as I am trying to
reconstruct my life. I sit on the floor of my study with
journals all around me. I open them and feathers fall from
their pages, sand cracks their spines, and sprigs of sage
pressed between passages of pain heighten my sense of
smell—and I remember the country I come from and how
it informs my life.

Most of the women in my family are dead. Cancer. At
thirty-four, I became the matriarch of my family. The losses
I encountered at the Bear River Migratory Bird Refuge as

Great Salt Lake was rising helped me to face the losses within my family. When most people had given up on the Refuge, saying the birds were gone, I was drawn further into its essence. In the same way that when someone is dying many retreat, I chose to stay.

Last night, I dreamed I was walking along the shores of Great Salt Lake. I noticed a purple bird floating in the waters, the waves rocking it gently. I entered the lake and, with cupped hands, picked up the bird and returned it to shore. The purple bird turned gold, dropped its tail, and began digging a burrow in the white sand, where it retreated and sealed itself inside with salt. I walked away. It was dusk. The next day, I returned to the lake shore. A wooden door frame, freestanding, became an arch I had to walk through. Suddenly, it was transformed into Athene's Temple. The bird was gone. I was left standing with my own memory.

In the next segment of the dream, I was in a doctor's office. He said, "You have cancer in your blood and you have nine months to heal yourself." I awoke puzzled and frightened.

Perhaps, I am telling this story in an attempt to heal myself, to confront what I do not know, to create a path for myself with the idea that "memory is the only way home."

I have been in retreat. This story is my return.

TTW
JULY 4, 1990

rrrrrrrr

BURROWING OWLS

rrrrrrrr

lake level: 4204.70'

Great Salt Lake is about twenty-five minutes from our home. From the mouth of Emigration Canyon where we live, I drive west past Brigham Young standing on top of "This Is the Place" monument. When I reach Foothill Drive, I turn right, pass the University of Utah and make another right, heading east until I meet South Temple, which requires a left-hand turn. I arrive a few miles later at Eagle Gate, a bronze arch that spans State Street. I turn right once more. One block later, I turn left on North Temple and pass the Mormon Tabernacle on Temple Square. From here, I simply follow the gulls west, past the Salt Lake City International Airport.

Great Salt Lake: wilderness adjacent to a city; a shifting shoreline that plays havoc with highways; islands too stark, too remote to inhabit; water in the desert that no one can drink. It is the liquid lie of the West.

I recall an experiment from school: we filled a cup with water—the surface area of the contents was only a few square inches. Then we poured the same amount of water into a large, shallow dinner plate—it covered nearly a square foot. Most lakes in the world are like cups of water. Great Salt Lake, with its average depth measuring only thirteen feet, is like the dinner plate. We then added two or three tablespoons of salt to the cup of water for the right amount of salinity to complete the analogue.

The experiment continued: we let the plate and cup of water stand side by side on the window sill. As they evaporated, we watched the plate of water dry up becoming encrusted with salt long before the cup. The crystals were beautiful.

Because Great Salt Lake lies on the bottom of the Great Basin, the largest closed system in North America, it is a terminal lake with no outlet to the sea.

The water level of Great Salt Lake fluctuates wildly in response to climatic changes. The sun bears down on the lake an average of about 70 percent of the time. The water frequently reaches ninety degrees Fahrenheit, absorbing enough energy to evaporate almost four feet of water annually. If rainfall exceeds the evaporation rate, Great Salt Lake rises. If rainfall drops below the evaporation rate, the lake recedes. Add the enormous volume of stream inflow from the high Wasatch and Uinta Mountains in the east, and one begins to see a portrait of change.

Great Salt Lake is cyclic. At winter's end, the lake level rises with mountain runoff. By late spring, it begins to decline when the weather becomes hot enough that loss of water by evaporation from the surface is greater than the combined inflow from streams, ground water, and precipitation. The lake begins to rise again in the autumn, when the temperature decreases, and the loss of water by evaporation is exceeded by the inflow.

Since Captain Howard Stansbury's *Exploration and Survey of the Great Salt Lake, 1852,* the water level has varied by as much as twenty feet, altering the shoreline in some places by as much as fifteen miles. Great Salt Lake is surrounded by salt flats, sage plains, and farmland; a slight rise in the water level extends its area considerably. In the past twenty years, Great Salt Lake's surface area has fluctuated from fifteen hundred square miles to its present twenty-five hundred square miles. Great Salt Lake is now approximately the size of Delaware and Rhode Island. It has been estimated that a ten foot rise in Great Salt Lake would cover an additional two hundred forty square miles.

To understand the relationship that exists at Great Salt Lake between area and volume, imagine pouring one inch of water into the bottom of a paper cone. It doesn't take much water to raise an inch. However, if you wanted to raise the water level one inch at the top of the cone, the volume of water added would have to increase considerably. The lake bed of Great Salt Lake is cone-shaped. It takes more water to raise the lake an inch when it is at high-level, and less water to raise it in low-level years.

Natives of the Great Basin, of the Salt Lake Valley in particular, speak about Great Salt Lake in the shorthand of lake levels. For example, in 1963, Great Salt Lake retreated to its historic low of 4191'. Ten years later, Great Salt Lake reached its historic mean, 4200'—about the same level explorers John Fremont and Howard Stansbury encountered in the 1840s and 50s.

On September 18, 1982, Great Salt Lake began to rise because of a series of storms that occurred earlier in the month. The precipitation of 7.04 inches for the month (compared to an annual average of about fifteen inches from 1875 to 1982) made it the wettest September on record for Salt Lake City. The lake continued to rise for the next ten months as a result of greater-than-average snowfall during the winter and spring of 1982–83, and unseasonably cool weather (thus little evaporation) during the spring of 1983. The rise from September 18, 1982 to June 30, 1983, was 5.1′, the greatest seasonal rise ever recorded.

During these years, talk on the streets of Salt Lake City has centered around the lake: 4204′ and rising. It is no longer just a backdrop for spectacular sunsets. It is the play of urban drama. Everyone has their interests. 4211.6′ was the historic high recorded in the 1870's. City officials knew the Salt Lake City International Airport would be underwater if the Great Salt Lake rose to 4220′. Developments along the lakeshore were sunk at 4208′. Farmers whose land was being flooded in daily increments were trying desperately to dike or sell. And the Southern Pacific Railroad labors to maintain their tracks above water, twenty-four hours a day, three hundred sixty-five days a year, and has been doing so since 1959.

My interest lay at 4206′, the level which, according to my topographical map, meant the flooding of the Bear River Migratory Bird Refuge.

There are those birds you gauge your life by. The burrowing owls five miles from the entrance to the Bear River Migratory Bird Refuge are mine. Sentries. Each year, they alert me to the regularities of the land. In spring, I find them nesting, in summer they forage with their young, and by winter they abandon the Refuge for a place more comfortable.

What is distinctive about these owls is their home. It rises from the alkaline flats like a clay-covered fist. If you were to peek inside the tightly clenched fingers, you would find a dark-holed entrance.

"*Tttss! Tttss! Tttss!*"

That is no rattlesnake. Those are the distress cries of the burrowing owl's young.

Adult burrowing owls will stand on top of the mound with their prey before them, usually small rodents, birds, or insects. The entrance is littered with bones and feathers. I recall finding a swatch of yellow feathers like a doormat across the threshold—meadowlark, maybe. These small owls pursue their prey religiously at dusk.

Burrowing owls are part of the desert community, taking advantage of the abandoned burrows of prairie dogs. Historically, bison would move across the American Plains, followed by prairie dog towns which would aerate the soil after the weight of stampeding hooves. Black-footed ferrets, rattlesnakes, and burrowing owls inhabited the edges, finding an abundant food source in the communal rodents.

With the loss of desert lands, a decline in prairie dog populations is inevitable. And so go the ferret and burrowing owl. Rattlesnakes are more adaptable.

In Utah, prairie dogs and black-footed ferrets are endangered species, with ferrets almost extinct. The burrowing owl is defined as "threatened," a political step away from endangered status. Each year, the burrowing owls near the Refuge become more blessed.

The owls had staked their territory just beyond one of the bends in the Bear River. Whenever I drove to the Bird Refuge, I stopped at their place first and sat on the edge of the road and watched. They would fly around me, their wings sometimes spanning two feet. Undulating from post to post, they would distract me from their nest. Just under a foot long, they have a body of feathers the color of wheat, balanced on two long, spindly legs. They can burn grasses with their stare. Yellow eyes magnifying light. The protective hissing of baby burrowing owls is an adaptive memory of their close association with prairie rattlers. Snake or owl? Who wants to risk finding out. In the summer of 1983, I worried about the burrowing owls, wondering if the rising waters of Great Salt Lake had flooded their home, too. I was relieved to find not only their mound intact, but four owlets standing on its threshold. One of the Refuge managers stopped on the road and commented on what a good year it had been for them.

"Good news," I replied. "The lake didn't take everything."

That was late August when huge concentrations of shorebirds were still feeding between submerged shadescale.

A few months later, a friend of mine, Sandy Lopez, was visiting from Oregon. We had spoken of the Bird Refuge many times. The whistling swans had arrived, and it seemed like a perfect day for the marsh.

To drive to the Bear River Migratory Bird Refuge from Salt Lake City takes a little over one hour. I have discovered the conversation that finds its way into the car often manifests itself later on the land.

We spoke of rage. Of women and landscape. How our bodies and the body of the earth have been mined.

"It has everything to do with intimacy," I said. "Men define intimacy through their bodies. It is physical. They define intimacy with the land in the same way."

"Many men have forgotten what they are connected to," my friend added. "Subjugation of women and nature may be a loss of intimacy within themselves."

She paused, then looked at me.

"Do you feel rage?"

I didn't answer for some time.

"I feel sadness. I feel powerless at times. But I'm not certain what rage really means."

Several miles passed.

"Do you?" I asked.

She looked out the window. "Yes. Perhaps your generation, one behind mine, is a step removed from the pain."

We reached the access road to the Refuge and both took out our binoculars, ready for the birds. Most of the waterfowl had migrated, but a few ruddy ducks, redheads, and shovelers remained. The marsh glistened like cut topaz.

As we turned west about five miles from the Refuge, a mile or so from the burrowing owl's mound, I began to speak of them, *Athene cunicularia.* I told Sandy about the time when my grandmother and I first discovered them. It was in 1960, the same year she gave me my Peterson's *Field Guide to Western Birds.* I know because I dated their picture. We have come back every year since to pay our respects. Generations of burrowing owls have been raised here. I turned to my friend and explained how four owlets had survived the flood.

We anticipated them.

About a half mile away, I could not see the mound. I took my foot off the gas pedal and coasted. It was as though I was in unfamiliar country.

The mound was gone. Erased. In its place, fifty feet back, stood a cinderblock building with a sign, CANADIAN GOOSE GUN CLUB. A new fence crushed the grasses with a handwritten note posted: KEEP OUT.

We got out of the car and walked to where the mound had been for as long as I had a memory. Gone. Not a pellet to be found.

A blue pickup pulled alongside us.

"Howdy." They tipped their ball caps. "What y'all lookin' for?"

I said nothing. Sandy said nothing. My eyes narrowed.

"We didn't kill 'em. Those boys from the highway department came and graveled the place. Two bits, they did it. I mean, you gotta admit those ground owls are messy little bastards. They'll shit all over hell if ya let 'em. And try and sleep with 'em hollering at ya all night long. They had to go. Anyway, we got bets with the county they'll pop up someplace around here next year."

The three men in the front seat looked up at us, tipped their caps again. And drove off.

Restraint is the steel partition between a rational mind and a violent one. I knew rage. It was fire in my stomach with no place to go.

I drove out to the Refuge on another day. I suppose I wanted to see the mound back in place with the family of owls bobbing on top. Of course, they were not.

I sat on the gravel and threw stones.

By chance, the same blue pickup with the same three men pulled alongside: the self-appointed proprietors of the newly erected Canadian Goose Gun Club.

"Howdy, ma'am. Still lookin' for them owls, or was it sparrows?"

One winked.

Suddenly in perfect detail, I pictured the burrowing owls' mound—that clay-covered fist rising from the alkaline flats. The exact one these beergut-over-beltbuckled men had leveled.

I walked calmly over to their truck and leaned my stomach against their door. I held up my fist a few inches from the driver's face and slowly lifted my middle finger to the sky.

"This is for you—from the owls and me."

My mother was appalled—not so much over the loss of the burrowing owls, although it saddened her, but by my behavior. Women did not deliver obscene gestures to men, regardless. She shook her head, saying she had no idea where I came from.

In Mormon culture, that is one of the things you do know—history and genealogy. I come from a family with deep roots in the American West. When the expense of outfitting several thousand immigrants to Utah was becoming too great for the newly established church, leaders decided to furnish the pioneers with small two-wheeled carts about the size of those used by apple peddlers, which could be pulled by hand from Missouri to the Salt Lake Valley. My ancestors were part of these original "handcart companies" in the 1850s. With faith, they would endure. They came with few provisions over the twelve-hundred-mile trail. It was a small sacrifice in the name of religious freedom. Almost one hundred and fifty years later, we are still here.

I am the oldest child in our family, a daughter with three younger brothers: Steve, Dan, and Hank.

My parents, John Henry Tempest, III, and Diane Dixon Tempest, were married in the Mormon Temple in Salt Lake City on September 18, 1953. My husband, Brooke Williams, and I followed the same tradition and were married on June 2, 1975. I was nineteen years old.

Our extended family includes both maternal and paternal grandparents: Lettie Romney Dixon and Donald "Sanky" Dixon, Kathryn Blackett Tempest and John Henry Tempest, Jr.

Aunts, uncles, and cousins are many, extending familial ties all across the state of Utah. If I ever wonder who I am, I simply attend a Romney family reunion and find myself in the eyes of everyone I meet. It is comforting and disturbing, at once.

I have known five of my great-grandparents intimately. They tutored me in stories with a belief that lineage mattered. Genealogy is in our blood. As a people and as a family, we have a sense of history. And our history is tied to land.

I was raised to believe in a spirit world, that life exists before the earth and will continue to exist afterward, that each human being, bird, and bulrush, along with all other life forms had a spirit life before it came to dwell physically on the earth. Each occupied an assigned sphere of influence, each has a place and a purpose.

It made sense to a child. And if the natural world was assigned spiritual values, then those days spent in wildness were sacred. We learned at an early age that God can be found wherever you are, especially outside. Family worship was not just relegated to Sunday in a chapel.

Our weekends were spent camped alongside a small stream in the Great Basin, in the Stansbury Mountains or Deep Creeks. My father would take the boys rabbit hunting while Mother and I would sit on a log in an aspen grove and talk. She would tell me stories of how when she was a girl she would paint red lips on the trunks of trees to practice kissing. Or how she would lie in her grandmother's lucerne patch and watch clouds.

"I have never known my full capacity for solitude," she would say.

"Solitude?" I asked.

"The gift of being alone. I can never get enough."

The men would return anxious for dinner. Mother would cook over a green Coleman stove as Dad told stories from his childhood—like the time his father took away his BB gun for a year because he shot off the heads of every red tulip in his mother's garden, row after row after row. He laughed. We laughed. And then it was time to bless the food.

After supper, we would spread out our sleeping bags in a circle, heads pointing to the center like a covey of quail, and watch the Great Basin sky fill with stars. Our attachment to the land was our attachment to each other.

The days I loved most were the days at Bear River. The Bird Refuge was a sanctuary for my grandmother and me. I call her "Mimi." We would walk along the road with binoculars around our necks and simply watch birds. Hundreds of birds. Birds so exotic to a desert child it forced the imagination to be still. The imagined was real at Bear River.

I recall one bird in particular. It wore a feathered robe of cinnamon, white, and black. Its body rested on long, thin legs. Blue legs. On the edge of the marsh, it gracefully lowered its head and began sweeping the water side to side with its delicate, upturned bill.

"Plee-ek! Plee-ek! Plee-ek!"

Three more landed. My grandmother placed her hand gently on my shoulder and whispered, "avocets." I was nine years old.

At ten, Mimi thought I was old enough to join the

Audubon Society on a special outing to the wetlands surrounding Great Salt Lake. We boarded a greyhound bus in downtown Salt Lake and drove north on U.S. Highway 91, paralleling the Wasatch Mountains on our right and Great Salt Lake on our left. Once relaxed and out of the city, we were handed an official checklist of birds at the Bear River Migratory Bird Refuge.

"All members are encouraged to take copious notes and keep scrupulous records of birds seen," proclaimed the gray-haired, ponytailed woman passing out cards.

"What do copious and scrupulous mean?" I asked my grandmother.

"It means pay attention," she said. I pulled out my notebook and drew pictures of the backs of birdwatchers' heads.

Off the highway, the bus drove through the small town of Brigham City with its sycamore-lined streets. It's like most Utah settlements with its Mormon layout: a chapel for weekly worship, a tabernacle for communal events, and a temple nearby (in this case Logan) where sacred rites are performed. Lawns are well groomed and neighborhoods are immaculate. But the banner arched over Main Street makes this town unique. In neon lights it reads, BRIGHAM CITY: GATEWAY TO THE WORLD'S GREATEST GAME BIRD REFUGE. So welded to the local color of this community, I daresay no one sees the sign anymore, except newcomers and perhaps the birds that fly under it.

A small, elderly man with wire-rimmed glasses and a worn golf cap, stood at the front of the bus and began speaking into the handheld microphone: "Ladies and gentlemen, in approximately ten miles we will be entering the Bear River Migratory Bird Refuge, America's first waterfowl sanctuary, established by a special act of Congress on April 23, 1928."

I was confused. I thought the marsh had been created in

the spirit world first and on earth second. I never made the connection that God and Congress were in cahoots. Mimi said she would explain the situation later.

The man went on to say that the Bird Refuge was located at the delta of the Bear River, which poured into the Great Salt Lake. This I understood.

"People, this bus is a clock. Eyes forward, please. Straight ahead is twelve o'clock; to the rear is six. Three o'clock is on your right. Any bird identified from this point on will be noted accordingly."

The bus became a bird dog, a labrador on wheels, which decided where high noon would be simply by pointing in that direction. What time would it be if a bird decided to fly from nine o'clock to three o'clock? Did that make the bird half past nine or quarter to three? Even more worrisome to me was the possibility of a flock of birds flying between four and five o'clock. Would you say, "Twenty birds after four? Four-thirty? Or simply move the hands of the clock forward to five? I decided not to bother my grandmother with these particulars and, instead, retreated to my unindexed field guide and turned to the color plates of ducks.

"Ibises at two o'clock!"

The brakes squeaked the bus to a halt. The doors opened like bellows and we all filed out. And there they were, dozens of white-faced glossy ibises grazing in the field. Their feathers on first glance were chestnut, but with the slightest turn they flashed irridescences of pink, purple, and green.

Another flock landed nearby. And another. And another. They coasted in diagonal lines with their heads and necks extended, their long legs trailing behind them, seeming to fall forward on hinges the second before they touched ground. By now, we must have been watching close to a hundred ibises probing the farmlands adjacent to the marsh.

Our leader told us they were eating earthworms and insects.

"Good eyes," I thought, as I could only see their decurved bills like scythes disappearing behind the grasses. I watched the wind turn each feather as the birds turned the soil.

Mimi whispered to me how ibises are the companions of gods. "Ibis escorts Thoth, the Egyptian god of wisdom and magic, who is the guardian of the Moon Gates in heaven. And there are two colors of ibis—one black and one white. The dark bird is believed to be associated with death, the white bird a celebration of birth."

I looked out over the fields of black ibis.

"When an ibis tucks its head underwing to sleep, it resembles a heart. The ibis knows empathy," my grandmother said. "Remember that, alongside the fact it eats worms."

She also told me that if I could learn a new way to tell time, I could also learn a new way to measure distance.

"The stride of an ibis was a measurement used in building the great temples of the Nile."

I sat down by the rear wheels of the bus and pondered the relationship between an ibis at Bear River and an ibis foraging on the banks of the Nile. In my young mind, it had something to do with the magic of birds, how they bridge cultures and continents with their wings, how they mediate between heaven and earth.

Back on the bus and moving, I wrote in my notebook "one hundred white-faced glossy ibises—companions of the gods."

Mimi was pleased. "We could go home now," she said. "The ibis makes the day."

But there were more birds. Many, many more. Within the next few miles, ducks, geese, and shorebirds were sighted around "the clock." The bus drove past all of them. With my arms out the window, I tried to touch the wings of avocets and stilts. I knew these birds from our private trips to the Refuge. They had become relatives.

As the black-necked stilts flew alongside the silver bus, their long legs trailed behind them like red streamers.

"Ip-ip-ip! Ip-ip-ip!"

Their bills were not flattened and upturned like avocets, but straight as darning needles.

The wind massaged my face. I closed my eyes and sat back in my seat.

Mimi and I got out of the bus and ate our lunch on the riverbank. Two western grebes, ruby-eyed and serpentine, fished, diving at good prospects. They surfaced with silver minnows struggling between sharp mandibles. Violet-green swallows skimmed the water for midges as a snowy egret stood on the edge of the spillway.

With a crab sandwich in one hand and binoculars in the other, Mimi explained why the Bird Refuge had in fact, been created.

"Maybe the best way to understand it," she said, "is to realize the original wetlands were recreated. It was the deterioration of the marshes at Bear River Bay that led to the establishment of a sanctuary."

"How?" I asked.

"The marshes were declining for several reasons: the diversion of water from the Bear River for irrigation, the backing-up of brine from Great Salt Lake during high-water periods, excessive hunting, and a dramatic rise in botulism, a disease known then as 'western duck disease.'

"The creation of the Bear River Migratory Bird Refuge helped to preserve the freshwater character of the marsh. Dikes were built to hold the water from the Bear River to stabilize, manage, and control water levels within the marsh.

This helped to control botulism and at the same time keep out the brine. Meanwhile, the birds flourished."

After lunch, I climbed the observation tower at the Refuge headquarters. Any fear of heights I may have had moving up the endless flights of steel stairs was replaced by the bird's eye view before me. The marsh appeared as a green and blue mosaic where birds remained in a fluid landscape.

In the afternoon, we drove the twenty-two-mile loop around the Refuge. The roads capped the dikes which were bordered by deep channels of water with bulrush and teasel. We saw ruddy ducks (the man sitting behind us called them "blue bills"), shovelers, teals, and wigeons. We watched herons and egrets and rails. Red-wing blackbirds poised on cattails sang with long-billed marsh wrens as muskrats swam inside shadows created by clouds. Large families of Canada geese occupied the open water, while ravens flushed the edges for unprotected nests with eggs.

The marsh reflected health as concentric circles rippled outward from a mallard feeding "bottoms up."

By the end of the day, Mimi and I had marked sixty-seven species on our checklist, many of which I had never seen before. A short-eared owl hovered over the cattails. It was the last bird we saw as we left the Refuge.

I fell asleep on my grandmother's lap. Her strong, square hands resting on my forehead shielded the sun from my eyes. I dreamed of water and cattails and all that is hidden.

When we returned home, my family was seated around the dinner table.

"What did you see?" Mother asked. My father and three brothers looked up.

"Birds . . ." I said as I closed my eyes and stretched my arms like wings.

"Hundreds of birds at the marsh."

rrrrrrrr

WHIMBRELS

rrrrrrrr

lake level: 4203.25'

The Bird Refuge has remained a constant. It is a landscape so familiar to me, there have been times I have felt a species long before I saw it. The long-billed curlews that foraged the grasslands seven miles outside the Refuge were trustworthy. I can count on them year after year. And when six whimbrels joined them—whimbrel entered my mind as an idea. Before I ever saw them mingling with curlews, I recognized them as a new thought in familiar country.

The birds and I share a natural history. It is a matter of rootedness, of living inside a place for so long that the mind and imagination fuse.

Maybe it's the expanse of sky above and water below that soothes my soul. Or maybe it's the anticipation of seeing something new. Whatever the magic of Bear River is—I appreciate this corner of northern Utah, where the numbers of ducks and geese I find resemble those found by early explorers.

Of the 208 species of birds who use the Refuge, sixty-two are known to nest here. Such nesting species include eared, western, and pied-billed grebes, great blue herons, snowy egrets, white-faced ibises, American avocets, black-necked stilts, and Wilson's phalaropes. Also nesting at Bear River are Canada geese, mallards, gadwalls, pintails, green-winged, blue-winged, and cinnamon teals, redheads, and ruddy ducks. It is a fertile community where the hope of each day rides on the backs of migrating birds.

These wetlands, emeralds around Great Salt Lake, provide critical habitat for North American waterfowl and shorebirds, supporting hundreds of thousands, even millions of individuals during spring and autumn migrations. The long-legged birds with their eyes focused down transform a seemingly sterile world into a fecund one. It is here in the marshes with the birds that I seal my relationship to Great Salt Lake.

I could never have anticipated its rise.

My mother was aware of a rise on the left side of her abdomen. I was deep in dream. This particular episode found me hiding beneath my grandmother's bed as eight black helicopters flew toward the house. I knew we were in danger.

The phone rang and everything changed.

"Good morning," I answered.

"Good morning, dear," my mother replied.

This is how my days always began. Mother and I checking in—a long extension cord on the telephone lets me talk and eat breakfast at the same time.

"You're back. So how was the river trip?" I asked, pouring myself a glass of orange juice.

"It was wonderful," she answered. "I loved the river and I loved the people. The Grand Canyon is a . . ."

There was a break in her voice. I set my glass on the counter.

She paused. "I didn't want to do this, Terry."

I think I knew what she was going to say before she said it. The same way, twelve years before, I knew something was wrong when I walked into our house after school and Mother was gone. In 1971, it had been breast cancer.

With my back against the kitchen wall, I slowly sank to the floor and stared at the yellow flowered wallpaper I had always intended to change.

"What I was going to say is that the Grand Canyon is a perfect place to heal—I've found a tumor, a fairly large mass in my lower abdomen. I was wondering if you could go with me to the hospital. John has to work. I'm scheduled for an ultrasound this afternoon."

I closed my eyes. "Of course."

Another pause.

"How long have you known about this?"

"I discovered it about a month ago."

I found myself getting angry until she answered the next obvious question.

"I needed time to live with it, to think about it—and more than anything else, I wanted to float down the Colorado River. This was the trip John and I had been dreaming about for years. I knew the days in the canyon would give me peace. And Terry, they did."

I sat on the white linoleum floor in my nightgown with my knees pulled in toward my chest, my head bowed.

"Maybe it's nothing, Mother. Maybe it's only a cyst. It could be benign, you know."

She did not answer.

"How do you feel?" I asked.

"I feel fine," she said. "But I would like to go shopping for a robe before my appointment at one."

We agreed to meet at eleven.

"I'm glad you're home." I said.

"So am I."

She hung up. The dial tone returned. I listened to the line until it became clear I had heard what I heard.

It's strange to feel change coming. It's easy to ignore. An underlying restlessness seems to accompany it like birds flocking before a storm. We go about our business with the usual alacrity, while in the pit of our stomach there is a sense of something tenuous.

These moments of peripheral perceptions are short, sharp flashes of insight we tend to discount like seeing the movement of an animal from the corner of our eye. We turn and there is nothing there. They are the strong and subtle impressions we allow to slip away.

I had been feeling fey for months.

Mother and I drove downtown, parked the car, and walked into Nordstrom's. I recalled the last department store we were in when the only agenda was which lipstick to choose.

We rode the escalator up two floors to sleepwear. Mother appeared to have nothing else on her mind but a beautiful piece of lingerie.

"What do you think about this one?" she asked as she held a navy blue satin robe up to her in the mirror.

"It's stunning," I answered. "I love the tiny white stars—"
"So do I. It's quite dramatic." She turned to the clerk. "I'll
take this, please." and handed her the robe.

"Would you like this gift wrapped?" asked the sales-
woman.

I started to say no. Mother said yes. "Thank you, that
would be very nice."

My mother's flair for drama always caught me off guard.
Her love of spontaneity made the most mundane enterprise
an occasion. She entered a room, mystery followed her. She
left and her presence lingered.

I thought of the last time we were in New York together.
We slept late, rising mid-morning to partake of steaming
hot blueberry muffins downtown in a sidewalk café. It was
my mother's sacrament. We shopped in the finest stores and
twirled in front of mirrors. We lived in the museums.
Having overspent our allotment of time at the Met in the
Caravaggio exhibit, we opted for a quick make-over at
Bloomingdale's to revive us for the theatre. The brass and
glass of the department store's first floor was blinding until
we finally bumped into the Lancôme counter.

"It's wonderful to be in a place where no one knows
you," Mother said as she sat in the chair reserved for customers. "I would never do this at home."

The salesclerk acquainted her with options. She looked at
my mother's hazel eyes, the structure of her face, her dark
hair cut short.

"Great bones," the make-up artist said. "For you, less is
more."

I watched the woman sweep blush across my mother's
cheekbones. A hint of brown eyeshadow deepened her eyes
as framboise was painted across her lips.

"How do I look?" she said.

"Dazzling," I answered.

Mother gave me her chair. The Lancôme woman looked at my face and shook her head.

"Do you spend a lot of time in the wind?"

The hospital doors seemed heavy as I pushed them open against the air trapped inside the vestibule. Once inside, it reeked of disease white-washed with antiseptics. A trip to the hospital is always a descent into the macabre. I have never trusted a place with shiny floors.

We found our way to the lab through the maze of hallways by following the color-coded tape on the floors. Mother was given instructions to change into the hospital's blue and white seersucker robe. They say the gowns are for convenience, so they can do what they have to do fast. But their robes seem more like socialistic wraps that let you know that you belong to the fraternity of the ill waiting patiently in rooms all across America.

"Diane Tempest."

She looked too beautiful to be sick. Wearing their white foam slippers, she disappeared down the hall into a room with closed doors.

I waited.

My eyes studied each person in the room. Why were they there and what were they facing? They all seemed to share an unnatural color. I checked my hands against theirs. I tried to pick up snippets of conversation that pieced together their stories. But voices were soft and words were few.

I could not read the expression on Mother's face when she came out of X-ray. She changed into her clothes and we walked out of the hospital to the car.

"It doesn't look good," she said. "It's about the size of a

grapefruit, filled with fluid. They are calling in the results to the doctor. We need to go to his office to find out what to do next."

There was little emotion in her face. This was a time for details. Pragmatism replaced sentiment.

At Krehl Smith's office, the future was drawn on an 8½ by 11 inch pad of yellow paper. The doctor (her obstetrician who had delivered two of her four babies) proceeded to draw the tumor in relationship to her ovaries. He stumbled over his own words, not having the adequate vocabulary to tell a patient who was also a friend that she most likely had ovarian cancer.

We got the picture. There was an awkward silence.

"So what are my options?" Mother asked.

"A hysterectomy as soon as you are ready. If it is ovarian cancer then we'll follow it up with chemotherapy and go from there . . ."

"I'll make that decision," she said.

The tears I had wanted to remain hidden splashed down on the notes I was taking, blurring the ink.

Arrangements were made for surgery on Monday morning. Mother wanted to prepare the family over the weekend. Dr. Smith suggested that two oncologists be called in on the case; Gary Smith and Gary Johnson. Mother agreed, requesting that she be able to meet with them before the operation for questions.

There was another awkward silence. Details done. Mother stood up from the straight back chair.

"Thank you, Krehl."

Their eyes met. She turned to walk out the door, when Krehl Smith put his arm through hers. "I'm so sorry, Diane. I know what you went through before. I wish I had more encouraging news."

"So do I," she said. "So do I."

Mother and I got into the car. It started to rain. In a peculiar sort of way, the weather gave us permission to cry. Driving home, Mother stared out her window. "You know, I hear the words on the outside, that I might have ovarian cancer, but they don't register on the inside. I keep saying to myself, this isn't happening to me, but then why shouldn't it? I am facing my own mortality—again—something I thought I had already done twelve years ago. Do you know how strange it is to know your days are limited? To have no future?"

Home. The family gathered in the living room. Mother had her legs on Dad's lap. Dad had his left arm around her, his right hand rubbing her knees and thighs. My brothers, Steve, Dan, and Hank were seated across the room. I sat on the hearth. A fire was burning, so were candles. Twelve years ago, we had been too young to see beyond our own pain; children of four, eight, twelve, and fifteen. Dad was thirty-seven, in shock from the thought of losing his wife. We did not do well. She did. Things were different now. We would do it together. We made promises that we would be here for her this time, that she would not have to carry us.

The conversation shifted to mountain climbing, the men's desire to climb the Grand Teton in the summer, then on to tales of scaling Mount Everest without oxygen—it could be done.

Mother said she would like to work in the garden if the weather cleared. We said we would all help.

"That's funny," she said. "No one has ever offered to help me before."

She then asked that we respect her decisions, that this was

her body and her life, not ours, and that if the tumor was malignant, she would choose not to have chemotherapy. We said nothing.

She went on to explain why she had waited a month before going to the doctor.

"In the long run I didn't think one month would matter. In the short run, it mattered a great deal. The heat of the sandstone penetrated my skin as I laid on the red rocks. Desert light bathed my soul. And traveling through the inner gorge of Vishnu schist, the oldest exposed rock in the West, gave me a perspective that will carry me through whatever I must face. Those days on the river were a meditation, a renewal. I found my strength in its solitude. It is with me now."

She looked at Dad, "Lava Falls, John. We've got some white water ahead."

I know the solitude my mother speaks of. It is what sustains me and protects me from my mind. It renders me fully present. I am desert. I am mountains. I am Great Salt Lake. There are other languages being spoken by wind, water, and wings. There are other lives to consider; avocets, stilts, and stones. Peace is the perspective found in patterns. When I see ring-billed gulls picking on the flesh of decaying carp, I am less afraid of death. We are no more and no less than the life that surrounds us. My fears surface in my isolation. My serenity surfaces in my solitude.

It is raining. And it seems as though it has always been raining. Every day another quilted sky rolls in and covers us with water. Rain. Rain. More rain. The Great Basin is being filled.

It isn't just the clouds' doing. The depth of snowpack in the Wasatch Mountains is the highest on record. It begins to melt, and streams you could jump over become raging rivers with no place to go. Local canyons are splitting at their seams as saturated hillsides slide.

Great Salt Lake is rising.

Brooke and I opt for marriage maintenance and drive out to Black's Rock on the edge of the lake to watch birds. They'll be there in spite of the weather. And they are.

Avocets and black-necked stilts are knee-deep in water alongside Interstate 80. Flocks of California gulls stand on a disappearing beach. We pull over, get out of the car and begin walking up and over lakeside boulders. I inhale the salty air. It is like ocean, even the lake is steel-blue with whitecaps.

Brooke walks ahead while I sit down with my binoculars and watch grebes. Eared grebes. Their red eyes flash intensely on the water, and I am amazed by such buoyancy in small bodies. Scanning the horizon, all I can see is water. "Lake Bonneville," I think to myself.

It is easy to imagine this lake, born twenty-eight thousand years ago, in the Pleistocene Epoch, just one in the succession of bodies of water to inhabit the Bonneville Basin over the last fifteen million years. It inundated nearly twenty thousand square miles of western Utah, spilling into southern Utah and eastern Nevada—a liquid hand pressing against the landscape that measured 285 miles long and 140 miles wide, with an estimated depth of 1000'.

Across from where I sit, Stansbury Island looms. Distinct bench levels tell a story of old shorelines, a record of where Lake Bonneville paused in its wild fluctuations over the course of fifteen thousand years. Its rise was stalled about twenty-three thousand years ago when the lake's elevation was about 4500' above sea level; over the next three thousand years, it rose very little. The relentless erosion of wave against rock during this stable period cut a broad terrace known to geologists as the Stansbury Shoreline.

The lake began to swell again until it reached the 5090' level sixteen thousand years ago. And then for a millenium and a half, the lake carved the Bonneville Shoreline, the highest of the three main terraces. Great tongues of ice occupied canyons in the Wasatch Mountains to the east, while herds of musk oxen, mammoths, and saber-tooth cats frequented the forested shores of Lake Bonneville. Schools of Bonneville cutthroat trout flashed through these waters (remnants of which still cling to existence in the refuge of small ponds in isolated desert mountains of the Great Basin). Fossil records suggest birds similar to red-tail hawk, sage grouse, mallard, and teal lived here. And packs of dire wolves called up the moon.

About 14,500 years ago, Lake Bonneville spilled over the rim of the Great Basin near Red Rock Pass in southeastern Idaho. Suddenly, the waters broke the Basin breaching the sediments down to bedrock, releasing a flood so spectacular it is estimated the maximum discharge of water was thirty-three million cubic feet per second. This event, known today as the Bonneville Flood, dropped the lake about 350', to 4740'. When the outlet channel was eroded to resistant rock, the lake stabilized once again and the Provo Shoreline was formed.

As the climate warmed drawing moisture from the inland sea, the lake began to shrink, until, eleven thousand years ago, it had fallen to present day levels of about 4200'. This trend toward warmer and drier conditions signified the end of the Ice Age.

A millenium later, the lake rose slightly to an elevation

of about 4250′, forming the Gilbert Shoreline, but soon receded. This marked the end of Lake Bonneville and the birth of its successor, Great Salt Lake.

As children, it was easy to accommodate the idea of Lake Bonneville. The Provo Shoreline looks like a huge bathtub ring around the Salt Lake Valley. It is a bench I know well, because we lived on it. It is the ledge that supported my neighborhood above Salt Lake City. Daily hikes in the foothills of the Wasatch yielded vast harvests of shells.

"Lake Bonneville . . ." we would say as we pocketed them. Never mind that they were the dried shells of land snails. We would sit on the benches of this ancient lake, stringing white shells into necklaces. We would look west to Great Salt Lake and imagine.

That was in 1963. I was eight years old. Great Salt Lake was a puddle, having retreated to a record low surface elevation of 4191.35′. Local papers ran headlines that read, GREAT SALT LAKE DISAPPEARING? and INLAND SEA SHRINKS.

My mother decided Great Salt Lake was something we should see before it vanished. And so, my brothers and I, with friends from the neighborhood, boarded our red Ford station wagon and headed west.

It was a long ride past the airport, industrial complexes, and municipal dumps. It was also hot. The backs of our thighs stuck to the Naugahyde seats. Our towels were wrapped around us. We were ready to swim.

Mother pulled into the Silver Sands Beach. The smell should have been our first clue, noxious hydrogen sulphide gas rising from the brine.

"Phew!" we all complained as we walked toward the beach, brine flies following us. "Smells like rotten eggs."

"You'll get used to it," Mother said. "Now go play. See if you can float."

We were dubious at best. Our second clue should have

been the fact that Mother did not bring her bathing suit, but rather chose to sit on the sand in her sunsuit with a thick novel in hand.

The ritual was always the same. Run into the lake, scream, and run back out. The salt seeped into the sores of our scraped knees and lingered. And if the stinging sensation didn't bring you to tears, the brine flies did.

We huddled around Mother, the old Saltair Pavilion was visible behind her, vibrating behind a screen of heatwaves. We begged her to take us home, pleading for dry towels. Total time at the lake: five minutes. She was unsympathetic.

"We're here for the afternoon, kids," she said, and then brought down her sunglasses a bit so we could see her eyes. "I didn't see anyone floating."

She had given us a dare. One by one, we slowly entered Great Salt Lake. Gradually, we would lean backward into the hands of the cool water and find ourselves being held by the very lake that minutes before had betrayed us. For hours we floated on our backs, imprinting on Great Basin skies. It was in these moments of childhood that Great Salt Lake flooded my psyche.

Driving home, Mother asked each of us what we thought of the lake. None of us said much. We were too preoccupied with our discomfort: sunburned and salty, we looked like red gumdrops. Our hair felt like steel wool, and we smelled. With the lake so low and salinity around 26 percent, one pound of salt to every four pounds of water (half a gallon), another hour of floating in Great Salt Lake and we might have risked being pickled and cured.

Brooke brought me back a handful of feathers and sat behind me. I leaned back into his arms. Three more days until Mother's surgery.

The family spontaneously gathered at Mother's and Dad's; children, spouses, grandparents, and cousins. We sat on the lawn, some talked, others played gin rummy, while Mother planted marigolds in her garden.

Mother and I talked.

"I don't want you to be disappointed, Terry."

"I won't be," I said softly. My hands patted the earth around each flower she planted.

"It's funny how the tears finally leave you," she said, turning her trowel in the soil. "I think I've experienced every possible emotion this week."

"And how do you feel now?" I asked.

She looked out at the lake, wiped her forehead with the back of her gardening glove, and removed more marigolds from the flat.

"I'll be glad to have the operation behind me. I'm ready to get on with my life."

Dad mowed the lawn between clumps of relatives. It felt good to be outside, to feel the heat, and to hear the sounds of neighborhoods on Saturdays in the spring.

The sun set behind Antelope Island. Great Salt Lake was a mirror on the valley floor. One had the sense of water being in this country now, as the quality of light was different lending a high gloss to the foothills.

At dusk, we moved inside to the living room and created a family circle. Mother sat on a chair in the center. As the eldest son, Steve annointed Mother with consecrated olive oil to seal the blessing. The men who held the Melchizedek Priesthood, the highest order of authority bestowed upon Mormon males, gathered around her, placing their hands on the crown of her head. My father prayed in a low, humble voice, asking that she might be the receptacle of her family's love, that she might know of her influence in our lives and be blessed with strength and courage and peace of mind.

Kneeling next to my grandmother, Mimi, I felt her strength and the generational history of belief Mormon ritual holds. We can heal ourselves, I thought, and we can heal each other.

"These things we pray for in the name of Jesus Christ, amen."

Mother opened her eyes. "Thank you . . ."

My sister-in-law, Ann, and I slipped into the kitchen to prepare dinner.

Some things don't change. After everyone had eaten, attention shifted to the weather report on the ten o'clock news, a Western ritual, especially when your livelihood depends on it as ours does. A family construction business, now in its fourth generation, has taught me to look up before I look down. You can't lay pipe when the ground is frozen, neither can you have crews digging trenches in mud.

The weatherman not only promised good weather, but announced that most of the planet would be clear tomorrow according to the satellite projection—a powerful omen in itself.

After everyone left, I asked Mother if I could feel the tumor. She lay down on the carpet in the family room and placed my hand on her abdomen. With her help, I found the strange rise on the left side and palpated my fingers around its perimeter.

With my hands on my mother's belly, I prayed.

We wait. Our family is pacing the hall. Other families are pacing other halls. Each tragedy has its own territory. A Tongan family in the room next to Mother's sings mourning songs for the dying. Their melancholy sweeps over us like the shadow of a raven. What songs would we sing, I wonder. Two doors down, a nurse calls for assistance in turning a patient over on a bed of ice. Minutes later, I hear the groaning of the chilled woman.

It has been almost four hours. For most of the time, I have been sitting with my mother's parents. My grandmother, Lettie, is in a wheelchair. She suffers from Parkinson's disease. Her delicate hands tremble as she strokes my hair. I am leaning against the side of her knee. She and my grandfather, Sanky, are heartsick. Mother is their only daughter; one of their two sons is dead. Mother has always cared for her parents. Now that she needs their help, Lettie feels the pain of a mother unable to physically attend to her daughter.

The three doctors appear: Smith, Smith, and Johnson, green-robed and capped. Dad meets them halfway, cowboy boots toe-to-toe with surgical papered shoes. I try to read lips as he receives the bad news followed by the good news.

"Yes, it was malignant. No we didn't get it all, but with the chemotherapy we have to offer, there is reason to be hopeful." The doctors say they will meet with us in a couple of days when they get the pathology report back, then they will go over specific details and options with Mother and the family.

Dad—tall, rugged, and direct—asks one question. "What's the bottom line—how much time do we have?"

The doctors meet his narrow blue eyes. Gary Smith shakes his head. "We can't tell you that. No one can."

The curse and charisma of cancer: the knowledge that from this point forward, all you have is the day at hand.

Dad turned around defeated, frustrated. "I'd like to get some answers." His impatience became his stride as he walked back down the hall.

Bad news is miraculously accommodated. With one hope dashed—the tumor was malignant (an easier word to stomach than cancer)—another hope is adopted: the chemotherapy will cure. Now all we had to do was convince Mother. We made a pact among ourselves that we would not discuss anything with her until the next morning. We wanted her to rest.

Two orderlies wheel Mother back into her room. The tubes, bags, blood, and lines dangling from four directions did not foster the hope we were trying to sustain. Our faith faltered in the presence of her face—white, wan, and weakened. Dad whispered that she looked like a skinned deer.

Mother opened her eyes and faintly chuckled, "That bad, uh?"

No one else laughed. We just looked at one another. We were awkward and ill-prepared.

Dad took Mother's hand and spoke to her reassuringly. He tried stroking her arm but quickly became frustrated and frightened by all the tubing connected to her veins. He sat with her as long as he could maintain his composure and then retreated to the hall where his parents, Mimi and Jack, were standing by.

Steve, Dan, and Hank took over, each one nursing her in his own way.

"Don't worry about fixing dinner for Dad, tonight, Mom, we'll take care of him," said Steve.

Dan walked out of the room and came back with a cup of ice chips. "Would you like to suck on these, Mother? Your mouth looks dry."

Hank, sixteen, stood in the corner and watched. Mother looked at him and extended her hand. He walked toward her and took it.

"Love you, Mom."

"I love you, too, dear," she whispered.

My brothers left the room. I stood at the foot of her bed, "How are you feeling, Mother?"

It was a hollow question, I knew, but words don't count when words don't matter. I moved to her side and stroked her forehead. Her eyes pierced mine.

"Did they get it all?"

I blinked and looked away.

"Did they, Terry? Tell me." She grabbed my hand.

I shook my head. "No, Mother."

She closed her eyes and I watched the muscles in her jaw tighten.

"How bad is it?"

Dad walked in and saw the tears streaming down my cheeks. "What happened?"

I shook my head again, left the room and walked down the hall. He followed me and took hold of my shoulder.

"You didn't tell her, did you?"

I turned around, still crying, and faced him. "Yes."

"Why? Why, when we agreed not to say anything until tomorrow? It wasn't your place." His anger flared like the corona of an eclipsed sun.

"I told her because she asked me, and I could not lie."

The pathologist's report defined Mother's tumor as Stage III epithelial ovarian cancer. It had metastasized to the abdominal cavity. Nevertheless, Dr. Gary Smith believes Mother has a very good chance against this type of cancer, given the treatment available. He is recommending one year of chemotherapy using the agents Cytoxan and cisplatin.

Before surgery, Mother said no chemotherapy.

Today, I walked into her room, the blinds were closed.

"Terry," she said through the darkness. "Will you help

me? I told myself I would not let them poison me. But now I am afraid not to. I want to live."

I sat down by her bed.

"Perhaps you can help me visualize a river—I can imagine the chemotherapy to be a river running through me, flushing the cancer cells out. Which river, Terry?"

"How about the Colorado?" I said.

It was the first time in weeks I had seen my mother smile.

June 1, 1983. Mayor Ted Wilson has ordered the channeling of three mountain streams, Red Butte, Emigration, and Parley's, into a holding pond at Liberty Park near the center of town. From Liberty Park, the water will be funneled into the Jordan River, which will eventually pour into Great Salt Lake.

Normally, these three Wasatch Front rivers converge underground in an eighty-inch pipe, but when the pipe gets too full, it blows all the manhole covers sky high, causing massive flooding on the streets. It's called "Project Earthworks."

Yesterday's temperature was sixty-two degrees Fahrenheit. Today it is ninety-two. All hell is about to break loose in the mountains. A quick thaw is a quick flood.

Ten days have passed and, between all of us, we have kept vigil. Mother's strength is returning and with typical wit, she hinted that a bit of privacy might be nice. I took her cue and drove out to the Bird Refuge.

It looked like any other spring. Western kingbirds lined the fences, their yellow bellies flashing bright above the barbed wire. Avocets and stilts were still occupying the same shallow ponds they had always inhabited, and the white-faced glossy ibises six miles from the Refuge were meticulously separating the grasses with their decurved bills. Closer in, the alkaline flats, usually dry, stark, and vacant, were wet. A quarter mile out, they were flooded.

The Bear River Migratory Bird Refuge at an elevation of 4206', was two feet from being inundated. I walked out as far as I could. It had been a long time since I had heard the liquid songs of red-wing blackbirds.

"*Konk-la-ree! Konk-la-ree! Konk-la-ree!*"

The marsh was flooding. The tips of cattails looked like snorkels jutting a few inches above water. Coots' nests floated. They would fare well. With my binoculars, I could see snowy egrets fishing the small cascades that were breaking over the road's asphalt shoulders.

I could not separate the Bird Refuge from my family. Devastation respects no boundaries. The landscape of my childhood and the landscape of my family, the two things I had always regarded as bedrock, were now subject to change. Quicksand.

Looking out over the water, now an ocean, I felt foolish for standing in the middle of what little road was left. Better to have brought a canoe. But I rolled up my pantlegs over the tops of my rubber boots and continued to walk. I knew my ground.

Up ahead, two dozen white pelicans were creating a spiral staircase as they flew. It looked like a feathered DNA molecule. Their wings reflected the sun. The light shifted, and they disappeared. It shifted again and I found form. Escher's inspiration. The pelicans rose higher and higher on black-tipped wings until they straightened themselves into an arrow pointing west to Gunnison Island.

To my left, long-billed dowitchers, stout and mottled

birds, pattered and probed, pattered and probed, perforating the mud in masses. In an instant, they flew, sweeping the sky as one great bird. Flock consciousness.

I turned away from the water and walked east toward the mountains. Foxtails by the roadside gathered light and held it. Dry stalks of rumex, russet from last year's fall, drew hunger pangs—the innocence of those days.

Before leaving, I noticed sago pondweed screening shallow water near the edge of the road. Tiny green circles of cholorophyll were converting sunlight to sugar. I knelt down and scooped up a handful. Microscopic animals and a myriad of larvae drained from my hands. Within seconds, the marsh in microcosm slipped through my fingers.

I was not prepared for the loneliness that followed.

ʳʳʳʳʳʳʳʳ

SNOWY EGRETS

ʳʳʳʳʳʳʳʳ

lake level: 4204.05'

I caught myself staring out the window again. Last time I looked at the clock it was 11:20 A.M. Now it is 12:30.

From my third floor office at the Utah Museum of Natural History, I look out over roofs and watch a pair of kestrels flying in and out of the cottonwoods. They have a nest nearby. Beyond them loom the Wasatch. Their peaks still hold the snow in early summer.

I was with Mother yesterday during her first chemotherapy treatment. Her fear and resistance did not help. Resignation, I suppose, would be worse. I held her forehead as she writhed, wretched, and heaved. She would cry and I would cry with her. I just kept saying, "Let it go, Mother, we'll get through this." At one point, after the nurses left, I got into bed with her and held her close to absorb her trembling body. She was so cold, even blankets wouldn't help. Dr.

Smith said the first treatment is the most severe, especially since she is still recovering from the surgery.

My desk is heaped high with papers, pink notes, and mail; bureaucratic accumulation from my "vacation time."

The phone rings—I don't answer.

My eyes focus on a plate full of shells I brought home from Mexico. It has sat on my windowsill during the three years I have worked here. I pull the crab claw out from under the pink murex. It repulses me. This is cancer, my mother's process, not mine.

The disengaged limb holds me, haunts me. I can't let it go. There is something in my resistance that warrants attention.

Cancer. The word has infinite power. It kills us with its name first, because we have allowed it to become synonymous with death.

The Oxford English Dictionary defines cancer as "anything that frets, corrodes, corrupts, or consumes slowly and secretly."

A person who is told she has cancer faces a hideous recognition that something monstrous is happening within her own body.

Cancer becomes a disease of shame, one that encourages secrets and lies, to protect as well as to conceal.

And then suddenly, within the rooms of secrecy, patient, doctor, and family find themselves engaged in war. Once again, medical language is loaded, this time with military metaphors: the fight, the battle, enemy infiltration, and defense strategies. I wonder if this kind of aggression waged against our own bodies is counterproductive to healing? Can we be at war with ourselves and still find peace?

How can we rethink cancer?

It begins slowly and is largely hidden. One cell divides

into two; two cells divide into four; four cells divide into sixteen . . . normal cells are consumed by abnormal ones. Over time, they congeal, consolidate, make themselves known. Call it a mass, call it a tumor. It surfaces and demands our attention. We can surgically remove it. We can shrink it with radiation. We can poison it with drugs. Whatever we choose, though, we view the tumor as foreign, something outside ourselves. It is however, our own creation. The creation we fear.

The cancer process is not unlike the creative process. Ideas emerge slowly, quietly, invisibly at first. They are most often abnormal thoughts, thoughts that disrupt the quotidian, the accustomed. They divide and multiply, become invasive. With time, they congeal, consolidate, and make themselves conscious. An idea surfaces and demands total attention. I take it from my body and give it away.

I pick up the crab claw and put it in my pocket. I can hardly wait to tell Mother.

The phone rings again—this time I answer.

"Museum education, may I help you?"

It is someone calling about the upcoming film series entitled, "The Gentle Earth," a confirmation that Toby McLeod's film, "Four Corners: A National Sacrifice Area," is available for a Salt Lake City premiere.

Good news. But I need to convince our director that it is in the museum's best interest to sponsor a film about uranium tailings in Navajoland. Nothing inspires me more than a little controversy. We are in the business of waking people up to their surroundings. A museum is a good place to be quietly subversive on behalf of the land.

I close my door and begin to plot my strategy.

Downtown; the North Temple storm pipe that handles City Creek had been doing fine all week in spite of an increase in water flow from an average 50 cubic feet per second to 375 cubic feet per second. (The previous record was 90 cubic feet per second.) But the rocks, silt, and debris had caused a dense mass like concrete to form. The water had had no place to go and, consequently, it was backing up onto city streets. Mountain Bell Communications Systems and the LDS Church Office Building were in immediate danger of flooding.

The mayor, Ted Wilson, telephoned President Gordon B. Hinckley, an apostle of the Mormon Church. His request: "Empty the ward houses."

"But it's the sabbath—" Hinckley replied.

"We need your help. City Creek has literally become unglued, and a two foot wall of water is charging through Memory Grove. The Church Office Building could be next. Your genealogical records . . ." Within ten minutes, Mormon chapels across the Salt Lake Valley were vacated.

"Go home and change your clothes—we've got a flood on our hands . . ." was the message given over the pulpit.

Mayor Wilson received a call back from Hinckley.

"The ox is in the mire."

By 2:00 P.M., thousands of volunteers with their shirtsleeves rolled up, Mormons and non-Mormons alike, lined State Street which runs north and south for miles down the heart of the city. Within hours, State Street was transformed into a river.

Ted Wilson, on the news last night called it "a victimless war." Mother and I watched it from her hospital room, knowing the men in our family were part of the community throng building the three-foot walls of sandbags.

Dad, Steve, Dan, and Hank wandered in around ten o'clock in muddy Levi's and great spirits. Brooke followed shortly after.

"You should see it on the streets, Diane!" Dad said. "It's incredible. Sandbags were delivered. The city engineers had envisioned the plan, but there was no midlevel management to execute it."

"So, let me guess," I chided Dad, "you became General Patton."

"Not exactly," he said smiling, "but sort of. We made a line from the truck to the street, sandbags being passed left to right, left to right, for what seemed like hours. Then the volunteers on the street judged the grade and built the banks accordingly. Everyone brought their own ideas how it could best be done. There was total cooperation."

Dad sat down on the edge of Mother's hospital bed. "When the water was finally released from City Creek and began flowing down State Street, you should have heard the cheers. Cries from the crowd followed the water block after block like a wave."

"All I know," said Hank, "is that it was a great way to get out of church."

The sandbag banks held City Creek for almost three miles. In some places, the water was three feet deep. Where cars once drove, fish swam. Where pedestrians once crossed, bridges now spanned. A car bridge between the city blocks of 500 and 600 South was erected for the price of seventy thousand dollars—no small risk financially, for a mayor who saw his town being truncated, cut in half by flooding and not having a clue how long it might last. But his hunch paid off. The city kept moving in spite of the floods. And the State Street River kept flowing.

The flooding of Salt Lake City lifted everyone's spirits. People went fishing. Signs saying YOU CATCH 'EM—WE'LL

COOK 'EM were posted in front of State Street restaurants. A few trout were caught and fried.

A bride and groom exchanged vows on the bridge. They later walked arm in arm into the Alta Club for their wedding breakfast. A crowd followed them throwing rice.

My favorite innovations were made by the kayakers who complained about having to portage around the city-block bridges and made local officials promise to build rialtos next time with appropriate clearance. Class-three rapids were reported between South Temple and 100 South.

July 1, 1983. Great Salt Lake has risen 5.1' since September 18, 1982, the greatest seasonal rise ever recorded. And it's still rising. Hal Cannon, a folklorist, and I drive out to the Bird Refuge to see how the marsh is faring. We decided to swap expertise: he would fill me in on noteworthy collectibles at the Deseret Industries, a Mormon thrift shop, if I would take him to Bear River to watch birds.

Inside the Deseret Industries in Brigham City, Hal looks for glass grapes, any color. They were made by every Mormon woman in Relief Society, the women's auxiliary organization, during the 1960s (my own mother included). Boxes of glass balls were set on top of banquet tables in the Cultural Hall. You could pick your color and size. Turquoise, amber, red, and purple seemed to be the most popular. And then you could choose from dainty glass grapes to balls the size of silver dollars. Each woman was provided with a stick, which served as the bunch stalk. These were painted brown, then shellacked. Green leaves made out of silk were added next with copper wires curled into tendrils. The last step was to glue the glass balls together until you had your bunch of grapes. This seemed to be where the women ran into problems—they didn't know when to stop. Some of the glass ball masterpieces flowed halfway down the tables, looking like mutant clusters of salmon eggs. The women ended up carrying them to their cars in both hands. Coffee tables at home were in danger of collapsing under their weight. Every home had one, whether the women liked them or not. It was a symbol of craft adeptness, an important tenet of Mormonism.

Mother wasn't great with crafts. Her grapes, an amber cluster with glass balls the size of quarters, were modest.

"I was just glad to get it done so I could go home," I remember her saying. Nevertheless, they stayed on the bookshelf in the kitchen for years, until dust finally obscured their luster and a new fad like gingham geese replaced them.

No grapes are to be found on this trip to the Deseret Industries. Instead, Hal finds an old tweed coat that fits his husky frame like a glove. I splurge and buy a pink cashmere sweater with pearled and sequined flowers down the front for five dollars.

We drive through the flooding Bird Refuge in Hal's turquoise Comet convertible. It is the perfect birdwatching vehicle. Dozens of avocets and stilts fly over us, flocks of ibises fly alongside. Gulls are everywhere. I love seeing their bellies. (Hal reminds me we need umbrellas.) We are in an avian parade traveling west. I threaten to crawl into the back seat, perch on top of the trunk, and make figure eight waves to all the marsh like a float queen.

Luckily, two snowy egrets fly over us and distract me. My dignity is preserved.

We park the car and walk to the edge of cattails. Hunkering down, we separate the stalks with our fingers and find the egrets. I nudge Hal. One egret spears a small frog. A blink and we would have missed it. We watch them walk along the periphery of the pond in their "golden slippers." Snowy egrets have yellow feet.

We have lost track of time in a birdwatchers' trance. Egret plumes like French lace billow in the breeze and underscore their amorous play. One egret rises, the other follows. Their steps are light and buoyant. Hal leans toward me and softly hums an Irish folktune. The two egrets stagger their leaps—one lifts, one lands, one lifts, one lands—and the dance continues.

BARN SWALLOWS

lake level: 4204.75'

What is it about the relationship of a mother that can heal or hurt us? Her womb is the first landscape we inhabit. It is here we learn to respond—to move, to listen, to be nourished and grow. In her body we grow to be human as our tails disappear and our gills turn to lungs. Our maternal environment is perfectly safe—dark, warm, and wet. It is a residency inside the Feminine.

When we outgrow our mother's body, our cramps become her own. We move. She labors. Our body turns upside down in hers as we journey through the birth canal. She pushes in pain. We emerge, a head. She pushes one more time, and we slide out like a fish. Slapped on the back by the doctor, we breathe. The umbilical cord is cut—not at our request. Separation is immediate. A mother reclaims her body, for her own life. Not ours. Minutes old, our first death is our own birth.

Mother and I are in Wyoming. The quaking aspens are ablaze like the bright light of a burning match. We walk along the Gros Ventre River with the Tetons behind us. She gave me my birth story: what she experienced during her pregnancy, what the birthing was like, and how she felt when she held me for the first time.

"I don't ever remember being so happy, Terry. Having a child completed something for me. I can't explain it. It's something you feel as a woman connected to other women."

She paused.

I asked her if she thought my life was selfish without children.

"Yes," she said. "But I'm not saying that's bad. By being selfish a woman ultimately has more to give in the long run, because she has a self to give away."

"Do you think I should have a child?" I asked.

"I can't answer that for you," she said. "All I can tell you is that it was the right choice for me."

Across the river, Mother and I watch two elk. Bulls in the midst of their harems. She says they are eating. I say their antlers are locked and they are sparring.

"You have the most vivid imagination," she says. "Let me see your binoculars." She pulls them up to her eyes. "Okay, I'll give you this one."

Walking back to our family's place, we are seized by the alpenglow, a cradle of pink light. The willows are rust and maroon, the mountains purple. Trumpeter swans float above their reflections on the river. A pair of bald eagles fly across the face of the Tetons. Their heads seemed brighter than the promise of snow.

The next day, we awake at dawn and travel once again down to the river bottoms. We watch a herd of pronghorn antelope grazing on the moraine. A buck flares his fanny at us.

"Am I imagining this one?" I turn to Mother and hand her the binoculars.

"Do you blame him?" Mother replies. "We are beautiful women."

This afternoon, I have found quiet hours alone picking tomatoes. As my fingers find ripe tomatoes, red and firm, through the labyrinth of leaves, I am absorbed into the present. My garden asks nothing more of me than I am able to give. I pull tomatoes, gently placing them in the copper colander. Pulling tomatoes. Pulling tomatoes. Some come easily.

Tonight I watched the sun sink behind the lake. The clouds looked like rainbow trout swimming in a lapis sky. I can honor its beauty or resent the smog in this valley which makes it possible. Either way, I am deceiving myself.

Mother has completed her sixth month of chemotherapy. In some ways, it is easy to become complacent, to take life for granted all over again. I welcome this luxury. I have the feeling Mother is living in the heart of each day. I am not.

Buddha says there are two kinds of suffering: the kind that leads to more suffering and the kind that brings an end to suffering.

I recall a barn swallow who had somehow wrapped his tiny leg around the top rung of a barbed-wire fence. I was walking the dikes at Bear River. When I saw the bird, my first instinct was to stop and help. But then, I thought, no, there is nothing I can do, the swallow is going to die. But I could not leave the bird. I finally took it in my hands and unwrapped it from the wire. Its heart was racing against my fingers. The swallow had exhausted itself. I placed it among the blades of grass and sat a few feet away. With each breath, it threw back its head, until the breaths grew fainter and fainter. The tiny chest became still. Its eyes were half closed. The barn swallow was dead.

Suffering shows us what we are attached to—perhaps the umbilical cord between Mother and me has never been cut. Dying doesn't cause suffering. Resistance to dying does.

rrrrrrrr

PEREGRINE FALCON

rrrrrrrr

lake level: 4205.40'

Not far from Great Salt Lake is the municipal dump. Acres of trash heaped high. Depending on your frame of mind, it is either an olfactory fright show or a sociological gold mine. Either way, it is best to visit in winter.

For the past few years, when the Christmas Bird Count comes around, I seem to be relegated to the landfill. The local Audubon hierarchy tell me I am sent there because I know gulls. The truth lies deeper. It's an under-the-table favor. I am sent to the dump because secretly they know I like it.

As far as birding goes, there's often no place better. Our urban wastelands are becoming wildlife's last stand. The great frontier. We've moved them out of town like all other "low-income tenants."

The dump where I count birds for Christmas used to have cattails—but I can't remember them. A few have popped up

54

below the hill again, in spite of the bulldozers, providing critical cover for coots, mallards, and a variety of other waterfowl. I've seen herons standing by and once a snowy egret, but for the most part, the habitat now is garbage, perfect for starlings and gulls.

I like to sit on the piles of unbroken Hefties, black bubbles of sanitation. It provides comfort with a view. Thousands of starlings cover refuse with their feet. Everywhere I look—feathered trash.

The starlings gorge themselves, bumping into each other like drunks. They are not discretionary. They'll eat anything, just like us. Three starlings picked a turkey carcass clean. Afterward, they crawled inside and wore it as a helmet. A carcass with six legs walking around—you have to be sharp counting birds at the dump.

I admire starlings' remarkable adaptability. Home is everywhere. I've seen them nesting under awnings on New York's Fifth Avenue, as well as inside aspen trunks in the Teton wilderness. Over 50 percent of their diet is insects. They are the most effective predators against the clover weevil in America.

Starlings are also quite beautiful if looked at with beginner's eyes. In autumn and winter, their plumage appears speckled, unkempt. But by spring, the lighter tips of their feathers have been worn away, leaving them with a black, glossy plumage, glistening with irridescences.

Inevitably, students at the museum will describe an elegant, black bird with flashes of green, pink, and purple. "About this big," they say (holding their hands about seven inches apart vertically). "With a bright yellow bill. What is it?"

"A starling," I answer.

What follows is a dejected look flushed with embarrassment.

"Is that all?"

The name precedes the bird.

I understand it. When I'm out at the dump with starlings, I don't want to like them. They are common. They are aggressive, and they behave poorly, crowding out other birds. When a harrier happens to cross-over from the marsh, they swarm him. He disappears. They want their trash to themselves.

Perhaps we project on to starlings that which we deplore in ourselves: our numbers, our aggression, our greed, and our cruelty. Like starlings, we are taking over the world.

The parallels continue. Starlings forage by day in open country competing with native species such as bluebirds for food. They drive them out. In late afternoon, they return in small groups to nest elsewhere, competing with cavity nesters such as flickers, martins, tree swallows, and chickadees. Once again, they move in on other birds' territories.

Starlings are sophisticated mimics singing songs of bobwhites, killdeer, flickers, and phoebes. Their flocks drape bare branches in spring with choruses of chatters, creeks, and coos. Like any good impostor, they confuse the boundaries. They lie.

What is the impact of such a species on the land? Quite simply, a loss of diversity.

What makes our relationship to starlings even more curious is that we loathe them, calling in exterminators because we fear disease, yet we do everything within our power to encourage them as we systematically erase the specialized habitats of specialized birds. I have yet to see a snowy egret spearing a bagel.

The man who wanted Shakespeare's birds flying in Central Park and altruistically brought starlings to America from England, is not to blame. We are—for creating more and more habitat for a bird we despise. Perhaps the only value in the multitudes of starlings we have garnished is that in some small way they allow us to comprehend what vast flocks of birds must have felt like.

The symmetry of starling flocks takes my breath away, I lose track of time and space. At the dump, all it takes is the sweep of my hand. They rise. Hundreds of starlings. They wheel and turn, twist and glide, with no apparent leader. They are the collective. A flight of frenzy. They are black stars against a blue sky. I watch them above the dump, expanding and contracting along the meridian of a winged universe.

Suddenly, the flock pulls together like a winced eye, then opens in an explosion of feathers. A peregrine falcon is expelled, but not without its prey. With folded wings he strikes a starling and plucks its body from mid-air. The flock blinks again and the starlings disperse, one by one, returning to the landfill.

The starlings at the Salt Lake City municipal dump give us numbers that look good on our Christmas Bird Count, thousands, but they become faceless when compared to one peregrine falcon. A century ago, he would have seized a teal.

I will continue to count birds at the dump, hoping for under-the-table favors, but don't mistake my motives. I am not contemplating starlings. It is the falcon I wait for—the duckhawk with a memory for birds that once blotted out the sun.

rrrrrrrr

WILSON'S PHALAROPE

rrrrrrrr

lake level: 4206.15'

In 1975, the Utah State Legislature passed a law stating Great
Salt Lake could not exceed 4202'. Almost ten years later, at
lake level 4206.15', Great Salt Lake is above the law. What
lasso can you use to corral the West's latest outlaw?
      The State of Utah is reviewing its options to control the
lake. They have come up with five alternatives:

Option One: Breaching the Causeway
The Southern Pacific Railroad Causeway, built in 1957,
divides Great Salt Lake in two, running west from Promontory Point to the Lakeside Mountains. The rock-fill structure spans almost thirteen miles. All freshwater inflow to the
lake enters the lake's south arm. Two fifteen-foot culverts
located in the middle of the causeway allow water to move
from the south arm to the north, but at a rate much lower
than that of the inflow to the south arm. Consequently, the

elevation difference between the south and north arms of Great Salt Lake is almost four feet. For this reason, the salinity in the southern arm is less, which affects the brine shrimp and algae populations.

If the causeway were breached with a larger opening, the level of the south arm could be reduced by one foot, buying enough time for the wet weather to subside and allowing Great Salt Lake to flow back to its original shape. Estimated cost: $3,000,000.

Option Two: Store the Water
If there's a water problem in the West, build a dam. The Bear River is the largest tributary of Great Salt Lake, responsible for 60 percent of all stream inflow. Dam it. Store it. Create a reservoir. Nine different reservoir sites are under consideration, but preliminary studies reveal the maximum possible storage would yield only three hundred thousand acre-feet and have only a minimal effect on the flooding problems.
Estimated cost: $100,000,000 plus.

Option Three: Divert the Water
Since the Bear River was diverted from the Snake River by a volcanic dam some twenty million years ago, why not simply reroute it back to its original path? Never mind the politics of water rights, state boundaries, and engineering logistics.
Estimated cost: $200,000,000.

Option Four: Diking
Protective diking along the shore of Great Salt Lake, from the town of Corrine in Box Elder County to the north to Interstate 80 in Tooele County south, seems like a logical solution.

Estimated cost: $500,000,000 plus.

Selective diking as a short-term solution to protect critical public facilities such as wastewater treatment plants, interstate highways, and the airport, is already underway.

A second diking concept, a long-term solution, would build a dike on the existing causeway which connects the northern end of Antelope Island to the mainland community of Syracuse. A second dike would connect the southern tip of Antelope Island to Interstate 80.

Other interisland diking would include linking Promontory Point, Fremont Island, and Antelope Island—a dot-to-dot exercise in hydrological engineering.

This project would require a large pumping plant to remove the inflows of the Bear, Weber, and Jordan Rivers so that an acceptable water level in the impounded areas could be maintained.

Estimated cost: $250,000,000.

Option Five: West Desert Pumping Project
Originally Brigham Young's idea, it was first investigated by the Army Corps of Engineers in 1976. The plan proposed to dike off Great Salt Lake near Lakeside (on the western shore of the lake) and pump water over the dike, letting it flow naturally into the West Desert. It was determined unfeasible because it threatened the United States Air Force bombing range. We would be flooding a critical national defense facility.

A pumping project would have to be devised that would lift lake water over Hogup Mountain Ridge, to the desert west of the Newfoundland Mountains, causing only minimal effect on the bombing reservation. This would mean the water would have to be released high enough so that it could flow by gravity into an evaporation pond, and then back again to Great Salt Lake.

The West Desert Pumping Project is being looked at as an extreme measure in the event that the unprecedented wet period continues into subsequent years.

Estimated cost: $90,000,000.

It was decided after much debate on the House and Senate floors that breaching the Southern Pacific Railroad Causeway would give the most immediate relief for the least money.

House Bill 30 was passed, which provides $3.5 million dollars to construct a three-hundred-foot opening. The contract is to begin immediately.

Evidently, to do nothing is not an option.

"If someone would have told me one year ago, I would be going through eleven months of chemotherapy for ovarian cancer, I would never have believed them," Mother said. "Now that it's over—I don't know how I managed. It's funny what you can will yourself through when you have to."

We were on our way to lunch to celebrate Mother's birthday. March 7, 1932. She was fifty-two years old.

"What would you tell your children of me?" Mother asked after we had seated ourselves in the restaurant at Hotel Utah.

I unfolded my napkin and placed it on my lap. I didn't want to think about such things.

"I'll let you tell them for yourself," I answered, taking a sip of water.

She paused and placed her napkin on her lap.

"Tell them I am the bird's nest behind the waterfall. Yes, tell them that."

Great Salt Lake has swallowed the causeway that led to Antelope Island. Gone. The road has been erased. Gentle waves cross over each other from north and south. A sign half submerged reads, SPEED LIMIT 45 m.p.h. It must apply to the birds. Thirty miles to the north, the Bird Refuge is underwater.

Three men from Parks and Recreation are removing the last boat slips. I'm sitting on one of the wooden frames as the crane removes others to my left. I ask if I am bothering them.

"You can sit out here as long as you like, lady," the foreman replies. "In fact, you can walk out as far as your heart will carry you . . ."

The men go about their business. Ten white pelicans glide over us, their wing beats slow and deliberate. One of the men looks up, looks at me, and raises his eyebrows. Five avocets fly north. A cluster of cinnamon teals wing south. Coots, eared grebes, and gulls float on the lake as Wilson's phalaropes pirouette in the water.

The men look bored with their work. One of the workers looks over his shoulder and throws two wooden slats in a pile.

"Did you know the female phalarope, the small bird between the slips, wears the bright plumage in this species, not the male?"

No one comments.

"And see how she keeps twirling around?" I say, keeping my binoculars on the birds.

"Let me guess . . ." replies one of the men. "Someone wound it up and lost the key."

"It's their way of stirring up food lodged at the bottom of the lake. Their feet create a whirlwind from which they can feed. They don't take in as much salt that way either.

I've even heard that some phalaropes have been seen spinning at sixty revolutions per minute."

"Who in the hell would spend their life counting how many times a bird spins around?" one of the men asks incredulously.

"That's what I do for a living," I said.

The three men stop working and stare. This time, I laugh.

"So, what do you really think about the governor wanting to build a new causeway to the island?"

"Me?" asks the employee who noticed the pelicans. "I just work here."

I tell him his eyes don't look like he just works here. He grins. He reminds me of my brothers.

"Between you and me, they ought to just let the lake do its thing. It will anyway. It always does. They can come back and rebuild the road on another day when it decides to recede."

He pauses after he finishes breaking up some asphalt. "It changes out here every week. In January, you could drive out to the island. Now, well, now you need a boat or wings," he says, chuckling. "And where you stand today, a week from now you would be knee-deep in brine."

Another worker agrees. "Last month, this was all mud-flats. Every day we watch the lake eat another chunk of the road."

"You never know what you're going to find out here. Last February, I saw icebergs the size of pickup trucks floating on the lake. And a month from now, it'll be a buggy nightmare. This lake attracts flies like a magnet attracts iron shavings. Best to go home, it's so hot and miserable."

The foreman adds, "Boil or freeze at the Great Salt Lake, and if the weather doesn't kill you, the brine flies will . . ." He hums to a tune from the Grateful Dead.

Two curlews fly overhead. Three great blue herons,

evenly spaced, fish along shore. As I hop down from the boat slip, the lake laps around my ankles.

The foreman turns around. "No matter what they tell you on the news, the lake's still risin'."

I leave the men at the causeway and find a more remote vantage point of Antelope Island on dry ground. From where I sit now, it looks like a large buckskinned animal sleeping on its side. The rural country with monarchs on milkweeds on the eastern shore of Great Salt Lake almost allows me to believe this is a calm and predictable place.

I watch the island intently with my binoculars, scanning the shoreline, noticing where beaches end and outcroppings of stone begin. The island appears still and serene, but I know better. Buffalo live here. I have also seen deer and coyotes. Vultures clip the ridgelines in search of carrion, and the wind is always present. But in spite of the human hand, Antelope Island remains remarkably pristine. A state park claims the northern tip with a few facilities for tourists, but with the causeway submerged, it becomes wild and uninterrupted country once again.

The pulse of Great Salt Lake, surging along Antelope Island's shores, becomes the force wearing against my mother's body. And when I watch flocks of phalaropes wing their way toward quiet bays on the island, I recall watching Mother sleep, imagining the dreams that were encircling her, wondering what she knows that I must learn for myself. The light changes, Antelope Island is blue. Mother awakened and I looked away.

Antelope Island is no longer accessible to me. It is my mother's body floating in uncertainty.

Mother goes in tomorrow for a "second look" to see if the chemotherapy has been effective. When the original tumor was removed a year ago, cancer cells peppered her small intestines. Dr. Smith will perform a laparoscopy and take tissue samples to biopsy, hoping all is clear. He feels very positive and is looking toward a possible cure.

Mother's strength is returning. She and Dad have been playing tennis again. He tells us her serve is as wicked as ever.

We are all anxious, except Mother. She says it doesn't matter what they find, all we have is now.

rrrrrrrr

CALIFORNIA GULLS

rrrrrrrr

lake level: 4207.75'

"Everything looks good," said Dr. Smith, practically dancing as he entered the room. "All I saw was healthy pink tissue. I couldn't be more pleased."

"So you really think Diane might be clear?" Dad asked. "No more cancer?"

"We won't know for sure until we get the pathology report back on Wednesday, but let's just say, I'm cautiously optimistic. They should be bringing her back to the room any time now. I'll see you on Wednesday."

Dr. Smith left.

Dad and I looked at each other.

"We heard it right? Didn't we?"

"Can you believe it?" I cried. "It's a miracle. I knew Mother could do it. I'll call Steve. Dan and Hank are still in school."

"I'll call the rest of the family," Dad said.

We hurried back to Mother's room to greet her just as the orderlies were wheeling her back. We met her in the hall with thumbs up.

"Honest?" she said a bit drowsy. "Do you promise? Everything was clear?"

Dad and I nodded with tears. She took Dad's hand and wouldn't let it go. At this moment, I realized how badly Mother wanted to live.

She looked at us again. "Really? Everything looked good?"

"Healthy pink tissue," I answer. "Dr. Smith said he couldn't be more pleased."

"I'm almost afraid to believe you—to let myself go," she said. "I just want to sleep, I am so tired. I want to sleep and dream and relax, something I have not allowed myself to do." She took a deep breath, closed her eyes, and sighed. "I can't believe it."

The nurse brought in a bouquet of spring flowers sent by friends.

"Oh, have you ever seen flowers so beautiful?" Mother said. "Bring them closer, please."

Dr. Smith requested that our family meet with him in Mother's room this afternoon. It didn't make sense.

"Why would he want us all here just to reiterate the good news?" Steve asked.

Dad was pacing the halls. "Something's gone wrong," he said. "I can feel it."

Mother was quiet. "I can't believe I was so stupid," she said. "I should never have allowed myself to believe you."

I was sick to my stomach. Dan and Hank sat on the edge of Mother's bed. Brooke leaned against the wall.

Dr. Smith walked into the room with the pathology report.

We all stood. Everyone but Mother. He looked around the room and sat down on an empty chair by the door.

"The pathology report was not as good as I had hoped."

Dad walked out of the room and back in again.

"It's not all bad, but it's not all good either. We found microscopic cells in three of the fifteen biopsies. I'm sorry, Diane, everything looked so good. I was premature in my judgments. It's just that we all wanted it so badly. You've worked so hard and done so well . . ."

Mother shook her head. She was furious. She turned abruptly to Dad and me. "I could have handled this, why couldn't you?"

Dr. Smith tried to continue reassuringly. "There is still a fairly good chance for cure. With six weeks of radiation therapy . . ."

"I don't even want to hear it," Mother said sobbing. "It's over. I'm tired of fighting. Just leave me alone, all of you. Go, please. I need time to myself." She rolled on to her side and faced the wall.

We left. I was heartsick. I had betrayed her. I felt as though I had killed her with my optimism and I was strapped with guilt. Why couldn't I have respected her belief that the outcome mattered less than the gift of each day. We had wanted everything back to its original shape. We had wanted a cure for Mother for ourselves, so we could get on with our lives. What we had forgotten was that she was living hers.

I fled for Bear River, for the birds, wishing someone would rescue me.

The California gulls rescued the Mormons in 1848 from losing their crops to crickets. The gull has become folklore. It is a story we know well. As word of Great Salt Lake's nasty disposition filtered through the westering grapevine in the 1840's, the appeal of the Great Basin was tainted. The Mormons were an exception. They saw it as Holy Land.

Brigham Young raised his hands above the Salt Lake Valley and said, "This is a good place to make Saints, and it is a good place for Saints to live; it is the place that the Lord has appointed, and we shall stay here until He tells us to go somewhere else."

God's country. Isolation and a landscape of grit were just what the Mormons were looking for. A land that no one else wanted meant religious freedom and community-building without persecution. It was an environment perfectly suited for a people unafraid of what only their hands could yield. They were a people motivated by the dream of Zion. They had found their Dead Sea and the River Jordan. The Great Basin desert was familiar to them if not by sight, at least by story.

But it wasn't easy. Winter quarters for the poorly provisioned families who had just arrived proved difficult. Their livestock had been decimated by wolves and Indian raids. Untended animals grazed down their crops and the harvest of 1847 consisted of only a few "marble-size potatoes." The starving pioneers were reduced to eating "crows, wolf meat, tree bark, thistle tops, sego lily bulbs, and hawks."

One member describes in his journal, "I would dig until I grew weak and faint and sit down and eat a root, and then I would begin again."

The harvest of 1848 looked more promising and the Saints' spirits were buoyed. But just when a full pantry for each family seemed assured, hordes of crickets invaded their wheat fields. The crickets were described as "wingless, dumpy, black, and swollen-headed creatures, with bulging eyes in cases like goggles, mounted upon legs of steel wire . . . a cross between a spider and a buffalo."

The pioneers fought them with brooms, shovels, pitchforks, and fire. Nothing seemed to halt their invasion. In desperation, the farmers and their families fell to their knees with exhaustion and prayed to the Lord for help.

> Upon looking up, I beheld what appeared like a vast flock of pigeons coming from the northwest. It was about three o'clock in the afternoon . . . there must have been thousands of them; their coming was like a great cloud; and when they passed between us and the sun, a shadow covered the field. I could see the gulls settling for more than a mile around us. They were very tame, coming within four or five rods of us.
>
> At first, we thought that they also were after the wheat and this fact added to our terror; but we soon discovered that they devoured only the crickets. Needless to say, we quit fighting and gave our gentle visitors the possession of the fields.

Their prayers had been answered. Their crops had been saved.

Over one hundred years later, Mormons still gather to tell the story of how the gulls freed them from the crickets. How the white angels ate as many crickets as their bellies would hold, flew to the shores of Great Salt Lake and regurgitated them, then returned to the field for more. We honor them as Utah's state bird.

While sitting on the edge of Great Salt Lake, I noticed the gulls flying in one direction. From four o'clock until dusk, with their slow, steady wing beats, they flew southwest. I pocketed this information like a small stone.

The next day, I returned and witnessed the same pilgrimage. After all these years of cohabitation, the gulls had finally, seized my imagination.

I had to follow.

The gulls were flying to their nesting colonies on the islands of Great Salt Lake. What they gain in remoteness (abeyance from predators and human interference) they sacrifice in food supply. Because of its high salinity, Great Salt Lake yields no fish. With the exception of brine shrimp, which make up a meager percentage of the gull's total diet, the water is sterile. Consequently, gulls must fly great distances between island nesting sites and foraging grounds. Round trips between fifty to one hundred miles are made from Hat and Gunnison Islands to the Bear River Migratory Bird Refuge. Daily. White pelicans, double-crested cormorants, and great blue herons, also colony nesters, must make these same migrations to the surrounding marshes of Great Salt Lake.

The population of colony-nesting birds on the islands fluctuates with the lake level and human disturbances. Herons, cormorants, and pelicans are much more sensitive to these pressures than gulls. One striking difference between the species is their territoriality. Herons are wary, skittish. Pelicans and cormorants are shy. If disturbed, great blue herons leave the island first, followed by the pelicans and cormorants. The gulls never leave. They just fly around in circles screaming at the intruders.

The populations of herons, cormorants, and pelicans are decreasing on the islands of Great Salt Lake, whereas evidence shows gull communities on the rise. Gulls are more resilient to change and less vulnerable than other birds to environmental stresses.

Refuge

William H. Behle, curator of ornithology at the Utah Museum of Natural History, in his classic study on the birds of Great Salt Lake, reported sixty thousand adult California gulls nesting on Gunnison Island on June 29, 1932. This was the highest gull concentration ever known on Great Salt Lake.

Since the flooding, most of the islands have either been abandoned by colony nesters or their populations have been greatly reduced. This seems to have happened for three reasons: lack of nesting space due to rising waters, increased human visitation to the islands, and, most important, lack of food due to the submerged marshes.

In drought conditions, bird populations also decline but for different reasons. In low water, most of the islands are attached to the mainland, making the birds more vulnerable to predators and human interference. Food supply is also threatened as the marshes shrink.

The balance between colony-nesting birds, the fluctuating Great Salt Lake, and its wetlands is a delicate one.

In 1958, Dr. Behle wrote prophetically,

> If present trends continue, there is danger that the islands of Great Salt Lake will be entirely abandoned by colonial birds. Herons already have abandoned all their historic nesting sites on the lake. Cormorants persist at Egg Island but are barely holding their own from year to year. Pelicans faced a critical condition in 1935 and seem to be slowly recovering but their existence is precarious. The gulls are moving to man-made dikes and the islands of the refuges on the east side of the lake.

For now, any remembrance of Great Salt Lake hosting an island archipelago of birds is limited to the journals of early explorers. Captain Howard Stansbury wrote on April 9, 1850,

Rounding the northern point of Antelope Island, we came to a small rocky islet, about a mile west of it, which was destitute of vegetation of any kind, not even a blade of grass being found upon it. It was literally covered with wild waterfowl: ducks, white brandt, blue herons, cormorants, and innumerable flocks of gulls, which had congregated here to build their nests. We found great numbers of these, built of sticks and rushes, in the crevices of the rock, and supplied ourselves without scruple, with as many eggs as we needed, primarily those of the heron, it being too early in the season for most of the other waterfowl.

And on May 8 of the same year:

The neck and shores on both of the little bays were occupied by immense flocks of pelicans and gulls, disturbed now for the first time, probably by the intrusion of man. They literally darkened the air as they rose upon the wing, and hovering over our heads, caused the surrounding rocks to re-echo with their discordant screams. The ground was thickly strewn with their nests, of which there must have been thousands.

I have seen hundreds of gulls nesting not on the islands of Great Salt Lake, but on the old P-dike at the Bear River Migratory Bird Refuge. To wander through a gull colony is disorienting. In the midst of shrieking gulls, you begin to speak, but your voice is silenced. They pull the clouds around you as you walk on eggshells. You quickly realize that you do not belong.

Hundreds of gulls hovered inches above my head, making their shrill repetitive cries, *"Halp! Halp! Halp!"* Several wing tips struck my forehead, a warning that I was too close to their nests. There were so many nests, I didn't know where to step, much less how to behave. Finally, I just stood in one place and watched.

Refuge

A California gull's nest is a shallow depression on the ground. They gather nesting material and line the hollow. The gull settles down, usually female, and with her body and sometimes the aid of her feet and bill, neatly arranges the feathers, grasses, and twigs into a cup-shaped nest. Depending on the resources available, they can range from simple to elaborate.

The nests at Bear River were simple. Bones from gulls and other animals were woven into their fabric, making them look like death wreathes. Clutches of umber eggs splotched with brown lay in their centers.

Most of the gulls I watched at the Bird Refuge were incubating eggs, an activity which takes from twenty-three to twenty-eight days. Both sexes share in the responsibility.

I wondered in the midst of so many gulls and so many eggs, how the birds could differentiate between them. They do. Parental recognition. The subtle distinctions in patterning and coloration among individual egg clutches test my eye for discrimination. Each brood bears its own coat of arms.

Young gulls are precocial, which means they are relatively well developed at hatching. They are covered with a thick coat of natal down, can leave the nest soon after they hatch, and can feed themselves within a short time. Precocial young are typical to most waterfowl, an adaptation against predators of ground-dwelling birds.

In contrast, altricial young are those birds born helpless, usually naked and with closed eyes, completely dependent on their parents for a sustained period after hatching. Altricial young are more common to passerine birds, which have the advantage of tree nesting. They can afford to be helpless.

It is tempting to pick up a baby gull. I must confess I have tried, but only got as far as its fierce little beak would let me. They come into life as speckled warriors, waving egg teeth on the tips of their upper mandibles. Their battle with the eggshell is tireless as they struggle anywhere from twenty minutes to ten hours. They stand in wet armor ready to face the world. All around me, eggs were moving, cracking, and breaking open. I would stoop a few feet from a nest and find myself staring eye-to-eye with a chick.

A month from now, in June, the young will be in juvenile plumage, looking like gulls who ventured too close to a campfire. Smoked feathers. They will stretch and beat their wings wildly until one day their own force will surprise them, lifting them a foot or two off the ground. Gradually, with a few running steps, their wings will carry them. In a matter of weeks, adolescent gulls will be agile fliers.

By July, the California gulls will prepare to leave their breeding grounds, taking their young with them. Banding records from Bear River indicate that most of the Great Salt Lake population winters along the Pacific coast from northern Washington to southern California.

I love to watch gulls soar over the Great Basin. It is another trick of the lake to lure gulls inland. On days such as this, when my soul has been wrenched, the simplicity of flight and form above the lake untangles my grief.

"*Glide*" the gulls write in the sky—and, for a few brief moments, I do.

I go to the lake for a compass reading, to orient myself once again in the midst of change. Each trip is unique. The lake is different. I am different. But the gulls are always here, ordinary—black, white, and gray.

I have refused to believe that Mother will die. And by

denying her cancer, even her death, I deny her life. Denial stops us from listening. I cannot hear what Mother is saying. I can only hear what I want.

But denial lies. It protects us from the potency of a truth we cannot yet bear to accept. It takes our hands and leads us to places of comfort. Denial flourishes in the familiar. It seduces us with our own desires and cleverly constructs walls around us to keep us safe.

I want the walls down. Mother's rage over our inability to face her illness has burned away my defenses. I am left with guilt, guilt I cannot tolerate because it has no courage. I hurt Mother through my own desire to be cured.

I continue to watch the gulls. Their pilgrimage from salt water to fresh becomes my own.

RAVENS

lake level: 4209.10'

Mother began her radiation treatment this morning. They tattooed her abdomen with black dots and drew a grid over her belly with a blue magic marker.

"After the technicians had turned my body into their bull's-eye," Mother said, "the radiologist casually walked in, read my report, and said, "You realize, Mrs. Tempest, you have less than a 40 percent chance of surviving this cancer."

"What did you say to him?" I asked as we were driving home.

"I honestly don't remember if I said anything. He rearranged the machinery above me, rearranged my body on the stainless steel slab, and then walked out of the room to zap me and protect himself."

"How do you feel, Mother?" I asked.

She folded her arms across her midriff.

"I feel abused."

This afternoon, I coaxed Mother into going swimming at Great Salt Lake, something we have not done for years. On our backs, we floated, staring up at the sky—the cool water held us—in spite of the light, harsh and blinding. I heard the whisperings of brine shrimp, felt their orange feathered bodies brushing against my own. I showed them to Mother. She shuddered.

We drifted for hours. Merging with salt water and sky so completely, we were resolved, dissolved, in peace.

We returned with salt crystals in our hair and sand in our navels to remind us we had not been dreaming.

The Southern Pacific Railroad Causeway was breached today. Water from the south arm of Great Salt Lake shot through the three-hundred-foot opening into Gunnison Bay like a wave of pent-up emotion.

I envy the release.

Governor Scott Matheson anticipates that the disparate water levels of the south and north arm of Great Salt Lake will equalize within the next couple of months. There will be a mixing of brine as the salt loads within Gunnison and Gilbert Bays redistribute themselves with new bidirectional flow.

A small piece of Great Salt Lake's integrity has been restored.

We celebrated Dad's birthday, July 26, 1933. He reminded Mother, as he does each year, that he is one year younger than she. She retorts each year with the same remark:

"That explains your immaturity and my wisdom." She

reminds him that this year she will be kind to him, that she has not wrapped his presents in black like she did when he turned forty.

Steve, Dan, Hank, and I, together with Brooke and Ann, presented Dad with a large cake flaming with candles.

He is fifty-one.

Great Salt Lake shimmered in the background. It rose another 5' from September 25, 1983, to July 1, 1984, the second-largest rise ever recorded for the lake. The net rise from September 18, 1982 to July 1, 1984 was 9.6'. In comparison, the previously recorded maximum net rise of Great Salt Lake between any two years was 4.7', during 1970 and 1972.

"Make a wish . . ." said Mother. And we watched him blow out the candles.

I wish old Saltair were still standing guard over Great Salt Lake. The magnificent Moorish pavilion built on a wooden trestle reigned supreme during the early 1900s. Its image captures the romance of another era for Utah residents.

Today, I walked where Saltair once stood. A few charred posts from the pier still stand, looking like ravens.

In 1962, Herk Harvey released the film, *Carnival of Souls,* which has become a cult classic, a precursor to *The Night of the Living Dead.* The heroine of the film, Miss Henry, comes to Utah to play the organ. The pastor introduces her to his parish by saying, "We have an organist capable of stirring the soul."

Miss Henry's affinity with the dead leads her in a trance-like journey to Great Salt Lake. As she stands on the board-walk of Saltair she watches the dead, one after another, emerge from the lake, zombies dripping with salt water. She follows the gaunt, dark-eyed corpses dressed in black into the dance hall, where she is moved to play the organ for them.

For my generation, Saltair had become a sinister piece of abandoned architecture. This was not the case for my grandparents. I recall having dinner with Lettie and Sanky shortly after Saltair had burned to the ground.

My grandfather and grandmother fell in love on moonlit nights at Saltair.

"I remember the way her chiffon dress would blow in the breeze as we stood on the boardwalk looking over the lake. And I remember a kiss or two before we went back inside . . ." he said.

My grandmother smiled as she described the particular peach-colored dress with an asymmetrical hem that she frequently wore to the resort so popular in the 1920s. She boasted about my grandfather's agility at the games of chance, how he would always win a Kewpie doll for her.

They described boarding the open-air train from Salt Lake City to Saltair, how everyone would sing songs in the early evening as the train delivered them to the pavilion perched on wooden trestles above Great Salt Lake.

The dance pavilion catered to some of the great bands of the day: Harry James, Wayne King, Bob Crosby, and Guy Lombardo.

"The dance floor was suspended on springs," my grandmother said. "The ceiling was decorated with huge lighted balls made of mirrors that cast starlight reflections on the dance hall."

"And the band would play until midnight," my grandfather added. "That's when the last train would go back to the city."

Saltair never regained its pre-Depression glamour after it was destroyed by fire the first time on April 12, 1925. Even after it reopened in 1929, the public was becoming more mobile, and the novelty of traveling by train to the lake was beginning to wane. It officially closed in 1968; two years later it was burned down by arsonists.

In 1981, developer Wally Wright, tried to reincarnate Saltair. He purchased a hanger from Hill Air Force Base to use as the structure and then tried to resurrect the Moorish architecture with concrete. It wasn't the same. Somehow, water slides, bumper boats, and fast-food shops didn't hold the integrity of the times. Even so, Wally Wright never got a fair shot at his own concession. Great Salt Lake rose before he completed construction.

And so another abandoned monument to the lake stands. There are ghosts at the Great Salt Lake who still dance on starlit nights.

＜＜＜＜＜＜＜＜

PINK FLAMINGOS

＜＜＜＜＜＜＜＜

lake level: 4208.00'

Great Salt Lake has dropped 1.35' during the summer of
1984. About one half of the decline was due to normal
evaporation, but the remainder resulted from the breach in
the causeway. Perhaps we were given a reprieve.

Mother came over this morning.

"Do you have a minute?" she asked. "Tamra Crocker
Pulsifer was operated on yesterday for a brain tumor and I
want to send her this letter. May I read it to you? I want
it to be right."

We walked into the living room. I opened the drapes and
we settled on the couch. Mother paused, then began:

Dearest Tammy:

When I heard of your surgery yesterday, I felt my
heart would break. I kept thinking that you are so young
to have to go through this. I would gladly, Tammy, take

this cross you are to bear upon my own shoulders if I could. I know what you are going through right now and I want you to know my prayers and love are with you. I wish I could talk to you in person and we could just cry together and share each other's feelings. There are times, however, when you must go through certain things alone and this is one of them. When I say alone, I mean you have to address this illness yourself. You have to decide how you are going to deal with it. I know what a strong young woman you are and what a fighter you can be.

When I was told I had cancer thirteen years ago, I experienced many different reactions. The night before my surgery, I was given a blessing and in the blessing I was told that I would not have cancer, that the lump would be benign, and that I would be fine.

During the surgery, I had a spiritual experience that changed my life. Just before I awakened in the recovery room, I was literally in the arms of my Heavenly Father. I could feel His love for me and how sorry He was that He couldn't keep this from me. What He could and did give me was far greater than not having cancer. He gave me the gifts of faith, hope, strength, love, and a joy and peace I had never felt before. These gifts were my miracle. I know that it is not the trials we are given but how we react to these trials that matters.

I am sending you a book, "The Healing Heart," by Norman Cousins. He has had two incurable illnesses and survived both of them. This book helped me last year more than anything I have ever read. It helped me to realize that I can help in the recuperative process of my own body. That we can help ourselves through positive thinking. In his book, he says, "Death is not the enemy; living in constant fear of it is."

I want to live and think as actively and creatively as it is physically possible for me to do. This year, with two major surgeries, one year of chemotherapy, and six weeks of radiation, has been the most difficult year of my life, and also the most beautiful. It has enabled me to sense and see things I never did before. It brings life into focus one day at a time. You live each moment and when you see the sunset at the end of the day, you are so grateful to be part of that experience.

Don't be so strong, Tammy, that you won't cry when you want to. Let people help you and love you. I can't tell you how important it was for me to let people do things for me. I resisted at first, but I don't know how I would have gotten through the radiation without the six beautiful women who picked me up each day and took me to my treatments. It gave me something to look forward to—each with a different friend—and I appreciated the love and support they gave me.

While I was taking chemotherapy and radiation, I took a lot of vitamins and I believe they helped me stay strong. Let me know if there is anything I can do to help you.

May the Lord bless you, Tammy, with His gifts. You are a very special young woman and I want you to know how much I admire all that you are.

Love, Diane

Mother finished reading the letter. A long silence followed. She looked over to me. "Do you think it's all right to send?"

Our correspondences show us where our intimacies lie. There is something very sensual about a letter. The physical contact of pen to paper, the time set aside to focus thoughts, the folding of the paper into the envelope, licking it closed, addressing it, a chosen stamp, and then the release of the letter to the mailbox—are all acts of tenderness.

And it doesn't stop there. Our correspondences have wings—paper birds that fly from my house to yours— flocks of ideas crisscrossing the country. Once opened, a connection is made. We are not alone in the world.

But how do we correspond with the land when paper and

ink won't do? How do we empathize with the Earth when so much is ravaging her?

The heartbeats I felt in the womb—two heartbeats, at once, my mother's and my own—are heartbeats of the land. All of life drums and beats, at once, sustaining a rhythm audible only to the spirit. I can drum my heartbeat back into the Earth, beating, hearts beating, my hands on the Earth— like a ruffed grouse on a log, beating, hearts beating—like a bittern in the marsh, beating, hearts beating. My hands on the Earth beating, hearts beating. I drum back my return.

"A mirage is created when the air next to the earth becomes warmer than the air immediately above it," said Brooke.

We were walking the salt flats east of Wendover, Nevada, beyond the rock graffiti, where stone signatures have been left in the sand by locals. It was Sunday, mid-September, and hot. A line of quicksilver danced ahead of us.

"I don't believe it's a mirage," I said. It looks like another finger of Great Salt Lake."

"It's a mirage, Ter," Brooke continued. "The lake we are seeing is actually the sand's surface appearing wet, due to the hot air immediately above it and the cooler air above that. See how we are standing on slightly higher ground looking down?"

"So?"

"So, as a result of bending light rays, the image of the sky is turned upside down. It just looks like a lake."

"I think it is the lake."

We keep walking toward it, to prove ourselves right. But the body of water that I see and Brooke distrusts keeps flowing farther away from us. We retreat and the lake seems to follow.

I concede.

Brooke grins. I forget he is a biologist with an analytic mind.

"It's all an illusion. Nothing is as it appears. The air is refracting the sun's rays, transforming sand into water; make sense?"

I look at him and nod. "I think it's about hope on a hot day."

Mother and Dad are in Switzerland. I received a letter from her today. It reads:

Dear Terry:

More and more, I am realizing the natural world is my connection to myself. Landscape brings me simplicity. I can shed the multiplicity of things at home and take one duffle bag wherever I go. How wonderful to shed clothes and be free of choices, to feel the sun on my back and the wind on my face. I find my peace, my solitude, in the time I am alone in nature.

John has been my guide, Terry. His nature is not to just sit back and be an observer to the land, but an active participant. When we went to Hawaii for the first time, nineteen years ago, we ran and embraced it all. We didn't just look at the ocean, we dove into the waves and tasted salt water on our lips. We greeted the sunrise on the crater in Maui and looked out over thousands of miles. I'll swear we saw the curvature of the earth. We celebrated each day by walking along the beach, picking up shells. We ran into the wind and fell on to the sand, watching the tiny sandpipers dart back and forth.

And we are doing that now. We are hiking up and down the Alps together, walking farther than I ever thought possible. We have slept on the grass next to cows with bells around their necks. We have walked thigh-high in wildflowers. The natural world is a third party in our marriage. It holds us close and lets us revel in the intimacy of all that is real.

I think of you constantly. Please give our love to everyone at home.

We love you,
Mother

I fold the letter back into its envelope and call Brooke. Perhaps, we can go south this weekend.

I love to make lists. Maybe it's my background in beehives and breadmaking, the whole business of being industrious and frugal (of which I am neither) that a list promotes. Or maybe it's the power that comes when you can cross something off a list. Done. Finished. Move on to the next chore. I can see in a very tangible form what I have accomplished in a day. Or perhaps it's the democratic nature of lists that I find so attractive. Each task is of equal importance on paper. So "pick up fresh flowers" carries the same weight as "do the laundry." It's the line slashed through the words that counts. Never mind that the pleasurable items are crossed off by noon and the difficult ones, meant for procrastination anyway, get moved to the next day's agenda. The point is that my intentions are honorable. My lists will defend me.

The life list of a birdwatcher is of a different order. It's not what you cross off that counts, but what you add. It is a tally of all the species of birds seen within a lifetime. A bird seen for the first time is called "a lifer."

The life list can be a private accounting of birds seen, a scrapbook of sorts, of places visited and birds watched. It provides the pleasure of traditional list-making (in this case, adding something new to the list instead of crossing something off). Those who use their bird list in this manner usually have no idea of their total sum of species. And it is done at random—when a person thinks about it.

At the end of each day, I write down the names of all birds seen and read them out loud, regardless of who is there. It's like throwing a party and afterwards talking about who came. There are always those you can count on and those who will surprise you. And, once in a blue moon, an accidental guest will arrive.

Within every checklist there are those birds listed as "accidentals," one species, or at best a few, that have wandered far from their normal range. They are flukes in a flock of predictable migrants. They are loners in an unfamiliar territory.

William H. Behle, author of *Utah Birds,* defines an accidental as "a species seen only one or two times since 1920 or one or two times in the last fifty years or another fifty year interval provided that species is just as likely to occur now as then." Accidental birds in Utah are substantiated by at least one recorded specimen.

On July 25, 1962, Don Neilson, manager of the Clear Lake Refuge, observed an American flamingo in Millard County, Utah. It stayed in the area through Columbus Day. He has color photographs to prove it.

Another sighting occurred on August 3, 1966, by W. E. Ritter and Reuben Dietz who saw a flamingo at Buffalo Bay, on the northeast shore of Antelope Island. The bird was washed out and pale, thousands of miles from its homeland, so they inferred it must be an escapee from Tracy Aviary or Hogle Zoo in Salt Lake City. Calls were made, but all captive flamingos were accounted for.

Then, in the summer of 1971, a third flamingo was seen at the Bear River Migratory Bird Refuge from early June through September 29. Once again, photographs verified the sighting.

I personally have seen flamingos throughout the state of Utah perched proudly on lawns and in the gravel gardens of trailer courts. These flamingos, of course, are not *Phoenicopterus ruber,* but pink, plastic flamingos that can easily be purchased at any hardware store.

It is curious that we need to create an environment foreign from our own. In 1985, over 450,000 plastic flamingos were purchased in the United States. And the number is rising.

Pink flamingos teetering on suburban lawns—our unnatural link to the natural world.

The flocks of flamingos that Louis Agassiz Fuertes lovingly painted in the American tropics are no longer accessible to us. We have lost the imagination to place them in a dignified world. And when they do grace the landscapes around us, they are considered "accidental." We no longer believe in the possibility of such things.

There have been other accidentals in Utah.

On July 2, 1919, a flock of five roseate spoonbills flew over the Barnes Ranch near Wendover, Nevada. Mr. Barnes, having never seen such a bird, shot one and kept it inside his house for years as a conversation piece. It strangely disappeared and was subsequently found; it now rests at the Utah Museum of Natural History.

The flamingo and roseate spoonbill are not the only rarities to visit the wetlands surrounding Great Salt Lake. Other accidentals include the European wigeon seen at the Bear River Bird Refuge on October 19, 1955, and another one sighted by Bill Pingree at the Lakefront Gun Club on December 15, 1963.

And when there is a species whose occurrence is open to question largely by virtue of the absence of a record specimen (a bird in the hand), but where the competence of the observer or observers constitutes "sufficient evidence to justify the inclusion of the species in the checklist," these birds are listed as "hypothetical." Hypothetical species sighted around Great Salt Lake include the red-necked grebe, reddish egret, Louisiana heron, harlequin duck, black scoter, black oystercatcher, wandering tattler, stilt sandpiper, bar-tailed godwit, parakeet auklet, northern parula warbler, and a palm warbler.

How can hope be denied when there is always the possibility of an American flamingo or a roseate spoonbill floating down from the sky like pink rose petals?

How can we rely solely on the statistical evidence and percentages that would shackle our lives when red-necked grebes, bar-tailed godwits, and wandering tattlers come into our country?

When Emily Dickinson writes, "Hope is a thing with feathers that lights upon our soul," she reminds us, as the birds do, of the liberation and pragmatism of belief.

ʃʃʃʃʃʃʃʃ

SNOW BUNTINGS

ʃʃʃʃʃʃʃʃ

lake level: 4209.15'

The eastern shore of Great Salt Lake is frozen, and for as far as I can see it translates into isolation. Desolation. The fog hangs low, with little delineation between earth and sky. A few ravens. A few eagles. And the implacable wind.

Snow crystals stand on the land like the raised hackles of wolves. Broken reeds and cattails are encased in ice. Great Salt Lake has not only entered the marsh, it has taken over.

Because of the high water level and the drop in salinity, Great Salt Lake can freeze and does. The transparent ice along the lake's edge is filled with bubbles of air trapped inside like the sustained notes of a soprano.

I walk these open spaces in silence, relishing the monotony of the Refuge in winter.

Perhaps I am here because of last night's dream, when I stood on the frozen lake before a kayak made of sealskin.

I walked on the ice toward the boat and picked up a handful of shredded hide and guts. An old Eskimo man said, "You have much to work with." Suddenly, the kayak was stripped of its skin. It was a rib cage of willow. It was the skeleton of a fish.

I want to see it for myself, wild exposure, in January, when this desert is most severe. The lake is like steel. I wrap my alpaca shawl tight around my face until only my eyes are exposed. I must keep walking to stay warm. Even the land is frozen. There is no give beneath my feet.

I want to see the lake as Woman, as myself, in her refusal to be tamed. The State of Utah may try to dike her, divert her waters, build roads across her shores, but ultimately, it won't matter. She will survive us. I recognize her as a wilderness, raw and self-defined. Great Salt Lake strips me of contrivances and conditioning, saying, "I am not what you see. Question me. Stand by your own impressions."

We are taught not to trust our own experiences. Great Salt Lake teaches me experience is all we have.

One month ago, this was frozen country, monochromatic and still. This morning, spring has moved in. Constellations of ducks: pintails, mallards, wigeons, and teals are flying in from the south and southwest. The air is wild with voices, avian dialects are being spoken from every direction. The sky vibrates with wings.

Mother does not share my affection for birds. This is her first trip to Bear River. We watch a dozen herons fishing. Beyond them, I spot a carcass, a scattering of blue feathers and bones. I lure Mother forward. Up ahead, we find a wing.

"A great blue heron . . ." I say, picking it up. The primary feathers are attached to the ulna like the teeth of a comb.

Blood stains the snow. More bones. More feathers. And tracks.

We bend down to get a closer look.

"What do you think?" asks Mother.

"Fox, maybe. I don't know. I think they're too small for a coyote."

The wind blows, scattering the white, downy feathers in all directions. They rest momentarily on the snow, rocking back and forth like cradles, until they cartwheel off in the breeze.

We walk west. Blue-winged teals congregate between greasewood. On the road adjacent to Great Salt Lake, huge compression fractures have heaved diagonal chunks of ice this way and that. I try to climb over them, but they present too great an obstacle. They line the shore for miles. We can travel no farther.

"I want you to read 'God Sees the Truth, but Waits,' " said Mother. "Tolstoy writes about a man, wrongly accused of a murder, who spends the rest of his life in a prison camp. Twenty-six years later, as a convict in Siberia, he meets the true murderer and has an opportunity to free himself, but chooses not to. His longing for home leaves him and he dies."

I ask Mother why this story matters to her.

"Each of us must face our own Siberia," she says. "We must come to peace within our own isolation. No one can rescue us. My cancer is my Siberia."

Suddenly, two white birds about the size of finches, dart in front of us and land on the snow.

"I don't know these birds, Mother. They're something new."

She hands me the bird book. I rapidly thumb through the pages. She is looking through the binoculars.

"They are white with black on their backs and I see a flash

of rust on their heads." She pauses, "I can also see black on the tips of their wings when they fly . . ."

I quickly look up to see if they are gone, but they have just moved a few feet up the road. I flip back through the book again.

"I found them! Here they are—page 412, snow buntings!" I hand the field guide to Mother, focusing on the birds to make sure.

"I can't believe it!" I look at her, then back to the buntings. "These are rare to the Refuge. I've never seen them before."

We watch them forage around the edges of melting snow. Mother brings down her binoculars.

"Where are they usually found?" she asks, looking at the field guide once again, this time at the map illustrating bunting distribution.

"Snow buntings are circumpolar, nesting in the Arctic on the tundra."

Mother watches the birds carefully. "Tell me, Terry, are these birds Tolstoy may have known?"

Mother and I met for lunch today. She shared with me the letter she received from Tamra Crocker Pulsifer. It read:

Dear Diane:

I feel you understand maybe far more than I do about this difficult time. The letter you sent was so good for me, and your constant support of myself and my mother have been deeply needed.

I've learned a most valuable lesson. Sometimes you have to totally rely on the arms, tears, and loving hearts of others, that this is truly where God's love lies, in the support of family and friends.

I appreciate the book you sent me and the lotion for

my twenty-ninth birthday. I really thought I would go gray before I went bald.

I wish at times, I had more options and choices. But right now, I am torn between the excitement of what I am learning about life and the sorrow I feel in that I have no future.

My beliefs are carrying me, Diane. How exciting to know that we will possibly be able to design and create our own worlds as our Father and Mother in Heaven have.

I am starting to forget days now, and often around 4:00 a.m., my pillow becomes wet with the challenges ahead. I am crying, Diane. Please, I say, help me laugh and mean it, help me find gratitude for the small things I have left.

Diane, never before have I wanted to do more, but I can't concentrate on what it is that I want. And then when I remember, I don't have the strength or the energy to carry it out. I know I want Adrian, Canace, Christian, Jeneva, and mostly, just to nurse and hold sweet Adrea. How can we be so sad and so full at the same time?

I say to myself, be cheerful and don't worry about the future. But then I think about my children. I wish I could interview one hundred robust women who had hearts of gold, to raise my family and teach them what I have taught them. Maybe next week, I'll place an ad in the newspaper.

Thank you for your example, Diane. I will love you forever.

Tamra

The eye of the cormorant is emerald. The eye of the eagle is amber. The eye of the grebe is ruby. The eye of the ibis is sapphire. Four gemstones mirror the minds of birds, birds who mediate between heaven and earth.

We miss the eyes of birds, focusing only on feathers.

rrrrrrrr

WHITE PELICANS

rrrrrrrr

lake level: 4209.90'

The Refuge is subdued, unusually quiet. The spring frenzy of courtship and nesting is absent, because there is little food and habitat available. Although the species count remains about the same, individual numbers are down. Way down. This afternoon, I watched a white-faced ibis nest float alongside a drowned cottonwood tree. Three eggs had been abandoned. I did not see the adults.

A colony-nesting bird survey has been initiated this spring by the Utah Division of Wildlife Resources to monitor changes in population and habitat use of selected species affected by the rising Great Salt Lake.

The historical nesting grounds on the islands of Great Salt Lake are gone, with the exception of a California gull colony on Antelope Island and the white pelicans on Gunnison. This means colony nesters are now dependent upon the vegetation surrounding the lake for their livelihood.

Great blue herons, snowy egrets, cattle egrets, and double-crested cormorants use trees, tall shrubs, or man-made structures for nesting.

Franklin gulls, black-crowned night herons, and white-faced ibises nest in emergent vegetation such as bulrushes and cattails.

American avocets, black-necked stilts, and other shorebirds are ground nesters who usually scrape together a few sticks around clumps of low-lying vegetation such as salt grass and pickleweed.

Don Paul, waterfowl biologist for the Division of Wildlife Resources, anticipates that the white-faced ibis and Franklin gull populations will be the hardest hit by the flood.

"Look around and tell me how many stands of bulrush you see?" He waves his hand over the Refuge. "It's gone, and I suspect, so are they. We should have our data compiled by the end of the summer."

I turn around three hundred and sixty degrees: water as far as I can see. The echo of Lake Bonneville lapping against the mountains returns.

The birds of Bear River have been displaced; so have I.

Nothing is familiar to me any more. I just returned home from the hospital, having had a small cyst removed from my right breast. Second time. It was benign. But I suffered the uncertainty of not knowing for days. My scars portend my lineage. I look at Mother and I see myself. Is cancer my path, too?

As a child, I was aware that my grandmother, Lettie, had only one breast. It was not a shocking sight. It was her body. She loved to soak in steaming, hot baths, and I would sit beside the tub and read her my favorite fairy tales.

"One more," she would say, completely relaxed. "You read so well."

What I remember is my grandmother's beauty—her moist, translucent skin, the way her body responded to the slow squeeze of her sponge, which sent hot water trickling over her shoulders. And I loved how she smelled like lavender.

Seeing Mother's scar did not surprise me either. It was not radical like her mother's. Her skin was stretched smooth and taut across her chest, with the muscles intact.

"It is an inconvenience," Mother said. "That's all."

When I look in the mirror and Brooke stands behind me and kisses my neck, I whisper in his ear, "Hold my breasts."

Hundreds of white pelicans stand shoulder to shoulder on an asphalt spit that eventually disappears into Great Salt Lake. They do not look displaced as they engage in head-bobbing, bill-snapping, and panting; their large, orange gular sacs fanning back and forth act as a cooling device. Some preen. Some pump their wings. Others stand, take a few steps forward, tip their bodies down, and then slide into the water, popping up like corks. Their immaculate white forms with carrotlike bills render them surreal in a desert landscape.

Home to the American white pelicans of Great Salt Lake is Gunnison Island, one hundred sixty-four acres of bare-boned terrain. Located in the northwest arm of the lake, it is nearly one mile long and a half-mile wide, rising approximately two hundred seventy-eight feet above the water.

So far, the flooding of Great Salt Lake has favored pelicans. The railroad trestle connecting the southern tip of the Promontory peninsula with the eastern shore of the lake slowed the rate of salt water intrusion into Bear River Bay.

The high levels of stream inflow help to keep much of Bear River Bay fresh, so fish populations are flourishing. So are the pelicans.

Like the California gulls, the pelicans of Gunnison Island must make daily pilgrimages to freshwater sites to forage on carp or chub. Many pelican colonies fly by day and forage by night, to take advantage of desert thermals. The isolation of Gunnison Island offers protection to young pelicans, because there are no predators aside from heat and relentless gulls. Bear River Bay remains their only feeding site on Great Salt Lake.

So are their social skills. White pelicans are gregarious. What one does, they all do. Take fishing for example: four, five, six, as many as a dozen or more forage as a group, forming a circle to corral and then to herd fish, almost like a cattle drive, toward shallower water where they can more efficiently scoop them up in their pouches.

Cooperative fishing has advantages. It concentrates their food source, conserves their energy, and yields results: the pelicans eat. They return to Gunnison Island with fish in their bellies (not in their pouches) and invite their young to reach deep inside their throats as they regurgitate morsels of fish.

It's not a bad model, cooperation in the name of community. Brigham Young tried it. He called it the United Order.

The United Order was a heavenly scheme for a totally self-sufficient society based on the framework of the Mormon Church. It was a seed of socialism planted by a conservative people. So committed was this "American Moses" to the local production of every needful thing that he even initiated a silkworm industry to wean the Saints from their dependence on the Orient for fine cloth.

Brigham Young, the pragmatist, received his inspiration

for the United Order not so much from God as from Lorenzo Snow, a Mormon apostle, who in 1864 established a mercantile cooperative in the northern Utah community named after the prophet. Brigham City became the model of people working on behalf of one another.

The town, situated on Box Elder Creek at the base of the Wasatch Mountains, sixty miles north of Salt Lake City, was founded in 1851. It consisted of some six families until 1854, when Lorenzo Snow moved to Brigham City with fifty additional families. He had been called by Brother Brigham to settle there and preside over the Latter-day Saints in that region.

The families that settled Brigham City were carefully chosen by the church leadership. Members included a schoolteacher, a mason, carpenter, blacksmith, shoemaker, and other skilled craftsmen and tradesmen who would ensure the economic and social vitality of the community.

Lorenzo Snow was creating a community based on an ecological model: cooperation among individuals within a set of defined interactions. Each person was operating within their own "ecological niche," strengthening and sustaining the overall structure or "ecosystem."

Apostle Snow, with a population of almost sixteen hundred inhabitants to provide for, organized a cooperative general store. Mormon historian Leonard J. Arrington explains, "It was his intention to use this mercantile cooperative as the basis for the organization of the entire economic life of the community and the development of the industries needed to make the community self-sufficient."

A tannery, a dairy, a woolen factory, sheep herds, and hogs were added to the Brigham City Cooperative. Other enterprises included a tin shop, rope factory, cooperage, greenhouse and nursery, brush factory, and a wagon and carriage repair shop. An education department supervised the school and seminary.

The community even made provisions for transients, declaring a "tramp department" which enlisted their labor for chopping wood in exchange for a good meal. After the Brigham City Cooperative was incorporated into Brigham Young's United Order, members were told,

> If brethren should be so unfortunate as to have any of their property destroyed by fire, or otherwise, the United Order will rebuild or replace such property for them. When these brethren, or any other members of the United Order die, the directors become the guardians of the family, caring for the interests and inheritances of the deceased for the benefit and maintenance of the wives and children, and when the sons are married, giving them a house and stewardship, as the father would have done for them. Like care will be taken of their interests if they are sent on missions or taken sick.

By 1874, the entire economic life of this community of four hundred families was owned and directed by the cooperative association. There was no other store in town. Fifteen departments (later to expand to forty) produced the goods and services needed by the community; each household obtained its food, clothing, furniture, and other necessities from these sources.

In 1877, the secretary of the association filed the total capital stock as $191,000 held among 585 shareholders. The total income paid by the various departments to some 340 employees was in excess of $260,000.

Brigham Young's ideal society where "all members would be tending to their own specialty" appeared to be in full bloom. The Brigham City Cooperative even caught the eye of British social reformer, Brontier O'Brien. He noted that the Mormons had "created a soul under the ribs of death." Edward Bellamy spent a week in Brigham City researching *Looking Backward,* a Utopian novel prophesying a new social and economic order.

Home industry was proving to be solid economics.

But signs of inevitable decay began to show. A descendant of a Brigham City man told Arrington that his grandfather formed a partnership with another prominent Brigham City citizen in the late 1860s. Their haberdashery was the only place in town where material other than homespun could be purchased. When they succeeded beyond their dreams, they were asked to join the association. They declined, and townfolk were immediately instructed not to trade with them. When some of the community persisted in trading with these men, despite orders from Church officials, members of the Church were placed at the door of the shop to record the names of all persons who did business inside, even though the men in partnership were Mormons in good standing. As a result of this tactic, the business soon failed and the men were forced to set up shop elsewhere.

The ecological model of the Brigham City Cooperative began to crumble. They were forgetting one critical component: diversity.

The United Order of Minutes, taken on July 20, 1880, states, "It was moved and carried unanimously that the council disapprove discountenance, and disfellowship all persons who would start an opposition store or who would assist to erect a building for that purpose."

History has shown us that exclusivity in the name of empire building eventually fails. Fear of discord undermines creativity. And creativity lies at the heart of adaptive evolution.

Lorenzo Snow's fears that the Brigham City Cooperative would not adapt and respond quickly enough to the needs of a growing population, materialized. Fire, debt, taxes, and fines befell the Order. In 1885, Apostle Snow was indicted on a charge of unlawful cohabitation (polygamy). He served eleven months in the Utah State Penitentiary before his conviction was set aside by the United States Supreme Court. Finally, as a result of the 1890s depression, the cooperative store went bankrupt. By 1896, all that remained of Brigham City's hive of industry was the unused honey stored on the shelves of the new general store.

Fifteen years of United Order graced Brigham City, Utah. A model for community cooperation? In part. But there is an organic difference between a system of self-sufficiency and a self-sustaining system. One precludes diversity, the other necessitates it. Brigham Young's United Order wanted to be independent from the outside world. The Infinite Order of Pelicans suggests there is no such thing.

"Can you count?" Don Paul asks me one morning at the Ogden airport.

"1, 2, 3 . . ." I joke.

"Get in, you'll do fine."

We board *Skywagon II* for Gunnison Island for the Division of Wildlife Resources annual count of breeding pelicans.

We are cleared and begin taxiing down the runway. In a few seconds we are airborne, flying over farmlands. The checkerboards of crops, so familiar to rural communities become submerged and suddenly, we are flying over water. To see how much Great Salt Lake dominates the landscape from the air is to adopt a radical respect for its geography.

"I had no idea . . ." I mused.

"Nobody does," answers Don. "Except for the birds."

Images of the Utah poet, Alfred Lambourne come to mind as we look out over his "inland sea."

> In outline the sea is peculiar, resembling somewhat a human hand. The fingers are pressed together and point north, northwest. The stretch of water forming the thumb is known as Bear River Bay, and the dividing mountains between thumb and fingers is Promontory Range. In the palm of the hand are four large islands— Stansbury, Antelope, Carrington, and Fremont. Three which are smaller lie away to the north—Strong's Knob, Gunnison, and Dolphin.

While Lorenzo Snow was maintaining the United Order, Lambourne was living out his own order of solitude on Gunnison Island. Lambourne inhabited the island for one year in 1895, with the hope of homesteading seventy-five acres. But his application was denied, the rationale being that the island was more suitable for mineral interests than agriculture. Given the Mormon Church's religious doctrine against the drinking of alcohol, his carefully tended vineyard did not do much to bolster his request for residency.

I can see the flooded offices of the Bear River Migratory Bird Refuge on my right. Herons and cormorants are nesting on the roofs. Fremont Island, on our left, looks like a piece of worked flint.

"No colony nesters down there," says Paul. "No native grasses. No nothing. Only Welsh ponies and sheep. That island has been beaten to death. It's privately owned now. Kit Carson painted a cross on one of those rock outcrops, but darned if I can find it. I've tried."

The pilot, Val, banks the plane to the left. Three more islands come into view.

"There's Stansbury, Carrington, and the tiny island beyond is Hat, formerly known as Bird Island. It used to be covered with nesting pelicans, herons, gulls, terns, and

cormorants. As you can see now, it's almost underwater."
Below us, rust ribbons of brine oscillate with the currents.
Gulls, grebes, and phalaropes feed along the shrimp lines.
"There's practically no brine left in the south arm," Paul
says. "As a result, most of the phalaropes and grebes have
moved up here."
"Up ahead, Gunnison Island," the pilot reports. Lam-
bourne's description is accurate:

> It is a rock, a rising of the partially submerged Desert
> Range of mountains, a summit of black limestone with
> longitudinal traversements of coarse conglomerates.

The plane circles the island rounding the west shore. The
pilot banks hard to the right so Don Paul can get a solid
counting. He begins charting the nesting pelicans. The island
is beaded with them.
"Most of these birds are young," He explains. "The adults
are feeding at Bear River Bay. I saw them feeding as we
flew over."
We circle the island once again, while he continues
counting, marking dots on his map of Gunnison.
"The colonies look like they're all synchros."
"Synchros?" I ask.
The plane crosses over to the east shore, which appears
rockier. I see no pelicans nesting on this side.
"The reproductive activities of pelicans within a specific
colony are highly synchronized. Egg-laying, hatching, and
fledging of chicks in any given colony usually occurs within
a five- to nine-day period."
We swing around the west shore of the island. He asks
Val to bank right again and fly as low as he can.
"But the interesting part of this environmental story is
that the reproductive activities of the pelican population on
Gunnison Island as a whole is asynchronous. The reproduc-

tive-cycle stages between colonies may differ by as much as four to eight weeks."

"What's the advantage?" I ask.

"Scientists hypothesize that coloniality increases an individual's chance of successfully finding food, either by an exchange of information within a colony about where food is particularly abundant, or by enabling pelicans to form groups, leaving the island in flocks so they can take advantage of the thermals. Then, when they find their foraging grounds, they will fish cooperatively."

"Colonial economics," Don Paul continues, "would not be advantageous if every colony was on the same breeding and feeding schedule. The competition for food would not only diminish the resource but also result in pelican mortality. Whereas, a month later, it's a different ballgame: there's plenty of food to go around. The staggering of intercolony development on Gunnison Island makes good ecological sense."

"We'll catch them one more time," says Val. "There's the triangulation post set up by Stansbury in 1850."

I can see three sticks on top of one of the peaks. I try to locate Lambourne's cabin but can only find the guano miner's shack. As we circle the island for the last time, I recognize the northern cliffs, which Lambourne describes as "a conchant lion. His massive head turned eastward, his monstrous paws rest on the lower shelves."

Not much has changed.

"That's it . . ." says Paul as *Skywagon II* levels and straightens for home.

"And the count?" I ask.

Don Paul looks over his papers. "Ten thousand breeding adults."

Water. Rock. Bird. I don't know if Brigham Young ever ventured to Gunnison Island or observed the finely tuned society of pelicans. But had his attention been focused more on Earth than "heaven on earth" his vision for managing the Saints in the Great Basin might have been altered.

ϝϝϝϝϝϝϝϝ

YELLOW=HEADED
BLACKBIRDS

ϝϝϝϝϝϝϝϝ

lake level: 4209.55'

Mother's health seems to be stable.

Great Salt Lake seems to be stable. I've waited a long time to see Fish Springs National Wildlife Refuge. Now seems to be a good time. It is another oasis in the desert, adjacent to the Deseret Test Center, one of the many military bombing ranges in the Great Basin.

I follow the old Pony Express Trail through miles of sagebrush. It's a four-hour drive west from Salt Lake City. Eye-squinting country. A thin green line appears on the horizon. Bulrushes. The liquid, lambent stage for birds.

They are all here: avocets, stilts, waterfowl galore, great blues, night herons, bitterns and blackbirds, willets, ibises, marsh hawks, and terns. I sit on the edge of the springs, my eyes unable to focus, as a black-and-white-winged dragonfly is snapped by the mandibles of a snowy egret.

Dusk is approaching. Meadowlarks and yellow-headed

blackbirds sing the shadows longer. Lake Bonneville has left its mark. Bathtub rings rim the Great Basin. Tonight these mountains are lavender with blue creases that fall like chintz. First stars appear. A crescent moon. I throw down my sleeping bag. The stillness of the desert instructs me like a trail of light over water.

There are dunes beyond Fish Springs. Secrets hidden from interstate travelers. They are the armatures of animals. Wind swirls around the sand and ribs appear. There is musculature in dunes.

And they are female. Sensuous curves—the small of a woman's back. Breasts. Buttocks. Hips and pelvis. They are the natural shapes of Earth. Let me lie naked and disappear. Crypsis.

The wind rolls over me. Particles of sand skitter across my skin, fill my ears and nose. I am aware only of breathing. The workings of my lungs are amplified. The wind picks up. I hold my breath. It massages me. A raven lands inches away. I exhale. The raven flies.

Things happen quickly in the desert.

ʃʃʃʃʃʃʃʃ

REDHEADS

ʃʃʃʃʃʃʃʃ

lake level: 4208.50'

September, 1985. Don Paul's study is out. The recent population and habitat studies performed by the Utah Division of Wildlife Resources shows that colony nesting species around Great Salt Lake have been affected by the rise in lake level. Some are adapting and some are not. The data collection was funded by Los Angeles City Power and Light, which was recently sued by the National Wildlife Federation for drawing down the water levels of Mono Lake.

Great blue herons, egrets, and cormorants, all tree nesters, have been aided by the flooding of the wetlands, as waterfowl management areas have become inaccessible to man and arboreal predators. Their preferred habitat for nesting: dead trees. Suddenly, there's lots of them, killed by the rising salt water. The cottonwoods and box elders that once provided shade and cover for songbirds have become barebranched rookeries for herons and cormorants.

They have not been without their problems, however. In some instances, where they had used the low tamarisk shrubs to nest in, eggs and young were drowned as the waters rose over a few weeks.

As was expected, white-faced ibises and Franklin gulls, both dependent on hard-stem bulrushes for nesting, have suffered the most. With 80 percent of the world's population of white-faced ibises nesting in Utah, these losses become significant.

In 1979, the Utah ibis population was estimated at 8690 pairs. The 1985 colony-nesting survey recorded 3438 pairs. The decline in Franklin gulls is even more radical: a late 1970s survey showed a thousand breeding pairs, compared to the fifty-one nests counted this year.

It is hoped that many breeding adult ibises and Franklin gulls have survived and moved on to more stable marshes in the Great Basin. Breeding numbers are reported higher at Fish Springs and at the Ruby Marshes in Nevada. The Cutler and Bear Lake marshes northeast of Bear River also show an increase in ibis and gull populations.

The avocets and stilts, along with other ground nesters around Great Salt Lake, have been completely displaced. Their nesting sites have been usurped by water, with mud-flats almost nonexistent. Some pairs of avocets have been seen nesting just off the interstate on gravel shoulders.

California has lost 95 percent of its wetlands over the past one hundred years. Eighty-five percent of Utah's wetlands have been lost in the last two. When wetlands are destroyed, many species go with them, and not just the birds that nest there. In Utah's case, tiger salamanders, leopard frogs, orchids, buttercups, myriads of insects and rodents, plus the birds and mammals that prey on them, are vanishing.

Marshes are among the most productive ecosystems on the planet. They are also among the most threatened.

Nationwide, seventy-six endangered species are dependent upon wetlands. Marshes all across the country are disappearing without fanfare, leaving the earth devoid of birdsong. The long-billed curlews who lose their broods to floods become a generation that much more precious to their species' survival. Whether it's because of drought, as is the case in the prairie pothole region to the north, or levels of high toxicity in California's central valley, or just plain development—our wetlands are disappearing.

Wetlands are one more paradox of Great Salt Lake. The marshes here are disappearing naturally. It's not the harsh winter or yearly spillover that threatens Utah's wetland birds and animals. It is lack of land. In the normal cycle of a rising Great Salt Lake, the birds would simply move up. New habitat would be found. New habitat would be created. They don't have those options today, as they find themselves flush against freeways and a rapidly expanding airport.

Refugees.

Before the rise of Great Salt Lake, thousands of whistling swans (now called "tundra swans" by the American Ornithologists' Union) descended on Bear River Bay each autumn. As many as sixty thousand swans have been counted at the Bear River Migratory Bird Refuge during mid-October and mid-November, making it the single largest concentration of migrating swans in North America.

In November, 1984, only two hundred fifty-nine whistling swans were counted at the Refuge. One year later: three.

Birds are opportunistic by nature, but resourcefulness fails in the presence of high-speed traffic and asphalt.

This year, the Utah State Legislature appropriated $98

million for flood control. The alternatives state waterfowl managers are reviewing are: wait for the lake to recede, as it inevitably will; try to acquire more habitat, especially newly created wetlands; or reduce the level of the lake.

Tim Provan, the waterfowl biologist for the Division of Wildlife Resources in Salt Lake City, points out that "The marshes don't produce young. They never have. They hold the birds during migration. The marshes let them rest and feed for extended periods—two, three, four months at a time. The seven to eight hundred thousand ducks we did produce have dropped 85 percent since the flood."

He goes on to say, "The Great Salt Lake marshes had one of the strongest populations of redheads, but they are extremely susceptible to high water. They have been hit the hardest. They are not producing young. Their population is down 60 to 80 percent. We have found a direct statistical relationship between loss of habitat and rate of production: 70 percent loss of habitat, 70 percent loss of young. Our redheads are going other places where they are less successful breeders and more subject to predation." He stares out his office window. "I've seen redheads, canvasbacks, shovelers, and teals just laying dormant in the water as though they were in shock."

"How long before the marshes of Bear River will return?" I ask him.

"It will be three to seven years after the lake recedes before it even begins to take a significant turn, because the soil is so saturated with salts. The recycling of nutrients, the reseeding of plants—that will be a fifteen- to twenty-year turnaround."

"The truth is, the system isn't out there to replace. No other system on the continent can replace or absorb this wetland complex. There is a certain threshold that once crossed, we can never recover. When the death rates exceed the birth rates, we are in trouble. Nobody knows the answers. We are working with the questions."

rrrrrrrr

KILLDEER

rrrrrrrr

lake level: 4208.40'

Mimi and Mother and I had our astrology done. It seemed like a reasonable thing to do. As Mimi said, "If it sheds light on all the confusion, why not?"

We decided to have a picnic by Great Salt Lake to discuss our charts. We sat on its edge where large boulders had been brought in to secure the shore. Each of us found our own niche in the sun. Three women: a Leo, a Pisces, a Virgo. A grandmother, mother, and daughter.

It was beautiful and it was hot. We saw six ruddy ducks, one pair of redheads, avocets and stilts, flocks of Franklin gulls, young shrikes on greasewood, and meadowlarks.

Mimi and I engaged in our birding ritual: locate, focus, observe, and identify. After the bird flies, we pore over the field guide and debate over which species we have just seen.

Mother was amused, saying she wished she liked birds as much as we did, but she had never recovered from Alfred

Hitchcock's film, "The Birds." She could see herself all too well as Tippi Hedren fleeing from the wrath of gulls, regardless of whether they were ring-billed or California.

"So what do you believe?" Mother asked.

"I believe every woman should own at least one pair of red shoes . . ." I answered.

Mother grinned, "I'm being serious."

"So am I."

"When I was a young woman with four children, I was always living ahead of myself," she said. "Everything I was doing was projected toward the future, and I was so busy, busy, busy, preparing for tomorrow, for the next week, for the next month. Then one day, it all changed. At thirty-eight years old, I found I had breast cancer. I can remember asking my doctor what I should plan for in my future. He said, 'Diane, my advice to you is to live each day as richly as you can.' As I lay in my bed after he left, I thought, will I be alive next year to take my son to first grade? Will I see my children marry? And will I know the joy of holding my grandchildren?" She looked out over the water, barefoot, her legs outstretched; a white visor held down her short, black hair. "For the first time in my life, I started to be fully present in the day I was living. I was alive. My goals were no longer long-range plans, they were daily goals, much more meaningful to me because at the end of each day, I could evaluate what I had done."

A flock of sandpipers wheeled in front of us.

"I believe that when we are fully present, we not only live well, we live well for others."

Mimi questioned her, "Why is it then, Diane, that we are so willing to give up our own authority?"

"It's easier," I interjected. "We don't have to think. The

responsibility belongs to someone else. Why are we so afraid of being selfish? And why do we distract and excuse ourselves from our own creativity?"

"Same reason," Mother replied. "It's easier. We haven't figured out that time for ourselves is ultimately time for our families. You can't be constantly giving without depleting the source. Somehow, somewhere, we must replenish ourselves."

"But that's antithetical to the culture we belong to," Mimi said. "We are taught to sacrifice, support, and endure. There are other virtues I am more interested in cultivating," she said, smiling.

"I have a joke." I said. "How does a man honor a woman?"

"I don't know—" Mother answered.

"He puts her on a pedestal and then asks her to get down on it."

Mimi laughed. Mother tried not to.

"That's terrible, Terry."

"Oh, Mother, loosen up. There's nobody spying on us—unless these rocks are bugged." I picked one up and looked underneath.

"We haven't touched our astrology charts," Mimi said, pulling out hers.

Mother and I found ours. We read each other's. We had already listened to the individual taped sessions.

"I liked the part about Terry being neat and meticulous," teased Mother. "I remember standing in the middle of your bedroom when you were about thirteen years old. Everything in your closet was on the floor, art and school papers were piled high on your desk. I remember thinking, I have two choices here—I can harp on her every day of her life, making certain her room is straight—or I can close the door and preserve our relationship."

"Thank you for choosing the latter," I said. "Brooke may feel otherwise."

"The thing that struck me about your chart, Diane," said Mimi, "was the tension in your life between your need for privacy and the obligation you feel toward your family."

"And I think I have paid a price physically," Mother said. She looked out over the lake, then back to me, "Did anything surprise you about your chart, Terry?"

"I think the part that helped me the most was recognizing that I operate with three minds. Remember when she said I can look at a teacup and say, 'Isn't this lovely, notice the pink roses on the white bone china,' or 'Isn't this fascinating, consider the cup in human history,' or 'Look at this teacup, the coffee stains and chip on its rim'—What about you, Mimi?" I asked.

"At seventy-nine, what did I learn? It was more an affirmation of what I already know. I am aware of my intense curiosity, my compulsion to understand the world around me. I value intelligence. I listened hard to those traits I have to watch. I realize I am a very frank, strong personality as a Leo, but I hope I can evolve to be a Leo with wisdom—

"I believe we must do things in our lives for the right reasons, because we enjoy doing them, with no expectation of getting something back in return. Otherwise, we are constantly being disappointed." She moved her turquoise bracelet back and forth on her wrist. "So I had two sons, John and Richard, because I wanted to, not because I thought they would rescue me in old age. I got out of all social organizations and clubs in my fifties so I could spend time with my grandchildren, not because they would give something back to Jack and me later on, but because that was what I wanted to do—and I have loved doing it. Believe me, these have been selfish decisions."

Silence followed.

Mimi looked at me. "And you, Terry?"

"I believe in facing life directly, to not be afraid of risking oneself for fear of losing too much." I paused. Here was my mother standing outside the shadow of cancer and my grandmother standing inside the threshold of old age. These were the women who had seen me through birth. These were the women I would see through death.

The three of us stared out at the lake, the color of Chinese porcelain, and were hypnotized by the waves.

"How do you find refuge in change?" I asked quietly. Mimi put her broad hand on mine. "I don't know . . ." she whispered. "You just go with it."

A killdeer landed a few feet from where we were sitting.

"*Kill-deer! Kill-deer! Kill-deer!*"

"What bird is that?" Mother asked.

"A killdeer," Mimi answered, picking up her binoculars.

I stood up to get a better look. All at once, it began to feign a broken wing, dragging it around the sand in a circle.

"Is it hurt?" Mother asked.

"No," I said. "We must be close to its nest. She's trying to distract us. It's a protective device."

"We're not so different," Mimi said, her silver hair shining in the sun. "Shall we go?"

As we got up to leave, Mother turned to me, "I'm so glad you wore your red shoes . . ."

r r r r r r r r r

WHISTLING SWAN

r r r r r r r r r

The snow continues to fall. Red apples cling to bare
branches.

I just returned from Tamra Crocker Pulsifer's funeral. It
was a reunion of childhood friends and family. Our neighborhood sat on wooden benches row after row in the chapel.
I sat next to Mother and wondered how much time we had
left together.

Walking the wrackline of Great Salt Lake after a
storm is quite different from walking along the seashore
after high tide. There are no shells, no popping kelp or crabs.
What remains is a bleached narrative of feathers, bones,
occasional birds encrusted in salt and deep piles of brine
among the scattered driftwood. There is little human debris
among the remote beaches of Great Salt Lake, except for the

shotgun shells that wash up after the duck-hunting season. Yesterday, I walked along the north shore of Stansbury Island. Great Salt Lake mirrored the plumage of immature gulls as they skimmed its surface. It was cold and windy. Small waves hissed each time they broke on shore. Up ahead, I noticed a large, white mound a few feet from where the lake was breaking.

It was a dead swan. Its body lay contorted on the beach like an abandoned lover. I looked at the bird for a long time. There was no blood on its feathers, no sight of gunshot. Most likely, a late migrant from the north slapped silly by a ravenous Great Salt Lake. The swan may have drowned.

I knelt beside the bird, took off my deerskin gloves, and began smoothing feathers. Its body was still limp—the swan had not been dead long. I lifted both wings out from under its belly and spread them on the sand. Untangling the long neck which was wrapped around itself was more difficult, but finally I was able to straighten it, resting the swan's chin flat against the shore.

The small dark eyes had sunk behind the yellow lores. It was a whistling swan. I looked for two black stones, found them, and placed them over the eyes like coins. They held. And, using my own saliva as my mother and grandmother had done to wash my face, I washed the swan's black bill and feet until they shone like patent leather.

I have no idea of the amount of time that passed in the preparation of the swan. What I remember most is lying next to its body and imagining the great white bird in flight.

I imagined the great heart that propelled the bird forward day after day, night after night. Imagined the deep breaths taken as it lifted from the arctic tundra, the camaraderie within the flock. I imagined the stars seen and recognized on clear autumn nights as they navigated south. Imagined their silhouettes passing in front of the full face of the harvest moon. And I imagined the shimmering Great Salt Lake calling the swans down like a mother, the suddenness of the storm, the anguish of its separation.

And I tried to listen to the stillness of its body.

At dusk, I left the swan like a crucifix on the sand. I did not look back.

ʄʄʄʄʄʄʄʄ

GREAT HORNED OWL

ʄʄʄʄʄʄʄʄ

lake level: 4208.45'

"It was a perfect archetype," Mimi said of Thanksgiving in
Milburn, Utah. "A log cabin in the woods with turkey on
the table and four generations gathered together to pray. It
couldn't be more American."

She was right. We had flocked to my aunt and uncle's
place in a small, rural community. Rich and Ruth invited
the entire Tempest tribe down for Thanksgiving. Twenty-
six relatives arrived throughout the day.

While Mimi, Mother, and Ruth were in the kitchen
preparing the feast, we children were allowed to be children
again.

"Your time will come . . ." Mimi warned.

We bolted outside, seven boys and two girls, more like
brothers and sisters than cousins. My cousin Lynne and I

walked along the creek as our brothers went looking for deer.

"How's Diane?" she asked.

"Good," I replied. "I think it's been an adjustment for her to realize the doctors have done all they can do. The chemotherapy and radiation are over. But you can't live by your prognosis. Mom has this uncanny ability to get on with her life. I honestly think she's fine." I reached down and picked up a feather.

"Great horned owl," I said, handing it to Lynne. "Maybe tonight we can go owling. It's a full moon, you know."

We returned and joined our fathers and grandfather on the porch.

"Find anything?" Rich asked.

Lynne showed him the feather.

"Great horned," he said. He pointed to the one tucked in the band of his cowboy hat.

Lynne and I smiled. Jack took it and ran it through his fingers. "Beautiful . . ." he said, passing it on to Dad. They continued discussing state politics, Dad using the feather to accentuate his points.

"They're letting in too many out-of-state contractors," he said passionately. "There's not enough work to go around."

"And the bidding has turned into a free-for-all," added Rich.

Inside, Mimi finished making the gravy—the same recipe her grandmother had used—and announced dinner was ready. Ruth opened the back door and rang the triangle. We each found our place around the huge pine table. My uncle prayed in his deepest voice, giving thanks for all that brought us together.

"Amen," we said in unison. The platters of food were passed.

After dinner, my cousin Bob built a fire. The men stretched out on the floor and slept. Other relatives were scattered throughout the cabin.

Mother and I were washing dishes. Mimi and Ruth checked the turkey for any last filaments of meat while Lynne divided up leftovers.

"Here you go, Diane," Ruth said, handing Mother the wishbone.

Mother took the wishbone and wiped it with her towel. "Should we let it dry or do it now?" she asked.

"Let's do it now," Lynne said.

Mother handed me the wishbone, knowing my end would break.

"Pull," she said with a mischievous smile.

ƎƎƎƎƎƎƎƎ

ROADRUNNER

ƎƎƎƎƎƎƎƎ

lake level: 4210.90'

I asked Mother if she would accompany me to the West Desert to check out a particular site where I was to lead a field trip for the museum. I have traded my position as curator of education for naturalist-in-residence, which means more time in the field, more time to write, and more time with Mother.

We drove west on Interstate 80 toward Nevada making fishtails on the flooded highway in Mother's Saab. Phalaropes were spinning where the median strip once was. With the sunroof open, I watched gulls. A large green sign on its way to being underwater read, GREAT SALT LAKE TEN MILES.

"I've never seen anything like this," said Mother. "What are they going to do with all this water?"

"Pump it away," I answered.

About seventy miles later, we saw where the dikes were

to be built on the salt flats. The whole country looked like a mirage against the purple backdrop of the Silver Island Range.

Only this time, it *was* the lake.

We were approaching a nine-story concrete structure, the newly erected, "Tree of Utah." Its brightly colored spheres (leaves?) resembled enormous tennis balls, thirteen feet in diameter, poised on top of an eighty-three foot lightning rod. We pulled off the freeway, got out of the car, and walked to its base.

I jumped onto the platform and read the plaque out loud: " 'Metaphor,' by Karl Momen."

We both looked at the steel tree and then at each other. This was the work of a European architect who saw the West Desert as "a large white canvas with nothing on it." This was his attempt "to put something out there to break the monotony."

With the light of morning, it cast a shadow across the salt flats like a mushroom cloud.

"Another roadside attraction in the West . . . " Mother said.

Another car stopped. We returned to ours and drove on. In the rearview mirror, the man-made tree rose from the salt flats like a small phallus dwarfed by the open space that surrounded it.

We checked into the Stateline Casino for the night. Wendover, Nevada, is to Salt Lake City what Las Vegas is to Los Angeles. Mother and I were given complimentary tickets redeemable for ten dollars worth of nickels. Mother agreed a night in front of a slot machine would be more entertaining than a movie. After settling into our room, we descended upon the casino.

We let our eyes adjust to the neon-induced darkness, the

black walls and gilded ceilings, the chaos of blips and bloops
from the adjacent video arcade, and the constant ringing of
bells, falling of coins, and ebullient cries of winners.

We sat at two adjacent red stools and began inserting nickels and pulling down levers. Almost instantly, Mother began
winning—cherries, bells, single bars, and doubles. I inched
my stool closer to the machine. Things started picking up. I
didn't take my eyes off the flashing cherries. Fast and furious,
we pulled the levers—simultaneously. Mother winning. Me
winning. Nickels were hitting our silver trays like heavy
rain. By now, my left foot was up on the counter between
our two machines for leverage. Five nickels in, pull the arm
down; spin, spin, spin; bar, bar, bar; nickels rain down.

A small crowd gathered.

"These women are hot!" someone yelled.

Three sevens. That's what we needed.

Five nickels in, pull the lever down, cherries roll back,
forward and stop. I was communing with sevens. I could see
them in my mind. *Concentrate,* I kept telling myself as I
whispered to the machine, "Let go . . . let go . . ." All evening,
I had been putting in five nickels for the big one-hundred-
dollar pot. My eyes were glazed and my arm was loose. Five
nickels, pull the lever; five nickels pull the lever; five nickels,
pull the lever; one nickel, pull the lever. . .

7–7–7. Mother looked. I looked. The pit manager slapped
his thigh and groaned. Two hundred nickels began dropping
into the tray. Twenty bucks. It could have been one hundred
and the release of two thousand nickels. But on that particular whirl, I played it safe.

The pit manager offered his condolences. Mother and I
laughed until we cried. Her mascara was running down her
cheeks.

"There's got to be a lesson here," I said, my foot still
resting on the side of the machine.

Mother pulled out her handkerchief, still laughing, and began wiping her eyes. "Oh, Terry please, just this one time, let it be bad luck!"

I received a letter from Mimi today. They are spending the winter in St. George, Utah. It reads:

Dearest Terry,

Jack has checked the mailbox every day for a week. As he was asleep this afternoon, I decided to do it myself, and there it was—clean, large, and white—your letter.

It is wonderful to hear from you. I'm so glad your time is your own now, even though there will be adjustments in your change of job.

I awakened this morning at 4:00 a.m. to see Halley's Comet. I tried to put on my slippers, robe, and jacket quietly, when suddenly I heard this voice ask, "And what may I ask are you up to?"

Jack decided to get up, too. We couldn't see the southern horizon from the porch so we decided to search for it. We were out the door by 5:15 a.m. The problem was where to go.

We tried the road to Bloomington Hills until we hit Black Road. It was a perfect view, but by this time, it was 6:00 a.m.—too late.

But what a morning. To watch the light slowly appear in the east—the colors changing moment to moment; the peach and pink of the sunrise, the deep purples, blues, and grays—I wasn't going to see Halley's Comet, but the beauty of the sky and earth were worth the effort.

I feel I need to make every effort to see "Halley." The writer Loren Eiseley made it come alive for me through his description. He had seen it as a child and hoped to see it again as a man. He died a few years ago. I feel I have to see it for him. Thank goodness, I saw the little there was in November. I have until March 22, after that the moon will be too bright.

It is in the eastern/southeastern sky, a little south of Capricornus. It's heading for the teapot. Find Aquarius, and then look directly east past Sagittarius—specifically the two stars that make up his tail. I hope you can find it. Look for both of us. And I'll do the same. In April, the comet will be very low on the horizon and difficult to see in the Mountain West.

I talked to Diane on the phone yesterday. She sounds good, busy as usual.

Terry, I think of you many times each day. Are you dreaming, dear? Send me some. It is helpful to write them down. We can discuss them over the phone if you wish.

Jack and I are feeling great and enjoying each other. After fifty-five years we understand each other so well. A fight is great now and then. It peps things up.

I'm looking forward to the Bird Refuge when we get home.

All our love,
Mimi

I saw it! Faintly above the southeastern horizon, just before dawn. Halley's Comet. A dusting of celestial particles. With my binoculars I thought I could even see its tail. It hung in the sky like a tear.

As the morning light leached into darkness, the comet vanished.

"One more time . . ." I kept whispering under my breath. "Let me see it one more time."

4210' and rising. The governor's office is once again considering pumping Great Salt Lake into the West Desert. The hopes that the breached Southern Pacific Causeway would reduce the lake and buy time until the weather subsided have been dashed.

The Utah legislature appropriated funds to conduct a required environmental impact study and develop final designs for the West Desert Pumping Project. The cost estimate revealed that the higher water level, among other factors, had increased the cost to nearly $90 million. The project involves pumping water into a canal at the Hogup Mountain Ridge and introducing it into the salt desert, where it would spread out over a five-hundred-square-mile evaporation pond on the western side of the Newfoundland Mountains. The water in the West Desert pond would be contained by two dikes: the Bonneville Dike, approximately twenty-five miles long, which would run from Floating Island south to Interstate 80 and then along I-80 for another twelve miles; and Dike Number Two, which would extend from the southern end of the Newfoundland Mountains and run seven miles in a southeasterly direction. This dike would contain an overflow weir, which would allow the heavy, concentrated brine to flow back into the north arm of Great Salt Lake, allowing the elevation of the western pond to be varied as a means of maximizing evaporation.

The heavy brine would be allowed to flow back into the lake for two reasons. First, the evaporation rate decreases rapidly with increased salinity concentrations (the main function of the project is to evaporate water); second, the salts settling to the bottom of the evaporation pond would decrease its storage capacity and eventually decrease the viability of the project.

This month, the governor's office requested a review of the project, to determine ways to reduce the overall cost.

The new analysis reveals that a major reduction in the cost of the project could be realized by taking water from the north arm instead of constructing the diversion structure and twelve mile canal to take water from the south arm.

This would be feasible because the salinity of the north arm has decreased from 22 percent in 1984, prior to the breach in the causeway, to 15 percent.

An additional reduction in cost has been found by assuming that the Bonneville Dike can be built at a lower elevation—and risking that the dike, under certain circumstances, would be overtopped.

These design changes reduce the overall price tag of the project from $90 million to $60 million. It is now called the "bare bones" of the West Desert Pumping Project.

"I thought the marsh would be here forever," I said to Mimi standing on the edge of the flooded Bird Refuge.

Her eyes scanned Great Salt Lake.

"Things change," she said.

Afterwards, we ate lunch at the Idle Isle. Country fare in the form of mashed potatoes and gravy, pot roast, corn, and two soft dinner rolls that pull apart. It is good comfort food where nothing is complicated except the decision after the meal as to which chocolates to take home.

Mimi talked about Mother, how at fifty, women wonder what they have done with their lives. What do they believe? What is of value? What should they do with the new freedom that is theirs now that their children are for the most part, grown?

"It's a wonderful time in a woman's life to really explore the possibilities. Your mother has changed a great deal over the years." Mimi said. "And I think her cancer had a lot to do with it. During the early 1970s when many women were rethinking their roles within the home and confronting their own independence, I saw Diane focusing on her health, living, surviving, so she could raise you children. Along the way, she became much more philosophical. I admire how she protects her energy and understands her limitations."

"What was it like when your mother passed away?" I asked Mimi.

"I was twenty-eight years old. I had just given birth to John when I found out Mother had died from a stomach ulcer. A sudden infection. She had just made plans to come from Washington, D.C. to see him."

She paused.

"I'll never forget the telegram, my sister Marion sent. I couldn't believe it. It was so final. Suddenly, the world seemed very dark. I couldn't imagine how I was going to live without her and I grieved deeply that she was never able to see her first grandchild. But I will tell you, Terry, you do get along. It isn't easy. The void is always with you. But you will get by without your mother just fine and I promise you, you will become stronger and stronger each day."

Mother. She is preoccupied. Yesterday, on the telephone, she said she didn't think she could make the family backpacking trip in the Tetons scheduled for summer.

"I think I may have pulled some muscles in my stomach," she said.

I want to believe her.

It rains and rains. Great Salt Lake continues to rise.

Eudora Welty, when asked what causes she would support, replied, "Peace, education, conservation, and quiet."

Mother, Mimi and Jack, and I are seeking quiet in St. George, Utah.

Early this morning, we decided against our planned hike to Beaver Dam Wash in the Mojave Desert. At dawn, another nuclear bomb was being detonated underground at the Nevada Test Site.

Mimi and I were in the living room reading, Jack was outside, when Mother exclaimed from the kitchen, "They're here!"

We ran out on the balcony. It was a slow-moving river, hundreds of people walking on behalf of nuclear disarmament. The Great Peace March. We left the house to greet them.

Up the hill toward Green Valley, they walked by us—a procession of children, parents, and grandparents.

"I could join them," Mother said under her breath as we clapped for them.

A song rose up from the activists:

We are a gentle, loving people
and we are walking, walking for our lives—

We walked with them. It was the first time I had ever heard Mother and Mimi sing outside of church.

From the corner of my eye, I saw a roadrunner poised on the desert. I have never considered them to be a patriotic bird, but with its patch of red, white, and blue skin painted like a flag on the side of its head, I looked at him differently.

MAGPIES

lake level: 4211.30'

The Mormon Church declared Sunday, May 5, 1986, a day of prayer on behalf of the weather; that the rains might be stopped. The "Citizens for the Return of Lake Bonneville" also declared it a day of prayer; that the rains might continue. Each organization viewed the other as a cult.
Monday, it rained.

Flocks of magpies have descended on our yard. I cannot sleep for all their raucous behavior. Perched on weathered fences, their green-black tails long as rulers, wave up and down, reprimanding me for all I have not done.
I have done nothing for weeks. I have no work. I don't want to see anyone much less talk. All I want to do is sleep.
Monday, I hit rock-bottom, different from bedrock which is solid, expansive, full of light and originality.

Rock-bottom is the bottom of the rock, the underbelly that rarely gets turned over; but when it does, I am the spider that scurries from daylight to find another place to hide.

Today, I feel stronger, learning to live within the natural cycles of a day and to not expect so much from myself. As women, we hold the moon in our bellies. It is too much to ask to operate on full-moon energy three hundred and sixty-five days a year. I am in a crescent phase. And the energy we expend emotionally belongs to the hidden side of the moon.

Mother called from St. George. Yesterday, she hiked alone in Zion National Park. Finally, she has her solitude. Her voice was radiant. "Until you go through this process of facing death, or the probability of it—no one can ever know there is something that takes its place. It goes beyond hope."

Mother's whole being is accelerated. I see her insatiable curiosity intensify. Her desire to absorb everything that is fresh and natural and alive is magnified. She is the bird touching both heaven and earth, flying with newfound knowledge of what it means to live. She is reading Zen, Krishnamurti, and Jung, asking herself questions she has never had the courage to explore. Suddenly, the shackles which have bound her are beginning to snap, as personal revelation replaces orthodoxy.

"When I get home, we'll have a chaparral tea party," Mother said. "It's supposed to strengthen your immune system. I'm drinking some now. It looks like a drug stash."

Her inner retreat of the past few months has momentarily been replaced by openness.

"It's all inside," she said. "I just needed to get away, to be reminded by the desert of who I am and who I am not. The exposed geologic layers in the redrock mirror the depths within myself."

She paused over the phone.

"Remember when I asked you what you believed in?" I nodded and took her bait. "Yes," I said. "So what do you believe, Mother?"

"I believe in me."

Last night, I spoke at one of the Circle Meetings of the Baptist Church. Afterward, a Kenyan friend, Wangari Waigwa-Stone, and I spoke about darkness and stars.

"I was raised under an African sky," she said. "Darkness was never something I was afraid of. The clarity, definition, and profusion of stars became maps as to how one navigates at night. I always knew where I was simply by looking up." She paused. "My sons do not have these guides. They have no relationship to darkness, nothing in their imagination tells them there are pathways in the night they can move through."

"I have a Norwegian friend who says, 'City lights are a conspiracy against higher thought,'" I added.

"Indeed," Wangari said, smiling, her rich, deep voice resonating. "I am Kikuyu. My people believe if you are close to the Earth, you are close to people."

"How so?" I asked.

"What an African woman nurtures in the soil will eventually feed her family. Likewise, what she nurtures in her relations will ultimately nurture her community. It is a matter of living the circle.

"Because we have forgotten our kinship with the land," she continued, "our kinship with each other has become pale. We shy away from accountability and involvement. We choose to be occupied, which is quite different from being engaged. In America, time is money. In Kenya, time is relationship. We look at investments differently."

"It all comes down to dollars and cents," Dad said over the phone this morning.

"I've got a tip from Mountain Fuel. It looks like the governor is going ahead with the West Desert Pumping Project. Thirty-seven miles of six-inch pipe will have to be laid for the natural-gas line to fuel the pumps. The line will run from a site near AMAX's plant to the pumping station near Hogup Ridge. If Mountain Fuel is awarded the $2.7 million contract to build the transmission line to supply power for the pumps, they'll open it for bidding within the next couple of months. I want to take a look at the country so it's in my mind before we actually start figuring footage. Do you want to drive out with me?"

I was delighted to get out.

No drive to the West Desert is simple, especially one to the west shore of Great Salt Lake. We took I-80, turned north toward Lakeside, and then bumped along dirt roads until Dad decided it was time to stretch out and walk.

"You've got to get a feel for the land before you can lay the pipe," he said. "Nothing is as it appears. What do you see?"

We stood on a ridge of the Hogup Mountains.

"I see miles and miles of salt flats and sage, greasewood, and shadescale."

"How does the digging look to you?"

"It looks fairly easy, not that much rock."

"That's where you'd get into trouble."

We hiked off the ridge toward the salt flats. Dad's pace was brisk. What appeared to be an easy walk took several hours. Dad began digging a test hole. The hole filled with water.

"The water table, of course." I mused.

"Exactly," he said. "Because of the lake level, these flats are saturated. You have to build that into your costs." He dug a few more. Same results.

"I'd love to get this job," he said, his eyes squinting from the sun. "It would be exciting to be part of this project, even though I think the whole concept is ridiculous. We'll pump the lake into places it had no intention of going . . . the lake will recede and then what will be left?"

"What would happen," I asked, "if the governor said, 'I've decided to do nothing. Great Salt Lake is cyclic. This is a natural phenomenon. Our roads are built on a flood plain. We will move them.'" I looked at my father.

"He'd be impeached," Dad said, laughing. "The lakeshore industry is hurting financially. The pumping project is a way to bail out the salt and mineral companies, Southern Pacific Railroad, and a political career as well."

"Or ruin one . . ." I said.

"Politicians don't understand that the land, the water, the air, all have minds of their own. I understand it because I work with the elements every day. Our livelihood depends on it. If it rains, we quit. If it's a hundred degrees outside, our men suffer. And when the ground freezes, we can't lay pipe. If we don't make adjustments with the environment, our company goes broke." He looked out over the huge body of water glistening with salt crystals. "Sure, this lake has a mind, but it cares nothing for ours."

A special session of the Utah legislature was called to authorize $60 million for the construction and operation of the West Desert Pumping Project. The okay was given, the funds released, with the first pump slated to begin its job of bailing Great Salt Lake out to the desert in February 1987.

A deep sadness washes over me for all that has been lost. The water level of Great Salt Lake is so high now that it recalls the memory and reality of Lake Bonneville. The Wasatch Mountains capped with snow seem to rise from a sparkling blue sea.

I am not adjusting. I keep dreaming the Refuge back to what I have known: rich, green bulrushes that border the wetlands, herons hidden behind cattails, concentric circles of ducks on ponds. I blow on these images like the last burning embers on a winter's night.

There is no one to blame, nothing to fight. No developer with a dream of condominiums. No toxic waste dump that would threaten the birds. Not even a single dam on the Bear River to oppose. Only a simple natural phenomenon: the rise of Great Salt Lake.

ｆｆｆｆｆｆｆｆ

LONG=BILLED
CURLEWS

ｆｆｆｆｆｆｆｆ

lake level: 4211.65'

It is snowing at Bear River in May. I can only drive out
three miles west of Brigham City. The lake stops me. Before
the flood, it was a fifteen mile trip. The waves of Great Salt
Lake are lapping just below where my car door opens. Gray
sky. Gray water. I have the sense that I am suspended in the
middle of the lake with pelicans, coots, and grebes. I keep
driving with the illusion that my old Peugeot station wagon
is really a boat. When the lake starts seeping into the floor-
boards, I come to my senses. I stop the car, carefully open
the door and climb on to the roof.

Today's storm has brought in the birds. Everywhere I
look, wind and wings. Swarms of swallows dip down at the
crest of each wave to feed. Ibises, avocets, and stilts forage
in the submerged grasses. Geese fly above them, and it is
unclear whether snowflakes fall or feathers. It is one of those
curious days when time and season are out of focus, when
what you know is hidden behind the weather.

I return the next day to find clear skies and fewer birds. Instead, it is midge heaven with dead carp heaved on the road by the waves of yesterday's storm. The smell is foul, but it doesn't seem to bother the fishermen. I have joined them with my low-rider lawn chair. We are evenly spaced like herons along the banks of the Bear River.

This is a heavily used area a few miles west of Brigham City, known to locals as "First River." It smells of stale fish eggs and trash. Broken slabs of concrete litter the ground. But it's the only place left near the Refuge to watch birds.

Unless you have a raft.

I watch two western grebes through my binoculars. Their eyes are rubies against white feathers. The male's black head-feathers are flared and flattened on top, so they resemble Grace Jones. The female is impressed as she swims alongside. All at once, they arch their backs, extend their necks, and dash across the flat water with great speed and grace. They sink back down. They rise up again, running across the water. They sink back down.

This is the western grebes' "water rush," their courtship dance that ensures the species. I brought along Julian Huxley's *The Courtship Habits of the Great Crested Grebe* to read by the river, just in case there were no birds.

After the grebes retreat into the bulrushes, I flip through the small book, stopping at Huxley's description of the "weed-trick ceremony."

> Taken all in all, the courtship is chiefly mutual and self-exhausting, the excitatory, sexual form of courtship such as weed offering or pure display serve not as exci-

tants to coition, as in most birds, but as excitants to some further act of courtship.

Although Huxley writes about European cousins to the western grebes, family characteristics are hard to shake. What great-crested grebes do, western grebes do also.

The two grebes I have been watching, white-throated and black-backed, begin circling one another and bobbing their heads. Between head-shakes, the male rolls his neck on to his back and seductively preens feathers.

Huxley describes this behavior to a tee: "The simplest form of courtship action is the bout of shaking . . ." Huxley elaborates:

> Shaking may take place either before or after courtship actions . . . it varies a certain amount in intensity and in length and also in the amount of habit-preening that takes place . . . each bird excites the other. One gently shakes its head under the force of rising emotional tension; the other bird had not quite got to that stage, but the sight of its mate shaking acts as a stimulus, and it too pricks up its head a little and gives a shake. This reacts on the first bird, and so the excitement is mutually increased and the process fulfills itself.

I am a voyeur. The fisherman to my left asks me if I have been here before. Without thinking, I turn to the man and shake my head in a rather grebelike way, then immediately blush, hoping he has not been watching the amorous birds and mistaken my behavior as flirtatious.

I decide to walk along the river's edge. I stir up clouds of midges. They rise in thick black columns that sound like the string section of an orchestra holding one note as the bow moves frantically back and forth across the bridge. I take a few steps, and the winged column narrows as they raise their pitch another octave.

Through my binoculars, in a continued scan, I spot three wrecked cars, one nose down in the cattails, a Pontiac with a great blue heron standing on its tail lights. There is a spray of gunshot. The heron flies. Three ibises spring up, then float back down into the grasses. I turn. Suddenly, I feel as vulnerable as the long-legged birds.

On my way home, I stop at a favorite pond to watch a pair of cinnamon teals. Barn swallows fly in and out from under the bridge. Dozens of nests are plastered with mud against the concrete beam. A barn swallow is busy lining its cuplike nest with white down feathers. It flies, returning seconds later, with another piece of down in its beak. I wonder where the cache is—most likely a goose nest.

The cliff swallows' nests are different from the barn swallows', although both are built beneath the bridge. Their nests are enclosed, with a small hole left open as an entrance. One pair, their nest barely a shelf, takes turns bringing back dabs of mud. Ten dabs of mud in five minutes. Within an hour, I watch them pack 120 beak-loads of mud onto their new residence. The swallows tirelessly fly to the mudflats on the edge of the pond, load up their bills, return to the construction site, vibrate their heads as they pour the mud onto the nest. Then they vigorously pat it and shape it around their nest. They alternate turns as the male flies from the nest to the mudflat, loads, while the female pats. He returns, she flies out. Over and over again, the same painstaking work, as their tiny feathered bodies quiver with purpose. The shelf slowly, steadily, becomes a closed dwelling.

The spinning of phalaropes. The courtship of grebes. The growth of a swallow's nest. Each—a natural history unfolding.

North of Promontory Point, where the golden spike commemorated the completion of the transcontinental railroad on May 10, 1869, there is a remote vale called Curlew Valley. It is the breeding ground of the long-billed curlew.

In recent years, the long-billed curlew, the largest North American shorebird, has been declining in number in the Great Basin, as it loses much of its breeding habitat to the plow and other land developments. In the midwest, it has been extirpated as a breeding species altogether.

The eskimo curlew is close to extinction. At the turn of the century, in its northward migrations a single flock covered forty to fifty acres in the grasslands of Nebraska. They were known as "prairie pigeons" or "dough birds." As wagonloads were shot and sold, they took the place of the passenger pigeon on the marketplace. Hunters followed the curlews' migration from state to state, literally making a killing. Those who remember the eskimo curlew's call, say it sounded like "the wind whistling through a ship's rigging."

If grasslands continue to shrink, the long-billed curlew could follow the same path as its relative. Its plaintive cry resounds like a warning.

Long-billed curlew, *Numenius americanus,* takes its genus from the Greek *neos,* meaning "new" and *mene,* "moon." The shape of its long bill was thought to resemble the curvature of the sliver moon.

If new moon is defined as no moon or dark moon, the curlew could be associated with destructive powers, for it was long believed that ghosts, goblins, and witches were at the peak of their power in the dark of the moon.

In folklore, this relationship between curlews and black magic stands. A prayer of the Scottish Highlands asks "to be saved from witches, warlocks, and aw lang-nebbed things." In Scotland, the word *whaup* is the name of both the curlew and a goblin with a long beak who moves about under the eaves of attics at night.

In *The Folklore of Birds,* Edward Armstrong writes, "Flocks of curlews, passing over at night and uttering their plaintive, musical calls have also been regarded as the Seven Whistlers, and in the north of England their voices were said to presage someone's death."

He goes on to say, "The curlew's low-pitched fluting is sufficiently near the range of human voice to arouse in the heart the sense of weirdness which we are apt to feel on hearing sounds which have some simulation to but do not really belong to the world of men."

Curlews have been seen as winged souls with foreboding messages. Curiosities of natural history have been defined by curlews. An old-timer of the moors once told a friend of mine there was always an accident after hearing "them long-billed curlews." He spoke of a flock passing overhead and, a few minutes later, their boat overturned. Seven men drowned.

But the flipside of darkness is light. The new moon is also the resurrected moon, soon to be crescent, quarter, then whole. It is the time in many cultures to sow seeds. During the waxing moon all those things that needed to grow are attended to.

In the dark of the moon there is growth. Plants do not flourish in the noonday sun, but rather in the privacy of the new moon.

Maybe it is not the darkness we fear most, but the silences contained within the darkness. Maybe it is not the absence of the moon that frightens us, but the absence of what we expect to be there. A wedge of long-billed curlews flying in the night punctuates the silences and their unexpected calls remind us the only thing we can expect is change.

I found the long-billed curlews at Curlew Valley. A dozen hovered over me like banshees,

"Cur-lee! Cur-lee! Cur-lee!"

I was in their territory and they did not like it. Because of their camouflage, those in the grasses were difficult to see. Movement was my only clue. I counted seven adults. Most were pecking and probing the overgrazed landscape, plucking out multitudes of grasshoppers in between the stubble. Others were contesting the boundaries of competing curlews as they chased each other with heads low in a running crouch. Two curlews faced each other, with necks extended, their long bills pointing toward the sky. They looked ready to fence. Tense gestures, until one bird backed down and flew. The triumphant curlew stepped forward and fluttered its strong, pointed wings above its head. Cinnamon underfeathers flashed like the bright slip of a Spanish dancer.

Female curlews, slightly larger than the males, were prostrate, their necks stretched outward from their bodies. I suspected they were on nests and did not disturb them.

Burr buttercups grew between the grasses like snares, and in prairie dogs' abandoned holes black widows, the size of succulent grapes, reigned.

The hostility of this landscape teaches me how to be quiet and unobtrusive, how to find grace among spiders with a poisonous bite. I sat on a lone boulder in the midst of the curlews. By now, they had grown accustomed to me. This too, I found encouraging—that in the face of stressful intrusions, we can eventually settle in. One begins to almost trust the intruder as a presence that demands greater intent toward life.

On a day like today when the air is dry and smells of salt, I have found my open space, my solitude, and sky. And I have found the birds who require it.

There is something unnerving about my solitary travels around the northern stretches of Great Salt Lake. I am never entirely at ease because I am aware of its will. Its mood can change in minutes. The heat alone reflecting off the salt is enough to drive me mad, but it is the glare that immobilizes me. Without sunglasses, I am blinded. My eyes quickly burn on Salt Well Flats. It occurs to me that I will return home with my green irises bleached white. If I return at all.

The understanding that I could die on the salt flats is no great epiphany. I could die anywhere. It's just that in the foresaken corners of Great Salt Lake there is no illusion of being safe. You stand in the throbbing silence of the Great Basin, exposed and alone. On these occasions, I keep tight reins on my imagination. The pearl-handed pistol I carry in my car lends me no protection. Only the land's mercy and a calm mind can save my soul. And it is here I find grace.

It's strange how deserts turn us into believers. I believe in walking in a landscape of mirages, because you learn humility. I believe in living in a land of little water because life is drawn together. And I believe in the gathering of bones as a testament to spirits that have moved on.

If the desert is holy, it is because it is a forgotten place that allows us to remember the sacred. Perhaps that is why every pilgrimage to the desert is a pilgrimage to the self. There is no place to hide, and so we are found.

In the severity of a salt desert, I am brought down to my knees by its beauty. My imagination is fired. My heart opens and my skin burns in the passion of these moments. I will have no other gods before me.

Wilderness courts our souls. When I sat in church throughout my growing years, I listened to teachings about Christ in the wilderness for forty days and forty nights, reclaiming his strength, where he was able to say to Satan, "Get thee hence." When I imagined Joseph Smith kneeling in a grove of trees as he received his vision to create a new religion, I believed their sojourns into nature were sacred. Are ours any less?

There is a Mormon scripture, from the Doctrine and Covenants section 88:44–47, that I carry with me:

> *The earth rolls upon her wings, and the sun giveth his light by day, and the moon giveth her light by night, and the stars also give their light, as they roll upon their wings in their glory, in the midst of the power of God.*
> *Unto what shall I liken these kingdoms that ye may understand?*
> *Behold all these are kingdoms and any man who hath seen any or the least of these hath seen God moving in his majesty and power.*

I pray to the birds.

I pray to the birds because I believe they will carry the messages of my heart upward. I pray to them because I believe in their existence, the way their songs begin and end each day—the invocations and benedictions of Earth. I pray to the birds because they remind me of what I love rather than what I fear. And at the end of my prayers, they teach me how to listen.

Hundreds of white pelicans appear—white against blue. They turn, disappear. Reappear, black against blue. They turn, disappear. Reappear, white against blue. Through my binoculars, I can see their bright orange bills, many with the characteristic knobs associated with courtship.

The grassy banks of Teal Spring are a welcome reprieve from the barren country I have come from. This is just one of the many small ponds at Locomotive Springs, ten miles from Curlew Valley. It is classified by the Utah Division of Wildlife Resources as a "first-magnitude marsh," which means a place with a stable water supply used by waterfowl for nesting, migration, and wintering. I would call it a first-magnitude marsh simply because it's green.

Brooke will come later this evening. Until then, I shall curl up in the grasses like a bedded animal and dream.

Marsh music. Red-wing blackbirds. Yellow-headed blackbirds. Song sparrows. Barn swallows snapping mosquitoes on the wing. Herons traversing the sky.

Brooke arrives and we walk.

The sign TEAL SPRING is silhouetted against a numinous sky. Its reflection in the pond looks like a black cross. We listen to the catcall of a redhead. Thousands of birds seem to be speaking behind us. We turn around and find only a fortress of greasewood.

Settling into our sleeping bag, I nestle into Brooke's body. We are safe. With our arms around each other, we watch ibis after ibis, heron after heron, teal after teal, fly over us. A few stars appear. We try counting them, until finally the sweet whimperings of shorebirds seduce us into sleep.

Sunrise. Teal Spring is transformed. The pinks and lavenders of the night before have been exchanged for the vitality of yellows and blues. Even the rushes, whose black reflection bled into the water twelve hours earlier, are golden. Instead of the stalks predominating, morning light has struck their flowering heads like a match. Small flames flicker on each tip.

To spend a night at the marsh is to wax and wane with birdsong. At sunset and for an hour or so afterward, the pitch and frenzy of birds is so high, so frantic, idle conversation is impossible. But after midnight, silence. The depth and stillness of Great Salt Lake comes over the wetlands like a mother's calming hand. Morning approaches slowly, until each voice in the marsh awakens.

Brooke and I walk miles across the northwestern wetlands and alkaline flats of the lake. Salt crystals attached to the mud look like blistered skin. The sun is searing and the black gnats are almost intolerable. Relief comes only through concentration, losing ourselves in the studied behavior of birds.

Marbled godwits forage the flats with avocets and stilts. It would be easy to confuse the godwits with curlews, except for their bicolored bills that point upward, not down. And I find their character very different from curlews—more trusting, more gentle, more calm. When a curlew is near, the air is stirred; they are anxious and aggressive. Godwits are serene. They demand little from you except the patience to observe. Curlews cause guilt. You are reminded of your intrusion, that you do not belong.

As we walked along an eroding dike, flush with the roaring lake, a blue heron flies off its nest leaving four large eggs. The nest is built of dried greasewood on an old weathered fence that fans out like an accordion. Two ravens hover with eyes on the eggs. We leave quickly, so the heron can return.

Walking back toward Teal Spring, we discover a dead curlew. Its body lies fixed, encrusted with salt. We kneel down and run our fingers down its long, curved bill. Brooke ponders over the genetic information a species is born with, the sophistication of cells and the memory held inside a gene pool. It is the embryology of a curlew that informs the stubby, straight beak of a chick to take a graceful curve down.

I say a silent prayer for the curlew, remembering the bond of two days before when I sat in their valley nurtured by solitude. I ask the curlew for cinnamon-barred feathers and take them.

They do not come easily.

rrrrrrr

WESTERN TANAGER

rrrrrrr

lake level: 4211.85'

4211.85'. Great Salt Lake has surpassed its historic high of
1873. The date is June 2, 1986. It is also our anniversary.
Eleven years.

Brooke and I vigorously shake a bottle of champagne,
pull the cork, and let it spray into the salty waters of the
south shore. With dripping hands, Brooke pours the champagne into the crystal goblets I hold.

"Don't worry about me in the coming months," he says.
"I know where you need to be."

We toast to marriage and the indomitable spirit of Great
Salt Lake.

I find that the time with Mother is spent in quiet
reflection, oftentimes, talking from our trips across the
desert.

Last weekend, we were driving home from St. George. As we were passing through Provo, Utah, the town where she was born, she turned to me.

"I just remembered the strangest thing from my childhood . . ."

"What is that?" I asked.

"I remember walking home from school one afternoon and seeing Mother and Dad standing in front of our house. I could see in their faces that something was wrong. As I walked up to the door, Dad said, 'Diane, Blackie was hit by a car.' They put their arms around me and cried. What they had just said to me did not seem real. I asked if I could see him. They told me they had buried him in the backyard while I was at school. Mother explained that she didn't want me to have to see my dog that way. In their minds, they had protected me from one of life's sorrows."

"That night, I remember sneaking out of the house in my nightgown, trying to find the place where they had buried my black lab. I found the disturbed soil, knelt on the damp grass and began digging with my bare hands to uncover him. I wanted to see his broken body. I wanted to cradle his bones and see for myself that he was dead. I wanted to cry over the death of my dog. But the hole was too deep and I never found him."

"Isn't that funny I would remember that incident after all these years?"

"Why do you think?" I asked.

"What are you saying?" Mother sounded puzzled.

"I don't know—maybe there is something in that story that you need right now, maybe that's why it surfaced."

Mother turned her head. From the corner of my eye, I saw her staring out her window.

"Maybe I have never been allowed to grieve. Maybe I have never allowed myself to grieve."

"There is no blockage as of now, Diane. We can try another type of chemotherapy called Leukeran, different from the cisplatin and Cytoxan you had two years ago. There's a chance it might shrink the tumor we've found."

"And if I do nothing?" Mother asked.

Dr. Smith looked over at me. I raised my eyebrows to indicate that I was simply a bystander.

"A blockage will occur. I don't know how soon, but you will not be able to eat. At that point, I think you will want the blockage removed—so there may be more surgery—but let's not get ahead of ourselves."

He paused. "You don't think you want to try Leukeran?"

"No," Mother said.

He paused again.

"I respect that. Let's just see how things go, then. Diane, I had hoped—"

"I know," she broke in. "I just want to be able to continue in the decision making. I'm not afraid of my own death, but I am afraid of the pain." She hesitated. "I hope I have the courage to face what's ahead."

"You do," he said. "Call me when you think I can help." Dr. Smith walked us to the door.

Mother turned to him and took his hand, "Thank you. You have been wonderful."

We left the clinic. I looked at Mother and asked how she could remain so strong.

"Tell me, Terry, what choice do I have?"

Mother has chosen not to say anything to Dad and the family until after Hank's birthday, not because she doesn't want them to know, but because she wants to protect herself.

"I don't want everyone hovering over me as though I have a day or two to live. Besides, this is terribly boring."

"I'm not sure I would use the word, boring . . ."

"Illness is boring," she said. "Take my word."

"You seem to have a different attitude, Mother. Is that true?"

"It feels good to finally be able to embrace my cancer. It's almost like a friend," she said. "For the first time, I feel like moving with it and not resisting what is ahead. Before, I always knew I had more time, that the disease was outside of myself. This time, I don't feel that way. The cancer is very much a part of me."

"Terry, I need you to help me through my death."

I laid my head on her lap and closed my eyes. I could not tell if it was my mother's fingers combing through my hair or the wind.

The Bear River Migratory Bird Refuge offices officially closed today, according to the U.S. Fish and Wildlife area supervisor in Denver.

"We have pretty well abandoned the sixty-five-thousand-acre refuge fourteen miles west of Brigham City, because it is impossible to second-guess the Great Salt Lake," said Phil Norton. He explained that the maintenance worker assigned to the Refuge is being transferred to the Fish Springs Refuge, near Dugway in Tooele County. Peter Smith, acting manager of Bear River, will be reassigned with the Denver district, and a part-time secretary will be looking for a job.

At its peak, Mr. Norton said, the Bird Refuge "employed eight full-time people and four seasonal workers." Refuge employees began preparing for high waters from the Great Salt Lake in 1983. The press release cited "most of the fourteen-mile-long blacktop road to the Refuge as under-water," and Box Elder county commissioner chairman, James W. White, said, "At today's prices, it will cost $1 million a mile to elevate and repair the road . . ."

During an inspection trip a month ago, Mr. Norton and Mr. Smith reported, "more than $150,000 worth of damage had been done to the government buildings as a result of the wind blowing large chunks of ice off the lake into the structures."

Bear River now belongs to the birds.

On July 1, 1986, I cooked my first turkey for Hank's twentieth birthday. Brooke came home from work last night and found it soaking in the bathtub. I had forgotten to take it out of the freezer. I wanted Mother to know I could carry out the family traditions, that Thanksgiving and Christmas would be in good hands. It didn't work. The turkey was terrible.

Even so, there was a warmth and closeness to the evening. No one else knew about Mother. We all knew. Sometimes it is appropriate to skate on surfaces.

Dawn to dusk. I have spent the entire day with Mother. Lying next to her. Rubbing her back. Holding her fevered hand close to my face. Stroking her hair. Keeping ice on the back of her neck. She is so uncomfortable. We are trying to work with the pain.

Her jaw tightens. She cramps. And then she breathes.

I am talking her through a visualization, asking her to imagine what the pain looks like, what color it is, to lean into the sensation rather than resisting it. We breathe through the meditation together.

The light begins to deepen. It is sunset. I open the shutters, so Mother can see the clouds. I return to her bedside. She takes my hand and whispers, "Will you give me a blessing?"

In Mormon religion, formal blessings of healing are given by men through the Priesthood of God. Women have no outward authority. But within the secrecy of sisterhood we have always bestowed benisons upon our families.

Mother sits up. I lay my hands upon her head and in the privacy of women, we pray.

It's the Fourth of July, and the family decides to celebrate in the Tetons. Mother says she is sick of lying in bed and needs a change of scenery. I wonder how far she can push herself.

Brooke and I, with Mother and Dad, hike to Taggart Lake.

The Taggart-Bradley fire of last fall has opened up the country. It is a garden of wildflowers with fireweed, spirea, harebell, lupine, and heart-leaf arnica shimmering against the charred bark of lodgepole pines.

I have never been aware of the creek's path until now. It feels good to be someplace lush. The salt desert is too stark for me now because my interior is bare.

We reach the lake, only a mile and a half away, but each step for Mother is a triumph of will. She rests on her favorite boulder, a piece of granite I have known since childhood. She leans into the shade of the woods and closes her eyes.

"This feels so good," she says as the wind circles her. "It feels so good to be cool. I feel like I'm burning up inside."

A western tanager, red, yellow, and black, flies to the low branch of a lodgepole.

"Look, Mother! A tanager!" I hand her my binoculars.

"You look for me . . ." she says.

ꞃꞃꞃꞃꞃꞃꞃꞃ

GRAY JAYS

ꞃꞃꞃꞃꞃꞃꞃꞃꞃ

lake level: 4211.40'

I am retreating into the Wasatch Mountains. I cannot travel west to Great Salt Lake. It is too exposed, too wicked and hot with one-hundred degree temperatures. The granite of Big Cottonwood Canyon invigorates me as I hike from Brighton to Lake Catherine. Glacier lilies blanket the meadows. Usually they are gone by now. I pick one and press it between the pages of my journal.

"For Mother—" I say to myself, rationalizing my act, when I know it is for me.

Hiking the narrow trail up the steep slope massages my lungs. I breathe deeply. Inhale. Exhale. Inhale. Exhale.

I climb up the last pass and break down into the cirque. My lungs and legs feel strong. I have the lake to myself. My ears begin to throb with the altitude. My eyes water in the wind. I take off my rucksack, pull out my windbreaker and lunch. I can see the rock I am going to sit on. I hike down a little further and settle in.

Peeling an orange is a good thing to do in the mountains. It slows you down. You bite into the tart rind, pull it back with your teeth and then let your fingers undress the citrus. Nothing else exists beyond or before this task. The naked fruit is in your hands waiting for sections to be separated. Halves. Quarters. And then the delicacy of breaking the orange down to its smallest smile.

I lay out these ten sections on the flat granite rock I am sitting on. The sun threatens to dry them. But I wait for the birds. Within minutes, Clark's nutcrackers and gray jays join me. I suck on oranges as the mountains begin to work on me.

This is why I always return. This is why I can always go home.

I brought the pressed glacier lily to Mother. I found her sitting in the chaise lounge on the porch with a glass of ice water in her hand. It has been almost a week since she has been able to eat.

Mother turned around. As she took the flower, she said, "Terry, what I have to do now goes beyond the family."

The Tempest clan met for a family portrait. Everyone: Mimi and Jack, Mother and Dad, Richard and Ruth, all nine grandchildren with spouses, plus two great-grandchildren. A large elm with ivy winding around its trunk stood regally in the background. It was all very formal. Nobody wanted to be there. It was my idea. I thought it would be a nice Christmas present for Mimi and Jack. The photographer framed us with his hands, then disappeared behind his black broadcloth.

"Smile!" he yelled. "You all look so somber. What's the matter, is somebody dying?"

We lost control. Laughs turned into tears into sidesplitting hysteria. Richard looked at Dad who looked at Mother who looked at Mimi who looked at Jack, and so on down the family.

The photographer stepped out from behind the camera and shook his head. "Did I say something funny?"

Mother is in surgery. Brooke brought us lunch. The men are talking politics. Dad is figuring a bid. Hank is writing. Steve and Dan are walking the halls. Again.

We wait.

I am suspended between the past and future, held by a spider's filament stretched across a river.

Five twenty-five P.M. My concentration snaps as the doctor enters.

"She's fine." he says. "We removed the blockage. It was at the very end of her small intestine, a much better situation than we anticipated. There is still a sprinkling of cancer cells, but we can work with them."

Dr. Smith looks at my father.

"Maybe a year . . ."

"You still don't understand, do you?" Mother said to me. "It doesn't matter how much time I have left. All we have is now. I wish you could all accept that and let go of your projections. Just let me live so I can die."

Her words cut through me like broken glass. This afternoon, she said, "Terry, to keep hoping for life in the midst of letting go is to rob me of the moment I am in."

We had a slide show in Mother's hospital room. Brooke projected all the different takes of the family portrait on the white wall. We needed Mother's help in deciding which image was the best of everyone. We also brought chocolate cake, ice cream, and balloons, because it was Dad's birthday. Mother wasn't interested.

We raised her bed so she could see the pictures. Finally, she asked to be returned to a horizontal position and simply said, "They all look fine."

The party ended early. Dad, Brooke, and I stayed. The men decided to take a walk outside the hospital. Mother was asleep, her breathing labored. I pulled a chair close to the side of her bed and began quietly breathing with her, emphasizing each exhale.

Almost an hour passed.

Dad and Brooke returned. I stood up and moved the chair back against the wall.

"She looks more relaxed," Dad said.

Brooke looked at me. We kissed her and left.

Mother does not seem to be getting better. Her spirit has turned inward. She has little energy for others. Even the gardenia by her bedside that once brought her great pleasure offers little solace.

Dad and I decide what Mother needs, after fourteen days in a small, square room with little light, is fresh air. Without asking the nurses' or doctor's permission, we sneak her out of the hospital. Gathering all the bottles, bags, and tubing necessary for transport, we wheeled her outside.

It was a glorious summer day with huge cumulus clouds towering over the Wasatch. We took Mother to some gardens of pansies and marigolds. The heat seemed to draw color back in to her pale cheeks and, for the first time in weeks, her eyes brightened.

Dad sat on the grass beside her wheelchair talking in soft tones about the beauty before her, tenderly rubbing her legs. She began to cry from the soles of her feet.

We sat in the sunshine for an hour or more, until she said she was ready to go back inside.

"Thank you."

Dad was wheeling Mother back toward the hospital when a large black dog appeared. We stopped. Mother put out her hand. The labrador licked her palm and then laid his head in her lap. She lifted her other hand from the armrest and gently stroked his head.

At last, my mother grieved.

Mother is home from the hospital. A neighbor who had seen the lights on in the bedroom at midnight brought over some hot, homemade custard. Dad took the glass bowl out to the balcony to cool and brought it back inside when it was comfortably warm. He fed it to Mother. She ate. We stood at the foot of their bed and watched. She had not been able to eat for almost four weeks, until now.

"Delicious . . ." Mother said cooing. "It's absolutely delicious."

These summer days have been relentless with emotional heat. I am exhausted and depleted. This afternoon when I was taking Mother her pain medication, the doorbell rang, and without thinking, I took the pill myself. Standing on the front porch were women from the Relief Society with dinner for the family. It wasn't until Mother asked moments later for the Percodan that I realized what I had done.

She is exhausted from the weeks of sustained pain, and tonight I realized it could be months. Every day is a crisis because our expectations make it so.

"When will I ever feel good again?" Mother asked.

That is the question we are all living.

Steve has been massaging her forehead between pain contractions that come in intervals as predictable as labor. Dan gives her a sponge bath with ice water on the hour to break the fever. I watch our family fight the undertow of grief.

When I left to kiss Mother good-bye for the day, I noticed she was wearing two strands of heishe and pipestone around her neck, not her customary pearls.

"Hank," she said with a grin. "He gave me his medicine beads when I got home yesterday."

"A little magic never hurts," I said.

Once home, I cried on the lawn with the sun sinking into the lake that appeared as a long silver blade across the horizon. But this time I was not crying for Mother. I was crying for me. I wanted my life back. I wanted my marriage back. I wanted my own time. But most of all I wanted the suffering for Mother to end. And then, in the midst of my sorrow, hope seeped in like another drug.

I wrestle with my optimism until the Percodan pins my shoulders to the bed.

I found Dad on his hands and knees, pulling small starts of scrub oak out of the garden.

"It's here—" he said as he looked up at me. "It's really here, isn't it?"

I shook my head and sat down beside him. "I don't know.

I think she'll get stronger. It's just so hard to see her in such pain when there's so little we can do."

Dad picked up his pile of seedlings and threw them in a bag. His tears were quickly absorbed into the soil. I moved closer and put my arm through his.

"I thought we would have more time—" he said, "I just thought we would have more time."

"You learn to relinquish," Mother said to me while I rubbed her back.

"You learn to be an open vessel and let life flow through you."

I do not understand.

"It's not that I am giving up," she said, "I am just going with it. It's as if I am moving into another channel of life that lets everything in. Suddenly, there is nothing more to fight."

How can I advocate fighting for life when I am in the tutelage of a woman who is teaching me how to let go?

This evening, August 6, 1986, we celebrated Mimi's eightieth birthday with the entire extended family. A thunderstorm exploded outside. Immediately, we all vacated the living room and sat on the front porch. With our backs against the house, we watched veins of lightning torch the sky.

"It's a dance," Mimi said.

Mother was home alone.

Mother has moved to Mimi and Jack's house.

"Anything to get rid of the monotony," she said.

We sat in the backyard under the sycamore tree, where

the hose was left running to simulate the sound of a stream.
"It's such a healing sound," Mimi said.

Mother rolled up her pantlegs and let the water run over her feet. She bent down gingerly and washed her face. "Don't tell John I'm playing in the sprinklers," she said. "He'll have me hiking Mount Olympus tomorrow. He's the only person I know who viewed having a hysterectomy as an advantage for backpacking—less weight to carry."

Nothing is working. Mother is writhing in pain.

"Something is terribly wrong," she said after Mimi and I tried to persuade her to eat. "I know my body."

"But the doctor says you are fine. It's just a very slow recovery process," I argued.

In my mind, I don't think she is trying hard enough. She has abandoned the pain medication and relaxation tapes.

In Mother's mind, we are not listening to what she is saying.

For the first time, Mimi is looking like an old woman. She is being worn down like the rest of us. Dad feels like a failure because Mother left home. Mimi and Jack feel like a failure because she is getting worse. I feel like a failure because I am losing my compassion.

We are spent.

I leave tomorrow for a week to participate in an archaeological dig in Boulder, Utah, at Anasazi State Park, sponsored by the museum.

"I'm glad you're leaving," Mother said.

So am I.

ffffffff

MEADOWLARKS

ffffffff

lake level: 4211.00'

A fresh drink of water. A cool breeze. And a swollen Escalante River after a thunderstorm. With the wind billowing my white cotton blouse, I breathe with a clarity of spirit I have not known for months. These expressive skies in constant motion, emotion, move me. Silence. Juniper green. Cottonwood green. Sage blue. Red earth. Burnished skin. Refuge once again, this time in the reverie of southern Utah.

Behind me is a panel of petroglyphs, three figures etched into the cliff by the Anasazi: a warrior, a woman, and a woman with child. They lived. They died. And something of their spirit remains.

A group of ten high school students and two instructors, of which I am one, have been excavating a site under the supervision of Larry Davis, chief ranger at Anasazi State Park.

Transects are measured, quadrants assigned, and the tedious process of removing top soil begins with small shovels and trowels. Each bit of dirt is screened over a wheelbarrow, potsherds are kept and cataloged, along with fragments of charcoal, bone, and worked stone. The afternoon sun beats down on our backs. We repeat these menial tasks over and over again until it becomes a meditation, of sorts. I am astonished by how much soil we have moved in a day. The site adjacent to ours has already been excavated. Larry informed us that they had uncovered a burial: an Anasazi woman, approximate date A.D. 1050–1200.

"But what was unusual about this site were the objects we found buried with her—three ollas, corrugated vessels used for carrying water, and several large balls of clay. You could still see the palm prints of the person who had made them." He paused. "She was wearing a turquoise pendant. We believe she was a potter."

"And where is she now?" I asked.

"We reburied her."

I feel like a potter trying to shape my life with the materials at hand. But my creation is internal. My vessel is my body, where I hold a space of healing for those I love. Each day becomes a firing, a further refinement of the potter's process.

I must also learn to hold a space for myself, to not give everything away. It reminds me of the Indian teachings of Samkhya:

> If you consciously hold within yourself three quarters of your power and use only one quarter to respond to any communication coming from others, you can stop

the automatic, immediate and thoughtless movement outwards, which leaves you with a feeling of emptiness, of having been consumed by life. This stopping of the movement outwards is not self-defense, but rather an effort to have the response come from within, from the deepest part of one's being.

In the middle of Larry Davis's demonstration on primitive technologies, I was handed a pink note by the ranger working at the desk.

"Call home. Brooke."

I got up from the sandstone boulder I was sitting on and felt my legs turn to jelly. It was a long half mile to the telephone.

"Diane's back in the hospital," Brooke said. "It looks like there may be another blockage."

My heart sank.

"Can they operate?" I asked.

"Tomorrow morning," he said. "Is there any way you can get home tonight? Dr. Smith thinks you should be here."

Boulder, Utah, is walled in by wilderness with Lake Powell to the south, Capital Reef to the east, Escalante canyons to the west, and the Boulder Mountains, north. No buses. No trains or planes. No vehicle of my own, just the university van we brought the students in . . .

"I'll find a ride," I said. "I can always hitchhike."

"Just be careful," he said. "I love you."

I hung up the receiver and picked it up again to call the LDS Hospital.

"1–321–1100," I knew it by heart. But the operator couldn't get me through to Mother's room. I put the phone down. I tried to call Dad, no answer. I called Mimi, no answer.

A woman ranger who couldn't help but overhear my conversation said, "I'll find you a ride."

Two hours later, I am in a black, windowless van, sandwiched between two men, one of whom is wearing a cut-off T-shirt which reads, HELMET LAWS SUCK.

"You've got a sick old lady, huh?"

"Yes," I reply, then put on my sunglasses. "I really appreciate the ride."

"No problemo," the dark-haired one answers. "Hope you don't mind, if we make a quick stop back at the homestead to pick up a few more folks."

We drive down a dirt road a few miles beyond Boulder until we reach a large white-washed log house. A shredded American flag is flying on a new painted pole. The van stops.

"Enjoy Sculptured Creek . . ." the other long-haired man says.

I sit by the tiny stream and make a bundle of sage. Sculptured Creek is littered with painted tires, iron peace signs, and other abstract metal objects. The men's names are Robert and Mike. They are artists, and we share a common background in our Mormon ancestry, like almost everyone else in the state of Utah.

Robert has just turned forty. Several women helped him celebrate. The long scar that runs up his right arm like a snake is a memento from Vietnam.

"How old are you?" he asks. "Sixteen? You probably never heard of the war."

I am not flattered. "I heard about it . . ."

He and Mike start whistling for someone. It must be a dog. It is a woman. She is running around the house giggling, chased by another man. I look twice—I think she has clothes on.

"What's that one's name?" Robert asks Mike. He shakes his head. They turn to me, "Stay cool, it won't be much longer, we just have to round everybody up."

Seconds later, two women in tube-tops and cutoffs stagger out from behind the bushes with a skinny man holding a shotgun. I can't decide if it's a throwback to Li'l Abner or the 1960s. The man, drunk, opens fire on the pasture.

"I love to make these little fillies run," he shouts as a horse and mule run in circles.

Home appears like a distant mirage in the rural wildness of Garfield County.

Robert and Mike decide I am a dud and ask if I mind riding in the other car. The tie-dyed blonde chooses to ride with them and, as she opens the back of the van, coos, "Oh, another mattress, how sweet!" She falls in face first. Mike slams the door behind her.

I get into the back seat of a lime green Pinto with another man and woman. As we begin to drive away from "the homestead," Robert walks up to the car and motions me to roll down my window.

"It's been a real pleasure, Terry. I hope everything turns out bitchin." He extends his hand.

I reach out to shake it, and he slips me something.

"I wouldn't want one of my girls traveling without protection . . ."

I roll up the window and discover he has handed me a condom. I fight back the tears.

Meanwhile, the couple who are French-kissing behind the steering wheel ask if I mind if we take a detour to look for her lost rock.

"I lost my pink rock with sparkles on it," she says. "I dropped it when I was riding the Harley."

Looking across the sandstoned desert, all I see are pink rocks.

After a good hour of dirt road driving with our heads out the window in search of the dropped rock, she gives up and settles for another.

"It's just not the same," she laments. "I was so attached to the other one."

"I'm sorry," I hear myself saying. "I know how hard it is to lose something you love."

We stop in the town of Scipio for gas, only to find the pumps not working. And who should come up in his bulging T-shirt and tight-crotched jeans but Robert.

"How'd you like my present?" he asks, trying to pin me against the car with his hips.

"Wrong brand." I remove his hand from my shoulder.

Five hours later, a little after ten, the couple drops me off in front of the hospital. I thank them. We exchange phone numbers. We had become friends. As it turned out, the woman's mother is a textile consultant for us at the museum.

If I had asked enough questions, I am certain I would have discovered Robert and I were related through polygamy several generations back. The dark side of residency.

The family is gathered in Mother's room. Brooke and I check in with our eyes. Lights are low. Mother appears calm, relieved to know there really was something wrong, that the pain wasn't imagined. The operation is scheduled for the morning.

I lean over and kiss her, handing her the bundle of sage wrapped in soft leather.

"I'm so glad you're here . . ." she whispers.

"So am I."

Mother has been in surgery for two and a half hours. Tension is like a shackled horse. Dad is reading. Steve and Grandpa are pacing the halls. I write.

We are pieces of clay being fired again.

One week has passed. Mother and I sit outside, by the hospital fountain and listen to meadowlarks sing, "Salt Lake City is a pretty, little place . . ." It is the song of my childhood. They will be migrating soon.

Mother is frightened, frightened to go on with her life when the future is so uncertain, wondering what her life will be from this point forward.

"I feel like I am on hold until the next thing happens." she says.

She is quiet and frail. She is weary from the physical torment.

"I will not come back here," she says. "I am done with this hospital."

Twenty pounds have been lost. Mother weighs one hundred. But it is her eyes that divulge her suffering. They are deep and dark and distant.

A person with cancer dies in increments, and a part of you slowly dies with them.

STORM PETREL

lake level: 4210.85'

For ten days, I have done nothing but watch whales quietly surfacing, diving deep, and surfacing.

Brooke and I are in Telegraph Cove, a quaint fishing village at the northern tip of Vancouver Island.

We are assisting Jeff Foott who is making a film on killer whales for Survival Anglia. Yesterday, we were out on Johnstone Strait in a twenty-foot Boston Whaler from six in the morning until nine at night.

I was stationed on a cliff with three biologists who had set up a hydrophone at a depth of fifteen feet in the water to record vocalizations. You hear the whales long before you see them. Even with my untrained ear, I could discern dialects—both by individuals, as well as pods.

Several times, a mother with calves would surface. Sleek black and white bodies wheeling through the sea, their dorsal fins appearing as flags. We listened to their tender murmurings. Some whales passed solitary and silent, while others came into the cove singing.

John Lilly suggests whales are a culture maintained by oral traditions. Stories. The experience of an individual whale is valuable to the survival of its community.

I think of my family stories—Mother's in particular— how much I need them now, how much I will need them later. It has been said when an individual dies, whole worlds die with them.

The same could be said of each passing whale.

We are enshrouded in fog, trolling toward a remote island off Johnstone Strait. A storm petrel led us here. We followed her through the mist until she disappeared. Perhaps she was an apparition.

An orange mask flashes on the cliff face. It is a pictograph with huge ominous eyebrows above open eyes and a gaping mouth. Reflections dance across the water, below the face.

Brooke and I jump off the boat and tie the Whaler to barnacle-encrusted boulders. It is low tide and the hissing of intertidal creatures reminds us we are not alone. We step carefully over kelp-covered rocks on to the lush island.

Spruce, hemlock, and giant cedars humble us. The dense undergrowth of alders and devil's club muffles our voices. It is cool and damp. Little light penetrates this ancient forest.

We walk single file for an hour or more. Suddenly, Brooke stops. At the base of a sheer granite wall are broken cedar boxes.

Three boxes. Three skulls. Inside one box are bones; a partial skeleton with crossed femurs wrapped in a woven mat made of cedar.

One skull is grimacing inside a small cavern beneath the cliff. And still another is wide-eyed staring with its detached jaw from the other twisted box. The bones are disintegrating faster than the textiles that hold them. Rope fragments scatter the clearing like small snakes.

Kwakiutl. These are the old ways of a Northwest Coast Indian people, to hang the dead in burial boxes of cedar over a cliff or suspended from a tree.

We do not stay long, nor do we disturb what we see.

Instead, we walk briskly back to the boat. Looking over my shoulder in the wake of the whaler, one would never know bones, human bones, were hidden in the heart of this island.

I feel like I am floating in salt water, completely at the mercy of currents. Mimi was operated on this morning for breast cancer, September 8, 1986, my birthday.

I accompanied Mimi and Jack to the doctor's for the biopsy report.

"Mrs. Tempest," the doctor said. "I have some good news and bad news. The bad news is the biopsy was malignant. You have a rare form of breast cancer known as Paget's Disease. The good news is at this stage, it is 90 percent curable."

The three of us sat across from his desk, numb.

"I recommend a simple mastectomy. It's an easy procedure, basically like cutting off a mole . . ."

Mimi leaned forward and put her elbows on his desk. "Young man, my breast is no mole."

He blinked. He became flustered.

"Of course not, Mrs. Tempest, I simply meant to say . . . "

"I know what you meant," she interrupted. "And I just want you to know what I mean. I may be eighty, but I am still a woman."

Today, I watched two orderlies wheel her out of surgery

on a stainless steel gurney. I followed them back to the room. They left and I closed the door.

"Damn," Mimi said as she covered her face with her hand.

Mother and I retreated to the lake for the afternoon and sat on a newly constructed dike. The beach had long since disappeared.

We did not discuss Mimi. Instead, we took off our shoes and dangled our feet in the water. I dipped my finger for a taste. I expected salt, it was fresh.

"Thirty-one years . . ." Mother said, smiling. "Happy birthday, dear."

She handed me a present wrapped in white paper with a turquoise ribbon. I unwrapped it carefully and opened the box. Inside was a round glass paperweight with gold and black swirls against a jade background.

I cradled the small globe of waves in my hands.

I remember Mimi asking me as a child to make a lens by curling my fingers around to my thumb. I closed one eye and, with the other, looked through my hand lens. I played with scale. Blades of grass were transformed into trees, a gravel bed became a boulder field. Small rivulets pouring over moss became the great rivers of our continent. My world was my own creation.

It still is.

Now if I take this lens and focus on Great Salt Lake, I see waves rolling in one after another: my mother, my grandmother, myself. I am adrift with no anchor to hold me in place.

A few months ago, this would have frightened me. Today, it does not.

R e f u g e

I am slowly, painfully discovering that my refuge is not found in my mother, my grandmother, or even the birds of Bear River. My refuge exists in my capacity to love. If I can learn to love death then I can begin to find refuge in change.

ꞏꞏꞏꞏꞏꞏꞏꞏ

GREATER YELLOWLEGS

ꞏꞏꞏꞏꞏꞏꞏꞏ

lake level: 4210.80'

"Put aside any romantic or spiritual notions of ancient desert people," said Kevin Jones, an archaeologist, as we walked out to Floating Island in the middle of the salt flats. "Believe me, there was no romance in people's lives ten thousand years ago. They pretty much acted as we would—assessing their situation and making decisions based on the choices at hand."

We hiked up the hill to the cave. Floating Island is an isolated outcrop of limestone separated from the Silver Island Range by at least a mile of salt flats. The cave measures ten meters across its mouth, twelve meters deep. It looks south over the West Desert with Great Salt Lake lapping east. No archaeological excavation has been conducted at this site until now.

The Silver Island Expedition, a project funded by the National Science Foundation, hopes to secure archaeological data and specimens before the West Desert Pumping Project floods the sites.

Excavation of Floating Island Cave was undertaken in order to mitigate the adverse effects of construction work on the island. W. W. Clyde, a Utah-based company, is cutting into the flanks of the island, using it as riprap for the building of the dikes.

Kevin empties a bucket of dirt from the cave on the swinging tray of the quadruped. My job is to screen the debris. I pick out a jasper chip and place it in a vial. I delight over the "tink" of rock in plastic.

I empty another bucket of dirt on the tray and shake it back and forth over the screen. I catalog objects in my mind: twigs, pinyon nuts and hulls, a horned lizard's crown, cedar berries, beads, bone beads; hundreds of tiny bones—femurs, tibias, fibulas, ulnas, scapulas, jaws, skulls (most likely from bats, small rodents, and rabbits); beetle carapaces, grasses, seeds, saltbush leaves, red flakes of jasper and obsidian, basket-pressed stone pottery, animal and human coprolites and dust. Lots of dust. My pores quickly become black.

A small arrowhead rolls back and forth across the screen. Kevin identifies it as a Rose Spring notched point.

I take a break and enter the cave to see what's going on.

Think about a people who made clay figurines with shuttered eyes, and then think about the Fremont, desert people who inhabited the eastern Great Basin and western Colorado Plateau from approximately 650 to 1250 A.D., roughly a thousand years ago. They planted corn, irrigated their fields, and used wild foods with ingenuity. In many ways, the Fremont correspond to the Anasazi to the south. But in many ways, they do not.

The Anasazi were a group of people attached to the Colorado Plateau with a complex social organization: clans, elaborate kivas, and road systems. In contrast, the Fremont were small bands of people, much more closely tied to their immediate environment. They were flexible, adaptive, and diverse.

Some archaeologists believe the Fremont developed from existing groups of hunters and gatherers of this region. They varied from large sedentary populations, villages, to highly mobile clans. An austere rock shelter above the salt flats, a verdant marsh on the edge of Great Salt Lake, and aspen hillsides in central Utah—all house the spirit of the Fremont.

Floating Island Cave looks out over Great Salt Lake. A deeply layered dry cave site helps archaeologists to interpret Fremont groups in two important ways. First, deposits at the sites represent repeated visits by hunting and gathering groups from more than ten thousand years ago to less than fifty years ago. The layer-by-layer excavation of these caves allows us to see how the Fremont developed from underlying hunting and gathering peoples, what their technologies were, and how they evolved. It also shows that although subsistence patterns varied with each cave, together they represent the mobile end of a wide range of subsistence and settlement patterns practiced by the Fremont.

The numerous caves in the limestone mountains of the Great Basin provided natural shelters and storage facilities for the hunter-gatherers who inhabited this region. The Fremont probably visited such caves as Danger, Hogup, Promontory, and Fish Springs in the late fall and winter. Most of these sites were located near spring-fed marshes, which provided edible bulrush runners in winter. Their diet was supplemented by stored foods such as pine nuts collected in the summer and early fall. Other cave sites such as Floating Island and Lakeside were visited for short periods of time where the Fremont collected pickleweed and grasshoppers.

"Hungry?" David Madsen, state archaeologist and director of the excavation, hands me some pickleweed seeds from the small bushes that hug the salt flats.

I taste them.

"Better these than grasshoppers," he says. "Two years ago, we were excavating Lakeside Cave on the western shore of Great Salt Lake when we discovered tens of thousands of grasshopper fragments within the deposits. Bits of insects pervaded every stratum we uncovered. We estimated that the cave contained remains from as many as five million grasshoppers. At first, we had no ready explanation for this phenomenon. Nor could we explain why the cave deposits were so evenly layered with sand from the nearby beach. Some two dozen specimens of dried human feces gave us our first clue: most consisted of grasshopper parts in a heavy matrix of sand. This told us that people ate the hoppers and suggested that the sand was somehow involved in processing them for consumption."

Kevin dumps two buckets of sediment on each of our trays. We continued screening as Madsen continued his discourse.

"Then, last year, practically by accident, we found enormous numbers of grasshoppers that had either flown or been blown into the salt water and had subsequently been washed up on shore, leaving neat windrows of salted, sun-dried grasshoppers stretched for miles along the beach. As a result of varying wave action, as many as five separate rows existed in places. They ranged from an inch wide to more than six feet wide and nine inches thick, and contained anywhere from five hundred to ten thousand grasshoppers per foot. The rows, well sorted by the waves, contained virtually nothing but hoppers coated with a thin veneer of sand."

"That's remarkable," I say as I place more bones in plastic vials.

"But what's interesting," he adds, "is up until then we had envisioned grasshopper collecting to be a tedious task. Now we realize that the hunter-gatherers at Lakeside Cave could have simply scooped up grasshoppers piled along the beaches and consumed them directly."

"And how do they taste?"

"Like desert lobster."

These sites along Bear River Bay with their artifacts reveal that the people were not strictly dependent upon corn. They thrived on the rich resources of the wetlands surrounding Great Salt Lake. Molluscs, fish, waterfowl, muskrats, antelope, deer, and bison, were taken as food. The fibers of bulrushes, cattails, and milkweed were woven into baskets and clothing.

From eight hundred to twelve hundred years ago and, again, from three hundred to five hundred years past, Fremont life flourished on the edges of Great Salt Lake.

The Fremont oscillated with the lake levels. As Great Salt Lake rose, they retreated. As the lake retreated, they were drawn back. Theirs was not a fixed society like ours. They followed the expanding and receding shorelines. It was the ebb and flow of their lives.

In many ways, the Fremont had more options than we have. What do we do when faced with a rising Great Salt Lake? Pump it west. What did the Fremont do? Move. They accommodated change where, so often, we are immobilized by it.

I wonder how, among the Fremont, mothers and daughters shared their world. Did they walk side by side along the lake edge? What stories did they tell while weaving strips of bulrush into baskets? How did daughters bury their mothers and exercise their grief? What were the secret rituals of women? I feel certain they must have been tied to birds.

I return to my screening chores. Hoist another bucket on to the tray, empty it, and spread the sediments. The sands blown across Lake Bonneville ten thousand years ago are now blowing across my hands. I keep screening the strata: bones, tufa, sheep and pack-rat scat, a grass fragment here and there. I put them in vials, put the vials in sacks, the sacks in boxes, and the boxes in the back of the truck, which will transport them to the basement of the Utah Historical Society. It is our uncanny ability to catalog and interpolate, pigeonhole and store, our past. Each day in the field translates to one month in the lab.

"Why do you do this, David?" I ask as we pack the pickup.

"Because I want to know how these people coped with the fluctuations. What attracts me to Great Basin archaeology is putting all the pieces together, the complexity of the parts creating the whole. Artifacts alone have never interested me. It's the stratigraphy that speaks. The human stories are told within the layers of sediments."

He turns around, okays the truck, and returns to the trench inside Danger Cave. A shaft of afternoon light strikes the column of sediments. The definitions are stunning.

"You're looking at almost continual habitation in this cave for the last ten thousand years," he says, still profiling the lake gravels. "And the Fremont story is an unfinished one."

Driving back to camp, he elaborates, "During the fifteen

hundred years that the Fremont can be distinguished, they produced an archaeological record as rich, yet as enigmatic, as any in the world. The record of how they lived, reacted, and responded to the changing world around them is a mirror of ourselves—of all peoples at all times in all places."

I look out the window at this seemingly bleak landscape wondering what it means to be human in arid country.

A blast on the salt flats startles me.

"They're preparing for the dikes," Madsen says.

Settled in camp. Solar-heated showers were taken and fresh clothes put on. We hardly recognized one another as we "dressed for dinner."

"We've got an hour before supper," Kevin said, saddling an all-terrain vehicle. "I want to show you something."

I threw my left leg over the black leather seat and put my arms around his waist. Before I could blink, we were speeding up a dirt road in Silver Reef Canyon.

"This is not the best way to see birds," I screamed in his ear.

"We're not here to see birds," he yelled back.

We flew over a hill and bounced on sand. We drove a mile or two further on a seldom-used road with junipers on either side. All I could see in front of me was Kevin's back. He turned off the motor and got off the vehicle.

"Here we are . . ."

Before us was a fleet of plywood tanks, each with a flag painted on the side: Japan, Britain, Russia, and France. In the case of Germany, a swastika decorated the one-dimensional machine.

"What in the world are these?" I asked, walking around each one. "This is like an army theater, and we're miles from anywhere . . ."

"Military targets." Kevin answered while throwing rocks at them. "We found them by accident the other day when we were looking for cave sites."

Horned larks fluttered around the tanks.

"They probably nest inside," I said.

We toyed with the idea of moving the tanks down canyon, marching single file toward camp, but decided against it.

"How long do you suppose they have been here?"

"No idea. But if you look straight up, that's not blue sky you see—that's military airspace. Tomorrow count how many sonic booms you hear."

The crew is back on the road to Floating Island.

Last night the temperatures dropped below freezing, not unusual for mid-October in the Basin. We are blowing into our hands to warm them. It is an orange sky in a purple landscape. Great Salt Lake has advanced since yesterday. The salt flats are on fire, ablaze with morning light.

Rock wrens pierce the silence as we hike up to the cave. Yesterday the crew uncovered some cordage. Kevin says today we should be able to excavate deep enough to see if it leads to anything. The artifact appears to be circular.

The slow, tedious work continues.

By midafternoon, David Zeanah whisks away the last sediments from the cordage with his brush and uncovers a bird's foot. He follows the string to another leg. He gently lifts it out of the site.

"A bird-foot necklace?" I ask.

The two legs, maybe four inches long severed at the hypotarsus, dangle from the circle. The reticulation on the legs mirrors the twisting of the cordage. Holes have been drilled at the top, with the sinew drawn through and wrapped twice around them. The feet and toes are long and slender, pointing down.

I look at the necklace more closely. My guess: greater yellowlegs, a common transient to Great Salt Lake as they pass southward in the autumn and northward in the spring—an elegant shorebird.

I imagine this necklace being worn in a spring ritual by a clan of bird people to celebrate the fecundity of the marshes.

"Can't you see them dancing around the fire?" I say to Kevin, "dressed in feathers with the shrill cry of bone whistles in the air?"

Kevin rolls back his eyes. "And where are these feathered robes and bone whistles now?"

That night before dinner, I'm helping Jimmy Kirkman, the cook.

"What do you think about the bird-foot necklace?" I ask him.

"I think secretly everyone wants to try it on, but nobody dares to admit their fantasies."

"Unprofessional?" I ask.

"Against an archaeologist's religion," he replies. "It's science, not art."

We both look at each other and immediately have the same idea. We sneak into the supply tent, grab a box of plastic forks and dental floss and together make twelve bird-foot necklaces. Twisted cordage is replaced by spearmint-flavored floss. Where shorebird's legs hung, white forks dangle. We hide them inside my pack.

Dinner is served: linguini with clam sauce. A bonfire is built with wood brought from home. We all move our lawn chairs closer in. Jimmy throws a handful of his magic dust igniting purple and turquoise flames, careful not to reveal his recipe of copper sulphite. I stare at the seductive flames through a kaleidescope someone has brought, while a strange brew of "archaeologist's cider" is passed around in a well-seasoned gin bottle. Madsen begins telling stories. His gestures become larger. More logs are placed on the fire. More magic dust. Flames rise.

Jimmy winks at me and together, we present each archaeologist with their own bird-foot necklace.

Without hesitation, they rip off their coats, put the necklaces over their heads and dance. They dance wildly like tribesmen around the fire, singing songs I have never heard before.

I returned to the museum directly from the field. One of the security guards asked me if I had been to a picnic.

"No, why?" I asked.

"I thought that's why you might be wearing forks around your neck?"

"Oh, these . . ." I replied, looking down at my bird-foot necklace. "New-wave jewelry."

Thousands of objects associated with Fremont culture are cataloged in the museum's collection. We preserve the past for the future. Contact with the artifacts is restricted: white cotton gloves and lab coats, dim lighting, cool room temperatures. Each artifact is numbered. Site recognition is immediate. Any object can be recalled and detailed on the museum's computerized data base known as MIMS (museum inventory management system). It is a controlled environment.

But sometimes the objects run away with you. They seize your imagination and begin to sing songs of another day, when bone whistles called blue-winged teals down to the wetlands of Bear River. You hear them. You turn around. You are alone. Suddenly, the single mitten made of deer hide moves and you see a cold hand shivering inside Promontory Cave. It waves from the distance of a thousand years. Artifacts are alive. Each has a voice. They remind us what it means to be human—that it is our nature to survive, to create works of beauty, to be resourceful, to be attentive to the world we live in. A necklace of olivella shells worn by a Fremont man or woman celebrates our instinctive desire for adornment, even power and prestige. A polished stone ball, incised bones, and stone tablets court the mysteries of private lives, communal lives, lives rooted in ritual and ceremony.

And sometimes you recognize images from your own experience. I recall looking at a Great Salt Lake gray variant potsherd. A design had been pecked on its surface. It was infinitely familiar, and then it came to me—shorebirds standing in water, long-legged birds, the dazzling light from the lake reflected on feathers. This was a picture I had seen a thousand times on the shores of Great Salt Lake: godwits, curlews, avocets, and stilts—birds the Fremont knew well.

One night, a full moon watched over me like a mother. In the blue light of the Basin, I saw a petroglyph on a large boulder. It was a spiral. I placed the tip of my finger on the center and began tracing the coil around and around. It spun off the rock. My finger kept circling the land, the lake, the sky. The spiral became larger and larger until it became a halo of stars in the night sky above

Stansbury Island. A meteor flashed and as quickly disappeared. The waves continued to hiss and retreat, hiss and retreat.

In the West Desert of the Great Basin, I was not alone.

ffffffff

CANADA GEESE

ffffffff

lake level: 4210.95'

Seventeen monks dressed in white robes were singing vespers before dusk. Mother and I sat inside the Abbey of Our Lady of the Holy Trinity on wooden pews. The light was translucent, the music transcendent. The English translation of what they were singing was "Bring me back home." It was as though we were inside the chamber of a shell. As the chants took on the monotony of waves, we bowed our heads in supplication.

After the vespers, we walked beneath the canopy of cottonwoods that lined the country lane. They were golden. An autumn breeze blew the leaves around us. Mother slipped her arm through mine. She was weak and increasingly fragile. Dressed in a midcalf denim skirt and blouse, with a tweed jacket draped over her shoulders, she was quietly walking with the present. I knew she was tired. I also knew the power of this October afternoon. In another time, this moment would surface and carry me over rugged terrain. It would become one reservoir of strength.

I saw in Mother's face the mature beauty that a woman in her fifties has earned. I also recognized her weight loss not so much as disease, but as a shedding of that which was no longer necessary. She was letting go. So was I. The only clue I had of her pain was in the forthrightness of her voice.

"I used to think the life of a monk was a selfish life," she said. "I don't believe that anymore."

We find a grassy knoll to sit on, as several flocks of Canada geese graze in an amber meadow. Acres of sunflowers bloom against a blue sky. Great Salt Lake, only a few miles west of the Huntington Monastery, is visible through the corridor of Ogden Canyon.

Small families of geese congregate before their migration. The shadows of clouds crossing over the fields bring them in and out of light.

"Wild geese are my favorite birds," Mother says. "They seem to know where they are from and where they are going."

One can think of migration as merely a mechanical movement from point A to point B, and back to point A, explain it in purely physiological terms: in the fall, the photoperiod is lessened, it correlates with a drop in temperature. Food becomes scarce. Birds eat more. They overeat, put on fat, become restless, and along comes an environmental cue, such as wind, a change in barometric pressure, a cold front—and a rush of flight! Birds migrate.

Alongside the biological facts, could migration be an ancestral memory, an archetype that dreams birds thousands of miles to their homeland? A highly refined intelligence that emerges as intuition, the only true guide in life? Could it be that a family of Canada geese journey south not out of a genetic predisposition, but out of a desire for a shared vision of a species? They travel in flocks as they position themselves in an inverted V formation, the white feathers that separate their black rumps from their tails appear as a crescent moon, reminding them once again that they are participating in another cycle.

We usually recognize a beginning. Endings are more difficult to detect. Most often, they are realized only after reflection. Silence. We are seldom conscious when silence begins—it is only afterward that we realize what we have been a part of. In the night journeys of Canada geese, it is the silence that propels them.

Thomas Merton writes, "Silence is the strength of our interior life. . . . If we fill our lives with silence, then we will live in hope."

Mother and I break bread for the geese. We leave small offerings throughout the meadow. It is bread made by the monks from stone-ground grain. She puts her arm back through mine as we walk shoulder-high in sunflowers.

BALD EAGLES

lake level: 4211.10'

Rooted. Brooke and I have moved to Emigration Canyon, right smack on the trail that Brigham Young and the Latter-day Saints walked down on their way into the Salt Lake Valley.

We planted four Colorado blue spruces today. House-warming gifts from Mother and Dad. I held the root ball of each tree and blessed them in this supple soil (so unusual for a wintry day), that they might become the guardians of our home.

Dad and Brooke waited impatiently as they leaned on their shovels.

"I'm sorry, Brooke." Dad said. "All this hocus-pocus did not come from me."

I looked at my father as I stood up and clapped the dirt from my hands. "Who are you kidding, Dad? You are the man who taught us as children about divining for water with sticks, taking us out to a job where you had hired a man as a waterwitch to find where a well might be dug."

"Come on, Terry."

"The way I look at it, John," Brooke said. "We're never going to figure it all out, so we might as well acknowledge the intangibles. Who knows, maybe these trees do have souls."

Mormon religion has roots firmly planted in a magical worldview. Divining rods, seer stones, astrology, and visions were all part of the experience of the founding Prophet, Joseph Smith.

Dowsing was viewed negatively by some clergymen, "not because it leads to treasure, but because it leads to information."

Divining rods were understood by many to be instruments of revelation, used not just to locate veins of water or minerals, but to shepherd answers to questions. In folk magic, a nod up meant yes, a lack of movement meant no. Joseph Smith was not only familiar with this tradition, he and his family were practitioners of it—along with use of seer stones, which they used for treasure seeking.

Critics of Mormonism have used this to cast doubt on the origins and faith of this American religion. They dismiss Joseph Smith's discovery of "the golden plates" buried near Palmyra, New York—which contained the holy doctrine translated in the Book of Mormon—as simply an extension of the treasure-hunting days of his youth.

Others claim that Smith's sensitivity to matters of the occult heightened his shamanistic gifts and contributed to his developing spirituality.

For me, it renders my religion human. I love knowing that Joseph Smith was a mystic who ascribed magical properties to animals and married his wives according to the astrological "mansions of the moon."

To acknowledge that which we cannot see, to give definition to that which we do not know, to create divine order out of chaos, is the religious dance.

I have been raised in a culture that believes in personal revelation, that it is not something buried and lost with ancient prophets of the Old Testament. In the early days of the Mormon Church, authority was found within the individual, not outside.

In 1971, when Mother was diagnosed with breast cancer, the doctors said she had less than a 20 percent chance of surviving two years. Mother did not know this. Dad did. I found out only because I overheard the conversation between my father and the doctors.

Months passed. Mother was healing. It was stake conference, a regional gathering of church members that meets two times a year. My father was a member of the stake high council, a group of high priests who direct the membership on both organizational and spiritual matters. President Thomas S. Monson, one of the Twelve Apostles, directly beneath the Prophet, who at that time was Joseph Fielding Smith, was conducting interviews for the position of stake president.

Before conference, President Monson met with my father privately, as he did with all councilmen. He asked him, if called, would he serve as stake president? My father's reply was no. In a religion that believes all leadership positions are decided by God, this was an unorthodox response.

"Brother Tempest, would you like to explain?"

My father simply said it would be inappropriate to spend time away from his wife when she had so little time left.

President Monson stood and said, "You are a man whose priorities are intact."

After conference, my father was returning to his car. He heard his name called, ignored it at first, until he heard it for a second time. He turned to find President Monson, who put his hand on Dad's shoulder.

"Brother Tempest, I feel compelled to tell you your wife will be well for many years to come. I would like to invite you and your family to kneel together in the privacy of your home at noon on Thursday. The Brethren will be meeting in the holy chambers of the Temple, where we will enter your wife's name among those to be healed."

Back home, our family was seated around the dinner table. Dad was late. Mother was furious. I'll never forget the look on his face when he opened the door. He walked over to Mother and held her tightly in his arms. He wept.

"What's happened, John?" Mother asked.

That Thursday, my brothers and I came home from school to pray. We knelt in the living room together as a family. No words were uttered. But in the quiet of that room, I felt the presence of angels.

"What would you have me know?" I asked. "Faith," my great-grandmother Vilate said to me. Mother and my grandmother Lettie and I were helping to pack up her apartment. She was moving herself to a retirement center. "Faith, my child. It is the first and sweetest principle of the gospel."

At the time, I did not appreciate her answer. Faith, to a college coed, was a denouncement of knowledge, a passive act more akin to resignation than resolve.

"Where would faith in the Vietnam War have gotten us? Or faith in the preservation of endangered species without legislation?" I argued.

"My darling, faith without works is dead."

That is all I remember of our discussion. But, today, the idea of faith returns to me. Faith defies logic and propels us beyond hope because it is not attached to our desires. Faith is the centerpiece of a connected life. It allows us to live by the grace of invisible strands. It is a belief in a wisdom superior to our own. Faith becomes a teacher in the absence of fact.

The four trees we planted will grow in the absence of my mother. Faith holds their roots, the roots I can no longer see.

"I can't believe this is my body," Mother said, as she looked in the mirror of the dressing room at Nordstrom's. "I could never have imagined myself this thin . . . and these scars . . ." She shuddered.

I took the red suede chemise, size 6, off the padded hanger and handed it to her. She stepped into the dress and put one arm in, then the other, then buttoned the front and turned the collar up.

"It's perfect, isn't it?" she said, turning sideways to see how it hung in the back.

"Perfect," I replied. "You look absolutely beautiful."

She turned to me, her eyes radiant. "Right now, at this moment, I can honestly tell you, I feel wonderful! John will love this, even if it is extravagant."

She gave me back the dress. "I'll take it," she said, quickly putting on her black skirt and sweater. I held her emerald green jacket behind her as she slipped her arms through.

"Thank you," she said as she picked up her purse. "Shall we move on to our Christmas list?"

The rest of the day was spent in a shopping frenzy: three Christian Dior nightgowns for aunts; a shirt and tie for Steve; a reindeer sweater for Brooke; books for Dan; guitar

strings for Hank; a ceramic crèche for Ann; a silver vase for
a niece; pistachios for neighbors; a dozen narcissus bulbs; a
pair of black patent-leather pumps to go with her new red
dress; and two Madame Alexander dolls for her granddaughters, Callie and Sara.

Waiting for all the packages to be wrapped—she stood
while I sat. Mother's energy and quick pace was back. I
followed three steps behind.

We had lunch at Hotel Utah: poached salmon. We
laughed and chatted over absolute trivia.

"Let's make this an annual affair," Mother said.

We both believed it.

By the time we picked up our packages, it was late
afternoon. I drove her home. As she got out of the car, she
screamed, "Oh, Terry, look!"

The sun was a scarlet ball shimmering above the lake.
Mother put down her shopping bags and applauded.

"How much should I tell her?" Dr. Smith asked me
in his office. Mother was in the examining room.

"Tell her the truth," I said. "As you have always done."
I could feel the tears well up in my eyes. I was trying to
be brave.

"You can't be surprised, Terry. I thought you had accepted this last summer."

"We did. I mean I had, but hope can be more powerful
and deceptive than love."

"Her weight loss of eight more pounds is not a result of
flu. It's the cancer. She doesn't have much time," he said. He
walked out and opened the door to Mother's room.

After the examination, he came back out and said things
looked better than he thought, that the tumor he had felt
in June was gone, and that the others felt smaller.

Mother was very quiet. On the way home, I asked, "What do you think?"

"It doesn't really matter, does it?" she said. "Let's just take one day at a time."

I had the sense that she wanted to cry. And I thought of her mother, how once in the nursing home, after we had been crying together, I said, "Oh, Grandmother, doesn't it feel good to cry?" and she replied, "Only if you know there is an end to your tears."

Mother and I returned to my new house in the canyon. I fixed some chamomile tea.

"This tastes so good," she said with her hands cupped around the mug. "I can't seem to keep warm."

Mother asked for some more tea. We both settled on the couch. I gave her a mohair shawl to wrap around her shoulders.

"I think I have denied having cancer for years. It's a survival skill. You put it out of your mind and you get on with your life." She paused. "I mean, you have momentary flashes of what is real in periods of crisis and you face them, but then your mind seems to leap over the illness. You forget you were ever sick, much less that you are living with a life-threatening disease. The curious thing in all of this is that I have never acknowledged my anger over losing my breast as a young woman. Isn't that strange? Why would that come up now, after almost sixteen years? I'm angry, Terry."

Mother broke down. We both cried.

"I guess I'm giving up all sense of who I am," she said. "Last month, when John and I were at the beach in Laguna, all I could do was stare out at the waves."

The lake is frozen. Because of the ice, you can travel further west—if you dare.

My friend Roz Newmark and I drove out to the Bird Refuge in my trustworthy station wagon—as far as we could, until the lake stopped us.

It was a dreamscape where the will of the land overtakes you. I felt as though we were standing under the wing of a great blue heron.

As we walked, each step brought about a wheezing from the ice. The ice was thin, showing asphalt below. Off the road, it was a magnifying glass for objects arrested in motion. A floating feather—when was it caught in the clamps of ice? The quill tapered off, bleeding into darkness.

The ice became thinner and thinner, until each step of our boots sounded like vertebrae popping. We stopped. Beneath the icy veneer was a stream of suspended detritus: two snail shells, root fragments and reeds, a Canada goose feather, down, burrs, a piece of styrofoam, a small hollowed corn cob, pebbles, decaying insects, fish bones, carp scales, a woman's shoe.

Farther out, the ice looked solid. It was milky and dense. Roz and I dared each other to go first. Finally, we took hold of each other's gloved hands and began skating away from the road. I held my breath, as though it would make us lighter.

We lasted until moans, groans, and squeaks of the ice sent us back in a hurry. Roz, a dancer by profession, had an enormous advantage as she threw her head back, swung her arms in the air and leaped across the ice. I settled for short shuffles. Back on the road, we swung each other Western-style, joyous in our bravado.

We wandered a mile or two west, still on the same road. Two ravens flew across, their caws like chatter in a cathedral. The quiet returned. Twelve bald eagles stood on the ice of Great Salt Lake, looking like white-hooded monks. From November through March, they grace northern Utah. When the ice disappears, so do they.

Eagles on ice, cleaning up carp: beak to flesh. Flesh to bone. They whittle carrion down to a sculpture, exhibited in a bleak and lonely landscape.

Ice can immobilize, but on Great Salt Lake it creates habitat. I pluck the edge of the ice—it rings with the character of crystal. Ice that supports eagles is of finest quality.

Where the Bear River bends and flows south, the eagles flew. They appeared as small thoughts against the Wasatch Range.

Roz was sitting on her heels, wondering how life goes on in the river beneath the ice. Taking off her gloves, she ran her hand back and forth across its surface. Trapped bubbles, resembling clusters of fish eggs, were a reminder that fish swam below.

"It is comforting to know this," she said.

Mother was dead. I sat up startled and leaned against the pine headboard of our bed. Mother was alive. I wrapped my arms around myself to stop shaking from the nightmare.

But the feeling I could not purge from my soul was that without a mother, one no longer has the luxury of being a child.

I have never felt so alone.

December 16, 1986. Mother and I made a pact that we would no longer discuss how she was feeling physically, unless she wanted to.

"Good," she said. "So what do you have going on today?"

"Grading papers for my class on 'Women and Nature,' at the university," I said. "And I've got a few things to take care of at the museum. What about you?"

"John and I have the annual Church Christmas Ball at Hotel Utah, tonight."

"Are you excited about going?" I asked.

"Very."

"Your red dress?"

"My red dress," she answered.

I saw Mother asking my father to dance before anyone dared step onto the parquet floor.

RED=SHAFTED FLICKER

lake level: 4211.15′

This morning, a red-shafted flicker hammers above the window, offering me a wake-up call. The flicker peers through the glass. I delight in the red flash of feathers on his cheek. Later, I hear another sound. A backhoe. I open the sliding glass doors and watch the shovel's silver jaw rip into the Earth. Mother Earth. Another house is to be built. I see them digging my mother's grave.

I ring the doorbell. No one comes. The doors are locked. I know Mother is home. I walk to the neighbors and borrow their key. I open the front door. Mother is sitting on the stairs, her hands cradling her head between her knees.

"Are you all right, Mother?" I ask.

She slowly lifts her head. "I am too weak to get to the door."

She starts to sob. I hold her as we rock back and forth on the stairwell.

Another gray day. Fierce winds continue to pull the temperatures down.

Mother gives in. Dad and I take her to Dr. Smith's office. Walking down the long corridor of the medical building, I realize how much I hate this place. The smells, the color of the paint, the wallpaper, the claustrophobia of rooms with no windows. It is 1983. It is 1984. It is June, July, and August, 1986. It is Christmas.

Dr. Smith leads us into the back room, where he carefully inserts an IV into Mother's arm. The air of illness suffocates me. I am nauseous.

For the next two hours, as glucose drips into Mother's veins (to give her strength through Christmas), Dr. Smith unfolds the truth to her like a red rose: petal by petal, he tells her of her limited time, that she is within weeks of her death.

"You mean I have cancer?" she asks.

Dad and I look at each other in disbelief. Dr. Smith looks to us, then to Mother. He takes her hand and gently outlines what she can expect.

Mother's expression is stoic. She looks at Dad, whose tears stream down his cheeks.

"I heard the words but was incapable of internalizing them," Mother said once we were home. "I only saw John's face and could feel what was being said through him."

Tonight we gathered as a family and grieved openly.

"I want it all out by tomorrow so we can enjoy Christmas Eve," Mother said.

Dad and I rolled back our eyes at the woman who must be in control. It was funny. We remain true to our character even in death.

We each spoke of our love for Mother, and she gently said, "I am sorry I cannot be with your feelings. It is very different for me." She did not elaborate.

Steve spoke about being children in 1971, how we would set the timer every fifteen minutes to pray for her to be well.

Dan told of Dad's ritual of always asking Mother to "fix me"—which meant tucking in his shirt before he zipped up his pants.

I recalled a time when I was eight years old: I came home from school, devastated because friends on the playground were making fun of my naturally curly hair when straight hair was in. They called me a "witch." Mother took me by the hand into the bathroom and sat me in front of the mirror.

"Tell me what you see," she said.

I couldn't look.

She took her hand and lifted my chin. "Tell me what you see."

I looked in the mirror, and she said, "I see a beautiful girl with green eyes. I want you to stay here until you see her too."

Dad told the story of being in Hawaii in 1973 with the family. "I don't even remember what we were arguing about but, suddenly, Diane stood up in the middle of the restaurant, pulled the tablecloth off the table and said, 'That's it! I am no longer your slave! From now on, I'm doing what pleases me!' That was the beginning of women's liberation in this family."

Hank sat on the hearth next to me, the fire massaging our backs. I thought to myself, here is the child who, since memory, has lived with the fear that his mother may die. He could not speak.

Dad gave Mother a blessing, to which she added—as the men in the family gathered around her to place their hands on her head—"Someday, I hope Terry and Ann and my granddaughters will be able to stand in the circle . . ." We held hands as Dad spoke of our desire to help Mother through this—that this was our time to care for her as she had cared for us.

"Help us to not be afraid," Dad prayed.

Afterward, he placed his hands on her sculptured face, kissed her, and said, "Diane, we can do this. And I want you to be home. No more hospitals, sweetheart, just home."

Mother strikes a match and lights a white candle in the middle of the pine boughs on the glass table in the living room. She wears her blue satin robe with stars. It is Christmas Eve. We sit in a circle, each holding a silver cup of cranberry juice. Brooke has prepared a toast:

> Last week, Steve and Ann hosted the family Christmas party. It was so natural that what was almost imperceptible was the the fact that a new generation had volunteered to make the leap from guest to host. The torch has been passed. This party symbolized the beginning of the changing of the guard.
>
> What are we guarding?
>
> We are guarding the moments given to each of us as members of a family, small bits of time where the family becomes not just a mirror, but a clear, still pond, which each of us can gaze into for glimpses of our real being.
>
> We are guarding the very ideal of family, the bond, the web connecting us all, which gives rise to an energy,

a lust for life lost to those for whom family has lost significance.

We are guarding the buffer of protection that a family's love gives to each of its members, unconditionally, a magic covering that we can wear as armor but not notice its weight.

Let us toast twice. First, to the older generation: May your days come to be many, full of comfort and understanding. May they be spent knowing that those days past have held a completeness uncommon and unknown to many, and that every detail of your beings continues in the lives of those who follow.

To the younger: May we accept these gifts, knowing that they are of this tradition, of this old-fashioned courage, of ethics, and that they can be carried along forever like rusting relics or they can be worn as wings.

Let us wear them as wings.

We raised our cups and drank. Gifts were exchanged among the extended family. Four generations. The presents were opened one at a time.

We arrived for Christmas brunch. Mother greeted us in the foyer. The dining-room table was set as it had always been on Christmas morning: the white damask tablecloth; the Spode dishes, each plate with a tin candleholder from Mexico clipped to its side holding a small red candle; the sterling; the crystal; and a centerpiece of poinsettias and pine boughs.

The naïveté or self-centeredness of children did not recognize this gesture as heroic. There are those things in life that tradition allows you to count on. Of course, we would have Christmas brunch and, of course, Mother would prepare it.

We stood in the buffet line, each of us dishing the traditional food on to our plates: egg and sausage casserole, fresh fruit cocktail, and warbread—a raisin cake passed down by Mamie Comstock Tempest, my great-grandmother, who made this for her family when provisions were scarce during World War I.

We seated ourselves around the table. Everything appeared normal—knives and forks ticking on china, platters of food being passed, more water being poured—until one by one, we noticed Mother was not eating.

I watched her look at Dad, who squeezed her hand under the table. This was the last thing my mother had to do.

ffffffff

DARK=EYED JUNCO

ffffffff

Since Christmas we have traveled a thousand miles.

Mother is in bed. I have just come from the kitchen after putting a drop of opium tincture into a glass of water. She drinks the age-old potion for pain.

What does pain prepare us for? Emily Dickenson says, "Pain prepares us for peace."

Thump!

A bird has hit the bedroom window as I write in my journal. I quietly get up from the chaise, open the door, and find a junco stunned on the snow. Its white outer tail feathers are splayed. I want to hold the bird, to bring it inside and save it. But I don't. Instead, I smooth its tiny feathers behind its neck, close the door, and return to Mother.

"What I have learned through all this," Mother says, "is that you just pick yourself up and go on."

I rub her back while she talks.

"I have fought for so long and I have worked so hard to live through this summer, this fall, Christmas—and every minute has been worth it. And now, it feels good to give in. I am ready to go."

"Terry, you have accepted this, haven't you?"

"My soul has—but my mind has not."

A few hours later, Mother hands me a slip of paper, handwritten on both sides.

"I want you to call Dr. Smith with this list of questions, so I know what to expect. I didn't hear a thing the day we were in his office."

I read down the list. "How long is the process of starvation? What happens? Is there pain? Do I force myself to keep eating when nothing stays down? What about liquids? Does that help stay dehydration? What about the opium? Should I continue to take one drop three times a day? Is there anything that will quell the nausea and make me feel more comfortable? Do we need a nurse?"

I call Gary Smith at home with Mother's questions. We go through them systematically, one by one. After about an hour, I return to Mother's bedside.

"Tell me everything," she says.

Suddenly, my clinical self with notebook in hand dissolves.

It is New Year's Eve. Instead of talking about resolutions we wish to make, we are talking about funeral arrangements we must plan. Mother is in the next room, breathing.

Hank thinks our discussion is a betrayal. "I want nothing to do with it," he says. "She is alive."

Dad is becoming increasingly tense, pacing around the family room, unable to make decisions.

I suggest some ideas.

"If you think you have all the answers, Terry, then why don't you just go ahead and plan the whole thing," he snaps.

Dad's fuse is getting shorter as his fears become greater. He disappears into the bedroom.

Steve and Dan agree to choose the plot and casket. Ann offers to make a burial dress.

Dad returns after checking on Mother. "She is asleep. See what you think. I think she could go tonight."

The only light in the room is from the hallway. One by one, we enter, kiss her good-night, then leave.

It is 1987.

Dad closes the door and sleeps next to her.

I hate to go to sleep tonight, knowing another day has passed. At least it has snowed. Finally, the edges of this winter have softened.

None of us are sleeping. Steve and I anticipate the phone call from Dad. Dad cannot sleep, fearing that Mother will die. Dan and Hank cannot sleep because the house is unnaturally still.

The Soleri chimes on our porch keep ringing in the wind. Another storm is brewing.

We wait. We wait for Mother to die. The laziness of grief has us moving slowly.

Each day, I make a decision to dress colorfully: reds, purples, and blues, something to entertain Mother.

This morning she says softly, "You are changing my scenery. I appreciate you dressing up for me. I look forward to your costumes."

It continues to snow. Mother continues to weaken. Somehow, having the world soft and white makes this easier to bear.

Mother has not eaten for weeks. I look into her dark eyes, which widen each day as they retreat farther into her skull. Nothing seems real. The family is insulated from the outside world by the walls of this house. It feels holy. Friends and neighbors are respecting Mother's privacy.

"This is my death," I can hear Mother say, "nobody else's. It belongs to me in the intimacy of my family."

My days are immersed in the pragmatic details of care. And I love caring for her, we all do, even though there are times when horror splashes our skin like scalding water as we watch her writhe with nausea and pain.

And the other side, always the other side, is as tender as the pain is severe—bathing her, washing her hair, rubbing her body with fine French creams, feeding her ice chips, stroking her hair, her hands, and her forehead.

It is sacred time.

Mother's voice still speaks with her spirited and inquisitive nature. These things don't change. Life in the face of death is merely compressed into grist.

This afternoon, Dan and I came into Mother's room while she was watching Julia Child prepare a chicken dish on television.

"Oh," she sighed with rapt attention. "I would have loved to have tried that . . ."

The household is like a stretched rubber band. Dad left early this morning, angry. He is helpless, unable to save his wife or protect his children. Our calm only fuels his fire.

Mother is increasingly uncomfortable, so weak she can barely stand. I help her onto the scale. Eighty pounds.

"I don't know how to die," she said to me. "My mind won't allow me to rest. You are losing me. I am losing all of you."

It is her restlessness that weighs on me now. Her anguish over us—the living watching the dying,—the dying watching the living. She is still the peacemaker trying to create a calm in the midst of her death. And there is nothing she can do to ameliorate the situation.

"I feel the tensions, Terry."

An individual doesn't get cancer, a family does.

I feel calm, having just returned from a brisk walk along the base of the foothills. The balm of fresh air; Great Salt Lake glistened on the horizon. The valley is in sharp focus, crystal clear. I am reminded that what I adore, admire, and draw from Mother is inherent in the Earth. My mother's spirit can be recalled simply by placing my hands on the black humus of mountains or the lean sands of desert. Her love, her warmth, and her breath, even her arms around me—are the waves, the wind, sunlight, and water.

She is resting. The nurse came and gave her a shot of Demerol. All Mother said today is how much she wants to sleep, "to not think or feel, just sleep."

I never imagined we would walk to the place where what we hoped for was death. Sleep. It is the same thing.

I read Mother a poem this afternoon by Wendell Berry:

THE PEACE OF WILD THINGS

When despair for the world grows in me
and I wake in the night at the least sound
in fear of what my life and my children's lives may be,
I go and lie down where the wood drake
rests in his beauty on the water, and the great heron
  feeds.
I come into the peace of wild things
who do not tax their lives with forethought
of grief. I come into the presence of still water.
And I feel above me the day-blind stars
waiting with their light. For a time
I rest in the grace of the world, and am free.

"Read it again," she said. "Slowly."

The rubber band holding the household together snapped. Dad feared he was having a heart attack. I am so anxious, water glasses slip from my hands and break. Every day we think it can't get any worse. But it does. We don't know what to do for Mother.

Dr. Smith is coming up to the house to install a morphine drip, which will take the edge off Mother's pain.

Mother was greatly relieved to see him.

"How are you feeling, Diane?" he asked empathetically. He seemed to know what we couldn't. He sat on the edge of the bed. We gave them their privacy.

A few minutes later, Dr. Smith opened the door and asked for our help. I assisted him as he put an IV just under Mother's clavicle. The blood, our mother's blood, was spurting on the Marimekko sheets. We watched him search for a vein with the needle. Each time he tried, we winced, until the sickle-shaped needle finally hooked skin to tie down the line. Mother's face was strained. As the doctor shot the blood back into her vein she pulled back her neck. It was pale blood, not the deep red I had once seen. Dr. Smith set up the drip, the bag of glucose and morphine on a mechanical tree that held the pump. He showed us how to operate it, how to mix the right formula, the shot of Heparin needed to keep her veins open. I tried it and immediately pricked my finger with the needle while trying to put the cap back on the syringe.

I tried it again and got it right.

In the living room, Dr. Smith told us it was going to get a lot worse before Mother would die. That news seemed unfathomable. He said Mother would be comfortable now, and that he would check in often, whenever we needed him. He took his coat, walked out the door and I faced my father, with no answers.

His eyes were red with rage. His voice pinned me against the wall. "My home will not be turned into a hospital. Enough is enough. I can't take it anymore. It's so easy for you to play Pollyanna and say what a wonderful experience this is, but you don't have to live here. You can go home to Brooke, to the peace of another house and forget what is happening. I can't."

"As soon as Diane lapses into a coma, I'm taking her to the hospital. Someone else can deal with it—by then it won't matter, anyway."

rrrrrrrr

SANDERLINGS

rrrrrrrr

lake level: 4211.35'

"Close the door," Mother said. "I've been waiting for you
to get here all morning." She held out her hand. "Something
wonderful is happening. I'm so happy. Always remember,
it is here, in this moment, and I had it."

I didn't understand.

"Something extraordinary is happening to me. The only
way I can describe it to you is that I am moving into a realm
of pure feeling. Pure color."

I took off my coat and folded it over the chair. Sitting
down beside her, I replied, "Maybe that's what this business
of eternal life is . . ."

She took my hand again. "No, no, you're missing it—it's
right here, right now . . ."

This afternoon, as I was lying next to Mother, she took hold of my arm.

"Terry, this isn't a joke, is it? I mean is there any possibility?"

"Of what, Mother?"

"Is there any possibility I may not die?"

I paused, not knowing what she was asking or what she wanted to hear.

"I don't think so . . ."

She closed her eyes and sighed. "Ah, I'm so glad."

Ten breaths per minute. The morphine pump purrs. Mother floats. She is relaxed. No more anxious days of nausea and wondering how to die. It is easier on her, harder on us. I miss the conscious edge between acceptance and struggle.

Mother sleeps.

I watch her skeleton push through skin, emerging bone by bone, rib by rib, until her vertebrae have become the ladder my fingers climb as I rub her back. Her face is a death mask, skin stretched over skull. The bone from her ear to her eye is like a bridge, and the orbital structures that protect her eyes look like spectacles. Nothing is hidden. She sees with dark, wide sunken eyes.

Her hands remain unchanged, becoming more beautiful, more expressive each day. Her fingers seem to lengthen and her nails grow long. We hold each other's hands, and I see and feel the years of my mother's nurturing: the hands that cradled me, cuddled me, stroked my head at birth; the hands that bathed me, disciplined me, and combed my hair as a child; the hands that called me, prepared my food, wrote me letters, and loved my father's body; the hands that worked in the garden on long summer days planting marigolds for fall.

These hands even at death are beautiful.

Death is no longer what I imagined it to be. Death is earthy like birth, like sex, full of smells and sounds and bodily fluids. It is a confluence of evanescence and flesh.

It is so peaceful lying next to Mother. I am not afraid. We listen to Chopin and talk some. Mother finally says, "I just want to listen to the silence with you by my side."

The fullness of silence. I am learning what this means. Mother and I have grown so used to simply being, at times I find it difficult to speak.

Yesterday she said, "Terry, talk to me about something . . ." I panicked. I didn't know how to respond. I stood up and quickly said, "Okay, just a minute, let me get you some ice chips." While in the kitchen, I leaned against the counter, my mind throbbing with ideas that I have been yearning to explore with her, but by the time I returned, the moment had passed. Words had once again, lost their urgency. Silence. That ringing silence. I sat by the side of her bed and held her hand while the ice chips melted.

Mother mentioned this morning how much the morphine pump sounds like helicopters coming over the rise. Her comment stopped me, and then I remembered my own association with helicopters, the dream I had four years earlier, when all this began.

Changing the morphine drip frightens me. I draw out 5 cc's of morphine from the amber bottle. It measures 75 milligrams. I could draw out more. Mother watches me. We are both thinking the same thing. I shoot the morphine through the blue target of the 5 percent dextrose bag instead of her vein. Then I draw up 1 cc of Heparin. Inject that into the bag and massage the fluid so it is sufficiently mixed. I puncture the bottom of the bag with the IV apparatus and hang it on the mechanical tree like a hummingbird feeder. I clear the line of air bubbles, then turn off the pump. Next, I clamp the active IV line, pull out the needle from the Heparin lock near Mother's clavicle and insert the new IV. I unwind the old tubing and rethread the new. Turn the pump back on. Reset the monitor and I am finished. After I see all is in order, I tape the Heparin lock to Mother's skin and spiral the white tape around the tubing for security. I have another few hours to relax.

Touch is more important than ever. I notice Mother holds my hand tighter than usual.

"It feels so good to hold your hand," she said. "I don't feel so disconnected."

Steve arrives. We trade places.

After dinner, Dad took the men of the family to the basketball game at the University of Utah. Mother felt like talking.

"What I leave with you, Terry, is this: Follow your feelings. I have followed mine."

I asked her once again if she thought Brooke and I should have a child.

"I would hate to see you miss out on the most beautiful experience life has to offer. What are you afraid of?"

"I am afraid of losing my solitude, my time to retreat and

my time to create. Brooke is as ambivalent as I am. My ideas, Mother, are my children."

"I would rather hold you in my arms than one of your books." She paused.

"You asked my opinion, and I have given it to you."

"And I will follow my feelings."

She rubbed my back. "I love you so. We don't need words, do we? Do you know how wonderful it is to be perfectly honest with your daughter? Do you know how rich you have made my life? I am seeing circles, circles of love."

She took her hand off my back and turned the other way. "I need to be alone, dear."

"Come as soon as you can," Dad said over the telephone. "The night nurse doesn't think she will last the day."

When I arrived, Mother was disoriented, saying she felt like she was off her center, that she didn't know what was happening to her.

"I keep dreaming about elephants and melting ice," she said faintly.

I timed her breaths. Four per minute.

Dad came in. We both held her hands. She closed her eyes. We became incredibly frightened, not knowing if this was it, feeling she might slip away at any moment.

Suddenly, sunlight streamed into the room, striking Mother's face. It was as though God's hands were reaching out to her. And then the light shifted to another part of the room. Her breathing stabilized. Dad left.

Steve, Dan, and Hank entered. I stayed. We began to tease Mother about who was the cutest baby. Mother came alive.

"You all looked the same, little clones. That's why we didn't have any more."

Her laugh, unexpected and free, caught us off guard. We read from a book of local poetry, which a neighbor had sent:

> *A sad-eyed dog*
> *with tired feet*
> *stepped out*
> *into the busy street*
> *without a warning*
> *he was gone,*
> *a crumpled heap*
> *upon our lawn.*

The five of us became hysterical, laughing so hard we were sobbing. Dad ran into the room in a panic. Steve read the poem for a second time and we doubled over in laughter again—this time with Dad crying. In the middle of this bedlam, Mimi walked into the room carrying a tray of barbequed chicken wings. We read it a third time and she had to put the tray down.

The mood of the day was set. Mother was ebullient. We forgot our grief. The doorbell rang. Friends visited. Dr. Krehl Smith walked in. Mother asked him how his trip to Myrtle Beach had been, how his wife, Beverly, was feeling, and so on and so on, until finally he said, "No more of this, I'm onto you. What about Diane?"

"Who?" she asked.

"Diane—you."

"At peace," she said, "and not in pain."

Krehl left. Another dear friend arrived. They talked incessantly. Mother drooled over ward gossip, news from the neighborhood, and current affairs.

"Delicious," Mother said, with a Cheshire cat grin. "I feel like I'm back in the world."

Her friend having held her composure walked out the

front door and broke into tears. Mimi held her, saying, "We all feel that way."

Brooke came in with his ski clothes still on, looking more like a wildman from the north than a son-in-law. He kissed Mother. His cheeks were still cold and red from the windburn.

"I'm glad somebody in this family looks healthy," she said. "Give me your blue eyes and blond hair."

We were exhausted. Mother kept going. "Today has been wonderful," she said, stretching her arms. "I'm so happy. I didn't want the end of my life to become boring."

It is 8:00 P.M. Mother has ordered the family out of her bedroom. She is watching *Gone With The Wind*.

"I will ring the bell you gave me, if I need you."

She was sitting upright with the lights out, her glasses on, and pillows propping her back.

An hour later, she rings the bell. She is lying on her back. I carry another pan of green-black bile that Mother has vomited into the bathroom. As I pour it into the toilet, I throw up, too. It is the stench of death. The movie is still playing. I hear Melanie say on her death bed, "Do not squander time, that is the stuff life is made of."

Mother asked if I noticed a difference in her today. I told her, yes, that she was more restful. I then asked her if she felt any different.

"No," she said. "I don't feel different, but I sense you are all treating me differently."

I count her breaths. They have the intensity and fullness of a surfacing whale's.

Refuge

A week ago, Mother asked me to write her a story. Today, I read it to her:

A long, long time ago, when stray shells on a beach were as common as gulls, there sat a silver-haired woman on a silver-stained log. Driftwood. You could say they both had become driftwood. And there she sat, staring at the waves.

"Stallions," she thought. "The waves break as stallions."

She rocked back and forth on the log, digging her heels in the sand. She saw seven crows. Black on white. Or was it white on black? She could hardly tell, as the sea foam swirled around the dark forms. They didn't move. Perhaps they were not birds at all, but stones. She knew her eyes could not be trusted. They were clouding over with age. Even so, she would ask the next person to tell her what she had seen.

But no one came.

The old woman stiffened with each gust of sea breeze, and she began to smell of salt. The crows, the stones (she still couldn't be sure), seemed to be moving closer—or she to them. She felt herself tiring and closed her eyes.

She dreamed of the way things are seen, as she continued to rock back and forth on the log, clutching her elbows for warmth.

It was new moon, and the tides were changing. Two men were walking. One remarked to the other how solitary the beach had become. They passed by the woman. They passed by the driftwood. They passed by the birds.

"What is that?" one asked. "Over there—"

He pointed toward the tethered figure on the weathered wood.

"Just crows," the other replied. "Just crows."

And they continued to walk along the edge of the surf without getting their feet wet.

The old woman in her reverie had heard what she hoped might be true.

"Just crows . . ."

And the seven crows she had almost mistaken as stones stayed near her.

"Trust your feelings . . . I have trusted mine." Mother's words echo in my heart.

I get up to wash my hands. In the mirror, I see my mother's face.

January 15, 1987. It is 2:00 P.M. The wind continues. The large bedroom windows rattle with each gust. I fear they will shatter. The house is cold. I am alone with Mother as she is dying. And for the first time in weeks, I am afraid. The child in me, which lives as long as she does, wishes that the doorbell would ring, that Mimi or Grandmother or my aunts or anyone, would be there to help me.

Mother is restless. As she breathes, her throat rattles. Her neck is swollen. I worry that she is uncomfortable. I moisten her lips with a pink sponge swab. She appears to be talking with someone in this room, someone I cannot see. All at once, she rises and says, "I'm ready to go," and begins walking out the door. The morphine pump wavers, ready to tip over, as the morphine line threatens to snap.

I leap to my feet and grab her waist before she collapses on the floor. I lay her gently back on the bed and pull the covers around her. She looks to the corner of the room and points. "Can't you see?"

I look but see nothing.

Mother falls back into a deep sleep, and silence returns to the room.

I am left trembling, frightened by all I don't know, all

I can't see. I leave Mother, close the door, and escape into the living room. Through the windows, my eyes focus on Great Salt Lake. It's still there, mirroring the sky. I collapse under the weight of grief and cry. I curl up on the floor in a fetal position. I am sick of death. I want life. I want to surround myself with flocks of white pelicans in full summer sun. I want to dance naked on sand dunes. I yearn to have someone hold me and save me from this pain.

And then it hits me—I still have a mother. She is in the room next door and deserves to know how I feel, to see the underside of my heart. Dad keeps telling me she no longer understands what we are saying, that she is in a coma. I don't believe him.

I walk back into her room, kneel at her bedside, and with bowed head and folded arms, I sob. I tell her I can no longer be strong in her presence. I tell her how agonizing this has been, how helpless I have felt, how much I hurt for her, for all that she has had to endure. I tell her how much I love her and how desperately I will miss her, that she has not only given me a reverence for life, but a reverence for death.

I cry out from my soul, burying my head in the quilt that covers her.

I feel my mother's hand gently stroking the top of my head.

Five P.M. The doorbell rings. It is Dr. Gary Smith. He walks into the bedroom to check on Mother.

"It's very close," he says. "I'm sorry John isn't here. Please tell him he has done well." He places his hand on Mother's wrist. "Good-night, Diane." He looks around the room, unplugs the telephone near her bed, and, as quickly as he was in, he is out.

Steve and I are getting ready to change the morphine

drip. Our aunt, Ruth, has arrived with a pot of beef stew and is stirring it over the stove in the kitchen. Dan is asleep downstairs. Hank is at work.

Suddenly, Dad roars up the driveway only to find it blocked by cars. His tires screech as he races back down and parks on the street. The door slams. Thundering up the stairs, he yells, "Get out! All of you! This is my house and my wife!"

Ruth leaves immediately, simply saying, "John, your dinner is on the stove."

Dad walks into the bedroom, finds us there and suddenly, the alarm on the pump goes off.

"Beep . . . beep . . . beep . . . beep . . . beep . . ."

It has never done this before. Steve and I look at each other. Mother's eyes are closed.

"I want you out of here, now!" he says. "I'll take care of this."

"Dad, we have a problem here, let's get it under control and then we'll leave." Steve says very rationally.

As our father's rage is unleashed, my brother and I feed him line, plastic tubing to untangle. For twenty minutes, we keep feeding him line. We have less than one hour to fix the malfunction before Mother's veins will close. I pull the IV needle from her clavicle, shoot her with Heparin and return to the pump to figure out what is wrong. All the while, the alarm keeps beeping, flashing its red light into the room.

Lights in the room are flickering as the storm wages its own battle outside. Tempest. Our father is honoring his name.

"Go. Now."

"Just help us untangle the line, Dad. We've almost got it," Steve keeps saying calmly, as we continue threading plastic tubing through his quivering hands. After almost forty minutes, the alarm stops. Silence. The problem is solved.

I hook the IV needle back into Mother's vein. Prepare another 5 cc's of morphine. Steve brings in the bag. I look at my father, whose eyes are like a rabid dog's. Shoot the morphine into the glucose bag, massage it, hang it from the tree, set the gauge, bend down and kiss Mother's forehead. With her eyes closed, she whispers, "Thank you." "I love you." I whisper back. And walk out the door.

Dad follows us to our cars. I look at him but I have no words. It is a blizzard. Driving up the canyon, almost sliding off the road, I harbor my own rage and wonder if the wind will ever stop.

I am home. Our power is out. Brooke is lighting candles. My brothers and I agree we will not return to the house. Our father is the one who needs to be with Mother when she goes.

I understand him, but I don't have to forgive him. Not tonight.

I'm sitting on our chaise longue, sipping tea. This morning a pink light outlined the eastern skyline. The sky is blue. The mountain before me is crisp in detail. All of the wildness of last night is gone.

Mimi and Grandmother called. They let me talk out my feelings. They both said the same thing, "Trust life. Understanding is love."

For years, I have imagined being by Mother's side at the moment of her death. I have to let that go—she has taught me there is no one moment of death. It is a process. Besides, she—
Dad just called—he wants us there . . .

It is the third day since Mother's death. A candle is lit.
Let me begin.

Friday, January 16, 1987

Dad called around noon.

"I'm sorry," he said. "I behaved poorly last night. I
just wanted to be alone with Diane and I wanted to
protect you kids from the burden I was afraid to carry.
I realized last night that I cannot save Diane from death
or shelter you from seeing it. I realized a relationship is
not made in the last thirty days or twenty-four hours,
but in a lifetime. And Diane and I have lived well." He
paused. "She's going quickly. Please come. I need all of
you. I don't want to be alone."

Walking into her room, I could see death was immi-
nent, and I was surprised to see the physical changes from
the night before. Her color had changed—especially
around the mouth and nose. Her face was waxen. Her
feet were cold. It was as though dying moves from your
toes upward.

Mother's breathing was regular, but strained as she
exhaled. So much going out. So little coming in. I knelt
at the foot of her bed with the soles of her feet pressed
against my forehead. It was the only place I could feel
her pulse. I rubbed her legs under the mohair blanket.
They were like ice. Dad paced the room, occasionally
sitting next to her to hold her hand. We took turns.

From one until four in the afternoon, we sat near her.
A meditation. Her breaths could now be heard as moans.
Her eyes were haunting, open, and clear. Time was
suspended like watching a fire. Gradually, Mother's
breaths became a mantra and the death mask we feared
was removed.

Dad spoke of what it had meant to him to "take care
of our own." In these hours, we began to realize the
magnitude of these past weeks, months, years. Talk of
everyday life crept in, basketball scores, the day's news,

even laughter, and there, Mother lay dying. I never doubted her presence.

The light in the room deepened. It occurred to me that Mother would wait until after sunset. And it was an exquisite one. An apricot aura radiated above the purple Oquirrh Range. I told Mother what a beautiful sunset it was—I recalled her applause.

We turned a small lamp on. Mother's color looked better and, for a moment, we believed she would never die. The belabored breathing continued. We took turns holding her hands. Rubbing her forehead. Moistening her mouth. And we could feel the cold moving up her body. Her head was turned now, and with each breath her head drew back, reminding me of the swallow I beheld at Bear River, moments before it died.

Dad began to get nervous. He worried that Mother could go on for a few more days, that we had kept vigil too many times. He smiled with an anxious grin, saying, "Diane, you may outlast me yet . . ."

He wanted to be there when she died, and yet he didn't. He was afraid he would not be able to survive it. After a few minutes of wavering, Dad decided to pick up his car downtown. Brooke said he would drive him.

Dan left. Steve and Ann disappeared to other parts of the house. Hank was gone. I was alone with Mother.

Our eyes met. Death eyes. I looked into them, eyes wide with knowledge, unblinking, objective eyes. Eyes detached from the soul. Eyes turned inward. I moved from the chaise across the room and sat cross-legged on the bed next to her. I took her right hand in mine and whispered, "Okay, Mother, let's do it . . ."

I began breathing with her. It began simply as a mirroring of her breath, taking the exertion of her exhale, "ah . . . ," and reflecting back a more peaceful expression, "awe . . ." Mother and I became one. One breathing organism. Everything we had ever shared in our lives manifested itself in this moment, in each breath. Here and now.

I was stunned by the way her eyes fixed on mine—

the duet we were engaged in. At other times, I just closed my eyes and merged with her, whispering once again, "Let go, Mother, let go . . ." But mostly, it was just breath . . . slowing down, quieting down, until only the sweetest, faintest expressions of breath remained.

Steve and Ann walk into the room. They can feel her spirit: Mother's wisps of breath creating an atmosphere of peace.
   I feel joy. I feel love. I feel her love for me, for all of us, for her life and her birth, the rebirth of her soul.
   I say to Steve, "She's going . . . she's going . . ."
   He sits next to her and takes her other hand. Faint breaths. Soft breaths. In my heart I say, "Let go . . . let go . . . follow the light . . ." There is a crescendo of movement, like walking up a pyramid of light. And it is sexual, the concentration of love, of being fully present. Pure feeling. Pure color. I can feel her spirit rising through the top of her head. Her eyes focus on mine with total joy—a fullness that transcends words.
   Just then, we hear the garage door open. Dad and Brooke are home. A few more breaths . . . one last breath—Dad walks into the room. Mother turns to him. Their eyes meet. She smiles. And she goes.
   He kneels by her side, takes her hand, and says, "Diane, finally you are at peace."

7:56 P.M. I stood by Brooke. I felt as though I had been midwife to my mother's birth.

We knelt around her body. Dan held Mother's head in his lap. Our father offered a prayer for the release of her spirit and gave thanks for her life of courage, of beauty, and for her generosity, which enabled us to be part of her journey. He asked that her love might always be with us, as our love will forever be with her. And with great humility, he acknowledged the power of family.
   In the privacy of one another's company, we openly celebrated and grieved Mother's passing. A flock of sanderlings wheeling over the waves of grief.

Erich Fromm writes: "The whole life of the individual is nothing but the process of giving birth to himself; indeed, we should be fully born when we die."

A full moon hung in a starlit sky. It was Mother's face illumined.

rrrrrrr

BIRDS=OF=PARADISE

rrrrrrrr

lake level: 4211.65'

Mother was buried yesterday.

These days at home have been a meditation as I have
scoured sinks and tubs, picked up week-worn clothes, and
vacuumed.

I have washed and wiped each dish by hand, dusted tables,
even under the feet of figurines.

I notice my mother's hairbrush resting on the counter.
Pulling out the nest of short, black hairs, I suddenly remember the birds.

I quietly open the glass doors, walk across the snow and
spread the mesh of my mother's hair over the tips of young
cottonwood trees—

For the birds—

For their nests—

In the spring.

"Wait here, I want to show you something . . ." My friend, who runs a trading post in Salt Lake City, disappeared into the back room and returned with a pair of moccasins.

They took my breath away. The moccasins were ankle-high and fully beaded, including the soles, which were an intricate design of snakes. Cut glass beads: red, blue, and green, hand-sewn on white deerskin. As I carefully turned them, I wondered how anyone on earth could wear these. To walk in these moccasins would destroy the exquisite handwork.

An Indian woman who had been browsing, smelling the baskets of sweet grass, quietly walked over to the counter.

"Those are burial moccasins," she said. I handed one to her, but she would not touch it. "You won't see many of these."

My friend looked at the woman and then at me. "She's right. A Shoshone woman from Grantsville, ten miles south of Great Salt Lake, brought them in yesterday. They had just buried her grandmother in Skull Valley with the best they had: a buffalo robe, pendleton blankets, jewelry, a beaded dress of buckskin, and the moccasins. The granddaughter made two pairs."

The Indian woman in the trading post identified herself as Cherokee. She explained how, among her people, they sew only one bead on the soles of their burial moccasins.

I thought of the Mormon rituals that surround our dead: the care Mimi and I took in preparing Mother's body with essential oils and perfumes, the way we dressed her in the burial dress Ann had made of white French cotton; the high collar that disguised her weight loss, the delicate tucks from the neck down, the simple elegance of its lines. I recalled the silk stockings; the satin slippers; and the green satin apron, embroidered with leaves, symbolic of Eve and associated with sacred covenants made in the Mormon temple, that we tied around her waist—how it had been hand-sewn by my great-grandmother's sister at the turn of the century. A gift from Mimi. And then I remembered the white veil which framed Mother's face.

I tried to forget my encounter with the mortician in the hallway of the mortuary prior to the dressing, the way he led me down two flights of stairs, through the maze of coffins, and then abruptly drew the maroon velvet curtains that revealed Mother's body, now a carapace, naked, cold, and stiff, on a stainless steel table. Her face had been painted orange. I asked him to remove the make-up. He told me it was not possible, that it would bruise the skin tissues. I told him I wanted it off if I had to remove it myself. The mortician left in disgust and returned with a rag drenched in turpentine. He reluctantly handed me the cloth and for one hour, I wiped my mother's face clean.

I remember arriving at the chapel early, so I could check on the flowers and have some meditative time with Mother's body before the funeral. The face paint was back on. I stood at the side of my mother's casket, enraged at our inability to let the dead be dead. And I wept over the hollowness of our rituals.

The same funeral director put his hand on my shoulder. I turned.

"I'm sorry, Mrs. Williams, she did not pass our inspection. We felt she had to have some color."

"Won't you sit down." he said. "Death is most difficult on the living."

"I'll stand, thank you." I said taking my handkerchief to Mother's face once again.

One by one, family members entered the room, walked to the open coffin and paid their respects. This was the first time my grandmother Lettie had seen her daughter since

Christmas Eve. Confined to a wheelchair in a nursing home, her only contact had been by phone. My grandfather Sanky stood behind her with his hands on her shoulders. She mourned like no other.

As is customary in Mormon tradition, Steve and I brought the white veil down over Mother's face and tied the bow beneath her chin. I had hidden sprigs of forsythia down by her feet. The casket was closed. Dan and Hank placed the large bouquet of tulips, lilacs, roses, and lilies, across the top. Dad stood back, frozen with protocol.

Friends came to call. The line grew longer and longer. We became public greeters, entertaining their sorrow as we put aside our own.

I cannot escape these flashbacks. Some haunt. Some heal.

Today is Mother's birthday. March 7, 1987. She would be fifty-five. I lay one bird-of-paradise across her grave.

In a dugout canoe, Brooke and I paddle through a narrow channel of mangroves. A four-foot tiger heron peers out with golden eyes, more mysterious, perhaps, than any bird I have ever seen. The canal widens and we find ourselves in a salt water bay reminiscent of home.

We are in Rio Lagartos, Mexico.

Row upon row of flamingos are dancing with the current. It is a ballet. The flamingos closest to shore step confidently, heads down as they filter small molluscs, crustaceans, and algae through their bills before the water is expelled through either side. These are not quiet birds.

Behind the feeders, a corp de ballet tiptoes in line, flowing in the opposite direction like a feathered river. They too, are nodding their heads, twittering, gliding with the black portion of their bills pointing upward. They move with remarkable syncopation.

American flamingos. Gray. White. Fuchsia and pink.

They span the red spectrum. Feathers float in the water. Delicately. Brooke leans over the gunnels of the canoe and retrieves one. It contracts out of water. He blows it dry.

The birds are a pink brushstroke against the dark green mangroves. A flock flies over us, their necks extended with their long legs trailing behind them. Pure exotica. In the afternoon light, they become flames against a cloudless blue sky. Early taxonomists must have had the same impression: the Latin family name assigned to flamingos is *Phoenicopteridae,* derived from the phoenix, which rose from its ashes to live again.

There is a holy place in the salt desert, where egrets hover like angels. It is a cave near the lake where water bubbles up from inside the earth. I am hidden and saved from the outside world. Leaning against the back wall of the cave, the curve of the rock supports the curve of my spine. I listen:

Drip. Drip-drip. Drip. Drip. Drip-drip.

My skin draws moisture from the rocks as my eyes adjust to the darkness.

Ancient murals of ceremonial art bleed from the cavern walls. Pictographs of waterbirds decorate the interior of the cave. Herons, egrets, and cranes. Tadpoles and serpents stain the walls red. Human figures dance wildly, backs arched, hips thrust forward. A spear-thrower lunges toward fish. Beyond him, stands a water-jug maiden faintly painted above ferns. So lucent are these forms on the weeping rocks, they could be smeared without thought.

I kneel at the spring and drink.

This is the secret den of my healing, where I come to whittle down my losses. I carve chevrons, the simple image of birds, on rabbit bones cleaned by eagles. And I sing without the embarrassment of being heard.

The men in my family have migrated south for one year to lay pipe in southern Utah.

My keening is for my family, fractured and displaced.

PINTAILS, MALLARDS, AND TEALS

ʄʄʄʄʄʄʄʄ

lake level: 4211.85'

April 1, 1987. Great Salt Lake has peaked for the second time at 4211.85'.

The birds have abandoned the lake. Borders are fluid, not fixed. There is no point even driving out to the Refuge. For now it is ocean. I hardly know where I am.

Since Mother's death, I have been liberated from my optimism. I have nothing to hope for because what I hoped for is gone.

There are no mirages.

A Sunday morning in April. It is General Conference. A gathering of Saints. Mormons from all over the world convene on Temple Square to sit on the wooden pews of the tabernacle (pews stained to look like oak, even though they are pine) to hear the latest council and doctrine from the Brethren.

I drive by the cast-iron gates, heading west with the gulls. Red light on North Temple. I stop. With my windows rolled down, I can hear the Tabernacle Choir singing "Abide With Me, 'Tis Eventide," as it is being broadcast throughout the grounds.

Abide: to wait for; to endure without yielding; to bear patiently; to accept without objection; to remain in a stable or fixed state; to continue in a place. "Abide with me," I have sung this song all my life.

Once out at the lake, I am free. Native. Wind and waves are like African drums driving the rhythm home. I am spun, supported, and possessed by the spirit who dwells here. Great Salt Lake is a spiritual magnet that will not let me go. Dogma doesn't hold me. Wildness does. A spiral of emotion. It is ecstacy without adrenaline. My hair is tossed, curls are blown across my face and eyes, much like the whitecaps cresting over waves.

Wind and waves. Wind and waves. The smell of brine is burning in my lungs. I can taste it on my lips. I want more brine, more salt. Wet hands. I lick my fingers, until I am sucking them dry. I close my eyes. The smell and taste combined reminds me of making love in the Basin; flesh slippery with sweat in the heat of the desert. Wind and waves. A sigh and a surge.

I pull away from the lake, pause, and rest easily in the sanctuary of sage.

Ten miles east, General Conference is adjourned.

In Mormon theology, the Holy Trinity is comprised of God the Father, Jesus Christ the Son, and the Holy Ghost. We call this the Godhead.

Where is the Motherbody?

We are far too conciliatory. If we as Mormon women believe in God the Father and in his son, Jesus Christ, it is only logical that a Mother-in-Heaven balances the sacred triangle. I believe the Holy Ghost is female, although she has remained hidden, invisible, deprived of a body, she is the spirit that seeps into our hearts and directs us to the well. The "still, small voice" I was taught to listen to as a child was "the gift of the Holy Ghost." Today I choose to recognize this presence as holy intuition, the gift of the Mother. My prayers no longer bear the "proper" masculine salutation. I include both Father and Mother in Heaven. If we could introduce the Motherbody as a spiritual counterpoint to the Godhead, perhaps our inspiration and devotion would no longer be directed to the stars, but our worship could return to the Earth.

My physical mother is gone. My spiritual mother remains. I am a woman rewriting my genealogy.

On the west shore of the lake, across from Dolphin Island, I cover myself with hot, white sand. Oolitic sands. These sands, perfectly round, have a nucleus of quartz or the fecal pellet of the tiny brine shrimp. An outer shell is then built around the core in concentric layers of aragonite. Great Salt Lake pearls. I wear them. My secret dowry of wealth.

A blank spot on the map translates into empty space, space devoid of people, a wasteland perfect for nerve gas, weteye bombs, and toxic waste.

The army believes that the Great Salt Lake Desert is an ideal place to experiment with biological warfare.

An official from the Atomic Energy Commission had one

comment regarding the desert between St. George, Utah, and Las Vegas, Nevada: "It's a good place to throw used razor blades."

A woman from the Department of Energy, who had mapped the proposed nuclear-waste repository in Lavender Canyon, adjacent to Canyonlands National Park, flew into Moab, Utah, from Washington, D.C., to check her calculations and witness this "blank spot." She was greeted by a local, who drove her directly to the site. Once there, she got out of the vehicle, stared into the vast, redrock wilderness and shook her head slowly, delivering four words:

"I had no idea."

Brooke and I with a few good friends decided to go south for the Fourth of July: Dark Canyon, a remote area in southeastern Utah. On the map it appears without character. For years, I had dreamed of entering this primitive area where one can walk barefoot on slickrock for days, finding cool, midday soaks in hidden potholes. But first, we had to descend into Black Steer Canyon. Dark Canyon was one day away.

Somewhere between thoughts of rattlesnakes and finding the safest route down the steep, talus slope, I lost my footing. Skin, bone to stone, my head hit on rock and with my hands in my pockets I tumbled down the cliff until I was caught and saved by an old, juniper tree.

One of my companions, who was hiking behind me, yelled to see if I was all right. I answered yes, but as soon as I rolled over and tried to stand up with my pack still on, he said, "No, Terry—you're not all right. Lie down."

The river of blood that dyed my white shirt red was not from a nosebleed, but rather a long, deep pressure wound on my forehead, which had popped open like a peach hitting pavement. Lying down on the scree slope, I couldn't get two thoughts out of my mind: How badly am I hurt? And who will take care of me? I could feel myself losing consciousness.

Brooke hiked back up the slope to reach me. Fortunately, one of the members of our group was an emergency medical technician, with a well-supplied first-aid kit. I looked into her eyes as she was trying to stop the bleeding and asked, "Am I going to die?"

"Yes," she said. "But not today."

I relaxed.

"All you need when you get home," one friend teased, "is some long bangs."

We had been joking all morning about how the only good part of this hike down Black Steer Canyon (now christened "Bum Steer Canyon") was that we wouldn't have to climb back out. The good news that I was going to live was now dampened by the view of the cliff before me. I was the only person who could carry me out.

With a tightly bandaged head, after some water and a twenty-minute rest, Brooke and I climbed out of the canyon. Once atop the mesa, where the going was flat, we traversed the desert in hundred-degree heat, pushed on by a shot of energy only adrenaline can produce. It took us four hours to get back to the car.

Ten hours later, we arrived in Salt Lake City and met Brooke's brother-in-law, a plastic surgeon at the LDS Hospital. He reopened the cut, which I saw with a mirror in hand, for the first time. It ran from my widow's peak straight down my forehead across the bridge of my nose down my cheek to the edge of my jaw. I saw the boney plate of my skull. Bedrock.

I have been marked by the desert. The scar meanders down the center of my forehead like a red, clay river. A natural feature on a map. I see the land and myself in context.

A blank spot on the map is an invitation to encounter the natural world, where one's character will be shaped by the landscape. To enter wilderness is to court risk, and risk favors the senses, enabling one to live well.

The landscapes we know and return to become places of solace. We are drawn to them because of the stories they tell, because of the memories they hold, or simply because of the sheer beauty that calls us back again and again.

I will return to Dark Canyon.

The unknown Utah that some see as a home for used razor blades, toxins, and biological warfare, is a landscape of the imagination, a secret we tell to those who will keep it.

"It's no secret among traditional peoples," Mimi said sitting next to me on my bed. "Many native cultures participate in scarification rituals. It's a sign that denotes change. The person who is scarred has undergone some kind of transformation."

The next time I looked into the mirror, I saw a woman with green eyes and a red scar painted down the center of her forehead.

"She had felt labor pains all night long," my cousin Lynne explained. "And then in the middle of the night, she woke up, walked into the bathroom, and gave birth to a tumor. She reached into the toilet bowl, pulled out the bloody mass and set it in the sink. She walked into the kitchen, opened the cupboard, returned with a plastic bag, and placed the tumor inside it, ziplocked it shut, put it in the refrigerator and went back to sleep. The next morning, Mimi called the doctor."

"And what did the doctor say?" I asked.

"The doctor said she would meet her at the hospital for some tests. Mimi arrived with Jack, pulled the plastic bag out of her purse and handed it to the nurse behind the desk. Gary Smith, who was on call at the time, looked at the mass, looked at Mimi and said, 'Mrs. Tempest, I can tell you right now, this is a cancerous growth.' The biopsy later was diagnosed as a mixed müllerian sarcoma."

"When did they perform the hysterectomy?"

"That afternoon," Lynne said.

"And what's the prognosis?"

"It doesn't look good."

"You Americans, why is death always such a surprise to you? Don't you understand the dance and the struggle are the same?"

The voice of a Zimbabwean woman comes back to me. We had met in Kenya a few years back. I had walked out on a film on famine in Ethiopia. I could not bear the suffering. She followed me out, grabbed my arm, and brought me back in.

Same hospital. Same floor. Different room. My legs barely support me. I walk in with a bouquet of miniature roses from our garden. Mimi is sitting up in bed, reading *Omni* magazine.

"I'm home . . ." I say smiling, having just returned from leading a ten-day pack trip into the Tetons sponsored by the museum. Dad had come as a participant.

I handed her the flowers trying to keep up my persona.

"I'm so grateful you and John were away, Terry. I can't bear to put you all through this again. At least they have me down the hall from where Diane was."

We both laughed, then cried.

"Mimi, what happened?"

"I let go of my conditioning . . . I could only say this to you. But when I looked into the water closet and saw what my body had expelled, the first thought that came into my mind was 'Finally, I am rid of the orthodoxy.' My advice to you, dear, is do it consciously."

Consciously? Here is the woman who, at eighty, dragged her two granddaughters to a weekend symposium entitled "The Way of the Dream," where for two days we watched twenty hours of film of Jungian dream analyst Marie-Louise von Franz, lecture on archetypal language. (The second day we showed up in our nightgowns.)

Consciously? Here is the woman who, when I was twelve years old, told me that the dream I had of letting a great horned owl fly into my house and perch on my shoulder was an omen that I would menstruate soon.

Consciously? Here is the woman who had seriously considered taking LSD under the supervision of a medical doctor so she could have "a mind-altering experience," who had read herself straight out of Mormonism and into Eastern religious thought—but refused to replace one dogma with another.

How could she have been more conscious?

And then I looked at this woman I loved, my spiritual mentor, who was a charter member of Greenpeace before I had ever heard of them, who was a financial supporter of every conservation group in America—and then I remembered; she also voted for Ronald Reagan twice.

"What are we going to do?" I asked.

"I promise you, Terry, I am fine. Cancer at eighty is very different from cancer at forty. You must get on with your life and I will get on with mine. We will just go with it."

"We've harnessed the lake!" exclaims Governor Norm Bangerter. "We are finally in control."

One hundred Utah Republicans wave their ceremonial cowboy hats in the air and cheer. These are the party donors known as "The Elephant Club," who are eager to see for themselves the completed West Desert Pumping Station, which will extend millions of acre-feet of water into a salt desert.

David Grant, from the Bureau of Economic Development, explains to the crowd, "The sixty-million-dollar expenditure for this project is an insurance premium. It will benefit the state of Utah in a specific set of circumstances. If Mother Nature goes beserk on us and we have massive flooding—well, that's one set of circumstances the policy won't cover. And if we find ourselves in a drought—the policy wasn't needed. But let me ask how many of us cash in our homeowner's insurance? We're just happy to have it."

The pumping project will return Great Salt Lake to elevation 4208.00'. It aids AMAX, Southern and Union Pacific Railroads, and Great Salt Lake Minerals. There has already been a $240 million loss to railroads, transportation, mineral, wildlife, recreation, and residential interests. The potential loss to the south shore—where the Salt Lake City International Airport, I-80, I-15, and railroads reside— would be close to $1 billion.

Railroads and industry surrounding Great Salt Lake understand—so much so that Great Salt Lake Minerals contributed $200,000.00 to the pumping project study, and Southern Pacific put up $7 million out of its own pocket to help finance the $23 million bid to restore the ten-mile causeway access to the pumps.

I hear my father's voice: "It all comes down to dollars and cents . . ."

AMAX isn't complaining. After the water is diverted from Great Salt Lake to the West Desert, the mineral content increases from 7 to 15 percent because of evaporation. A canal from the holding pond to the magnesium plant has been designed and is one step away from being implemented. AMAX expects increased revenues of $30 million and the creation of two hundred jobs.

The state of Utah may deny subsidizing industry at the taxpayer's expense, but they cannot deny that they are an administration with great imagination. Consider some of the alternatives in state files for managing Great Salt Lake:

University of Utah professor of geology William Lee Stokes advocated nuking the lake, creating a cavern by the atomic explosion that would drain the water to the center of the earth.

Another idea: Dye the lake purple. Certain colors enhance evaporation 10 to 15 percent. Dark purple is one of them. The problem with dying Great Salt Lake purple was not purple pelicans and gulls, but rather that the dye would only penetrate 30 inches of the water's surface. They would have to paint the lake repeatedly and with the volume of water in Great Salt Lake at 30 million acre-feet, this management scheme would have depleted the world's source of purple dye, not to mention a cost of $300 million to taxpayers.

But all is not lost. Utah has itself a major tourist attraction. According to Lt. Governor Val Oveson, "This could become a tourist destination on an international scale—

pumping salt water into a desert—there's nothing like it in the world!"

Public tours are already in motion. Thousands of curious individuals and a fair share of skeptics have boarded buses across the causeway from Lakeside, Utah, to witness the pumps. Public officials believe if they charge a dollar per person, they can recover the state's $60 million with that many tourists.

Ron Ollis, public affairs officer for the division of water resources, who conducted the first half-dozen tours last month, said, "While we were conducting tours, the air force was dropping bombs. The military did maneuvers above the causeway for the public. Some of the pilots stormed the railroad tracks in 'Top Gun' fashion. Everybody loved it."

The Utah Test and Training Range, adjacent to the pumping station, is property of the United States Air Force. The contractor who put the natural-gas line into the station had to sign a document absolving the air force of liability in the event that his workers hit a mine or a fused bomb.

When you enter the project you are given a pair of Decidamp hearing protectors. You squeeze-roll them between your fingers, put them in your ears, and wait for them to expand.

The world is silenced.

In the pumping station everything is explained:

AIR COMPRESSOR, 300 KW GENERATORS, NATURAL-GAS LINE, AIR CLUTCH, RIGHT ANGLE GEAR DRIVE, ENGINE COOLANT LINE, HEAT EXCHANGES, ENGINE COOLANT SURG TANK.

A banner over the engine reads, DRESSER–RAND NATURAL GAS–FUELED ENGINE, BUILT WITH PRIDE IN PAINTED POST, NEW YORK. THREE OF THESE ENGINES WILL DRIVE INGERSOLL–RAND FIFTY-FOOT-TALL PUMPS THAT WILL DROP THE LEVEL OF GREAT SALT LAKE BY 1,300,000 GALLONS PER MINUTE FOR THE DE-PARTMENT OF WATER RESOURCES OF UTAH.

I hold on to the railing to see the pump shaft below. It resembles the beater in my mixmaster.

Outside, I take off my ear plugs and put them in my skirt pocket. Salt foam swirls beneath the bridge. Phalaropes spin near shore. The four-mile canal that shuttles Great Salt Lake to the desert has been named "Rio Buena Vista," after the mythical river that the Spaniards supposed drained the salt lake into the Pacific Ocean.

"How long will these pumps last, given the corrosive effect of the salt?" I ask Ollis.

"The pumps are made of aluminum-bronze alloy. They have a fifty-year life expectancy."

"How much does it take to operate these pumps?"

"$2.3 million per year. The $100,000 per month needed to pay for the natural gas that fuels the pumps is included in the $60 million appropriated by the state legislature."

"And how long will it take for the lake to go down with the pumps operating?"

"We anticipate a drop of one foot the first year, that's 2.2 million acre-feet of water displaced—or 325,000 acres or 500 square miles of water in the West Desert, depending on how you choose to look at it. Keep in mind, right now there are 30.2 million acre-feet of water in Great Salt Lake, roughly 2400 square miles."

Ron Ollis is sharp. I like him.

Governor Bangerter addresses the Elephant Club: "This was the kind of decision you wish you didn't have to make, but when the lake is lapping at your doorstep, you do what you have to do to solve the problem. And let me tell you, Great Salt Lake is a big problem."

The Governor cuts the red ribbon. There is a round of cheers. The West Desert Pumping Station is officially christened. One of the Elephant Club members turns to Great Salt Lake before getting back on to the bus. "Now, what I'd really like to see us do is pump the salt out of the lake completely."

I sit on the banks of the new "Rio Buena Vista" and watch a vein of Great Salt Lake flow west. A rattlesnake stretched across a boulder stops my eyes. The head and rattles have been cut off by a trophy hunter. I walk over to the snake and lift its body, which still articulates between each delicate rib. Forty-two diamonds run down its back. It must measure over three feet long. I wrap the snake around my neck, leave the pumping station, and set out across the desert.

Poet Robert Hass writes, "You hear pain singing in the nerves of things; it is not a song."

My father no longer hunts. Neither do my brothers.

"I can no longer participate in the killing," Dad said. "When I see the deer, I see Diane."

Hank put his gun down years ago. So did Dan. Steve carries his rifle into the hills, but he has not shot a deer since 1983.

"I see the buck in my scope but I can't find a good enough reason to pull the trigger."

For the men in my family, their grief has become their compassion.

This afternoon, I walked along the shores of Farmington Bay. Four California gulls, three pintails, a blue-winged teal, one Canada goose, two mallards, a western grebe, and an American merganser—dead—individual birds, randomly shot. Their limp bodies were strewn along the beach.

R e f u g e

I realize months afterward that my grief is much larger than I could ever have imagined. The headless snake without its rattles, the slaughtered birds, even the pumped lake and the flooded desert, become extensions of my family. Grief dares us to love once more.

BITTERNS

lake level: 4210.20'

I found the birds! Malheur National Wildlife Refuge in southeastern Oregon, has adopted and absorbed the flocks of Great Salt Lake. Not all of them, of course. But many. Especially the colony-nesters. Thousands of white-faced ibises, double-crested cormorants, and snowy egrets circle the lake. Huge cumulus clouds looking like clipper ships sail across the sky. The air is quivering with pintails, mallards, and teals. It is like coming home.

Bitterns stand their ground in the camouflage of cattails, their bills pointing toward heaven. A short-eared owl flaps over fields that are back-lit in pink. Even curlews are dancing on the uplands. These are my wetlands that sparkle and sing. It has been so long since my lungs have been filled with the musky scent of the marsh.

I sit on the rich, moist earth, green earth and draw my knees to my chest. All is not lost. The birds have simply moved on. They give me the courage to do the same.

Refuge

These wetlands held in the spacious arms of the Great Basin are refuges—Malheur in Oregon; Stillwater and the Ruby Marshes in Nevada; Fish Springs and Bear River in Utah; sapphires in the desert bordered by birds.

ffffffff

SNOWY PLOVERS

ffffffff

lake level: 4209.10'

The day the pumps were turned on, the lake did an about-face on its own. Great Salt Lake is receding, having dropped more than two feet from last year's lake level high of 4211.85'.

Where the water has pulled back, the land looks as though it is recovering from a long illness. Barbed-wire fences act as strainers. Sheets of algae and rotting vegetation hang like handmade paper and bobs of tangled hair.

A "bomb catcher" is being built in the West Desert. It is the newest component of the West Desert Pumping Project.

The United States Air Force has disclosed information from their own environmental assessment report: although most bombs exploded on impact during training missions conducted since World War II, some did not. There is a fear that unexploded bombs, including some in watertight containers embedded in the salt flats might be dislodged by the pond water and float toward Great Salt Lake.

"Imagine a giant comb about eleven hundred feet long," says Brent S. Bingham, president of Bingham Engineering, Inc., the Salt Lake City company that has designed the bomb catcher. "It consists of twenty-two hundred fiberglass bars, five feet tall and six inches apart, that will span the spillway, preventing bombs from being carried into the lake by the stream of water pouring out of the new holding pond west of the Newfoundland Mountains."

Mr. Bingham told newpaper reporters today that no bombs have been seen floating in the pond, which is two and a half to three feet deep, but state officials don't want to take any chances.

Dee Hansen, Director of the Utah Department of Natural Resources says, "The bomb catcher is not for major bombs. It's for phosphorous bombs and different types of bombs in canvas bags . . . The Air Force experimented with a bunch of stuff out there. Most of it has probably deteriorated if it didn't explode. But the Air Force is pretty cautious, and we want them to be." He adds, "An explosive ordnance disposal unit from Hill Air Force Base inspected the corridor for the twelve-mile-long Newfoundland Dike before construction began. They found some unexploded ordnance in the area, which were retrieved."

All I can see are thousands upon thousands of tumbleweeds cartwheeling over the surface of the water, beating the floating bombs to the strainer.

The West Desert Pumping Project is one of thirteen engineering efforts nominated for the 1988 Outstanding Civil Engineering Achievement Award presented by the American Society of Civil Engineers.

"The award recognizes engineering projects that demonstrate the greatest engineering skills and represent the greatest contribution to civil engineering progress and mankind," said Sheila Brand, spokesperson for the society.

We had several calls at the museum today from people who wanted to know if there had been an earthquake. According to the seismology station on campus, there had been no tremors.

It turns out the rattling vibrations were in the air, not the ground.

Atmospheric shock waves were generated when the air force exploded twenty-five thousand pounds of munitions near Great Salt Lake at 2:30 P.M.

Airman First Class Jay Joerz, with Hill Air Force Base public affairs, said, "Munitions are disposed of on a regular basis at the test and training range just west of the lake. Weather conditions must have been just right for the shock wave to carry so far. Yesterday we had another twenty-five thousand–pound explosion and nobody noticed."

Snowy plovers have shown a 50 percent decline in abundance on the California, Oregon, and Washington coasts since the 1960s, due to the loss of coastal habitats. The National Audubon Society petitioned the U.S. Fish and Wildlife Service in March, 1988, to list the coastal population of the western snowy plover as a threatened species. The present population estimate for the western United States, excluding Utah, is ten thousand adult snowy plovers, rising to thirteen thousand individuals after breeding season. A knowledge of inland population numbers and distribution is essential to our understanding of the status of the species as a whole. That's why we are counting them in Utah.

I have been combing the salt flats north of Crocodile Mountain for them since early morning. So far, my count is zero.

Margy Halpin, a non-game biologist leading the survey for the Utah Division of Wildlife Resources, and I are walking parallel to each other, maybe a half-mile apart. The distance between us feels greater than it is because of the intense heat and glare of the alkaline terrain.

I walk slowly, following the western shoreline of Great Salt Lake. Clay bluffs along the water's edge resemble Normandy: they have eroded into fantastic shapes, alcoves, and tunnels from past wave action. There are no footprints here.

Windrows of brine flies and ladybug carcasses twist along the beach. Otherwise, it is littered with limestone chips, which clamor like coins when walked upon. The heat is brutal. I pause to dip my scarf in the lake and tie it back around my forehead.

I turn west away from the lake and walk back across the salt flats. Another hour passes. I see movement. Two snowy plovers skitter ahead. Margy also has them in view—we motion each other simultaneously waving with our right hands. If they were not dashing across the white-brocaded landscape, they would be impossible to see. They are perfectly camouflaged.

Margy and I join each other and sit on the salt to watch them. I have to squint through my binoculars to shut out the light reflecting off the flats. Heatwaves blur the plovers.

They appear to be foraging on half-inch golden beetles. We pick up one of the insects close to us for a better examination of what the plovers are eating. The golden carapace is translucent, gemlike. We set the creature back on its course, and it skeeters away.

Snowy plovers are the scribes of the salt flats. Their tracks are cursive writing, cabalistic messages for the bird-watcher who cares enough to follow their eccentric wanderings.

We spot two more adult plovers with chicks. Two chicks. Margy and I check with each other to make sure.

"Ku-wheet! Ku-wheet! Ku-wheet!"

On this day, their calls are the only dialogue in the desert.

The snowy plover is considered to be an uncommon summer resident around the shores of Great Salt Lake, so our total count of six on June 11, 1988, is no surprise. They are listed as common residents of Pyramid Lake in Nevada and Mono Lake in California. Long-term distribution records show that snowy plover populations rise as Great Salt Lake retreats. More habitat supports more birds.

What intrigues me about these "pale ghost-birds of the beaches" is how they manage their lives in such a forbidding landscape. The only shade on the salt flats is the shadow they cast. There is little fresh water, if any. And their diet consists of insects indigenous to alkaline habitats—brine flies and beetles.

Fred Ryser explains, in *Birds of the Great Basin,* how this "wet food, even during the driest and hottest time of year, contains much water of succulence . . . with each mouthful of food, the plover drinks."

To cool off, the snowy plover stands in the salt water and lets the brackish water evaporate from its body.

Another question rises with the heat of the salt desert. Why don't their eggs bake?

Snowy plovers nest in shallow scrapes, open and exposed. Some plovers will use brine fly pupal carcasses for a nesting bed, and then line them with small pebbles and shells. Both male and female snowys incubate the eggs; on hot days, such as today, they trade places frequently, alternating from sitting to standing (not so unlike us). Parenting plovers have been seen to soak in salt water and, upon returning to their clutch of eggs, will ruffle their wet feathers, sprinkling the eggs with water. An average clutch size is three eggs. Research suggests half the broods in Utah might fledge two young.

Margy and I share drinks from her canteen. I have a throbbing headache, which tells me I have been ignoring my own need for water. I fear I may be suffering from heatstroke and begin to worry about getting home. Too much exposure.

Before walking back through shoulder-high greasewood, I take a quick swim in the lake. The silky waters of Great Salt Lake cool my parched skin, even though the salt burns. This offers a momentary reprieve from my nausea. I lick my swollen lips and am careful not to rub my eyes.

I catch up with Margy and follow her through the maze of greasewood. We hear rattles and stop. It is the driest sound on earth. We take another path and walk briskly toward Crocodile Mountain.

Driving home alone on the solitary dirt road that winds around the lake, I am struck with delirium. I stop the car. Nothing looks familiar. I get out and heave violently behind the sagebrush.

The next thing I remember is waking up in a dark motel room in Tremonton, Utah. I call Brooke to see if he can tell me what happened. He is not home. Snowy plovers come to mind. They can teach me how to survive.

November 15, 1988. Lettie Romney Dixon passed away at noon from lingering illness. My grandfather, Sanky, has not left her side for months. Last night, I sat with them all night long. He held her hand and I held his. Mother felt near. Death has become a familiar landscape. I can smell it.

We prepare my grandmother's body. Her tiny arms stiff around her chest are like chicken wings because of Parkinson's Disease. They have not been able to hold those she loved for years. This was the pain I could not embrace. Her blue eyes did. And now they are closed.

My uncle Don, from out of town, walks into the room. We hug. I see my mother's face in his and do not hear a word he says.

Once home, I split open a ripe pomegranate. Red juice trickles over my hand and spills on to my lap as I eat the tart, succulent seeds.

Mothers. Daughters. Granddaughters. The myth of Demeter and Persephone lives through us.

"This cannot be a coincidence, can it?" I ask my cousin Lynne, over the telephone. "Three women in one family unrelated by blood, all contract cancer within months of each other?"

"I have no idea, Terry. All I know is that my mother has breast cancer and her surgery is tomorrow."

"Is there a pattern here, Lynne, that we are not seeing?"

Lynne's voice breaks. "What I do know," she says, "is that I resent so much being asked of the women and so little being asked of the men." There is a long pause. "I'm scared, Terry. I'm scared for you and me."

"So am I. So am I."

Something is wrong and I can't figure it out—the egg collection at the Museum of Natural History. On first appearance, these clutches of eggs arranged in a nest of cotton move me. The size range and color differentiation is stunning, from the pink and brown splotching of a peregrine falcon's eggs to the perfectly white, perfectly round eggs of a great horned owl. And the smaller birds' eggs are individual works of art, canvases on calcium spheres—some spotted, some striped.

But when I hold one of these eggs, there is no gravity in my hand. A weightless shell. Life has literally been blown out through a pinhole.

It dawns on me, eggs are not meant to be seen. This collection is a sacrilege, the exposed medicine bundles of a tribe. These eggs are the hidden wealth of a species, tenderly guarded beneath the warm, bare brood patch of a female bird.

Secrets were housed inside these shells, enough avian lives to repopulate a marsh, even Bear River. But we have sacrificed them in the name of biology to substantiate the obvious, that we know where each bird comes from. These hollow eggs are our stockpile of evidence.

On my way home, I drop by to visit Mimi. She is painting on her easel in the dining room. She rinses her brushes and we sit in her turquoise study.

"What's on your mind?" she asks.

"Tell me what eggs symbolize?"

She runs her hand through her short gray hair. "For me, it is where life originates. In mythic times, the Cosmic Egg was believed to be held within the pelvis of the ancient Bird Goddess. Why do you ask?"

I describe my encounter with the egg collection at the museum, how disturbing it was.

"The hollow eggs translated into hollow wombs. The

Earth is not well and neither are we. I saw the health of the planet as our own."

Mimi listened intently. She stood and turned sideways to switch on the lamp. It was dusk. I could not help but notice her distended belly, pregnant with tumor.

"It's all related," she said. "I feel certain."

"The total number of snowy plovers counted around Great Salt Lake was 487, with 26 young in 11 broods," I tell Mimi as we drive out to Stansbury Island. "Biologists figure we may have two thousand breeding pairs in Utah." She wanted to get out of the house for a change of view. Her strength is holding in spite of the cancer.

We had just seen four snowys scurrying between clumps of pickleweed.

Just outside Grantsville, thousands of Wilson's phalaropes and eared grebes were feeding in the median ponds adjacent to the freeway. No doubt a migratory stop.

In recognition of Great Salt Lake's critical role as a migrational mirror reflecting ducks, geese, swans, and shorebirds down for food and rest, the Western Hemisphere Shorebird Reserve Network has identified the lake as a crucial link in the chain of primary migratory, breeding, and wintering sites along the great shorebird flyways that extend from the Arctic to the southern tip of South America.

By becoming part of the network, Great Salt Lake could gain international support for local conservation efforts and wetlands management. It has been nominated by the Utah Division of Wildlife Resources, the U.S. Fish and Wildlife Service, and Bureau of Land Management. And just recently, the Utah Division of Parks and Recreation, along with the Division of State Lands and Forestry, endorsed the nomination.

To qualify, a site must entertain in excess of 250,000 birds a year, or more than 30 percent of a species' flyway population.

Great Salt Lake qualifies. It hosts millions of birds in a season. Don Paul points out, however, that the lake qualifies on the basis of Wilson's phalaropes alone—flocks of 500,000 to 1,000,000 are not uncommon during July and August, when they are en route to South America.

The Western Hemisphere Reserve Shorebird Network has paired Great Salt Lake with Laguna Del Mar Chiquita, the salt lake in the Cordoba region of Argentina where the phalaropes winter. They are sister reserves.

"Think about one phalarope flying those distances," Mimi said, looking through her binoculars. "And then think about flocks of phalaropes, millions of individuals being driven on their collective journey. We go about our lives giving little thought, if any, to such miracles."

There is a chorus of wings navigating the planet. Twenty million shorebirds migrate through the United States each year to arctic breeding grounds in the spring and back to their wintering sites in South America. One bird may cover as many as fifteen thousand miles in a year.

Great Salt Lake is a refuge for these migrants. And there are certainly other strategic sites along the migratory path, essential to the health and well-being of those birds dependent upon wetlands. The Copper River Delta in Alaska, Canada's Bay of Fundy, Grays Harbor in Washington, the Cheyenne Bottoms of Kansas, and Delaware Bay in New Jersey are just a few of the oases that nurture hundreds of thousands of shorebirds.

Without these places of refuge, successful migrations would cease for millions of birds. None of these sites are secure. Conservation laws are only as strong as the people who support them. We look away and they are in danger of being overturned, compromised, and weakened.

Wetlands have a long history of being dredged, drained, and filled, or regarded as wastelands on the periphery of our towns. Already in Utah, there are those who envision a salt-free Great Salt Lake. A proposal has been drafted for the Utah State Legislature to introduce the concept of "Lake Wasatch." The Lake Wasatch Coalition would impound freshwater flowing into Great Salt Lake from the Bear, Weber, Ogden, and Jordan Rivers and other tributaries, by means of more than eighteen miles of inter-island dikes stretching through four counties between Interstate-80, Antelope Island, Fremont Island, and Promontory Point.

They see Lake Wasatch as fifty-two miles long and twelve and a half miles wide—three times the size of Lake Powell in southern Utah and northwest Arizona.

With 192 miles of shoreline, which unlike Lake Powell, is mostly under private ownership, there would be opportunities for unlimited lakeside development. Promotors already have plans for Antelope Island. They see it as an ideal site for a theme park with high-rise hotels and condominiums.

Lake Wasatch is a chamber of commerce dream. Finally, the Great Salt Lake would be worth something.

What about the birds?

Mimi turns to me, her legs outstretched on the sands of Half-Moon Bay.

"How do you place a value on inspiration? How do you quantify the wildness of birds, when for the most part, they lead secret and anonymous lives?"

GREAT BLUE HERON

lake level: 4207.05'

A heron stands on the edge of the lake, solitary and serene. The wind shinnies up her back, raising a few feathers, but her focus remains steady. This is a bird who knows how to protect herself. She has weathered the changes well. Throughout the high water and now its retreat, the true blue heron has stayed home. Perhaps this is a generational stance, the legacy of her lineage.

I would like to believe she is reclusive at heart, in spite of the communal nesting of her species. I would like to wade along the edges with her, this great blue heron. She belongs to the meditation of water.

But then this is another paradox of mine—wanting to be a bird when I am human.

The Gnostics teach me:

> For what is inside of you is what is outside of you and
> the one who fashions you on the outside is the one who
> shaped the inside of you And what you see outside of
> you, you see inside of you, it is visible and it is your
> garment.

Refuge is not a place outside myself. Like the lone heron
who walks the shores of Great Salt Lake, I am adapting as
the world is adapting.

Mimi and I are on a Great Basin pilgrimage. It was
a trip I wanted to take while she still had her strength.

"Now tell me where we are going?" she asks.

"All I will tell you is that what Stonehenge is to England,
'Sun Tunnels' are to the Great Basin. At least, that's how I
choose to look at them."

We turn north off the interstate and eventually find our
way on a dirt road, which meanders through an endless sea
of sage. I explain how artist Nancy Holt spent three years,
from 1973 through 1976, creating "Sun Tunnels," how it is
a sculpture built on forty acres she bought in the West
Desert, specifically as a site for the work.

Mimi puts on her glasses, opens an article on the sculpture
that I brought for her, and reads Nancy Holt's words out
loud:

> Sun Tunnels marks the yearly extreme positions of the
> sun on the horizon—the tunnels being aligned with the
> angles of the rising and the setting of the sun on the days
> of the solstices around June 21 and December 21. On
> those days the sun is centered through the tunnels, and
> is nearly centered for about ten days before and after the
> solstices.
>
> The four concrete tunnels are laid out on the desert

in an open X configuration eighty-six feet long on the diagonal. Each tunnel is eighteen feet long, and has an outside diameter of nine and a half feet and an inside diameter of eight feet, with a wall thickness of seven and a quarter inches.

Cut through the wall in the upper half of each tunnel are holes of four different sizes—seven, eight, nine, and ten inches in diameter. Each tunnel has a different configuration of holes corresponding to stars in four different constellations—Draco, Perseus, Columba, and Capricorn. The sizes of the holes vary relative to the magnitude of the stars to which they correspond. During the day, the sun shines through the holes, casting a changing pattern of pointed ellipses and circles of light on the bottom half of each tunnel. On nights when the moon is more than a quarter full, moonlight shines through the holes, casting its own paler pattern. The shapes and positions of the light cast differ from hour to hour, day to day, and season to season, relative to the positions of the sun and moon in the sky.

Each tunnel weighs twenty-two tons and rests on a buried concrete foundation. Due to the density, shape, and thickness of the concrete, the temperature is fifteen to twenty degrees cooler inside the tunnels in the heat of the day. There is also a considerable echo in the tunnels."

Mimi put down the article. "I can't wait to see them."

"I visited with Nancy Holt when I was in New York," I tell Mimi. "During our conversation, she talked about the process she personally underwent while conceiving them. She camped at the site for ten days and, at the time, wondered if she could stay in the desert that long. After a few days, she located a particular sound within the land and began to chant. This song became her connection to the Great Salt Lake desert. She told me she fluctuated from feeling very small to feeling very expansive. I remember her words, 'I became like the ebb and flow of light inside the tunnels.' "

"I understand that," Mimi says. "I remember going into my last surgery with two syllables in my mind, 'Ah, om,' 'Ah, om . . .' I closed my eyes and hummed those two words over and over until I was perfectly calm."

I stopped the car. "We're here."

Mimi looked out her window. "This is it? You mean these four pieces of conduit pipe? This looks like a job site of the Tempest Company!"

In Nancy Holt's "Sun Tunnels," the Great Basin landscape is framed within circles and we remember the shape of our planet, the shape of our eyes, our mouth in song and in prayer. These tunnels breathe as the ellipses expand and contract with the fickle light.

Smooth walls trick me into headstands, cartwheels, and somersaults. The sun hides and I want to say something—anything. The tunnels give import to my voice. It echoes. I laugh and chide and flirt with the gods until I find myself flat on my back with spots of sunlight covering my body—and I burst into tears, knowing it is only a matter of time until I am burned like paper beneath a magnifying glass. By morning, I will be left, frozen on the salt flats—forgotten forever were it not for my bones—bones that become whistles for the wind to blow through.

Mimi and I have not spoken for hours, each of us comfortable in our silences. A harrier hovers over the sage. Black, white, and gray. Male. I find a stash of feathers beneath the shadescale, I suspect horned lark. As I separate the brittle branches, sure enough, I find the foot with an extended hallax, lark for certain. A black beetle crosses the clay. One, two, three . . . seven, eight, nine mountain ranges are visible underneath this dome of sky.

I return to the east tunnel and fall asleep. When I awaken, I see Mimi standing in the center of the four "Sun Tunnels." She is turning slowly, looking outward in each direction.

ʄʄʄʄʄʄʄʄ

SCREECH OWLS

ʄʄʄʄʄʄʄʄ

lake level: 4206.00'

Mimi passed away this morning at 5:10 A.M., June 27, 1989.

One week ago, she said to me, "You know, Terry, it's the strangest thing, I keep expecting to see an owl one morning."

"Have you ever seen an owl here?" I asked, looking out her bedroom window through the trees.

"No," she said.

"Have you ever heard one?"

"No, but I just keep thinking that one morning I will wake up to see an owl."

Four days later, I was lying next to her. We were talking. I took hold of her broad, square hand.

"Mimi, when you die, if there really is something beyond death will you send me a sign, so I will know you are fine?"

She looked at me with her eyes that always squinted when she smiled and laughed.

"It doesn't work. I asked my father the same thing and he never came back."

Jack sits quietly beside Mimi's body. A single candle burns on her dressing table. Reflected in the mirror, it appears as two.

Dad and Richard leave to call the family. I walk outside.

The sky is electric blue, the sycamore and horse chestnut trees are silhouetted in black. I walk down the porch, past the bedroom windows, to the privacy of the backyard. I think I hear the cooing of mourning doves above the lilacs. I look up to see them, but they are not doves at all.

They are owls. Two owls are circling each other on top of the telephone pole.

"Dance. Dance. Dance," I hear Mimi say.

I stand below them. One screech owl turns, faces me, then flies. The other owl turns. We stare. It lifts its wings over its head, flutters them, then disappears in the direction of the other.

> *Ah, not to be cut off,*
> *not through the slightest partition*
> *shut out from the law of the stars.*
> *The inner—what is it?*
> *if not intensified sky,*
> *hurled through with birds and deep*
> *with the winds of homecoming.*
> > *—Rainer Maria Rilke*

Lying in my hammock at home, the wind rocks me back and forth. It is all that is left to comfort me.

Mimi and I shared a clandestine vision of things. I could afford to dream because she could interpret the story. We spoke through the shorthand of symbols: an egg, an owl. And most of what we shared was secret, much like the migrations of birds.

If I am to survive, I must let my secrets out like white doves held captive too long. I am a woman with wings.

With Mother I buried my innocence. With Mimi I will bury my haven.

Auden echoes from the open grave, "Our dreams of safety must disappear."

The Division of Water Resources has officially turned off the pumps. Great Salt Lake is on its own. The flood is over.

The Bear River Migratory Bird Refuge is able to breathe once again at lake level: 4206.00′.

AVOCETS AND STILTS

lake level: 4204.70'

The way to the Bird Refuge is clear for the first time in seven years. Great Salt Lake has retreated from sight, except for the faint line of silver on the horizon.

Refuge headquarters is unrecognizable. The buildings have been leveled. An old exhibit panel with a silhouette of a redhead flying over cattails stands akimbo among the wreckage. A partial title remains, reading simply, HISTORY—

Climbing over the rubble, spiders are everywhere. They are reinhabiting the Refuge. Their gossamer threads are binding it all together. Within minutes, I am draped with them. Even the avocets with their long, thin legs, sky blue, have silken strands trailing behind them.

The smell from the newly exposed land is ripe. Stilts walk on the cracked mud with skirts of brine flies around their red legs. Only a thin vein of water flows through the old canal, but volunteers have secured the banks. The U.S. Fish and Wildlife Service has promised $23 million toward the restoration of the Bear River Migratory Bird Refuge.

I turn. All at once, a thousand avocets take flight. More. Tens of thousands. A white and black flurry of birds circles me. The soft whistling of wings fills both time and space. I can no longer see the sky—above me, before me and behind me, avocets and stilts flock.

Oh, blessed wings.

In this moment, I realize how little I have hung on to for so long.

Brooke and I slip our red canoe into Half-Moon Bay. Great Salt Lake accepts us like a lover. We dip our wooden paddles into the icy waters and make strong, rapid strokes, north. The canoe powers gracefully ahead.

For two hours we paddle forward, toward the heart of the lake.

At the bow of the boat, I face the wind. Small waves take us up and down, up and down. The water, now bottle green, becomes a seesaw. We keep paddling.

The past seven years are with me. Mother and Mimi are present. The relationships continue—something I did not anticipate.

Flocks of pintails, mallards, and teals fly over us. There are other flocks behind them, undulating strands of birds like hieroglyphics that constantly rewrite themselves. Spring migration has begun.

We keep paddling. I have a turquoise and black shawl wrapped around me, protecting my face from the cold. This shawl is from Mexico, a gift to myself from the Day of the Dead.

I recall the impulse in me that said, "Go." I needed a ritual, a celebration to move me from death to life. I wore red for eight days—a simple cotton dress, drop-waisted and loose. I wanted no restrictions.

And when I entered the village of Tepotzlán, I bought flowers: gardenias, calla lilies, and lavender. But it was the marigolds that moved me. They were flames in the marketplace. Villagers plant seeds in May to be harvested for this occasion. They call them *cempaxuchil,* the flower with one thousand blossoms.

In the *mercado,* there was a man purchasing masks of jaguar, frog, and deer. I watched him. He knew something. I bought a mask for myself, an owl made of papier-maché. The man left. I followed him through the market. He bought loaves of bread, chicken, molé, tomatoes, clumps of cilantro, basil, and thyme. In the plaza, he stopped and abruptly turned around. Our eyes met. I pretended to be buying incense.

"Habla usted Inglés?" I asked.

"Sí," he said shifting his rucksack of food and masks to his other shoulder. (It turned out he was a North American who left the States in 1969. He had not been back since).

"What should I know about *el Día de los Muertos?"*

He looked at me long and hard.

"What do you want from them?" he asked.

"From whom?" I responded.

"From your Dead."

I looked away.

"There is a small adobe up the hill. Look for a turquoise door. The Dead will be there—five o'clock tomorrow afternoon, the eve of *el Día de los Muertos.* If you are to find it . . . you will."

I found the turquoise door. A white gauze curtain was billowing from the doorway. As soon as I walked in, an old woman with a long gray braid running down her back led me out, behind the adobe, and baptized me in lime water. Once inside, I sat down on one of the four white pews with a dozen or more villagers. It was a small white room. A woman knelt in front of a white altar, reciting prayers. Candles were burning. Thirteen candles. The shrine was smothered with white gladiolas. From one corner to the other, white crepe-paper flowers were strung between straws. The villagers prayed out loud with the kneeling woman.

I folded my hands across my lap and bowed my head. I was filled with gratitude for the graciousness of these people, that I could sit with them. A wave of emotion crested in me and broke. I wept silently for all I had lost. I reentered my own landscape of grief with perfect recall.

Songs were sung. More prayers were offered. And slowly my individual sorrow was absorbed into a sea of collective tears. We all wept.

Two women and a child, all dressed in white, sat in straight-backed chairs adjacent to the altar. In time, each one rose, trembling uncontrollably, sucking air through clenched teeth. They were in a quivering, hissing trance. I watched the Dead enter their bodies. They became taller, more robust, and confident. One by one, I listened to their stories. I watched their hands gesture the past as a mother spoke through her daughter, a sister spoke through her sister, and a mother spoke through her son.

After each account, the trembling and hissing returned, until the spirits slipped out through the storytellers' teeth and the peasants were returned to themselves. They collapsed, exhausted, back into the large white chairs.

Their stories were not so unlike my own. It was the reverberation of tone I recognized, like a piece of music you return to again and again that awakens the soul. The voices of my Dead came back to me.

Wearing my owl mask, I danced in the cobblestone streets. Bonfires lit every corner. Townsfolk circled them warming their hands. Tequila poured through the gutters. In one glance, I saw both lovers and murderers kissing and knifing each other against doors. Puppet shows were performed in the plaza as firecrackers exploded at our feet. Costumed children paraded through the village, carrying illuminated gourds as lanterns. All night long there is the relentless clamouring of bells, and the baying of dogs.

Carrying a lit candle, I entered the procession of masked individuals walking toward the cemetery. We followed the pathway of petals—marigold petals sprinkled so the Dead could follow.

The iron gates were open. Hundreds of candles were flickering as families left offerings on the graves of their kin: photographs, flowers, and food; calaveras—sugared skulls among them. Men and women washed the blue-tiled tombs that rose from the ground like altars, while other relatives cut back the vines that obscured the names of their loved ones. There were no tears here.

A crescent moon rose above the mountains, a blood-red sickle.

"*¿Porque està aquí?*" asked an old woman whose arms were wide with marigolds.

I looked up and stood. "*Mi madre está muerta.*"

She points down. "*¿Aquí?*"

"*No, no aquí*"—not here. I try to explain in poor Spanish.

"She is buried back home, *Los Estados Unidos,* but this is a good place to remember her."

We both pause.

The woman motions me to another place in the cemetery. I follow her until she turns around. She slowly sweeps her hand across five or six graves. *"Mi familia,"* she says smiling. *"Mi esposo, mi madre y padre, mis niños."* Then her hand moves up as she recklessly waves to the sky. *"Muy bonito . . . está cielo arriba . . . con las nubes como las rosas . . . los Muertos están conmigos."* I translate her words. "Very beautiful—this sky above us . . . with clouds like roses . . . the Dead are among us."

She hands me a marigold.

"Gracias," I say to her. "This is the flower my mother planted each spring."

My mind returns to the lake. Our paddling has become a meditation. We are miles from shore. In sight are four blue islands: Stansbury Island on our right, Carrigan Island to our left, and straight ahead we can see Antelope Island and Fremont.

My hands are numb. We bring in our paddles and allow ourselves to float. Brooke pulls out a thermos from his pack and pours two cups of hot chocolate. I spread cream cheese over poppyseed bagels. We eat.

There is no place on earth I would rather be. Our red canoe becomes a piece of driftwood in the current. Swirls of brine shrimp eggs cloud the water. I dip my empty cup into the lake. It fills with them, tiny pink spherical eggs. They are a mystery to me. I return them. I lean into the bow of the canoe. Brooke leans into the stern. We are balanced in the lake. For what seems like hours, we float, simply staring at the sky, watching clouds, watching birds, and breathing.

A ring-billed gull flies over us, then another. I sit up and carefully take out a pouch from my pocket, untying the leather thong that has kept the delicate contents safe. Brooke sits up and leans forward. I shake petals into his hands and then into my own. Together we sprinkle marigold petals into Great Salt Lake.

My basin of tears.

My refuge.

r r r r r r r r

THE CLAN OF ONE=BREASTED WOMEN

r r r r r r r r

Epilogue

I belong to a Clan of One-Breasted Women. My mother, my grandmothers, and six aunts have all had mastectomies. Seven are dead. The two who survive have just completed rounds of chemotherapy and radiation.

I've had my own problems: two biopsies for breast cancer and a small tumor between my ribs diagnosed as a "borderline malignancy."

This is my family history.

Most statistics tell us breast cancer is genetic, hereditary, with rising percentages attached to fatty diets, childlessness, or becoming pregnant after thirty. What they don't say is living in Utah may be the greatest hazard of all.

We are a Mormon family with roots in Utah since 1847. The "word of wisdom" in my family aligned us with good foods—no coffee, no tea, tobacco, or alcohol. For the most part, our women were finished having their babies by the time they were thirty. And only one faced breast cancer prior to 1960. Traditionally, as a group of people, Mormons have a low rate of cancer.

Is our family a cultural anomaly? The truth is, we didn't think about it. Those who did, usually the men, simply said, "bad genes." The women's attitude was stoic. Cancer was part of life. On February 16, 1971, the eve of my mother's surgery, I accidently picked up the telephone and overheard her ask my grandmother what she could expect.

"Diane, it is one of the most spiritual experiences you will ever encounter."

I quietly put down the receiver.

Two days later, my father took my brothers and me to the hospital to visit her. She met us in the lobby in a wheelchair. No bandages were visible. I'll never forget her radiance, the way she held herself in a purple velvet robe, and how she gathered us around her.

"Children, I am fine. I want you to know I felt the arms of God around me."

We believed her. My father cried. Our mother, his wife, was thirty-eight years old.

A little over a year after Mother's death, Dad and I were having dinner together. He had just returned from St. George, where the Tempest Company was completing the gas lines that would service southern Utah. He spoke of his love for the country, the sandstoned landscape, bare-boned and beautiful. He had just finished hiking the Kolob trail in Zion National Park. We got caught up in reminiscing, recalling with fondness our walk up Angel's Landing on his fiftieth birthday and the years our family had vacationed there.

Over dessert, I shared a recurring dream of mine. I told my father that for years, as long as I could remember, I saw this flash of light in the night in the desert—that this image had so permeated my being that I could not venture south without seeing it again, on the horizon, illuminating buttes and mesas.

"You did see it," he said.

"Saw what?"

"The bomb. The cloud. We were driving home from Riverside, California. You were sitting on Diane's lap. She was pregnant. In fact, I remember the day, September 7, 1957. We had just gotten out of the Service. We were driving north, past Las Vegas. It was an hour or so before dawn, when this explosion went off. We not only heard it, but felt it. I thought the oil tanker in front of us had blown up. We pulled over and suddenly, rising from the desert floor, we saw it, clearly, this golden-stemmed cloud, the mushroom. The sky seemed to vibrate with an eerie pink glow. Within a few minutes, a light ash was raining on the car."

I stared at my father.

"I thought you knew that," he said. "It was a common occurrence in the fifties."

It was at this moment that I realized the deceit I had been living under. Children growing up in the American Southwest, drinking contaminated milk from contaminated cows, even from the contaminated breasts of their mothers, my mother—members, years later, of the Clan of One-Breasted Women.

It is a well-known story in the Desert West, "The Day We Bombed Utah," or more accurately, the years we bombed Utah: above ground atomic testing in Nevada took place from January 27, 1951 through July 11, 1962. Not only were the winds blowing north covering "low-use segments of the population" with fallout and leaving sheep dead in their tracks, but the climate was right. The United States of the 1950s was red, white, and blue. The Korean War was raging. McCarthyism was rampant. Ike was it, and the cold war was hot. If you were against nuclear testing, you were for a communist regime.

Much has been written about this "American nuclear tragedy." Public health was secondary to national security. The Atomic Energy Commissioner, Thomas Murray, said, "Gentlemen, we must not let anything interfere with this series of tests, nothing."

Again and again, the American public was told by its government, in spite of burns, blisters, and nausea, "It has been found that the tests may be conducted with adequate assurance of safety under conditions prevailing at the bombing reservations." Assuaging public fears was simply a matter of public relations. "Your best action," an Atomic Energy Commission booklet read, "is not to be worried about fallout." A news release typical of the times stated, "We find no basis for concluding that harm to any individual has resulted from radioactive fallout."

On August 30, 1979, during Jimmy Carter's presidency, a suit was filed, *Irene Allen v. The United States of America.* Mrs. Allen's case was the first on an alphabetical list of twenty-four test cases, representative of nearly twelve hundred plaintiffs seeking compensation from the United States government for cancers caused by nuclear testing in Nevada.

Irene Allen lived in Hurricane, Utah. She was the mother of five children and had been widowed twice. Her first husband, with their two oldest boys, had watched the tests from the roof of the local high school. He died of leukemia in 1956. Her second husband died of pancreatic cancer in 1978.

In a town meeting conducted by Utah Senator Orrin Hatch, shortly before the suit was filed, Mrs. Allen said, "I am not blaming the government, I want you to know that, Senator Hatch. But I thought if my testimony could help in any way so this wouldn't happen again to any of the generations coming up after us . . . I am happy to be here this day to bear testimony of this."

God-fearing people. This is just one story in an anthology of thousands.

On May 10, 1984, Judge Bruce S. Jenkins handed down his opinion. Ten of the plaintiffs were awarded damages. It was the first time a federal court had determined that nuclear tests had been the cause of cancers. For the remaining fourteen test cases, the proof of causation was not sufficient. In spite of the split decision, it was considered a landmark ruling. It was not to remain so for long.

In April, 1987, the Tenth Circuit Court of Appeals overturned Judge Jenkins's ruling on the ground that the United States was protected from suit by the legal doctrine of sovereign immunity, a centuries-old idea from England in the days of absolute monarchs.

In January, 1988, the Supreme Court refused to review the Appeals Court decision. To our court system it does not matter whether the United States government was irresponsible, whether it lied to its citizens, or even that citizens died from the fallout of nuclear testing. What matters is that our government is immune: "The King can do no wrong."

In Mormon culture, authority is respected, obedience is revered, and independent thinking is not. I was taught as a young girl not to "make waves" or "rock the boat."

"Just let it go," Mother would say. "You know how you feel, that's what counts."

For many years, I have done just that—listened, observed, and quietly formed my own opinions, in a culture that rarely asks questions because it has all the answers. But one by one, I have watched the women in my family die common, heroic deaths. We sat in waiting rooms hoping for good news, but always receiving the bad. I cared for them, bathed their scarred bodies, and kept their secrets. I watched beautiful women become bald as Cytoxan, cisplatin, and Adriamycin were injected into their veins. I held their foreheads as they vomited green-black bile, and I shot them with morphine when the pain became inhuman. In the end, I witnessed their last peaceful breaths, becoming a midwife to the rebirth of their souls.

The price of obedience has become too high.

The fear and inability to question authority that ultimately killed rural communities in Utah during atmospheric testing of atomic weapons is the same fear I saw in my mother's body. Sheep. Dead sheep. The evidence is buried.

I cannot prove that my mother, Diane Dixon Tempest, or my grandmothers, Lettie Romney Dixon and Kathryn Blackett Tempest, along with my aunts developed cancer from nuclear fallout in Utah. But I can't prove they didn't.

My father's memory was correct. The September blast we drove through in 1957 was part of Operation Plumbbob, one of the most intensive series of bomb tests to be initiated. The flash of light in the night in the desert, which I had always thought was a dream, developed into a family nightmare. It took fourteen years, from 1957 to 1971, for cancer to manifest in my mother—the same time, Howard L. Andrews, an authority in radioactive fallout at the National Institute of Health, says radiation cancer requires to become evident. The more I learn about what it means to be a "downwinder," the more questions I drown in.

What I do know, however, is that as a Mormon woman of the fifth generation of Latter-day Saints, I must question everything, even if it means losing my faith, even if it means becoming a member of a border tribe among my own people. Tolerating blind obedience in the name of patriotism or religion ultimately takes our lives.

When the Atomic Energy Commission described the country north of the Nevada Test Site as "virtually uninhabited desert terrain," my family and the birds at Great Salt Lake were some of the "virtual uninhabitants."

One night, I dreamed women from all over the world circled a blazing fire in the desert. They spoke of change, how they hold the moon in their bellies and wax and wane with its phases. They mocked the presumption of even-tempered beings and made promises that they would never fear the witch inside themselves. The women danced wildly as sparks broke away from the flames and entered the night sky as stars.

And they sang a song given to them by Shoshone grandmothers:

Ah ne nah, nah	Consider the rabbits
nin nah nah—	How gently they walk on the earth—
ah ne nah, nah	Consider the rabbits
nin nah nah—	How gently they walk on the earth—
Nyaga mutzi	We remember them
oh ne nay—	We can walk gently also—
Nyaga mutzi	We remember them
oh ne nay—	We can walk gently also—

The women danced and drummed and sang for weeks, preparing themselves for what was to come. They would reclaim the desert for the sake of their children, for the sake of the land.

A few miles downwind from the fire circle, bombs were being tested. Rabbits felt the tremors. Their soft leather pads on paws and feet recognized the shaking sands, while the roots of mesquite and sage were smoldering. Rocks were hot from the inside out and dust devils hummed unnaturally. And each time there was another nuclear test, ravens watched the desert heave. Stretch marks appeared. The land was losing its muscle.

The women couldn't bear it any longer. They were mothers. They had suffered labor pains but always under the promise of birth. The red hot pains beneath the desert promised death only, as each bomb became a stillborn. A contract had been made and broken between human beings and the land. A new contract was being drawn by the women, who understood the fate of the earth as their own.

Under the cover of darkness, ten women slipped under a barbed-wire fence and entered the contaminated country. They were trespassing. They walked toward the town of Mercury, in moonlight, taking their cues from coyote, kit fox, antelope squirrel, and quail. They moved quietly and deliberately through the maze of Joshua trees. When a hint of daylight appeared they rested, drinking tea and sharing their rations of food. The women closed their eyes. The time had come to protest with the heart, that to deny one's genealogy with the earth was to commit treason against one's soul.

At dawn, the women draped themselves in mylar, wrapping long streamers of silver plastic around their arms to blow in the breeze. They wore clear masks, that became the faces of humanity. And when they arrived at the edge of Mercury, they carried all the butterflies of a summer day in their wombs. They paused to allow their courage to settle.

The town that forbids pregnant women and children to enter because of radiation risks was asleep. The women moved through the streets as winged messengers, twirling around each other in slow motion, peeking inside homes and watching the easy sleep of men and women. They were astonished by such stillness and periodically would utter a shrill note or low cry just to verify life.

The residents finally awoke to these strange apparitions. Some simply stared. Others called authorities, and in time, the women were apprehended by wary soldiers dressed in desert fatigues. They were taken to a white, square building on the other edge of Mercury. When asked who they were and why they were there, the women replied, "We are mothers and we have come to reclaim the desert for our children."

The soldiers arrested them. As the ten women were blindfolded and handcuffed, they began singing:

> *You can't forbid us everything*
> *You can't forbid us to think—*
> *You can't forbid our tears to flow*
> *And you can't stop the songs that we sing.*

The women continued to sing louder and louder, until they heard the voices of their sisters moving across the mesa:

> *Ah ne nah, nah*
> *nin nah nah—*
> *Ah ne nah, nah*
> *nin nah nah—*
> *Nyaga mutzi*
> *oh ne nay—*
> *Nyaga mutzi*
> *oh ne nay—*

"Call for reinforcements," one soldier said.

"We have," interrupted one woman, "we have—and you have no idea of our numbers."

I crossed the line at the Nevada Test Site and was arrested with nine other Utahns for trespassing on military lands. They are still conducting nuclear tests in the desert. Ours was an act of civil disobedience. But as I walked toward the town of Mercury, it was more than a gesture of peace. It was a gesture on behalf of the Clan of One-Breasted Women.

As one officer cinched the handcuffs around my wrists, another frisked my body. She did not find my scars.

We were booked under an afternoon sun and bused to Tonopah, Nevada. It was a two-hour ride. This was familiar country. The Joshua trees standing their ground had been named by my ancestors, who believed they looked like prophets pointing west to the Promised Land. These were the same trees that bloomed each spring, flowers appearing like white flames in the Mojave. And I recalled a full moon in May, when Mother and I had walked among them, flushing out mourning doves and owls.

The bus stopped short of town. We were released.

The officials thought it was a cruel joke to leave us stranded in the desert with no way to get home. What they didn't realize was that we were home, soul-centered and strong, women who recognized the sweet smell of sage as fuel for our spirits.

ACKNOWLEDGEMENTS

First and foremost, I must honor my father, John Henry Tempest, III. He is a proud and private man. I thank him for understanding and respecting my desire to tell this story. He read each draft, edited and discussed the scaffolding of ideas built around a tender, and often-times painful, chronology. I have relied on his courage and vulnerability in trying to tell the truth. *Refuge* has been a collaborative project including my brothers and sister-in-law: Stephen Dixon Tempest, Daniel Dixon Tempest, and William Henry Tempest, each one for different reasons; Steve for his gravity, Dan for his perceptions, and Hank for feeling it all; my sister-in-law, Ann Peterson Tempest, invited me to be at the birth of their third daughter, Diane Kathryn Tempest, on January 27, 1990. Her gift was my healing. She intuitively knew I needed to see life coming in.

I thank my nieces, Callie and Sara, for their sweet companionship.

My grandfathers, Jack Tempest and Sanky Dixon quietly hold our family together. Both men in their mid-eighties continue to teach us about adaptability. For their wisdom, for their joy, I am indebted.

Ruth and Richard Tempest, Bob, Lynne, Michael, Matthew, and David; Steve Earl and Elizabeth Hansen Tempest. We are one tribe. Diane and Don Dixon, Debbie and Skip McWhorter, Shelley and Lee Johnson, Cami and Scott Dixon, Sean and Kerry Dixon; we share my mother's blood.

Marion Blackett, Norinne Tempest, Bea Berg, Ann Williams, and Natalie McCullough are relatives who offered insight into family matters. Extended family members; Blacketts, Bullens, Romneys, and Dixons offered physical and spiritual support. I thank them all for their web of concern.

Rex and Rosemary Williams have been a constant source of love in my life. To them, my debt is great.

To these friends, I owe a garden of flowers: Martha Moench, Jan Dalebout, Nancy Roberts, Roz Newmark, Hal Cannon, Meg Brady, Joan and Ted Major, Jack Turner, Med Bennett, Jan and Joey Williams, Becky and Dave Thomas, Nan and Steve Hasler, Amy and Tom Williams, David Brewer, Mary Beth Raynes, Sue and Thayer Christensen, Wangari Waigwa-Stone, Jeff Giese, Margy and Chris Noble, Glen Lathrop, Bruce Hucko, June Pace, Lynn Berryhill, Jeffrey Montague, Annick Smith, Bill Kittredge, Pauline Weggeland, Emma Lou Thayne, Margo and Fred Silvester, Shelley and Rich Fenton, Gene Hoopes, Sally Smith, Beth Sundstrom, Greta DeJong, Flo Krall, Gwen Webster, Melissa and Scott Wood, Rich Wandschneider, P. K. Price, Donna Land Maldonado, Darci Cummins, Ron Barness, Betsy Burton, Patrick DeFrietas, Chet Morris, Steve Wilcox, Sam Weller, Steve Ashley, Karma Armstrong, Tom Lyon, G. Barnes, Kim Stafford, Sharon and Bill Loya, Marilyn Ellingson, Steve Casimiro, Liz Montague, Story and Bill Resor, Victoria Smith, Jim Harrison, and Barry Lopez.

My neighbors, John and Anne Milliken, offered food, friendship, and retreat. Anne was my editor across the fence whose daily inquiries about my mother moved the manuscript forward. Conversations over tea turned into paragraphs.

I want to acknowledge the generosity of Lorna Miller and Don Albrecht at the Crescent H Ranch in Wilson, Wyoming; Heather Burgess of the Ucross Foundation; Jane and Ken Sleight at Pack Creek Ranch in Moab, Utah; and Deborah Meier in New York City. They offered places of solitude in which to write.

Don Hague and the staff at the Utah Museum of Natural History are a remarkable community of individuals. They have my love and most sincere regard. Mary Gesicki in her grace, is responsible for creating an atmosphere I could work in.

Many individuals provided invaluable information that contributed greatly to the body of the manuscript and my understanding of Great Salt Lake. To their research and writings, I am indebted. Dr. William H. Behle laid the foundation for all studies of birds at Great Salt Lake. *The Birds of Great Salt Lake* (University of Utah Press, 1958) remains a classic. I have used his research and knowledge extensively. His friendship and passion for ornithology have inspired me—first as his student and, later, as an instructor at the Utah Museum of Natural History. My love is his. Margy Halpin's bright soul inspired wonderful encounters and discussions surrounding the birds of Bear River, snowy plovers in particular. I acknowledge her expertise in reviewing the checklist. Peter Paton of Utah State University along with Anne Wallace provided biological data on snowy plovers and white pelicans. Sally Jackson and John A. Kadlec's work, "Recent Flooding of Wetlands Around Great Salt Lake, Utah," from the Department of Fisheries and Wildlife in Logan, Utah, was key in my understanding this complex ecosystem. A. Lee Foote offered insight into wetland ecology and plants important to waterfowl. The Utah Division of Wildlife Resources has been supportive of my work through the years: Tim Provan, Don Paul, Joel Huener, Tom Aldrich, Susan Aune, and Brenda Schussman, in particular. Their research, pacific flyway reports, and companionship in the field have been very valuable. Jim Barnes and Clayton White of Brigham Young University have offered ecological perspectives that altered my own, regarding birds and place. Emmett A. Alford, also of Brigham Young University, contributed to my understanding of white-faced ibises, their nesting behavior in relationship to habitat. Fred Ryser's work, *Birds of the Great Basin* (University of Nevada Press, 1985) has been central. Ella Sorenson has been my guide through taxonomy, along with Eric Reichart, curator of birds and mammals at the University of Utah. "Utah Birds: A Revised Checklist" by William H. Behle, Ella D. Sorenson, and Clayton M. White, *Occasional Publication,* No. 4 (Utah Museum of Natural History, 1985) has been my bible.

In the field of anthropology, David Madsen has been a star. His sense of Great Basin archaeology has broadened any understanding of what it means to live in arid country. Kevin Jones has been a companion in my understanding of Fremont culture. Our days at Floating Island were a gift. I have relied on their research and findings outlined in "The Silver Island Expedition, 1988," University of Utah Anthropological Papers (in press). Liz Manion's "Partial Analysis of Lakeside Cave Fauna" delivered at the XXI Great Basin Anthropology Conference was also helpful. Jim Kirkman, at the Utah State Antiquities office, took me through the excavation material and explained the meticulous process of cataloging. His knowledge is magic. Larry Davis and Dee Dee O'Brien were guides and helpmates down south in Anasazi country. Ann Hanniball, curator of collections at the Utah Museum of Natural History, has been my anchor through the maze of information always offering human insight into scientific equations. To her, I am indebted.

In my quest to understand Great Salt Lake, its fluctuations and geomorphology, I must thank especially Genevieve Atwood, Don R. Mabey, and Donald R. Currey. Their Map 73, "Major Levels of Great Salt Lake and Lake Bonneville" produced by the Utah Geological and Mineral Survey; Department of Natural Resources, gave me my blueprint of study. Without them, I am afraid I would have been reduced to metaphor. Standing out in the Basin, listening to them describe the lake levels, made Lake Bonneville a tangible presence. I found myself standing underwater. I also acknowledge their predecessors, G. K. Gilbert and R. J. Spencer. Ted Arnow, from the U.S. Geological Survey, also played a substantial role in my understanding. I am indebted to his work on water-level and water-quality changes in Great Salt Lake from 1843 to 1985. Ron Ollis of the Utah Division of Water Resources was extremely generous in his explanations of the West Desert Pumping Project. I used his research and public information materials heavily. Gode Davis and Cliff Nielsen wrote provocatively about the rise of Great Salt Lake in *Utah Holiday* (March 1987). I must thank them for their fine work which took me further into the politics of place associated with the flooding. Former Salt Lake City mayor Ted Wilson gave me a day of stories in the chronology of the State Street flood. His wit, wisdom, and savvy of Utah politics inspired me. Mark Rosenfeld reminded me that brine shrimp and brine fly larvae are not the only inhabitants of Great Salt Lake. I appreciate his sharing knowledge of fish in the Great Basin. Frank DeCourten, museum curator, was the individual who walked me through each lake level of Lake Bonneville and painted pictures of the Ice Age so vivid that I could no longer hold the Pleistocene Epoch as an abstraction. Lake levels have been supplied by the Utah Division of Water Resources. In the medical profession, my heart goes to Dr. Gary Smith. He literally carried Mother and our family through death. And he did it with honesty, dignity, and compassion. He is family. Gary Johnson and Krehl Smith walked alongside. To these men, I am most grateful. The nursing staff at the LDS Hospital on 8 East was a beautiful expression of care, in particular, Faye Harder and Rolene Thompson. These were the women that Mother could be honest with and confide in without having to protect family. Dirk Noyes, Vicki Macy, and William F. Reilly were the caretakers of Mimi. They never lied. Hal Bourne, Hank Duffy, Howie Garber, and Steve Prescott also offered medical assurance along the way.

Natalie Clausen, Carol Mercereau, Marlisa DeJong, and Ann Kreilkamp, along with Rachel Bassett, were my healers. Women of great spirit.

The Mormon community we are a part of also healed us. I wish to acknowledge members of the Monument Park 11th and 14th wards. Bishop Craig Carman and Bishop Frank Nelson, in particular. Elder Hugh Pinnock brought both prayer and humor into our home through friendship. Beth Lords, Darlene Nilson, Joan James, and Diane Tonnesen provided daily rituals which Mother relied on. Each one of her friends can write their name in here. Aenona and LaMar Crocker shared common ground. As a family, we honor the life of their daughter, Tamra Crocker Pulsifer. Neighbors extend the notion of family. We were fed by them. Thank you.

Leonard Arrington has taught me about my own people. His thorough and thoughtful research into the history of Mormonism conveyed in *Brigham Young: American Moses* (Knopf, 1984) and *The Mormon Experience,* coauthored with Davis Bitton (Knopf, 1979), served as a catalyst and source for my discussions of the United Order, Brigham Young, and the early plight of the Latter-day Saints. I am grateful for his integrity in telling our history straight. He is trustworthy. Insights into Joseph Smith were gleaned from Michael Quinn's book, *Early Mormonism and the Magic World View* (Signature Books, 1987). Dale Morgan's vision of Great Salt Lake and its history has served as a baseline commentary. *Great Salt Lake—A Scientific, Historical and Economic Overview,* edited by J. Wallace Gwynn, Ph.D. (Utah Geological and Mineral Survey, 1980), has been an essential text.

Nancy Holt, the creator of "Sun Tunnels," was generous enough to let me into her home in New York where we shared our love of the Great Basin. I am indebted to her sense of place and her gift of forty acres outside Lucin, Utah. Most of the information used in this book came from direct quotations used in *Artforum* (April, 1977). Out of respect for her privacy, I chose not to draw from our personal conversations, rather allowed them to feed my own understanding of her work. Katie Nelson, a friend and art critic, first took me to the "Sun Tunnels." Once again, she drew me into the unseen world.

Regarding the epilogue, "The Clan of One-Breasted Women," I wish to thank Nini Rich who accompanied me to the Nevada Test Site in 1988. The stunning narrative of Philip L. Fradkin in his book *Fallout* (University of Arizona Press, 1989), provided the factual background of the essay. John G. Fuller also contributed to my understanding of nuclear politics in his book *The Day We Bombed Utah* (New American Library, 1984). To both men, I owe thanks. The Shoshone women who I was fortunate enough to cross the line with gave me their song. Carole Gallagher pushed me through my own denial. Her stories of radiation victims are eloquently photographed in *Nuclear Towns: The Secret War in the American West* (Doubleday, 1991). Senator Orrin Hatch and Congressman Wayne Owens from Utah have been fierce advocates of downwinders, passing a compensation bill in the fall of 1990. Bless them. Don Snow and Deb Clow, editors of *Northern Lights,* are responsible for the original essay. They are alchemists. Howard Berkes of National Public Radio and Karen Rathe of the *Seattle Times* took the message further. Charles F. Wilkinson provided astute comments and legal guidance on the various court cases. His encouragement to be bold moved me beyond fear.

Tenia Holland and Kara Edwards provided perspective on the manuscript as they worked with me on nuts and bolts. Linda Rawlins provided accurate Spanish translations in my discussion of the Day of the Dead.

I have had dear traveling companions who have walked repeatedly with me through this country: Lyn Dalebout, Dru Weggeland Brewer, Christopher Merrill, Geoff Foote, Laura Simms, Sandy Lopez (who kept me in flowers), and Ann Zwinger, who is my mentor. Lynne Ann Tempest shared my grief. Doug Peacock never fled from conversations of death. He stayed with me. To these friends, I express my devotion.

My family in New York, to whom I owe my professional life, were unwavering in their belief and support. Laurie Graham Schieffelin through her friendship and editorial wisdom crafted the manuscript. Linda Asher stretched the ideas and encouraged precision of language. Carl Brandt held the vision of *Refuge* when I became weary. He never lost faith. He gave me mine. I especially want to thank Dan Frank, my editor at Pantheon, for his lack of sentimentality in his insistence that I tell the right story. We have traveled far together.

Lastly, I wish to express my deepest gratitude to my husband, Brooke Williams, who is fearless and wise in his capacity to love. He is bedrock.

BIRDS
ASSOCIATED WITH
GREAT SALT LAKE

Great Salt Lake supports a rich diversity and abundance of breeding, migrating, and wintering birds. Rare but regular species are included in this list, which has been organized according to the phylogenetic order used in common field guides.

Common Loon   *Gavia immer*
Pied-billed Grebe   *Podilymbus podiceps*
Horned Grebe   *Podiceps auritus*
Eared Grebe   *Podiceps nigricollis*
Clark's Grebe   *Aechmophorus clarkii*
Western Grebe   *Aechmophorus occidentalis*
American White Pelican   *Pelecanus erythrorhynchos*
Double-crested Cormorant   *Phalacrocorax auritus*
American Bittern   *Botaurus lentiginosus*
Least Bittern   *Ixobrychus exilis*
Great Blue Heron   *Ardea herodias*
Great Egret   *Casmerodius albus*
Snowy Egret   *Egretta thula*
Cattle Egret   *Bubulcus ibis*

Refuge

Green-backed Heron   *Butorides striatus*
Black-crowned Night Heron   *Nycticorax nycticorax*
White-faced Ibis   *Plegadis chihi*
Tundra Swan (Whistling Swan)   *Cygnus columbianus*
Trumpeter Swan   *Cygnus buccinator*
Greater White-fronted Goose   *Anser albifrons*
Snow Goose   *Chen caerulescens*
Ross's Goose   *Chen rossii*
Brant   *Branta bernicla*
Canada Goose   *Branta canadensis*
Green-winged Teal   *Anas crecca*
Mallard   *Anas platyrhynchos*
Northern Pintail   *Anas acuta*
Blue-winged Teal   *Anas discors*
Cinnamon Teal   *Anas cyanoptera*
Northern Shoveler   *Anas clypeata*
Gadwall   *Anas strepera*
American Wigeon   *Anas americana*
Canvasback   *Aythya valisineria*
Redhead   *Aythya americana*
Ring-necked Duck   *Aythya collaris*
Greater Scaup   *Aythya marila*
Lesser Scaup   *Aythya affinis*
Oldsquaw   *Clangula hyemalis*
Surf Scoter   *Melanitta perspicillata*
White-winged Scoter   *Melanitta fusca*
Common Goldeneye   *Bucephala clangula*
Barrow's Goldeneye   *Bucephala islandica*
Bufflehead   *Bucephala albeola*
Hooded Merganser   *Lophodytes cucullatus*
Common Merganser   *Mergus merganser*
Red-breasted Merganser   *Mergus serrator*
Ruddy Duck   *Oxyura jamaicensis*
Turkey Vulture   *Cathartes aura*
Osprey   *Pandion haliaetus*
Bald Eagle   *Haliaeetus leucocephalus*
Northern Harrier (Marsh Hawk)   *Circus cyaneus*
Sharp-shinned Hawk   *Accipiter striatus*
Cooper's Hawk   *Accipiter cooperii*

Northern Goshawk   *Accipiter gentilis*
Swainson's Hawk   *Buteo swainsoni*
Red-tailed Hawk   *Buteo jamaicensis*
Ferruginous Hawk   *Buteo regalis*
Rough-legged Hawk   *Buteo lagopus*
Golden Eagle   *Aquila chrysaetos*
American Kestrel   *Falco sparverius*
Merlin   *Falco columbarius*
Peregrine Falcon   *Falco peregrinus*
Prairie Falcon   *Falco mexicanus*
Chukar   *Alectoris chukar*
Ring-necked Pheasant   *Phasianus colchicus*
Sage Grouse   *Centrocercus urophasianus*
Virginia Rail   *Rallus limicola*
Sora   *Porzana carolina*
Common Moorhen   *Gallinula chloropus*
Sandhill Crane   *Grus canadensis*
Black-bellied Plover   *Pluvialis squatarola*
Lesser Golden Plover   *Pluvialis dominica*
Snowy Plover   *Charadrius alexandrinus*
Semipalmated Plover   *Charadrius semipalmatus*
Killdeer   *Charadrius vociferus*
Black-necked Stilt   *Himantopus mexicanus*
American Avocet   *Recurvirostra americana*
Greater Yellowlegs   *Tringa melanoleuca*
Lesser Yellowlegs   *Tringa flavipes*
Solitary Sandpiper   *Tringa solitaria*
Willet   *Catoptrophorus semipalmatus*
Spotted Sandpiper   *Actitis macularia*
Whimbrel   *Numenius phaeopus*
Long-billed Curlew   *Numenius americanus*
Marbled Godwit   *Limosa fedoa*
Red Knot   *Calidris canutus*
Sanderling   *Calidris alba*
Semipalmated Sandpiper   *Calidris pusilla*
Western Sandpiper   *Calidris mauri*
Least Sandpiper   *Calidris minutilla*
Baird's Sandpiper   *Calidris bairdii*
Pectoral Sandpiper   *Calidris melanotos*

Refuge

Dunlin   *Calidris alpina*
Stilt Sandpiper   *Calidris himantopus*
Long-billed Dowitcher   *Limnodromus scolopaceus*
Common Snipe   *Gallinago gallinago*
Wilson's Phalarope   *Phalaropus tricolor*
Red-necked Phalarope   *Phalaropus lobatus*
Franklin's Gull   *Larus pipixcan*
Bonaparte's Gull   *Larus philadelphia*
Ring-billed Gull   *Larus delawarensis*
California Gull   *Larus californicus*
Herring Gull   *Larus argentatus*
Thayer's Gull   *Larus thayeri*
Glaucous Gull   *Larus hyperboreus*
Caspian Tern   *Sterna caspia*
Common Tern   *Sterna hirundo*
Forster's Tern   *Sterna forsteri*
Black Tern   *Chlidonias niger*
Rock Dove   *Columba livia*
Mourning Dove   *Zenaida macroura*
Common Barn Owl   *Tyto alba*
Western Screech Owl   *Otus kennicottii*
Great Horned Owl   *Bubo virginianus*
Burrowing Owl   *Athene cunicularia*
Long-eared Owl   *Asio otus*
Short-eared Owl   *Asio flammeus*
Common Nighthawk   *Chordeiles minor*
Common Poorwill   *Phalaenoptilus nuttallii*
Black-chinned Hummingbird   *Archilochus alexandri*
Broad-tailed Hummingbird   *Selasphorus platycercus*
Rufous Hummingbird   *Selasphorus rufus*
Belted Kingfisher   *Ceryle alcyon*
Downy Woodpecker   *Picoides pubescens*
Hairy Woodpecker   *Picoides villosus*
Northern Flicker   *Colaptes auratus*
Western Wood Pewee   *Contopus sordidulus*
Willow Flycatcher   *Empidonax traillii*
Hammond's Flycatcher   *Empidonax hammondii*
Dusky Flycatcher   *Empidonax oberholseri*
Gray Flycatcher   *Empidonax wrightii*

Cordilleran Flycatcher *Empidonax occidentalis*
Say's Phoebe *Sayornis saya*
Western Kingbird *Tyrannus verticalis*
Horned Lark *Eremophila alpestris*
Tree Swallow *Tachycineta bicolor*
Violet-green Swallow *Tachycineta thalassina*
Northern Rough-winged Swallow *Stelgidopteryx serripennis*
Bank Swallow *Riparia riparia*
Cliff Swallow *Hirundo pyrrhonota*
Barn Swallow *Hirundo rustica*
Scrub Jay *Aphelocoma coerulescens*
Black-billed Magpie *Pica pica*
Common Raven *Corvus corax*
Black-capped Chickadee *Parus atricapillus*
Red-breasted Nuthatch *Sitta canadensis*
Brown Creeper *Certhia americana*
Rock Wren *Salpinctes obsoletus*
House Wren *Troglodytes aedon*
Marsh Wren *Cistothorus palustris*
Golden-crowned Kinglet *Regulus satrapa*
Ruby-crowned Kinglet *Regulus calendula*
Mountain Bluebird *Sialia currucoides*
Townsend's Solitaire *Myadestes townsendi*
Hermit Thrush *Catharus guttatus*
American Robin *Turdus migratorius*
Gray Catbird *Dumetella carolinensis*
Northern Mockingbird *Mimus polyglottos*
Sage Thrasher *Oreoscoptes montanus*
American Pipit *Anthus rubescens*
Bohemian Waxwing *Bombycilla garrulus*
Cedar Waxwing *Bombycilla cedrorum*
Northern Shrike *Lanius excubitor*
Loggerhead Shrike *Lanius ludovicianus*
European Starling *Sturnus vulgaris*
Solitary Vireo *Vireo solitarius*
Warbling Vireo *Vireo gilvus*
Orange-crowned Warbler *Vermivora celata*
Virginia's Warbler *Vermivora virginiae*
Yellow Warbler *Dendroica petechia*

Yellow-rumped Warbler   *Dendroica coronata*
Black-throated Gray Warbler   *Dendroica nigrescens*
Townsend's Warbler   *Dendroica townsendi*
MacGillivray's Warbler   *Oporornis tolmiei*
Common Yellowthroat   *Geothylpis trichas*
Wilson's Warbler   *Wilsonia pusilla*
Yellow-breasted Chat   *Icteria virens*
Western Tanager   *Piranga ludoviciana*
Black-headed Grosbeak   *Pheucticus melanocephalus*
Lazuli Bunting   *Passerina amoena*
Green-tailed Towhee   *Pipilo chlorurus*
Rufous-sided Towhee   *Pipilo erythrophthalmus*
American Tree Sparrow   *Spizella arborea*
Chipping Sparrow   *Spizella passerina*
Brewer's Sparrow   *Spizella breweri*
Vesper Sparrow   *Pooecetes gramineus*
Lark Sparrow   *Chondestes grammacus*
Sage Sparrow   *Amphispiza belli*
Savannah Sparrow   *Passerculus sandwichensis*
Grasshopper Sparrow   *Ammodramus savannarum*
Song Sparrow   *Melospiza melodia*
Lincoln's Sparrow   *Melospiza lincolnii*
White-crowned Sparrow   *Zonotrichia leucophrys*
Dark-eyed Junco   *Junco hyemalis*
Red-winged Blackbird   *Agelaius phoeniceus*
Western Meadowlark   *Sturnella neglecta*
Yellow-headed Blackbird   *Xanthocephalus xanthocephalus*
Brewer's Blackbird   *Euphagus cyanocephalus*
Brown-headed Cowbird   *Molothrus ater*
Northern Oriole   *Icterus galbula*
Cassin's Finch   *Carpodacus cassinii*
House Finch   *Carpodacus mexicanus*
Pine Siskin   *Carduelis pinus*
American Goldfinch   *Carduelis tristis*
Evening Grosbeak   *Coccothraustes vespertinus*
House Sparrow   *Passer domesticus*

ABOUT THE AUTHOR

Terry Tempest Williams is Naturalist-in-Residence at the Utah Museum of Natural History in Salt Lake City. Her first Book, *Pieces of White Shell: A Journey to Navajoland* (1984), based on her experiences as a teacher among the Navajo, is a personal retelling and exploration of Native American myths. It received the 1984 Southwest Book Award. The link between story and landscape is further explored in *Coyote's Canyon* (1989), personal narratives of Southern Utah's desert canyons accompanied by the photographs of John Telford. Terry Tempest Williams lives in Salt Lake City. She serves on the Governing Council of the Wilderness Society in Washington, D.C.

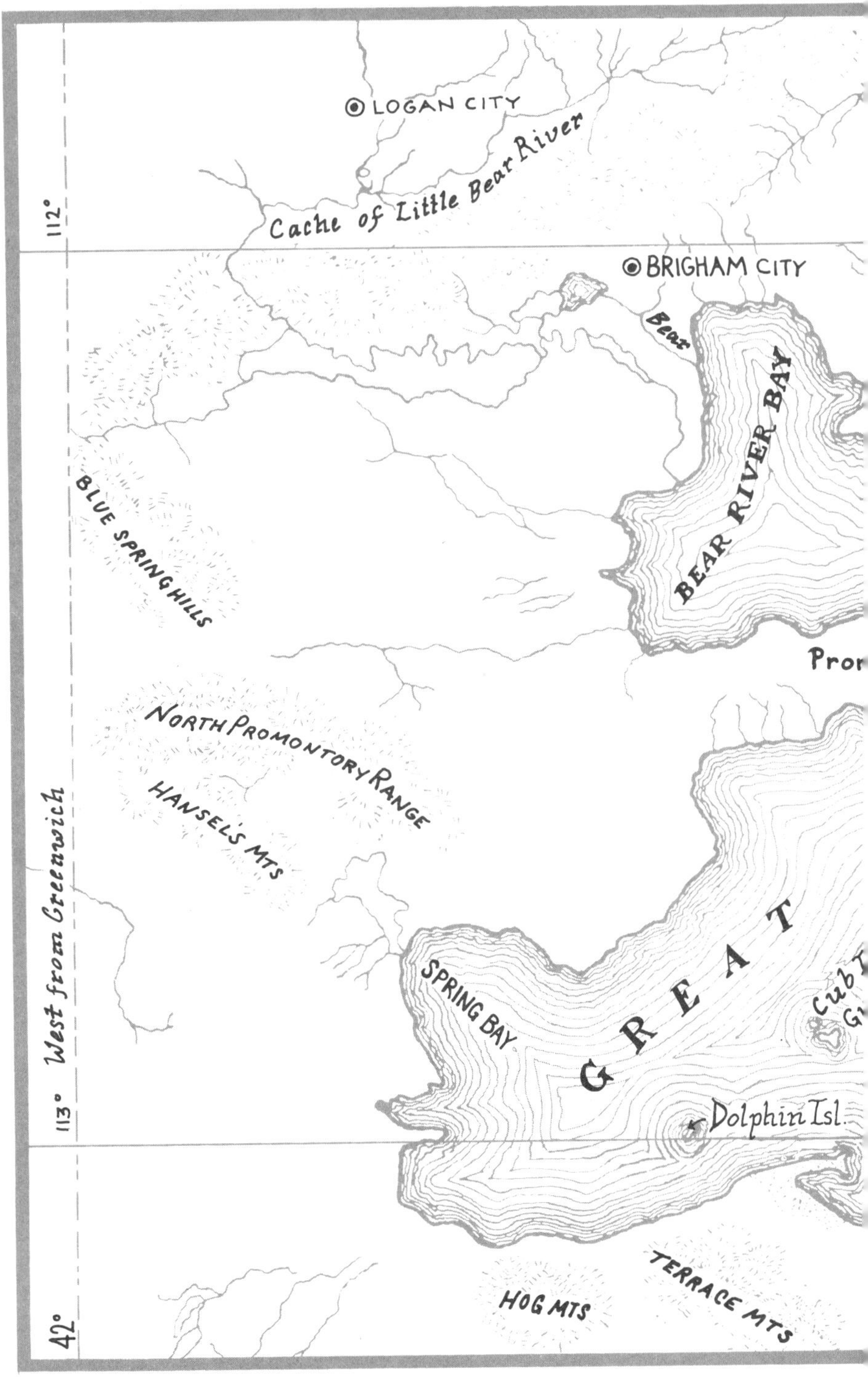

LOGAN CITY
Cache of Little Bear River
112°
BRIGHAM CITY
Bear
BEAR RIVER BAY
BLUE SPRING HILLS
Pror
NORTH PROMONTORY RANGE
HANSEL'S MTS
West from Greenwich
SPRING BAY
GREAT
Dolphin Isl.
113°
HOG MTS
TERRACE MTS
42°